Power Systems Modelling and Fault Analysis

Power Systems Modelling and Fault Analysis

Theory and Practice

Nasser D. Tleis

BSc, MSc, PhD, CEng, FIEE

AMSTERDAM · BOSTON · HEIDELBERG · LONDON · NEW YORK · OXFORD
PARIS · SAN DIEGO · SAN FRANCISCO · SINGAPORE · SYDNEY · TOKYO

Newnes is an imprint of Elsevier

ELSEVIER

Newnes

Newnes is an imprint of Elsevier
Linacre House, Jordan Hill, Oxford OX2 8DP, UK
30 Corporate Drive, Suite 400, Burlington, MA 01803, USA

First published 2008

Notice
No responsibility is assumed by the publisher for any injury and/or damage to persons
or property as a matter of products liability, negligence or otherwise, or from any use
or operation of any methods, products, instructions or ideas contained in the material
herein. Because of rapid advances in the medical sciences, in particular, independent
verification of diagnoses and drug dosages should be made

British Library Cataloguing in Publication Data
A catalogue record for this book is available from the British Library

Library of Congress Cataloging-in-Publication Data
A catalog record for this book is available from the Library of Congress

ISBN-13: 978-0-7506-8074-5

For information on all Newnes publications
visit our web site at http://books.elsevier.com

Typeset by Charon Tec Ltd (A Macmillan Company), Chennai, India
www.charontec.com
Printed and bound in Hungary

08 09 10 11 12   10 9 8 7 6 5 4 3 2 1

Dedicated to the late Mr Rafik Hariri
former Prime Minister of Lebanon
without whom this book would not have been written

Contents

List of Electrical Symbols   xvii

Foreword   xix

Preface   xxi

Biography   xxiv

1	**Introduction to power system faults**	**1**
1.1	General	1
1.2	Structure of power systems	1
1.3	Need for power system fault analysis	2
	1.3.1 General	2
	1.3.2 Health and safety considerations	3
	1.3.3 Design, operation and protection of power systems	3
	1.3.4 Design of power system equipment	4
1.4	Characteristics of power system faults	4
	1.4.1 Nature of faults	4
	1.4.2 Types of faults	4
	1.4.3 Causes of faults	5
	1.4.4 Characterisation of faults	6
1.5	Terminology of short-circuit current waveform and current interruption	8
1.6	Effects of short-circuit currents on equipment	12
	1.6.1 Thermal effects	12
	1.6.2 Mechanical effects	12
1.7	Per-unit analysis of power systems	15
	1.7.1 General	15
	1.7.2 Single-phase systems	15
	1.7.3 Change of base quantities	18
	1.7.4 Three-phase systems	19
	1.7.5 Mutually coupled systems having different operating voltages	20
	1.7.6 Examples	25

**2   Theory of symmetrical components and connection of
     phase sequence networks during faults**   **28**
   2.1   General   28
   2.2   Symmetrical components of a three-phase power system   29
         2.2.1   Balanced three-phase voltage and current phasors   29
         2.2.2   Symmetrical components of unbalanced voltage or
                 current phasors   31
         2.2.3   Apparent power in symmetrical component terms   34
         2.2.4   Definition of phase sequence component networks   34
         2.2.5   Sequence components of unbalanced three-phase
                 impedances   36
         2.2.6   Sequence components of balanced three-phase
                 impedances   39
         2.2.7   Advantages of symmetrical components frame of
                 reference   40
         2.2.8   Examples   40
   2.3   Analysis of balanced and unbalanced faults in the sequence
         reference frame   43
         2.3.1   General   43
         2.3.2   Balanced three-phase to earth short-circuit faults   43
         2.3.3   Balanced three-phase clear of earth short-circuit faults   45
         2.3.4   Unbalanced one-phase to earth short-circuit faults   47
         2.3.5   Unbalanced phase-to-phase or two-phase short-circuit
                 faults   49
         2.3.6   Unbalanced two-phase to earth short-circuit faults   51
         2.3.7   Unbalanced one-phase open-circuit faults   55
         2.3.8   Unbalanced two-phase open-circuit faults   56
         2.3.9   Example   58
   2.4   Fault analysis and choice of reference frame   59
         2.4.1   General   59
         2.4.2   One-phase to earth short-circuit faults   60
         2.4.3   Two-phase to earth short-circuit faults   61
   2.5   Analysis of simultaneous faults   63
         2.5.1   General   63
         2.5.2   Simultaneous short-circuit faults at the same location   63
         2.5.3   Cross-country faults or simultaneous faults at different
                 locations   65
         2.5.4   Simultaneous open-circuit and short-circuit faults at
                 the same location   66
         2.5.5   Simultaneous faults caused by broken and fallen to
                 earth conductors   68
         2.5.6   Simultaneous short-circuit and open-circuit faults on
                 distribution transformers   69
   Further reading   73

3 Modelling of multi-conductor overhead lines and cables   **74**
  3.1   General   74
  3.2   Phase and sequence modelling of three-phase
       overhead lines   74
       3.2.1 Background   74
       3.2.2 Overview of the calculation of overhead line
              parameters   76
       3.2.3 Untransposed single-circuit three-phase lines with and
              without earth wires   89
       3.2.4 Transposition of single-circuit three-phase lines   96
       3.2.5 Untransposed double-circuit lines with earth wires   102
       3.2.6 Transposition of double-circuit overhead lines   108
       3.2.7 Untransposed and transposed multiple-circuit lines   123
       3.2.8 Examples   127
  3.3   Phase and sequence modelling of three-phase cables   140
       3.3.1 Background   140
       3.3.2 Cable sheath bonding and earthing arrangements   142
       3.3.3 Overview of the calculation of cable parameters   145
       3.3.4 Series phase and sequence impedance matrices of
              single-circuit cables   154
       3.3.5 Shunt phase and sequence susceptance matrices of
              single-circuit cables   164
       3.3.6 Three-phase double-circuit cables   168
       3.3.7 Examples   170
  3.4   Sequence π models of single-circuit and double-circuit
       overhead lines and cables   173
       3.4.1 Background   173
       3.4.2 Sequence π models of single-circuit overhead
              lines and cables   175
       3.4.3 Sequence π models of double-circuit overhead
              lines   177
       3.4.4 Sequence π models of double-circuit cables   180
  3.5   Sequence π models of three-circuit overhead lines   180
  3.6   Three-phase modelling of overhead lines and cables
       (phase frame of reference)   182
       3.6.1 Background   182
       3.6.2 Single-circuit overhead lines and cables   183
       3.6.3 Double-circuit overhead lines and cables   184
  3.7   Computer calculations and measurements of overhead line and
       cable parameters   186
       3.7.1 Computer calculations of overhead line and cable
              parameters   186
       3.7.2 Measurement of overhead line parameters   187
       3.7.3 Measurement of cable parameters   193

3.8   Practical aspects of phase and sequence parameters of
      overhead lines and cables   197
      3.8.1   Overhead lines   197
      3.8.2   Cables   197
Further reading   198

4 Modelling of transformers, static power plant and static load   200
4.1   General   200
4.2   Sequence modelling of transformers   200
      4.2.1   Background   200
      4.2.2   Single-phase two-winding transformers   202
      4.2.3   Three-phase two-winding transformers   213
      4.2.4   Three-phase three-winding transformers   224
      4.2.5   Three-phase autotransformers with and without
              tertiary windings   230
      4.2.6   Three-phase earthing or zig-zag transformers   242
      4.2.7   Single-phase traction transformers connected to
              three-phase systems   243
      4.2.8   Variation of transformer's PPS leakage impedance with
              tap position   245
      4.2.9   Practical aspects of ZPS impedances of transformers 246
      4.2.10 Measurement of sequence impedances of three-phase
              transformers   249
      4.2.11 Examples   254
4.3   Sequence modelling of QBs and PS transformers   261
      4.3.1   Background   261
      4.3.2   PPS, NPS and ZPS modelling of QBs and PSs   263
      4.3.3   Measurement of QB and PS sequence impedances   268
4.4   Sequence modelling of series and shunt reactors and
      capacitors   272
      4.4.1   Background   272
      4.4.2   Modelling of series reactors   273
      4.4.3   Modelling of shunt reactors and capacitors   275
      4.4.4   Modelling of series capacitors   278
4.5   Sequence modelling of static variable compensators   283
      4.5.1   Background   283
      4.5.2   PPS, NPS and ZPS modelling   284
4.6   Sequence modelling of static power system load   285
      4.6.1   Background   285
      4.6.2   PPS, NPS and ZPS modelling   286
4.7   Three-phase modelling of static power plant and load in the
      phase frame of reference   286
      4.7.1   Background   286
      4.7.2   Three-phase modelling of reactors and capacitors   286
      4.7.3   Three-phase modelling of transformers   287

	4.7.4	Three-phase modelling of QBs and PSs	297
	4.7.5	Three-phase modelling of static load	299
	Further reading		300
5	**Modelling of ac rotating machines**		**301**
	5.1	General	301
	5.2	Overview of synchronous machine modelling in the phase frame of reference	302
	5.3	Synchronous machine modelling in the $dq0$ frame of reference	304
		5.3.1 Transformation from phase ryb to $dq0$ frame of reference	304
		5.3.2 Machine $dq0$ equations in per unit	306
		5.3.3 Machine operator reactance analysis	308
		5.3.4 Machine parameters: subtransient and transient reactances and time constants	310
	5.4	Synchronous machine behaviour under short-circuit faults and modelling in the sequence reference frame	314
		5.4.1 Synchronous machine sequence equivalent circuits	314
		5.4.2 Three-phase short-circuit faults	315
		5.4.3 Unbalanced two-phase (phase-to-phase) short-circuit faults	324
		5.4.4 Unbalanced single-phase to earth short-circuit faults	328
		5.4.5 Unbalanced two-phase to earth short-circuit faults	332
		5.4.6 Modelling the effect of initial machine loading	337
		5.4.7 Effect of AVRs on short-circuit currents	339
		5.4.8 Modelling of synchronous motors/compensators/ condensers	342
		5.4.9 Examples	343
	5.5	Determination of synchronous machines parameters from measurements	348
		5.5.1 Measurement of PPS reactances, PPS resistance and d-axis short-circuit time constants	348
		5.5.2 Measurement of NPS impedance	352
		5.5.3 Measurement of ZPS impedance	353
		5.5.4 Example	353
	5.6	Modelling of induction motors in the phase frame of reference	357
		5.6.1 General	357
		5.6.2 Overview of induction motor modelling in the phase frame of reference	358
	5.7	Modelling of induction motors in the dq frame of reference	362
		5.7.1 Transformation to dq axes	362
		5.7.2 Complex form of induction motor equations	363
		5.7.3 Operator reactance and parameters of a single-winding rotor	363

| | | 5.7.4 | Operator reactance and parameters of double-cage or deep-bar rotor | 364 |

5.7.4 Operator reactance and parameters of double-cage or
      deep-bar rotor   364
5.8 Induction motor behaviour under short-circuit faults and
    modelling in the sequence reference frame   368
    5.8.1 Three-phase short-circuit faults   368
    5.8.2 Unbalanced single-phase to earth short-circuit faults   375
    5.8.3 Modelling the effect of initial motor loading   377
    5.8.4 Determination of motor's electrical parameters
          from tests   378
    5.8.5 Examples   383
5.9 Modelling of wind turbine generators in short-circuit
    analysis   385
    5.9.1 Types of wind turbine generator technologies   385
    5.9.2 Modelling of fixed speed induction generators   388
    5.9.3 Modelling of small speed range wound rotor induction
          generators   388
    5.9.4 Modelling of doubly fed induction generators   389
    5.9.5 Modelling of series converter-connected generators   393
Further reading   396

6 Short-circuit analysis techniques in ac power systems   **397**
6.1 General   397
6.2 Application of Thévenin's and superposition's theorems to the
    simulation of short-circuit and open-circuit faults   398
    6.2.1 Simulation of short-circuit faults   398
    6.2.2 Simulation of open-circuit faults   400
6.3 Fixed impedance short-circuit analysis techniques   402
    6.3.1 Background   402
    6.3.2 Passive short-circuit analysis techniques   402
    6.3.3 The ac short-circuit analysis techniques   403
    6.3.4 Estimation of dc short-circuit current component
          variation with time   403
    6.3.5 Estimation of ac short-circuit current component
          variation with time   404
6.4 Time domain short-circuit analysis techniques in large-scale
    power systems   404
6.5 Analysis of the time variation of ac and dc short-circuit current
    components   405
    6.5.1 Single short-circuit source connected by a
          radial network   405
    6.5.2 Parallel independent short-circuit sources connected
          by radial networks   408
    6.5.3 Multiple short-circuit sources in interconnected
          networks   412

6.6   Fixed impedance short-circuit analysis of large-scale power
       systems   417
       6.6.1   Background   417
       6.6.2   General analysis of balanced three-phase short-circuit
                faults   417
       6.6.3   General analysis of unbalanced short-circuit faults   428
       6.6.4   General analysis of open-circuit faults   435
6.7   Three-phase short-circuit analysis of large-scale power
       systems in the phase frame of reference   438
       6.7.1   Background   438
       6.7.2   Three-phase models of synchronous and induction
                machines   438
       6.7.3   Three-phase analysis of ac current in the phase frame
                of reference   441
       6.7.4   Three-phase analysis and estimation of X/R ratio of
                fault current   445
       6.7.5   Example   448
Further reading   450

7   International standards for short-circuit analysis in
     ac power systems   451
     7.1   General   451
     7.2   International Electro-technical Commission 60909-0
            Standard   451
            7.2.1   Background   451
            7.2.2   Analysis technique and voltage source at the
                     short-circuit location   452
            7.2.3   Impedance correction factors   453
            7.2.4   Asynchronous motors and static converter drives 456
            7.2.5   Calculated short-circuit currents   458
            7.2.6   Example   462
     7.3   UK Engineering Recommendation ER G7/4   463
            7.3.1   Background   463
            7.3.2   Representation of machines and passive load   464
            7.3.3   Analysis technique   465
            7.3.4   Calculated short-circuit currents   466
            7.3.5   Implementation of ER G7/4 in the UK   467
     7.4   American IEEE C37.010 Standard   469
            7.4.1   Background   469
            7.4.2   Representation of system and equipment   469
            7.4.3   Analysis technique   470
            7.4.4   Calculated short-circuit currents   471
     7.5   Example calculations using IEC 60909, UK ER G7/4 and
            IEEE C37.010   473

7.6   IEC 62271-100-2001 and IEEE C37.04-1999 circuit-breaker
       standards   479
       7.6.1   Short-circuit ratings   479
       7.6.2   Assessment of circuit-breakers short-circuit duties
               against ratings   481
Further reading   483

**8 Network equivalents and practical short-circuit current
   assessments in large-scale ac power systems**   **485**
8.1   General   485
8.2   Power system equivalents for large-scale system studies   485
       8.2.1   Theory of static network reduction   485
       8.2.2   Need for power system equivalents   487
       8.2.3   Mathematical derivation of power system equivalents   489
8.3   Representation of power systems in large-scale studies   496
       8.3.1   Representation of power generating stations   496
       8.3.2   Representation of transmission, distribution and
               industrial networks   497
8.4   Practical analysis to maximise short-circuit current predictions   498
       8.4.1   Superposition analysis and initial ac loadflow
               operating conditions   498
       8.4.2   Effect of mutual coupling between overhead line
               circuits   499
       8.4.3   Severity of fault types and substation configuration   503
8.5   Uncertainties in short-circuit current calculations: precision
       versus accuracy   504
8.6   Probabilistic short-circuit analysis   507
       8.6.1   Background   507
       8.6.2   Probabilistic analysis of ac short-circuit current
               component   507
       8.6.3   Probabilistic analysis of dc short-circuit current
               component   509
       8.6.4   Example   515
8.7   Risk assessment and safety considerations   516
       8.7.1   Background   516
       8.7.2   Relevant UK legislation   517
       8.7.3   Theory of quantified risk assessment   517
       8.7.4   Methodology of quantified risk assessment   518
Further reading   519

9 Control and limitation of high short-circuit currents   **520**
9.1   General   520
9.2   Limitation of short-circuit currents in power system
       operation   520
       9.2.1   Background   520
       9.2.2   Re-certification of existing plant short-circuit rating   521

9.2.3   Substation splitting and use of circuit-breaker
autoclosing   521

9.2.4   Network splitting and reduced system parallelism   523

9.2.5   Sequential disconnection of healthy then faulted
equipment   524

9.2.6   Increasing short-circuit fault clearance time   524

9.2.7   De-loading circuits   525

9.2.8   Last resort generation disconnection   525

9.2.9   Example   525

9.3   Limitation of short-circuit currents in power system design
and planning   527

9.3.1   Background   527

9.3.2   Opening of unloaded delta-connected transformer
tertiary windings   527

9.3.3   Specifying higher leakage impedance for new
transformers   528

9.3.4   Upgrading to higher nominal system voltage levels   528

9.3.5   Uprating and replacement of switchgear and other
substation equipment   529

9.3.6   Wholesale replacement of switchgear and other
substation equipment   529

9.3.7   Use of short-circuit fault current limiters   529

9.3.8   Examples   529

9.4   Types of short-circuit fault current limiters   531

9.4.1   Background   531

9.4.2   Earthing resistor or reactor connected to
transformer neutral   531

9.4.3   Pyrotechnic-based fault current limiters   532

9.4.4   Permanently inserted current limiting series reactor   533

9.4.5   Series resonant current limiters using a bypass switch   534

9.4.6   Limiters using magnetically coupled circuits   534

9.4.7   Saturable reactor limiters   536

9.4.8   Passive damped resonant limiter   536

9.4.9   Solid state limiters using power electronic switches   538

9.4.10   Superconducting fault current limiters   539

9.4.11   The ideal fault current limiter   543

9.4.12   Applications of fault current limiters   543

9.4.13   Examples   546

Further reading   549

**10 An introduction to the analysis of short-circuit earth return
current, rise of earth potential and electrical interference   550**

10.1   Background   550

10.2   Electric shock and tolerance of the human body to
ac currents   551

10.2.1   Step, touch, mesh and transferred potentials   551

	10.2.2	Electrical resistance of the human body	552
	10.2.3	Effects of ac current on the human body	553
10.3		Substation earth electrode system	555
	10.3.1	Functions of substation earth electrode system	555
	10.3.2	Equivalent resistance to remote earth	555
10.4		Overhead line earthing network	561
	10.4.1	Overhead line earth wire and towers earthing network	561
	10.4.2	Equivalent earthing network impedance of an infinite overhead line	561
10.5		Analysis of earth fault ZPS current distribution in overhead line earth wire, towers and in earth	563
10.6		Cable earthing system impedance	567
10.7		Overall substation earthing system and its equivalent impedance	567
10.8		Effect of system earthing methods on earth fault current magnitude	568
10.9		Screening factors for overhead lines	569
10.10		Screening factors for cables	571
	10.10.1	General	571
	10.10.2	Single-phase cable with metallic sheath	571
	10.10.3	Three-phase cable with metallic sheaths	573
10.11		Analysis of earth return currents for short-circuits in substations	576
10.12		Analysis of earth return currents for short circuits on overhead line towers	577
10.13		Calculation of rise of earth potential	579
10.14		Examples	580
10.15		Electrical interference from overhead power lines to metal pipelines	584
	10.15.1	Background	584
	10.15.2	Electrostatic or capacitive coupling from power lines to pipelines	585
	10.15.3	Electromagnetic or inductive coupling from power lines to pipelines	588
	10.15.4	Resistive or conductive coupling from power systems to pipelines	595
	10.15.5	Examples	595
Further reading			603

Appendices			**605**
A.1		Theory and analysis of distributed multi-conductor lines and cables	605
A.2		Typical data of power system equipment	608
	A.2.1	General	608
	A.2.2	Data	609

| **Index** | | | **619** |

List of electrical symbols

Resistor

Resistor with non-linear current/voltage characteristics

Varistor or surge arrester with non-linear current/voltage characteristics

Inductor

Iron-cored inductor

Capacitor

Impedance

Reactor

Short-circuit fault

Normally closed circuit-breaker

Normally open circuit breaker

Normally closed disconnector

Normally open disconnector

Fast closing switch

Fast opening switch

Fuse

Two-winding transformers of various winding connection arrangements

Variable tap ratio ideal transformer

Autotransformers

Interconnected star or zig-zag earthing transformer

Three-winding transformers

Autotransformers with a delta-connected tertiary winding

Neutral solidly earthed

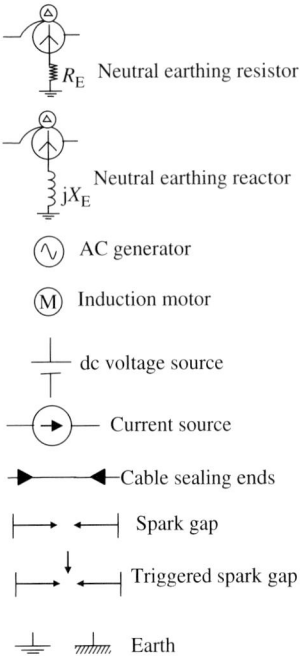

Neutral earthing resistor

Neutral earthing reactor

AC generator

Induction motor

dc voltage source

Current source

Cable sealing ends

Spark gap

Triggered spark gap

Earth

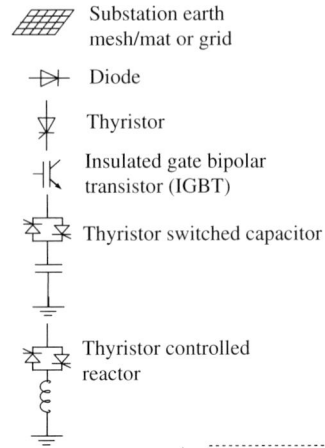

Substation earth mesh/mat or grid

Diode

Thyristor

Insulated gate bipolar transistor (IGBT)

Thyristor switched capacitor

Thyristor controlled reactor

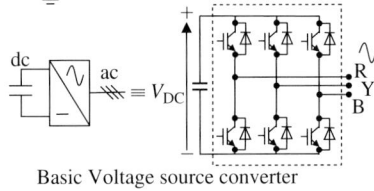

Basic Voltage source converter

Foreword

This is a new era for the electricity sector. The challenges we face in the near future are greater than at any time since the major network development programmes of the mid 20th Century. Thankfully, power transmission technology and its control and protection has made enormous leaps enabling better utilisation of assets, greater efficiency and improved quality of supply. This will help us meet the challenges ahead.

From a technology perspective we are now seeing the construction of new national networks, the formation or integration of regional networks and major network renewal programmes. There is also the need to develop and integrate new generation technologies and implement new control and power electronics solutions in more active and integrated transmission and distribution networks. The technology problem is therefore becoming richer and more complex – and demanding of novel solutions; it also requires a greater understanding of the characteristics and performance of the systems we need to build.

But technology is not the only development over the last few decades. We have also seen the development of a more competitive market place for electricity with second and third generation market models now being implemented, the unbundling of utility companies all providing benefits for consumers.

And crucially we are now understanding the impact of human activity on the environment and seeking to reduce emissions and develop more sustainable networks. This creates new pressures to incorporate the new greener technologies and meet planning and amenity constraints.

From a social perspective we know that electricity has entwined itself into the very heart and veins of society and all the services we now take for granted. We have learnt this lesson very keenly in the opening of the 21st century with rude reminders on what can happen when electricity supplies are lost.

Academia and the industry need to help the next generations of engineers to rise to these challenges. I believe that now is the time for the resurgence in engineering and electrical engineering disciplines; in particular the power generation, transmission and distribution sectors. It is vital that we develop and equip engineers with the verve, excitement, knowledge and talent they require to serve society's needs.

This book fills a major gap in providing the tools for this generation of engineers. It carefully targets the knowledge required by practitioners as well as academics in understanding power systems and their characteristics and how this can be modelled and incorporated into the development of the networks of the future.

Nasser Tleis is distinguished as an academic and a senior manager in the industry. Nasser has been at the forefront in developing academic capability as well as building generations of engineers capable of taking forward this knowledge and experience in the practical application of the techniques. This book captures Nasser's unique blend of the theoretical and applied knowledge to become a reference text and work book for our academics and engineers.

It gives me great pleasure to write the Forward to this book. It comes at a time where its contents are most relevant and I am confident it will bridge a gap between academic treatments and the very real need for application to power systems for the future by a new breed of practitioners.

Nick Winser
Director of Transmission, National Grid
Warwick, United Kingdom

Preface

The objective of this book is to present a practical treatment of modelling of electrical power systems, and the theory and practice of power system fault analysis. The treatment is designed to be sufficiently in-depth and generally adequate to serve the needs of practising electrical power engineers. Practical knowledge of power systems modelling and analysis techniques is essential for power system engineers working in the planning, design, operation, protection and incident analysis of generation, transmission, distribution and industrial power systems.

In many universities, undergraduate levels cover very little electrical engineering and even at postgraduate levels, course contents have become more fundamental, theoretical and basic. Nowadays, many undergraduate and postgraduate university teachers have no or very little practical industrial experience. This book is intended to provide a practical source of material for postgraduate students, researchers and university teachers in electrical power engineering. Further, over the last 20 years or so, the ongoing liberalisation and restructuring of electricity supply industries have been accompanied by significant loss of experienced electrical power engineers, mostly to retirement. Many new engineers entering industry are neither adequately equipped academically nor are they finding many experienced engineers to train them! Technical learned society papers are necessarily concise and specialised. Though not necessarily brief, books on power system modelling and fault analysis generally tend to follow a highly theoretical treatment and lack sufficient practical information and knowledge that leaves the reader with inadequate understanding. In writing this book, one of my aims has been to attempt to bridge a gap between those theoretical books and the specialised technical papers.

The aim of this book is to present practical power system modelling and analysis techniques as applied in modern industry practices. Therefore, strict academic and basic fundamental theories have largely been omitted to save valuable space. Basic knowledge is presumed in the following areas: analysis of three-phase alternating current electrical circuits; theory of electrostatic and electromagnetic fields; calculation of resistance, inductance and capacitance of lines; basic theories of electromagnetic transformers and ac rotating machines, complex phasor algebra, matrix algebra, linear differential equations and Laplace transforms. These basic topics are well covered in many power systems and mathematics textbooks.

In support of the in-depth material I present in this book, I have included a comprehensive and most relevant list of technical references.

I have used SI units throughout and I hope this is not seen as a disadvantage where non-SI units are still in use.

Chapter 1 discusses the nature, causes and effects of faults in power systems, presents fundamental concepts and definitions of short-circuit currents and circuit-breaker interruption as well as a practical treatment of per-unit system of analysis. Chapter 2 presents the theory of symmetrical components and a practical and detailed treatment of the connection of sequence networks under various fault conditions including simultaneous faults. Chapter 3 is concerned with the advanced modelling and analysis of practical multi-conductor overhead lines and cables in the phase coordinates and sequence reference frames. Chapter 4 presents the modelling, in the phase coordinates and sequence reference frames, of transformers, quadrature boosters, phase shifters, series and shunt reactors, series and shunt capacitors, static variable compensators and power system load. Chapter 5 presents the modelling in the phase coordinates, $dq0$ and sequence reference frames of synchronous generators and induction motors. Modern wind turbine generators such as squirrel-cage induction generators, wound rotor doubly fed induction generators and generators connected to the ac grid through power electronics converters are also covered. In Chapters 3–5, practical measurement techniques of the electrical parameters of various power system equipment are presented. The models presented in these three chapters can be used in various power system analysis applications including positive phase sequence (PPS) load flow and PPS transient stability, multiphase unbalanced load flow and multiphase fault analysis, etc. Chapter 6 presents methods for the simulation of short-circuit and open-circuit faults as well as static and dynamic short-circuit analysis techniques in ac power systems. New and strong emphasis is given to the analysis of the time variation of the ac and dc components of short-circuit current. This emphasis is important in generation, transmission and industrial power systems. In addition, the expansion in the connection of small-scale distributed generation in distribution systems is exacerbating short-circuit current problems where switchgear are traditionally not designed either for make duties or for significant dc short-circuit current component.

The emphasis on the analysis of the dc short-circuit current component also reflects the increase in X/R ratios of power system equipment due to the use of higher system voltages and/or more efficient and lower loss transformers. An introduction to modern short-circuit analysis in the phase coordinates frame of reference is given. Chapter 7 describes and highlights the differences among the three international approaches to the analysis of short-circuit currents: the International Electro-technical Commission IEC 60909 Standard, the UK Engineering Recommendation ER G7/4 and the American IEEE C37.010 Standard. Chapter 8 presents the formulation of power system equivalents by network reduction techniques. It discusses uncertainties present in short-circuit analysis, gives an introduction to probabilistic short-circuit analysis and to the theory of quantified risk assessment including safety considerations. Chapter 9 presents practical methods for the control and limitation of high short-circuit currents in power system design, operational planning and real-time operation. In addition, the various technologies of existing and some future short-circuit fault current limiters are described including their applications. Chapter 10 describes the effects of ac currents on the human body and its electrical resistance. It describes the components that make up extended substation earthing systems, and gives an introduction to the analysis of short-circuit earth or ground return current and rise of earth potential. The phenomenon of electrical interference from power lines is discussed and analysis techniques of induced voltages are presented with a particular focus on coupling interference from overhead power lines to metallic pipelines. Two Appendices are included: the first presents the analysis of multi-conductor lines and cables and the second presents typical data for power system equipment.

I have used actual power system equipment data and solved practical examples representing some of the type of problems faced by practising power engineers. I have solved or shown how to solve the examples using hand calculations and electronic calculators. I believe in the 'feel' and unique insight that hand calculations provide which can serve as a good foundation for the power engineer to adequately specify, model, analyse and interpret complex and large power system analysis results.

Many colleagues in National Grid have given me help and encouragement. Mr Andy Stevenson who gave his and National Grid's support at the start of this project, Mr Tony Johnson, Mr Tom Fairey and Dr Andrew Dixon for reviewing various chapters and preparing their figures. I am indebted to Dr Zia Emin for reviewing Chapter 3, for the high quality figures prepared and for our many useful discussions.

Writing this book whilst full-time employed and raising a young family has been difficult. I would like to thank my wife Hanadi for her patience, unfailing support and encouragement throughout. I also want to thank my daughters Serene and Lara for never complaining why I ignored them for such a long time!

Like most books that contain material of a reference nature, the published work is only the visible part of a huge iceberg. It is hoped that the nature of the material and list of references included may give a 'taster' of the size of the invisible iceberg!

Finally, the book may contain errors of a typographical nature or otherwise. Should a second edition be required, I would be grateful to receive your comments and any suggestions for improvements on any aspect of this book on Nasser.Tleis@uk.ngrid.com.

Nasser Tleis
England, UK
June 2007

Biography

Dr Nasser Tleis obtained his PhD degree in Electrical Power Engineering from The University of Manchester Institute of Science and Technology, England, UK in 1989. He joined the Central Electricity Generating Board in London, then the newly formed National Grid Company on privatisation in 1990. He has extensive experience in electrical power engineering in the following areas: power plant performance specification including steady state and transient analysis, substation insulation coordination, planning and design of transmission systems, voltage control strategies and reactive compensation planning, power systems thermal and loadflow analysis, voltage stability, frequency control and stability, transient and dynamic stability including long-term dynamics, faults and short-circuit analysis, multiphase power flow and unbalance analysis, analysis and measurement of harmonic distortion and voltage flicker.

He has worked on the development and validation of dynamic models of generation plant, their excitation and governor control systems, dynamic models for transmission plant control systems as well as developing new power systems analysis techniques. For the last few years, he has led the development of the Great Britain Grid Code specifying the technical connection requirements for new non-synchronous generation technologies, particularly wind turbine generators. He is currently involved in the assessment of the changing technical behaviour of power systems that include a mix of synchronous and non-synchronous generation technologies including the development of steady state, transient and dynamic models. He is a Chartered Engineer and a Fellow of the Institution of Engineering and Technology (formerly Institution of Electrical Engineers).

1

Introduction to power system faults

1.1 General

In this introductory chapter, we introduce the important terminology of fault current waveform, discuss the need for power system fault analysis and the effects of fault currents in power systems. Per-unit analysis concept of single-phase and three-phase power systems is presented including the base and per-unit equations of self and mutual impedances and admittances.

1.2 Structure of power systems

Electrical ac power systems consist of three-phase generation systems, transmission and distribution networks, and loads. The networks supply large three-phase industrial loads at various distribution and transmission voltages as well as single-phase residential and commercial loads. In some countries, e.g. North America, the term subtransmission is used to denote networks with voltage classes between transmission and distribution. Distribution voltages are typically 10–60 kV, subtransmission voltages are typically 66–138 kV and transmission voltages are typically above 138 kV. Generated voltages are up to 35 kV for generators used in large electrical power stations. Power station auxiliary supply systems and industrial power systems supply a significant amount of induction motor load. Residential and commercial loads include a significant amount of single-phase induction motor loads.

For over a century, electric power systems used synchronous machines for the generation of electricity. However, in the twenty-first century, the generation of electricity from renewable energy sources such as wind has begun to expand at a large pace. Generally, such generation systems use a variety of asynchronous

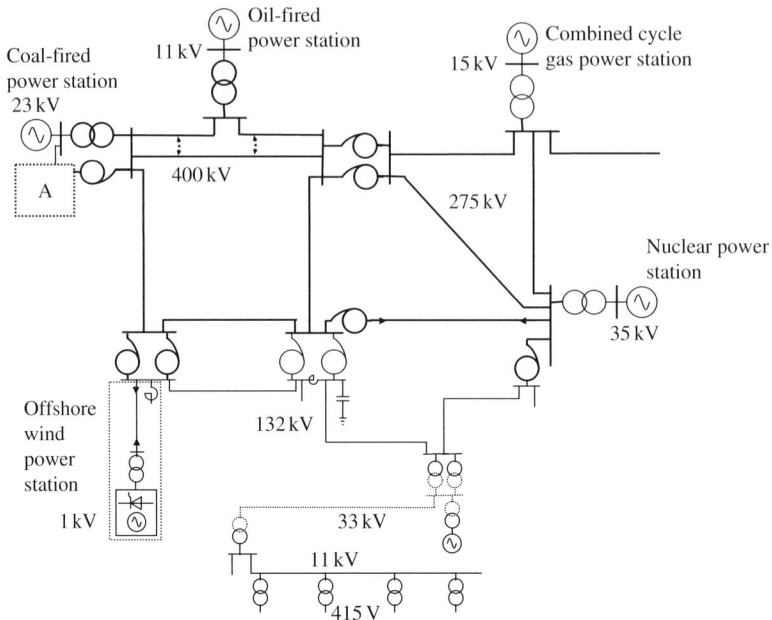

Figure 1.1 Typical structure and components of a generation, transmission and distribution power system

machines as well as machines interfaced to the three-phase network through a low voltage direct current link or a power electronics converter. Typical ratings of wind turbine generators are currently up to 5 MW and typical generated voltage range from 0.4 to 5 kV. The mix of synchronous, asynchronous and converter isolated electrical generation systems is expected to change the behaviour of three-phase power systems following disturbances such as short-circuit faults. Figure 1.1 illustrates a typical structure and components of a generation, transmission and distribution power system and Figure 1.2 illustrates a typical auxiliary electrical supply system for a large power station representing box A in Figure 1.1.

1.3 Need for power system fault analysis

1.3.1 General

Short-circuit analysis is carried out in electrical power utility systems, industrial power systems, commercial power systems and power station auxiliary systems. Other special applications are in concentrated power system installations on board military and commercial ships and aircraft. Short-circuit calculations are generally performed for a number of reasons. These are briefly described in the next sections.

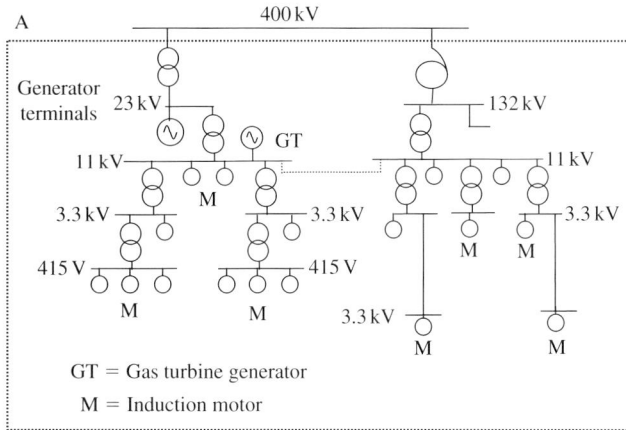

Figure 1.2   Typical structure and components of a power station auxiliary electrical supply system

1.3.2 Health and safety considerations

Short-circuit fault analysis is carried out to ensure the safety of workers as well as the general public. Power system equipment such as circuit-breakers can fail catastrophically if they are subjected to fault duties that exceed their rating. Other equipment such as busbars, transformers and cables can fail thermally or mechanically if subjected to fault currents in excess of ratings. In addition, to ensure safety, short-circuit fault analysis is carried out and used in the calculation of rise of earth potential at substations and overhead line towers. Other areas where fault analysis is carried out are for the calculation of induced voltages on adjacent communication circuits, pipelines, fences and other metallic objects.

1.3.3 Design, operation and protection of power systems

Short-circuit current calculations are made at the system design stage to determine the short-circuit ratings of new switchgear and substation infrastructure equipment to be procured and installed. System reinforcements may be triggered by network expansion and/or the connection of new generating plant to the power system. Routine calculations are also made to check the continued adequacy of existing equipment as system operating configurations are modified. In addition, calculations of minimum short-circuit currents are made and these are used in the calculation of protection relay settings to ensure accurate and coordinated relay operations. In transmission systems, short-circuit currents must be quickly cleared to avoid loss of synchronism of generation plant and major power system black-outs. Maximum short-circuit current calculations are carried out for the design of substation earth electrode systems. Short-circuit analysis is also carried out as part of initial power quality assessments for the connection of disturbing loads to electrical power networks. These assessments include voltage flicker, harmonic analysis and voltage unbalance. Other areas where short-circuit analysis is carried out is in the modification of an existing system or at the design stage of new electrical power installations such as a new offshore oil platform, new petrochemical process plant or the auxiliary electrical power system of a new power station. The aim is to determine the short-circuit rating of new switchgear and other substation infrastructure equipment that will be procured and installed.

1.3.4 Design of power system equipment

Switchgear manufacturers design their circuit-breakers to ensure that they are capable of making, breaking and carrying, for a short time, the specified short-circuit current. Equipment with standardised short-circuit ratings are designed and produced by manufacturers. Also, manufacturers of substation infrastructure equipment and other power system plant, e.g. transformers and cables, use the short-circuit current ratings specified by their customers to ensure that the equipment is designed to safely withstand the passage of these currents for the duration specified.

1.4 Characteristics of power system faults

1.4.1 Nature of faults

A fault on a power system is an abnormal condition that involves an electrical failure of power system equipment operating at one of the primary voltages within the system. Generally, two types of failure can occur. The first is an insulation failure that results in a short-circuit fault and can occur as a result of overstressing and degradation of the insulation over time or due to a sudden overvoltage condition. The second is a failure that results in a cessation of current flow or an open-circuit fault.

1.4.2 Types of faults

Short-circuit faults can occur between phases, or between phases and earth, or both. Short circuits may be one-phase to earth, phase to phase, two-phase to earth, three-phase clear of earth and three-phase to earth. The three-phase fault that symmetrically affects the three phases of a three-phase circuit is the only balanced fault whereas all the other faults are unbalanced. Simultaneous faults are a combination of two or more faults that occur at the same time. They may be of the same or different types and may occur at the same or at different locations. A broken overhead line conductor that falls to earth is a simultaneous one-phase open-circuit and one-phase short-circuit fault at one location. A short-circuit fault occurring at the same time on each circuit of a double-circuit overhead line, where the two circuits are strung on the same tower, is a simultaneous fault condition. A one-phase to earth short-circuit fault in a high impedance earthed distribution system may cause a sufficient voltage rise on a healthy phase elsewhere in the system that a flashover and short-circuit fault occurs. This is known as a cross-country fault. Most faults do not change in type during the fault period but some faults do change and evolve from say a one-phase to earth short circuit to engulf a second phase where it changes to a two-phase to earth short circuit fault. This can occur on overhead lines or in substations where the flashover arc of the faulted phase spreads to other healthy phases. Internal short circuits to earth and open-circuit faults can also occur on windings of transformers, reactors and machines as well as faults between a number of winding turns of the same phase.

1.4.3 Causes of faults

Open-circuit faults may be caused by the failure of joints on cables or overhead lines or the failure of all the three phases of a circuit-breaker or disconnector to open or close. For example, two phases of a circuit-breaker may close and latch but not the third phase or two phases may properly open but the third remains stuck in the closed position. Except on mainly underground systems, the vast majority of short-circuit faults are weather related followed by equipment failure. The weather factors that usually cause short-circuit faults are: lightning strikes, accumulation of snow or ice, heavy rain, strong winds or gales, salt pollution depositing on insulators on overhead lines and in substations, floods and fires adjacent to electrical equipment, e.g. beneath overhead lines. Vandalism may also be a cause of short-circuit faults as well as contact with or breach of minimum clearances between overhead lines and trees due to current overload.

Lightning strikes may discharge currents in the range of a few kiloamps up to 100 or 200 kA for a duration of several microseconds. If the strike hits an overhead line or its earth wire, the voltage produced across the insulator may be so large that a back-flashover and short circuit occurs. This may involve one or all three phases of a three-phase electrical circuit and as a result a transient power frequency short-circuit current flows. For example, consider a 132 kV overhead three-phase transmission line of steel tower construction and one earth wire. The surge impedances of the tower, line phase conductors and earth wire are given as 220, 350 and 400 Ω, respectively, and the tower's earthing or footing resistance is 50 Ω. The line's rated lightning impulse withstand voltage to earth is 650 kV peak phase to earth (nominal peak phase to earth voltage is $(132 \text{ kV} \times \sqrt{2})/\sqrt{3} = 107.8 \text{ kV}$). A lightning strike with a reasonably modest peak current of 10 kA hits a tower on the line. The voltage produced across the line's insulator is approximately equal to 10 kA $\times$ 105 Ω = 1050 kV. This significantly exceeds the line's insulation strength causing back-flashover on all three phases of the line and a three-phase to earth short-circuit fault. If the shielding of the earth wire fails and the lightning strike hits one of the phase conductors near a tower, then the voltage produced across the line's insulator is approximately equal to 10 kA $\times$ 175 Ω = 1750 kV.

In this case, a smaller lightning current in the order of 3.8 kA would be sufficient to cause a back-flashover and hence a short-circuit fault. On lower voltage distribution lines, even 'indirect' lightning strikes, i.e. those that hit nearby objects to the line may produce a sufficiently large voltage on the line to cause an insulator flashover and a short-circuit fault. Other causes of short-circuit faults are fires. The smoke of fires beneath overhead lines consists of small particles which encourage the breakdown of the air that is subjected to the intense electric field of a high voltage power line. The hot air in the flames of a fire has a much lower insulation strength than air at ambient temperature. A flashover across an insulator to earth or from a phase conductor to a tree may occur.

Equipment failure, e.g. machines, transformers, reactors, cables, etc., cause many short-circuit faults. These may be caused by failure of internal insulation due to ageing and degradation, breakdown due to high switching or lightning over-voltages, by mechanical incidents or by inappropriate installation. An example is a breakdown of a cable's polymer insulation due to ageing or to the creation of voids within the insulation caused by an external mechanical force being accidentally applied on the cable.

Short-circuit faults may also be caused by human error. A classical example is one where maintenance staffs inadvertently leave isolated equipment connected through safety earth clamps when maintenance work is completed. A three-phase to earth short-circuit fault occurs when the equipment is reenergised to return it to service.

On mainly overhead line systems, the majority of short-circuit faults, typically 80–90%, tend to occur on overhead lines and the rest on substation equipment and busbars combined. Typically, on a high voltage transmission system with overhead line steel tower construction, such as the England and Wales transmission system, long-term average short-circuit fault statistics show that around 300 short-circuit faults occur per annum. Of these, 67% are one-phase to earth, 25% are phase to phase, 5% are three-phase to earth and three-phase clear of earth, and 3% are two-phase to earth. About 77% of one-phase to earth faults are caused by lightning strikes followed by wind and gales then salt pollution on insulators. Although lightning can cause some phase-to-phase faults, by far the most common causes of these faults are snow/ice followed by wind/gales that cause two line conductors to clash. The majority of three-phase to earth and two-phase to earth faults in England and Wales are caused by lightning then wind and gales. On wood-pole overhead lines, e.g. some 132 kV and lower in England and Wales, between 50% and 67% of short-circuit faults are two-phase and three-phase faults.

1.4.4 Characterisation of faults

Because they are unbalanced, one-phase open-circuit and two-phase open-circuit faults are characterised by the negative and zero phase sequence voltages and currents they generate at the fault location and elsewhere in the power system particularly at substations where electrical machines are connected. Machines are vulnerable to damage by overheating due to excessive negative phase sequence currents flowing into them. Short-circuit faults are characterised by the short-circuit current and its components. These are the ac or symmetrical root mean square (rms) short-circuit current, dc short-circuit current or dc time constant or X/R ratio, and the overall asymmetrical short-circuit current. These are described in the next section.

To characterise the short-circuit infeed from one short-circuit source, a group of such sources or an entire system, the concept of short-circuit fault level is useful:

$$\text{Short-circuit fault level}_{(MVA)} = \sqrt{3} V_{\text{Phase–Phase(kV)}} I_{\text{rms(kA)}} \qquad (1.1)$$

where $I_{\text{rms(kA)}}$ is the rms short-circuit current infeed at the point of fault and $V_{\text{Phase–Phase(kV)}}$ is the prefault phase-to-phase voltage at the point of fault. For example, for an rms short-circuit current of 54 kA and a 404 kV prefault voltage in a 400 kV system, the short-circuit fault level is equal to 37.786 GVA. The system short-circuit fault level or infeed gives a measure of the strength or weakness of the system at the point of fault. For a given system short-circuit fault level or $\text{MVA}_{\text{Infeed}}$ at a busbar, the equivalent system impedance seen at the busbar in per unit on MVA_{Base} and phase-to-phase $V_{\text{kV}}^{\text{Prefault}}$ is given by

$$Z_{S(pu)} = \frac{\text{MVA}_{\text{Base}}}{\text{MVA}_{\text{Infeed}}} \times \frac{V_{\text{Prefault(kV)}}^2}{V_{\text{Base(kV)}}^2} \qquad (1.2)$$

where the definition of base quantities is presented in Section 1.7. Where the prefault and base voltages are equal, we have

$$Z_{S(pu)} = \frac{\text{MVA}_{\text{Base}}}{\text{MVA}_{\text{Infeed}}} \qquad (1.3)$$

High system strength is characterised by a high short-circuit fault level infeed and thus low system impedance, and vice versa. Z_S is also equal to the Thévenin's impedance.

Sometimes, an MVA figure is used to describe the short-circuit rating of a circuit-breaker in MVA. This practice is discouraged as it can easily lead to confusion and errors. For example, consider a 5000 MVA, 132 kV circuit-breaker that is to be used in a 110 kV system. The 5000 MVA rating becomes a meaningless and the correct figure should be

$$\sqrt{3} \times 110\,\text{kV} \times \frac{5000\,\text{MVA}}{\sqrt{3} \times 132\,\text{kV}} = 4166.7\,\text{MVA}$$

Another important characteristic of a short-circuit fault in the case of earth faults is its fault impedance. In general, in the event of a flashover, the fault impedance consists of the flashover arc resistance and the earthing impedance of the object to which the flashover occurs. For example, where this object is an overhead line's steel tower, the earthing impedance is that of the tower footing resistance in parallel with the line's earth wire impedance, where it exists (this is discussed in detail in Chapter 10). The fault impedance is generally neglected in higher voltage systems when calculating maximum short-circuit currents to obtain conservative results. However, earthing impedances are taken into account when calculating rise of earth potential for short-circuit faults in substations and on overhead lines' towers. In lower voltage distribution systems, many short-circuit faults occur due to a flashover from a line to a tree which may present a significant earthing impedance depending on earth resistivity. The resistance of the flashover arc can be estimated using the following empirical formula:

$$R_{\text{Arc}} = 1.81 \times \frac{\ell}{I_{\text{rms}}^{1.4}} \ \Omega \tag{1.4}$$

where ℓ is the length of the arc in m in still air and I_{rms} is the rms short-circuit current in kA. For example, for $\ell = 7.75$ m on a typical 400 kV overhead line and $I_{\text{rms}} = 30$ kA, we obtain $R_{\text{Arc}} = 0.12 \ \Omega$.

1.5 Terminology of short-circuit current waveform and current interruption

In order to calculate short-circuit current duties on power system equipment, it is important to define the terminology used in characterising the short-circuit current waveform. Figure 1.3 shows a simple balanced three-phase electric circuit where L and R are the circuit inductance and resistance for each phase, and L_e and R_e are the earth return path inductance and resistance, respectively.

The balanced three-phase voltage sources are given by

$$v_i(t) = \sqrt{2}V_{\text{rms}} \sin(\omega t + \varphi_i) \quad i = \text{r, y, b} \tag{1.5}$$

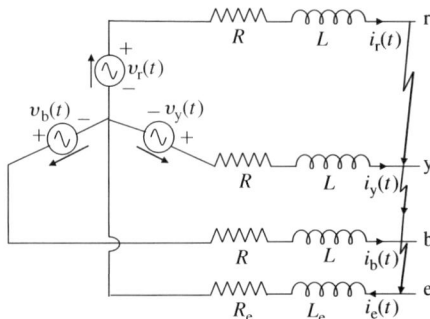

Figure 1.3   Basic balanced three-phase electric circuit with earth return

where V_{rms} is rms voltage magnitude, $\omega = 2\pi f$ in rad/s, f is power frequency in Hz and φ_i is voltage phase angle in rad given by

$$\varphi_y = \varphi_r - 2\pi/3 \quad \varphi_b = \varphi_r + 2\pi/3 \tag{1.6}$$

If a solid three-phase to earth connection or short-circuit fault is made simultaneously between phases r, y, b and earth e at $t = 0$, we can write

$$L\frac{di_i(t)}{dt} + Ri_i(t) + L_e\frac{di_e(t)}{dt} + R_e i_e(t) = v_i(t) \quad i = r, y, b \tag{1.7}$$

Substituting $i = r$, y, b in Equation (1.7) and adding the three equations, we obtain

$$L\frac{d}{dt}[i_r(t) + i_y(t) + i_b(t)] + R[i_r(t) + i_y(t) + i_b(t)]$$

$$+ 3L_e\frac{di_e(t)}{dt} + 3R_e i_e(t) = v_r(t) + v_y(t) + v_b(t) \tag{1.8}$$

Since the three-phase voltage sources are balanced we have

$$v_r(t) + v_y(t) + v_b(t) = 0 \tag{1.9}$$

Also, from Figure 1.3, we have

$$i_r(t) + i_y(t) + i_b(t) = i_e(t) \tag{1.10}$$

Therefore, substituting Equations (1.9) and (1.10) into Equation (1.8), we obtain

$$(L + 3L_e)\frac{di_e(t)}{dt} + (R + 3R_e)i_e(t) = 0 \tag{1.11}$$

The solution of Equation (1.11) is given by

$$i_e(t) = K \times \exp\left[\frac{-t}{\left(\frac{L+3L_e}{R+3R_e}\right)}\right] \tag{1.12}$$

where K is a constant that satisfies the initial conditions. Since the three-phase system is symmetrical and balanced, $i_e(t = 0) = 0$. Thus, Equation (1.12) gives $K = 0$ and $i_e(t) = 0$. That is, following a simultaneous three-phase short circuit, no current will flow in the earth return connection and the three fault currents $i_i(t)$ will flow independently as in single-phase circuits. Therefore, with $i_e(t) = 0$, the solution of Equation (1.7) is given by

$$i_i(t) = \sqrt{2}I_{rms}\left\{\sin\left[\omega t + \varphi_i - \tan^{-1}\left(\frac{\omega L}{R}\right)\right]\right.$$

$$\left. - \sin\left[\varphi_i - \tan^{-1}\left(\frac{\omega L}{R}\right)\right] \times \exp\left[\frac{-t}{(L/R)}\right]\right\} \tag{1.13}$$

where

$$I_{\text{rms}} = \frac{V_{\text{rms}}}{\sqrt{R^2 + (\omega L)^2}} \tag{1.14}$$

Equation (1.13) can be written as the sum of an ac component and a unidirectional dc component as follows:

$$i_i(t) = i_{i(ac)}(t) + i_{i(dc)}(t) \quad i = r, y, b \tag{1.15}$$

where

$$i_{i(ac)}(t) = \sqrt{2}I_{\text{rms}} \sin\left[\omega t + \varphi_i - \tan^{-1}\left(\frac{\omega L}{R}\right)\right] \tag{1.16}$$

and

$$i_{i(dc)}(t) = -\sqrt{2}I_{\text{rms}} \sin\left[\varphi_i - \tan^{-1}\left(\frac{\omega L}{R}\right)\right] \times \exp\left[\frac{-t}{(L/R)}\right] \tag{1.17}$$

In this analysis, the magnitude of the ac current component is constant because we assumed that the source inductance L is constant or time independent. This assumption is only generally valid if the location of short-circuit fault is electrically remote from electrical machines feeding short-circuit current into the fault. This aspect is discussed in detail later in this book where practical time-dependent source inductance is considered. The initial magnitude of the dc current component in any phase depends on the instant on the voltage waveform when the short circuit occurs, i.e. on φ_r and on the magnitude of the ac current component I_{rms}. The rate of decay of the dc current component in the three phases depends on the circuit time constant L/R or circuit X/R ratio where $X/R = \omega L/R$. Again, the assumption of a constant L results in a time-independent X/R ratio or constant rate of decay.

Short-circuit currents are detected by protection relays which initiate the interruption of these currents by circuit-breakers. Figure 1.4 shows a general asymmetrical short-circuit current waveform and the terminology used to describe the various current components as well as the short-circuit current interruption.

From Figure 1.4, the following quantities and terms are defined:

t_F = Instant of short-circuit fault.
Δt_1 = Protection relay time.
t_A = Instant of 'initial peak' of short-circuit current.
t_1 = Instant of energisation of circuit-breaker trip circuit.
Δt_2 = Circuit-breaker opening time.
t_2 = Instant of circuit-breaker contact separation = instant of arc initiation.
Δt_3 = Circuit-breaker current arcing time.
t_3 = Instant of final arc extinction = instant of short-circuit current interruption.
t_B = Instant of peak of major current loop just before current interruption.
$2\sqrt{2}I_k'' = 2.828I_k''$ = Theoretical current at the instant of short-circuit fault t_F where I_k'' is the rms short-circuit current at $t = t_F$.

(a)   Typical short-circuit current waveform

Time sequence to circuit-breaker current interruption

(b)

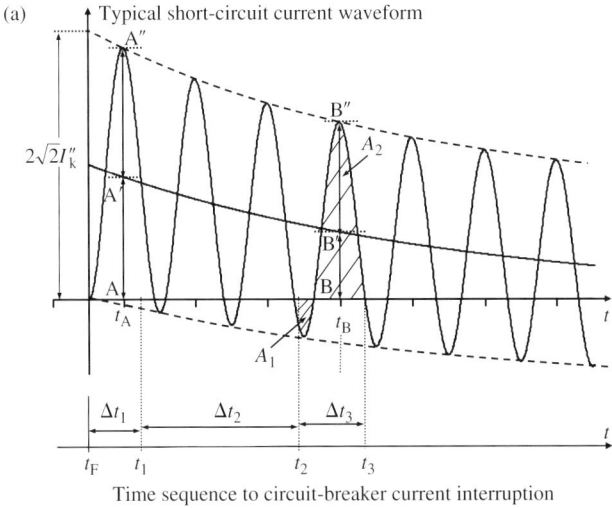

Contacts and arc-quenching mechanism

Figure 1.4 (a) Asymmetrical short-circuit current waveform and current interruption and (b) modern 420 kV SF6 circuit breaker

$AA'' = $ 'Initial peak' short-circuit current at $t = t_A$ and is denoted i_p. i_p is also termed '½ cycle peak' current, 'peak make' current or 'making' current. $AA'' = AA' + A'A''$ where $AA' = $ magnitude of dc current component at $t = t_A$ and $A'A'' = $ peak ac current component at $t = t_A$ and is equal to $\sqrt{2} \times$ rms current at $t = t_A$.

$BB'' = $ Peak short-circuit current at $t = t_B$ and is denoted i_b. $BB'' = BB' + B'B''$ where $BB' = $ magnitude of dc current component at $t = t_B$ and $B'B'' = $ peak ac current component at $t = t_B$ and is equal to $\sqrt{2} \times I_b$, where I_b is rms current at $t = t_B$.

Percentage dc current component at $t = t_B$ given by $(\text{BB}' \times 100)/\text{B}'\text{B}''$.
$A_1 + A_2 =$ Area that corresponds to the arc energy measured from instant of contact
separation or arc initiation t_2 to instant of final arc extinction or current
interruption t_3.

1.6 Effects of short-circuit currents on equipment

1.6.1 Thermal effects

Short-circuit currents flowing through the conductors of various power system equipment create thermal effects on conductors and equipment due to heating and excess energy input over time as measured by I^2T where I is the short-circuit current magnitude and T is the short-circuit current duration. Because the short-circuit duration, including short circuits cleared by protection in back-up clearance times, is quite short, the heat loss from conductors during the short circuit is usually very low. Generally, both ac and dc components of the short-circuit current contribute to the thermal heating of conductors. Extreme values of the time constant of the dc short-circuit current component may be up to 150 ms so that by 500 ms from the instant of short circuit, the dc component is nearly vanished and nearly all its generated heat will have dissipated by around 1 s. For the ac current component, the heat dissipation depends on the ratio of the initial subtransient rms current to steady state rms current. For a typical ratio of about 2.5, the amount of conductor heat dissipation will be around 50% at 1 s and nearly 65% at 3 s. The three-phase short-circuit fault normally gives rise to the highest thermal effect on equipment. For cables of larger conductor sizes, the thermal limit is usually imposed by the sheath/armour as opposed to the core conductor under short-circuit earth fault conditions. During the making of high short-circuit currents by inadequately rated circuit-breakers, their contacts may weld together encouraged by a prestriking flashover arc and possibly contact popping.

1.6.2 Mechanical effects

Short-circuit currents flowing through the conductors of various power system equipment create electromagnetic forces and mechanical stresses on equipment such as overhead line conductors, busbars and their supports, cables and transformer windings. Mechanical forces on transformer windings are both radial and axial. The radial force is a repulsion force between the inner and outer windings and tends to crush the inner winding towards the transformer core and burst the outer winding outwards. The axial force tends to displace the windings, or part of the winding, with respect to each other. The transformer windings must be designed to withstand the mechanical forces created by the short-circuit currents. The electromagnetic forces produced by short-circuit currents in three-core unarmoured cables tend to repel the cores from each other and could burst the

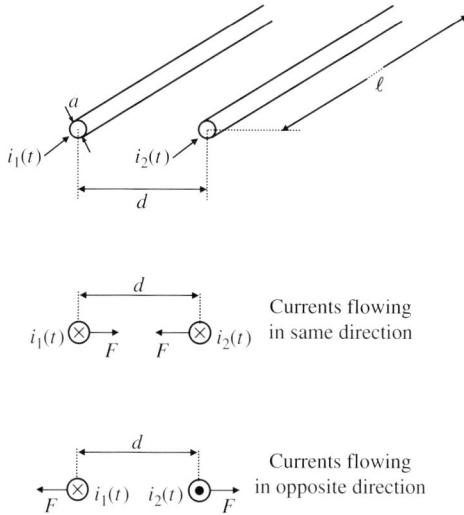

Figure 1.5 Electromagnetic forces on two parallel round current carrying conductors

cable altogether leading to insulation damage if the cores are not adequately bound.

The electromagnetic forces acting on two parallel, round conductors carrying currents i_1 and i_2, as shown in Figure 1.5, are given by

$$F(t) = \frac{\mu_o i_1(t) i_2(t)}{2\pi} \left[\sqrt{\left(\frac{\ell}{d}\right)^2 + 1} - 1 \right] \text{ N} \tag{1.18}$$

where d is the distance between the centres of the conductors in m and ℓ is the conductor length between supports in m. $\mu_o = 4\pi 10^{-7}$ H/m is the permeability of vacuum. Since $i_1(t)$ and $i_2(t)$ are instantaneous currents in amps as given in Equation (1.13), the electromagnetic force $F(t)$ is clearly time dependent. In the general case, $F(t)$ will contain a dc component, a power frequency component and a double-power frequency component, i.e. a 100 Hz component in a 50 Hz system. In practical installations, $\ell \gg d$ and $d \gg a$ where a is the conductor diameter in m. Therefore, Equation (1.18) reduces to

$$F(t) = \frac{\mu_o i_1(t) i_2(t)}{2\pi} \frac{\ell}{d} \text{ N} \tag{1.19}$$

Where the currents flow in the two conductors in the same direction, the forces are compressive, i.e. pulling the conductors together but if the currents are flowing in opposite directions, the forces would repel the conductors away from each other. This is illustrated in Figure 1.5. To illustrate how this equation may be

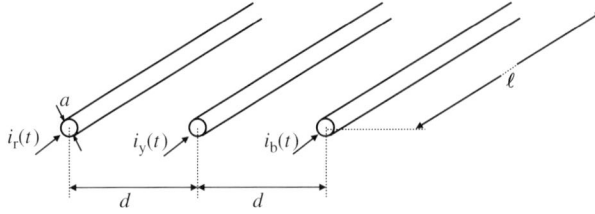

Figure 1.6   Electromagnetic forces on three-phase parallel round conductors

used, consider an overhead line with a twin conductor bundle per phase (overhead lines are discussed in detail in Chapter 3) and $d = 0.5$ m. The short-circuit current satisfies Equation (1.13) and has an rms value of 20 kA per phase and an X/R ratio equal to 15. The maximum attraction forces on each conductor per metre produced by the first peak of the asymmetrical short-circuit current are to be calculated. Individual conductor rms current is $20/2 = 10$ kA. Using Equation (1.13), the maximum value of the initial peak current is calculated assuming maximum dc current offset which corresponds to a short-circuit fault that occurs at a voltage phase angle $\varphi = \tan^{-1}\left(\frac{X}{R}\right) - \frac{\pi}{2}$. Therefore, using Equation (1.13), the initial peak current is equal to

$$i_{p1} = i_{p2} = \sqrt{2} \times 10 \times \left[1 + \exp\left(\frac{-\pi}{15}\right)\right] = 2.56 \times 10 = 25.6\,\text{kA}$$

The attraction forces on each conductor are calculated as

$$F = \frac{4\pi 10^{-7}(25.6 \times 10^3)^2}{2\pi}\frac{1}{0.5} = 262\,\text{N/m}$$

The electromagnetic forces acting on three rigid, parallel and round conductors, shown in Figure 1.6, under a balanced three-phase short-circuit fault condition can be calculated approximately using the following equations:

$$F_r(t) = \frac{\mu_o \ell}{4\pi d}[2i_r(t)i_y(t) + i_r(t)i_b(t)]\,\text{N} \qquad (1.20)$$

$$F_y(t) = \frac{\mu_o \ell}{4\pi d}[2i_y(t)i_b(t) - 2i_y(t)i_r(t)]\,\text{N} \qquad (1.21)$$

$$F_b(t) = \frac{-\mu_o \ell}{4\pi d}[i_b(t)i_r(t) + 2i_b(t)i_y(t)]\,\text{N} \qquad (1.22)$$

In Equations (1.20)–(1.22), $i_r(t)$, $i_y(t)$ and $i_b(t)$ are the instantaneous currents given in Equation (1.13) in amps. Substituting these currents into Equations (1.20)–(1.22), it can be shown that the maximum force occurs on the middle conductor and is given by

$$F_y = \frac{\sqrt{3}\mu_0 \ell}{4\pi d} i_p^2 \, \text{N} \qquad (1.23)$$

where i_p is the initial peak short-circuit current that corresponds to AA″ in Figure 1.4. During a phase-to-phase short-circuit fault, it can be shown that the maximum electromagnetic force between the two faulted conductors is given by

$$F_y = \frac{\mu_0 \ell}{2\pi d} i_p^2 \, \text{N} \qquad (1.24)$$

where i_p is the initial peak short-circuit current.

Where the conductors are not round, e.g. rectangular, d in all the electromagnetic force equations is replaced by an effective distance d_{eff} equal to the effective geometric mean distance between the rectangular conductors.

1.7 Per-unit analysis of power systems

1.7.1 General

Steady state network analysis on initially balanced three-phase networks generally employ complex phasors and involve the calculation of active and reactive power flows, voltages and currents in the network. Usually, such calculations are carried out using per-unit values of actual physical quantities such as voltages, currents or impedances. The advantages of per-unit systems of analysis are generally covered in most introductory power systems textbooks and will not be repeated. However, as we will discuss later in this book, the use of actual physical units such as volts, amps, ohms and Siemens is more advantageous in multiphase network analysis.

1.7.2 Single-phase systems

The per-unit system of analysis is based on the application of Ohm's law to a single impedance or admittance as illustrated in Figure 1.7(a).

Using actual physical units of kilovolts, kiloamps, ohms and Siemens, the voltage drop across the impedance and injected current into the admittance are given by

$$V_{\text{Actual,kV}} = Z_{\text{Actual},\Omega} \times I_{\text{Actual,kA}} \qquad (1.25a)$$

$$I_{\text{Actual,kA}} = Y_{\text{Actual,S}} \times V_{\text{Actual,kV}} \qquad (1.25b)$$

where V, I, Z and Y are complex phasors representing actual physical quantities of voltage, current, impedance and admittance, respectively. To calculate a per-unit value for each of these quantities, a corresponding base quantity must be defined

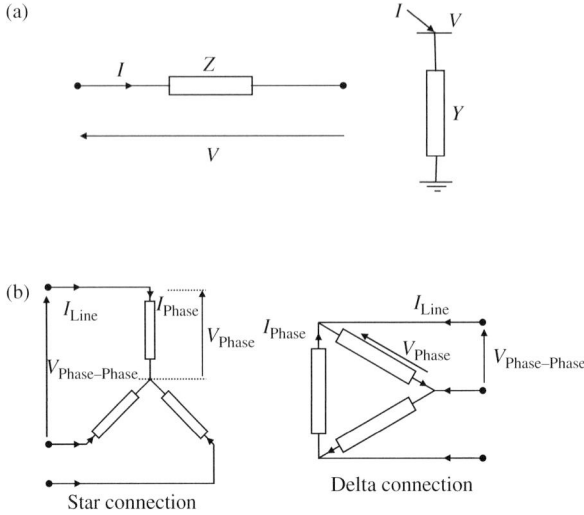

Figure 1.7   Per-unit analysis of (a) single-phase impedance and admittance and (b) three-phase connections

first. Let $V_{Base,kV}$ and $I_{Base,kA}$ be the base voltage and base current, respectively. Therefore, Equations (1.25a) and (1.25b) can be written as follows:

$$\frac{V_{Actual,kV}}{V_{Base,kV}} = \frac{Z_{Actual,\Omega}}{\left(\frac{V_{Base,kV}}{I_{Base,kA}}\right)} \times \frac{I_{Actual,kA}}{I_{Base,kA}}$$

and

$$\frac{I_{Actual,kA}}{I_{Base,kA}} = \frac{Y_{Actual,S}}{\left(\frac{I_{Base,kA}}{V_{Base,kV}}\right)} \times \frac{V_{Actual,kV}}{V_{Base,kV}}$$

or

$$V_{pu} = Z_{pu} \times I_{pu} \qquad\qquad (1.26a)$$

and

$$I_{pu} = Y_{pu} \times V_{pu} \qquad\qquad (1.26b)$$

where

$$V_{pu} = \frac{V_{Actual,kV}}{V_{Base,kV}} \qquad I_{pu} = \frac{I_{Actual,kA}}{I_{Base,kA}} \qquad (1.27)$$

and

$$Z_{\text{pu}} = \frac{Z_{\text{Actual},\Omega}}{Z_{\text{Base},\Omega}} \quad Z_{\text{Base},\Omega} = \frac{V_{\text{Base,kV}}}{I_{\text{Base,kA}}} \tag{1.28}$$

$$Y_{\text{pu}} = \frac{Y_{\text{Actual,S}}}{Y_{\text{Base,S}}} \quad Y_{\text{Base,S}} = \frac{I_{\text{Base,kA}}}{V_{\text{Base,kV}}} = \frac{1}{Z_{\text{Base},\Omega}} \tag{1.29}$$

By defining $V_{\text{Base,kV}}$ and $I_{\text{Base,kA}}$, $Z_{\text{Base},\Omega}$ and $Y_{\text{Base,S}}$ are also defined as shown in Equations (1.28) and (1.29), and the per-unit values of the actual voltage and current are also defined in Equation (1.27). It should be noted that the base quantities $V_{\text{Base,kV}}$ and $I_{\text{Base,kA}}$ are defined as real numbers so that the phase angles of V_{pu} and I_{pu} remain unchanged from $V_{\text{Actual,kV}}$ and $I_{\text{Actual,kA}}$, respectively. Using Equation (1.27), we have

$$V_{\text{Actual,kV}} \times I_{\text{Actual,kA}} = V_{\text{Base,kV}} V_{\text{pu}} \times I_{\text{Base,kA}} I_{\text{pu}}$$

or

$$V_{\text{pu}} \times I_{\text{pu}} = \frac{V_{\text{Actual,kV}} \times I_{\text{Actual,kA}}}{V_{\text{Base,kV}} \times I_{\text{Base,kA}}}$$

or

$$\text{per-unit MVA} = \frac{\text{MVA}_{\text{Actual}}}{\text{MVA}_{\text{Base}}} \tag{1.30a}$$

where

$$\text{MVA}_{\text{Actual}} = V_{\text{Actual,kV}} \times I_{\text{Actual,kA}}$$

$$\text{MVA}_{\text{Base}} = V_{\text{Base,kV}} \times I_{\text{Base,kA}} \quad \text{and} \quad \text{per-unit MVA} = V_{\text{pu}} \times I_{\text{pu}} \tag{1.30b}$$

We note that by defining $V_{\text{Base,kV}}$ and $I_{\text{Base,kA}}$, MVA_{Base} is also defined according to Equation (1.30b). In practical power system analysis, it is more convenient to define or choose MVA_{Base} and $V_{\text{Base,kV}}$ and calculate $I_{\text{Base,kV}}$ if required. Therefore, using Equations (1.27) and (1.30b), we have

$$I_{\text{pu}} = \frac{I_{\text{Actual,kA}}}{\left(\dfrac{\text{MVA}_{\text{Base}}}{V_{\text{Base,kV}}}\right)} \quad \text{and} \quad V_{\text{pu}} = \frac{V_{\text{Actual,kA}}}{\left(\dfrac{\text{MVA}_{\text{Base}}}{I_{\text{Base,kA}}}\right)} \tag{1.31}$$

Also, using Equations (1.28) and (1.30b), we can write

$$Z_{\text{Base},\Omega} = \frac{V_{\text{Base,kV}}}{I_{\text{Base,kA}}} \times \frac{V_{\text{Base,kV}}}{V_{\text{Base,kV}}} = \frac{V_{\text{Base,kV}}^2}{V_{\text{Base,kV}} I_{\text{Base,kA}}}$$

and

$$Y_{\text{Base,S}} = \frac{1}{Z_{\text{Base},\Omega}} = \frac{V_{\text{Base,kV}} I_{\text{Base,kA}}}{V_{\text{Base,kV}}^2}$$

or using Equation (1.30b), we have

$$Z_{Base,\Omega} = \frac{V^2_{Base,kV}}{MVA_{Base}} \quad \text{and} \quad Y_{Base,S} = \frac{MVA_{Base}}{V^2_{Base,kV}} \tag{1.32}$$

Substituting Equation (1.32) in Equations (1.28) and (1.29), we obtain

$$Z_{pu} = \frac{Z_{Actual,\Omega}}{\left(\frac{V^2_{Base,kV}}{MVA_{Base}}\right)} \quad \text{and} \quad Y_{pu} = \frac{Y_{Actual,S}}{\left(\frac{MVA_{Base}}{V^2_{Base,kV}}\right)} \tag{1.33}$$

To convert per-unit values to per cent, the per-unit values are multiplied by 100.

1.7.3 Change of base quantities

The per-unit values that are calculated by generator manufacturers and sometimes transformer manufacturers are usually based on the rated voltage and rated apparent power of the equipment. However, transmission and distribution system analysis using per-unit values are based on a single MVA base of typically 100 MVA. In industrial power systems and power station auxiliary supply systems, 10 MVA base is generally found more convenient. For transformers, it is almost always the case that the rated voltage of at least one of the windings is not equal to the base voltage of the network to which the winding is connected. This is discussed in detail in Chapter 4. To determine how, in general, per-unit values are calculated using new base quantities, let the two base quantities be MVA_{Base-1}, $V_{Base-1,kV}$ and MVA_{Base-2}, $V_{Base-2,kV}$, and it is required to convert from Base-1 to the new Base-2. Using Equation (1.27), we can write

$$V_{pu-1} = \frac{V_{Actual,kV}}{V_{Base-1,kV}} \quad \text{and} \quad V_{pu-2} = \frac{V_{Actual,kV}}{V_{Base-2,kV}}$$

giving

$$V_{pu-2} = \frac{V_{Base-1,kV}}{V_{Base-2,kV}} \times V_{pu-1} \tag{1.34}$$

Using Equation (1.31), we have

$$I_{pu-1} = \frac{I_{Actual,kA}}{\left(\frac{MVA_{Base-1}}{V_{Base-1,kV}}\right)} \quad \text{and} \quad I_{pu-2} = \frac{I_{Actual,kA}}{\left(\frac{MVA_{Base-2}}{V_{Base-2,kV}}\right)}$$

giving

$$I_{pu-2} = \frac{V_{Base-2,kV}}{V_{Base-1,kV}} \times \frac{MVA_{Base-1}}{MVA_{Base-2}} \times I_{pu-1} \tag{1.35a}$$

or using Equation (1.30b), we have

$$I_{pu\text{-}2} = \frac{I_{Base\text{-}1,kA}}{I_{Base\text{-}2,kA}} \times I_{pu\text{-}1} \tag{1.35b}$$

From Equation (1.33), we have

$$Z_{pu\text{-}1} = \frac{Z_{Actual,\Omega}}{\left(\dfrac{V_{Base\text{-}1,kV}^2}{MVA_{Base\text{-}1}}\right)} \quad \text{and} \quad Z_{pu\text{-}2} = \frac{Z_{Actual,\Omega}}{\left(\dfrac{V_{Base\text{-}2,kV}^2}{MVA_{Base\text{-}2}}\right)}$$

or

$$Z_{pu\text{-}2} = \left(\frac{V_{Base\text{-}1,kV}^2}{V_{Base\text{-}2,kV}^2}\right)\left(\frac{MVA_{Base\text{-}2}}{MVA_{Base\text{-}1}}\right) Z_{pu\text{-}1} \tag{1.36}$$

and similarly for the per-unit admittance

$$Y_{pu\text{-}2} = \left(\frac{V_{Base\text{-}2,kV}^2}{V_{Base\text{-}1,kV}^2}\right)\left(\frac{MVA_{Base\text{-}1}}{MVA_{Base\text{-}2}}\right) Y_{pu\text{-}1} \tag{1.37}$$

1.7.4 Three-phase systems

The above analysis of single-phase systems can be easily extended to three-phase systems or equipment using the following basic relations that apply irrespective of whether the equipment is star or delta connected

$$MVA_{3\text{-}Phase} = 3 \times MVA_{1\text{-}Phase} \tag{1.38}$$

and

$$MVA_{3\text{-}Phase} = \sqrt{3} \times V_{Phase\text{--}Phase,kV} \times I_{Line,kA} \tag{1.39}$$

where $MVA_{3\text{-}Phase}$ is three-phase base apparent power, $MVA_{1\text{-}Phase}$ is one-phase base apparent power, $V_{Phase\text{--}Phase,kV}$ is phase-to-phase base voltage and $I_{Line,kA}$ is base line current. The relationships between base voltages and base currents for star- and delta-connected three-phase equipment is shown in Figure 1.7(b):

$$\text{Star: } V_{Phase\text{--}Phase,kV} = \sqrt{3} \times V_{Phase,kV} \quad \text{and} \quad I_{Line,kA} = I_{Phase,kA} \tag{1.40}$$

$$\text{Delta: } V_{Phase\text{--}Phase,kV} = V_{Phase,kV} \quad \text{and} \quad I_{Line,kA} = \sqrt{3} \times I_{Phase,kA} \tag{1.41}$$

where $V_{Phase,kV}$ and $I_{Phase,kA}$ correspond to the base voltage and current quantities as used in Section 1.7.2 for a single-phase system. Using Equations (1.38) and

(1.40) in Equations (1.31), (1.33), (1.34), (1.35b), (1.36) and (1.37), we obtain

$$Z_{pu} = \frac{Z_{Actual,\Omega}}{\left(\dfrac{V^2_{(Phase-Phase)Base,kV}}{MVA_{(3-Phase)Base}}\right)} \qquad Y_{pu} = \frac{Y_{Actual,S}}{\left(\dfrac{MVA_{(3-Phase)Base}}{V^2_{(Phase-Phase)Base,kV}}\right)} \qquad (1.42)$$

$$V_{pu} = \frac{V_{(Phase-Phase)Actual,kV}}{\left(\dfrac{MVA_{(3-Phase)Base}}{\sqrt{3}I_{(Line)Base,kA}}\right)} \qquad I_{pu} = \frac{I_{(Line)Actual,kA}}{\left(\dfrac{MVA_{(3-Phase)Base}}{\sqrt{3}V_{(Phase-Phase)Base,kV}}\right)} \qquad (1.43)$$

$$V_{pu-2} = \frac{V_{(Phase-Phase)Base-1,kV}}{V_{(Phase-Phase)Base-2,kV}} \times V_{pu-1} \qquad I_{pu-2} = \frac{I_{Line(Base-1),kA}}{I_{Line(Base)-2,kA}} \times I_{pu-1}$$

$$(1.44)$$

$$Z_{pu-2} = \left(\frac{V^2_{(Phase-Phase)Base-1,kV}}{V^2_{(Phase-Phase)Base-2,kV}}\right)\left(\frac{MVA_{(3-Phase)Base-2}}{MVA_{(3-Phase)Base-1}}\right)Z_{pu-1} \qquad (1.45)$$

$$Y_{pu-2} = \left(\frac{V^2_{(Phase-Phase)Base-2,kV}}{V^2_{(Phase-Phase)Base-1,kV}}\right)\left(\frac{MVA_{(3-Phase)Base-1}}{MVA_{(3-Phase)Base-2}}\right)Y_{pu-1} \qquad (1.46)$$

1.7.5 Mutually coupled systems having different operating voltages

In many electrical power systems, double-circuit overhead lines are used for the transport of bulk power. The modelling of these lines is presented in detail in Chapter 3. For now, we state that these lines consist of two three-phase circuits strung on the same tower. In this section, we present the general case of the derivation of the per-unit values of the mutual impedance and mutual susceptance of two mutually coupled circuits operating at different voltages. The analysis is general and applies also to mutually coupled cable circuits and mutually coupled transformer windings.

Base and per-unit values of mutual inductive impedance

Figure 1.8 shows two mutually coupled inductive circuits having different design and operating voltages. The series voltage drop across each circuit is given by

$$V_1 - V'_1 = Z_{S1}I_1 + Z_M I_2 \qquad (1.47a)$$
$$V_2 - V'_2 = Z_{S2}I_2 + Z_M I_1 \qquad (1.47b)$$

where V is in volts, I is in amps, Z_{S1} and Z_{S2} are the self-impedances of circuits 1 and 2, respectively, in ohms, and Z_M is the mutual impedance between the two

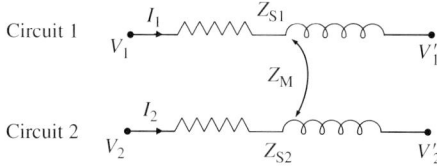

Figure 1.8 Per-unit analysis of two mutually coupled inductive circuits having different design and operating voltages

circuits in ohms. Using the same base apparent power for both circuits, let the base quantities of circuits 1 and 2 be defined as follows:

$$V_{1(\text{pu})} = \frac{V_1}{V_{1(\text{Base})}} \quad I_{1(\text{pu})} = \frac{I_1}{I_{1(\text{Base})}} \quad Z_{S1(\text{Base})} = \frac{V_{1(\text{Base})}}{I_{1(\text{Base})}} \quad Z_{S1(\text{pu})} = \frac{Z_{S1}}{Z_{S1(\text{Base})}}$$

$$V_{2(\text{pu})} = \frac{V_2}{V_{2(\text{Base})}} \quad I_{2(\text{pu})} = \frac{I_2}{I_{2(\text{Base})}} \quad Z_{S2(\text{Base})} = \frac{V_{2(\text{Base})}}{I_{2(\text{Base})}} \quad Z_{S2(\text{pu})} = \frac{Z_{S2}}{Z_{S2(\text{Base})}}$$

$$S_{(\text{Base})} = V_{1(\text{Base})}I_{1(\text{Base})} = V_{2(\text{Base})}I_{2(\text{Base})} \tag{1.48}$$

We note that for now, we have not yet defined the base or per-unit values for Z_M. Using Equation (1.48) and substituting the actual quantities, e.g. $V_1 = V_{1(\text{pu})} \times V_{1(\text{Base})}$, into Equation (1.47), we obtain

$$V_{1(\text{Base})}[V_{1(\text{pu})} - V'_{1(\text{pu})}] = Z_{S1(\text{pu})} \frac{V_{1(\text{Base})}}{I_{1(\text{Base})}} I_{1(\text{Base})}I_{1(\text{pu})} + Z_M I_{2(\text{Base})}I_{2(\text{pu})} \tag{1.49a}$$

and

$$V_{2(\text{Base})}[V_{2(\text{pu})} - V'_{2(\text{pu})}] = Z_{S2(\text{pu})} \frac{V_{2(\text{Base})}}{I_{2(\text{Base})}} I_{2(\text{Base})}I_{2(\text{pu})} + Z_M I_{1(\text{Base})}I_{1(\text{pu})} \tag{1.49b}$$

Dividing Equation (1.49a) by $V_{1(\text{Base})}$ and Equation (1.49b) by $V_{2(\text{Base})}$, we obtain

$$V_{1(\text{pu})} - V'_{1(\text{pu})} = Z_{S1(\text{pu})}I_{1(\text{pu})} + Z_M \frac{I_{2(\text{Base})}}{V_{1(\text{Base})}} I_{2(\text{pu})} \tag{1.50a}$$

and

$$V_{2(\text{pu})} - V'_{2(\text{pu})} = Z_{S2(\text{pu})}I_{2(\text{pu})} + Z_M \frac{I_{1(\text{Base})}}{V_{2(\text{Base})}} I_{1(\text{pu})} \tag{1.50b}$$

or

$$V_{1(\text{pu})} - V'_{1(\text{pu})} = Z_{S1(\text{pu})}I_{1(\text{pu})} + Z_{M(\text{pu})}I_{2(\text{pu})} \tag{1.51a}$$

$$V_{2(\text{pu})} - V'_{2(\text{pu})} = Z_{S2(\text{pu})}I_{2(\text{pu})} + Z_{M(\text{pu})}I_{1(\text{pu})} \tag{1.51b}$$

From Equations (1.50) and (1.51), the per-unit value of Z_M is defined as

$$Z_{M(pu)} = \frac{Z_M}{Z_{M(Base)}} = \frac{Z_M}{\left(\frac{V_{1(Base)}}{I_{2(Base)}}\right)} = \frac{Z_M}{\left(\frac{V_{2(Base)}}{I_{1(Base)}}\right)} \tag{1.52a}$$

And the base impedance of Z_M is defined as

$$Z_{M(Base)} = \frac{V_{1(Base)}}{I_{2(Base)}} = \frac{V_{2(Base)}}{I_{1(Base)}} \tag{1.52b}$$

From Equation (1.48), $S_{(Base)} = V_{1(Base)}I_{1(Base)} = V_{2(Base)}I_{2(Base)}$ or

$$I_{2(Base)} = \frac{S_{(Base)}}{V_{2(Base)}} \quad \text{and} \quad I_{1(Base)} = \frac{S_{(Base)}}{V_{1(Base)}} \tag{1.53}$$

Substituting Equation (1.53) into Equation (1.52b), and using $S_{(Base)}$ in MVA, $V_{1(Base)}$ and $V_{2(Base)}$ in kV, we obtain

$$Z_{M(Base)-\Omega} = \frac{V_{1(Base)-kV} V_{2(Base)-kV}}{S_{(Base)-MVA}} \tag{1.54}$$

Substituting Equation (1.54) into Equation (1.52a), we have

$$Z_{M(pu)} = \frac{Z_{M(\Omega)}}{\left(\frac{V_{1(Base)-kV} V_{2(Base)-kV}}{S_{(Base)-MVA}}\right)} \tag{1.55}$$

In the above equations, if the voltages are phase to earth then the apparent power is single-phase power but if the voltages used are phase to phase, then the apparent power to be used is the three-phase value.

An equivalent alternative approach is to define a per-unit mutual impedance on the base of each circuit as follows:

$$Z_{M1(pu)} = \frac{Z_{M(\Omega)}}{\left(\frac{V_{1(Base)-kV}}{I_{1(Base)-kA}}\right)} \quad \text{and} \quad Z_{M2(pu)} = \frac{Z_{M(\Omega)}}{\left(\frac{V_{2(Base)-kV}}{I_{2(Base)-kA}}\right)} \tag{1.56}$$

Now

$$Z_{M1(pu)} \times Z_{M2(pu)} = \frac{Z_{M(\Omega)}}{\left(\frac{V_{1(Base)-kV}}{I_{1(Base)-kA}}\right)} \times \frac{Z_{M(\Omega)}}{\left(\frac{V_{2(Base)-kV}}{I_{2(Base)-kA}}\right)} = \frac{Z_{M(\Omega)}}{\left(\frac{V_{1(Base)-kV}}{I_{2(Base)-kA}}\right)} \times \frac{Z_{M(\Omega)}}{\left(\frac{V_{2(Base)-kV}}{I_{1(Base)-kA}}\right)}$$

and using Equation (1.52b), we obtain $Z_{M1(pu)} \times Z_{M2(pu)} = Z_{M(pu)}Z_{M(pu)} = Z^2_{M(pu)}$ or

$$Z_{M(pu)} = \sqrt{Z_{M1(pu)} \times Z_{M2(pu)}} \tag{1.57}$$

Equations (1.55) and (1.57) are identical.

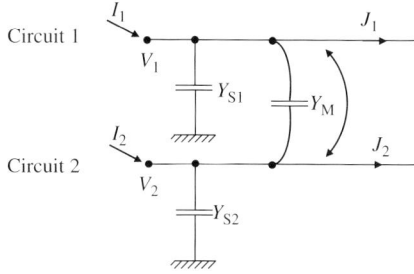

Figure 1.9 Per-unit analysis of two mutually coupled capacitive circuits having different design and operating voltages

Base and per-unit values of mutual capacitive admittance

Figure 1.9 shows two mutually coupled capacitive circuits having different design and operating voltages. Using self and mutual admittance terms to maintain generality, the injected current into each circuit at one end of the circuit is given by

$$I_1 = Y_{S1}V_1 + Y_M(V_1 - V_2) + J_1 \tag{1.58a}$$

$$I_2 = Y_{S2}V_2 + Y_M(V_2 - V_1) + J_2 \tag{1.58b}$$

where V is in volts, I is in amps, Y_{S1} and Y_{S2} are the self-admittances of circuits 1 and 2 in Siemens, respectively, and Y_M is the mutual admittance between the two circuits in Siemens. Using the same base apparent power for both circuits, let the base quantities of circuits 1 and 2 be defined as follows:

$$V_{1(pu)} = \frac{V_1}{V_{1(Base)}} \quad I_{1(pu)} = \frac{I_1}{I_{1(Base)}} \quad Y_{S1(Base)} = \frac{I_{1(Base)}}{V_{1(Base)}} \quad Y_{S1(pu)} = \frac{Y_{S1}}{Y_{S1(Base)}}$$

$$V_{2(pu)} = \frac{V_2}{V_{2(Base)}} \quad I_{2(pu)} = \frac{I_2}{I_{2(Base)}} \quad Y_{S2(Base)} = \frac{I_{2(Base)}}{V_{2(Base)}} \quad Y_{S2(pu)} = \frac{Y_{S2}}{Y_{S2(Base)}}$$

$$S_{(Base)} = V_{1(Base)}I_{1(Base)} = V_{2(Base)}I_{2(Base)} \tag{1.59}$$

We note that for now, we have not yet defined the base or per-unit values for Y_M. Using Equation (1.59) in Equation (1.58), we obtain

$$I_{1(Base)}I_{1(pu)} = [Y_{S1(pu)}Y_{S1(Base)} + Y_M]V_{1(pu)}V_{1(Base)}$$
$$- Y_M V_{2(pu)}V_{2(Base)} + I_{1(Base)}J_{1(pu)} \tag{1.60a}$$

and

$$I_{2(Base)}I_{2(pu)} = [Y_{S2(pu)}Y_{S2(Base)} + Y_M]V_{2(pu)}V_{2(Base)}$$
$$- Y_M V_{1(pu)}V_{1(Base)} + I_{2(Base)}J_{2(pu)} \tag{1.60b}$$

Dividing Equation (1.60a) by $I_{1(\text{Base})}$ and Equation (1.60b) by $I_{2(\text{Base})}$, we obtain

$$I_{1(\text{pu})} = \left[Y_{S1(\text{pu})} + \frac{Y_M}{\left(\frac{I_{1(\text{Base})}}{V_{1(\text{Base})}} \right)} \right] V_{1(\text{pu})} - \frac{Y_M}{\left(\frac{I_{1(\text{Base})}}{V_{2(\text{Base})}} \right)} V_{2(\text{pu})} + J_{1(\text{pu})} \tag{1.61a}$$

and

$$I_{2(\text{pu})} = \left[Y_{S2(\text{pu})} + \frac{Y_M}{\left(\frac{I_{2(\text{Base})}}{V_{2(\text{Base})}} \right)} \right] V_{2(\text{pu})} - \frac{Y_M}{\left(\frac{I_{2(\text{Base})}}{V_{1(\text{Base})}} \right)} V_{1(\text{pu})} + J_{2(\text{pu})} \tag{1.61b}$$

or

$$I_{1(\text{pu})} = [Y_{S1(\text{pu})} + Y_{M1(\text{pu})}] V_{1(\text{pu})} - Y_{M(\text{pu})} V_{2(\text{pu})} + J_{1(\text{pu})} \tag{1.62a}$$

and

$$I_{2(\text{pu})} = [Y_{S2(\text{pu})} + Y_{M2(\text{pu})}] V_{2(\text{pu})} - Y_{M(\text{pu})} V_{2(\text{pu})} + J_{2(\text{pu})} \tag{1.62b}$$

where

$$Y_{M(\text{pu})} = \frac{Y_M}{\left(\frac{I_{1(\text{Base})}}{V_{2(\text{Base})}} \right)} = \frac{Y_M}{\left(\frac{I_{2(\text{Base})}}{V_{1(\text{Base})}} \right)} = \frac{Y_M}{Y_{M(\text{Base})}} \tag{1.63a}$$

and the base admittance of Y_M is given by

$$Y_{M(\text{Base})} = \frac{I_{1(\text{Base})}}{V_{2(\text{Base})}} = \frac{I_{2(\text{Base})}}{V_{1(\text{Base})}} \tag{1.63b}$$

Also

$$Y_{M1(\text{pu})} = \frac{Y_M}{\left(\frac{I_{1(\text{Base})}}{V_{1(\text{Base})}} \right)} \qquad Y_{M2(\text{pu})} = \frac{Y_M}{\left(\frac{I_{2(\text{Base})}}{V_{2(\text{Base})}} \right)} \tag{1.64}$$

Using $S_{\text{Base}} = V_{1(\text{Base})} I_{1(\text{Base})} = V_{2(\text{Base})} I_{2(\text{Base})}$ in Equation (1.63b), $S_{(\text{Base})}$ in MVA, $V_{1(\text{Base})}$ and $V_{2(\text{Base})}$ in kV, we obtain

$$Y_{M(\text{Base})\text{-}S} = \frac{S_{(\text{Base})\text{-MVA}}}{V_{1(\text{Base})\text{-kV}} V_{2(\text{Base})\text{-kV}}} \tag{1.65}$$

Substituting Equation (1.65) in Equation (1.63a), we obtain

$$Y_{M(\text{pu})} = \frac{Y_{M(S)}}{\left(\frac{S_{(\text{Base})\text{-MVA}}}{V_{1(\text{Base})\text{-kV}} V_{2(\text{Base})\text{-kV}}} \right)} \tag{1.66}$$

In the above equations, if the voltages are phase to earth then the apparent power is single-phase power but if the voltages used are phase to phase, then the apparent power to be used is the three-phase value. From Equation (1.64), we have

$$Y_{M1(pu)} \times Y_{M2(pu)} = \frac{Y_M}{\left(\dfrac{I_{1(Base)-kA}}{V_{2(Base)-kV}}\right)} \times \frac{Y_M}{\left(\dfrac{I_{2(Base)-kV}}{V_{1(Base)-kA}}\right)} = Y_{M(pu)}Y_{M(pu)} = Y^2_{M(pu)}$$

or

$$Y_{M(pu)} = \sqrt{Y_{M1(pu)} \times Y_{M2(pu)}} \qquad (1.67)$$

For practical overhead line analysis at power frequency, e.g. 50 or 60 Hz, $Y_{M(pu)} = jB_{M(pu)}$, $Y_{M1(pu)} = jB_{M1(pu)}$ and $Y_{M2(pu)} = jB_{M2(pu)}$ where B is shunt susceptance. Therefore, Equation (1.67) reduces to

$$jB_{M(pu)} = j\sqrt{B_{M1(pu)} \times B_{M2(pu)}} \qquad (1.68)$$

1.7.6 Examples

Example 1.1   Figure 1.10 shows two inductively coupled circuits having operating (and design) voltages of 275 and 400 kV phase to phase, carrying currents of 1 and 2 kA, and have actual voltage magnitudes at end A of 270 and 406 kV, respectively. The self and mutual impedances of the circuits are shown on the figure. Calculate all quantities shown on the figure in per unit using a 100 MVA base, 275 and 400 kV base voltages for the circuits. In addition, calculate the induced voltages in each circuit in actual and in per-unit values.

The per-unit voltages at end A are 270 kV/275 kV = 0.9818 pu and 406 kV/400 kV = 1.015 pu. Circuit 1 base current is 100 MVA/($\sqrt{3} \times 275$ kV) = 0.21 kA and circuit 1 current in per unit is 1 kA/0.21 kA = 4.76 pu. Circuit 2 base current is 100 MVA/($\sqrt{3} \times 400$ kV) = 0.1443 kA and circuit 2 current in per unit is 2 kA/0.1443 kA = 13.86 pu.

Base impedance of circuit 1 is $(275$ kV$)^2$/100 MVA = 756.25 Ω. Per-unit value of circuit 1 self-impedance is $(6+j30)$Ω/756.25 Ω = $(0.00793 + j0.0396)$pu or $(0.793 + j3.96)\%$.

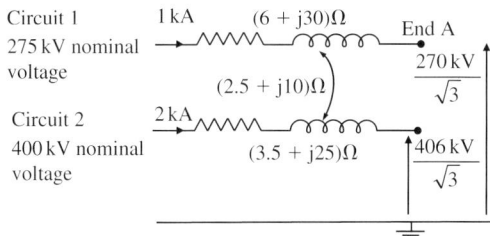

Figure 1.10   Two inductively coupled circuits having different operating (and design) voltages for use in Example 1.1

Base impedance of circuit 2 is $(400\,\mathrm{kV})^2/100\,\mathrm{MVA} = 1600\,\Omega$. Per-unit value of circuit 2 self-impedance is $(3.5 + \mathrm{j}25)\Omega/1600\,\Omega = (0.0022 + \mathrm{j}0.0156)\mathrm{pu}$ or $(0.22 + \mathrm{j}1.56)\%$.

Base impedance of mutual impedance is $(400\,\mathrm{kV} \times 275\,\mathrm{kV})/100\,\mathrm{MVA} = 1100\,\Omega$ and per-unit value of mutual impedance is $(2.5 + \mathrm{j}10)\Omega/1100\,\Omega = (0.00227 + \mathrm{j}0.00909)\mathrm{pu}$ or $(0.227 + \mathrm{j}0.909)\%$. Alternatively,

$$Z_{M1(pu)} = \frac{(2.5 + \mathrm{j}10)}{\left(\dfrac{275\,\mathrm{kV}}{\sqrt{3} \times 0.1433\,\mathrm{kA}}\right)} \quad\text{and}\quad Z_{M2(pu)} = \frac{(2.5 + \mathrm{j}10)}{\left(\dfrac{400\,\mathrm{kV}}{\sqrt{3} \times 0.21\,\mathrm{kA}}\right)}$$

and

$$Z_{M(pu)} = \sqrt{Z_{M1(pu)} \times Z_{M2(pu)}}$$
$$= (0.00227 + \mathrm{j}0.00909)\mathrm{pu} \text{ or } (0.227 + \mathrm{j}0.909)\%$$

Induced voltage in circuit 2 due to current in circuit 1 is $(2.5 + \mathrm{j}10)\Omega \times 1\,\mathrm{kA} = (2.5 + \mathrm{j}10)\mathrm{kV}$ or in per unit $(2.5 + \mathrm{j}10)\mathrm{kV}/(400\,\mathrm{kV}/\sqrt{3}) = (0.0108 + \mathrm{j}0.0433)\mathrm{pu}$. Alternatively, it can be calculated as $(0.00227 + \mathrm{j}0.00909)\mathrm{pu} \times 4.76\,\mathrm{pu} = (0.0108 + \mathrm{j}0.0433)\mathrm{pu}$. The induced per-unit voltage in circuit 1 due to circuit 2 current can be similarly calculated.

Example 1.2   Figure 1.11 shows two 400 and 275 kV capacitively coupled circuits. The actual voltages at end A, the self susceptance of each circuit and the mutual susceptance between the two circuits are shown on the figure. Calculate all quantities shown on the figure in per-unit using a 100 MVA base. In addition, calculate the charging current of each circuit at end A including the current flowing through the mutual susceptance. All calculations are to be carried out in physical units and in per-unit.

The per-unit voltages at end A are $283\,\mathrm{kV}/275\,\mathrm{kV} = 1.029$ pu and $404\,\mathrm{kV}/400\,\mathrm{kV} = 1.01$ pu. Circuit 1 base current is 0.21 kA and circuit 2 base current is 0.1443 kA.

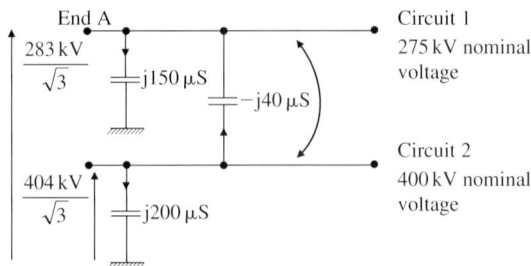

Figure 1.11   Two capacitively coupled circuits having different operating (and design) voltages for use in Example 1.2

Base admittance of circuit 1 is $100\,\text{MVA}/(275\,\text{kV})^2 = 0.00132\,\text{S}$ and per-unit value of self-susceptance is $j150 \times 10^{-6}\,\text{S}/0.00132\,\text{S} = j0.1136$ pu or $j11.36\%$. Base admittance of circuit 2 is $100\,\text{MVA}/(400\,\text{kV})^2 = 6.25 \times 10^{-4}\,\text{S}$ and per-unit value of self-susceptance is $j200 \times 10^{-6}\,\text{S}/(6.25 \times 10^{-4}\,\text{S}) = j0.32$ pu or $j32\%$.

Base value of mutual susceptance is $100\,\text{MVA}/(400\,\text{kV} \times 275\,\text{kV}) = 9.09 \times 10^{-4}\,\text{S}$ and per-unit value of mutual susceptance is $j40 \times 10^{-6}\,\text{S}/(9.09 \times 10^{-4}\,\text{S}) = j0.044$ pu or $j4.4\%$. Alternatively,

$$Y_{M1(pu)} = \frac{j40 \times 10^{-6}\,\text{S}}{0.00132\,\text{S}} = j0.0303\,\text{pu}$$

and

$$Y_{M2(pu)} = \frac{j40 \times 10^{-6}\,\text{S}}{6.25 \times 10^{-4}\,\text{S}} = j0.064\,\text{pu}$$

thus $Y_{M(pu)} = \sqrt{j0.0303 \times j0.064} = j0.044$ pu or $j4.4\%$.

Circuit 1 charging current due to self susceptance is equal to $j150 \times 10^{-6}\,\text{S} \times (283 \times 10^3\,\text{V}/\sqrt{3}) = j24.51\,\text{A}$ or $j24.51\,\text{A}/210\,\text{A} = j0.1167$ pu. Similarly, circuit 2 charging current due to self susceptance is equal to $j46.65\,\text{A}$ or $j46.65\,\text{A}/144.3\,\text{A} = j0.3230$ pu.

Current flowing through the mutual susceptance is equal to $j40 \times 10^{-6}\,\text{S} \times (404 \times 10^3 - 283 \times 10^3)\text{V}/\sqrt{3} = j2.79\,\text{A}$.

Charging current of circuit 1 is equal to $j24.51\,\text{A} - j2.79\,\text{A} = j21.72\,\text{A}$ or $j21.72\,\text{A}/210\,\text{A} = j0.103$ pu.

Charging current of circuit 2 is equal to $j46.65\,\text{A} + j2.79\,\text{A} = j49.44\,\text{A}$ or $j49.44\,\text{A}/144.3\,\text{A} = j0.3426$ pu.

Theory of symmetrical components and connection of phase sequence networks during faults

2.1 General

The analysis of three-phase ac power systems is greatly simplified if the system is assumed perfectly balanced. The three-phase system can then be modelled and analysed on a per-phase basis because knowledge of voltages and currents in one phase allows us to calculate those in the other two phases where they differ by $\pm 120°$ phase displacement from those of the known phase. However, in practical three-phase power systems, there are a number of sources of unbalance which can be divided into internal and external sources. Internal sources of unbalance are those due to the inherent small differences in the phase impedances and susceptances of three-phase power plant such as overhead lines and transformers. Three-phase generators, however, produce, by design, a balanced set of three-phase voltages. External sources of unbalance are those that impose an unbalanced condition on the power system network such as the connection of unbalanced three-phase loads, or single-phase loads, to a three-phase power system, the occurrence of unbalanced short-circuit and open-circuit faults. These external conditions create unbalanced voltages and currents in the three-phase network even if this network is balanced in terms of impedance and susceptance elements. The modelling and analysis of external balanced and unbalanced faults are the core topic of this chapter.

The analysis of unbalanced short-circuit and open-circuit faults in practical power systems makes extensive use of the theory of symmetrical components. Fortescue proposed this general theory in a famous paper in 1918. It applies to a system of N unbalanced phasors, including the case of three unbalanced phasors representing three-phase power systems. The theory enables the transformation of three unbalanced phasors into a three set of balanced phasors called the positive phase sequence (PPS), negative phase sequence (NPS), and zero-phase sequence (ZPS) phasors. This property presents an extremely powerful analysis tool that enables the formation of three separate and uncoupled equivalent networks called the PPS, NPS and ZPS networks provided that the three-phase power system network in physical phase terms is internally balanced. The types of external unbalanced conditions imposed on the actual three-phase network determine the methods of connection of these sequence networks. This chapter introduces the theory of symmetrical components, the definitions of sequence networks, the methods of connecting these networks for different fault types, and the analysis of sequence and phase currents and voltages for various conditions of external unbalances that may be imposed on the three-phase power system network.

2.2 Symmetrical components of a three-phase power system

2.2.1 Balanced three-phase voltage and current phasors

Mathematically, balanced three-phase voltages (or currents) can be defined as complex instantaneous, real instantaneous or complex phasor quantities. Denoting the three phases as R, Y and B, three-phase complex instantaneous voltages are defined as follows:

$$V_R(t) = \sqrt{2}V_{rms} \exp[j(\omega t + \phi)] = (\sqrt{2}V_{rms}e^{j\phi})e^{j\omega t} = \mathbf{V}_R e^{j\omega t} \tag{2.1a}$$

$$V_Y(t) = \sqrt{2}V_{rms} \exp[j(\omega t + \phi - 2\pi/3)] = (\sqrt{2}V_{rms}e^{j(\phi-2\pi/3)})e^{j\omega t} = \mathbf{V}_Y e^{j\omega t} \tag{2.1b}$$

$$V_B(t) = \sqrt{2}V_{rms} \exp[j(\omega t + \phi + 2\pi/3)] = (\sqrt{2}V_{rms}e^{j(\phi+2\pi/3)})e^{j\omega t} = \mathbf{V}_B e^{j\omega t} \tag{2.1c}$$

where V_{rms} is the root mean square (rms) value of any of the three-phase voltage waveforms and ϕ is the phase angle that determines the magnitude of the three-phase voltages at $t = 0$. From Equation (2.1), the three-phase real instantaneous voltages are defined as follows:

$$v_R(t) = \text{Real}[V_R(t)] = \sqrt{2}V_{\text{rms}}\cos(\omega t + \phi) \tag{2.2a}$$

$$v_Y(t) = \text{Real}[V_Y(t)] = \sqrt{2}V_{\text{rms}}\cos(\omega t + \phi - 2\pi/3) \tag{2.2b}$$

$$v_B(t) = \text{Real}[V_B(t)] = \sqrt{2}V_{\text{rms}}\cos(\omega t + \phi + 2\pi/3) \tag{2.2c}$$

and the three-phase complex phasor voltages are defined as follows:

$$V_R = \sqrt{2}V_{\text{rms}}e^{j\phi} \tag{2.3a}$$

$$V_Y = \sqrt{2}V_{\text{rms}}e^{j(\phi-2\pi/3)} \tag{2.3b}$$

$$V_B = \sqrt{2}V_{\text{rms}}e^{j(\phi+2\pi/3)} \tag{2.3c}$$

and the peak phase voltages are given by $\hat{V}_R = \hat{V}_Y = \hat{V}_B = \sqrt{2}V_{\text{rms}}$.

From Equation (2.3), it is observed that for a balanced three-phase set of voltage or current phasors R, Y and B, all three phasors are equal in magnitude and are displaced from each other by 120° in phase.

Figure 2.1 illustrates Equations (2.1)–(2.3) in their instantaneous and phasor forms. We now define a new complex number h that has a magnitude of unity and a phase angle of 120° such that the effect of multiplying any phasor by h is to advance or rotate the original phasor counter clockwise by 120° whilst keeping its

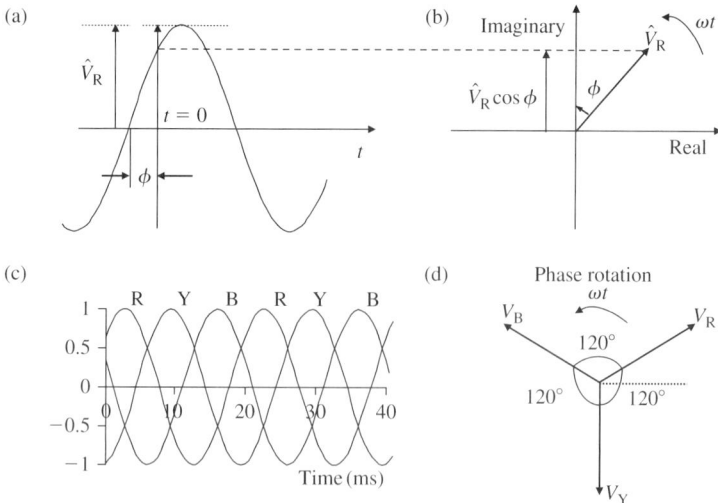

Figure 2.1 Instantaneous and phasor definitions of balanced three-phase voltages: (a) instantaneous; (b) phasor at $t = 0$; (c) balanced set of three-phase voltages or currents with phase rotation RYB, YBR, BRY, etc. and (d) balanced set of three phasors at $t = 0$

magnitude unchanged. h is thus known as an operator and is given by

$$h = e^{j2\pi/3} = -0.5 + j0.866 \quad j = \sqrt{-1} \tag{2.4a}$$

$$h^2 = e^{-j2\pi/3} = -0.5 - j0.866 \tag{2.4b}$$

$$h^3 = e^{j2\pi} = 1$$

and

$$1 + h + h^2 = 0 \tag{2.4c}$$

Using the operator h, and phase R as the reference phasor, Equations (2.3b) and (2.3c) can be written as $V_Y = h^2 V_R$ and $V_B = h V_R$. Therefore, we can write the balanced three-phase voltages in vector form as follows:

$$\mathbf{V}_{RYB} = \begin{bmatrix} V_R \\ V_Y \\ V_B \end{bmatrix} = \begin{bmatrix} V_R \\ h^2 V_R \\ h V_R \end{bmatrix} \tag{2.5}$$

where $\mathbf{V}_{RYB}$ denotes a column vector. The above equation equally applies to three-phase balanced currents, i.e.

$$\mathbf{I}_{RYB} = \begin{bmatrix} I_R \\ I_Y \\ I_B \end{bmatrix} = \begin{bmatrix} I_R \\ h^2 I_R \\ h I_R \end{bmatrix} \tag{2.6}$$

2.2.2 Symmetrical components of unbalanced voltage or current phasors

Consider a set of three-phase unbalanced voltage phasors V_R, V_Y and V_B that are unbalanced in both magnitude and phase as shown in Figure 2.2(a). According to the symmetrical components theory, each one of these phasors can be decomposed into three balanced phasors known as the PPS, NPS and ZPS sets as shown in Figure 2.2(b)–(d). The three PPS phasors, shown in Figure 2.2(b), are equal in magnitude, displaced from each other by an equal phase of 120° and have the same phase sequence or rotation as the original unbalanced set, i.e. RYB, YBR, BRY, etc. The three NPS phasors, shown in Figure 2.2(c), are equal in magnitude, displaced from each other by an equal phase of 120° and have a phase sequence or rotation that is opposite to that of the original unbalanced set, i.e. RBY, BYR, YRB, etc. The three ZPS phasors, shown in Figure 2.2(d), are equal in magnitude and are all in phase with each other.

We choose the letters P, N, Z as superscripts to denote PPS, NPS and ZPS quantities, respectively. Therefore, the three unbalanced phasors can be written in terms of their three PPS, NPS and ZPS symmetrical components as follows:

$$V_R = V_R^P + V_R^N + V_R^Z \tag{2.7a}$$

$$V_Y = V_Y^P + V_Y^N + V_Y^Z \tag{2.7b}$$

$$V_B = V_B^P + V_B^N + V_B^Z \tag{2.7c}$$

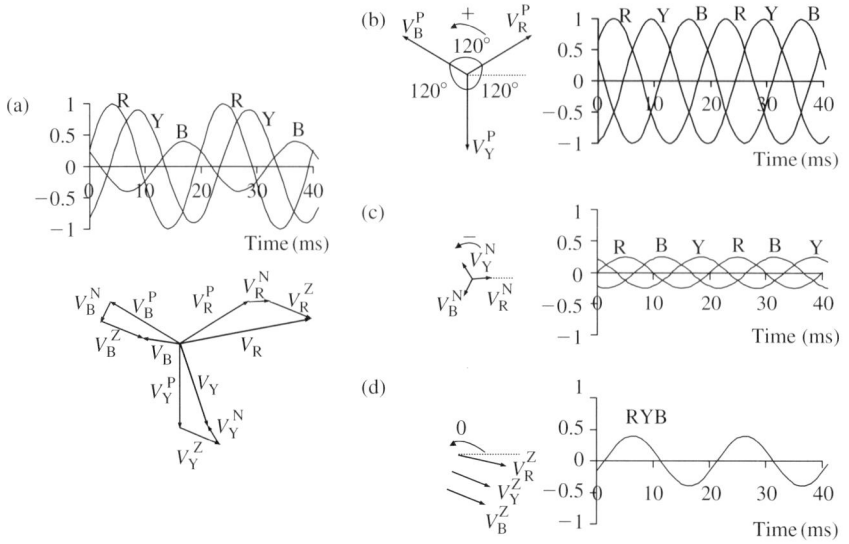

Figure 2.2   Unbalanced three-phase voltages and their symmetrical components: (a) unbalanced instant-aneous voltages and their phasors; (b) balanced PPS phasors; (c) balanced NPS phasors and (d) ZPS phasors

Using phase R as the reference phase, the h operator, Figures 2.2(b)–(d), the equations following can be written:

For phase Y   *For phase B*

$$V_Y^P = h^2 V_R^P \qquad V_B^P = h V_R^P \tag{2.8a}$$

$$V_Y^N = h V_R^N \qquad V_B^N = h^2 V_R^N \tag{2.8b}$$

$$V_Y^Z = V_R^Z \qquad V_B^Z = V_R^Z \tag{2.8c}$$

Substituting Equations (2.8) in Equations (2.7b) and (2.7c), we obtain

$$V_R = V_R^P + V_R^N + V_R^Z \tag{2.9a}$$

$$V_Y = h^2 V_R^P + h V_R^N + V_R^Z \tag{2.9b}$$

$$V_B = h V_R^P + h^2 V_R^N + V_R^Z \tag{2.9c}$$

or in matrix form

$$\begin{bmatrix} V_R \\ V_Y \\ V_B \end{bmatrix} = \begin{bmatrix} 1 & 1 & 1 \\ h^2 & h & 1 \\ h & h^2 & 1 \end{bmatrix} \begin{bmatrix} V_R^P \\ V_R^N \\ V_R^Z \end{bmatrix} = \mathbf{H} \begin{bmatrix} V_R^P \\ V_R^N \\ V_R^Z \end{bmatrix} \tag{2.10a}$$

where

$$\mathbf{H} = \begin{bmatrix} 1 & 1 & 1 \\ h^2 & h & 1 \\ h & h^2 & 1 \end{bmatrix} \tag{2.10b}$$

or

$$\mathbf{V}_{RYB} = \mathbf{H}\mathbf{V}_R^{PNZ} \tag{2.11}$$

H is the transformation matrix that transforms PPS, NPS and ZPS quantities into their corresponding phase quantities. Conversely, if the unbalanced phase quantities are known, the phase R sequence quantities can be calculated from

$$\mathbf{V}_R^{PNZ} = \mathbf{H}^{-1}\mathbf{V}_{RYB} \tag{2.12a}$$

where

$$\mathbf{H}^{-1} = \frac{1}{3}\begin{bmatrix} 1 & h & h^2 \\ 1 & h^2 & h \\ 1 & 1 & 1 \end{bmatrix} \tag{2.12b}$$

It is interesting to note that

$$\mathbf{H}^{-1} = \frac{1}{3}(\mathbf{H}^*)^t \tag{2.13}$$

Expanding Equation (2.12a) using Equation (2.12b), we obtain

$$V_R^P = \frac{1}{3}(V_R + hV_Y + h^2V_B) \tag{2.14a}$$

$$V_R^N = \frac{1}{3}(V_R + h^2V_Y + hV_B) \tag{2.14b}$$

$$V_R^Z = \frac{1}{3}(V_R + V_Y + V_B) \tag{2.14c}$$

The PPS, NPS and ZPS components of phases Y and B phasors can be calculated using Equation (2.8).

The above analysis applies to an arbitrary unbalanced three-phase set of phasors so that currents as well as voltage phasors could have been used. When current phasors are used, Equations (2.11) and (2.12a) can be written as

$$\mathbf{I}_{RYB} = \mathbf{H}\mathbf{I}_R^{PNZ} \tag{2.15a}$$

$$\mathbf{I}_R^{PNZ} = \mathbf{H}^{-1}\mathbf{I}_{RYB} \tag{2.15b}$$

2.2.3 Apparent power in symmetrical component terms

In a three-phase power system, the total apparent power can be expressed in terms of actual phase voltages and currents as follows:

$$S_{\text{3-Phase}} = P_{\text{3-Phase}} + jQ_{\text{3-Phase}} = [V_R \ V_Y \ V_B] \times \begin{bmatrix} I_R^* \\ I_Y^* \\ I_B^* \end{bmatrix} = \mathbf{V}_{RYB}^t \times \mathbf{I}_{RYB}^* \quad (2.16)$$

Substituting Equations (2.11) and (2.15a) into Equation (2.16), we obtain

$$
\begin{aligned}
S_{\text{3-Phase}} = P_{\text{3-Phase}} + jQ_{\text{3-Phase}} &= (\mathbf{H} \times \mathbf{V}_R^{PNZ})^t \times (\mathbf{H} \times \mathbf{I}_R^{PNZ})^* \\
&= (\mathbf{V}_R^{PNZ})^t \mathbf{H}^t \times \mathbf{H}^* (\mathbf{I}_R^{PNZ})^*
\end{aligned}
\quad (2.17)
$$

It can readily be shown that $\mathbf{H}^t \mathbf{H}^* = 3\mathbf{U}$, where $\mathbf{U}$ is the unit matrix, therefore, Equation (2.17) can be written as

$$
\begin{aligned}
S_{\text{3-Phase}} = P_{\text{3-Phase}} + jQ_{\text{3-Phase}} &= 3(\mathbf{V}_R^{PNZ})^t (\mathbf{I}_R^{PNZ})^* \\
&= 3V_R^P (I_R^P)^* + 3V_R^N (I_R^N)^* + 3V_R^Z (I_R^Z)^* \\
&= 3(S_R^P + S_R^N + S_R^Z)
\end{aligned}
\quad (2.18)
$$

The total apparent power in the three-phase unbalanced system can be calculated as three times the sum of the PPS, NPS and ZPS apparent powers. It is interesting to note that only like sequence terms are involved in each multiplication and that no cross-sequence terms appear.

2.2.4 Definition of phase sequence component networks

We have already established that the symmetrical component theory allows us to replace a three-phase set of unbalanced voltages (or currents) with a three separate sets of balanced voltages (or currents) defined as the PPS, NPS and ZPS sets. When an external unbalanced condition, such as a single-phase short-circuit fault, is imposed on the network, PPS, NPS and ZPS voltages and currents appear on the network at the point of fault. A phase sequence network is one that carries currents and voltages of one particular phase sequence such as PPS, NPS or ZPS sequence. It should be remembered that because the actual three-phase network is assumed perfectly balanced, the PPS, NPS and ZPS networks are separate, i.e. there is no intersequence mutual coupling between them, and they are only connected at the point of unbalance in the system, as will be seen later. In addition, the assumption of a perfectly balanced three-phase network means that PPS voltage sources and hence PPS currents exist in the PPS network only. Although the NPS and ZPS networks still exist and can still be artificially constructed, they are totally redundant because they carry no NPS or ZPS voltages or currents.

Each sequence network contains terms of one sequence only and hence each one can be considered as an independent single-phase network. That is the active PPS network contains only PPS voltages, PPS currents and PPS impedances/susceptances. The passive NPS network contains only NPS voltages, NPS currents and NPS impedances/susceptances. The passive ZPS network contains only ZPS voltages, ZPS currents and ZPS impedances/susceptances. The NPS and ZPS voltages in the NPS and ZPS networks appear as a result of the unbalanced condition imposed on the actual three-phase network; they should not be confused with the voltage sources that exist in the PPS network.

Let us consider first the type of unbalanced conditions that can appear in the three-phase network at a point F relative to the network neutral point N as shown in Figure 2.3(a). The three sequence networks can then be constructed from the actual three-phase network components and network topology, and illustrated as shown in Figure 2.3(b). It should be noted that the derivation of sequence models of power system components is the subject of later chapters. For now, it suffices to state that the entire PPS, NPS and ZPS networks can each be reduced using Thévenin's theorem. The reduction results in a single equivalent voltage source at point F and a single Thévenin's equivalent impedance seen looking back into the relevant network from point F as illustrated in Figure 2.3(b). Remembering that only PPS voltages existed in the network prior to the occurrence of the unbalanced condition, the resultant PPS, NPS and ZPS equivalent circuits are illustrated in Figure 2.3(c).

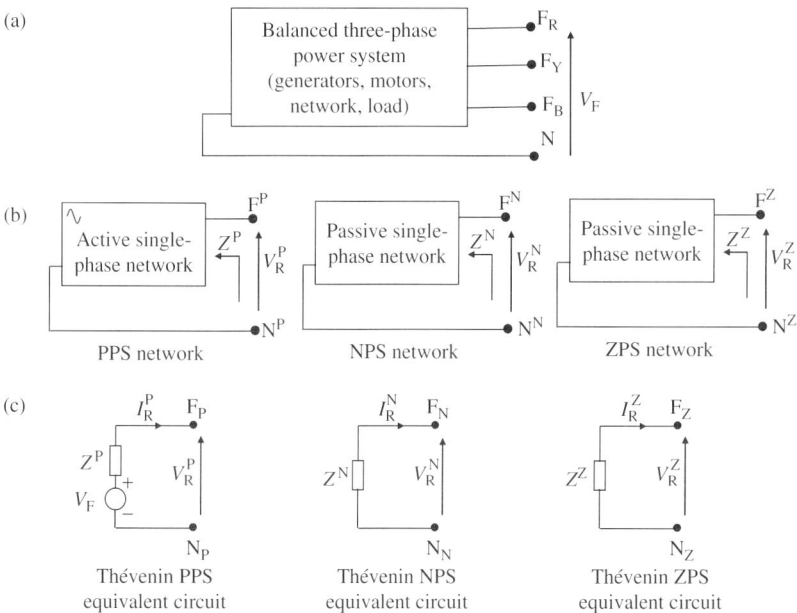

Figure 2.3 Definition of Thévenin's PPS, NPS and ZPS sequence networks

From Figure 2.3(c), sequence voltage and current relations at the point of unbalance F can be written for each sequence network. For the active PPS network:

$$V_R^P = V_F - Z^P I_R^P \tag{2.19}$$

where V_F is the PPS voltage at the point of fault F immediately before the unbalance condition is applied, V_R^P is the resultant PPS voltage at the point of fault, and I_R^P is the PPS current flowing out of the PPS network into the point of fault.

For the passive NPS network:

$$V_R^N = -Z^N I_R^N \tag{2.20}$$

where V_R^N is the NPS voltage at the point of fault and I_R^N is the NPS current flowing out of the NPS network into the point of fault.

For the passive ZPS network:

$$V_R^Z = -Z^Z I_R^Z \tag{2.21}$$

where V_R^Z is the ZPS voltage at the point of fault and I_R^Z is the ZPS current flowing out of the ZPS network into the point of fault.

As will be seen in later chapters, the PPS and NPS impedances are the same for most static power system plant but are generally different for rotating machines. In a ZPS network, because ZPS voltages, and ZPS currents, are co-phasal, the ZPS currents can only flow if there is a return connection, through either a neutral or earth wire or the general body of the earth. This is a very important consideration when determining the ZPS equivalent circuits for transformers and the analysis of earth return currents as will be seen in Chapters 4 and 10, respectively.

2.2.5 Sequence components of unbalanced three-phase impedances

Figure 2.4 shows a system of static unbalanced three-phase mutually coupled impedance elements.

The voltage drop equations across phases R, Y and B, in matrix form, are given by

$$
\begin{array}{ccc} & R & Y & B \end{array}
$$
$$
\begin{bmatrix} V_R \\ V_Y \\ V_B \end{bmatrix} = \begin{array}{c} R \\ Y \\ B \end{array} \begin{bmatrix} Z_{RR} & Z_{RY} & Z_{RB} \\ Z_{YR} & Z_{YY} & Z_{YB} \\ Z_{BR} & Z_{BY} & Z_{BB} \end{bmatrix} \begin{bmatrix} I_R \\ I_Y \\ I_B \end{bmatrix} \tag{2.22a}
$$

or in concise matrix notation form:

$$\mathbf{V}_{RYB} = \mathbf{Z}_{RYB} \mathbf{I}_{RYB} \tag{2.22b}$$

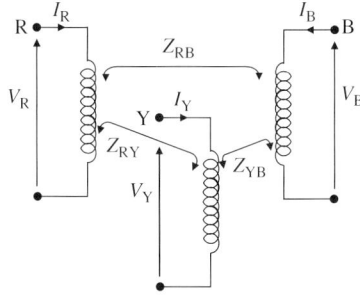

Figure 2.4 Unbalanced static mutually coupled three-phase impedances

where $\mathbf{Z_{RYB}}$ is defined as the phase impedance matrix of the unbalanced three-phase mutually coupled system shown in Figure 2.4. The diagonal elements are the self-impedances of each phase and the off-diagonal elements are the mutual impedances between the phases. It is important to note that this is the impedance matrix in the actual or physical phase frame of reference, RYB, where each impedance is an actual physical element.

What are the PPS, NPS and ZPS impedances of this system and what light do they shed on the use of the symmetrical components' theory in practice? To answer this question, we substitute Equations (2.11) and (2.15a) into Equation (2.22b), we obtain

$$\mathbf{HV_R^{PNZ}} = \mathbf{Z_{RYB}HI_R^{PNZ}}$$

and pre-multiplying by $\mathbf{H^{-1}}$, we obtain

$$\mathbf{V_R^{PNZ}} = \mathbf{H^{-1}Z_{RYB}HI_R^{PNZ}} = \mathbf{Z^{PNZ}I_R^{PNZ}} \qquad (2.23)$$

where $\mathbf{Z^{PNZ}}$ is defined as the sequence impedance matrix of the unbalanced three-phase system shown in Figure 2.4 and is given by

$$\mathbf{Z^{PNZ}} = \mathbf{H^{-1}Z_{RYB}H} \qquad (2.24)$$

It is important to note that this is the impedance matrix in the sequence frame of reference obtained by transforming the phase impedance matrix according to Equation (2.24) using the transformation matrices shown.

Assuming that the mutual phase impedances are bilateral or reciprocal that is $Z_{RY} = Z_{YR}$, etc., Equation (2.24) becomes

$$\mathbf{Z^{PNZ}} = \begin{array}{c} \\ P \\ N \\ Z \end{array} \begin{array}{ccc} P & N & Z \\ \left[\begin{array}{ccc} Z^{PP} & Z^{PN} & Z^{PZ} \\ Z^{NP} & Z^{NN} & Z^{NZ} \\ Z^{ZP} & Z^{ZN} & Z^{ZZ} \end{array} \right] \end{array} \qquad (2.25)$$

where

$$Z^{PP} = Z^{NN} = \frac{1}{3}[(Z_{RR} + Z_{YY} + Z_{BB}) - (Z_{RY} + Z_{RB} + Z_{YB})]$$

$$= S^{PP} - M^{PP} \tag{2.26a}$$

$$Z^{PN} = \frac{1}{3}[Z_{RR} + h^2 Z_{YY} + h Z_{BB}) + 2(h^2 Z_{RB} + h Z_{RY} + Z_{YB})]$$

$$= S^{PN} + 2M^{PN} \tag{2.26b}$$

$$Z^{PZ} = Z^{ZN} = \frac{1}{3}[(Z_{RR} + h Z_{YY} + h^2 Z_{BB}) - (h^2 Z_{RY} + h Z_{RB} + Z_{YB})]$$

$$= S^{PZ} - M^{PZ} \tag{2.26c}$$

$$Z^{NP} = \frac{1}{3}[(Z_{RR} + h Z_{YY} + h^2 Z_{BB}) + 2(h^2 Z_{RY} + h Z_{RB} + Z_{YB})]$$

$$= S^{PZ} + 2M^{PZ} \tag{2.26d}$$

$$Z^{NZ} = Z^{ZP} = \frac{1}{3}[(Z_{RR} + h^2 Z_{YY} + h Z_{BB}) - (h^2 Z_{RB} + h Z_{RY} + Z_{YB})]$$

$$= S^{PN} - M^{PN} \tag{2.26e}$$

$$Z^{ZZ} = \frac{1}{3}[(Z_{RR} + Z_{YY} + Z_{BB}) + 2(Z_{RY} + Z_{RB} + Z_{YB})]$$

$$= S^{PP} + 2M^{PP} \tag{2.26f}$$

Dropping the R notation from the current vector $\mathbf{I}_R^{PNZ}$ for convenience and using Equations (2.23), (2.24) and (2.25), we can write

$$\begin{bmatrix} V^P \\ V^N \\ V^Z \end{bmatrix} = \begin{bmatrix} S^{PP} - M^{PP} & S^{PN} + 2M^{PN} & S^{PZ} - M^{PZ} \\ S^{PZ} + 2M^{PZ} & S^{PP} - M^{PP} & S^{PN} - M^{PN} \\ S^{PN} - M^{PN} & S^{PZ} - M^{PZ} & S^{PP} + 2M^{PP} \end{bmatrix} \begin{bmatrix} I^P \\ I^N \\ I^Z \end{bmatrix} \tag{2.27}$$

which can be written as

$$\begin{bmatrix} V^P \\ V^N \\ V^Z \end{bmatrix} = \begin{bmatrix} a^{PP} & b^{PN} & c^{PZ} \\ d^{NP} & a^{PP} & e^{NZ} \\ e^{NZ} & c^{PZ} & f^{ZZ} \end{bmatrix} \begin{bmatrix} I^P \\ I^N \\ I^Z \end{bmatrix} \tag{2.28}$$

Equation (2.28) is important and requires a physical explanation. The sequence impedance matrix is full, non-diagonal and non-symmetric. A non-symmetric matrix is exemplified by the different mutual coupling terms b and d (also c and e) in the matrix. A non-diagonal matrix means that mutual coupling between the sequence circuits exist. For example, taking the PPS circuit, the total PPS voltage drop in this circuit consists of three components as follows:

$$V^P = a^{PP} I^P + b^{PN} I^N + c^{PZ} I^Z \tag{2.29}$$

The first term on the right is the PPS voltage drop induced by the flow of PPS current in the PPS circuit. The second term is an additional PPS voltage drop caused by the flow of NPS current in the NPS circuit acting on the mutual impedance from the PPS to the NPS circuit. The third term is an additional PPS voltage drop caused by the flow of ZPS current in the ZPS circuit acting on the mutual impedance from the PPS to the ZPS circuit.

We have calculated the sequence impedance matrix of a general three-phase mutually coupled system and this turns out to be non-diagonal and non-symmetric. What does this mean in practice? Let us remember that the basic principle of the symmetrical component theory rests on the fact that the PPS, NPS and ZPS sequence circuits are separate with no mutual coupling between them. In mathematical terms, this means that the sequence impedance matrix has to be diagonal so that a PPS voltage drop is produced by the flow of PPS current only, and likewise for the NPS and ZPS circuits. The transformation of the phase impedance matrix of Equation (2.22a) from the phase frame of reference to the sequence frame of reference, Equation (2.28), produced no advantages whatsoever. In such situations, the symmetrical component theory may not be used and analysis in the phase frame of reference is more advantageous.

2.2.6 Sequence components of balanced three-phase impedances

The application of the symmetrical component theory in practical power system short-circuit analysis requires the PPS, NPS and ZPS circuits to be separate and uncoupled. This is achieved by the removal of the off-diagonal mutual elements in the sequence impedance matrix of Equation (2.28) so that the matrix becomes diagonal. This can be accomplished if the physical phase impedances of the original three-phase system shown in Figure 2.4 are assumed to be balanced, i.e. the self-impedances are assumed to be equal, and the mutual impedances are also assumed to be equal, i.e.

$$Z_{RR} = Z_{YY} = Z_{BB} = Z_{Self} \tag{2.30a}$$

$$Z_{RY} = Z_{YB} = Z_{BR} = Z_{Mutual} \tag{2.30b}$$

Substituting Equation (2.30) into Equations (2.26) and (2.27), we obtain

$$\begin{bmatrix} V^P \\ V^N \\ V^Z \end{bmatrix} = \begin{bmatrix} Z_{Self} - Z_{Mutual} & 0 & 0 \\ 0 & Z_{Self} - Z_{Mutual} & 0 \\ 0 & 0 & Z_{Self} + 2 \times Z_{Mutual} \end{bmatrix} \begin{bmatrix} I^P \\ I^N \\ I^Z \end{bmatrix}$$

$$\tag{2.31}$$

where,

$$\text{PPS impedance} = \text{NPS impedance} = Z^P = Z^N = Z_{Self} - Z_{Mutual} \tag{2.32a}$$

$$\text{ZPS impedance} = Z^Z = Z_{\text{Self}} + 2 \times Z_{\text{Mutual}} \qquad (2.32b)$$

Also, if the PPS, NPS and ZPS impedances are known, the balanced self and mutual phase impedances can be calculated as follows:

$$Z_{\text{Self}} = \frac{1}{3}(Z^Z + 2Z^P) \qquad (2.33a)$$

$$Z_{\text{Mutual}} = \frac{1}{3}(Z^Z - Z^P) \qquad (2.33b)$$

2.2.7 Advantages of symmetrical components frame of reference

The significant advantage of the symmetrical component theory is that it provides a mathematical method that allows us to decompose the original complex mutually coupled three-phase system equations into three separate sets of equations. The PPS, NPS and ZPS sets of currents and voltages are each calculated from the PPS, NPS and ZPS circuits, respectively, i.e. in the sequence frame of reference. The sequence of currents and voltages are then easily transformed back into the phase frame of reference. However, the creation of three uncoupled single-phase sequence circuits is based on the assumption that the original three-phase system is perfectly balanced. In reality, this is not the case, as will be seen in later chapters. Nonetheless, reasonable assumptions of balance can be made to allow the use of the symmetrical component theory in practical applications. Nowadays, almost all large-scale short-circuit analysis in practical power systems is carried out in the sequence frame of reference with transformation to the phase frame of reference carried out as a final step in the analysis. There are, nonetheless, specialised short-circuit applications where analysis in the phase frame of reference is used. This is discussed in Chapter 6.

2.2.8 Examples

Example 2.1   Prove that a balanced set of three-phase voltages contains only a PPS voltage component.

Substituting Equations (2.5) and (2.6) into Equation (2.14), we obtain

$$V_R^P = \frac{1}{3}(V_R + hh^2 V_R + h^2 h V_R) = V_R \quad \text{since } h^3 = 1$$

$$V_R^N = \frac{1}{3}(V_R + h^2 h^2 V_R + hh V_R) = 0 \quad \text{since } 1 + h + h^2 = 0$$

$$V_R^Z = \frac{1}{3}(V_R + h^2 V_R + h V_R) = 0$$

Example 2.2   What is the phase sequence nature of MW and MVAr power flows in a balanced three-phase power system?

In a balanced three-phase system, NPS and ZPS voltages and currents are zeroes and only PPS voltages and currents exist. Therefore, and according to Equation (2.18), only PPS apparent power exists and hence the MW and MVAr power flows in a balanced three-phase system are PPS quantities.

Example 2.3   What is the relationship between the ZPS current and the neutral current in a three-phase power system with a neutral wire or neutral connection to earth.

From Equation (2.7), rewritten for phase currents, we have

$$I_R = I_R^P + I_R^N + I_R^Z \quad I_Y = I_Y^P + I_Y^N + I_Y^Z \quad I_B = I_B^P + I_B^N + I_B^Z$$

The neutral current is the sum of the currents in phases R, Y and B, thus

$$I_{Neutral} = I_R + I_Y + I_B = (I_R^P + I_Y^P + I_B^P) + (I_R^N + I_Y^N + I_B^N) + (I_R^Z + I_Y^Z + I_B^Z)$$

or

$$I_{Neutral} = 0 + 0 + (I_R^Z + I_Y^Z + I_B^Z) = 3I_R^Z$$

Since the PPS (and NPS) currents are balanced, by definition, their sum is zero and thus neither produce any neutral current. Since the ZPS currents are equal and in phase, the neutral current is the sum of the ZPS currents in each phase and this is equal to three times the ZPS current in phase R.

Example 2.4   The three voltages of a three-phase system are balanced in phase but not in magnitude and are

$$V_R = 1\angle 0 \, pu \quad V_Y = 0.97\angle{-}120 \, pu \quad V_B = 1.01\angle 120 \, pu$$

Calculate the PPS, NPS and ZPS voltages of phase R voltage. Express the NPS and ZPS voltage magnitudes as percentages of the PPS voltage magnitude.

Using Equation (2.14), we obtain

$$V_R^P = \frac{1}{3}(1 + 1\angle 120 \times 0.97\angle{-}120 + 1\angle{-}120 \times 1.01\angle 120) = 0.99 \, pu$$

$$V_R^N = \frac{1}{3}(1 + 1\angle{-}120 \times 0.97\angle{-}120 + 1\angle 120 \times 1.01\angle 120)$$

$$= 0.012\angle{-}73.89 \, pu$$

$$V_R^Z = \frac{1}{3}(1 + 0.97\angle-120 + 1.01\angle120) = 0.012\angle73.89\,\text{pu}$$

$$\frac{|V_R^N|}{|V_R^P|} = \frac{|V_R^Z|}{|V_R^P|} = \frac{0.012}{0.99} = 0.01212 \text{ or } 1.212\%$$

Example 2.5   The three voltages of a three-phase system have equal magnitudes of 1 pu but phase Y lags phase R by 115° and the phase displacement between phases Y and B is 120°. Calculate the PPS, NPS and ZPS voltages of phase R. Express the NPS and ZPS voltage magnitudes as percentages of the PPS voltage magnitude.

Taking phase R as an arbitrary reference, the phase voltages can be written as

$$V_R = 1\angle0\,\text{pu} \quad V_Y = 1\angle-115\,\text{pu} \quad V_B = 1\angle125\,\text{pu}$$

Thus, the three voltages are balanced in magnitude but not in phase. Using Equation (2.14), we obtain

$$V_R^P = \frac{1}{3}(1 + 1\angle120 \times 1\angle-115 + 1\angle-120 \times 1\angle125) = 0.9991\angle3.3\,\text{pu}$$

$$V_R^N = \frac{1}{3}(1 + 1\angle-120 \times 1\angle-115 + 1\angle120 \times 1\angle125) = 0.029\angle-87.5\,\text{pu}$$

$$V_R^Z = \frac{1}{3}(1 + 1\angle-115 + 1\angle125) = 0.029\angle-87.5\,\text{pu}$$

$$\frac{|V_R^N|}{|V_R^P|} = \frac{|V_R^Z|}{|V_R^P|} = \frac{0.029}{0.999} = 0.02903 \text{ or } 2.9\%$$

Example 2.6   The three unbalanced voltages of a three-phase system under a very large unbalanced condition are

$$V_R = 0\angle0\,\text{pu} \quad V_Y = 1.1\angle-140\,\text{pu} \quad V_B = 1.1\angle140\,\text{pu}.$$

Calculate the PPS, NPS and ZPS voltages of phase R. Express the NPS and ZPS voltage magnitudes as percentages of the PPS voltage magnitude.

Using Equation (2.14), we obtain

$$V_R^P = \frac{1}{3}(0 + 1\angle120 \times 1.1\angle-140 + 1\angle-120 \times 1.1\angle140) = 0.689\angle0\,\text{pu}$$

$$V_R^N = \frac{1}{3}(0 + 1\angle-120 \times 1.1\angle-140 + 1\angle120 \times 1.1\angle140) = -0.1273\angle0\,\text{pu}$$

$$V_R^Z = \frac{1}{3}(0 + 1.1\angle -140 + 1.1\angle 140) = -0.5617\angle 0 \text{ pu}$$

$$\frac{|V_R^N|}{|V_R^P|} = \frac{0.1273}{0.689} = 0.185 \text{ or } 18.5\% \qquad \frac{|V_R^Z|}{|V_R^P|} = \frac{0.5617}{0.689} = 0.815 \text{ or } 81.5\%$$

2.3 Analysis of balanced and unbalanced faults in the sequence reference frame

2.3.1 General

In general, there are three types of faults in three-phase power systems. These are short-circuit faults between one or more phases which may or may not involve earth, open-circuit faults on one or two phases and simultaneous faults where more than one fault occurs at the same time at the same or at different locations in the network. Short-circuit faults are sometimes referred to as shunt faults whereas open-circuit faults as series faults. When a short-circuit fault occurs at a point F on the three-phase network, the conditions imposed by the fault at F must be observed between the relevant phase(s) and the neutral point in the sequence networks. However, when an open-circuit fault occurs at a point F in the network, the conditions imposed by the fault must be observed between the two sides of the open circuit, say points F and F', in the sequence networks. In this section, we will derive methods of connecting the three PPS, NPS and ZPS networks for various fault types that can occur in power systems. The unbalanced fault condition applied in the three-phase network is arranged to be symmetrical with respect to phase R which is taken as the reference phase. This results in simpler mathematical derivation as will be shown later.

2.3.2 Balanced three-phase to earth short-circuit faults

A three-phase to earth short-circuit fault at a point in a three-phase system is a balanced or symmetrical fault that can still be analysed using the symmetrical components theory. Figure 2.5(a) shows the representation of this fault. F_R, F_Y and F_B are points in the three-phase system where the three-phase fault is assumed to occur through the balanced fault impedances Z_F, and F_R', F_Y' and F_B' are the true points of fault.

From Figure 2.5(a), the voltages at point F are given by

$$V_R = Z_F I_R \quad V_Y = Z_F I_Y \quad V_B = Z_F I_B \tag{2.34}$$

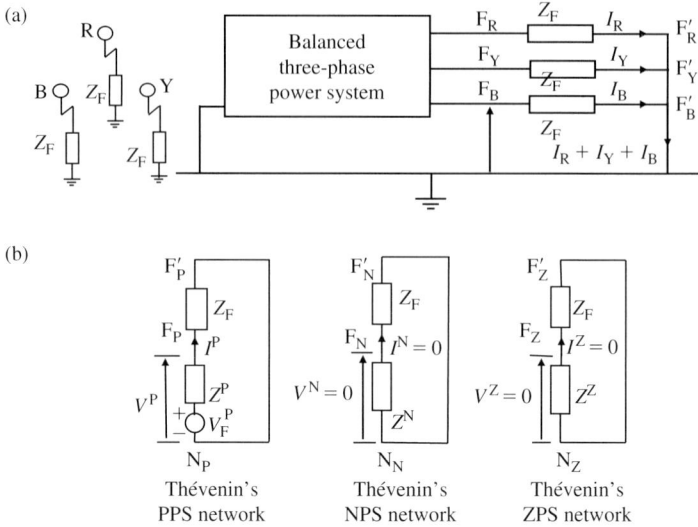

Figure 2.5   Balanced three-phase to earth short-circuit fault through a fault impedance Z_F

Using Equations (2.6), (2.14a) and (2.19), it can be shown that the sequence voltages and currents at the fault point are given by

$$V_R^P = Z_F I_R^P = V_F^P - Z^P I_R^P \quad V_R^N = 0 \quad V_R^Z = 0 \tag{2.35}$$

and

$$I_R^P = \frac{V_F^P}{Z^P + Z_F} \quad I_R^N = 0 \quad I_R^Z = 0 \tag{2.36}$$

Therefore, the three-phase power system remains balanced and symmetrical after the occurrence of such a fault because the fault impedances are equal in the three phases. Therefore, only PPS voltages exist and only PPS currents can flow.

Since phase R is used as the reference, it is convenient from now on to drop the R notation in the PPS, NPS and ZPS voltage and current equations whilst always remembering that these sequence quantities are those of phase R.

Using Equations (2.15a) and (2.10b), the phase fault currents are given by

$$I_R = I^P = \frac{V_F^P}{Z^P + Z_F} \tag{2.37a}$$

$$I_y = h^2 I_R = \frac{h^2 V_F^P}{Z^P + Z_F} \tag{2.37b}$$

$$I_B = hI_R = \frac{hV_F^P}{Z^P + Z_F} \qquad (2.37c)$$

and, using Equation (2.34), the phase fault voltages at the fault point F are given by

$$V_R = I^P Z_F = \frac{V_F^P}{Z^P + Z_F} Z_F \qquad (2.38a)$$

$$V_Y = h^2 V_R = \frac{h^2 V_F^P}{Z^P + Z_F} Z_F \qquad (2.38b)$$

$$V_B = h V_R = \frac{h V_F^P}{Z^P + Z_F} Z_F \qquad (2.38c)$$

As expected for a balanced and symmetrical short-circuit fault, the sum of the three-phase currents $I_R + I_Y + I_B$ is equal to zero hence the net fault current flowing into earth is zero. Similarly, the sum of the three-phase voltages $V_R + V_Y + V_B$ is equal to zero. Figure 2.5(b) shows the connections of the Thévenin's PPS, NPS and ZPS equivalent circuits that satisfy Equations (2.35) and (2.36).

The case of a solid or bolted three-phase to earth short-circuit fault is obtained by setting $Z_F = 0$.

2.3.3 Balanced three-phase clear of earth short-circuit faults

A three-phase short-circuit fault clear of earth at a point F in a three-phase system is represented by connecting an equal fault impedance between each pair of phases as shown in Figure 2.6(a), i.e., as a delta connection.

In order to calculate the equivalent fault impedance 'seen' in each phase, a transformation of delta to star with isolated neutral is needed as follows:

$$Z_R = Z_Y = Z_B = \frac{Z_F \times Z_F}{Z_F + Z_F + Z_F} = \frac{1}{3} Z_F \qquad (2.39)$$

The fault representation is illustrated in Figure 2.6(b).

The phase fault currents and voltages are given by

$$I_R + I_Y + I_B = 0 \qquad (2.40a)$$

and

$$V_R = V_Y = V_B = 0 \quad \text{at point F}' \qquad (2.40b)$$

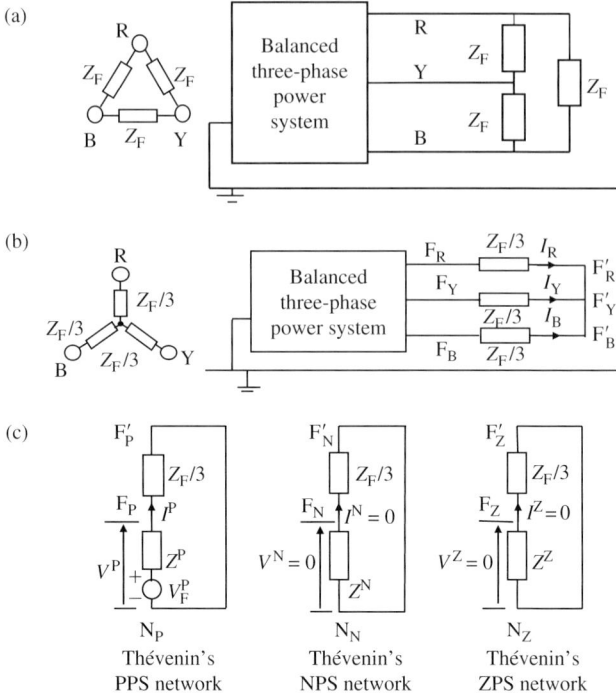

Figure 2.6   Balanced three-phase short-circuit fault clear of earth through a fault impedance Z_F

and

$$V_R = \frac{1}{3}Z_F I_R \quad V_Y = \frac{1}{3}Z_F I_Y \quad V_B = \frac{1}{3}Z_F I_B \qquad (2.40c)$$

Using Equations (2.14), (2.19), (2.20) and (2.21), we obtain

$$V^P = \frac{1}{3}Z_F I^P = V_F^P - Z^P I^P \quad V^N = 0 \quad V^Z = 0 \qquad (2.41a)$$

$$I^P = \frac{V_F^P}{Z^P + \frac{1}{3}Z_F} \quad I^N = 0 \quad I^Z = 0 \qquad (2.41b)$$

The phase fault currents are given by

$$I_R = I^P = \frac{V_F^P}{Z^P + \frac{1}{3}Z_F} \qquad (2.42a)$$

$$I_Y = h^2 I_R = \frac{h^2 V_F^P}{Z^P + \frac{1}{3}Z_F} \qquad (2.42b)$$

$$I_B = hI_R = \frac{hV_F^P}{Z^P + \frac{1}{3}Z_F} \qquad (2.42c)$$

The phase fault voltages at point F are given by

$$V_R = I^P \frac{1}{3} Z_F = \frac{V_F^P}{1 + \frac{3Z^P}{Z_F}} \qquad (2.43a)$$

$$V_Y = h^2 V_R = \frac{h^2 V_F^P}{1 + \frac{3Z^P}{Z_F}} \qquad (2.43b)$$

$$V_Y = hV_R = \frac{hV_F^P}{1 + \frac{3Z^P}{Z_F}} \qquad (2.43c)$$

Figure 2.6(c) shows the connections of the PPS, NPS and ZPS equivalent networks that satisfy the fault condition at the fault point F. The case of a solid or bolted three-phase clear of earth short-circuit fault is obtained by setting $Z_F = 0$.

2.3.4 Unbalanced one-phase to earth short-circuit faults

Figure 2.7(a) shows a representation of an unbalanced one-phase to earth fault on phase R through a fault impedance Z_F.

For ease of notation in the rest of this chapter, we will replace V_F^P with V_F. The conditions at the point of fault are

$$I_Y = I_B = 0 \qquad (2.44a)$$

and

$$V_R = Z_F I_R \qquad (2.44b)$$

Using Equation (2.14) for currents instead of voltages, we obtain

$$I^P = I^N = I^Z = \frac{1}{3} I_R \quad \text{or} \quad I_R = 3 \times I^Z \qquad (2.45)$$

Substituting Equations (2.19), (2.20) and (2.21) into Equation (2.9a) and using Equation (2.45), we obtain

$$V_R = Z_F \times 3I^P = V^P + V^N + V^Z = V_F - (Z^P + Z^N + Z^Z)I^P$$

or

$$V_F = (Z^P + Z_F)I^P + (Z^N + Z_F)I^N + (Z^Z + Z_F)I^Z \qquad (2.46)$$

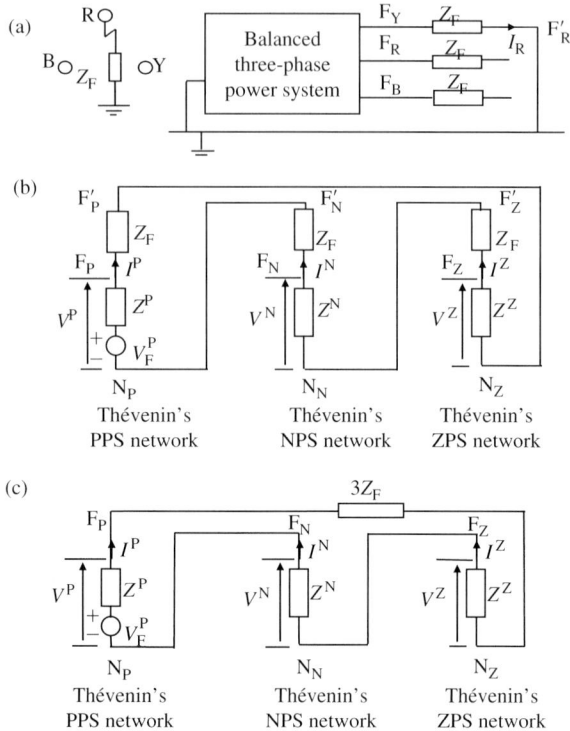

Figure 2.7 Unbalanced one-phase to earth short-circuit fault through a fault impedance Z_F

Thus, the sequence fault currents are given by

$$I^P = I^N = I^Z = \frac{V_F}{[(Z^P + Z_F) + (Z^N + Z_F) + (Z^Z + Z_F)]}$$

$$= \frac{V_F}{(Z^P + Z^N + Z^Z + 3Z_F)} \tag{2.47}$$

Equations (2.45) and (2.46) show that the PPS, NPS and ZPS networks should be connected in series as shown in Figure 2.7(b) with the fault impedance Z_F appearing as an external impedance in series with each sequence network.

The PPS, NPS and ZPS voltages at the fault point are calculated using Figure 2.7(b) or (c) giving

$$V^P = \frac{(Z^N + Z^Z + 3Z_F)}{(Z^P + Z^N + Z^Z + 3Z_F)} V_F \tag{2.48}$$

$$V^N = \frac{-Z^N}{(Z^P + Z^N + Z^Z + 3Z_F)} V_F \tag{2.49}$$

$$V^Z = \frac{-Z^Z}{(Z^P + Z^N + Z^Z + 3Z_F)} V_F \tag{2.50}$$

The phase fault current is calculated using Equations (2.45) and (2.47) giving

$$I_R = \frac{3V_F}{(Z^P + Z^N + Z^Z + 3Z_F)} \tag{2.51}$$

The phase fault voltage is calculated from Equation (2.44) giving

$$V_R = \frac{3Z_F}{(Z^P + Z^N + Z^Z + 3Z_F)} V_F \tag{2.52}$$

The phase voltages on the healthy or unfaulted phases at the point of fault can be calculated using Equation (2.9) and Equations (2.48), (2.49) and (2.50) giving

$$V_Y = K_{E1} \times h^2 V_F \tag{2.53}$$

where

$$K_{E1} = 1 - \frac{(Z^P + h^2 Z^N + h Z^Z)}{(Z^P + Z^N + Z^Z + 3Z_F)} \tag{2.54}$$

and

$$V_B = K_{E2} \times h V_F \tag{2.55}$$

where

$$K_{E2} = 1 - \frac{(Z^P + h Z^N + h^2 Z^Z)}{(Z^P + Z^N + Z^Z + 3Z_F)} \tag{2.56}$$

K_{E1} and K_{E2} are termed the earth fault factors and have a magnitude that typically range from 1 to 1.8 depending on the method of system earthing used. This factor is the ratio of the rms phase to earth voltage at the fault point during the fault to the rms phase to earth voltage without a fault. This factor determines the extent of voltage rise on the healthy phases during the fault. Systems that are defined as 'effectively earthed' are those where the earth fault factor is less than or equal to 1.4.

The case of a solid or bolted single-phase to earth short-circuit fault is obtained by setting $Z_F = 0$.

2.3.5 Unbalanced phase-to-phase or two-phase short-circuit faults

Figure 2.8(a) shows a representation of an unbalanced two-phase fault on phases Y and B through a fault impedance Z_F.

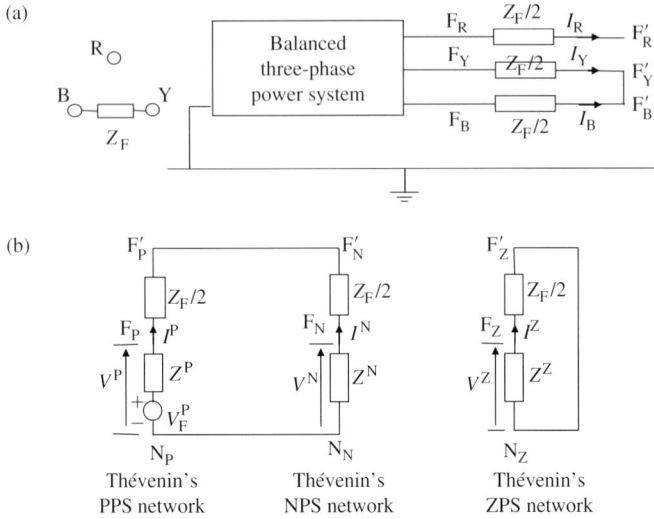

Figure 2.8   Unbalanced two-phase short-circuit fault through a fault impedance Z_F

The conditions at the point of fault are

$$I_R = 0 \quad I_Y = -I_B \quad V_Y - V_B = Z_F I_Y \tag{2.57a}$$

Substituting the above phase currents into Equation (2.15b), we obtain

$$I^Z = 0 \quad \text{and} \quad I^P = -I^N \tag{2.57b}$$

Substituting $I^Z = 0$ into Equation (2.21) gives $V^Z = 0$ and using Equations (2.7) and (2.8), we can write

$$V_Y - V_B = (h^2 - h)(V^P - V^N) \tag{2.58a}$$

Also, using Equations (2.7b) and (2.8) for phase currents, we obtain

$$Z_F I_Y = Z_F(h^2 I^P + h I^N) \tag{2.58b}$$

Equating $V_Y - V_B$ from Equations (2.57a) and (2.58a), using Equation (2.58b) and $I^P = -I^N$, we obtain

$$V^P - V^N = Z_F I^P \tag{2.59a}$$

Substituting Equations (2.19) and (2.20) into Equation (2.59a), and using $I^P = -I^N$, we obtain

$$V_F - \left(Z^P + \frac{1}{2}Z_F\right)I^P = -\left(Z^N + \frac{1}{2}Z_F\right)I^N \tag{2.59b}$$

The sequence fault currents are calculated from Equation (2.59b) using $I^P = -I^N$ giving

$$I^P = -I^N = \frac{V_F}{[(Z^P + \frac{1}{2}Z_F) + (Z^N + \frac{1}{2}Z_F)]} = \frac{V_F}{(Z^P + Z^N + Z_F)} \tag{2.60}$$

Equations (2.57b) and (2.59b) show that the PPS and NPS networks should be connected in parallel as shown in Figure 2.8(b) with half the fault impedance, $\frac{1}{2}Z_F$, appearing as an external impedance in series with the PPS and NPS networks. The ZPS currents and voltages are zero.

The phase fault currents are calculated using Equation (2.15a) giving

$$I_Y = -I_B = -j\sqrt{3}I^P = \frac{-j\sqrt{3}V_F}{(Z^P + Z^N + Z_F)} \tag{2.61}$$

The phase voltages on the faulted phases Y and B are calculated using Equations (2.9b), (2.9c), (2.19), (2.20), (2.21) and (2.60) giving

$$V_Y = \left(\frac{h^2 Z_F - Z^N}{Z^P + Z^N + Z_F} \right) V_F \tag{2.62a}$$

$$V_B = \left(\frac{h Z_F - Z^N}{Z^P + Z^N + Z_F} \right) V_F \tag{2.62b}$$

The phase voltage on the healthy or unfaulted phase R can be calculated using Equation (2.9a) giving

$$V_R = \left(\frac{2Z^N + Z_F}{Z^P + Z^N + Z_F} \right) V_F \tag{2.63}$$

The case of a solid or bolted two-phase short-circuit fault is obtained by setting $Z_F = 0$.

2.3.6 Unbalanced two-phase to earth short-circuit faults

Figure 2.9(a) shows a representation of an unbalanced two-phase to earth fault on phases Y and B through a fault impedance Z_F.

The conditions at the point of fault are given by

$$I_R = 0 \quad V_Y = Z_F I_Y \quad V_B = Z_F I_B \tag{2.64}$$

From Equation (2.7a) for currents, and $I_R = 0$, we obtain

$$I^P + I^N + I^Z = 0 \quad \text{or} \quad I^Z = -(I^P + I^N) \tag{2.65}$$

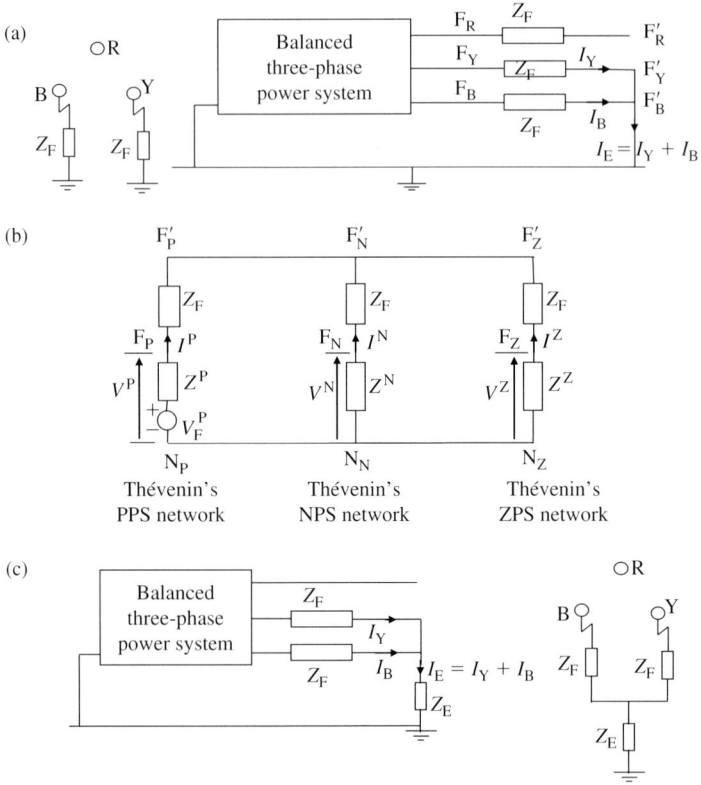

Figure 2.9 Unbalanced two-phase to earth short-circuit fault through a fault impedance Z_F

Rewriting Equations (2.9b) and (2.9c), and using Equation (2.64), we have

$$V_Y = h^2 V^P + hV^N + V^Z = Z_F I_Y = Z_F(h^2 I^P + hI^N + I^Z) \quad (2.66a)$$

$$V_B = hV^P + h^2 V^N + V^Z = Z_F I_B = Z_F(hI^P + h^2 I^N + I^Z) \quad (2.66b)$$

Subtracting Equation (2.66b) from Equation (2.66a), we obtain

$$V_Y - V_B = (h^2 - h)V^P + (h - h^2)V^N = Z_F[(h^2 - h)I^P + (h - h^2)I^N]$$

which, using $h^2 - h = -j\sqrt{3}$, reduces to

$$V^P - Z_F I^P = V^N - Z_F I^N \quad (2.67)$$

Now, substituting Equations (2.19) and (2.20) into Equation (2.67), we obtain

$$-(Z^N + Z_F)I^N = V_F - (Z^P + Z_F)I^P \quad (2.68)$$

Adding Equations (2.66a) and (2.66b), we obtain

$$V_Y + V_B = (h^2 + h)V^P + (h^2 + h)V^N + 2V^Z = Z_F[(h^2 + h)I^P + (h^2 + h)I^N + 2I^Z]$$

which, using $h^2 + h = -1$ and Equations (2.20), (2.21) and (2.67), reduces to

$$V^N - Z_F I^N = V^Z - Z_F I^Z \tag{2.69a}$$

and

$$-(Z^Z + Z_F)I^Z = -(Z^N + Z_F)I^N = V_F - (Z^P + Z_F)I^P \tag{2.69b}$$

Equations (2.65) and (2.69) show that the PPS, NPS and ZPS networks should be connected in parallel, as shown in Figure 2.9(b), with the fault impedance Z_F appearing as an external impedance in each sequence network.

The sequence fault currents are calculated using Equations (2.65) and (2.69), or alternatively from Figure 2.9(b). The use of the equations is illustrated for deriving the sequence currents. Using Equation (2.69) and expressing I^N and I^Z in terms of I^P then substituting into Equation (2.65), we obtain

$$I^P + \frac{(Z^P + Z_F)I^P - V_F}{(Z^N + Z_F)} + \frac{(Z^P + Z_F)I^P - V_F}{(Z^Z + Z_F)} = 0$$

which after a little algebra gives

$$I^P = \frac{[(Z^N + Z_F) + (Z^Z + Z_F)]V_F}{(Z^N + Z_F)(Z^Z + Z_F) + (Z^N + Z_F)(Z^P + Z_F) + (Z^P + Z_F)(Z^Z + Z_F)} \tag{2.70a}$$

$$I^N = \frac{-(Z^Z + Z_F)V_F}{(Z^N + Z_F)(Z^Z + Z_F) + (Z^N + Z_F)(Z^P + Z_F) + (Z^P + Z_F)(Z^Z + Z_F)} \tag{2.70b}$$

and

$$I^Z = \frac{-(Z^N + Z_F)V_F}{(Z^N + Z_F)(Z^Z + Z_F) + (Z^N + Z_F)(Z^P + Z_F) + (Z^P + Z_F)(Z^Z + Z_F)} \tag{2.70c}$$

The phase fault currents are calculated from the sequence currents using Equation (2.15a) giving

$$I_Y = \frac{-j\sqrt{3}V_F[(Z^Z + Z_F) - h(Z^N + Z_F)]}{(Z^N + Z_F)(Z^Z + Z_F) + (Z^N + Z_F)(Z^P + Z_F) + (Z^P + Z_F)(Z^Z + Z_F)} \tag{2.71a}$$

and

$$I_B = \frac{j\sqrt{3}V_F[(Z^Z + Z_F) - h^2(Z^N + Z_F)]}{(Z^N + Z_F)(Z^Z + Z_F) + (Z^N + Z_F)(Z^P + Z_F) + (Z^P + Z_F)(Z^Z + Z_F)}$$

(2.71b)

The total fault current flowing into earth at the point of fault is the sum of the phase fault currents I_Y and I_B giving

$$I_E = I_Y + I_B$$

$$= 3I^Z = \frac{-3(Z^N + Z_F)V_F}{(Z^N + Z_F)(Z^Z + Z_F) + (Z^N + Z_F)(Z^P + Z_F) + (Z^P + Z_F)(Z^Z + Z_F)}$$

(2.72)

The phase voltage on the healthy or unfaulted phase R can be calculated using Equation (2.9a) giving

$$V_R = \frac{3V_F[Z^N Z^Z + Z_F(Z^N + Z^Z + Z_F)]}{(Z^N + Z_F)(Z^Z + Z_F) + (Z^N + Z_F)(Z^P + Z_F) + (Z^P + Z_F)(Z^Z + Z_F)}$$

(2.73)

The phase voltages on the faulted phases Y and B can be calculated using Equation (2.64) giving

$$V_Y = \frac{-j\sqrt{3}V_F Z_F[(Z^Z + Z_F) - h(Z^N + Z_F)]}{(Z^N + Z_F)(Z^Z + Z_F) + (Z^N + Z_F)(Z^P + Z_F) + (Z^P + Z_F)(Z^Z + Z_F)}$$

(2.74a)

$$V_B = \frac{j\sqrt{3}V_F Z_F[(Z^Z + Z_F) - h^2(Z^N + Z_F)]}{(Z^N + Z_F)(Z^Z + Z_F) + (Z^N + Z_F)(Z^P + Z_F) + (Z^P + Z_F)(Z^Z + Z_F)}$$

(2.74b)

The case of a solid or bolted two-phase to earth short-circuit fault is obtained by setting $Z_F = 0$. Also, in the case where an earth impedance Z_E is present in the common connection to earth, as shown in Figure 2.9(c), it can be shown that Z^Z in Figure 2.9(b) must be replaced by $(Z^Z + 3Z_E)$. The reader is encouraged to prove this statement.

2.3.7 Unbalanced one-phase open-circuit faults

Figure 2.10(a) shows the representation of an unbalanced one-phase open-circuit fault occurring on phase R and creating points F and F′ in a balanced three-phase power system.

The conditions imposed by the fault are

$$I_R = 0 \quad \text{hence } I^P + I^N + I^Z = 0 \tag{2.75}$$

Equation (2.75) represents the sequence currents of the phase current I_R from points F to F′. In addition, since these two points are still connected together on phases Y and B, we have

$$V_Y^{FF'} = 0 \quad \text{and} \quad V_B^{FF'} = 0 \tag{2.76}$$

The sequence components of the voltages in Equation (2.76) can be calculated using Equation (2.14) giving

$$V^P = V^N = V^Z = \frac{1}{3} V_R^{FF'} \tag{2.77}$$

Equations (2.75) and (2.77) are satisfied by connecting the PPS, NPS and ZPS equivalent networks in parallel at points F to F′ as shown in Figure 2.10(b). It should be noted that Z^P, Z^N, and Z^Z are the PPS, NPS and ZPS equivalent Thévenin impedances, respectively, as 'seen' looking back into the respective sequence network from between the points F and F′. The voltage source in the PPS equivalent network in Figure 2.10(b) has not been derived yet. This voltage source is the open-circuit PPS voltage appearing across F and F′. This voltage is calculated as the multiplication of the prefault load current in phase R, i.e. the current flowing before

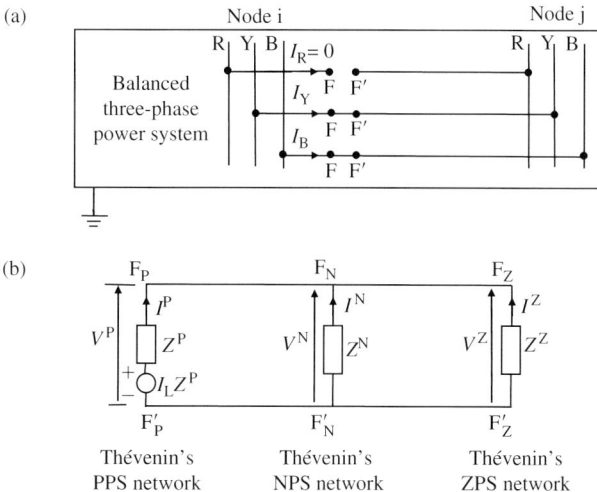

Figure 2.10   Unbalanced one-phase open-circuit fault

the open circuit occurs (note that this is a PPS current because the three-phase power system is balanced), and Z^P, i.e.

$$V_R^{FF'} = I_L Z^P \tag{2.78}$$

From Figure 2.10(b), the PPS current is given by

$$I^P = \frac{I_L Z^P}{\left(Z^P + \frac{Z^N Z^Z}{Z^N + Z^Z}\right)} = \frac{(Z^N + Z^Z)}{(Z^P Z^N + Z^N Z^Z + Z^Z Z^P)} I_L Z^P \tag{2.79}$$

The NPS current is given by

$$I^N = -I^P \frac{Z^Z}{(Z^N + Z^Z)}$$

or

$$I^N = \frac{-Z^Z}{(Z^P Z^N + Z^N Z^Z + Z^Z Z^P)} I_L Z^P \tag{2.80}$$

The ZPS current is given by

$$I^Z = -I^P \frac{Z^N}{(Z^N + Z^Z)}$$

or

$$I^Z = \frac{-Z^N}{(Z^P Z^N + Z^N Z^Z + Z^Z Z^P)} I_L Z^P \tag{2.81}$$

The sequence voltages are given by

$$V^P = V^N = V^Z = \frac{Z^N Z^Z}{(Z^P Z^N + Z^N Z^Z + Z^Z Z^P)} I_L Z^P \tag{2.82}$$

and the phase voltage across the open circuit is given by

$$V_R = \frac{Z^N Z^Z}{(Z^P Z^N + Z^N Z^Z + Z^Z Z^P)} (3 I_L Z^P) \tag{2.83}$$

2.3.8 Unbalanced two-phase open-circuit faults

Figure 2.11(a) shows a representation of an unbalanced two-phase open-circuit fault occurring on phases Y and B and creating points F and F' in a balanced three-phase power system.

The conditions imposed by the fault are given by

$$I_Y = 0 \quad \text{and} \quad I_B = 0 \tag{2.84a}$$

(a)

(b)

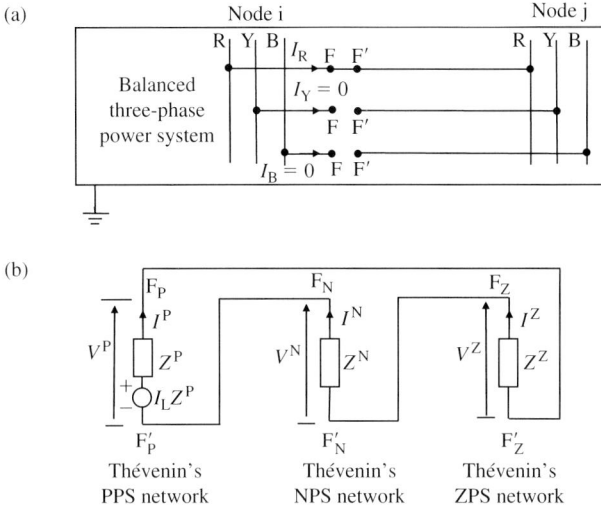

Figure 2.11   Unbalanced two-phase open-circuit fault

The sequence current components in Equation (2.84a) are given by

$$I^P = I^N = I^Z = \frac{1}{3}I_R \qquad (2.84b)$$

In addition, since points F to F' are still connected together on phase R

$$V_R^{FF'} = V^P + V^N + V^Z = 0 \qquad (2.85)$$

Equations (2.84b) and (2.85) are satisfied by connecting the PPS, NPS and ZPS equivalent networks in series at the points F to F' as shown in Figure 2.11(b). From this, the sequence currents are given by

$$I^P = I^N = I^Z = \frac{I_L Z^P}{Z^P + Z^N + Z^Z} \qquad (2.86)$$

where I_L is the prefault current flowing in phase R between F and F' just before the open-circuit fault occurs on phases Y and B.

The sequence voltages are given by

$$V^P = I^P(Z^N + Z^Z) = \frac{(Z^N + Z^Z)}{(Z^P + Z^N + Z^Z)}I_L Z^P \qquad (2.87a)$$

$$V^N = -I^N Z^N = \frac{-Z^N}{(Z^P + Z^N + Z^Z)}I_L Z^P \qquad (2.87b)$$

$$V^Z = -I^Z Z^Z = \frac{-Z^Z}{(Z^P + Z^N + Z^Z)}I_L Z^P \qquad (2.87c)$$

The phase voltages on the faulted phases Y and B are calculated using Equations (2.9b) and (2.9c) giving

$$V_Y^{FF'} = \frac{j\sqrt{3}(hZ^Z - Z^N)}{(Z^P + Z^N + Z^Z)} I_L Z^P \tag{2.88a}$$

$$V_B^{FF'} = \frac{-j\sqrt{3}(h^2Z^Z - Z^N)}{(Z^P + Z^N + Z^Z)} I_L Z^P \tag{2.88b}$$

The case of a three-phase open-circuit fault does not require any consideration since this is not normally a fault but rather a normal switching operation in power systems such as the opening of the three phases of a circuit-breaker. Only PPS currents and voltages continue to exist in the resulting balanced power system. Such studies that involve the opening of the three phases of circuit-breakers, for example, to simulate the disconnection of a circuit and calculate the resultant currents, voltages active and reactive power flows, are known as PPS power flow studies.

2.3.9 Example

Example 2.7   In this example, we assume $Z_F = 0$ and $Z^P = Z^N$:

(a) Compare the relative magnitudes of a two-phase short-circuit fault and a three-phase short-circuit fault.
   From Equations (2.36) and (2.61), we have

$$|I_{3\phi}| = \frac{V_F}{Z^P} \quad \text{and} \quad |I_{2\phi}| = \frac{\sqrt{3}}{2}\frac{V_F}{Z^P} \quad \text{hence } |I_{2\phi}| = 0.866 \times |I_{3\phi}|$$

(b) For both one-phase to earth and two-phase to earth faults, derive general expressions for the earth fault currents in terms of three-phase fault currents and expressions for residual voltages. Comment on the effect of ZPS to PPS impedance ratio.

Let $K_{ZP} = Z^Z/Z^P$. The residual voltage is the sum of the three-phase voltage phasors. In actual three-phase power systems, both residual voltages and earth fault currents are measured using current and voltage transformers. These measurements are used for the detection of earth faults by protection relays.

One-phase short-circuit fault
The earth fault current is given by Equation (2.48). Hence,

$$I_{1\phi} = \frac{3}{2Z^P + Z^Z}\frac{V_F}{Z^P} = \frac{3}{2 + K_{ZP}}I_{3\phi} \quad \text{or} \quad \frac{I_{1\phi}}{I_{3\phi}} = \frac{3}{2 + K_{ZP}}$$

Using Equations (2.50) and (2.51), the residual voltage is equal to

$$V_{E(1\phi)} = V_R + V_Y + V_B = \frac{-3K_{ZP}}{2 + K_{ZP}} V_F \quad \text{or} \quad \frac{V_{E(1\phi)}}{V_F} = \frac{-3K_{ZP}}{2 + K_{ZP}}$$

Two-phase to earth short-circuit fault

The earth fault current is given by Equation (2.72). Hence,

$$I_{E(2\phi-E)} = \frac{-3}{1 + 2K_{ZP}} \frac{V_F}{Z^P} = \frac{-3}{1 + 2K_{ZP}} I_{3\phi} \quad \text{or} \quad \frac{I_{E(2\phi-E)}}{I_{3\phi}} = \frac{-3}{1 + 2K_{ZP}}$$

Using Equation (2.73), the residual voltage is equal to

$$V_{E(2\phi-E)} = V_R + V_Y + V_B = \frac{3K_{ZP}}{1 + 2K_{ZP}} V_F \quad \text{or} \quad \frac{V_{E(2\phi-E)}}{V_F} = \frac{3K_{ZP}}{1 + 2K_{ZP}}$$

$$K_{ZP} = \frac{Z^Z}{Z^P} = \frac{R^Z + jX^Z}{R^P + jX^P} \approx \frac{R^Z + jX^Z}{jX^P} = \frac{X^Z}{X^P} - j\frac{R^Z}{X^P}$$

The complex ratio K_{ZP} has been simplified assuming that in high voltage networks, the PPS X^P/R^P ratio is generally larger than 5. It is quite interesting to examine the variation of

$$\frac{I_{1\phi}}{I_{3\phi}} = \frac{3}{2 + K_{ZP}} \qquad \frac{I_{E(2\phi-E)}}{I_{3\phi}} = \frac{-3}{1 + 2K_{ZP}}$$

$$\frac{V_{E(1\phi)}}{V_F} = \frac{-3K_{ZP}}{2 + K_{ZP}} \qquad \frac{V_{E(2\phi-E)}}{V_F} = \frac{3K_{ZP}}{1 + 2K_{ZP}}$$

as functions of K_{ZP} where the latter varies from 0 to 5.

For $K_{ZP} = 0$, 1, 5 and ∞, $I_{1\phi}/I_{3\phi} = 1.5$, 1, 0.43 and 0, respectively, and $|I_{E(2\phi-E)}/I_{3\phi}| = 3$, 1, 0.27 and 0, respectively. In other words, for $K_{ZP} = (Z^Z/Z^P) < 1$, $I_{1\phi} > I_{3\phi}$ and $I_{E(2\phi-E)} > I_{3\phi}$, and $I_{E(2\phi-E)} > I_{1\phi}$, and vice versa. For $K_{ZP} = 0$, 1, 5 and ∞, $|V_{E(1\phi)}/V_F| = 0$, 1, 2.14 and 3, respectively, and $|V_{E(2\phi-E)}/V_F| = 0$, 1, 1.36 and 1.5, respectively.

2.4 Fault analysis and choice of reference frame

2.4.1 General

The mathematical equations derived in previous sections for unbalanced short-circuit and open-circuit faults were based on arranging the unbalance conditions to be symmetrical about phase R which was taken as the reference phase. In this section, we will show that whilst any phase could be chosen as the reference, the choice of phase R results in the simplest mathematical derivations.

2.4.2 One-phase to earth short-circuit faults

In the case of a one-phase to earth short circuit on phase R with phase R chosen as the reference, the symmetrical components of the faulted phase are given by

$$V_R^P + V_R^N + V_R^Z = 0 \quad \text{and} \quad I_R^P = I_R^N = I_R^Z \tag{2.89}$$

The above equations are represented in Figure 2.7.

However, for a one-phase to earth short circuit on phase Y with phase Y chosen as the reference, the symmetrical components of the faulted phase are given by

$$V_Y^P + V_Y^N + V_Y^Z = 0 \quad \text{and} \quad I_Y^P = I_Y^N = I_Y^Z \tag{2.90}$$

Expressing the sequence currents and voltages of Equation (2.90) in terms of phase R using Equation (2.8), we obtain

$$h^2 V_R^P + h V_R^N + V_R^Z = 0 \quad \text{and} \quad h^2 I_R^P = h I_R^N = I_R^Z \tag{2.91}$$

Similarly, for a one-phase to earth short circuit on phase B with phase B chosen as the reference, the symmetrical components of the faulted phase are given by

$$V_B^P + V_B^N + V_B^Z = 0 \quad \text{and} \quad I_B^P = I_B^N = I_B^Z \tag{2.92}$$

Expressing the sequence currents and voltages of Equation (2.92) in terms of phase R using Equation (2.8), we obtain

$$h V_R^P + h^2 V_R^N + V_R^Z = 0 \quad \text{and} \quad h I_R^P = h^2 I_R^N = I_R^Z \tag{2.93}$$

Equations (2.89), (2.91) and (2.93) can be represented by the connection of the PPS, NPS and ZPS networks as shown in Figure 2.12. The sequence networks are still connected in series but through three complex multipliers k^P, k^N and k^Z. These multipliers are applied to the sequence voltages and currents and are equal to either 1, h or h^2 according to Equations (2.89), (2.91) and (2.93) and as shown in Case 1 in Figure 2.12. The function of the multiplier is to apply the same phase shift to the relevant sequence current and voltage whilst keeping their magnitudes unchanged. In describing this multiplier, we have deliberately avoided the use of the term phase shifting transformer so as to avoid confusion with the property of physical transformers that transform voltages and currents by inverse ratios or multipliers.

Figure 2.12 shows that the phase shifts are applied to the PPS and NPS voltages and currents for faults on either phase Y or B. However, the ZPS current and voltage multiplier is always equal to unity.

For convenience in practical analysis, it is normal to avoid applying phase shifts to the active PPS network, which contains generating or voltage sources, and instead apply the phase shifts to the passive NPS and ZPS networks. This is easily

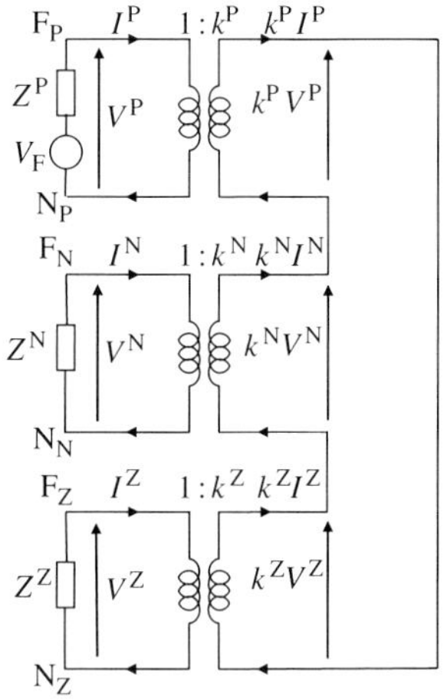

One-phase short circuit on phase:

	k^P	k^N	k^Z
(R–E R is reference)	1	1	1
(Y–E Y is reference)	h^2	h	1
(B–E B is reference)	h	h^2	1

Case 1

$h = e^{j2\pi/3}$

One-phase short circuit on phase:

	k^P	k^N	k^Z
(R–E R is reference)	1	1	1
(Y–E Y is reference)	1	h^2	h
(B–E B is reference)	1	h	h^2

Case 2

Figure 2.12 One-phase short-circuit fault on either phase R, Y or B

accomplished by dividing Equations (2.91) and (2.93) by h^2 and h, respectively, and the resulting multipliers k^P, k^N and k^Z are as shown in Case 2 in Figure 2.12.

In summary, in the case of a single-phase to earth short-circuit fault on any phase in a three-phase network, the assumption of the fault being on phase R results in the simplest mathematical equations because the three multipliers k^P, k^N and k^Z are all equal to unity.

2.4.3 Two-phase to earth short-circuit faults

For a two-phase to earth short-circuit fault, we have three cases as follows:

(a) Phase Y to B to earth short-circuit fault, phase R is the reference phase

$$V_R^P = V_R^N = V_R^Z \quad \text{and} \quad I_R^P + I_R^N + I_R^Z = 0 \qquad (2.94)$$

(b) Phase R to B to earth short-circuit fault, phase Y is the reference phase

$$V_Y^P = V_Y^N = V_Y^Z \quad \text{and} \quad I_Y^P + I_Y^N + I_Y^Z = 0 \qquad (2.95a)$$

or expressed in terms of phase R

$$h^2 V_R^P = h V_R^N = V_R^Z \quad \text{and} \quad h^2 I_R^P + h I_R^N + I_R^Z = 0 \qquad (2.95b)$$

(c) Phase R to Y to earth short-circuit fault, phase B is the reference phase

$$V_B^P = V_B^N = V_B^Z \quad \text{and} \quad I_B^P + I_B^N + I_B^Z = 0 \tag{2.96a}$$

or expressed in terms of phase R

$$hV_R^P = h^2 V_R^N = V_R^Z \quad \text{and} \quad hI_R^P + h^2 I_R^N + I_R^Z = 0 \tag{2.96b}$$

Equations (2.94), (2.95b) and (2.7b) can be represented by the connection of the PPS, NPS and ZPS networks as shown in Figure 2.13. The sequence networks are still connected in parallel but through three complex multipliers k^P, k^N and k^Z that are equal to either 1, h or h^2 according to Equations (2.95b) and (2.96b) as shown in Case 1 in Figure 2.13. Again, applying phase shifts to the active PPS network can be avoided by dividing Equations (2.95b) and (2.96b) by h^2 and h, respectively, and the resulting multipliers k^P, k^N and k^Z are as shown in Case 2 in Figure 2.13.

A comparison of the tables showing the complex multipliers k^P, k^N and k^Z in Figures 2.12 and 2.13 show that for the same reference phase, each multiplier, k^P, k^N or k^Z, has the same value irrespective of the fault type.

The methodology presented can be easily extended to any other unbalanced short-circuit or unbalanced open-circuit fault. The reader is encouraged to repeat the analysis for other fault conditions. These include: (a) a one-phase open-circuit fault on any phase, i.e. R or Y or B; (b) a two-phase open-circuit fault on any two phases, i.e. R–Y or Y–B or R–B; and (c) a two-phase short-circuit fault on any two phases, i.e. R–Y or Y–B or R–B.

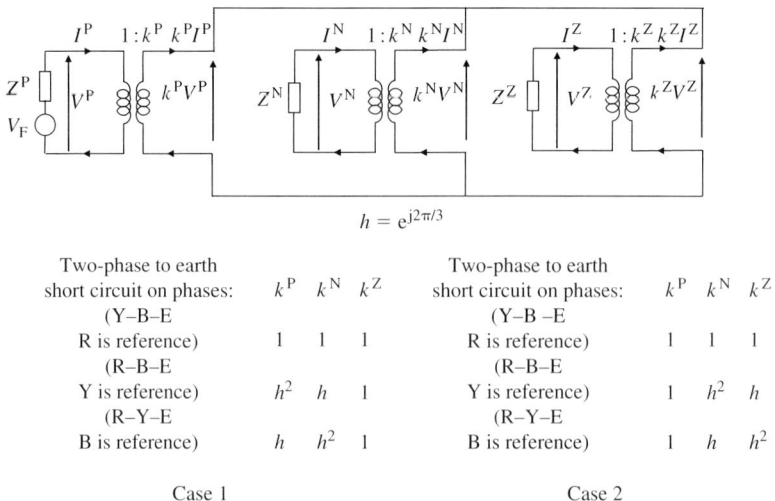

$$h = e^{j2\pi/3}$$

Two-phase to earth short circuit on phases:	k^P	k^N	k^Z		Two-phase to earth short circuit on phases:	k^P	k^N	k^Z
(Y–B–E R is reference)	1	1	1		(Y–B –E R is reference)	1	1	1
(R–B–E Y is reference)	h^2	h	1		(R–B–E Y is reference)	1	h^2	h
(R–Y–E B is reference)	h	h^2	1		(R–Y–E B is reference)	1	h	h^2
Case 1					Case 2			

Figure 2.13   Two-phase to earth short circuit on Y–B–E, R–B–E or R–Y–E

2.5 Analysis of simultaneous faults

2.5.1 General

Simultaneous faults are more than one fault that occur at the same time in a three-phase power system either at the same or at different locations. Because there is a very large theoretical combination of such faults, we will limit our attention to a representative number of cases that are of practical interest. These are three cases of two simultaneous faults and one case of three simultaneous faults. The analysis of simultaneous faults can be simplified by deriving the conditions at the fault locations with respect to the same reference phase, i.e. phase R. These conditions determine the method of connection of the PPS, NPS and ZPS networks. In this analysis, we assume that the full network has been reduced to an equivalent as seen from the faulted boundary locations and the latter, denoted as nodes J and L, have been retained. Network reduction is discussed in detail in Chapter 8.

2.5.2 Simultaneous short-circuit faults at the same location

The two-simultaneous faults we consider are illustrated in Figure 2.14(a) and consist of a solid one-phase to earth short-circuit fault and a solid two-phase short-circuit fault. Conceptually, since all three phases are faulted, the two faults can also be considered as a single three-phase unbalanced fault! Using phase R as the

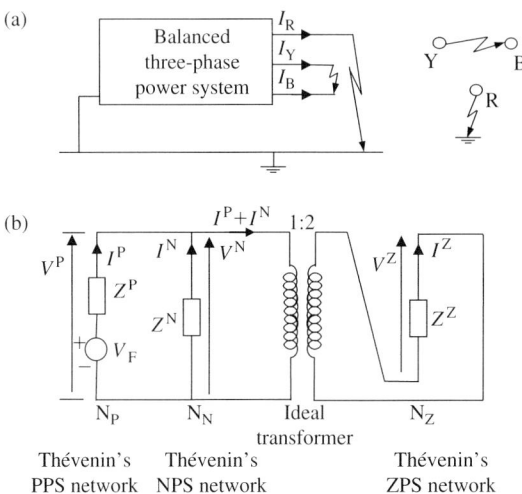

Figure 2.14 Simultaneous one-phase to earth and two-phase short-circuits at the same location, or three-phase unbalanced short-circuit fault

reference phase, the single-phase short circuit assumed to occur on phase R and the two-phase short circuit assumed to occur on phases Y and B, the conditions at the fault location are given by

$$V_R = 0 \quad I_Y + I_B = 0 \quad V_Y = V_B \tag{2.97}$$

Using Equations (2.9b) and (2.9c) for I_Y and I_B instead of voltages, we obtain

$$I^P + I^N = 2I^Z \tag{2.98}$$

Substituting Equations (2.19), (2.20) and (2.21) for PPS, NPS and ZPS voltages at the fault point into Equations (2.9), we have

$$V_R = (V_F - Z^P I^P) + (-Z^N I^N) + (-Z^Z I^Z) = 0 \tag{2.99a}$$

$$V_Y = h^2(V_F - Z^P I^P) + h(-Z^N I^N) + (-Z^Z I^Z) \tag{2.99b}$$

and

$$V_B = h(V_F - Z^P I^P) + h^2(-Z^N I^N) + (-Z^Z I^Z) \tag{2.99c}$$

Adding Equations (2.99), we obtain $V_R + V_Y + V_B = -3Z^Z I^Z = 2V_Y = 2V_B$ or

$$V_Y = V_B = \frac{-3Z^Z I^Z}{2} \tag{2.100}$$

Also, using Equations (2.99b) and (2.99c) as well as $V_Y = V_B$ from Equation (2.97), we have $V_Y - hV_B = (1 - h)V_Y = (1 - h)(Z^N I^N - Z^Z I^Z)$ or

$$V_Y = Z^N I^N - Z^Z I^Z \tag{2.101}$$

Equating Equations (2.100) and (2.101), we obtain

$$I^Z = \frac{-2Z^N}{Z^Z} I^N \tag{2.102}$$

Substituting Equation (2.102) into Equation (2.98), we obtain

$$I^N = \frac{Z^Z}{4Z^N + Z^Z} I^P \quad \text{and} \quad I^Z = \frac{2Z^N}{4Z^N + Z^Z} I^P \tag{2.103}$$

Substituting Equations (2.103) into Equation (2.99a), the sequence fault currents are

$$I^P = \frac{4Z^N + Z^Z}{4Z^P Z^N + Z^N Z^Z + Z^Z Z^P} V_F \qquad (2.104a)$$

$$I^N = \frac{-Z^Z}{4Z^P Z^N + Z^N Z^Z + Z^Z Z^P} V_F \qquad (2.104b)$$

$$I^Z = \frac{2Z^N}{4Z^P Z^N + Z^N Z^Z + Z^Z Z^P} V_F \qquad (2.104c)$$

Using Equation (2.9), the phase fault currents are given by

$$I_R = \frac{6Z^N}{4Z^P Z^N + Z^N Z^Z + Z^Z Z^P} V_F \qquad (2.105a)$$

$$I_Y = I_B = \frac{-j\sqrt{3}(2Z^N + Z^Z)}{4Z^P Z^N + Z^N Z^Z + Z^Z Z^P} V_F \qquad (2.105b)$$

Subtracting Equation (2.9c) from Equation (2.9b) and using $V_Y = V_B$, we obtain $V^P = V^N$. Using Equation (2.9a) with $V_R = 0$ and $V^P = V^N$, we obtain $V^Z = -2V^P$. Using Equation (2.20), we obtain

$$V^P = V^N = \frac{Z^N Z^Z}{4Z^P Z^N + Z^N Z^Z + Z^Z Z^P} V_F \qquad (2.106a)$$

and

$$V^Z = \frac{-2Z^N Z^Z}{4Z^P Z^N + Z^N Z^Z + Z^Z Z^P} V_F \qquad (2.106b)$$

Finally, using Equation (2.100), the phase voltages on phases Y and B are given by

$$V_Y = V_B = \frac{-3Z^N Z^Z}{4Z^P Z^N + Z^N Z^Z + Z^Z Z^P} V_F \qquad (2.106c)$$

Using Equation (2.98) or $I^Z = (I^P + I^N)/2$ and $V^Z = -2V^P = -2V^N$, the equivalent circuit of this simultaneous fault condition is shown in Figure 2.14(b).

It should be noted that the real transformation ratio acts on voltage and current as a normal ideal transformer does.

2.5.3 Cross-country faults or simultaneous faults at different locations

A cross-country fault is a condition where there are two simultaneous one-phase to earth short-circuit faults affecting the same circuit but at different locations and possibly involving different phases. Therefore, the fault conditions are a one-phase to earth, phase R–Earth (E) short-circuit fault at location J and one-phase to earth short-circuit fault on either R–E, Y–E or B–E at location L. The fault conditions at both fault locations are given as follows:

Location J: R–E short-circuit fault

$$V_J^P + V_J^N + V_J^Z = 0 \quad \text{and} \quad I_J^P = I_J^N = I_J^Z \tag{2.107}$$

Location L: R–E short-circuit fault

$$V_L^P + V_L^N + V_L^Z = 0 \quad \text{and} \quad I_L^P = I_L^N = I_L^Z \tag{2.108}$$

Location L: Y–E short-circuit fault

$$h^2 V_L^P + h V_L^N + V_L^Z = 0 \quad \text{and} \quad h^2 I_L^P = h I_L^N = I_L^Z$$

or

$$V_L^P + h^2 V_L^N + h V_L^Z = 0 \quad \text{and} \quad I_L^P = h^2 I_L^N = h I_L^Z \tag{2.109}$$

Location L: B–E short-circuit fault

$$h V_L^P + h^2 V_L^N + V_L^Z = 0 \quad \text{and} \quad h I_L^P = h^2 I_L^N = I_L^Z$$

or

$$V_L^P + h V_L^N + h^2 V_L^Z = 0 \quad \text{and} \quad I_L^P = h I_L^N = h^2 I_L^Z \tag{2.110}$$

Figure 2.15 shows the connection of the PPS, NPS and ZPS networks at the two fault locations J and L using the three sets of Equations (2.107) and (2.108), (2.107) and (2.109), and (2.107) and (2.110).

Straightforward analysis using Kirchoff's voltage and current laws can be carried out using Figure 2.15.

2.5.4 Simultaneous open-circuit and short-circuit faults at the same location

We now consider a one-phase open-circuit fault on phase R at location J and a one-phase to earth short-circuit fault R–E or Y–E or B–E at the same location.

The conditions created by the open-circuit fault on phase R at location J creating open-circuit J–J′ are given by

$$V_{JJ'}^P = V_{JJ'}^N = V_{JJ'}^Z \quad \text{and} \quad I_{JJ'}^P + I_{JJ'}^N + I_{JJ'}^Z = 0 \tag{2.111}$$

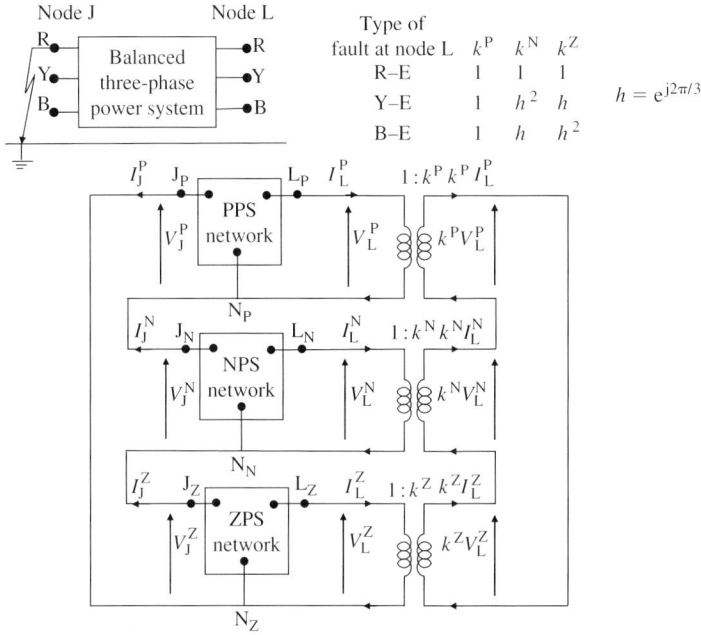

Figure 2.15 Cross-country simultaneous one-phase to earth short-circuit faults

The conditions created by a R–E short-circuit fault at location J are given by

$$V_J^P + V_J^N + V_J^Z = 0 \quad \text{and} \quad I_J^P = I_J^N = I_J^Z \qquad (2.112)$$

The conditions created by a Y–E short-circuit fault at location J are given by

$$h^2 V_J^P + h V_J^N + V_J^Z = 0 \quad \text{and} \quad h^2 I_J^P = h I_J^N = I_J^Z$$

or

$$V_J^P + h^2 V_J^N + h V_J^Z = 0 \quad \text{and} \quad I_J^P = h^2 I_J^N = h I_J^Z \qquad (2.113)$$

The conditions created by a B–E short-circuit fault at location J are given by

$$h V_J^P + h^2 V_J^N + V_J^Z = 0 \quad \text{and} \quad h I_J^P = h^2 I_J^N = I_J^Z$$

or

$$V_J^P + h V_J^N + h^2 V_J^Z = 0 \quad \text{and} \quad I_J^P = h I_J^N = h^2 I_J^Z \qquad (2.114)$$

Figure 2.16 shows the connection of the PPS, NPS and ZPS networks using the three sets of equations (2.111) and (2.112), (2.111) and (2.113), and (2.111) and (2.114).

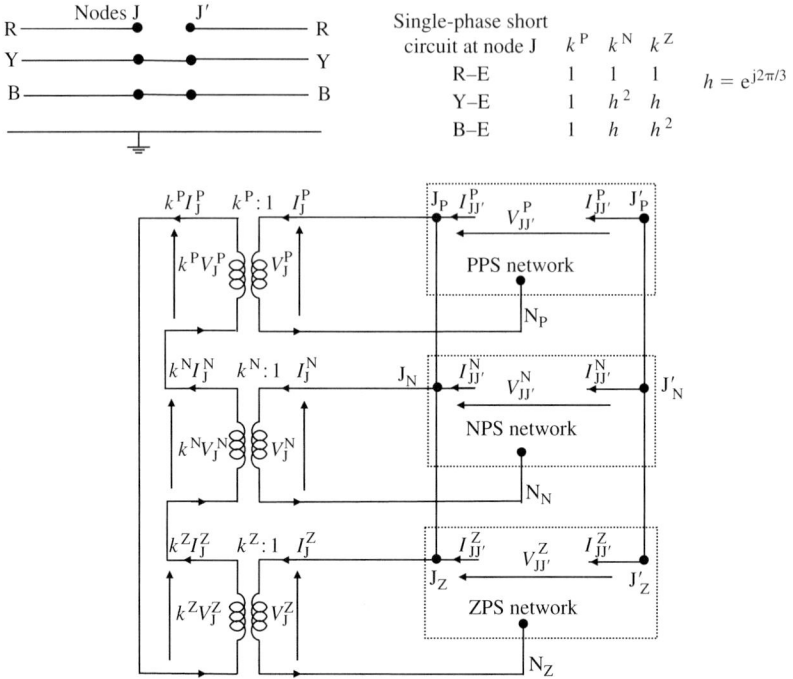

Figure 2.16 Simultaneous one-phase open-circuit and one-phase short-circuit faults

Straightforward analysis using Kirchoff's voltage and current laws can be carried out using Figure 2.16.

2.5.5 Simultaneous faults caused by broken and fallen to earth conductors

Figure 2.17 illustrates a one-phase open-circuit fault on phase R caused by a broken phase conductor. The conductors on both sides of the open circuit are assumed to fall to earth thus creating a one-phase to earth short-circuit fault on each side.

The conditions created by the open-circuit fault on phase R creating open-circuit J–J' are given by

$$V_{JJ'}^{P} = V_{JJ'}^{N} = V_{JJ'}^{Z} \quad \text{and} \quad I_{JJ'}^{P} + I_{JJ'}^{N} + I_{JJ'}^{Z} = 0 \qquad (2.115)$$

Short-circuit fault on phase R at location J

$$V_{J}^{P} + V_{J}^{N} + V_{J}^{Z} = 0 \quad \text{and} \quad I_{J}^{P} = I_{J}^{N} = I_{J}^{Z} \qquad (2.116)$$

Short-circuit fault on phase R at location J'

$$V_{J'}^{P} + V_{J'}^{N} + V_{J'}^{Z} = 0 \quad \text{and} \quad I_{J'}^{P} = I_{J'}^{N} = I_{J'}^{Z} \qquad (2.117)$$

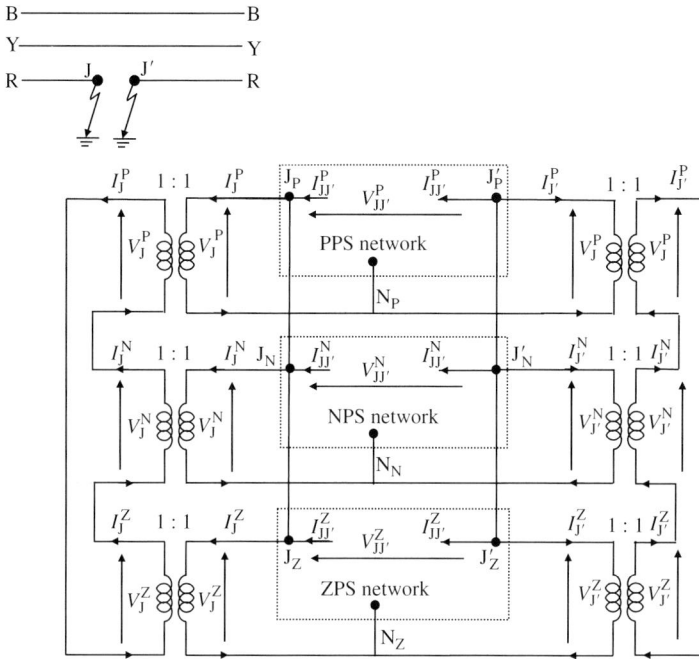

Figure 2.17   Three simultaneous faults caused by a broken and fallen to earth conductor

Figure 2.17 shows the connection of the PPS, NPS and ZPS networks using Equations (2.115), (2.116) and (2.117).

Straightforward analysis using Kirchoff's voltage and current laws can be carried out using Figure 2.17.

2.5.6 Simultaneous short-circuit and open-circuit faults on distribution transformers

We will now analyse in detail a simultaneous fault case that involves a two-winding distribution transformer that is extensively used in distribution substations. The high voltage winding is delta connected with fuses being used as incoming protection against high currents. The low voltage winding is star connected with the neutral solidly earthed and the transformer supplies a balanced three-phase load. The simultaneous faults may be created when a short-circuit fault occurs on the transformer low voltage side which causes one fuse to blow and clear before the other fuses, or a circuit-breaker upstream, open and clear the fault. This can lead to a situation where the short-circuit currents are too low to operate any further protection so that the prolonged duration of such currents may overstress or even damage power plant. The transformer is assumed to supply a balanced star-connected three-phase static load having an impedance per phase of Z_L and therefore the corresponding load PPS, NPS and ZPS impedances are all equal to Z_L. The detailed sequence modelling of transformers will be covered in Chapter 4, but we will present this case now because of its practical importance and relevance in this chapter. We will also make use of the transformer phase shifts introduced by the star–delta winding connections and presented in Chapter 4, and we will consider the transformer vector group to be D11yn or in accordance with American ANSI standard. This means that when stepping up from the low voltage to the high voltage side, the transformer's PPS currents (and voltages) are advanced by $30°$ whereas the NPS currents (and voltages) are retarded by $30°$. The simultaneous faults are illustrated in Figure 2.18(a) and the equivalent sequence network connections are shown in Figure 2.18(b).

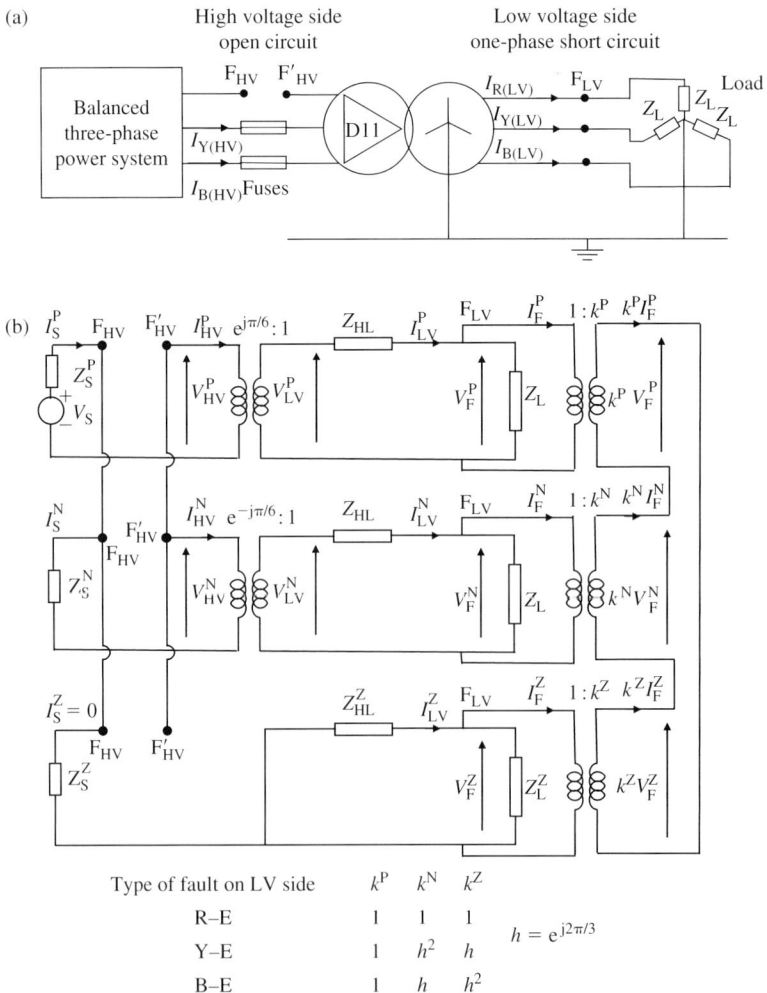

Type of fault on LV side	k^P	k^N	k^Z
R–E	1	1	1
Y–E	1	h^2	h
B–E	1	h	h^2

$$h = e^{j2\pi/3}$$

Figure 2.18 Simultaneous one-phase short circuit on transformer low voltage side and one-phase open circuit on transformer high voltage side

From Figure 2.18(b), the following relations can be written

$$I_S^P = I_{HV}^P = I_{LV}^P e^{j30°} \quad \text{and} \quad V_{HV}^P = V_{LV}^P e^{j30°} \tag{2.118}$$

$$I_S^N = I_{HV}^N = I_{LV}^N e^{-j30°} \quad \text{and} \quad V_{HV}^N = V_{LV}^N e^{-j30°} \tag{2.119}$$

$$I_S^P = -I_S^N \tag{2.120}$$

$$k^P I_F^P = k^N I_F^N = k^Z I_F^Z \quad \text{and} \quad k^P V_F^P + k^N V_F^N + k^Z V_F^Z = 0 \tag{2.121}$$

Therefore, from the above current equations, we obtain

$$I_{LV}^P e^{j30°} = -I_{LV}^N e^{-j30°} \quad I_{LV}^N = -I_{LV}^P e^{j60°} \tag{2.122}$$

and

$$I_F^P = k^N I_F^N \quad I_F^Z = \frac{k^N}{k^Z} I_F^N \quad k^P = 1 \tag{2.123}$$

Applying Kirchoff's voltage law to the high voltage side of the PPS network of Figure 2.18(b), we obtain $V_S = I_S^P Z_S^P - I_S^N Z_S^N + V_{HV}^P - V_{HV}^N$ or using Equations (2.118)–(2.121)

$$V_S = (Z_S^P + Z_S^N) I_{LV}^P e^{j30°} + V_{LV}^P e^{j30°} - V_{LV}^N e^{-j30°} \tag{2.124}$$

Applying Kirchoff's voltage law to the low voltage side of the PPS network of Figure 2.18(b), and after a little algebra, we obtain

$$V_{LV}^P = I_{LV}^P (Z_{HL} + Z_L) - I_F^P Z_L \tag{2.125}$$

Applying Kirchoff's voltage law to the low voltage side of the NPS network of Figure 2.18(b), and using Equation (2.122), we obtain

$$V_{LV}^N = -(Z_{HL} + Z_L) I_{LV}^P e^{j60°} - I_F^N Z_L \tag{2.126}$$

Substituting Equations (2.125) and (2.126) into Equation (2.124), and using Equation (2.123), we obtain, after a little algebra

$$V_S = [(Z_S^P + Z_S^N) + 2(Z_{HL} + Z_L)] I_{LV}^P e^{j30°} - Z_L I_F^N (k^N e^{j30°} - e^{-j30°}) \tag{2.127}$$

Using Equations (2.122) and (2.123) as well as Figure 2.18(b), the PPS, NPS and ZPS voltages at the short-circuit fault point are given by

$$V_F^P = Z_L (I_{LV}^P - k^N I_F^N) \tag{2.128}$$

$$V_F^N = -Z_L (I_{LV}^P e^{j60°} + I_F^N) \tag{2.129}$$

From the low voltage side of the ZPS network of Figure 2.18(b), we have $V_F^Z = Z_L^Z(I_{LV}^Z - I_F^Z) = -Z_{HL}^Z I_{LV}^Z$ which, using Equation (2.123) and Figure 2.18(b) becomes

$$V_F^Z = -\frac{k^N}{k^Z} \frac{Z_L^Z Z_{HL}^Z}{Z_L^Z + Z_{HL}^Z} I_F^N \tag{2.130}$$

Now, by substituting Equations (2.128), (2.129) and (2.130) into the sequence voltages of Equation (2.121), we can express I_F^N in terms of I_{LV}^P as

$$I_F^N = \frac{Z_L(1 - k^N e^{j60°})}{k^N \left(2Z_L + \frac{Z_L^Z Z_{HL}^Z}{Z_L^Z + Z_{HL}^Z}\right)} I_{LV}^P \tag{2.131}$$

Substituting Equation (2.131) into Equation (2.127) and solving for I_{LV}^P, we obtain

$$I_{LV}^P = \frac{V_S e^{-j30°}}{[(Z_S^P + Z_S^N) + 2(Z_{HL} + Z_L)] - \frac{Z_L^2[k^N(2 - k^N e^{j60°}) - e^{-j60°}]}{k^N \left(2Z_L + \frac{Z_L^Z Z_{HL}^Z}{Z_L^Z + Z_{HL}^Z}\right)}} \tag{2.132}$$

The calculation of I_{LV}^P enables us to calculate the required sequence currents and voltages in Figure (2.18b) using Equations (2.118) to (2.131) by back substitution. We will present in Chapter 6 that the voltage source V_S is calculated from the initial load flow solution just before the occurrence of the short-circuit fault as follows:

$$V_S = (Z_S^P + Z_{HL})I_{Load} + V_{Load} \tag{2.133}$$

The short-circuit fault current on the faulted phase can be calculated from Equation (2.123) as follows:

$$I_F = 3k^N I_F^N \tag{2.134}$$

The corresponding currents on the high voltage side can be calculated using Equations (2.15a), (2.118), (2.119) and (2.122) to give

$$I_Y = \sqrt{3} I_{LV}^P e^{-j60°} \tag{2.135}$$

and

$$I_B = \sqrt{3} I_{LV}^P e^{j120°} = -I_Y \tag{2.136}$$

The reader is encouraged to repeat the above analysis for a solid two-phase to earth short-circuit fault on the transformer low voltage side considering all three fault combinations of R–Y–E, R–B–E and Y–B–E.

Further reading

Books

[1] Wagner, C.F., *et al.*, *Symmetrical Components*, McGraw-Hill Book Company, Inc., 1933.

[2] Anderson, P.M., *Analysis of Faulted Power Systems*, Iowa State Press, Ames, IA, 1973.

[3] Grainger, J. and Stevenson, W.D., *Power System Analysis*, 1994, ISBN 0701133380.

[4] Blackburn, J.L., *Symmetrical Components for Power Systems Engineering*, 1993, ISBN 0824787676.

[5] Elgerd, O.I., *Electric Energy Systems Theory*, McGraw-Hill Int. Ed., 1983, ISBN 0-07-66273-8.

[6] Weedy, B.M., *Electric Power Systems*, John Wiley & Sons, 1967, ISBN 0-471-92445-8.

Paper

[7] Fortescue, C.L., Method of symmetrical coordinates applied to the solution of polyphase networks, *Transactions of AIEE*, Vol. 37, 1918, 1027–1140.

Modelling of multi-conductor overhead lines and cables

3.1 General

In this chapter, we present the modelling of multi-conductor overhead lines and cables both in the phase and sequence frames of reference. Calculations and measurement techniques of the electrical parameters, or constants, of lines and cables are described. Transposition analysis of single-circuit and multiple-circuit overhead lines, and sheaths and cores of cables are presented as well as their π models in the sequence and phase frames of reference.

3.2 Phase and sequence modelling of three-phase overhead lines

3.2.1 Background

The transmission and distribution of three-phase electrical power on overhead lines requires the use of at least three-phase conductors. Most low voltage lines use three-phase conductors forming a single three-phase circuit. Many higher voltage lines consist of a single three-phase circuit or two three-phase circuits strung or suspended from the same tower structure and usually called a double-circuit line. The two circuits may be strung in a variety of configurations such as vertical, horizontal or triangular configurations. Figure 3.1 illustrates typical single-circuit

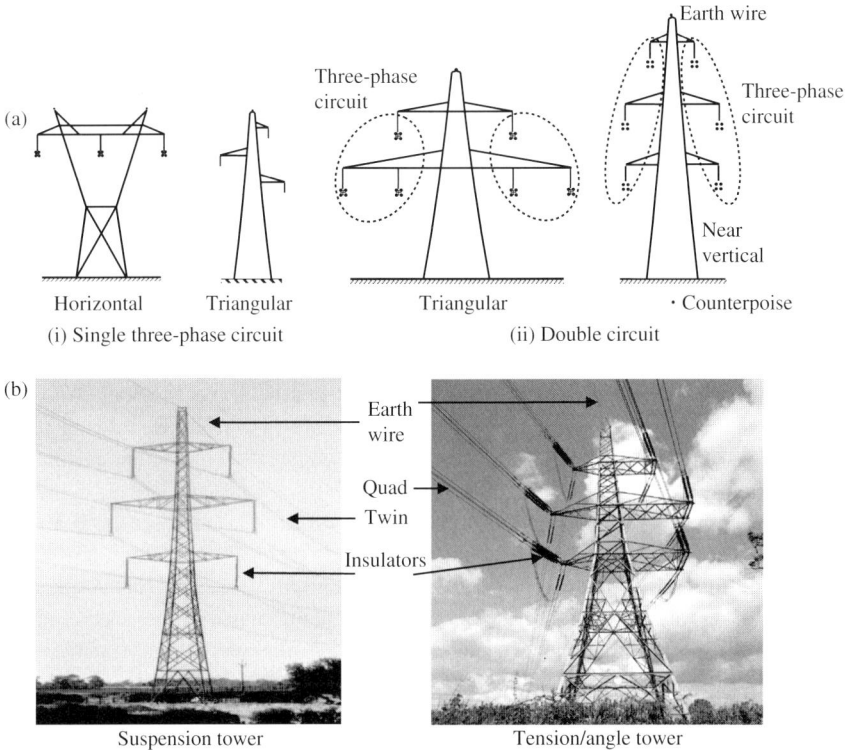

Figure 3.1 (a) Typical single-circuit and double-circuit overhead lines and (b) double-circuit overhead lines with one earth wire: twin bundle = 2 conductors per phase and quad bundle = 4 conductors per phase

lines and double-circuit lines in horizontal, triangular and vertical phase conductor arrangements. A line may also consist of two circuits running physically in parallel but on different towers. In addition, a few lines have been built with three, four or even six three-phase circuits strung on the same tower structure in various horizontal and/or triangular formations. In England and Wales, almost 99% of the 400 kV and 275 kV overhead transmission system consists of vertical or near vertical double-circuit line configurations.

In addition to the phase conductors, earth wire conductors may be strung to the tower top and normally bonded to the top of the earthed tower. Earth wires perform two important functions; shielding the phase conductors from direct lightning strikes and providing a low impedance path for the short-circuit fault current in the event of a back flashover from the phase conductors to the tower structure. The ground itself over which the line runs is an important additional lossy conductor having a complex and distributed electrical characteristics. In the case of high resistivity or lossy earths, it is usual to use a counterpoise, i.e. a wire buried underground beneath the tower base and connected to the footings of the towers. This serves to reduce the effective tower footing resistance. Where a metallic pipeline runs in close proximity to an overhead line, a counterpoise may also be used in parallel with the pipeline in order to reduce the induced voltage on the pipeline from the power line.

Therefore, a practical overhead transmission line is a complex arrangement of conductors all of which are mutually coupled not only to each other but also to earth. The mutual coupling is both electromagnetic (i.e. inductive) and electrostatic (i.e. capacitive). The asymmetrical positions of the phase conductors with respect to each other, the earth wire(s) and/or the surface of the earth cause some unbalance in the phase impedances, and hence currents and voltages. This is undesirable and in order to minimise the effect of line unbalance, it is possible to interchange the conductor positions at regular intervals along the line route, a practice known as transposition. The aim of this is to achieve some averaging of line parameters and hence balance for each phase. However, in practice, and in order to avoid the inconvenience, costs and delays, most lines are not transposed along their routes but transposition is carried out where it is physically convenient at the line terminals, i.e. at substations.

Bundled phase conductors are usually used on transmission lines at 220 kV and above. These are constructed with more than one conductor per phase separated at regular intervals along the span length between two towers by metal spacers. Conductor bundles of two, three, four, six and eight are in use in various countries and in Great Britain, two, three (triangle) and four (square or rectangle) conductor bundles are used at 275 and 400 kV. The purpose of bundled conductors is to reduce the voltage gradients at the surface of the conductors because the bundle appears as an equivalent conductor of much larger diameter than that of the component conductors. This minimises active losses due to corona, reduces noise generation, e.g. radio interference, reduces the inductive reactance and increases the capacitive susceptance or capacitance of the line. The latter two effects improve the steady state power transfer capability of the line. Figure 3.1(a)(ii) shows a typical 400 kV double-circuit line of vertical phase conductor arrangement having four bundled conductors per phase, one earth wire and one counterpoise wire. The total number of conductors in such a multi-conductor system is $(4 \times 3) \times 2 \mid 1 \mid 1 = 26$ conductors, all of which are mutually coupled to each other and to earth.

3.2.2 Overview of the calculation of overhead line parameters

General

A line is a static power plant that has electrical parameters distributed along its length. The basic parameters of the line are conductor series impedance and shunt admittance. Each conductor has a self-impedance and there is a mutual impedance between any two conductors. The impedance generally consists of a resistance and a reactance. The shunt admittance consists of the conductor's conductance to ground and the susceptance between conductors and between each conductor and earth. The conductance of the air path to earth represents the leakage current along the line insulators due to corona. This is negligibly small and is normally ignored in short circuit, power flow and transient stability analysis.

Practical calculations of multi-conductor line parameters with series impedance expressed in pu length (e.g. Ω/km) and shunt susceptance in μS/km are carried out using digital computer programs. These parameters are then used to form the line series impedance and shunt admittance matrices in the phase frame of reference as will be described later. The line capacitances or susceptances are calculated from the line potential coefficients which are essentially dependent on the line and tower physical dimensions and geometry. The calculations use the method of image conductors, assumes that the earth is a plane at a uniform zero potential and that the conductor radii are much smaller than the spacings among the conductors. The self and mutual impedances depend on the conductor material, construction, tower or line physical dimensions or geometry, and on the earth's resistivity. Figure 3.2 is a general illustration of overhead line tower physical dimensions and spacings of conductors above the earth's surface as well as conductor images below earth used for the calculation of line electrical parameters.

The fundamental theories used in the calculations of resistance, inductance and capacitance parameters of overhead lines are extensively covered in most basic power system textbooks and will not be repeated here. Practical digital computer

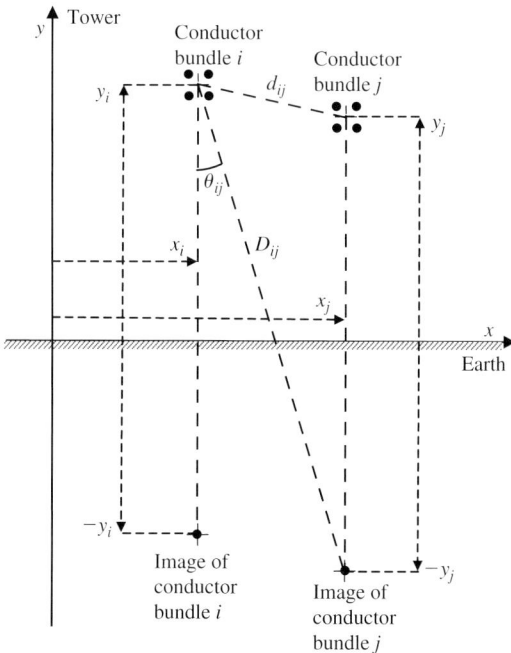

Figure 3.2 A general illustration of an overhead line physical dimensions and conductor spacings relative to tower centre and earth

based calculations used in industry consider the effect of earth as a lossy conducting medium. The equations used in the calculation of line parameters are presented below.

Potential coefficients, shunt capacitances and susceptances

Using Figure 3.2, given a set of N conductors, the potential V of conductor i due to conductor's own charge and charges on all other conductors is given by

$$V_i|_{i=1,\ldots,N} = \sum_{j=1}^{N} P_{ij}Q_j \ \text{V} \tag{3.1}$$

where P_{ij} is the Maxwell's potential coefficient expressed in km/F and Q_j is the charge in C/km. The equations assume an infinitely long perfectly horizontal conductors above earth whose effect is included using the method of electrostatic images. This method is generally valid up to a frequency of about 1 MHz. The potential of a conductor i above earth due to its own charge and an equal but negative charge on its own image enables us to calculate the self-potential coefficient of conductor i as follows:

$$P_{ii} = 17.975109 \times \log_e \left(\frac{2y_i}{r_i}\right) \text{km}/\mu\text{F} \tag{3.2a}$$

where y_i is the height of conductor i above earth in m and r_i is the radius of conductor i in m. Clearly y_i is much greater than r_i. The potential of conductor i due to a charge on conductor j and an equal but negative charge on the image of conductor j enables us to calculate the mutual potential coefficient between conductor i and conductor j as follows:

$$P_{ij} = 17.975109 \times \log_e \left(\frac{D_{ij}}{d_{ij}}\right) \quad \text{and} \quad P_{ji} = P_{ij} \ \text{km}/\mu\text{F} \tag{3.2b}$$

where D_{ij} is the distance between conductor i and the image beneath the earth's surface of conductor j in m, and d_{ij} is the distance between conductor i and conductor j in m. Equation (3.1) can be rewritten in matrix form as

$$\mathbf{V} = \mathbf{PQ} \ \text{V} \tag{3.3a}$$

where $\mathbf{P}$ is a potential coefficient matrix of $N \times N$ dimension. Multiplying both sides of Equation (3.3a) by $\mathbf{P}^{-1}$, we obtain

$$\mathbf{Q} = \mathbf{CV} \ \text{C} \tag{3.3b}$$

where

$$\mathbf{C} = \mathbf{P}^{-1} \ \mu\text{F/km} \tag{3.3c}$$

C is line's shunt capacitance matrix and is equal to the inverse of the potential coefficient matrix **P**. Under steady state conditions, the current and voltage vectors are phasors and are related by

$$\mathbf{I} = \mathbf{YV} = \mathbf{jBV} = \mathbf{j}\omega\mathbf{CV} \text{ amps} \tag{3.4a}$$

since the line's conductance is negligible at power frequency f. The line's shunt susceptance matrix is given by

$$\mathbf{B} = \omega\mathbf{C} = 2\pi f\mathbf{C}\,\mu\text{S/km} \tag{3.4b}$$

Series self and mutual impedances

Self impedance

The equations assume infinitely long and perfectly horizontal conductors above a homogeneous conducting earth having a uniform resistivity $\rho_e(\Omega\text{m})$ and a unit relative permeability. Proximity effect between conductors is neglected. Using Figure 3.2, the series voltage drop in V of each conductor due to current flowing in the conductor itself and currents flowing in all other conductors in the same direction is given by

$$V_i|_{i=1,\dots,N} = \sum_{j=1}^{N} Z_{ij}I_j \text{ V/km} \tag{3.5a}$$

where Z is the impedance expressed in Ω/km and I is the current in amps. The self-impedance of conductor i is given by

$$Z_{ii} = [R_{i(c)} + \mathbf{j}X_{i(c)}] + \mathbf{j}X_{i(g)} + [R_{i(e)} + \mathbf{j}X_{i(e)}]\,\Omega/\text{km} \tag{3.5b}$$

where subscript c represents the contribution of conductor i resistance and internal reactance, g represents a reactance contribution to conductor i due to its geometry, i.e. an external reactance contribution and e represents correction terms to conductor i resistance and reactance due to the contribution of the earth return path. If skin effect is neglected, i.e. assuming a direct current (dc) condition or zero frequency, the internal dc impedance of a solid magnetic round conductor, illustrated in Figure 3.3(a), is given by

$$Z_{i(c)} = R_{i(c)} + \mathbf{j}X_{i(c)} = \frac{1000\rho_c}{\pi r_c^2} + \mathbf{j}4\pi 10^{-4}f(\mu_r/4)\,\Omega/\text{km} \tag{3.6}$$

where μ_r and ρ_c are the relative permeability and resistivity of the conductor, respectively, and r_c is the conductor's radius.

Where skin effect is to be taken into account, the following exact equation for the internal impedance of a solid round conductor can be used

$$Z_{i(c)} = \frac{1000\rho_c}{2\pi r_c} \times (\sqrt{2}\delta^{-1}e^{\mathbf{j}\pi/4}) \times \frac{I_0[(\sqrt{2}\delta^{-1}r_c)e^{\mathbf{j}\pi/4}]}{I_1[(\sqrt{2}\delta^{-1}r_c)e^{\mathbf{j}\pi/4}]}\,\Omega/\text{km} \tag{3.7a}$$

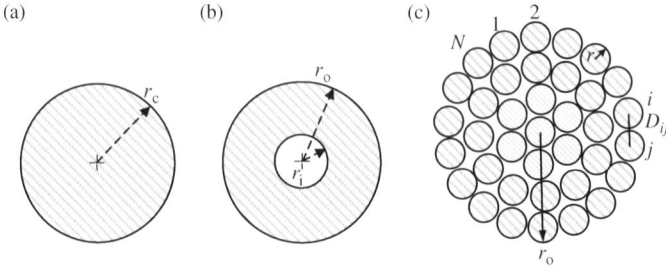

Figure 3.3 Illustration of some conductor types: cross-section of a (a) solid round conductor, (b) tubular round conductor and (c) 30/7 stranded conductor (30 Aluminium, 7 inner steel stands)

where

$$\sqrt{2}\,\delta^{-1}e^{j\pi/4} = \frac{1}{\delta} + j\frac{1}{\delta}$$

is defined as the complex propagation constant and δ is skin depth or depth of penetration into the conductor and is given by

$$\delta = 503.292 \times \sqrt{\frac{\rho_c}{f\mu_r}}\,\text{m} \qquad (3.7b)$$

and I_i are modified Bessel functions of the first kind of order i. For calculation of line parameters close to power frequency, the following alternative equation, suitable for hand calculations using electronic calculators, is found accurate up to about 200 Hz:

$$Z_{i(c)} = \frac{1000\rho_c}{\pi r_c^2}\left(1 + \frac{\pi^4 r_c^4 f^2 \mu_c^2}{3\times10^{14}\rho_c^2}\right) + j4\pi 10^{-4}f\frac{\mu_r}{4}\left(1 - \frac{\pi^4 r_c^4 f^2 \mu_c^2}{6\times10^{14}\rho_c^2}\right)\,\Omega/\text{km} \qquad (3.8)$$

Equation (3.8) shows that skin effect causes an increase in the conductor's effective ac resistance and a decrease in its effective ac internal reactance. Also, at $f = 0$, Equation (3.8) reduces to Equation (3.6).

In the case of a tubular or hollow conductor, illustrated in Figure 3.3(b), the dc internal impedance is given by

$$Z_{i(c)} = \frac{1000\rho_c}{\pi(r_o^2 - r_i^2)} + j4\pi 10^{-4}f\frac{\mu_r}{4}\left[1 - \frac{2r_i^2}{(r_o^2 - r_i^2)} + \frac{4r_i^4}{(r_o^2 - r_i^2)^2}\log_e\left(\frac{r_o}{r_i}\right)\right]\,\Omega/\text{km} \qquad (3.9)$$

Mathematically, the case of a solid round conductor is a special case of the hollow conductor since by setting $r_i = 0$, Equation (3.9) reduces to Equation (3.6). Aluminium conductor steel reinforced (ACSR) or modern gapped-type conductors can be represented as hollow conductors if the effect of steel saturation can be ignored.

Saturation may be caused by the flow of current through the helix formed by each Aluminium strand that produces a magnetic field within the steel.

Where skin effect of a hollow conductor is to be taken into account, the following exact equation for the internal impedance of such a conductor can be used

$$Z_{i(c)} = \frac{1000\rho_c}{2\pi(r_o^2 - r_i^2)}\left(1 - \frac{r_i^2}{r_o^2}\right)(\sqrt{2}\delta^{-1}r_o)j\frac{D_1}{D_2}\ \Omega/\text{km} \tag{3.10a}$$

$$\begin{aligned}D_2 &= I_0[(\sqrt{2}\delta^{-1}r_o)e^{j\pi/4}] \times K_1[(\sqrt{2}\delta^{-1}r_i)e^{j\pi/4}] + I_1[(\sqrt{2}\delta^{-1}r_i)e^{j\pi/4}]\\ &\quad \times K_0[(\sqrt{2}\delta^{-1}r_o)e^{j\pi/4}]\end{aligned}$$

$$\begin{aligned}D_2 &= I_1[(\sqrt{2}\delta^{-1}r_o)e^{j\pi/4}] \times K_1[(\sqrt{2}\delta^{-1}r_i)e^{j\pi/4}] - I_1[(\sqrt{2}\delta^{-1}r_i)e^{j\pi/4}]\\ &\quad \times K_1[(\sqrt{2}\delta^{-1}r_o)e^{j\pi/4}]\end{aligned} \tag{3.10b}$$

where I_i and K_i are modified Bessel functions of the first and second kind of order i, respectively. Mathematical solutions suitable for digital computations that provide sufficient accuracy are available in standard handbooks of mathematical functions and also in modern libraries of digital computer programs. Equation (3.7a) that represents the case of a solid conductor can be obtained from Equation (3.10a) by substituting $r_i = 0$.

The external reactance of conductor i due to its geometry is given by

$$X_{i(g)} = 4\pi 10^{-4}f \log_e\left(\frac{2y_i}{r_i}\right)\ \Omega/\text{km} \tag{3.11}$$

In practice, the internal reactance of a conductor is much smaller than its external reactance except in the case of very large conductors at high frequencies. Generally, precision improvements in the former would have very small effect on the total reactance.

The contribution of the earth's correction terms to the self-impedance of conductor i is presented after the next section.

Mutual impedances

The mutual impedance between conductor i and conductor j is given by

$$Z_{ij} = jX_{ij(g)} + [R_{ij(e)} + jX_{ij(e)}]\ \Omega/\text{km} \tag{3.12a}$$

and

$$X_{ij(g)} = 4\pi 10^{-4}f \log_e\left(\frac{D_{ij}}{d_{ij}}\right)\ \Omega/\text{km} \tag{3.12b}$$

Earth return path impedances

The contributions of correction terms to the self and mutual impedances, due to the earth return path, are generally given as infinite series. The resistance and reactance general correction terms are calculated in terms of two parameters m_{ij} and θ_{ij} as follows

$$R_{ij(e)} = 4\pi 10^{-4} f \times 2 \left[\frac{\pi}{8} - b_1 m_{ij} \cos\theta_{ij} + b_2 m_{ij}^2 \right.$$

$$\times \left\{ \log_e \left(\frac{e^{C2}}{m_{ij}} \right) \cos(2\theta_{ij}) + \theta_{ij} \sin(2\theta_{ij}) \right\} + b_3 m_{ij}^3 \cos(3\theta_{ij})$$

$$\left. - d_4 m_{ij}^4 \cos(4\theta_{ij}) - b_5 m_{ij}^5 \cos(5\theta_{ij}) + \cdots \right] \tag{3.13a}$$

$$X_{ij(e)} = 4\pi 10^{-4} f \times 2 \left[\frac{1}{2} \log_e \left(\frac{1.85138}{m_{ij}} \right) + b_1 m_{ij} \cos\theta_{ij} - d_2 m_{ij}^2 \cos(2\theta_{ij}) \right.$$

$$+ b_3 m_{ij}^3 \cos(3\theta_{ij}) - b_4 m_{ij}^4 \left\{ \log_e \left(\frac{e^{C4}}{m_{ij}} \right) \cos(4\theta_{ij}) \right.$$

$$\left. \left. + \theta_{ij} \sin(4\theta_{ij}) \right\} + b_5 m_{ij}^5 \cos(5\theta_{ij}) - \cdots \right] \tag{3.13b}$$

where

$$m_{ij} = \begin{cases} \sqrt{2} \times \dfrac{2y_i}{\delta} & \text{for the self-impedance terms or } i = j \\ \sqrt{2} \times \dfrac{D_{ij}}{\delta} & \text{for the mutual impedance terms or } i \neq j \end{cases} \tag{3.14a}$$

$$\theta_{ij} = \begin{cases} 0 & \text{for the self-impedance terms or } i = j \\ \cos^{-1}\left(\dfrac{y_i + y_j}{D_{ij}} \right) & \text{for the mutual impedance terms or } i \neq j \end{cases} \tag{3.14b}$$

and δ is the skin depth defined in Equation (3.7b). The coefficients used in Equation (3.13) are given by

$$b_1 = \frac{1}{3\sqrt{2}} \quad b_2 = \frac{1}{16} \quad b_n = b_{n-2} \frac{\text{sign}}{n(n+2)} \quad C_2 = 1.3659315$$

$$C_n = C_{n-2} + \frac{1}{n} + \frac{1}{n+2} \quad \text{and} \quad d_n = \frac{\pi}{4} b_n$$

The sign in b_n alternates every four terms that is sign $= +1$ for $n = 1, 2, 3, 4$ then sign $= -1$ for $n = 5, 6, 7, 8$ and so on.

Various forms of Equation (3.13) can be given depending on the value of m_{ij}. For short-circuit, power flow and transient stability analysis, the line parameters are calculated at power frequency, i.e. 50 or 60 Hz. For such calculations, m_{ij} is normally less than unity and generally one term in the series would be sufficient to give good accuracy. At higher frequencies, two cases are distinguished; the first is for $1 < m_{ij} \leq 5$, where the full series is usually used whereas for $m_{ij} > 5$, the series converges to an asymptotic form and simple finite expressions that provide acceptable accuracy may be used provided that $\theta_{ij} < 45°$. Generally, the number of correction terms required increases with frequency if sufficient accuracy is to be maintained.

Using the skin depth δ of Equation (3.7b), and an earth relative permeability of unity, the effect of earth return path is defined as an equivalent conductor at a depth given by

$$D_{\text{erc}} = 1.309125 \times \delta$$

or

$$D_{\text{erc}} = 658.87 \times \sqrt{\frac{\rho_e}{f}}\ \text{m} \tag{3.15}$$

Therefore, for the self-impedance, the resistance and reactance correction terms are given by

$$R_{i(e)} = \pi^2 10^{-4} f - \frac{f}{911.812 \times \dfrac{D_{\text{erc}}}{2y_i}} + \cdots \ \Omega/\text{km} \tag{3.16a}$$

$$X_{i(e)} = 4\pi 10^{-4} f \log_e \left(\frac{D_{\text{erc}}}{2y_i} \right) + \frac{f}{911.812 \times \dfrac{D_{\text{erc}}}{2y_i}} + \cdots \ \Omega/\text{km} \tag{3.16b}$$

and for the mutual impedance, the correction terms are given by

$$R_{ij(e)} = \pi^2 10^{-4} f - \frac{f}{911.812 \times \dfrac{D_{\text{erc}}}{y_i + y_j}} + \cdots \ \Omega/\text{km} \tag{3.17a}$$

and

$$X_{ij(e)} = 4\pi 10^{-4} f \log_e \left(\frac{D_{\text{erc}}}{D_{ij}} \right) + \frac{f}{911.812 \times \dfrac{D_{\text{erc}}}{y_i + y_j}} + \cdots \ \Omega/\text{km} \tag{3.17b}$$

Typical values of earth resistivity are: 1–20 Ωm for garden and marshy soil, 10–100 Ωm for loam and clay, 60–200 Ωm for farmland, 250–500 Ωm for sand, 300–1000 Ωm for pebbles, 1000–10 000 Ωm for rock and 10^9 Ωm for sandstone. Resistivity of sea water is typically 0.1–1 Ωm. Using an average earth resistivity of a reasonably wet soil of 20 Ωm, the depth of the equivalent earth return conductor D_{erc} given in Equation (3.15) is equal to 416.7 m at 50 Hz and 380.4 m at 60 Hz.

Summary of self and mutual impedances

Substituting Equations (3.11), (3.16a) and (3.16b) in Equation (3.5), the self-impedance of conductor i with earth return is given by

$$Z_{ii} = R_{i(c)} + \pi^2 10^{-4} f - \frac{f}{911.812 \times \dfrac{D_{erc}}{2y_i}} + \cdots$$

$$+ j \left[X_{i(c)} + 4\pi 10^{-4} f \log_e \left(\frac{2y_i}{r_i} \right) + 4\pi 10^{-4} f \log_e \left(\frac{D_{erc}}{2y_i} \right) \right.$$

$$\left. + \frac{f}{911.812 \times \dfrac{D_{erc}}{2y_i}} + \cdots \right] \Omega/\text{km} \qquad (3.18a)$$

Substituting Equations (3.12) and (3.17) in Equation (3.12), the mutual impedance between conductor i and conductor j with earth return is given by

$$Z_{ij} = \pi^2 10^{-4} f - \frac{f}{911.812 \times \dfrac{D_{erc}}{y_i + y_j}} + \cdots$$

$$+ j \left[4\pi 10^{-4} f \log_e \left(\frac{D_{ij}}{d_{ij}} \right) + 4\pi 10^{-4} f \log_e \left(\frac{D_{erc}}{D_{ij}} \right) \right.$$

$$\left. + \frac{f}{911.812 \times \dfrac{D_{erc}}{y_i + y_j}} + \cdots \right] \Omega/\text{km} \qquad (3.18b)$$

For non-digital computer calculations of impedances at power frequency, e.g. using electronic hand calculators, the use of the first earth correction term for resistance and reactance is usually sufficient. Assuming that the power frequency inductance of a general tubular conductor does not appreciably reduce below its dc value given in Equation (3.9), Equation (3.18a) reduces to

$$Z_{ii} = R_{i(c)} + \pi^2 10^{-4} f + j4\pi 10^{-4} f \left[\frac{\mu_r}{4} \times f(r_o, r_i) + \log_e \left(\frac{D_{erc}}{r_o} \right) \right] \Omega/\text{km} \qquad (3.19a)$$

where

$$f(r_o, r_i) = 1 - \frac{2r_i^2}{(r_o^2 - r_i^2)} + \frac{4r_i^4}{(r_o^2 - r_i^2)^2} \log_e \left(\frac{r_o}{r_i} \right) \qquad (3.19b)$$

For a non-magnetic solid conductor with $\mu_r = 1$ and $r_i = 0$, we obtain from Equation (3.19b), $f(r_o, r_i) = 1$ giving a value for the internal inductance in

Equation (3.19a) of 1/4. Combining this with the logarithmic term, the latter changes to $\log_e[D_{\text{erc}}/(0.7788 \times r_{\text{o}})]$ where $0.7788 \times r_{\text{o}}$ is known as the geometric mean radius of the conductor.

Similarly, combining its two logarithmic terms, Equation (3.18b) reduces to

$$Z_{ij} = \pi^2 10^{-4} f + j4\pi 10^{-4} f \log_e \left(\frac{D_{\text{erc}}}{d_{ij}} \right) \Omega/\text{km} \qquad (3.20a)$$

When calculating mutual impedances between conductors of circuits that are separated by a distance d, Equation (3.20a) is generally sufficiently accurate for

$$d_{ij} \leq \frac{D_{\text{erc}}}{7.32} \, \text{m} \qquad (3.20b)$$

For an earth resistivity of 20 Ωm and $f = 50$ Hz, $D_{\text{erc}} = 416.7$ m and $d_{ij} \leq 57$ m.

Stranded conductors

Overhead lines with stranded conductors are extensively used and their effect is important in calculating power frequency line parameters. Figure 3.3(c) illustrates a conductor that consists of N strands. Even a tightly packed bundle of a large number of strands will still leave small unfilled gaps between the strands. The total cross-sectional area of the strands is smaller than the equivalent cross-sectional area of a solid conductor. This reduction factor is the area of the conductors divided by the area of the equivalent solid conductor and can be used to increase the equivalent conductor effective resistivity whilst keeping its overall radius the same. However, for line parameter calculations, the equivalent geometric mean radius of the conductor is calculated and for a general conductor that consists of N strands each having a radius r, this is given by

$$\text{GMR}_c = \left(r^N \prod_{i=1}^{N-1} \prod_{j=i+1}^{N} D_{ij}^2 \right)^{N-2} \qquad (3.21)$$

The effect of stranding causes a reduction in the outer radius of the equivalent conductor of the bundle of strands. A reduction factor, termed the stranding factor, is defined as the ratio of the geometric mean radius, GMR_c, of the conductor to the conductor's outer radius and this can be calculated for any known stranding arrangement. For example, for a homogeneous conductor that consists of seven strands arranged on a hexagonal array, it can be shown that the area reduction factor is equal to 0.77778 and the stranding factor is equal to 0.72557.

Bundled phase conductors

Figure 3.4 illustrates a general case of a phase bundle consisting of N conductors per phase with asymmetrical spacings between the conductors. Practical examples

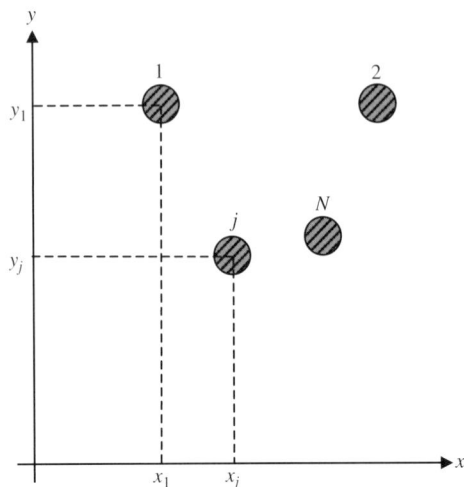

Figure 3.4   General asymmetrical bundle of phase conductors

of such asymmetrical arrangement are rectangular and non-equilateral triangular bundles. Asymmetrical conductor bundles of two or more conductors can be
represented by an equivalent single conductor. The radius of the equivalent conductor can be calculated by applying the GMR technique to an arbitrary set of axes
and origin as shown in Figure 3.4. The GMR of the equivalent conductor of the
entire bundle is given by

$$
\text{GMR}_{\text{Eq}} = \left\{ (\text{GMR}_c)^N \prod_{i=1}^{N-1} \prod_{j=i+1}^{N} \left[(x_i - x_j)^2 + (y_i - y_j)^2 \right] \right\}^{N-2}
\tag{3.22}
$$

where N is ≥ 2 and is the number of conductors in the bundle. When impedance
calculations are carried out, GMR_c is the geometric mean radius of one conductor
in the bundle. When potential coefficient calculations are carried out, GMR_c is
equal to the conductor's outer radius.

   Special cases of symmetrical conductor bundles are shown in Figure 3.5. This
shows bundles of two, three, four and six conductors, with circles drawn through
the centres of the individual conductors. It can be shown that the general Equation
(3.22) of the equivalent GMR_{Eq} reduces to

$$
\text{GMR}_{\text{Eq}} = \sqrt[N]{N \times \text{GMR}_c \times A^{N-1}}
\tag{3.23a}
$$

where A is the radius of a circle through the centres of the bundled conductors and
is given by

$$
A = \frac{d}{2 \times \sin\left(\frac{\pi}{N}\right)}
\tag{3.23b}
$$

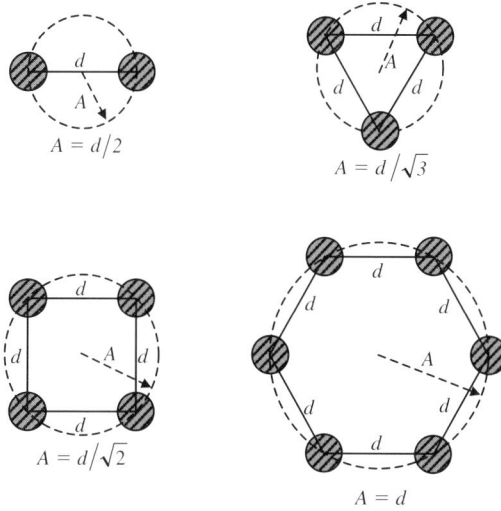

$A = d/2$

$A = d/\sqrt{3}$

$A = d/\sqrt{2}$

$A = d$

Figure 3.5 Typical symmetrical bundles of phase conductors

and d is the spacing between any two adjacent conductors in the symmetrical bundle.

It should be noted that for N subconductors in a bundle, the internal resistance and inductance included in Equation (3.19a) must be divided by N.

An alternative method of bundling phase conductors is to form the full phase impedance and susceptance matrices from the parameters of all conductors including bundled subconductors and earth wires. For example, for a single-circuit line with one earth wire and four conductors per phase, the dimension of the resultant phase impedance matrix is 13×13. The bundled conductors are effectively short-circuited by zero impedance so the voltages are equal for all the subconductors in the bundle and the sum of currents in all subconductors is equal to the equivalent phase current. For example, for phase R, the subconductor voltages are $V_1 = V_2 = V_3 = V_4 = V_R$ whereas for the subconductor currents, $I_1 + I_2 + I_3 + I_4 = I_R$. Using these voltage and current constraints, the bundle conductors can then be combined using standard matrix reduction techniques to produce a 4×4 matrix that represents three equivalent phase conductors and one earth wire.

Average height of conductor above earth

The calculation of line parameters is based on the assumption of perfectly horizontal conductors above the earth's plane. The average sag of phase conductors and earth wires between towers together with the height of the conductor at the tower can be used to calculate an average conductor height for use in the calculation of line's electrical parameters. Using Figure 3.6, it can be shown that the average

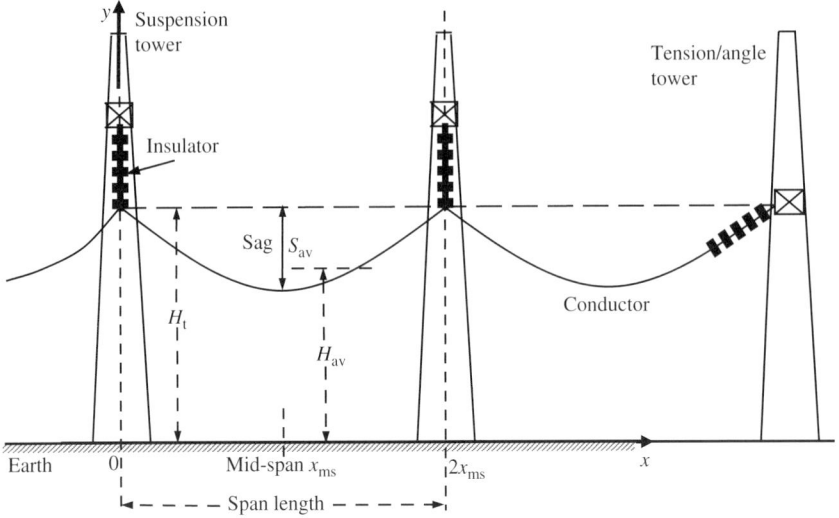

Figure 3.6   Average conductor height above earth between two towers

height of the earth wire conductor above ground is given by

$$H_{av} = \frac{\sqrt{(2H_t - S_{av})S_{av}}}{\log_e \left(\frac{H_t + \sqrt{(2H_t - S_{av})S_{av}}}{H_t - S_{av}} \right)} \text{ m} \tag{3.24}$$

where H_t is the conductor height at the tower in m and S_{av} is the average conductor sag in m measured at mid-span between two suspension towers and usually assumed to apply to the entire line length, i.e. ignoring the effect of angle or tension towers. For span lengths of up to 400 to 500 m, a simple formula can be derived by assuming that the variation of conductor height with distance between the two towers is a parabola, that is

$$y = H_t - \frac{2S_{av}}{x_{ms}}x + \frac{S_{av}}{x_{ms}^2}x^2 \text{ m} \tag{3.25}$$

where x_{ms} is half the span length. The average height above ground between the two towers is calculated by integrating Equation (3.25) over the span length, i.e.

$$H_{av} = y_{av} = \frac{1}{2x_{ms}} \times \int_0^{2x_{ms}} y \, dx$$

Thus,

$$H_{av} = H_t - \frac{2}{3}S_{av} \text{ m} \tag{3.26}$$

The calculation of average conductor height should take into account the different sags of phase and earth wire conductors.

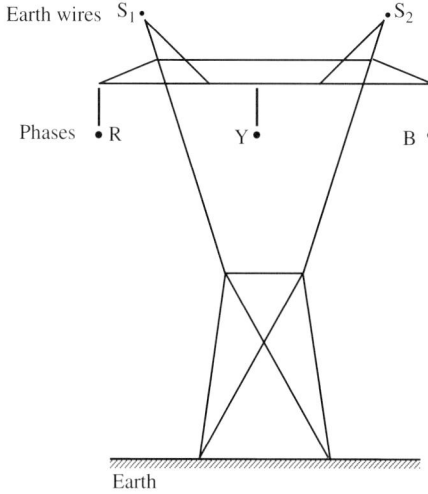

Figure 3.7   Multi-conductor single-circuit three-phase line with two earth wires

3.2.3 Untransposed single-circuit three-phase lines with and without earth wires

Consider a general case of a multi-conductor single-circuit three-phase overhead line with asymmetrical spacings between the conductors, two earth wires and with earth return as shown in Figure 3.7.

Phase and sequence series impedance matrices

Figure 3.8(a) shows the coupled series inductive circuit of Figure 3.7 where each phase and earth wire conductors, and earth are represented as equivalent self and mutually coupled impedances.

To derive a general formulation of the series phase impedance matrix for such a line, the phase R series impedance voltage drop equation can be written as

$$V_R - V_R' = \Delta V_R = Z_{RR}I_R + Z_{RY}I_Y + Z_{RB}I_B - Z_{RS1}I_{S1} - Z_{RS2}I_{S2}$$
$$- Z_{RE}I_E + V_E \tag{3.27a}$$

$$V_E = Z_{EE}I_E - Z_{ER}I_R - Z_{EY}I_Y - Z_{EB}I_B + Z_{ES1}I_{S1} + Z_{NS2}I_{S2} \tag{3.27b}$$

With the phases at the receiving end all earthed, $I_R + I_Y + I_B = I_E + I_{S1} + I_{S2}$ and using Equation (3.27b) in Equation (3.27a), we have

$$\Delta V_R = (Z_{RR} - 2Z_{RE} + Z_{EE})I_R + (Z_{RY} - Z_{RE} - Z_{YE} + Z_{EE})I_Y$$
$$+ (Z_{RB} - Z_{RE} - Z_{BE} + Z_{EE})I_B + (Z_{RE} - Z_{RS1} + Z_{S1E} - Z_{EE})I_{S1}$$
$$+ (Z_{RE} - Z_{RS2} + Z_{S2E} - Z_{EE})I_{S2}$$

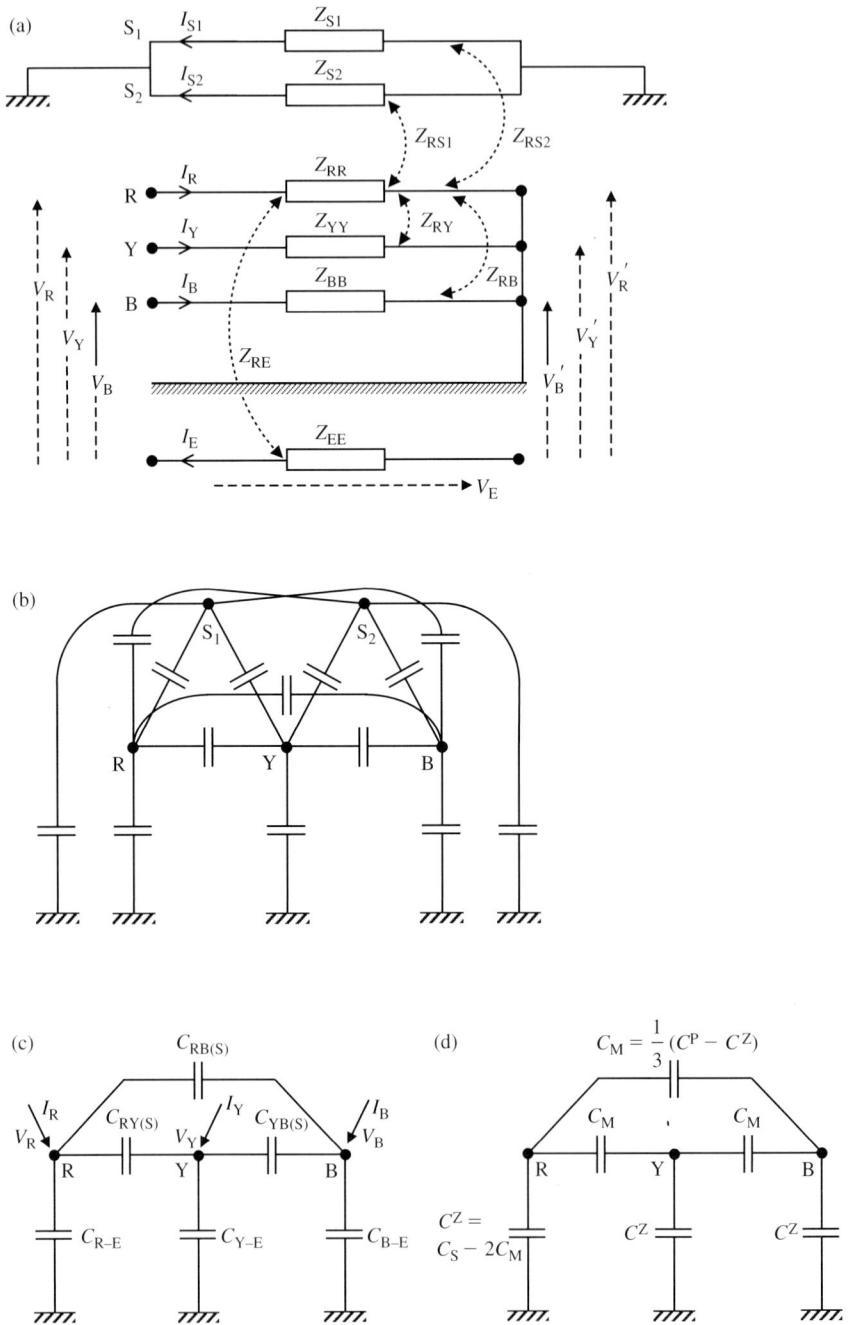

Figure 3.8 Series impedance and shunt susceptance circuits of Figure 3.7: (a) three-phase coupled series impedance circuit; (b) three-phase coupled shunt capacitance circuit; (c) reduced three-phase shunt capacitance circuit with earth wires eliminated and (d) three-phase shunt capacitance circuit in sequence terms

or

$$\Delta V_R = Z_{RR-E}I_R + Z_{RY-E}I_Y + Z_{RB-E}I_B + Z_{RS1-E}I_{S1} + Z_{RS2-E}I_{S2} \quad (3.28a)$$

where

$$
\begin{aligned}
Z_{RR-E} &= Z_{RR} - 2Z_{RE} + Z_{EE} & Z_{RY-E} &= Z_{RY} - (Z_{RE} + Z_{YE}) + Z_{EE} \\
Z_{RB-E} &= Z_{RB} - (Z_{RE} + Z_{BE}) + Z_{EE} & Z_{RS1-E} &= Z_{RE} - (Z_{RS1} - Z_{S1E}) - Z_{EE} \\
Z_{RS2-E} &= Z_{RE} - (Z_{RS2} - Z_{S2E}) - Z_{EE}
\end{aligned}
$$

$$(3.28b)$$

The self and mutual phase impedances defined in Equation (3.28b) as well as those of the earth wires include the effect of earth impedance Z_{EE}. We can write similar equations for ΔV_Y, ΔV_B, ΔV_{S1}, ΔV_{S2} and combine them all to obtain the following series voltage drop and partitioned series phase impedance matrix:

$$
\begin{bmatrix}
\Delta V_R \\
\Delta V_Y \\
\Delta V_B \\
\Delta V_{S1} \\
\Delta V_{S2}
\end{bmatrix}
=
\left[
\begin{array}{ccc|cc}
Z_{RR-E} & Z_{RY-E} & Z_{RB-E} & Z_{RS1-E} & Z_{RS2-E} \\
Z_{YR-E} & Z_{YY-E} & Z_{YB-E} & Z_{YS1-E} & Z_{YS2-E} \\
Z_{BR-E} & Z_{BY-E} & Z_{BB-E} & Z_{BS1-E} & Z_{BS2-E} \\
Z_{S1R-E} & Z_{S1Y-E} & Z_{S1B-E} & Z_{S1S1-E} & Z_{S1S2-E} \\
Z_{S2R-E} & Z_{S2Y-E} & Z_{S2B-E} & Z_{S2S1-E} & Z_{S2S2-E}
\end{array}
\right]
\begin{bmatrix}
I_R \\
I_Y \\
I_B \\
I_{S1} \\
I_{S2}
\end{bmatrix}
\quad (3.29)
$$

In large-scale power system short-circuit analysis, we are interested in the calculation of short-circuit currents on the faulted phases R, Y or B or a combination of these but generally not in the currents flowing in the earth wires. One exception is the calculation of earth fault return currents covered in Chapter 10. For the elimination of the earth wires, the partitioned Equation (3.29) is rewritten as follows:

$$
\begin{bmatrix}
\Delta \mathbf{V}_{RYB} \\
\Delta \mathbf{V}_{S1S2}
\end{bmatrix}
=
\begin{bmatrix}
\mathbf{Z}_{AA} & \mathbf{Z}_{AS} \\
\mathbf{Z}_{SA} & \mathbf{Z}_{SS}
\end{bmatrix}
\begin{bmatrix}
\mathbf{I}_{RYB} \\
\mathbf{I}_{S1S2}
\end{bmatrix}
\quad (3.30a)
$$

where

$$\Delta \mathbf{V}_{RYB} = [V_R \quad V_Y \quad V_B]^t \quad \mathbf{I}_{RYB} = [I_R \quad I_Y \quad I_B]^t$$

$$
\Delta \mathbf{V}_{S1S2} =
\begin{bmatrix}
\Delta V_{S1} \\
\Delta V_{S2}
\end{bmatrix}
\quad
\mathbf{I}_{S1S2} =
\begin{bmatrix}
I_{S1} \\
I_{S2}
\end{bmatrix}
$$

$$
\mathbf{Z}_{AA} =
\begin{bmatrix}
Z_{RR-E} & Z_{RY-E} & Z_{RB-E} \\
Z_{YR-E} & Z_{YY-E} & Z_{YB-E} \\
Z_{BR-E} & Z_{BY-E} & Z_{BB-E}
\end{bmatrix}
\quad
\mathbf{Z}_{AS} =
\begin{bmatrix}
Z_{RS1-E} & Z_{RS2-E} \\
Z_{YS1-E} & Z_{YS2-E} \\
Z_{BS1-E} & Z_{BS2-E}
\end{bmatrix}
$$

$$
\mathbf{Z}_{SA} =
\begin{bmatrix}
Z_{S1R-E} & Z_{S1Y-E} & Z_{S1B-E} \\
Z_{S2R-E} & Z_{S2Y-E} & Z_{S2B-E}
\end{bmatrix}
\quad
\mathbf{Z}_{SS} =
\begin{bmatrix}
Z_{S1-S1-E} & Z_{S1-S2-E} \\
Z_{S2-S1-E} & Z_{S2-S2-E}
\end{bmatrix}
$$

$$(3.30b)$$

$\mathbf{Z}_{AA}$ consists of the self and mutual impedances of phase conductors R, Y and B with earth return. $\mathbf{Z}_{AS}$ consists of the mutual impedances between the phase conductors R, Y and B, and earth wires S_1 and S_2, with earth return. $\mathbf{Z}_{SA} = \mathbf{Z}_{AS}^t$ noting that the individual impedances are symmetric, i.e. $Z_{RS1-E} = Z_{S1R-E}$. $\mathbf{Z}_{SS}$ consists of the self and mutual impedances of the earth wires S_1 and S_2 with earth return. The earth return effect is shown in Equation (3.28b). Expanding Equation (3.30a), we obtain

$$\Delta\mathbf{V}_{RYB} = \mathbf{Z}_{AA}\mathbf{I}_{RYB} + \mathbf{Z}_{AS}\mathbf{I}_{S1S2} \tag{3.31a}$$

$$\Delta\mathbf{V}_{S1S2} = \mathbf{Z}_{AS}^t\mathbf{I}_{RYB} + \mathbf{Z}_{SS}\mathbf{I}_{S1S2} \tag{3.31b}$$

In the majority of line installations, the earth wires are bonded to the tower tops and are only partially earthed by the footing resistance of each tower to which they are connected but solidly earthed at substations. However, it is usual to assume zero tower footing resistances and that the earth wires are at zero voltage at all points. Thus, using $\Delta\mathbf{V}_{S1S2} = 0$ in Equation (3.31b) and substituting the result in Equation (3.31a), we obtain

$$\Delta\mathbf{V}_{RYB} = \mathbf{Z}_{RYB(S-E)}\mathbf{I}_{RYB} \tag{3.32a}$$

where

$$\mathbf{Z}_{RYB(S-E)} = \mathbf{Z}_{AA} - \mathbf{Z}_{AS}\mathbf{Z}_{SS}^{-1}\mathbf{Z}_{AS}^t \tag{3.32b}$$

and

$$\mathbf{Z}_{RYB(S-E)} = \begin{bmatrix} Z_{RR(S-E)} & Z_{RY(S-E)} & Z_{RB(S-E)} \\ Z_{YR(S-E)} & Z_{YY(S-E)} & Z_{YB(S-E)} \\ Z_{BR(S-E)} & Z_{BY(S-E)} & Z_{BB(S-E)} \end{bmatrix} \tag{3.32c}$$

Equation (3.32b) shows that the self and mutual phase impedances of the phase conductors of matrix $\mathbf{Z}_{AA}$ are reduced by the presence of the earth wires. Equation (3.32c) indicates the elements of the phase impedance matrix of the single-circuit three-phase line with both earth wires eliminated and including the effect of the earth impedance. It is noted that the mathematical elimination of the earth wires only eliminates their presence from the full matrix but not their effects which are included in the modified elements of the reduced matrix, i.e. $Z_{RR(S-E)} \neq Z_{RR-E}$.

Although far less common, segmented earth wires may be used to prevent circulating currents in earth wires and associated I^2R losses. This is a 'T' arrangement where the earth wires are bonded to the top of the middle tower but insulated at the adjacent towers on either side. This is equivalent to $I_{S1} = I_{S2} = 0$ in Equation (3.29) and hence the reduced phase impedance matrix $\mathbf{Z}_{RYB(S-E)}$ is directly obtained from Equation (3.29) by deleting the last two rows and columns that correspond to the earth wires. This results in $\mathbf{Z}_{RYB(S-E)} = \mathbf{Z}_{AA}$. The impedance matrix of Equation (3.32c) is symmetric about the diagonal and in the case of asymmetrical spacings between the conductors, the self or diagonal terms are generally not equal to each other, and neither are the mutual or off-diagonal terms. Currents flowing in any one conductor will induce voltage drops in the other two conductors and these may be unequal even if the currents are balanced. This is because the mutual impedances, which are dependent on the physical spacings of the conductors, are unequal. Rewriting the voltage drop equation using Equation (3.32c) and dropping the S and E notation for convenience, we can write

$$
\begin{bmatrix} \Delta V_R \\ \Delta V_Y \\ \Delta V_B \end{bmatrix} = \begin{bmatrix} Z_{RR} & Z_{RY} & Z_{RB} \\ Z_{YR} & Z_{YY} & Z_{YB} \\ Z_{BR} & Z_{BY} & Z_{BB} \end{bmatrix} \begin{bmatrix} I_R \\ I_Y \\ I_B \end{bmatrix} \tag{3.33}
$$

Assuming balanced three-phase currents, i.e. $I_Y = h^2 I_R$ and $I_B = h I_R$ where $h = e^{j2\pi/3}$, we obtain from Equation (3.33)

$$
\Delta V_R = (Z_{RR} + h^2 Z_{RY} + h Z_{RB}) I_R \tag{3.34a}
$$

$$
\Delta V_Y = (Z_{YY} + h^2 Z_{YB} + h Z_{YR}) I_Y \tag{3.34b}
$$

$$
\Delta V_B = (Z_{BB} + h^2 Z_{BR} + h Z_{BY}) I_B \tag{3.34c}
$$

Equation (3.34) describes the per-phase or single-phase representation of the three-phase system when balanced currents flow. However, the three per-phase equivalent impedances are clearly unequal and a single per-phase representation cannot be used.

The sequence impedance matrix of the phase impedance matrix given in Equation (3.33) can be calculated, assuming a phase rotation of RYB, using $\mathbf{V}_{RYB} = \mathbf{H}\mathbf{V}^{PNZ}$ and $\mathbf{I}_{RYB} = \mathbf{H}\mathbf{I}^{PNZ}$ where $\mathbf{H}$ is the sequence to phase transformation matrix given in Chapter 2. Therefore,

$$
\mathbf{Z}^{PNZ} = \mathbf{H}^{-1}\mathbf{Z}_{RYB}\mathbf{H} = \begin{array}{c} \\ P \\ N \\ Z \end{array} \begin{array}{ccc} P & N & Z \\ \begin{bmatrix} Z^{PP} & Z^{PN} & Z^{PZ} \\ Z^{NP} & Z^{NN} & Z^{NZ} \\ Z^{ZP} & Z^{ZN} & Z^{ZZ} \end{bmatrix} \end{array} \tag{3.35}
$$

where the nine sequence impedance elements of this matrix are as given by Equation (2.26) in Chapter 2.

The conversion to the sequence reference frame still produces a full and even asymmetric sequence impedance matrix that includes intersequence mutual coupling. Where this intersequence coupling is to be eliminated, the circuit has to be perfectly transposed. Transposition is dealt with in Section 3.2.4.

Phase and sequence shunt susceptance matrices

Figure 3.8(b) shows the shunt capacitance circuit of Figure 3.7 involving phase conductors and earth. To derive the shunt phase susceptance matrix, we use the potential coefficients calculated from the line's dimensions. Thus, the voltage on each conductor to ground as a function of the electric charges on all the conductors is given by

$$
\begin{bmatrix}
V_R \\
V_Y \\
V_B \\
V_{S1} \\
V_{S2}
\end{bmatrix}
=
\left[
\begin{array}{ccc|cc}
P_{RR} & P_{R-Y} & Z_{R-B} & P_{R-S1} & P_{R-S2} \\
P_{Y-R} & P_{YY} & Z_{Y-B} & P_{Y-S1} & P_{Y-S2} \\
P_{B-R} & P_{B-Y} & Z_{BB} & P_{B-S1} & P_{B-S2} \\
\hline
P_{S1-R} & P_{S1-Y} & Z_{S1-B} & P_{S1S1} & P_{S1-S2} \\
P_{S2-R} & P_{S2-Y} & Z_{S2-B} & P_{S2-S1} & P_{S2S2}
\end{array}
\right]
\begin{bmatrix}
Q_R \\
Q_Y \\
Q_B \\
Q_{S1} \\
Q_{S2}
\end{bmatrix}
\qquad (3.36a)
$$

which can be written as

$$
\begin{bmatrix}
\mathbf{V}_{RYB} \\
\mathbf{V}_{S1S2}
\end{bmatrix}
=
\begin{bmatrix}
\mathbf{P}_{AA} & \mathbf{P}_{AS} \\
\mathbf{P}_{SA} & \mathbf{P}_{SS}
\end{bmatrix}
\begin{bmatrix}
\mathbf{Q}_{RYB} \\
\mathbf{Q}_{S1S2}
\end{bmatrix}
\qquad (3.36b)
$$

Again, with the earth wires at zero voltage, they are eliminated from Equation (3.36b) as follows

$$
\mathbf{V}_{RYB} = \mathbf{P}_{RYB(S)}\mathbf{Q}_{RYB} \qquad (3.37a)
$$

where

$$
\mathbf{P}_{RYB(S)} = \mathbf{P}_{AA} - \mathbf{P}_{AS}\mathbf{P}_{SS}^{-1}\mathbf{P}_{AS}^{t} \qquad (3.37b)
$$

Again, $\mathbf{P}_{RYB(S)}$ is the reduced potential coefficient matrix that includes the effects of the eliminated earth wires. Equation (3.37b) shows that the self and mutual potential coefficients of the phase conductors of the matrix $\mathbf{P}_{AA}$ are reduced by the presence of the earth wires. To derive the shunt phase capacitance matrix of the line, multiplying Equation (3.37a) by $\mathbf{P}_{RYB(S)}^{-1}$, we obtain

$$
\mathbf{Q}_{RYB} = \mathbf{C}_{RYB(S)}\mathbf{V}_{RYB} \qquad (3.38a)
$$

where

$$
\mathbf{C}_{RYB(S)} = \mathbf{P}_{RYB(S)}^{-1} \qquad (3.38b)
$$

and $\mathbf{C}_{RYB(S)}$ is the shunt phase capacitance matrix.

Expanding Equation (3.38b), and noting that the capacitance matrix elements include the effect of the eliminated earth wires, we have

$$
\mathbf{C}_{RYB(S)} =
\begin{bmatrix}
C_{RR-S} & -C_{RY(S)} & -C_{RB(S)} \\
-C_{YR(S)} & C_{YY-S} & -C_{YB(S)} \\
-C_{BR(S)} & -C_{BY(S)} & C_{BB(S)}
\end{bmatrix}
\qquad (3.38c)
$$

In Equation (3.38c), the elements of this capacitance matrix are increased by the presence of the earth wires which reduce the potential coefficients. Equation (3.38c) is illustrated in the reduced shunt capacitance equivalent shown in

Figure 3.8(c) after the elimination of the earth wires but not their effects. Using Equations (3.4b) and (3.38c) and dropping the S notation for convenience, the nodal admittance matrix of Figure 3.8(c) is given by

$$
\begin{bmatrix} I_R \\ I_Y \\ I_B \end{bmatrix} = \begin{bmatrix} jB_{RR} & -jB_{RY} & -jB_{RB} \\ -jB_{YR} & jB_{YY} & -jB_{YB} \\ -jB_{BR} & -jB_{BY} & jB_{BB} \end{bmatrix} \begin{bmatrix} V_R \\ V_Y \\ V_B \end{bmatrix} \tag{3.39a}
$$

or

$$
\mathbf{I}_{RYB} = j\mathbf{B}_{RYB}\mathbf{V}_{RYB} \tag{3.39b}
$$

The negative signs for the off-diagonal capacitance or susceptance terms are due to the matrices being in nodal form. For example, from Figure 3.8(c), and using susceptances instead of capacitances, the injected current into node R is given by

$$
\begin{aligned}
I_R &= jB_{R-E}V_R + jB_{RY}(V_R - V_Y) + jB_{RB}(V_R - V_B) \\
&= jB_{RR}V_R - jB_{RY}V_Y - jB_{RB}V_B
\end{aligned} \tag{3.40a}
$$

where

$$
B_{RR} = B_{R-E} + B_{RY} + B_{RB} \tag{3.40b}
$$

and similarly for I_Y and I_B. The off-diagonal terms represent shunt susceptances between two-phase conductors, e.g. R and Y, etc. The diagonal terms, e.g. that for conductor R, represent the sum of the shunt capacitances between conductor R and all other conductors including earth as shown in Figure 3.8(c).

The shunt susceptance matrix of Equation (3.39a) is symmetric about the diagonal but in the case of asymmetrical spacings between the conductors, the self that is diagonal terms are generally not equal to each other, and neither are the mutual or off-diagonal terms. Therefore, as for the series phase impedance matrix, the sequence shunt susceptance matrix is given by

$$
\mathbf{B}^{PNZ} = \mathbf{H}^{-1}\mathbf{B}_{RYB}\mathbf{H} = \begin{matrix} \\ P \\ N \\ Z \end{matrix} \begin{matrix} P & N & Z \\ \begin{bmatrix} B^{PP} & B^{PN} & B^{PZ} \\ B^{NP} & B^{NN} & B^{NZ} \\ B^{ZP} & B^{ZN} & B^{ZZ} \end{bmatrix} \end{matrix} \tag{3.41}
$$

where the nine sequence susceptance elements of this matrix are as given by Equation (2.26) in Chapter 2 but with Z replaced by B.

As for the series sequence impedance matrix, the intersequence mutual coupling present can be eliminated by assuming the line to be perfectly transposed. This is dealt with in the Section 3.2.4.

An alternative method to obtain the reduced 3×3 capacitance matrix of Equation (3.38c) is to calculate the inverse of the 5×5 potential coefficient matrix of Equation (3.36a) which gives a 5×5 shunt capacitance matrix. The 3×3 shunt capacitance matrix can then be directly obtained by simply deleting the last two rows and columns that correspond to the earth wires. The reader is encouraged to prove this statement.

We have presented in this section the general case of an untransposed single-circuit three-phase line with two earth wires. The cases where the line has only one earth wire or no earth wires become special cases from a mathematical viewpoint. The 5×5 matrices of Equations (3.29) and (3.36a) become 4×4 matrices in the case of one earth wire and 3×3 matrices in the case of no earth wires and the rest of the analysis is similar. If the earth wires are identical and are symmetrical with respect to the three-phase circuit, they can be initially analytically replaced by an equivalent single earth wire whose equivalent impedance is half the sum of the self-impedance of one earth wire and the mutual impedance between the earth wires. However, analytical calculations are not necessary because of the extensive use in industry of digital computer calculations of line parameters or constants.

3.2.4 Transposition of single-circuit three-phase lines

We have shown in Section 3.2.3 that the calculated sequence series impedance and shunt susceptance matrices include full intersequence mutual couplings. However, as presented in Chapter 2, sequence component reference frame analysis is based on separate positive phase sequence (PPS), negative phase sequence (NPS) and zero phase sequence (ZPS) circuits. To eliminate the intersequence mutual couplings, an assumption can be made that the line is perfectly transposed. The objective of transposition is to produce equal series self-impedances, and equal series mutual impedances in the phase frame of reference and similarly for the shunt self-susceptances and shunt mutual susceptances. Perfect phase transposition means that each phase conductor occupies successively the same physical positions as the other two conductors in two successive line sections as shown in Figure 3.9.

Figure 3.9 shows three sections of a transposed line. This represents a perfectly transposed line where the three sections have equal length. Perfect transposition results in the same total voltage drop for each phase conductor and hence equal average series self-impedances of each phase conductor. This effect also applies to the average series phase mutual impedances, average shunt phase self-susceptances and average shunt phase mutual susceptances. Figure 3.9(a) illustrates a complete forward transposition cycle, i.e. three transpositions where the line is divided into three sections and t, m and b are used to designate the conductor physical positions on the tower. If the three conductors of the circuit are designated C1, C2 and C3, then a forward transposition is defined as one where the conductor positions for the three sections are C1C2C3, C3C1C2 then C2C3C1 as shown in Figure 3.9(a).

Using Equation (3.33), the series voltage drops per-unit length across conductors C1, C2 and C3 for each section of the line can be calculated taking into account the changing positions of the three conductors on the tower and hence their changing

(a)

(b)

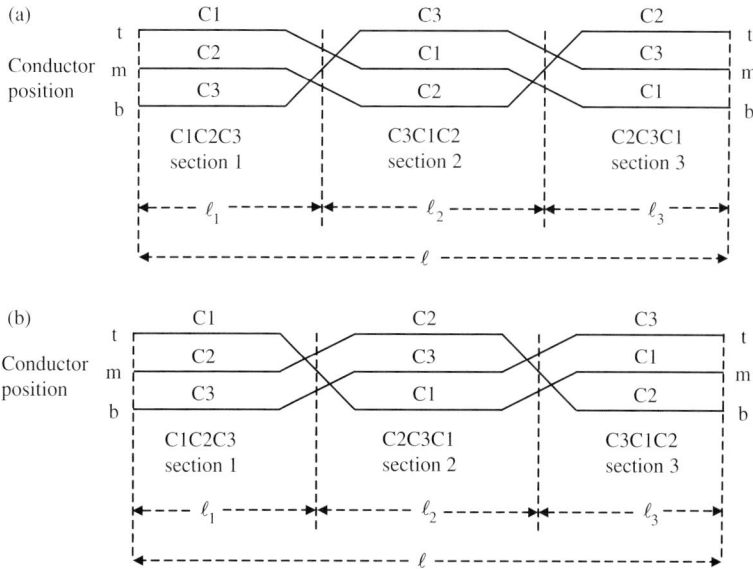

Figure 3.9 Transposition of a single-circuit three-phase line: (a) forward successive phase transpositions and (b) reverse successive phase transpositions

impedances. With equal section lengths, i.e. $\ell_1 = \ell_2 = \ell_3 = \frac{1}{3}\ell$, the voltage drops for each section of the line are given as

$$
\begin{bmatrix} V_{C1} \\ V_{C2} \\ V_{C3} \end{bmatrix}_{\text{Section-1}} = \frac{1}{3} \begin{matrix} C1 \\ C2 \\ C3 \end{matrix} \begin{bmatrix} Z_{tt} & Z_{tm} & Z_{tb} \\ Z_{mt} & Z_{mm} & Z_{mb} \\ Z_{bt} & Z_{bm} & Z_{bb} \end{bmatrix} \begin{bmatrix} I_{C1} \\ I_{C2} \\ I_{C3} \end{bmatrix} \quad \text{or} \quad \mathbf{V}_{C\text{Section-1}} = \frac{1}{3} \mathbf{Z}_{\text{Section-1}} \mathbf{I}_C
$$

(3.42a)

$$
\begin{bmatrix} V_{C1} \\ V_{C2} \\ V_{C3} \end{bmatrix}_{\text{Section-2}} = \frac{1}{3} \begin{matrix} C1 \\ C2 \\ C3 \end{matrix} \begin{bmatrix} Z_{mm} & Z_{mb} & Z_{mt} \\ Z_{bm} & Z_{bb} & Z_{bt} \\ Z_{tm} & Z_{tb} & Z_{tt} \end{bmatrix} \begin{bmatrix} I_{C1} \\ I_{C2} \\ I_{C3} \end{bmatrix} \quad \text{or} \quad \mathbf{V}_{C\text{Section-2}} = \frac{1}{3} \mathbf{Z}_{\text{Section-2}} \mathbf{I}_C
$$

(3.42b)

$$
\begin{bmatrix} V_{C1} \\ V_{C2} \\ V_{C3} \end{bmatrix}_{\text{Section-3}} = \frac{1}{3} \begin{matrix} C1 \\ C2 \\ C3 \end{matrix} \begin{bmatrix} Z_{bb} & Z_{bt} & Z_{bm} \\ Z_{tb} & Z_{tt} & Z_{tm} \\ Z_{mb} & Z_{mt} & Z_{mm} \end{bmatrix} \begin{bmatrix} I_{C1} \\ I_{C2} \\ I_{C3} \end{bmatrix} \quad \text{or} \quad \mathbf{V}_{C\text{Section-3}} = \frac{1}{3} \mathbf{Z}_{\text{Section-3}} \mathbf{I}_C
$$

(3.42c)

where Z is a total impedance of the line. Therefore, the voltage drop across each conductor of the line is given by

$$
\mathbf{V}_{\text{Total}} = \sum_{i=1}^{3} \mathbf{V}_{\text{Section-}i} = \frac{1}{3}(\mathbf{Z}_{\text{Section-1}} + \mathbf{Z}_{\text{Section-2}} + \mathbf{Z}_{\text{Section-3}})\mathbf{I}_C = \mathbf{Z}_{\text{Phase}} \mathbf{I}_C
$$

(3.43a)

where

$$\mathbf{Z}_{\text{Phase}} = \frac{1}{3}(\mathbf{Z}_{\text{Section-1}} + \mathbf{Z}_{\text{Section-2}} + \mathbf{Z}_{\text{Section-3}}) = \begin{bmatrix} Z_S & Z_M & Z_M \\ Z_M & Z_S & Z_M \\ Z_M & Z_M & Z_S \end{bmatrix} \quad (3.43\text{b})$$

and

$$Z_S = \frac{1}{3}(Z_{tt} + Z_{mm} + Z_{bb}) \quad Z_M = \frac{1}{3}(Z_{tm} + Z_{mb} + Z_{bt}) \quad (3.43\text{c})$$

$\mathbf{Z}_{\text{Phase}}$ is the phase impedance matrix of the perfectly transposed line, noting that individual conductor impedances are symmetric.

We have shown how to calculate the phase impedance matrix for each line transposition section from first principles. However, a general matrix analysis approach is more suitable for modern calculations by digital computers. Let us define a transposition matrix that has the following characteristics:

$$T = \begin{bmatrix} 0 & 0 & 1 \\ 1 & 0 & 0 \\ 0 & 1 & 0 \end{bmatrix} \quad \mathbf{T}^{-1} = \mathbf{T}^t = \mathbf{T}^2 = \begin{bmatrix} 0 & 1 & 0 \\ 0 & 0 & 1 \\ 1 & 0 & 0 \end{bmatrix} \quad \mathbf{T}^3 = \mathbf{U} \text{ the identity matrix}$$

$$(3.44)$$

Using Equation (3.42) and noting the successive changing positions of the three conductors in Figure 3.9(a), Equation (3.42) can be rewritten as

$$\begin{bmatrix} V_{C1} \\ V_{C2} \\ V_{C3} \end{bmatrix}_{\text{Section-1}} = \frac{1}{3} \begin{bmatrix} Z_{tt} & Z_{tm} & Z_{tb} \\ Z_{mt} & Z_{mm} & Z_{mb} \\ Z_{bt} & Z_{bm} & Z_{bb} \end{bmatrix} \begin{bmatrix} I_{C1} \\ I_{C2} \\ I_{C3} \end{bmatrix} = \frac{1}{3}\mathbf{Z}_{\text{Section-1}} \begin{bmatrix} I_{C1} \\ I_{C2} \\ I_{C3} \end{bmatrix}$$

$$\begin{bmatrix} V_{C3} \\ V_{C1} \\ V_{C2} \end{bmatrix}_{\text{Section-2}} = \frac{1}{3}\mathbf{Z}_{\text{Section-1}} \begin{bmatrix} I_{C3} \\ I_{C1} \\ I_{C2} \end{bmatrix} \quad \text{and} \quad \begin{bmatrix} V_{C2} \\ V_{C3} \\ V_{C1} \end{bmatrix}_{\text{Section-3}} = \frac{1}{3}\mathbf{Z}_{\text{Section-1}} \begin{bmatrix} I_{C2} \\ I_{C3} \\ I_{C1} \end{bmatrix}$$

$$(3.45)$$

Using the transposition matrix defined in Equation (3.44), we can write for Section 2

$$\begin{bmatrix} V_{C3} \\ V_{C1} \\ V_{C2} \end{bmatrix}_{\text{Section-2}} = \mathbf{T} \begin{bmatrix} V_{C1} \\ V_{C2} \\ V_{C3} \end{bmatrix}_{\text{Section-2}} = \frac{1}{3}\mathbf{Z}_{\text{Section-1}}\mathbf{T} \begin{bmatrix} I_{C1} \\ I_{C2} \\ I_{C3} \end{bmatrix}$$

or

$$\begin{bmatrix} V_{C1} \\ V_{C2} \\ V_{C3} \end{bmatrix}_{\text{Section-2}} = \mathbf{V}_{C(\text{Section-2})} = \frac{1}{3}\mathbf{T}^t\mathbf{Z}_{\text{Section-1}}\mathbf{T}\,\mathbf{I}_C = \frac{1}{3}\mathbf{Z}_{\text{Section-2}}\mathbf{I}_C$$

or

$$\mathbf{Z}_{\text{Section-2}} = \mathbf{T}^t \, \mathbf{Z}_{\text{Section-1}} \mathbf{T} \tag{3.46}$$

Similarly for Section 3, we can write

$$\begin{bmatrix} V_{C2} \\ V_{C3} \\ V_{C1} \end{bmatrix}_{\text{Section-3}} = \mathbf{T}^{-1} \begin{bmatrix} V_{C1} \\ V_{C2} \\ V_{C3} \end{bmatrix}_{\text{Section-2}} = \frac{1}{3}\mathbf{Z}_{\text{Section-1}}\mathbf{T}^{-1} \begin{bmatrix} I_{C1} \\ I_{C2} \\ I_{C3} \end{bmatrix}$$

or

$$\begin{bmatrix} V_{C1} \\ V_{C2} \\ V_{C3} \end{bmatrix}_{\text{Section-3}} = V_{C(\text{Section-3})} = \frac{1}{3}\mathbf{T}\,\mathbf{Z}_{\text{Section-1}}\mathbf{T}^t\,\mathbf{I}_C = \frac{1}{3}\mathbf{Z}_{\text{Section-3}}\mathbf{I}_C$$

or

$$\mathbf{Z}_{\text{Section-3}} = \mathbf{T}\,\mathbf{Z}_{\text{Section-1}}\mathbf{T}^t \tag{3.47}$$

In summary, the phase impedance matrices of the first, second and third line transposition sections are given by

$$\mathbf{Z}_{\text{Section-}i} = \begin{cases} \mathbf{Z}_{\text{Section-1}} & \text{for } i = 1 \\ \mathbf{T}^t \, \mathbf{Z}_{\text{Section-1}} \mathbf{T} & \text{for } i = 2 \\ \mathbf{T} \, \mathbf{Z}_{\text{Section-1}} \mathbf{T}^t & \text{for } i = 3 \end{cases} \tag{3.48a}$$

and the phase impedance matrix of the perfectly transposed line is given by

$$\mathbf{Z}_{\text{Phase}} = \frac{1}{3}\left[\mathbf{Z}_{\text{Section-1}} + \mathbf{T}^t\mathbf{Z}_{\text{Section-1}}\mathbf{T} + \mathbf{T}\,\mathbf{Z}_{\text{Section-1}}\mathbf{T}^t\right] \tag{3.48b}$$

This analysis approach is straightforward if the effect of the matrix $\mathbf{T}$ is recognised. The effect of pre-multiplying matrix $\mathbf{Z}_{\text{Section-1}}$ by matrix $\mathbf{T}$ is to shift its row 2 elements up to row 1, row 3 elements up to row 2 and row 1 elements to row 3. Also, the effect of post-multiplying matrix $\mathbf{Z}_{\text{Section-1}}$ by matrix $\mathbf{T}^t$ is to shift its column 2 elements to column 1, column 3 elements to column 2 and column 1 elements to column 3. A reverse successive transposition cycle could also be used to obtain the same result as illustrated in Figure 3.9(b). The reader is encouraged to show that the phase impedance matrix of each transposition section, using transposition matrix $\mathbf{T}$, is given by

$$\mathbf{Z}_{\text{Section-}i} = \begin{cases} \mathbf{Z}_{\text{Section-1}} & \text{for } i = 1 \\ \mathbf{T} \, \mathbf{Z}_{\text{Section-1}} \mathbf{T}^t & \text{for } i = 2 \\ \mathbf{T}^t\mathbf{Z}_{\text{Section-1}}\mathbf{T} & \text{for } i = 3 \end{cases} \tag{3.49a}$$

and the phase impedance is given by

$$\mathbf{Z}_{Phase} = \frac{1}{3} \left[\mathbf{Z}_{Section-1} + \mathbf{T} \, \mathbf{Z}_{Section-1} \mathbf{T}^t + \mathbf{T}^t \, \mathbf{Z}_{Section-1} \mathbf{T} \right] \tag{3.49b}$$

Equation (3.43b) is the phase impedance matrix of our balanced or perfectly transposed single-circuit line. Assuming R, Y, B is the electrical phase rotation of conductors 1, 2 and 3, respectively, the sequence impedance matrix, calculated as shown in Chapter 2, using $\mathbf{Z}^{PNZ} = \mathbf{H}^{-1}\mathbf{Z}_{Phase}\mathbf{H}$, is given by

$$\mathbf{Z}^{PNZ} = \begin{bmatrix} Z^P & 0 & 0 \\ 0 & Z^N & 0 \\ 0 & 0 & Z^Z \end{bmatrix} \tag{3.50a}$$

where

$$Z^P = Z^N = Z_S - Z_M = \frac{1}{3}[(Z_{tt} + Z_{mm} + Z_{bb}) - (Z_{tm} + Z_{mb} + Z_{bt})]$$

$$Z^Z = Z_S + 2Z_M = \frac{1}{3}[(Z_{tt} + Z_{mm} + Z_{bb}) + 2(Z_{tm} + Z_{mb} + Z_{bt})] \tag{3.50b}$$

Combining Equation (3.50a) with sequence voltage and current vectors, the sequence voltage drops are given by

$$\Delta V^P = Z^P I^P \quad \Delta V^N = Z^N I^N \quad \Delta V^Z = Z^Z I^Z \tag{3.51}$$

Equation (3.51) shows that the three sequence voltage drop equations are decoupled.

For the shunt phase susceptance matrix, we follow the same method as for the series phase impedance matrix. Therefore, using Figure 3.9(a), we can write

$$\mathbf{B}_{Section-i} = \begin{cases} \mathbf{B}_{Section-1} & \text{for } i = 1 \\ \mathbf{T}^t \mathbf{B}_{Section-1} \mathbf{T} & \text{for } i = 2 \\ \mathbf{T} \, \mathbf{B}_{Section-1} \mathbf{T}^t & \text{for } i = 3 \end{cases} \tag{3.52a}$$

and the transposed shunt phase susceptance matrix is given by

$$\mathbf{B}_{Phase} = \frac{1}{3}[\mathbf{B}_{Section-1} + \mathbf{T}^t \mathbf{B}_{Section-1}\mathbf{T} + \mathbf{T} \, \mathbf{B}_{Section-1}\mathbf{T}^t] \tag{3.52b}$$

The resultant shunt phase susceptance matrix of our perfectly transposed line is given by

$$\mathbf{B}_{Phase} = \begin{bmatrix} B_S & -B_M & -B_M \\ -B_M & B_S & -B_M \\ -B_M & -B_M & B_S \end{bmatrix} \tag{3.53a}$$

where

$$B_S = \frac{1}{3}(B_{tt} + B_{mm} + B_{bb}) \quad \text{and} \quad B_M = \frac{1}{3}(B_{tm} + B_{mb} + B_{bt}) \tag{3.53b}$$

Using $\mathbf{B}^{PNZ} = \mathbf{H}^{-1}\mathbf{B}_{Phase}\mathbf{H}$, the shunt sequence susceptance matrix is given by

$$\mathbf{B}^{PNZ} = \begin{bmatrix} B^P & 0 & 0 \\ 0 & B^N & 0 \\ 0 & 0 & B^Z \end{bmatrix} \tag{3.54a}$$

where

$$B^P = B^N = B_S + B_M = \frac{1}{3}[(B_{tt} + B_{mm} + B_{bb}) + (B_{tm} + B_{mb} + B_{bt})] \tag{3.54b}$$

and

$$B^Z = B_S - 2B_M = \frac{1}{3}[(B_{tt} + B_{mm} + B_{bb}) - 2(B_{tm} + B_{mb} + B_{bt})] \tag{3.54c}$$

Combining Equation (3.54a) with sequence voltages and currents, the nodal sequence currents are given by

$$I^P = jB^P V^P \quad I^N = jB^N V^N \quad I^Z = jB^Z V^Z \tag{3.55}$$

Equation (3.55) shows that the three sequence current equations are decoupled.

It is instructive to view the capacitance parameters of the three-phase reduced capacitance equivalent of Figure 3.8(c) in sequence terms noting that for the balanced line represented by Equation (3.53), the shunt and mutual terms of Figure 3.8(c) become $C_S - 2C_M$ and C_M, respectively. Using Equations (3.54b) and (3.54c) for capacitance, we obtain

$$C_M = \frac{C^P - C^Z}{3} \quad \text{and} \quad C_S - 2C_M = C^Z$$

The result is shown in Figure 3.8(d).

We have assumed in this section that the line is perfectly transposed that is each unit length is divided into three sections of equal lengths. In practice, lines are rarely perfectly transposed because of the expense and inconvenience. The general case where the line may be semi-transposed at one or two locations some distance(s) along the line route can be considered as follows. From Figure 3.9(a), we assume that the lengths of the three line sections are ℓ_1, ℓ_2 and ℓ_3 where $\ell_1 + \ell_2 + \ell_3 = \ell$ and ℓ is the total line unit length. Thus, from Equation (3.42), we can write

$$\mathbf{Z}_{Section-1} = \frac{\ell_1}{\ell} \begin{bmatrix} Z_{tt} & Z_{tm} & Z_{tb} \\ Z_{mt} & Z_{mm} & Z_{mb} \\ Z_{bt} & Z_{bm} & Z_{bb} \end{bmatrix} \quad \mathbf{Z}_{Section-2} = \frac{\ell_2}{\ell} \begin{bmatrix} Z_{mm} & Z_{mb} & Z_{mt} \\ Z_{bm} & Z_{bb} & Z_{bt} \\ Z_{tm} & Z_{tb} & Z_{tt} \end{bmatrix}$$

and

$$\mathbf{Z}_{Section-3} = \frac{\ell_3}{\ell} \begin{bmatrix} Z_{bb} & Z_{bt} & Z_{bm} \\ Z_{tb} & Z_{tt} & Z_{tm} \\ Z_{mb} & Z_{mt} & Z_{mm} \end{bmatrix} \tag{3.56a}$$

Therefore, the phase impedance matrix of the general case of a transposed line is

$$\mathbf{Z}_{\text{Phase}} = \frac{1}{\ell} \begin{bmatrix} \ell_1 Z_{\text{tt}} + \ell_2 Z_{\text{mm}} + \ell_3 Z_{\text{bb}} & \ell_1 Z_{\text{tm}} + \ell_2 Z_{\text{mb}} + \ell_3 Z_{\text{bt}} & \ell_1 Z_{\text{tb}} + \ell_2 Z_{\text{mt}} + \ell_3 Z_{\text{bm}} \\ \ell_1 Z_{\text{mt}} + \ell_2 Z_{\text{bm}} + \ell_3 Z_{\text{tb}} & \ell_1 Z_{\text{mm}} + \ell_2 Z_{\text{bb}} + \ell_3 Z_{\text{tt}} & \ell_1 Z_{\text{mb}} + \ell_2 Z_{\text{bt}} + \ell_3 Z_{\text{tm}} \\ \ell_1 Z_{\text{bt}} + \ell_2 Z_{\text{tm}} + \ell_3 Z_{\text{mb}} & \ell_1 Z_{\text{bm}} + \ell_2 Z_{\text{tb}} + \ell_3 Z_{\text{mt}} & \ell_1 Z_{\text{bb}} + \ell_2 Z_{\text{tt}} + \ell_3 Z_{\text{mm}} \end{bmatrix}$$

$$(3.56b)$$

or

$$\mathbf{Z}_{\text{Phase}} = \begin{bmatrix} Z_{\text{S1}} & Z_{\text{M1}} & Z_{\text{M2}} \\ Z_{\text{M1}} & Z_{\text{S2}} & Z_{\text{M3}} \\ Z_{\text{M2}} & Z_{\text{M3}} & Z_{\text{S3}} \end{bmatrix} \qquad (3.56c)$$

The matrix of Equation (3.56c) is symmetric but the self or diagonal terms are unequal to each other, and the mutual or off-diagonal terms on a given row or column are unequal to each other. Therefore, the corresponding sequence impedance matrix will be full with intersequence mutual coupling. This result should be expected as the line is no longer perfectly transposed since perfect transposition occurs only when $\ell_1 = \ell_2 = \ell_3 = \ell/3$. Similar analysis applies to the susceptance matrix.

3.2.5 Untransposed double-circuit lines with earth wires

For double-circuit three-phase lines with earth wires strung on the same tower, there is mutual coupling between the conductors of the two circuits besides that between the conductors within each circuit. The self-circuit and inter-circuit mutual coupling needs to be defined, together with its significance, in sequence component terms, for use in large-scale power frequency steady state analysis.

Phase and sequence series impedance matrices

Figure 3.10 illustrates a typical double-circuit line with two earth wires. For circuit A, the phase conductors are numbered 1, 2 and 3 and occupy positions t_1, m_1 and b_1. For circuit B, the phase conductors are numbered 4, 5 and 6 and occupy positions t_2, m_2, b_2. The earth wires are numbered 7 and 8.

The general formulation of the series phase impedance matrix for this line can be derived from the voltage drops across each conductor in a manner similar to that presented in Section 3.2.2. Let Z_{ii} be the self-impedance of conductor i with earth return and Z_{ij} be the mutual impedance between conductors i and j with earth return. The series voltage drops across all phase and earth wire conductors,

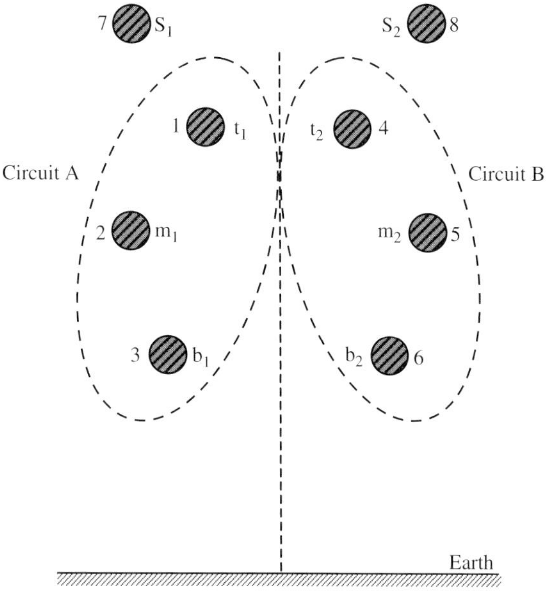

Figure 3.10 Typical double-circuit three-phase overhead line with two earth wires

denoted C1 to C8, are given by

$$
\begin{bmatrix} \Delta V_{1(A)} \\ \Delta V_{2(A)} \\ \Delta V_{3(A)} \\ \hline \Delta V_{4(B)} \\ \Delta V_{5(B)} \\ \Delta V_{6(B)} \\ \hline \Delta V_{7(S)} \\ \Delta V_{8(S)} \end{bmatrix}
=
\begin{array}{c} \\ \text{Circuit A} \\ \\ \\ \text{Circuit B} \\ \\ \\ \\ \end{array}
\begin{array}{c} \\ \text{C1} \\ \text{C2} \\ \text{C3} \\ \text{C4} \\ \text{C5} \\ \text{C6} \\ \text{C7} \\ \text{C8} \end{array}
\begin{array}{c} \quad\text{Circuit A} \qquad\qquad \text{Circuit B} \qquad\quad \text{Earth wires} \\
\begin{array}{ccc|ccc|cc} \text{C1} & \text{C2} & \text{C3} & \text{C4} & \text{C5} & \text{C6} & \text{C7} & \text{C8} \end{array} \\
\left[\begin{array}{ccc|ccc|cc}
Z_{11} & Z_{12} & Z_{13} & Z_{14} & Z_{15} & Z_{16} & Z_{17} & Z_{18} \\
Z_{21} & Z_{22} & Z_{23} & Z_{24} & Z_{25} & Z_{26} & Z_{27} & Z_{28} \\
Z_{31} & Z_{32} & Z_{33} & Z_{34} & Z_{35} & Z_{36} & Z_{37} & Z_{38} \\ \hline
Z_{41} & Z_{42} & Z_{43} & Z_{44} & Z_{45} & Z_{46} & Z_{47} & Z_{48} \\
Z_{51} & Z_{52} & Z_{53} & Z_{54} & Z_{55} & Z_{56} & Z_{57} & Z_{58} \\
Z_{61} & Z_{62} & Z_{63} & Z_{64} & Z_{65} & Z_{66} & Z_{67} & Z_{68} \\ \hline
Z_{71} & Z_{72} & Z_{73} & Z_{74} & Z_{75} & Z_{76} & Z_{77} & Z_{78} \\
Z_{81} & Z_{82} & Z_{83} & Z_{84} & Z_{85} & Z_{86} & Z_{87} & Z_{88}
\end{array}\right] \end{array}
\begin{bmatrix} I_{1(A)} \\ I_{2(A)} \\ I_{3(A)} \\ \hline I_{4(B)} \\ I_{5(B)} \\ I_{6(B)} \\ \hline I_{7(S)} \\ I_{8(S)} \end{bmatrix}
\tag{3.57a}
$$

or

$$
\begin{bmatrix} \Delta \mathbf{V}_A \\ \Delta \mathbf{V}_B \\ \Delta \mathbf{V}_S \end{bmatrix} = \begin{bmatrix} \mathbf{Z}_{AA} & \mathbf{Z}_{AB} & \mathbf{Z}_{AS} \\ \mathbf{Z}_{BA} & \mathbf{Z}_{BB} & \mathbf{Z}_{BS} \\ \mathbf{Z}_{SA} & \mathbf{Z}_{SB} & \mathbf{Z}_{SS} \end{bmatrix} \begin{bmatrix} \mathbf{I}_A \\ \mathbf{I}_B \\ \mathbf{I}_S \end{bmatrix} \tag{3.57b}
$$

where

$$
\Delta \mathbf{V}_A = \begin{bmatrix} \Delta V_{1(A)} \\ \Delta V_{2(A)} \\ \Delta V_{3(A)} \end{bmatrix} \quad \Delta \mathbf{V}_B = \begin{bmatrix} \Delta V_{4(B)} \\ \Delta V_{5(B)} \\ \Delta V_{6(B)} \end{bmatrix} \quad \Delta \mathbf{V}_S = \begin{bmatrix} \Delta V_{7(S)} \\ \Delta V_{8(S)} \end{bmatrix}
$$

$$\mathbf{I}_A = \begin{bmatrix} I_{1(A)} \\ I_{2(A)} \\ I_{3(A)} \end{bmatrix} \quad \mathbf{I}_B = \begin{bmatrix} I_{4(B)} \\ I_{5(B)} \\ I_{6(B)} \end{bmatrix} \quad \mathbf{I}_S = \begin{bmatrix} I_{7(S)} \\ I_{8(S)} \end{bmatrix} \quad \mathbf{Z}_{SS} = \begin{bmatrix} Z_{77} & Z_{78} \\ Z_{87} & Z_{88} \end{bmatrix}$$

$$\mathbf{Z}_{AA} = \begin{bmatrix} Z_{11} & Z_{12} & Z_{13} \\ Z_{21} & Z_{22} & Z_{23} \\ Z_{31} & Z_{23} & Z_{33} \end{bmatrix} \quad \mathbf{Z}_{AB} = \begin{bmatrix} Z_{14} & Z_{15} & Z_{16} \\ Z_{24} & Z_{25} & Z_{26} \\ Z_{34} & Z_{35} & Z_{36} \end{bmatrix} \quad \mathbf{Z}_{AS} = \mathbf{Z}_{SA}^t = \begin{bmatrix} Z_{17} & Z_{18} \\ Z_{27} & Z_{28} \\ Z_{37} & Z_{38} \end{bmatrix}$$

$$\mathbf{Z}_{BA} = \mathbf{Z}_{AB}^t = \begin{bmatrix} Z_{41} & Z_{42} & Z_{43} \\ Z_{51} & Z_{52} & Z_{53} \\ Z_{61} & Z_{62} & Z_{63} \end{bmatrix} \quad \mathbf{Z}_{BB} = \begin{bmatrix} Z_{44} & Z_{45} & Z_{46} \\ Z_{54} & Z_{55} & Z_{56} \\ Z_{64} & Z_{65} & Z_{66} \end{bmatrix} \quad \mathbf{Z}_{BS} = \mathbf{Z}_{SB}^t = \begin{bmatrix} Z_{47} & Z_{48} \\ Z_{57} & Z_{58} \\ Z_{67} & Z_{68} \end{bmatrix}$$

$$(3.57c)$$

$\mathbf{Z}_{AA}$ consists of the self and mutual impedances of circuit A phase conductors 1, 2 and 3. $\mathbf{Z}_{BB}$ consists of the self and mutual impedances of circuit B phase conductors 4, 5 and 6. $\mathbf{Z}_{BA} = \mathbf{Z}_{AB}^t$ consists of the mutual impedances between the phase conductors of circuit A and the phase conductors of circuit B. $\mathbf{Z}_{AS} = \mathbf{Z}_{SA}^t$ consists of the mutual impedances between circuit A phase conductors 1, 2 and 3 and the earth wires 7 and 8. $\mathbf{Z}_{BS} = \mathbf{Z}_{SB}^t$ consists of the mutual impedances between circuit B phase conductors 4, 5 and 6 and the earth wires 7 and 8. $\mathbf{Z}_{SS}$ consists of the self and mutual impedances of conductors 7 and 8 that represent the earth wires.

To eliminate the two earth wires, we set $\Delta \mathbf{V}_S = 0$ in Equation (3.57b), and after a little matrix algebra, we obtain

$$\begin{bmatrix} \Delta \mathbf{V}_A \\ \Delta \mathbf{V}_B \end{bmatrix} = \begin{bmatrix} \mathbf{Z}'_{AA} & \mathbf{Z}'_{AB} \\ \mathbf{Z}'_{BA} & \mathbf{Z}'_{BB} \end{bmatrix} \begin{bmatrix} \mathbf{I}_A \\ \mathbf{I}_B \end{bmatrix} = \mathbf{Z}_{Phase} \begin{bmatrix} \mathbf{I}_A \\ \mathbf{I}_B \end{bmatrix} \qquad (3.58a)$$

where

$$\mathbf{Z}'_{AA} = \mathbf{Z}_{AA} - \mathbf{Z}_{AS}\mathbf{Z}_{SS}^{-1}\mathbf{Z}_{AS}^t \quad \mathbf{Z}'_{AB} = \mathbf{Z}_{AB} - \mathbf{Z}_{AS}\mathbf{Z}_{SS}^{-1}\mathbf{Z}_{BS}^t$$

$$\mathbf{Z}'_{BA} = \mathbf{Z}_{BA} - \mathbf{Z}_{BS}\mathbf{Z}_{SS}^{-1}\mathbf{Z}_{AS}^t \quad \mathbf{Z}'_{BB} = \mathbf{Z}_{BB} - \mathbf{Z}_{BS}\mathbf{Z}_{SS}^{-1}\mathbf{Z}_{BS}^t$$

and

$$\mathbf{Z}_{Phase} = \begin{array}{c} \\ \\ \text{Circuit A} \\ \\ \\ \text{Circuit B} \end{array} \begin{array}{c} \text{C1} \\ \text{C2} \\ \text{C3} \\ \text{C4} \\ \text{C5} \\ \text{C6} \end{array} \overset{\displaystyle \overset{\text{Circuit A}}{\overbrace{\begin{array}{ccc} \text{C1} & \text{C2} & \text{C3} \end{array}}} \overset{\text{Circuit B}}{\overbrace{\begin{array}{ccc} \text{C4} & \text{C5} & \text{C6} \end{array}}}}{\left[\begin{array}{ccc|ccc} Z_{11-S} & Z_{12-S} & Z_{13-S} & Z_{14-S} & Z_{15-S} & Z_{16-S} \\ Z_{21-S} & Z_{22-S} & Z_{23-S} & Z_{24-S} & Z_{25-S} & Z_{26-S} \\ Z_{31-S} & Z_{32-S} & Z_{33-S} & Z_{34-S} & Z_{35-S} & Z_{36-S} \\ \hline Z_{41-S} & Z_{42-S} & Z_{43-S} & Z_{44-S} & Z_{45-S} & Z_{46-S} \\ Z_{51-S} & Z_{52-S} & Z_{53-S} & Z_{54-S} & Z_{55-S} & Z_{56-S} \\ Z_{61-S} & Z_{62-S} & Z_{63-S} & Z_{64-S} & Z_{65-S} & Z_{66-S} \end{array} \right]}$$

$$(3.58b)$$

Equation (3.58b) is the series phase impedance matrix of the double-circuit line containing the impedance elements of the two circuits A and B with earth return, and with both earth wires eliminated. In the general case of asymmetrical spacings between the conductors within each circuit and between the two circuits, the self-impedance matrices of each circuit, $\mathbf{Z}'_{AA}$ of circuit A and $\mathbf{Z}'_{BB}$ of circuit B, and the mutual impedance matrices between the two circuits, $\mathbf{Z}'_{AB}$ and $\mathbf{Z}'_{BA}$, are not balanced within themselves. Currents flowing in any one conductor will induce voltage drops in the five other conductors and these may be unequal even if the currents are balanced. The sequence impedance matrix of the 6×6 phase impedance matrix of Equation (3.58b) can be calculated by applying the phase-to-sequence transformation matrix to each voltage and current matrix vector in Equation (3.58a). Assuming an electrical phase sequence of R, Y, B for conductors 1, 2, 3 of circuit A and similarly for conductors 4, 5, 6 of circuit B, we can apply $\mathbf{V}_{RYB} = \mathbf{H}\,\mathbf{V}^{PNZ}$ and $\mathbf{I}_{RYB} = \mathbf{H}\,\mathbf{I}^{PNZ}$ to circuit A and circuit B voltage and current vectors. Therefore, the sequence voltage drop equation is given by

$$
\begin{bmatrix} \Delta\mathbf{V}_A^{PNZ} \\ \Delta\mathbf{V}_B^{PNZ} \end{bmatrix} = \begin{bmatrix} \mathbf{H}^{-1} & 0 \\ 0 & \mathbf{H}^{-1} \end{bmatrix} \begin{bmatrix} \mathbf{Z}'_{AA} & \mathbf{Z}'_{AB} \\ \mathbf{Z}'_{BA} & \mathbf{Z}'_{BB} \end{bmatrix} \begin{bmatrix} \mathbf{H} & 0 \\ 0 & \mathbf{H} \end{bmatrix} \begin{bmatrix} \mathbf{I}_A^{PNZ} \\ \mathbf{I}_B^{PNZ} \end{bmatrix} = \mathbf{Z}^{PNZ} \begin{bmatrix} \mathbf{I}_A^{PNZ} \\ \mathbf{I}_B^{PNZ} \end{bmatrix}
$$

(3.59a)

where

$$
\mathbf{Z}^{PNZ} = \begin{bmatrix} \mathbf{H}^{-1}\mathbf{Z}'_{AA}\mathbf{H} & \mathbf{H}^{-1}\mathbf{Z}'_{AB}\mathbf{H} \\ \mathbf{H}^{-1}\mathbf{Z}'_{BA}\mathbf{H} & \mathbf{H}^{-1}\mathbf{Z}'_{BB}\mathbf{H} \end{bmatrix}
$$

(3.59b)

Equation (3.59b) shows that both the self-impedance matrices of each circuit and the mutual impedance matrices between the two circuits must be pre-and-post multiplied by the appropriate transformation matrix viz $\mathbf{Z}^{PNZ} = \mathbf{H}^{-1}\mathbf{Z}'_{Phase}\mathbf{H}$. In each case of Equation (3.59b), this conversion will produce, in the general case of asymmetrical spacings, a full asymmetric sequence matrix. Even if the two circuits are identical, the inter-circuit sequence matrices will not be equal (the reader is encouraged to prove this statement). Consequently, in order to derive appropriate sequence impedances for a double-circuit line, including equal inter-circuit parameters, for use in large-scale power frequency steady state analysis, certain transposition assumptions need to be made. This is dealt with in Section 3.2.6.

Phase and sequence shunt susceptance matrices

We now derive the shunt phase susceptance matrix of our double-circuit line using the potential coefficients calculated from the line's physical dimensions or geometry. The voltage on each conductor to ground is a function of the electric charges on all conductors, thus

$$
\begin{bmatrix}
V_{1(A)} \\
V_{2(A)} \\
V_{3(A)} \\
V_{4(B)} \\
V_{5(B)} \\
V_{6(B)} \\
V_{7(S)} \\
V_{8(S)}
\end{bmatrix}
=
\begin{array}{c}
\text{Circuit A} \\
\text{Circuit B}
\end{array}
\begin{array}{c}
\text{C1} \\ \text{C2} \\ \text{C3} \\ \text{C4} \\ \text{C5} \\ \text{C6} \\ \text{C7} \\ \text{C8}
\end{array}
\begin{bmatrix}
P_{11} & P_{12} & P_{13} & P_{14} & P_{15} & P_{16} & P_{17} & P_{18} \\
P_{21} & P_{22} & P_{23} & P_{24} & P_{25} & P_{26} & P_{27} & P_{28} \\
P_{31} & P_{32} & P_{33} & P_{34} & P_{35} & P_{36} & P_{37} & P_{P38} \\
P_{41} & P_{42} & P_{43} & P_{44} & P_{45} & P_{46} & P_{47} & P_{48} \\
P_{51} & P_{52} & P_{53} & P_{54} & P_{55} & P_{56} & P_{57} & P_{58} \\
P_{61} & P_{62} & P_{63} & P_{64} & P_{65} & P_{66} & P_{67} & P_{68} \\
P_{71} & P_{72} & P_{73} & P_{74} & P_{75} & P_{76} & P_{77} & P_{78} \\
P_{81} & P_{82} & P_{83} & P_{84} & P_{85} & P_{86} & P_{87} & P_{88}
\end{bmatrix}
\begin{bmatrix}
Q_{1(A)} \\
Q_{2(A)} \\
Q_{3(A)} \\
Q_{4(B)} \\
Q_{5(B)} \\
Q_{6(B)} \\
Q_{7(S)} \\
Q_{8(S)}
\end{bmatrix}
$$

$$
\begin{array}{c}
\text{Circuit A} \quad\quad \text{Circuit B} \quad \text{Earth wires} \\
\text{C1} \ \text{C2} \ \text{C3} \quad \text{C4} \ \text{C5} \ \text{C6} \quad \text{C7} \quad \text{C8}
\end{array}
$$

(3.60a)

or

$$
\begin{bmatrix}
V_A \\
V_B \\
V_S
\end{bmatrix}
=
\begin{bmatrix}
P_{AA} & P_{AB} & P_{AS} \\
P_{BA} & P_{BB} & P_{BS} \\
P_{SA} & P_{SB} & P_{SS}
\end{bmatrix}
\begin{bmatrix}
Q_A \\
Q_B \\
Q_S
\end{bmatrix}
$$

(3.60b)

where

$$
P_{AA} =
\begin{bmatrix}
P_{11} & P_{12} & P_{13} \\
P_{21} & P_{22} & P_{23} \\
P_{31} & P_{23} & P_{33}
\end{bmatrix}
\quad
P_{AB} =
\begin{bmatrix}
P_{14} & P_{15} & P_{16} \\
P_{24} & P_{25} & P_{26} \\
P_{34} & P_{35} & P_{36}
\end{bmatrix}
\quad
P_{AS} = P_{SA}^t =
\begin{bmatrix}
P_{17} & P_{18} \\
P_{27} & P_{28} \\
P_{37} & P_{38}
\end{bmatrix}
$$

$$
P_{BA} = P_{AB}^t \quad
P_{BB} =
\begin{bmatrix}
P_{44} & P_{45} & P_{46} \\
P_{54} & P_{55} & P_{56} \\
P_{64} & P_{65} & P_{66}
\end{bmatrix}
\quad
P_{BS} = P_{SB}^t =
\begin{bmatrix}
P_{47} & P_{48} \\
P_{57} & P_{58} \\
P_{67} & P_{68}
\end{bmatrix}
$$

$$
P_{SS} =
\begin{bmatrix}
P_{77} & P_{78} \\
P_{87} & P_{88}
\end{bmatrix}
$$

(3.60c)

To eliminate the earth wires, we set $V_S = 0$ in Equation (3.60b), and after a little matrix algebra, we obtain

$$
\begin{bmatrix}
V_A \\
V_B
\end{bmatrix}
=
\begin{bmatrix}
P'_{AA} & P'_{AB} \\
P'_{BA} & P'_{BB}
\end{bmatrix}
\begin{bmatrix}
Q_A \\
Q_B
\end{bmatrix}
= P_{Phase}
\begin{bmatrix}
Q_A \\
Q_B
\end{bmatrix}
$$

(3.61a)

where

$$
P'_{AA} = P_{AA} - P_{AS}P_{SS}^{-1}P_{AS}^t \quad P'_{AB} = P_{AB} - P_{AS}P_{SS}^{-1}P_{BS}^t
$$

$$
P'_{BA} = P_{BA} - P_{BS}P_{SS}^{-1}P_{AS}^t \quad P'_{BB} = P_{BB} - P_{BS}P_{SS}^{-1}P_{BS}^t
$$

and

$$\mathbf{P}_{\text{Phase}} = \left[\begin{array}{ccc|ccc} P_{11-S} & P_{12-S} & P_{13-S} & P_{14-S} & P_{15-S} & P_{16-S} \\ P_{21-S} & P_{22-S} & P_{23-S} & P_{24-S} & P_{25-S} & P_{26-S} \\ P_{31-S} & P_{32-S} & P_{33-S} & P_{34-S} & P_{35-S} & P_{36-S} \\ \hline P_{41-S} & P_{42-S} & P_{43-S} & P_{44-S} & P_{45-S} & P_{46-S} \\ P_{51-S} & P_{52-S} & P_{53-S} & P_{54-S} & P_{55-S} & P_{56-S} \\ P_{61-S} & P_{62-S} & P_{63-S} & P_{64-S} & P_{65-S} & P_{66-S} \end{array} \right] \tag{3.61b}$$

The Maxwell's or capacitance coefficient matrix of the double-circuit line is given by $\mathbf{C}_{\text{Phase}} = \mathbf{P}_{\text{Phase}}^{-1}$. Dropping the S notation for convenience and remembering that this has the form of a nodal admittance matrix, we have

$$\mathbf{C}_{\text{Phase}} = \left[\begin{array}{ccc|ccc} C_{11} & -C_{12} & -C_{13} & -C_{14} & -C_{15} & -C_{16} \\ -C_{21} & C_{22} & -C_{23} & -C_{24} & -C_{25} & -C_{26} \\ -C_{31} & -C_{32} & C_{33} & -C_{34} & -C_{35} & -C_{36} \\ \hline -C_{41} & -C_{42} & -C_{43} & C_{44} & -C_{45} & -C_{46} \\ -C_{51} & -C_{52} & -C_{53} & -C_{54} & C_{55} & -C_{56} \\ -C_{61} & -C_{62} & -C_{63} & -C_{64} & -C_{65} & C_{66} \end{array} \right] \tag{3.62a}$$

or

$$\mathbf{C}_{\text{Phase}} = \left[\begin{array}{cc} \mathbf{C}_{\text{AA}} & \mathbf{C}_{\text{AB}} \\ \mathbf{C}_{\text{BA}} & \mathbf{C}_{\text{BB}} \end{array} \right] \tag{3.62b}$$

and using $\mathbf{B}_{\text{Phase}} = \omega \mathbf{C}_{\text{Phase}}$, the nodal shunt phase susceptance matrix is given by

$$\left[\begin{array}{c} \mathbf{I}_{\text{A}} \\ \mathbf{I}_{\text{B}} \end{array} \right] = \left[\begin{array}{cc} j\mathbf{B}_{\text{AA}} & j\mathbf{B}_{\text{AB}} \\ j\mathbf{B}_{\text{BA}} & j\mathbf{B}_{\text{BB}} \end{array} \right] \left[\begin{array}{c} \mathbf{V}_{\text{A}} \\ \mathbf{V}_{\text{B}} \end{array} \right] = j\mathbf{B}_{\text{Phase}} \left[\begin{array}{c} \mathbf{V}_{\text{A}} \\ \mathbf{V}_{\text{B}} \end{array} \right] \tag{3.62c}$$

Again, in the general case of asymmetrical spacings between the conductors within each circuit and between the two circuits, the self-susceptance matrices of each circuit, $\mathbf{B}_{\text{AA}}$ of circuit A and $\mathbf{B}_{\text{BB}}$ of circuit B, and the mutual susceptance matrices between the two circuits, $\mathbf{B}_{\text{AB}}$ and $\mathbf{B}_{\text{BA}}$, are not balanced within themselves. The sequence susceptance matrix of Equation (3.62c) is calculated by applying the phase-to-sequence transformation matrix to each voltage and current matrix vector. Thus, using $\mathbf{V}_{\text{RYB}} = \mathbf{H}\mathbf{V}^{\text{PNZ}}$ and $\mathbf{I}_{\text{RYB}} = \mathbf{H}\mathbf{I}^{\text{PNZ}}$, we obtain

$$\left[\begin{array}{c} \mathbf{I}_{\text{A}}^{\text{PNZ}} \\ \mathbf{I}_{\text{B}}^{\text{PNZ}} \end{array} \right] = \left[\begin{array}{cc} \mathbf{H}^{-1} & 0 \\ 0 & \mathbf{H}^{-1} \end{array} \right] \left[\begin{array}{cc} j\mathbf{B}_{\text{AA}} & j\mathbf{B}_{\text{AB}} \\ j\mathbf{B}_{\text{BA}} & j\mathbf{B}_{\text{BB}} \end{array} \right] \left[\begin{array}{cc} \mathbf{H} & 0 \\ 0 & \mathbf{H} \end{array} \right] \left[\begin{array}{c} \mathbf{V}_{\text{A}}^{\text{PNZ}} \\ \mathbf{V}_{\text{B}}^{\text{PNZ}} \end{array} \right] = \mathbf{B}^{\text{PNZ}} \left[\begin{array}{c} \mathbf{V}_{\text{A}}^{\text{PNZ}} \\ \mathbf{V}_{\text{B}}^{\text{PNZ}} \end{array} \right] \tag{3.63a}$$

where

$$\mathbf{B}^{PNZ} = \begin{bmatrix} \mathbf{H}^{-1}\mathbf{B}_{AA}\mathbf{H} & \mathbf{H}^{-1}\mathbf{B}_{AB}\mathbf{H} \\ \mathbf{H}^{-1}\mathbf{B}_{BA}\mathbf{H} & \mathbf{H}^{-1}\mathbf{B}_{BB}\mathbf{H} \end{bmatrix} \qquad (3.63b)$$

This conversion will produce, in each case, a full asymmetric sequence matrix and even if the two circuits are identical, the inter-circuit sequence matrices will not be equal. Consequently, in order to derive appropriate sequence susceptances for a double-circuit line, including equal inter-circuit parameters, certain transposition assumptions need to be made. This is dealt with in Section 3.2.6.

An alternative method to obtain the 6×6 capacitance or susceptance matrix is to calculate the inverse of the 8×8 potential coefficient matrix of Equation (3.60a) giving a 8×8 shunt capacitance matrix including, explicitly, the earth wires. The required 6×6 shunt phase capacitance matrix can be directly obtained by simply deleting the last two rows and columns that correspond to the earth wires. The reader is encouraged to prove this statement.

3.2.6 Transposition of double-circuit overhead lines

The transformation of a balanced 3×3 phase matrix of a single-circuit line to the sequence reference frame produces a diagonal sequence matrix with no inter-sequence mutual coupling. Therefore, when the perfect three-cycle transposition presented in Section 3.2.4 is applied to each circuit of a double-circuit line, we obtain balanced Z'_{AA} and Z'_{BB} in Equation (3.58a) and similarly for each circuit susceptance matrix. However, will such independent circuit transpositions produce balanced inter-circuit Z'_{AB} and Z'_{BA} matrices, as well as balanced inter-circuit susceptance matrices? If not, what should the transposition assumptions be in order to produce balanced inter-circuit phase impedance and susceptance matrices so that following transformation to the sequence reference frame, the inter-circuit sequence mutual coupling is either of like sequence or of zero sequence only? Like sequence coupling means that only PPS mutual coupling exists between the PPS impedances of the two circuits and similarly for the NPS and ZPS impedances. The answer can be illustrated by formulating the inter-circuit mutual impedance matrix for each section and obtaining the average for the total per-unit length of the line. Figure 3.11 shows independent circuit transpositions of a double-circuit line with either triangular or near vertical arrangements of phase conductors for each circuit.

In Figure 3.11, the conductors within each circuit are independently and perfectly transposed by rotating them in a forward direction as in the case of the single-circuit line of Figure 3.9(a). Let us designate t_1, m_1 and b_1 as the conductor positions of circuit A, and t_2, m_2 and b_2 as the conductor positions of circuit B. The self-matrices of each circuit and the inter-circuit matrices can be derived as in the case of a single-circuit line. The average self-phase impedance matrix of

Phase conductor arrangement	Transposition section 1		Transposition section 2		Transposition section 3	
	Circuit A	Circuit B	Circuit A	Circuit B	Circuit A	Circuit B
Near vertical or vertical	1⊙t_1 2⊙m_1 3⊙b_1	t_2⊙4 m_2⊙5 b_2⊙6	3⊙t_1 1⊙m_1 2⊙b_1	t_2⊙6 m_2⊙4 b_2⊙5	2⊙t_1 3⊙m_1 1⊙b_1	t_2⊙5 m_2⊙6 b_2⊙4
Triangular	⊙t_1 (1) ⊙m_1(2) ⊙b_1(3)	⊙t_2(4) ⊙m_2(5) ⊙b_2(6)	⊙t_1(3) ⊙m_1(1) ⊙b_1(2)	⊙t_2(6) ⊙m_2(4) ⊙b_2(5)	⊙t_1(2) ⊙m_1(3) ⊙b_1(1)	⊙t_2(5) ⊙m_2(6) ⊙b_2(4)

Circuit A: t_1—C1, m_1—C2, b_1—C3 ; C3, C1, C2 ; C2, C3, C1 → t, m, b

Circuit B: t_2—C4, m_2—C5, b_2—C6 ; C6, C4, C5 ; C5, C6, C4 → t, m, b

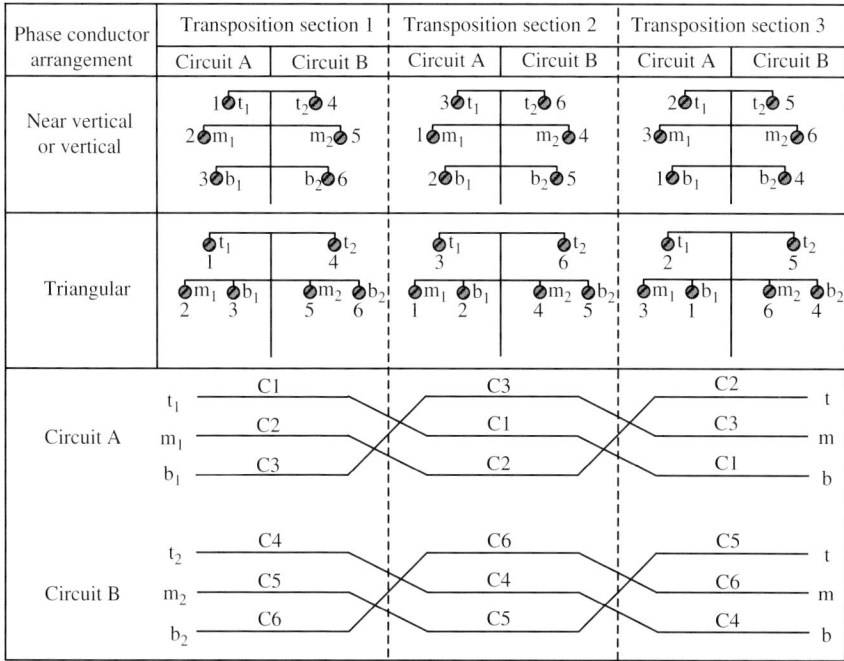

Figure 3.11 Typical double-circuit lines with perfect within circuit transpositions

circuit A per-unit length is given by

$$\mathbf{Z}_{AA} = \begin{bmatrix} Z_{S(A)} & Z_{M(A)} & Z_{M(A)} \\ Z_{M(A)} & Z_{S(A)} & Z_{M(A)} \\ Z_{M(A)} & Z_{M(A)} & Z_{S(A)} \end{bmatrix} \tag{3.64a}$$

where

$$Z_{S(A)} = \frac{1}{3}\left(Z_{t1t1} + Z_{m1m1} + Z_{b1b1}\right) \quad \text{and} \quad Z_{M(A)} = \frac{1}{3}\left(Z_{t1m1} + Z_{m1b1} + Z_{b1t1}\right) \tag{3.64b}$$

and similarly for the self-phase impedance matrix of circuit B except that suffices A and 1 change to B and 2, respectively. The inter-circuit series mutual phase impedance matrices in per-unit length for the three line sections can be written by inspection as follows:

$$\mathbf{Z}_{AB-1} = \begin{array}{c} \\ C1 \\ C2 \\ C3 \end{array} \begin{array}{ccc} C4 & C5 & C6 \\ \begin{bmatrix} Z_{t1t2} & Z_{t1m2} & Z_{t1b2} \\ Z_{m1t2} & Z_{m1m2} & Z_{m1b2} \\ Z_{b1t2} & Z_{b1m2} & Z_{b1b2} \end{bmatrix} \end{array}$$

$$\mathbf{Z}_{AB-2} = \begin{array}{c} \\ C1 \\ C2 \\ C3 \end{array} \begin{array}{ccc} C4 & C5 & C6 \\ \begin{bmatrix} Z_{m1m2} & Z_{m1b2} & Z_{m1t2} \\ Z_{b1m2} & Z_{b1b2} & Z_{b1t2} \\ Z_{t1m2} & Z_{t1b2} & Z_{t1t2} \end{bmatrix} \end{array}$$

and

$$
\mathbf{Z}_{AB-3} = \begin{array}{c} \\ C1 \\ C2 \\ C3 \end{array} \overset{\begin{array}{ccc} C4 & C5 & C6 \end{array}}{\left[\begin{array}{ccc} Z_{b1b2} & Z_{b1t2} & Z_{b1m2} \\ Z_{t1b2} & Z_{t1t2} & Z_{t1m2} \\ Z_{m1b2} & Z_{m1t2} & Z_{m1m2} \end{array} \right]} \tag{3.65a}
$$

Therefore, the average inter-circuit mutual impedance matrix per-unit length is given by

$$
\mathbf{Z}_{AB} = \frac{1}{3} \sum_{i=1}^{3} \mathbf{Z}_{AB-i}
$$

$$
= \frac{1}{3} \left[\begin{array}{ccc} Z_{t1t2} + Z_{m1m2} + Z_{b1b2} & Z_{t1m2} + Z_{m1b2} + Z_{b1t2} & Z_{t1b2} + Z_{m1t2} + Z_{b1m2} \\ Z_{m1t2} + Z_{b1m2} + Z_{t1b2} & Z_{m1m2} + Z_{b1b2} + Z_{t1t2} & Z_{m1b2} + Z_{b1t2} + Z_{t1m2} \\ Z_{b1t2} + Z_{t1m2} + Z_{m1b2} & Z_{b1m2} + Z_{t1b2} + Z_{m1t2} & Z_{b1b2} + Z_{t1t2} + Z_{m1m2} \end{array} \right]
$$

or

$$
\mathbf{Z}_{AB} = \begin{array}{c} \\ C1 \\ C2 \\ C3 \end{array} \overset{\begin{array}{ccc} C4 & C5 & C6 \end{array}}{\left[\begin{array}{ccc} Z_{S(AB)} & Z_{M(AB)} & Z_{N(AB)} \\ Z_{N(AB)} & Z_{S(AB)} & Z_{M(AB)} \\ Z_{M(AB)} & Z_{N(AB)} & Z_{S(AB)} \end{array} \right]} \quad \text{and} \quad \mathbf{Z}_{BA} = \mathbf{Z}_{AB}^{t} \tag{3.65b}
$$

Therefore, using Equations (3.64a) and (3.65b), the 6×6 series phase impedance matrix of the double-circuit line with three-phase transpositions shown in Figure 3.11 is

$$
\mathbf{Z}_{Phase} = \begin{array}{c} \\ \\ \text{Circuit A} \\ \\ \\ \text{Circuit B} \\ \\ \end{array} \overset{\begin{array}{cc} \text{Circuit A} & \text{Circuit B} \end{array}}{\left[\begin{array}{ccc|ccc} Z_{S(A)} & Z_{M(A)} & Z_{M(A)} & Z_{S(AB)} & Z_{M(AB)} & Z_{N(AB)} \\ Z_{M(A)} & Z_{S(A)} & Z_{M(A)} & Z_{N(AB)} & Z_{S(AB)} & Z_{M(AB)} \\ Z_{M(A)} & Z_{M(A)} & Z_{S(A)} & Z_{M(AB)} & Z_{N(AB)} & Z_{S(AB)} \\ \hline Z_{S(AB)} & Z_{N(AB)} & Z_{M(AB)} & Z_{S(B)} & Z_{M(B)} & Z_{M(B)} \\ Z_{M(AB)} & Z_{S(AB)} & Z_{N(AB)} & Z_{M(B)} & Z_{S(B)} & Z_{M(B)} \\ Z_{N(AB)} & Z_{M(AB)} & Z_{S(AB)} & Z_{M(B)} & Z_{M(B)} & Z_{S(B)} \end{array} \right]}
$$

$$\tag{3.66}$$

Examining the inter-circuit mutual phase impedance matrix, the self or diagonal terms are all equal, as expected, but the off-diagonal terms are generally not equal. However, these terms will be equal in the special case where circuits A and B are symmetrical with respect to each other and have vertical or near vertical phase conductor arrangement where the spacings within circuit A and B are equal. For the transpositions shown in Figure 3.11, the off-diagonal terms will not be equal in the case of triangular phase conductor arrangements even if the internal spacings of circuit A are equal to the corresponding ones of circuit B. See Example 3.5 for an alternative conductor numbering arrangement.

We will now consider the case where the inter-circuit matrix of Equation (3.65b) is balanced, i.e. $Z_{M(AB)} = Z_{N(AB)}$. The transformation of Equation (3.66) to the sequence reference frame requires knowledge of the electrical phasing of the conductors of both circuits. If conductors 1, 2, 3 of circuit A are phased R, Y, B, and conductors C4, C5, C6 of circuit B are similarly phased, the sequence matrix transformation can be calculated using Equation (3.59b). The result is given as

$$
\mathbf{Z}^{PNZ} =
\begin{array}{c}
\text{Circuit A} \\
\begin{array}{cc}
\\ \\ \text{Circuit A} \\ \\ \text{Circuit B} \\ \\
\end{array}
\begin{array}{c}
P \\ N \\ Z \\ P \\ N \\ Z
\end{array}
\begin{array}{|ccc|ccc|}
\hline
Z_A^P & 0 & 0 & Z_{AB}^P & 0 & 0 \\
0 & Z_A^N & 0 & 0 & Z_{AB}^N & 0 \\
0 & 0 & Z_A^Z & 0 & 0 & Z_{AB}^Z \\
\hline
Z_{AB}^P & 0 & 0 & Z_B^P & 0 & 0 \\
0 & Z_{AB}^N & 0 & 0 & Z_B^N & 0 \\
0 & 0 & Z_{AB}^Z & 0 & 0 & Z_B^Z \\
\hline
\end{array}
\end{array}
$$

$$\text{(3.67a)}$$

where

$$Z_A^P = Z_A^N = Z_{S(A)} - Z_{M(A)} = \frac{1}{3}[(Z_{t1t1} + Z_{m1m1} + Z_{b1b1}) - (Z_{t1m1} + Z_{m1b1} + Z_{b1t1})]$$

$$\text{(3.67b)}$$

$$Z_A^Z = Z_{S(A)} + 2Z_{M(A)} = \frac{1}{3}[(Z_{t1t1} + Z_{m1m1} + Z_{b1b1}) + 2(Z_{t1m1} + Z_{m1b1} + Z_{b1t1})]$$

$$\text{(3.67c)}$$

and similarly for circuit B except that the suffices A and 1 are replaced with B and 2, respectively. Also

$$Z_{AB}^P = Z_{AB}^N = Z_{S(AB)} - Z_{M(AB)} = \frac{1}{3}[(Z_{t1t2} + Z_{m1m2} + Z_{b1b2}) - (Z_{t1m2} + Z_{m1b2} + Z_{b1t2})]$$

$$\text{(3.68a)}$$

and

$$Z_{AB}^Z = Z_{S(AB)} + 2Z_{M(AB)} = \frac{1}{3}[(Z_{t1t2} + Z_{m1m2} + Z_{b1b2}) + 2(Z_{t1m2} + Z_{m1b2} + Z_{b1t2})]$$

$$\text{(3.68b)}$$

Equation (3.67a) requires a physical explanation. Each three-phase circuit has self-PPS/NPS and ZPS impedances. In addition, there is an equal mutual PPS impedance coupling between circuits A and B. Also, similar NPS and ZPS inter-circuit impedance coupling exists. There is no intersequence coupling between the two circuits. In other words, a PPS current flowing in circuit B will induce a PPS voltage only in circuit A and similarly for NPS and ZPS currents. Further, as will be presented in Section 3.4, the PPS, NPS and ZPS mutual impedances between the two circuits create a physical circuit for the double-circuit line that can be represented by an appropriate π equivalent circuit.

In the above analysis, circuits A and B were assumed to be similarly phased, i.e. RYB/RYB. However, will we obtain the same desired result of Equation (3.67a) if the conductors of circuit B were phased differently? In England and Wales 400 kV and 275 kV transmission system, double-circuit overhead lines are generally of vertical or near vertical construction sometimes with an offset middle arm tower. These lines are not transposed between substations and in order to reduce their degree of unbalance, measured in NPS and ZPS voltages and/or currents, it is standard British practice to use electrical phase transposition at substations. That is for RYB phasing of the top, middle and bottom conductors on circuit A, circuit B would be phase BYR of the top, middle and bottom conductors. This means that circuit B has the same middle phase as circuit A but the top and bottom phases are interchanged with respect to circuit A. In order to transform Equation (3.66) to the sequence frame, we need to determine the applicable transformation matrices. For circuit A, using RYB phase sequence of conductors 1, 2, 3, and for circuit B, using BYR phase sequence of conductors 4, 5, 6, we have

$$Circuit\ A: \mathbf{V}_{\mathrm{RYB}} = \mathbf{H}\mathbf{V}^{\mathrm{PNZ}} \quad and \quad \mathbf{I}_{\mathrm{RYB}} = \mathbf{H}\mathbf{I}^{\mathrm{PNZ}} \quad where\ \mathbf{H} = \begin{bmatrix} 1 & 1 & 1 \\ h^2 & h & 1 \\ h & h^2 & 1 \end{bmatrix}$$

(3.69a)

$$Circuit\ B: \mathbf{V}_{\mathrm{BYR}} = \mathbf{H}_1 \mathbf{V}^{\mathrm{PNZ}} \quad and \quad \mathbf{I}_{\mathrm{BYR}} = \mathbf{H}_1 \mathbf{I}^{\mathrm{PNZ}} \quad where\ \mathbf{H}_1 = \begin{bmatrix} h & h^2 & 1 \\ h^2 & h & 1 \\ 1 & 1 & 1 \end{bmatrix}$$

(3.69b)

Therefore, applying Equation (3.69) to Equation (3.58a), we obtain

$$\begin{bmatrix} \Delta \mathbf{V}_{\mathrm{A}}^{\mathrm{PNZ}} \\ \Delta \mathbf{V}_{\mathrm{B}}^{\mathrm{PNZ}} \end{bmatrix} = \begin{bmatrix} \mathbf{H}^{-1} & 0 \\ 0 & \mathbf{H}_1^{-1} \end{bmatrix} \begin{bmatrix} \mathbf{Z}_{\mathrm{AA}}' & \mathbf{Z}_{\mathrm{AB}}' \\ \mathbf{Z}_{\mathrm{BA}}' & \mathbf{Z}_{\mathrm{BB}}' \end{bmatrix} \begin{bmatrix} \mathbf{H} & 0 \\ 0 & \mathbf{H}_1 \end{bmatrix} \begin{bmatrix} \mathbf{I}_{\mathrm{A}}^{\mathrm{PNZ}} \\ \mathbf{I}_{\mathrm{B}}^{\mathrm{PNZ}} \end{bmatrix} = \mathbf{Z}^{\mathrm{PNZ}} \begin{bmatrix} \mathbf{I}_{\mathrm{A}}^{\mathrm{PNZ}} \\ \mathbf{I}_{\mathrm{B}}^{\mathrm{PNZ}} \end{bmatrix}$$

(3.70a)

where

$$\mathbf{Z}^{\mathrm{PNZ}} = \begin{bmatrix} \mathbf{H}^{-1}\mathbf{Z}_{\mathrm{AA}}'\mathbf{H} & \mathbf{H}^{-1}\mathbf{Z}_{\mathrm{AB}}'\mathbf{H}_1 \\ \mathbf{H}_1^{-1}\mathbf{Z}_{\mathrm{BA}}'\mathbf{H} & \mathbf{H}_1^{-1}\mathbf{Z}_{\mathrm{BB}}'\mathbf{H}_1 \end{bmatrix} = \begin{bmatrix} \mathbf{Z}_{\mathrm{AA}}^{\mathrm{PNZ}} & \mathbf{Z}_{\mathrm{AB}}^{\mathrm{PNZ}} \\ \mathbf{Z}_{\mathrm{BA}}^{\mathrm{PNZ}} & \mathbf{Z}_{\mathrm{BB}}^{\mathrm{PNZ}} \end{bmatrix}$$

(3.70b)

Applying Equation (3.70b) to Equation (3.66), we find that $\mathbf{Z}_{\mathrm{AA}}^{\mathrm{PNZ}}$ is clearly diagonal. $\mathbf{Z}_{\mathrm{BB}}^{\mathrm{PNZ}}$ is also diagonal of a form similar to $\mathbf{Z}_{\mathrm{AA}}^{\mathrm{PNZ}}$. $\mathbf{Z}_{\mathrm{AB}}^{\mathrm{PNZ}}$ and $\mathbf{Z}_{\mathrm{BA}}^{\mathrm{PNZ}}$ are obtained through multiplication by different transformation matrices.

Dropping the prime for convenience and with $Z_{M(AB)} = Z_{N(AB)}$, that is $Z_{BA} = Z_{AB}^{t} = Z_{AB}$, the sequence matrices of $\mathbf{Z}_{AB}$ and $\mathbf{Z}_{BA}$ are given by

$$
\mathbf{Z}_{AB}^{PNZ} = \mathbf{H}^{-1}\mathbf{Z}_{AB}\,\mathbf{H}_{1} =
\begin{array}{c}
 \\ P \\ N \\ Z
\end{array}
\begin{array}{ccc}
P & N & Z \\
\left[\begin{array}{ccc}
0 & Z_{AB}^{PN} & 0 \\
Z_{AB}^{NP} & 0 & 0 \\
0 & 0 & Z_{AB}^{ZZ}
\end{array}\right]
\end{array}
\tag{3.71a}
$$

where

$$
Z_{AB}^{PN} = h^{2}[Z_{S(AB)} - Z_{M(AB)}] \quad Z_{AB}^{NP} = h[Z_{S(AB)} - Z_{M(AB)}]
$$

$$
Z_{AB}^{ZZ} = Z_{S(AB)} + 2Z_{M(AB)}
\tag{3.71b}
$$

and

$$
\mathbf{Z}_{BA}^{PNZ} = \mathbf{Z}_{AB}^{PNZ}
\tag{3.71c}
$$

Therefore, the full sequence impedance matrix, including voltage and current vectors, is given by

$$
\begin{bmatrix}
\Delta V_{A}^{P} \\
\Delta V_{A}^{N} \\
\Delta V_{A}^{Z} \\
\hline
\Delta V_{B}^{P} \\
\Delta V_{B}^{N} \\
\Delta V_{B}^{Z}
\end{bmatrix}
=
\begin{array}{c}
\\
\text{Circuit A}\ \ \begin{array}{c} P \\ N \\ Z \end{array} \\
\\
\text{Circuit B}\ \ \begin{array}{c} P \\ N \\ Z \end{array}
\end{array}
\left[\begin{array}{ccc|ccc}
Z_{A}^{P} & 0 & 0 & 0 & Z_{AB}^{PN} & 0 \\
0 & Z_{A}^{N} & 0 & Z_{AB}^{NP} & 0 & 0 \\
0 & 0 & Z_{A}^{Z} & 0 & 0 & Z_{AB}^{ZZ} \\
\hline
0 & Z_{AB}^{PN} & 0 & Z_{B}^{P} & 0 & 0 \\
Z_{AB}^{NP} & 0 & 0 & 0 & Z_{B}^{N} & 0 \\
0 & 0 & Z_{AB}^{ZZ} & 0 & 0 & Z_{B}^{Z}
\end{array}\right]
\begin{bmatrix}
I_{A}^{P} \\
I_{A}^{N} \\
I_{A}^{Z} \\
\hline
I_{B}^{P} \\
I_{B}^{N} \\
I_{B}^{Z}
\end{bmatrix}
\tag{3.72}
$$

This undesirable result shows that there is PPS to NPS mutual coupling between circuit A and circuit B. This means that a NPS current in circuit B will produce a PPS voltage drop in circuit A by acting on the impedance Z_{AB}^{PN}. Similarly, a PPS current in circuit B will produce a NPS voltage drop in circuit A by acting on the impedance Z_{AB}^{NP}. It should be noted that $Z_{AB}^{PN} \neq Z_{AB}^{NP}$ as seen in Equation (3.71b). However, the two self-ZPS impedances of circuit A and B are coupled by the same inter-circuit ZPS mutual impedance and that this is equal to that where both circuits A and B had the same phasing rotation of RYB.

We have shown that three transpositions may be sufficient to produce balanced inter-circuit mutual phase impedance matrices in very special cases of tower geometry, conductor arrangements and conductor electrical phasing. However, in general, the transformation of Equation (3.65b) into the sequence reference frame, after three-phase transpositions, can produce either a full mutual sequence

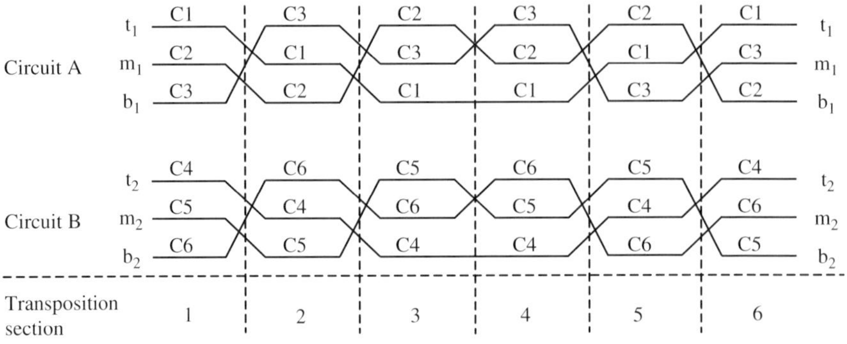

Figure 3.12 Double-circuit line with six-phase transpositions within each circuit; both circuits are phased RYB

impedance matrix between the two circuits or mutual coupling of unlike sequence terms, as shown in Equation (3.72).

Examination of Equation (3.65b) reveals that this unbalanced matrix can be made balanced if the off-diagonal terms $Z_{M(AB)}$ and $Z_{N(AB)}$ can each be changed to $Z_{M(AB)} + Z_{N(AB)}$. This can be achieved by using three further transpositions, i.e. a total of six transpositions as follows: circuit A fourth, fifth and sixth transpositions retain the same conductor sequence in the middle positions, i.e. C2, C1 and C3. Also, the conductors in the t and b positions of the fourth, fifth and sixth transpositions are obtained by interchanging the conductors of the first, second and third transpositions, respectively. Similar transpositions are also applied to circuit B fourth, fifth and sixth transpositions. The resultant six transpositions are shown in Figure 3.12.

The inter-circuit mutual impedance matrices for the first three transposition sections are given in Equation (3.65a) and those for the fourth, fifth and sixth transposition sections can be written by inspection using Figure 3.12 as follows

$$\mathbf{Z}_{AB-4} = \begin{matrix} \\ C1 \\ C2 \\ C3 \end{matrix} \begin{matrix} \begin{matrix} C4 & C5 & C6 \end{matrix} \\ \begin{bmatrix} Z_{b1b2} & Z_{b1m2} & Z_{b1t2} \\ Z_{m1b2} & Z_{m1m2} & Z_{m1t2} \\ Z_{t1b2} & Z_{t1m2} & Z_{t1t2} \end{bmatrix} \end{matrix} \qquad \mathbf{Z}_{AB-5} = \begin{matrix} \\ C1 \\ C2 \\ C3 \end{matrix} \begin{matrix} \begin{matrix} C4 & C5 & C6 \end{matrix} \\ \begin{bmatrix} Z_{m1m2} & Z_{m1t2} & Z_{m1b2} \\ Z_{t1m2} & Z_{t1t2} & Z_{t1b2} \\ Z_{b1m2} & Z_{b1t2} & Z_{b1b2} \end{bmatrix} \end{matrix}$$

and

$$\mathbf{Z}_{AB-6} = \begin{matrix} \\ C1 \\ C2 \\ C3 \end{matrix} \begin{matrix} \begin{matrix} C4 & C5 & C6 \end{matrix} \\ \begin{bmatrix} Z_{t1t2} & Z_{t1b2} & Z_{t1m2} \\ Z_{b1t2} & Z_{b1b2} & Z_{b1m2} \\ Z_{m1t2} & Z_{m1b2} & Z_{m1m2} \end{bmatrix} \end{matrix} \qquad (3.73)$$

Therefore, using Equations (3.65a) and (3.73), the series mutual phase impedance matrix per-unit length between circuits A and B for the six-line sections is given by

$$\mathbf{Z}_{AB} = \frac{1}{6}\sum_{i=1}^{6}\mathbf{Z}_{AB-i} = \begin{bmatrix} Z_{S(AB)} & Z_{M(AB)} & Z_{M(AB)} \\ Z_{M(AB)} & Z_{S(AB)} & Z_{M(AB)} \\ Z_{M(AB)} & Z_{M(AB)} & Z_{S(AB)} \end{bmatrix} \qquad (3.74a)$$

where

$$Z_{S(AB)} = \frac{1}{3}(Z_{t1t2} + Z_{m1m2} + Z_{b1b2}) \qquad (3.74b)$$

and

$$Z_{M(AB)} = \frac{1}{6}(Z_{t1m2} + Z_{t1b2} + Z_{m1t2} + Z_{m1b2} + Z_{b1t2} + Z_{b1m2}) \qquad (3.74c)$$

The 6×6 series phase impedance matrices of the double-circuit line with the six-phase transpositions shown in Figure 3.12 are given by

$$\mathbf{Z}_{Phase} = \begin{array}{c} \\ \text{Circuit A} \\ \\ \\ \text{Circuit B} \\ \\ \end{array} \begin{bmatrix} \overset{\text{Circuit A}}{} & & & \overset{\text{Circuit B}}{} & & \\ Z_{S(A)} & Z_{M(A)} & Z_{M(A)} & Z_{S(AB)} & Z_{M(AB)} & Z_{M(AB)} \\ Z_{M(A)} & Z_{S(A)} & Z_{M(A)} & Z_{M(AB)} & Z_{S(AB)} & Z_{M(AB)} \\ Z_{M(A)} & Z_{M(A)} & Z_{S(A)} & Z_{M(AB)} & Z_{M(AB)} & Z_{S(AB)} \\ Z_{S(AB)} & Z_{M(AB)} & Z_{M(AB)} & Z_{S(B)} & Z_{M(B)} & Z_{M(B)} \\ Z_{M(AB)} & Z_{S(AB)} & Z_{M(AB)} & Z_{M(B)} & Z_{S(B)} & Z_{M(B)} \\ Z_{M(AB)} & Z_{M(AB)} & Z_{S(AB)} & Z_{M(B)} & Z_{M(B)} & Z_{S(B)} \end{bmatrix}$$

$$(3.75a)$$

where

$$Z_{S(A)} = \frac{1}{3}(Z_{t1t1} + Z_{m1m1} + Z_{b1b1}) \qquad Z_{S(B)} = \frac{1}{3}(Z_{t2t2} + Z_{m2m2} + Z_{b2b2})$$

$$Z_{M(A)} = \frac{1}{3}(Z_{t1m1} + Z_{m1b1} + Z_{b1t1}) \qquad Z_{M(B)} = \frac{1}{3}(Z_{t2m2} + Z_{m2b2} + Z_{b2t2})$$

$$Z_{S(AB)} = \frac{1}{3}(Z_{t1t2} + Z_{m1m2} + Z_{b1b2})$$

$$Z_{M(AB)} = \frac{1}{6}(Z_{t1m2} + Z_{t1b2} + Z_{m1t2} + Z_{m1b2} + Z_{b1t2} + Z_{b1m2}) \qquad (3.75b)$$

The corresponding sequence impedance matrix with both circuits A and B having the same electrical phasing RYB is given by

$$
\mathbf{Z}^{PNZ} =
\begin{array}{c}
\text{Circuit A} \\
\quad
\end{array}
$$

$$
\mathbf{Z}^{PNZ} = \text{Circuit A} \quad \text{Circuit B}
$$

$$
\begin{array}{cc}
 & \begin{array}{ccc|ccc}
\text{P} & \text{N} & \text{Z} & \text{P} & \text{N} & \text{Z}
\end{array} \\
\begin{array}{c}
\text{Circuit A} \\
\\
\text{Circuit B}
\end{array}
\begin{array}{c}
\text{P}\\ \text{N}\\ \text{Z}\\ \text{P}\\ \text{N}\\ \text{Z}
\end{array}
&
\left[
\begin{array}{ccc|ccc}
Z_A^P & 0 & 0 & Z_{AB}^P & 0 & 0 \\
0 & Z_A^N & 0 & 0 & Z_{AB}^N & 0 \\
0 & 0 & Z_A^Z & 0 & 0 & Z_{AB}^Z \\
\hline
Z_{AB}^P & 0 & 0 & Z_B^P & 0 & 0 \\
0 & Z_{AB}^N & 0 & 0 & Z_B^N & 0 \\
0 & 0 & Z_{AB}^Z & 0 & 0 & Z_B^Z
\end{array}
\right]
\end{array}
$$

(3.76a)

where

$$
Z_A^P = Z_A^N = Z_{S(A)} - Z_{M(A)} = \frac{1}{3}[(Z_{t1t1} + Z_{m1m1} + Z_{b1b1}) - (Z_{t1m1} + Z_{m1b1} + Z_{b1t1})]
$$

$$
Z_A^Z = Z_{S(A)} + 2Z_{M(A)} = \frac{1}{3}[(Z_{t1t1} + Z_{m1m1} + Z_{b1b1}) + 2(Z_{t1m1} + Z_{m1b1} + Z_{b1t1})]
$$

(3.76b)

and similarly for circuit B except suffices A and 1 are replaced by B and 2, respectively. For the inter-circuit parameters

$$
Z_{AB}^P = Z_{AB}^N = Z_{S(AB)} - Z_{M(AB)}
$$

$$
= \frac{1}{3}(Z_{t1t2} + Z_{m1m2} + Z_{b1b2}) - \frac{1}{6}(Z_{t1m2} + Z_{t1b2} + Z_{m1t2} + Z_{m1b2} + Z_{b1t2} + Z_{b1m2})
$$

$$
Z_{AB}^Z = Z_{S(AB)} + 2Z_{M(AB)}
$$

$$
= \frac{1}{3}[(Z_{t1t2} + Z_{m1m2} + Z_{b1b2}) + (Z_{t1m2} + Z_{t1b2} + Z_{m1t2} + Z_{m1b2} + Z_{b1t2} + Z_{b1m2})]
$$

(3.76c)

Equation (3.76a) shows that each circuit is represented by a self-PPS/NPS impedance and a self-ZPS impedance but with no mutual sequence coupling within each circuit. In addition, there is like sequence PPS/NPS and ZPS mutual impedance coupling between the two circuits. It is important to note that this inter-circuit sequence mutual coupling is of the same sequence type that is only PPS coupling appears in the PPS circuits, NPS coupling in the NPS circuits and ZPS coupling in the ZPS circuits.

The reader is encouraged to show that for the six-phase transpositions shown in Figure 3.12, the transposed 6×6 shunt phase susceptance matrix can be obtained as

$$
\mathbf{B}_{\text{Phase}} =
\begin{array}{c}
\\
\text{Circuit A} \\
\\
\\
\text{Circuit B} \\
\\
\end{array}
\begin{bmatrix}
B_{S(A)} & -B_{M(A)} & -B_{M(A)} & -B_{S(AB)} & -B_{M(AB)} & -B_{M(AB)} \\
-B_{M(A)} & B_{S(A)} & -B_{M(A)} & -B_{M(AB)} & -B_{S(AB)} & -B_{M(AB)} \\
-B_{M(A)} & -B_{M(A)} & B_{S(A)} & -B_{M(AB)} & -B_{M(AB)} & -B_{S(AB)} \\
-B_{S(AB)} & -B_{M(AB)} & -B_{M(AB)} & -B_{S(B)} & -B_{M(B)} & -B_{M(B)} \\
-B_{M(AB)} & -B_{S(AB)} & -B_{M(AB)} & -B_{M(B)} & -B_{S(B)} & -B_{M(B)} \\
-B_{M(AB)} & -B_{M(AB)} & -B_{S(AB)} & -B_{M(B)} & -B_{M(B)} & B_{S(B)}
\end{bmatrix}
$$

with overhead labels **Circuit A** and **Circuit B**.

(3.77a)

where

$$
B_{S(A)} = \frac{1}{3}(B_{t1t1} + B_{m1m1} + B_{b1b1}) \qquad B_{S(B)} = \frac{1}{3}(B_{t2t2} + B_{m2m2} + B_{b2b2})
$$

$$
B_{M(A)} = \frac{1}{3}(B_{t1m1} + B_{m1b1} + B_{b1t1}) \qquad B_{M(B)} = \frac{1}{3}(B_{t2m2} + B_{m2b2} + B_{b2t2})
$$

$$
B_{S(AB)} = \frac{1}{3}(B_{t1t2} + B_{m1m2} + B_{b1b2})
$$

$$
B_{M(AB)} = \frac{1}{6}(B_{t1m2} + B_{t1b2} + B_{m1t2} + B_{m1b2} + B_{b1t2} + B_{b1m2}) \qquad (3.77b)
$$

and the corresponding shunt sequence susceptance matrix is given by

		Circuit A			Circuit B		
		P	N	Z	P	N	Z
Circuit A	P	B_A^P	0	0	B_{AB}^P	0	0
	N	0	B_A^N	0	0	B_{AB}^N	0
	Z	0	0	B_A^Z	0	0	B_{AB}^Z
Circuit B	P	B_{AB}^P	0	0	B_B^P	0	0
	N	0	B_{AB}^N	0	0	B_B^N	0
	Z	0	0	B_{AB}^Z	0	0	B_B^Z

$$
\mathbf{B}^{\text{PNZ}} =
$$

(3.78a)

where

$$
B_A^P = B_A^N = B_{S(A)} + B_{M(A)} = \frac{1}{3}[(B_{t1t1} + B_{m1m1} + B_{b1b1}) + (B_{t1m1} + B_{m1b1} + B_{b1t1})]
$$

$$
B_A^Z = B_{S(A)} - 2B_{M(A)} = \frac{1}{3}[(B_{t1t1} + B_{m1m1} + B_{b1b1}) - 2(B_{t1m1} + B_{m1b1} + B_{b1t1})]
$$

(3.78b)

and similarly for circuit B except suffices A and 1 are replaced by B and 2, respectively.

$$B_{AB}^P = B_{AB}^N = B_{S(AB)} + B_{M(AB)}$$

$$= \frac{1}{3}(B_{t1t2} + B_{m1m2} + B_{b1b2}) + \frac{1}{6}(B_{t1m2} + B_{t1b2} + B_{m1t2} + B_{m1b2} + B_{b1t2} + B_{b1m2})$$

$$B_{AB}^Z = B_{S(AB)} - 2B_{M(AB)} = \frac{1}{3}(B_{t1t2} + B_{m1m2} + B_{b1b2}) - \frac{1}{3}(B_{t1m2} + B_{t1b2}$$
$$+ B_{m1t2} + B_{m1b2} + B_{b1t2} + B_{b1m2}) \qquad (3.78c)$$

As for the series sequence impedance matrix, Equation (3.78a) shows that there is inter-circuit mutual sequence susceptance coupling of like sequence only.

We have shown that the six transpositions shown in Figure 3.12 result in diagonal sequence self and inter-circuit mutual impedance/susceptance matrices for the general case of asymmetrical spacing of conductors provided that the two circuits have the same electrical phasing. If one circuit has different phasing from the other, then the sequence inter-circuit mutual impedance/susceptance matrices will result in intersequence coupling as shown in the case of RYB/BYR phasing that resulted in Equation (3.72). Therefore, the six transpositions applied for a double-circuit line of general asymmetrical spacings required to produce like sequence coupling, or diagonal sequence inter-circuit impedance/susceptance matrices, are dependent on the actual electrical phasing used for both circuits. Using the common RYB/BYR phasing arrangement employed on over 90% of 400 and 275 kV double-circuit lines in England and Wales, Figure 3.13 shows the six transpositions required.

The general transposition analysis using the matrix analysis method presented for single-circuit lines will now be extended and applied to double-circuit lines.

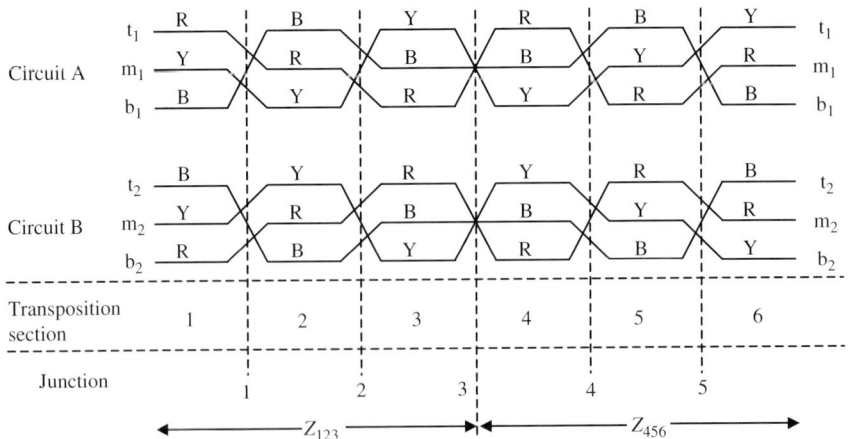

Figure 3.13 Double-circuit line with six-phase transpositions within each circuit; circuits are phased RYB/BYR

Using Figure 3.13, the series phase impedance matrix of Section 1 is given as

$$
\mathbf{Z}_{\text{Section-1}} =
\begin{array}{c}
\\
\text{Circuit A}
\\
\\
\text{Circuit B}
\end{array}
\begin{array}{c}
\\
\begin{array}{c} R \\ Y \\ B \end{array}
\\
\begin{array}{c} B \\ Y \\ R \end{array}
\end{array}
\overset{\begin{array}{ccc ccc} \quad\text{Circuit A} & & & & \text{Circuit B} & \\ R & Y & B & B & Y & R \end{array}}{
\left[
\begin{array}{ccc|ccc}
Z_{t1t1} & Z_{t1m1} & Z_{t1b1} & Z_{t1t2} & Z_{t1m2} & Z_{t1b2} \\
Z_{m1t1} & Z_{m1m1} & Z_{m1b1} & Z_{m1t2} & Z_{m1m2} & Z_{m1b2} \\
Z_{b1t1} & Z_{b1m1} & Z_{b1b1} & Z_{b1t2} & Z_{b1m2} & Z_{b1b2} \\
\hline
Z_{t2t1} & Z_{t2m1} & Z_{t2b1} & Z_{t2t2} & Z_{t2m2} & Z_{t2b2} \\
Z_{m2t1} & Z_{m2m1} & Z_{m2b1} & Z_{m2t2} & Z_{m2m2} & Z_{m2b2} \\
Z_{b2t1} & Z_{b2m1} & Z_{b2b1} & Z_{b2t2} & Z_{b2m2} & Z_{b2b2}
\end{array}
\right]}
$$

(3.79)

We note in Figure 3.13 the forward and reverse transpositions at the first junction of circuit A and circuit B, respectively. Therefore, using the transposition matrix **T** defined in Equation (3.44) for circuit A, and using its transpose for circuit B, we define a new 6×6 transposition matrix for our double-circuit line as follows:

$$
\mathbf{T}_{\text{AB}} =
\left[
\begin{array}{ccc|ccc}
0 & 1 & 0 & 0 & 0 & 0 \\
0 & 0 & 1 & 0 & 0 & 0 \\
1 & 0 & 0 & 0 & 0 & 0 \\
\hline
0 & 0 & 0 & 0 & 0 & 1 \\
0 & 0 & 0 & 1 & 0 & 0 \\
0 & 0 & 0 & 0 & 1 & 0
\end{array}
\right]
\begin{array}{l} \\ \text{Circuit A} \\ \\ \\ \text{Circuit B} \\ \\ \end{array}
\qquad
\mathbf{T}_{\text{AB}}^{-1} = \mathbf{T}_{\text{AB}}^{t} = \mathbf{T}_{\text{AB}}^{2}
$$

(3.80)

Applying the transposition matrix at the first junction in Figure 3.13, we obtain

$$
\mathbf{Z}_{\text{Section-2}} = \mathbf{T}_{\text{AB}} \mathbf{Z}_{\text{Section-1}} \mathbf{T}_{\text{AB}}^{t}
$$

(3.81a)

Applying the transposition matrix again at the second in Figure 3.13 junction, we obtain

$$
\mathbf{Z}_{\text{Section-3}} = \mathbf{T}_{\text{AB}} \mathbf{Z}_{\text{Section-2}} \mathbf{T}_{\text{AB}}^{t} = \mathbf{T}_{\text{AB}} (\mathbf{T}_{\text{AB}} \mathbf{Z}_{\text{Section-1}} \mathbf{T}_{\text{AB}}^{t}) \mathbf{T}_{\text{AB}}^{t}
$$

$$
= \mathbf{T}_{\text{AB}}^{2} \mathbf{Z}_{\text{Section-1}} (\mathbf{T}_{\text{AB}}^{t})^{2} = \mathbf{T}_{\text{AB}}^{t} \mathbf{Z}_{\text{Section-1}} \mathbf{T}_{\text{AB}}
$$

(3.81b)

The impedance matrices for Sections 4, 5 and 6 can be derived in terms of $\mathbf{Z}_{\text{Section-1}}$ noting that new transposition matrices need to be defined for junctions 3 and 4. Alternatively, the impedance matrix of one section can be derived in terms of the impedance matrix of the previous section. That is the impedance matrix of section n is derived as a function of that of the $(n-1)$ section and so on. The reader is encouraged to attempt both derivations and prove that they give the same result.

However, for us, we will follow a third alternative approach. We note that in the first three transpositions, circuit A is phased with a PPS order namely RYB, BRY and YBR whereas circuit B is phased with a NPS order BYR, YRB and RBY. Let the impedance matrix resulting from these first three transpositions be $\mathbf{Z}_{123}$. Therefore, using Equation (3.81), we have

$$\mathbf{Z}_{123} = \mathbf{Z}_{\text{Section-1}} + \mathbf{T}_{\text{AB}}\mathbf{Z}_{\text{Section-1}}\mathbf{T}_{\text{AB}}^{t} + \mathbf{T}_{\text{AB}}^{t}\mathbf{Z}_{\text{Section-1}}\mathbf{T}_{\text{AB}} \qquad (3.82)$$

To complete the six transpositions, the matrix $\mathbf{Z}_{123}$ is connected in series with a new matrix $\mathbf{Z}_{456}$ so that circuit A is now phased with a NPS order RBY, BYR and YRB whereas circuit B is now phased with a PPS order YBR, RYB and BRY. This means that matrix $\mathbf{Z}_{456}$ has the top and bottom conductors of circuit A, and the top and bottom conductors of circuit B, interchanged with respect to matrix $\mathbf{Z}_{123}$. For this, a new transposition matrix $\mathbf{I}$ is defined as follows:

$$\mathbf{I} = \begin{bmatrix} 0 & 0 & 1 & 0 & 0 & 0 \\ 0 & 1 & 0 & 0 & 0 & 0 \\ 1 & 0 & 0 & 0 & 0 & 0 \\ \hline 0 & 0 & 0 & 0 & 0 & 1 \\ 0 & 0 & 0 & 0 & 1 & 0 \\ 0 & 0 & 0 & 1 & 0 & 0 \end{bmatrix} \quad \text{where } \mathbf{I} = \mathbf{I}^{t} = \mathbf{I}^{-1} \text{ and } \mathbf{I}^{2} = \mathbf{U} \qquad (3.83)$$

To clarify this approach, we note that the effect of pre-multiplying a column matrix $\begin{bmatrix} C1 \\ C2 \\ C3 \end{bmatrix}$, or a 3×3 square matrix $\begin{bmatrix} Z_{11} & Z_{12} & Z_{13} \\ Z_{21} & Z_{22} & Z_{23} \\ Z_{31} & Z_{32} & Z_{33} \end{bmatrix}$, by the matrix $\begin{bmatrix} 0 & 0 & 1 \\ 0 & 1 & 0 \\ 1 & 0 & 0 \end{bmatrix}$ is to produce a new column matrix $\begin{bmatrix} C3 \\ C2 \\ C1 \end{bmatrix}$, or a new 3×3 square matrix $\begin{bmatrix} Z_{31} & Z_{32} & Z_{33} \\ Z_{21} & Z_{22} & Z_{23} \\ Z_{11} & Z_{12} & Z_{13} \end{bmatrix}$, so that the middle row retains its position but the top and bottom rows interchange their positions. Similarly, the effect of post-multiplying our 3×3 matrix by $\begin{bmatrix} 0 & 0 & 1 \\ 0 & 1 & 0 \\ 1 & 0 & 0 \end{bmatrix}$ is to produce a new 3×3 square matrix with the middle column retaining its position but columns 1 and 3 interchange their positions. Therefore, the new impedance matrix of the fourth, fifth and sixth transpositions is given by

$$\mathbf{Z}_{456} = \mathbf{I}\mathbf{Z}_{123}\mathbf{I} \qquad (3.84)$$

Finally, the series phase impedance matrix per-unit length of our double-circuit line with six transpositions and circuit A phased RYB and circuit B phased

BYR is given by

$$\mathbf{Z}_{\text{Phase}} = \frac{1}{6}(\mathbf{Z}_{123} + \mathbf{Z}_{456}) \tag{3.85}$$

After some algebra, it can be shown that $\mathbf{Z}_{\text{Phase}}$ is given by

$$\mathbf{Z}_{\text{Phase}} = \begin{array}{c} \\ \\ \text{Circuit A} \\ \\ \\ \\ \text{Circuit B} \\ \\ \end{array}
\begin{array}{c} \\ R \\ Y \\ B \\ \\ B \\ Y \\ R \end{array}
\overset{\overset{\displaystyle \text{Circuit A} \qquad\qquad \text{Circuit B}}{R \quad Y \quad B \quad B \quad Y \quad R}}{
\left[\begin{array}{ccc|ccc}
Z_{S(A)} & Z_{M(A)} & Z_{M(A)} & Z_{M(AB)} & Z_{M(AB)} & Z_{S(AB)} \\
Z_{M(A)} & Z_{S(A)} & Z_{M(A)} & Z_{M(AB)} & Z_{S(AB)} & Z_{M(AB)} \\
Z_{M(A)} & Z_{M(A)} & Z_{S(A)} & Z_{S(AB)} & Z_{M(AB)} & Z_{M(AB)} \\ \hline
Z_{M(AB)} & Z_{M(AB)} & Z_{S(AB)} & Z_{S(B)} & Z_{M(B)} & Z_{M(B)} \\
Z_{M(AB)} & Z_{S(AB)} & Z_{M(AB)} & Z_{M(B)} & Z_{S(B)} & Z_{M(B)} \\
Z_{S(AB)} & Z_{M(AB)} & Z_{M(AB)} & Z_{M(B)} & Z_{M(B)} & Z_{S(B)}
\end{array}\right]}
\tag{3.86}$$

where the elements of $\mathbf{Z}_{\text{Phase}}$ are as given by Equation (3.75b).

It is interesting to note the positions of the diagonal and off-diagonal elements of the inter-circuit mutual impedance matrix in comparison with those obtained in Equation (3.75a). The sequence impedance matrix of Equation (3.86) is calculated, noting the RYB/BYR phasing of circuits A and B, as follows:

$$\mathbf{Z}^{\text{PNZ}} = \mathbf{H}_{\text{dc}}^{-1}\mathbf{Z}_{\text{Phase}}\mathbf{H}_{\text{dc}} \tag{3.87a}$$

where

$$\mathbf{H}_{\text{dc}} = \begin{bmatrix} \mathbf{H} & 0 \\ 0 & \mathbf{H}_1 \end{bmatrix} =
\begin{array}{c} \\ \\ \text{Circuit A} \\ \\ \\ \\ \end{array}
\begin{array}{c} R \\ Y \\ B \\ B \\ Y \\ R \end{array}
\left[\begin{array}{ccc|ccc}
1 & 1 & 1 & 0 & 0 & 0 \\
h^2 & h & 1 & 0 & 0 & 0 \\
h & h^2 & 1 & 0 & 0 & 0 \\ \hline
0 & 0 & 0 & h & h^2 & 1 \\
0 & 0 & 0 & h^2 & h & 1 \\
0 & 0 & 0 & 1 & 1 & 1
\end{array}\right]
\begin{array}{l} \\ \\ \\ \\ \text{Circuit B} \\ \\ \end{array}
\tag{3.87b}$$

and

$$\mathbf{H}_{\text{dc}}^{-1} = \begin{bmatrix} \mathbf{H}^{-1} & 0 \\ 0 & \mathbf{H}_1^{-1} \end{bmatrix} =
\begin{array}{c} \text{Circuit A} \end{array}
\frac{1}{3}
\left[\begin{array}{ccc|ccc}
1 & h & h^2 & 0 & 0 & 0 \\
1 & h^2 & h & 0 & 0 & 0 \\
1 & 1 & 1 & 0 & 0 & 0 \\ \hline
0 & 0 & 0 & h^2 & h & 1 \\
0 & 0 & 0 & h & h^2 & 1 \\
0 & 0 & 0 & 1 & 1 & 1
\end{array}\right]
\begin{array}{l} \\ \\ \\ \\ \text{Circuit B} \\ \\ \end{array}
\tag{3.87c}$$

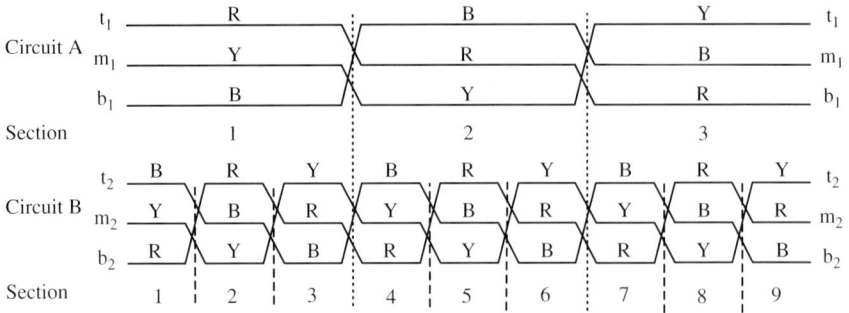

Figure 3.14   Double-circuit line ideal nine transpositions with ZPS inter-circuit mutual coupling only

H_{dc} is the transformation matrix for a double-circuit or two mutually coupled circuit line. The result takes the form of Equation (3.76a) with diagonal self-circuit and inter-circuit impedance matrices. The value of the sequence elements of the matrices are as given in Equation (3.76).

The above phase and sequence impedance matrix analysis for double-circuit lines applies equally to the line's shunt phase and sequence susceptance matrices.

We have shown that with the use of six transpositions, the effect of mutual sequence coupling between the two circuits is not entirely eliminated with like sequence coupling between the two circuits remaining and this may indeed be the desired result. However, if required, it is even possible to eliminate the PPS and NPS coupling and retain ZPS coupling only between the two circuits. Because there are two three-phase circuits of conductors, each phase conductor has to be transposed within its circuit and with respect to the parallel circuit. In other words, if one circuit is subject to three transpositions, then for each one of its sections, the other circuit should undergo full three section transpositions. This produces a total of nine transpositions for our double-circuit line as shown in Figure 3.14.

It can be shown analytically by inspection or by matrix analysis that the effect of this nine transposition assumption is to produce the following 6×6 phase impedance matrix:

$$\mathbf{Z}_{\text{Phase}} = \frac{1}{9}\sum_{i=1}^{9}\mathbf{Z}_{\text{Phase (Section-}i)} = \begin{array}{c} \text{Circuit A} \\ \text{Circuit B} \end{array} \overbrace{\begin{bmatrix} Z_{S(A)} & Z_{M(A)} & Z_{M(A)} \\ Z_{M(A)} & Z_{S(A)} & Z_{M(A)} \\ Z_{M(A)} & Z_{M(A)} & Z_{S(A)} \\ \hline Z_{AB} & Z_{AB} & Z_{AB} \\ Z_{AB} & Z_{AB} & Z_{AB} \\ Z_{AB} & Z_{AB} & Z_{AB} \end{bmatrix}}^{\text{Circuit A}} \overbrace{\begin{array}{|ccc} Z_{AB} & Z_{AB} & Z_{AB} \\ Z_{AB} & Z_{AB} & Z_{AB} \\ Z_{AB} & Z_{AB} & Z_{AB} \\ \hline Z_{S(B)} & Z_{M(B)} & Z_{M(B)} \\ Z_{M(B)} & Z_{S(B)} & Z_{M(B)} \\ Z_{M(B)} & Z_{M(B)} & Z_{S(B)} \end{array}}^{\text{Circuit B}}$$

$$(3.88a)$$

where $Z_{S(A)}$, $Z_{M(A)}$, $Z_{S(B)}$ and $Z_{M(B)}$ are given in Equation (3.75b) and Z_{AB} is given by

$$Z_{AB} = \frac{1}{9}(Z_{t1t2} + Z_{t1m2} + Z_{t1b2} + Z_{m1t2} + Z_{m1m2} + Z_{m1b2} + Z_{b1t2}$$
$$+ Z_{b1m2} + Z_{b1b2}) \tag{3.88b}$$

The elements of the inter-circuit mutual matrix are all equal to Z_{AB} given in Equation (3.88b). This is equal to the average of all nine mutual impedances between the six conductors of the two circuits. The corresponding sequence matrix of Equation (3.88a) calculated using any electrical phasing of circuits A and B, e.g. RYB/RYB or RYB/BYR, is given by

$$
\mathbf{Z}^{PNZ} =
\begin{array}{c}
\\
\text{Circuit A} \\
\\
\\
\text{Circuit B} \\
\\
\end{array}
\begin{array}{c}
\\
\text{P} \\
\text{N} \\
\text{Z} \\
\text{P} \\
\text{N} \\
\text{Z}
\end{array}
\overset{\displaystyle \begin{array}{cccccc} \text{Circuit A} & & & \text{Circuit B} & & \\ \text{P} & \text{N} & \text{Z} & \text{P} & \text{N} & \text{Z} \end{array}}{
\left[
\begin{array}{ccc|ccc}
Z_A^P & 0 & 0 & 0 & 0 & 0 \\
0 & Z_A^N & 0 & 0 & 0 & 0 \\
0 & 0 & Z_A^Z & 0 & 0 & Z_{AB}^Z \\
\hline
0 & 0 & 0 & Z_B^P & 0 & 0 \\
0 & 0 & 0 & 0 & Z_B^N & 0 \\
0 & 0 & Z_{AB}^Z & 0 & 0 & Z_B^Z
\end{array}
\right]}
\tag{3.89a}
$$

$$
Z_{AB}^Z = 3Z_{AB} = \frac{1}{3}\left(Z_{t1t2} + Z_{t1m2} + Z_{t1b2} + Z_{m1t2} + Z_{m1m2} + Z_{m1b2} + Z_{b1t2}\right.
$$
$$
\left. + Z_{b1m2} + Z_{b1b2}\right)
\tag{3.89b}
$$

Equation (3.88a) shows that the effect of this ultimate nine transposition assumption is to equalise all nine elements of the inter-circuit phase matrix. In the sequence reference frame, this eliminates the PPS and NPS mutual coupling between the two circuits but not the ZPS mutual coupling which will always be present. It is informative for the reader to derive the mutual phase impedance/susceptance matrices between the two circuits for all nine transpositions and show that the total has the form shown in Equation (3.88a).

3.2.7 Untransposed and transposed multiple-circuit lines

Multiple that is three or more three-phase circuits strung on the same tower or running in parallel in close proximity in a corridor are sometimes used in electrical power networks. Similar to double-circuit lines, mutual inductive and capacitive coupling exists between all conductors in such a complex multi-conductor system. Figure 3.15 illustrates typical three and four circuit tower arrangements. Two double-circuit lines may also run in close physical proximity to each other along the same route or right of way. The coupling between these lines is usually neglected in practice if the lines are electromagnetically coupled for a short distance only relative to the shortest circuit so that the effect on the overall electrical parameters of such a circuit is negligible. Where this is not the case the formulation of the phase impedance and susceptance matrices of the entire multi-conductor system and elimination of the earth wire(s) follows a similar approach to that presented in Section 3.2.5 for double-circuit lines. After the elimination of the earth wire(s), the

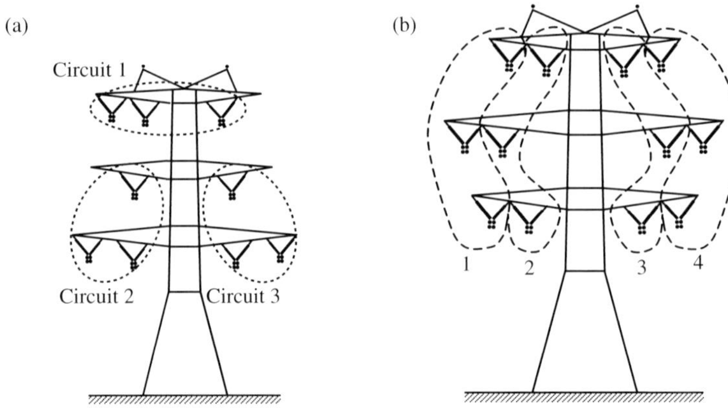

(a)   (b)

Figure 3.15   Typical (a) three and (b) four-circuit towers

dimension of the resultant phase matrix would be $3 \times N$ where N is the number of coupled three-phase circuits. As in the case of double-circuit lines, the sequence impedance and susceptance matrices can be derived by introducing appropriate transposition assumptions to retain like sequence PPS, NPS and ZPS inter-circuit coupling but not intersequence coupling. Alternatively, ideal transpositions may be chosen so as to retain ZPS inter-circuit mutual coupling only.

Lines with three coupled circuits

Consider a line with three mutually coupled circuits 1, 2 and 3 erected on the same tower as illustrated in Figure 3.15(a). The transposed and balanced 9×9 series phase impedance matrix that retains PPS, NPS and ZPS inter-circuit mutual coupling can be derived using the technique presented for double-circuit lines. Each circuit is perfectly transposed within itself and the inter-circuit mutual impedance matrices between any two circuits result in equal self and equal mutual impedances. However, the inter-circuit mutual coupling impedances between circuits 1 and 2, circuits 1 and 3, and circuits 2 and 3 are assumed unequal to maintain generality. It can be shown that the resultant balanced phase impedance matrix is given by

$$
Z_{\text{Phase}} =
\begin{array}{c}
\\ \\
\text{Circuit 1}
\\ \\ \\
\text{Circuit 2}
\\ \\ \\
\text{Circuit 3}
\\ \\
\end{array}
\begin{array}{c}
\\
\begin{array}{c} R \\ Y \\ B \end{array}
\\
\begin{array}{c} R \\ Y \\ B \end{array}
\\
\begin{array}{c} R \\ Y \\ B \end{array}
\end{array}
\left[
\begin{array}{ccc|ccc|ccc}
Z_{S1} & Z_{M1} & Z_{M1} & Z_{S12} & Z_{M12} & Z_{M12} & Z_{S13} & Z_{M13} & Z_{M13} \\
Z_{M1} & Z_{S1} & Z_{M1} & Z_{M12} & Z_{S12} & Z_{M12} & Z_{M13} & Z_{S13} & Z_{M13} \\
Z_{M1} & Z_{M1} & Z_{S1} & Z_{M12} & Z_{M12} & Z_{S12} & Z_{M13} & Z_{M13} & Z_{S13} \\ \hline
Z_{S12} & Z_{M12} & Z_{M12} & Z_{S2} & Z_{M2} & Z_{M2} & Z_{S23} & Z_{M23} & Z_{M23} \\
Z_{M12} & Z_{S12} & Z_{M12} & Z_{M2} & Z_{S2} & Z_{M2} & Z_{M23} & Z_{S23} & Z_{M23} \\
Z_{M12} & Z_{M12} & Z_{S12} & Z_{M2} & Z_{M2} & Z_{S2} & Z_{M23} & Z_{M23} & Z_{S23} \\ \hline
Z_{S13} & Z_{M13} & Z_{M13} & Z_{S23} & Z_{M23} & Z_{M23} & Z_{S3} & Z_{M3} & Z_{M3} \\
Z_{M13} & Z_{S13} & Z_{M13} & Z_{M23} & Z_{S23} & Z_{M23} & Z_{M3} & Z_{S3} & Z_{M3} \\
Z_{M13} & Z_{M13} & Z_{S13} & Z_{M23} & Z_{M23} & Z_{S23} & Z_{M3} & Z_{M3} & Z_{S3}
\end{array}
\right]
$$

$$
\begin{array}{ccc}
\text{Circuit 1} & \text{Circuit 2} & \text{Circuit 3} \\
R \quad Y \quad B & R \quad Y \quad B & R \quad Y \quad B
\end{array}
$$

(3.90)

The elements of this matrix can be calculated as for a double or two mutually coupled circuits given in Section 3.2.6 by averaging the appropriate self and mutual phase impedance terms of the original untransposed matrix.

For the ideal transposition the inter-circuit impedance matrices have all nine elements equal thereby retaining only ZPS inter-circuit mutual coupling. The ideal transposition assumption results in $Z_{S12} = Z_{M12} = A_{12}$ for circuits 1 and 2, $Z_{S13} = Z_{M13} = A_{13}$ for circuits 1 and 3, and $Z_{S23} = Z_{M23} = A_{23}$ for circuits 2 and 3. The sequence impedance matrix can be calculated based on knowledge of the electrical phasings of the three circuits. The above series impedance matrix assumes that all circuits are phased RYB and the corresponding sequence impedance matrix can be calculated using

$$
\mathbf{Z}^{PNZ} = \begin{bmatrix} \mathbf{H}^{-1} & 0 & 0 \\ 0 & \mathbf{H}^{-1} & 0 \\ 0 & 0 & \mathbf{H}^{-1} \end{bmatrix} \mathbf{Z}_{Phase} \begin{bmatrix} \mathbf{H} & 0 & 0 \\ 0 & \mathbf{H} & 0 \\ 0 & 0 & \mathbf{H} \end{bmatrix} \tag{3.91}
$$

giving

$$
\mathbf{Z}^{PNZ} =
\begin{array}{cc}
& \begin{array}{c}
\text{Circuit 1} \hspace{2em} \text{Circuit 2} \hspace{2em} \text{Circuit 3}
\end{array}
\end{array}
$$

		Circuit 1			Circuit 2			Circuit 3		
		P	N	Z	P	N	Z	P	N	Z
Circuit 1	P	Z_1^P	0	0	Z_{12}^P	0	0	Z_{13}^P	0	0
	N	0	Z_1^N	0	0	Z_{12}^N	0	0	Z_{13}^N	0
	Z	0	0	Z_1^Z	0	0	Z_{12}^Z	0	0	Z_{13}^Z
Circuit 2	P	Z_{12}^P	0	0	Z_2^P	0	0	Z_{23}^P	0	0
	N	0	Z_{12}^N	0	0	Z_2^N	0	0	Z_{23}^N	0
	Z	0	0	Z_{12}^Z	0	0	Z_2^Z	0	0	Z_{23}^Z
Circuit 3	P	Z_{13}^P	0	0	Z_{23}^P	0	0	Z_3^P	0	0
	N	0	Z_{13}^N	0	0	Z_{23}^N	0	0	Z_3^N	0
	Z	0	0	Z_{13}^Z	0	0	Z_{23}^Z	0	0	Z_3^Z

(3.92a)

where

$$
Z_i^P = Z_i^N = Z_{Si} - Z_{Mi} \quad \text{and} \quad Z_i^Z = Z_{Si} + 2Z_{Mi} \quad \text{for } i = 1 \text{ to } 3
$$

$$
Z_{12}^P = Z_{12}^N = Z_{S12} - Z_{M12} \quad Z_{13}^P = Z_{13}^N = Z_{S13} - Z_{M13}
$$

$$
Z_{23}^P = Z_{23}^N = Z_{S23} - Z_{M23}
$$

$$
Z_{12}^Z = Z_{S12} + 2Z_{M12} \quad Z_{13}^Z = Z_{S13} + 2Z_{M13}
$$

$$
Z_{23}^Z = Z_{S23} + 2Z_{M23} \tag{3.92b}
$$

In the ideal transposition case where only ZPS inter-circuit mutual coupling is retained, we have $Z_{12}^P = Z_{12}^N = Z_{13}^P = Z_{13}^N = Z_{23}^P = Z_{23}^N = 0$.

The 9 × 9 transposed and balanced shunt phase susceptance matrix that retains PPS, NPS and ZPS inter-circuit mutual coupling, or ZPS inter-circuit mutual coupling only, is similar in form to the series phase impedance matrix. Similarly, the sequence susceptance matrix is similar in form to the sequence impedance matrix.

Lines with four coupled circuits

We now briefly outline for the most interested of readers the balanced phase impedance matrix for an unusual case of identical four circuits or two double-circuit lines running in close proximity to each other. In deriving such a matrix, it is assumed that each circuit is perfectly transposed and hence represented by a self and a mutual impedance. The inter-circuit mutual impedance matrices between any two circuits are also assumed balanced reflecting a transposition assumption similar to that for double-circuit lines. Further, the inter-circuit mutual coupling matrices between any two pair of circuits are assumed unequal to maintain generality. It can be shown that the balanced series phase impedance matrix of such a complex multi-conductor system with transpositions that retain PPS/NPS and ZPS inter-circuit mutual coupling is given by

$$
\mathbf{Z}_{\text{Phase}} =
\begin{array}{c}
\begin{array}{ccccccccccccc}
 & & \text{Circuit 1} & & & \text{Circuit 2} & & & \text{Circuit 3} & & & \text{Circuit 4} & \\
 & R & Y & B & R & Y & B & R & Y & B & R & Y & B
\end{array} \\
\begin{array}{c}
\text{Circuit 1} \\ \\ \\
\text{Circuit 2} \\ \\ \\
\text{Circuit 3} \\ \\ \\
\text{Circuit 4} \\ \\
\end{array}
\left[
\begin{array}{ccc|ccc|ccc|ccc}
A_1 & B_1 & B_1 & C & D & D & E & F & F & G & H & H \\
B_1 & A_1 & B_1 & D & C & D & F & E & F & H & G & H \\
B_1 & B_1 & A_1 & D & D & C & F & F & E & H & H & G \\
\hline
C & D & D & A_2 & B_2 & B_2 & J & K & K & L & M & M \\
D & C & D & B_2 & A_2 & B_2 & K & J & K & M & L & M \\
D & D & C & B_2 & B_2 & A_2 & K & K & J & M & M & L \\
\hline
E & F & F & J & K & K & A_3 & B_3 & B_3 & U & W & W \\
F & E & F & K & J & K & B_3 & A_3 & B_3 & W & U & W \\
F & F & E & K & K & J & B_3 & B_3 & A_3 & W & W & U \\
\hline
G & H & H & L & M & M & U & W & W & A_4 & B_4 & B_4 \\
H & G & H & M & L & M & W & U & W & B_4 & A_4 & B_4 \\
H & H & G & M & M & L & W & W & U & B_4 & B_4 & A_4
\end{array}
\right]
\end{array}
$$

$$(3.93)$$

Alternatively, an ideal transposition assumption results in $C=D$, $E=F$, $G=H$, $J=K$, $L=M$ and $U=W$. The sequence impedance matrix that corresponds to Equation (3.93) can be calculated with all circuits phased RYB using

$$
\mathbf{Z}^{\text{PNZ}} =
\begin{bmatrix}
\mathbf{H}^{-1} & 0 & 0 & 0 \\
0 & \mathbf{H}^{-1} & 0 & 0 \\
0 & 0 & \mathbf{H}^{-1} & 0 \\
0 & 0 & 0 & \mathbf{H}^{-1}
\end{bmatrix}
\mathbf{Z}_{\text{Phase}}
\begin{bmatrix}
\mathbf{H} & 0 & 0 & 0 \\
0 & \mathbf{H} & 0 & 0 \\
0 & 0 & \mathbf{H} & 0 \\
0 & 0 & 0 & \mathbf{H}
\end{bmatrix}
$$

$$(3.94)$$

giving

$$\mathbf{Z}^{\text{PNZ}} =$$

		Circuit 1			Circuit 2			Circuit 3			Circuit 4		
		P	N	Z	P	N	Z	P	N	Z	P	N	Z
Circuit 1	P	Z_1^P	0	0	Z_{12}^P	0	0	Z_{13}^P	0	0	Z_{14}^P	0	0
	N	0	Z_1^N	0	0	Z_{12}^N	0	0	Z_{13}^N	0	0	Z_{14}^N	0
	Z	0	0	Z_1^Z	0	0	Z_{12}^Z	0	0	Z_{13}^Z	0	0	Z_{14}^Z
Circuit 2	P	Z_{12}^P	0	0	Z_2^P	0	0	Z_{23}^P	0	0	Z_{24}^P	0	0
	N	0	Z_{12}^N	0	0	Z_2^N	0	0	Z_{23}^N	0	0	Z_{24}^N	0
	Z	0	0	Z_{12}^Z	0	0	Z_2^Z	0	0	Z_{23}^Z	0	0	Z_{24}^Z
Circuit 3	P	Z_{13}^P	0	0	Z_{23}^P	0	0	Z_3^P	0	0	Z_{34}^P	0	0
	N	0	Z_{13}^N	0	0	Z_{23}^N	0	0	Z_3^N	0	0	Z_{34}^N	0
	Z	0	0	Z_{13}^Z	0	0	Z_{23}^Z	0	0	Z_3^Z	0	0	Z_{34}^Z
Circuit 4	P	Z_{14}^P	0	0	Z_{24}^P	0	0	Z_{34}^P	0	0	Z_4^P	0	0
	N	0	Z_{14}^N	0	0	Z_{24}^N	0	0	Z_{34}^N	0	0	Z_4^N	0
	Z	0	0	Z_{14}^Z	0	0	Z_{24}^Z	0	0	Z_{34}^Z	0	0	Z_4^Z

$$(3.95a)$$

where

$$Z_i^P = Z_i^N = A_i - B_i \quad \text{and} \quad Z_i^Z = A_i + 2B_i \quad \text{for } i = 1 \text{ to } 4$$

$$Z_{12}^P = Z_{12}^N = C - D \quad Z_{12}^Z = C + 2D \quad Z_{13}^P = Z_{13}^N = E - F \quad Z_{13}^Z = E + 2F$$

$$Z_{14}^P = Z_{14}^N = G - H \quad Z_{14}^Z = G + 2H \quad Z_{23}^P = Z_{23}^N = J - K \quad Z_{23}^Z = J + 2K$$

$$Z_{24}^P = Z_{24}^N = L - M \quad Z_{24}^Z = L + 2M \quad Z_{34}^P = Z_{34}^N = U - W \quad Z_{34}^Z = U + 2W$$

$$(3.95b)$$

In the ideal transposition case where only ZPS inter-circuit mutual coupling is retained, we have $Z_{12}^P = Z_{12}^N = Z_{13}^P = Z_{13}^N = Z_{14}^P = Z_{14}^N = Z_{23}^P = Z_{23}^N = Z_{24}^P = Z_{24}^N = 0$.

3.2.8 Examples

Example 3.1   Consider a single-circuit overhead line with solid non-magnetic conductors and no earth wire. Using the self and mutual impedance expressions for the phase conductors, and assuming the line is perfectly transposed, derive expressions for the PPS/NPS and ZPS impedances of the circuit. Ignore the conductor's internal inductance skin effect.

Using Equation (3.19), the self and mutual impedances with earth return are given by

$$Z_{ii} = R_{i(c)} + \pi^2 10^{-4} f + j4\pi 10^{-4} f \left[\frac{1}{4} + \log_e \left(\frac{D_{\text{erc}}}{r_o} \right) \right]$$

and

$$Z_{ij} = \pi^2 10^{-4} f + j4\pi 10^{-4} f \log_e \left(\frac{D_{\text{erc}}}{d_{ij}} \right)$$

The nine elements of the original phase impedance matrix are calculated using Equation (3.50). Therefore, the PPS/NPS impedance is given by

$$Z^P = Z^N = \frac{1}{3}[(Z_{11} + Z_{22} + Z_{33}) - (Z_{12} + Z_{23} + Z_{13})]$$

$$= R_{i(\text{c})} + j4\pi 10^{-4} f \log_e \left(\frac{\text{GMD}}{\text{GMR}} \right)$$

where $\text{GMD} = \sqrt[3]{d_{12} d_{23} d_{13}}$ and $\text{GMR} = 0.7788 \times r_{\text{o}}$. It is interesting to note that the earth return resistance and reactance present in the self and mutual impedances with earth return cancel out and are not present in the PPS/NPS impedance. Using Equation (3.50), the ZPS impedance is given by

$$Z^Z = \frac{1}{3}[(Z_{11} + Z_{22} + Z_{33}) + 2(Z_{12} + Z_{23} + Z_{13})]$$

$$= R_{i(\text{c})} + 3\pi^2 10^{-4} f + j4\pi 10^{-4} f \log_e \left(\frac{D_{\text{erc}}^3}{\text{GMR} \times \text{GMD}^2} \right)$$

Example 3.2 Consider the double-circuit line shown in Figure 3.16 having a triangular conductor arrangement within each circuit and symmetrical arrangement with respect to the tower.

For the conductor numbering and electrical phasing given, prove that a three-cycle reverse transposition of circuit A and a similar but forward transposition of circuit B is sufficient to produce a balanced inter-circuit phase matrix and a diagonal inter-circuit sequence matrix.

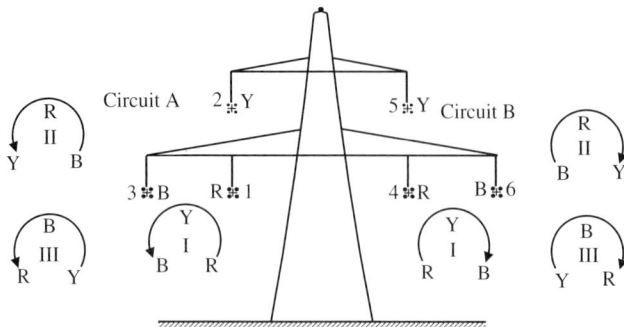

Figure 3.16 Double-circuit line configuration used in Example 3.2

From the geometry of the two circuits, the inter-circuit mutual impedance matrix of the three transposition sections can be written by inspection as follows:

$$
\mathbf{Z}_{AB-1} = \begin{array}{c} \\ R \\ Y \\ B \end{array}
\begin{array}{ccc} R & Y & B \\ \left[\begin{array}{ccc} Z_{14} & Z_{15} & Z_{16} \\ Z_{24} & Z_{25} & Z_{26} \\ Z_{34} & Z_{35} & Z_{36} \end{array}\right] \end{array}
\qquad
\mathbf{Z}_{AB-2} = \begin{array}{c} \\ R \\ Y \\ B \end{array}
\begin{array}{ccc} R & Y & B \\ \left[\begin{array}{ccc} Z_{25} & Z_{26} & Z_{24} \\ Z_{35} & Z_{36} & Z_{34} \\ Z_{15} & Z_{16} & Z_{14} \end{array}\right] \end{array}
$$

$$
\mathbf{Z}_{AB-3} = \begin{array}{c} \\ R \\ Y \\ B \end{array}
\begin{array}{ccc} R & Y & B \\ \left[\begin{array}{ccc} Z_{36} & Z_{34} & Z_{35} \\ Z_{16} & Z_{14} & Z_{15} \\ Z_{26} & Z_{24} & Z_{25} \end{array}\right] \end{array}
$$

$$
\mathbf{Z}_{AB} = \frac{1}{3}\sum_{i=1}^{3}\mathbf{Z}_{AB-i} = \frac{1}{3}\left[\begin{array}{ccc}
Z_{14}+Z_{25}+Z_{36} & Z_{15}+Z_{26}+Z_{34} & Z_{16}+Z_{24}+Z_{35} \\
Z_{24}+Z_{35}+Z_{16} & Z_{25}+Z_{36}+Z_{14} & Z_{26}+Z_{34}+Z_{15} \\
Z_{34}+Z_{15}+Z_{26} & Z_{35}+Z_{16}+Z_{24} & Z_{36}+Z_{14}+Z_{25}
\end{array}\right]
$$

Again, from the geometry of the circuits, we can write $Z_{24}=Z_{15}$, $Z_{26}=Z_{35}$, $Z_{34}=Z_{16}$ and $Z_{16}=Z_{34}$. Therefore, the inter-circuit phase impedance matrix is given by

$$
\mathbf{Z}_{AB} = \left[\begin{array}{ccc}
Z_S & Z_M & Z_M \\
Z_M & Z_S & Z_M \\
Z_M & Z_M & Z_S
\end{array}\right]
$$

and is balanced, and the corresponding diagonal sequence matrix is given by

$$
\mathbf{Z}_{AB}^{PNZ} = \left[\begin{array}{ccc}
Z_S - Z_M & 0 & 0 \\
0 & Z_S - Z_M & 0 \\
0 & 0 & Z_S + 2Z_M
\end{array}\right]
$$

Example 3.3   Consider a 275 kV single-circuit overhead line with a symmetrical horizontal phase conductor configuration and two earth wires. The spacing of the conductors relative to the centre of the tower and earth are shown in Figure 3.17. The crosses numbered 1–5 represent the mid span conductor positions due to conductor sag.

The physical data of conductors is as follows:

Phase conductors: $2 \times 175\,\text{mm}^2$ ACSR per phase, 30/7 strands (30 Aluminium, 7 Steel), conductor stranding factor $=0.82635$, conductor outer radius $=9.765\,\text{mm}$, conductor ac resistance $=0.1586\,\Omega/\text{km}$, height at tower $=19.86\,\text{m}$, average sag $=10.74\,\text{m}$.

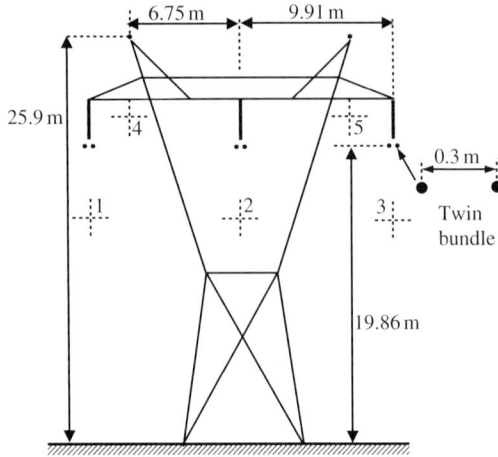

Figure 3.17   Single-circuit line used in Example 3.3

Earth wire conductor: $1 \times 175\,\text{mm}^2$ ACSR, 30/7 strands, outer radius $= 9.765\,\text{mm}$, ac resistance $= 0.1489\,\Omega/\text{km}$, height at tower $= 25.9\,\text{m}$, average sag $= 8.25\,\text{m}$.

Earth resistivity $= 20\,\Omega\text{m}$. Nominal frequency $f = 50\,\text{Hz}$.

Calculate the potential coefficient and capacitance matrices, phase susceptance matrix with earth wires eliminated, perfectly transposed phase susceptance matrix and corresponding sequence susceptance matrix. Calculate the full phase impedance matrix, reduced phase impedance matrix with earth wires eliminated, perfectly transposed balanced phase impedance matrix, and sequence impedance matrices for the untransposed and balanced phase impedance matrices.

Average height of phase conductors $= 19.86 - (2/3) \times 10.74 = 12.7\,\text{m}$
Average height of earth wire conductor $= 25.9 - (2/3) \times 8.25 = 20.4\,\text{m}$

Potential coefficients, capacitance, phase and sequence susceptance matrices
The GMR of each stranded conductor is equal to its radius i.e. $\text{GMR}_\text{C} = 9.765\,\text{mm}$. We have two subconductors per phase so $\text{GMR}_\text{Eq} = \sqrt{2 \times 9.765 \times 150} = 54.1248\,\text{mm}$ where the radius of a circle through the centres of the two conductors is equal to $300/2 = 150\,\text{mm}$.

$$P_{11} = 17.975109 \log_e \left(\frac{2 \times 12.7}{54.1248 \times 10^{-3}} \right) = 110.568\,\text{km}/\mu\text{F} = P_{22} = P_{33}$$

$$P_{44} = 17.975109 \log_e \left(\frac{2 \times 20.4}{9.765 \times 10^{-3}} \right) = 149.8698\,\text{km}/\mu\text{F} = P_{55}$$

$$P_{12} = 17.975109 \log_e \left(\frac{\sqrt{9.91^2 + (2 \times 12.7)^2}}{9.91} \right)$$

$$= 18.19\,\text{km}/\mu\text{F} = P_{21} = P_{23} = P_{32}$$

$$P_{13} = 17.975109 \log_e \left(\frac{\sqrt{(2 \times 9.91)^2 + (2 \times 12.7)^2}}{2 \times 9.91} \right) = 8.733 \, \text{km}/\mu\text{F} = P_{31}$$

$$P_{14} = 17.975109 \log_e \left(\frac{\sqrt{(20.4 + 12.7)^2 + (9.91 - 6.75)^2}}{\sqrt{(20.4 - 12.7)^2 + (9.91 - 6.75)^2}} \right)$$

$$= 24.896 \, \text{km}/\mu\text{F} = P_{41} = P_{35} = P_{53}$$

$$P_{15} = 17.975109 \log_e \left(\frac{\sqrt{(20.4 + 12.7)^2 + (9.91 + 6.75)^2}}{\sqrt{(20.4 - 12.7)^2 + (9.91 + 6.75)^2}} \right)$$

$$= 12.629 \, \text{km}/\mu\text{F} = P_{51} = P_{34} = P_{43}$$

$$P_{24} = 17.975109 \log_e \left(\frac{\sqrt{(20.4 + 12.7)^2 + 6.75^2}}{\sqrt{(20.4 - 12.7)^2 + 6.75^2}} \right)$$

$$= 21.4556 \, \text{km}/\mu\text{F} = P_{42} = P_{25} = P_{52}$$

$$P_{45} = 17.975109 \log_e \left(\frac{\sqrt{(2 \times 6.75)^2 + (2 \times 20.4)^2}}{2 \times 6.75} \right) = 20.814 \, \text{km}/\mu\text{F} = P_{54}.$$

Therefore, the potential coefficient matrix is equal to

$$\mathbf{P} = \begin{array}{c} 1 \\ 2 \\ 3 \\ 4 \\ 5 \end{array} \left[\begin{array}{ccc|cc} 110.568 & 18.19 & 8.733 & 24.896 & 12.629 \\ 18.19 & 110.568 & 18.19 & 21.455 & 21.455 \\ 8.733 & 18.19 & 110.568 & 12.629 & 24.896 \\ \hline 24.896 & 21.455 & 12.629 & 149.869 & 20.814 \\ 12.629 & 21.455 & 24.896 & 20.814 & 149.869 \end{array} \right] \text{km}/\mu\text{F}$$

and the full phase capacitance matrix is equal to

$$\mathbf{C} = \mathbf{P}^{-1} = \begin{array}{c} 1 \\ 2 \\ 3 \\ 4 \\ 5 \end{array} \left[\begin{array}{ccc|cc} 9.614\text{E}-3 & -1.191\text{E}-3 & -3.199\text{E}-4 & -1.344\text{E}-3 & -3.999\text{E}-4 \\ -1.191\text{E}-3 & 9.813\text{E}-3 & -1.191\text{E}-3 & -9.718\text{E}-4 & -9.718\text{E}-4 \\ -3.199\text{E}-4 & -1.191\text{E}-3 & 9.614\text{E}-3 & -3.999\text{E}-4 & -1.344\text{E}-3 \\ \hline -1.344\text{E}-3 & -9.718\text{E}-4 & -3.999\text{E}-4 & 7.162\text{E}-3 & -6.759\text{E}-4 \\ -3.999\text{E}-4 & -9.718\text{E}-4 & -1.344\text{E}-3 & -6.759\text{E}-4 & 7.162\text{E}-3 \end{array} \right] \mu\text{F}/\text{km}$$

Eliminating the two earth wires, the phase capacitance and susceptance matrices of the untransposed line are equal to

$$\mathbf{C}_{\text{Phase}} = \begin{array}{c} 1 \\ 2 \\ 3 \end{array} \left[\begin{array}{ccc} 9.614\text{E}-3 & -1.191\text{E}-3 & -3.199\text{E}-4 \\ -1.191\text{E}-3 & 9.813\text{E}-3 & -1.191\text{E}-3 \\ -3.199\text{E}-4 & -1.191\text{E}-3 & 9.614\text{E}-3 \end{array} \right] \mu\text{F}/\text{km}$$

and

$$\mathbf{B}_{\text{Phase}} = \omega \mathbf{C} = \begin{array}{c} 1 \\ 2 \\ 3 \end{array} \begin{bmatrix} 3.02 & -0.374 & -0.1 \\ -0.374 & 3.083 & -0.374 \\ -0.1 & -0.374 & 3.02 \end{bmatrix} \mu \text{S}/\text{km}$$

Assume phase conductors 1, 2 and 3 are phased R, Y and B, respectively,

$$\mathbf{B}^{\text{PNZ}} = \mathbf{H}^{-1} \mathbf{B}_{\text{Phase}} \mathbf{H} \quad \text{and} \quad \mathbf{H} = \begin{bmatrix} 1 & 1 & 1 \\ h^2 & h & 1 \\ h & h^2 & 1 \end{bmatrix}$$

Thus, the sequence susceptance matrix of the untransposed line is equal to

$$\mathbf{B}^{\text{PNZ}} = \begin{bmatrix} 3.324 & -0.102 - \text{j}0.176 & 0.035 - \text{j}0.061 \\ -0.102 + \text{j}0.176 & 3.324 & 0.035 + \text{j}0.061 \\ 0.035 + \text{j}0.061 & 0.035 - \text{j}0.061 & 2.475 \end{bmatrix} \mu \text{S}/\text{km}$$

The self and mutual elements of the susceptance matrix of the perfectly transposed line are equal to

$$B_{\text{Self}} = \frac{1}{3}(3.02 + 3.083 + 3.02) = 3.041$$

and

$$B_{\text{Mutual}} = \frac{1}{3}(-0.374 - 0.1 - 0.374) = -0.283$$

Thus, the balanced phase susceptance matrix is equal to

$$\mathbf{B}_{\text{Phase}} = \begin{bmatrix} 3.041 & -0.283 & -0.283 \\ -0.283 & 3.041 & -0.283 \\ -0.283 & -0.283 & 3.041 \end{bmatrix} \mu \text{S}/\text{km}$$

and the corresponding sequence susceptance matrix is equal to

$$\mathbf{B}^{\text{PNZ}} = \begin{bmatrix} 3.324 & 0 & 0 \\ 0 & 3.324 & 0 \\ 0 & 0 & 2.475 \end{bmatrix} \mu \text{S}/\text{km}$$

Phase and sequence impedance matrices
The GMR of each stranded conductor is equal to $\text{GMR}_{\text{c}} = 0.82635 \times 9.765 = 8.0693$ mm. We have two subconductors per phase so $\text{GMR}_{\text{Eq}} = \sqrt{2 \times 8.0693 \times 150} = 49.2015$ mm. The ac resistance per phase is equal to $0.1586/2 = 0.0793$ Ω/km. The depth of equivalent earth return conductor

$D_{erc} = 658.87 \times \sqrt{20/50} = 416.7$ m. The phase impedances with earth return are equal to

$$Z_{11} = 0.0793 + \pi^2 10^{-4} \times 50 + j4\pi 10^{-4} \times 50 \left[\frac{1}{4 \times 2} + \log_e \left(\frac{416.7}{49.2015 \times 10^{-3}} \right) \right]$$

$$= (0.12865 + j0.5761)\Omega/\text{km} = Z_{22} = Z_{33}$$

$$Z_{44} = 0.1489 + \pi^2 10^{-4} \times 50 + j4\pi 10^{-4} \times 50 \left[\frac{1}{4} + \log_e \left(\frac{416.7}{8.0693 \times 10^{-3}} \right) \right]$$

$$= (0.19825 + j0.6975)\Omega/\text{km} = Z_{55}$$

$$Z_{12} = \pi^2 10^{-4} \times 50 + j4\pi 10^{-4} \times 50 \log_e \left(\frac{416.7}{9.91} \right)$$

$$= (0.04935 + j0.2349)\Omega/\text{km} = Z_{21} = Z_{23} = Z_{32}$$

$$Z_{13} = 0.04935 + j4\pi 10^{-4} \times 50 \log_e \left(\frac{416.7}{2 \times 9.91} \right)$$

$$= (0.04935 + j0.19136)\Omega/\text{km} = Z_{31}$$

$$Z_{14} = 0.04935 + j4\pi 10^{-4} \times 50 \log_e \left(\frac{416.7}{\sqrt{(9.91 - 6.75)^2 + (20.4 - 12.7)^2}} \right)$$

$$= (0.04935 + j0.24588)\Omega/\text{km} = Z_{41} = Z_{35} = Z_{53}$$

$$Z_{15} = 0.04935 + j4\pi 10^{-4} \times 50 \log_e \left(\frac{416.7}{\sqrt{(20.4 - 12.7)^2 + (9.91 + 6.75)^2}} \right)$$

$$= (0.04935 + j0.1962)\Omega/\text{km} = Z_{51} = Z_{34} = Z_{43}$$

$$Z_{24} = 0.04935 + j4\pi 10^{-4} \times 50 \log_e \left(\frac{416.7}{\sqrt{(20.4 - 12.7)^2 + 6.75^2}} \right)$$

$$= (0.04935 + j0.23286)\Omega/\text{km} = Z_{42} = Z_{25} = Z_{52}$$

$$Z_{45} = 0.04935 + j4\pi 10^{-4} \times 50 \log_e \left(\frac{416.7}{2 \times 6.75} \right)$$

$$= (0.04935 + j0.2155)\Omega/\text{km} = Z_{54}$$

Therefore, the full phase impedance matrix is equal to

$$\mathbf{Z} = \begin{array}{c} 1 \\ 2 \\ 3 \\ 4 \\ 5 \end{array} \begin{bmatrix} 0.1286 + j0.584 & 0.0493 + j0.235 & 0.0493 + j0.191 & 0.0493 + j0.246 & 0.0493 + j0.196 \\ 0.0493 + j0.235 & 0.1286 + j0.584 & 0.0493 + j0.235 & 0.0493 + j0.233 & 0.0493 + j0.233 \\ 0.0493 + j0.191 & 0.0493 + j0.235 & 0.1286 + j0.584 & 0.0493 + j0.196 & 0.0493 + j0.246 \\ 0.0493 + j0.246 & 0.0493 + j0.233 & 0.0493 + j0.196 & 0.198 + j0.697 & 0.0493 + j0.215 \\ 0.0493 + j0.196 & 0.0493 + j0.233 & 0.0493 + j0.246 & 0.0493 + j0.215 & 0.198 + j0.697 \end{bmatrix} \Omega/\text{km}$$

Eliminating the two earth wires, 4 and 5, the untransposed phase impedance matrix is calculated as follows

$$\mathbf{Z}_{\text{Phase}} = \begin{bmatrix} 0.1286 + j0.584 & 0.0493 + j0.235 & 0.0493 + j0.191 \\ 0.0493 + j0.235 & 0.1286 + j0.584 & 0.0493 + j0.235 \\ 0.0493 + j0.191 & 0.0493 + j0.235 & 0.1286 + j0.584 \end{bmatrix}$$

$$- \begin{bmatrix} 0.0493 + j0.246 & 0.0493 + j0.196 \\ 0.0493 + j0.233 & 0.0493 + j0.233 \\ 0.0493 + j0.196 & 0.0493 + j0.246 \end{bmatrix} \times \begin{bmatrix} 0.198 + j0.697 & 0.0493 + j0.215 \\ 0.0493 + j0.215 & 0.198 + j0.697 \end{bmatrix}^{-1}$$

$$\times \begin{bmatrix} 0.0493 + j0.246 & 0.0493 + j0.233 & 0.0493 + j0.196 \\ 0.0493 + j0.196 & 0.0493 + j0.233 & 0.0493 + j0.246 \end{bmatrix}$$

$$\mathbf{Z}_{\text{Phase}} = \begin{bmatrix} 0.111 + j0.475 & 0.031 + j0.123 & 0.03 + j0.086 \\ 0.031 + j0.122 & 0.11 + j0.465 & 0.031 + j0.122 \\ 0.03 + j0.086 & 0.031 + j0.122 & 0.111 + j0.475 \end{bmatrix} \Omega/\text{km}$$

The corresponding sequence impedance matrix is equal to

$$\mathbf{Z}^{\text{PNZ}} = \begin{bmatrix} 0.08 + j0.362 & -0.023 + j0.014 & -0.008 - j0.004 \\ 0.024 + j0.013 & 0.08 + j0.362 & 0.007 - j0.004 \\ 0.008 - j0.005 & -0.008 - j0.004 & 0.171 + j0.692 \end{bmatrix} \Omega/\text{km}$$

The self and mutual elements of the impedance matrix of the perfectly transposed line are equal to

$$Z_{\text{Self}} = \frac{1}{3}(0.111 + j0.475 + 0.11 + j0.465 + 0.111 + j0.475)$$
$$= 0.11 + j0.472$$

and

$$Z_{\text{Mutual}} = \frac{1}{3}(0.031 + j0.123 + 0.03 + j0.086 + 0.031 + j0.122)$$
$$= 0.03 + j0.110$$

Thus balanced phase and corresponding sequence impedance matrices are

$$\mathbf{Z}_{\text{Phase}} = \begin{bmatrix} 0.11 + j0.472 & 0.03 + j0.110 & 0.03 + j0.110 \\ 0.03 + j0.110 & 0.11 + j0.472 & 0.03 + j0.110 \\ 0.03 + j0.110 & 0.03 + j0.110 & 0.11 + j0.472 \end{bmatrix} \Omega/\text{km}$$

and

$$\mathbf{Z}^{\text{PNZ}} = \begin{bmatrix} 0.08 + j0.362 & 0 & 0 \\ 0 & 0.08 + j0.362 & 0 \\ 0 & 0 & 0.171 + j0.692 \end{bmatrix} \Omega/\text{km}$$

Example 3.4   Consider a 400 kV double-circuit overhead line with a near ver-tical phase conductor configuration and one earth wire. The conductors are numbered 1–7 and their spacings including average conductor sag relative to the centre of the tower and earth are shown in Figure 3.18.

The conductors of the two circuits have the following physical data.

Phase conductors: $4 \times 400 \text{ mm}^2$ ACSR per phase, 54/7 strands (54 Aluminium, 7 Steel), conductor stranding factor $= 0.80987$, conductor outer radius $= 14.31$ mm, conductor ac resistance $= 0.0684 \ \Omega/\text{km}$, Figure 3.18 shows average height of the centre of phase conductor bundle including average sag.

Earth wire conductor: $1 \times 400 \text{ mm}^2$ ACSR, 54/7 strands, outer radius $=$ 9.765 mm, ac resistance $= 0.0643 \ \Omega/\text{km}$, Figure 3.18 shows average height of earth wire including average sag.

Earth resistivity $= 20 \ \Omega\text{m}$. Nominal frequency $f = 50$ Hz.

Calculate the 7×7 potential coefficient and capacitance matrices, 6×6 phase susceptance matrix with earth wire eliminated, the 6×6 phase suscep-tance matrix for six and ideal nine transpositions. Calculate the corresponding 6×6 sequence susceptance matrix in each case. Calculate the full 7×7 and reduced 6×6 phase impedance matrices with earth wire eliminated as well as the 6×6 phase matrices for six and ideal nine transpositions. Calculate the corresponding sequence impedance matrices in each case.

Potential coefficients, capacitance, phase and sequence susceptance matrices
For each stranded conductor, the GMR is equal to $\text{GMR}_C = 14.31$ mm. For the four conductor bundle, $\text{GMR}_{Eq} = \sqrt[4]{4 \times 14.31 \times (353.553)^3} = 224.267$ mm

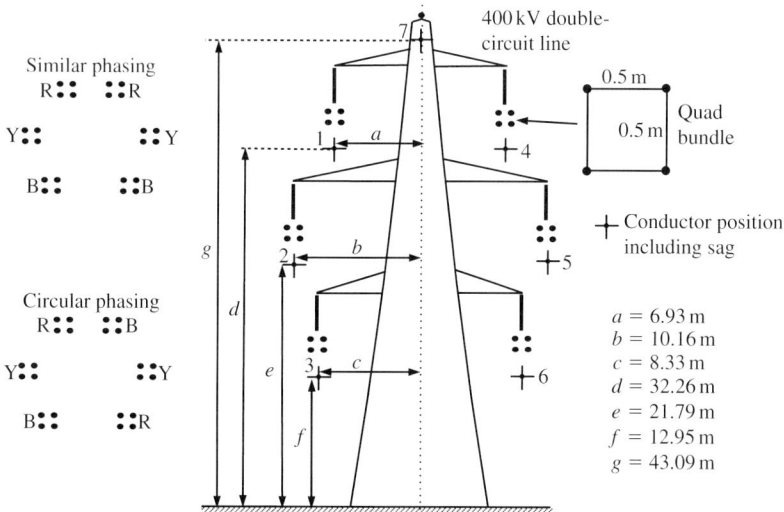

Figure 3.18   Double-circuit line used in Example 3.4

where the radius of a circle through the centres of the four conductors is equal to $R = 500 \, \text{mm}/\sqrt{2} = 353.553 \, \text{mm}$.

$$P_{11} = 17.975109 \log_e \left(\frac{2 \times 32.26}{224.267 \times 10^{-3}} \right) = 101.773 \, \text{km}/\mu\text{F} = P_{44}$$

$$P_{12} = 17.975109 \log_e \left(\frac{\sqrt{[2 \times 21.79 + (32.26 - 21.79)]^2 + (10.16 - 6.93)^2}}{\sqrt{(32.26 - 21.79)^2 + (10.16 - 6.93)^2}} \right)$$

$$= 28.719 \, \text{km}/\mu\text{F}$$

$$= P_{21} = P_{45} = P_{54}$$

The rest of the self and mutual potential coefficients are calculated similarly. Therefore, the potential coefficient matrix is equal to

$$
\mathbf{P} = \begin{array}{c} 1 \\ 2 \\ 3 \\ 4 \\ 5 \\ 6 \\ 7 \end{array}
\begin{bmatrix}
101.773 & 28.719 & 15.253 & 28.05 & 18.69 & 11.9 & 30.985 \\
28.719 & 94.72 & 22.165 & 18.69 & 15.48 & 11.727 & 18.0 \\
15.253 & 22.165 & 85.366 & 11.9 & 11.727 & 11.043 & 10.46 \\
28.05 & 18.69 & 11.9 & 101.773 & 28.719 & 15.253 & 30.985 \\
18.69 & 15.48 & 11.727 & 28.719 & 94.72 & 22.165 & 18.0 \\
11.9 & 11.727 & 11.043 & 15.253 & 22.165 & 85.366 & 10.46 \\
30.985 & 18.0 & 10.46 & 30.985 & 18.0 & 10.46 & 156.833
\end{bmatrix} \, \text{km}/\mu\text{F}
$$

and the full-phase capacitance matrix is equal to $\mathbf{C} = \mathbf{P}^{-1}$ or

$$
\mathbf{C} = \begin{array}{c} 1 \\ 2 \\ 3 \\ 4 \\ 5 \\ 6 \\ 7 \end{array}
\begin{bmatrix}
0.012 & -2.542\text{E}-3 & -8.323\text{E}-4 & -1.948\text{E}-3 & -8.449\text{E}-4 & -4.475\text{E}-4 & -1.485\text{E}-3 \\
-2.542\text{E}-3 & 0.012 & -2.394\text{E}-3 & -8.449\text{E}-4 & -7.262\text{E}-4 & -6.351\text{E}-4 & -4.617\text{E}-4 \\
-8.323\text{E}-4 & -2.394\text{E}-3 & 0.013 & -4.475\text{E}-4 & -6.351\text{E}-4 & -9.402\text{E}-4 & -1.891\text{E}-4 \\
-1.948\text{E}-3 & -8.449\text{E}-4 & -4.475\text{E}-4 & 0.012 & -2.542\text{E}-3 & -8.323\text{E}-4 & -1.485\text{E}-3 \\
-8.449\text{E}-4 & -7.262\text{E}-4 & -6.351\text{E}-4 & -2.542\text{E}-3 & 0.012 & -2.394\text{E}-3 & -4.617\text{E}-4 \\
-4.475\text{E}-4 & -6.351\text{E}-4 & -9.402\text{E}-4 & -8.323\text{E}-4 & -2.394\text{E}-3 & 0.013 & -1.891\text{E}-4 \\
-1.485\text{E}-3 & -4.617\text{E}-4 & -1.891\text{E}-4 & -1.485\text{E}-3 & -4.617\text{E}-4 & -1.891\text{E}-4 & 7.094\text{E}-3
\end{bmatrix} \, \mu\text{F}/\text{km}
$$

Eliminating the earth wire, the phase capacitance and susceptance matrices of the untransposed line are equal to

$$
\mathbf{C}_{\text{Phase}} = \begin{array}{c} 1 \\ 2 \\ 3 \\ 4 \\ 5 \\ 6 \end{array}
\begin{bmatrix}
0.012 & -2.542\text{E}-3 & -8.323\text{E}-4 & -1.948\text{E}-3 & -8.449\text{E}-4 & -4.475\text{E}-4 \\
-2.542\text{E}-3 & 0.012 & -2.394\text{E}-3 & -8.449\text{E}-4 & -7.262\text{E}-4 & -6.351\text{E}-4 \\
-8.323\text{E}-4 & -2.394\text{E}-3 & 0.013 & -4.475\text{E}-4 & -6.351\text{E}-4 & -9.402\text{E}-4 \\
-1.948\text{E}-3 & -8.449\text{E}-4 & -4.475\text{E}-4 & 0.012 & -2.542\text{E}-3 & -8.323\text{E}-4 \\
-8.449\text{E}-4 & -7.262\text{E}-4 & -6.351\text{E}-4 & -2.542\text{E}-3 & 0.012 & -2.394\text{E}-3 \\
-4.475\text{E}-4 & -6.351\text{E}-4 & -9.402\text{E}-4 & -8.323\text{E}-4 & -2.394\text{E}-3 & 0.013
\end{bmatrix} \, \mu\text{F}/\text{km}
$$

and

$$
\mathbf{B}_{\text{Phase}} =
\begin{array}{c}
1 \\ 2 \\ 3 \\ 4 \\ 5 \\ 6
\end{array}
\left[
\begin{array}{ccc|ccc}
3.77 & -0.799 & -0.261 & -0.612 & -0.265 & -0.141 \\
-0.799 & 3.77 & -0.752 & -0.265 & -0.228 & -0.2 \\
-0.261 & -0.752 & 4.084 & -0.141 & -0.2 & -0.295 \\
\hline
-0.612 & -0.265 & -0.141 & 3.77 & -0.799 & -0.261 \\
-0.265 & -0.228 & -0.2 & -0.799 & 3.77 & -0.752 \\
-0.141 & -0.2 & -0.295 & -0.261 & -0.752 & 4.084
\end{array}
\right] \mu\text{S/km}
$$

The corresponding sequence susceptance matrix is calculated using $\mathbf{B}^{\text{PNZ}} = \mathbf{H}_{\text{dc}}^{-1} \mathbf{B}_{\text{Phase}} \mathbf{H}_{\text{dc}}$. For the similar phasing of the two circuits, i.e. RYB/RYB, as illustrated in Figure 3.18, $\mathbf{H}_{\text{dc}}$ can be obtained by modifying Equation (3.87) appropriately giving

$$
\mathbf{B}^{\text{PNZ}} =
\begin{array}{c}
1 \\ 2 \\ 3 \\ 4 \\ 5 \\ 6
\end{array}
\left[
\begin{array}{ccc|ccc}
4.479 & -0.2 - j0.219 & 0.022 - j0.246 & -0.177 & -0.114 - j0.091 & -0.118 - j0.017 \\
-0.2 + j0.219 & 4.479 & 0.022 + j0.246 & -0.114 + j0.091 & -0.177 & -0.118 + j0.017 \\
0.022 + j0.246 & 0.022 - j0.246 & 2.667 & -0.118 + j0.017 & -0.118 - j0.017 & -0.782 \\
\hline
-0.177 & -0.114 - j0.091 & -0.118 - j0.017 & 4.479 & -0.2 - j0.219 & 0.022 - j0.246 \\
-0.114 + j0.091 & -0.177 & -0.118 + j0.017 & -0.2 + j0.219 & 4.479 & 0.022 + j0.246 \\
-0.118 + j0.017 & -0.118 - j0.017 & -0.782 & 0.022 + j0.246 & 0.022 - j0.246 & 2.667
\end{array}
\right]
$$

For the circular phasing of the two circuits of RYB/BYR, as illustrated in Figure 3.18, $\mathbf{H}_{\text{dc}}$ is given in Equation (3.87). Therefore,

$$
\mathbf{B}^{\text{PNZ}} =
\begin{array}{c}
1 \\ 2 \\ 3 \\ 4 \\ 5 \\ 6
\end{array}
\left[
\begin{array}{ccc|ccc}
4.479 & -0.2 - j0.219 & 0.022 - j0.246 & 0.136 - j0.053 & 0.088 + j0.153 & -0.118 - j0.017 \\
-0.2 + j0.219 & 4.479 & 0.022 + j0.246 & 0.088 - j0.153 & 0.136 + j0.053 & -0.118 + j0.017 \\
0.022 + j0.246 & 0.022 - j0.246 & 2.667 & 0.073 - j0.094 & 0.073 + j0.094 & -0.782 \\
\hline
0.136 + j0.053 & 0.088 + j0.153 & 0.073 + j0.094 & 4.479 & -0.09 - j0.283 & 0.202 - j0.142 \\
0.088 - j0.153 & 0.136 - j0.053 & 0.073 - j0.094 & -0.09 + j0.283 & 4.479 & 0.202 + j0.142 \\
-0.118 + j0.017 & -0.118 - j0.017 & -0.782 & 0.202 + j0.142 & 0.202 - j0.142 & 2.667
\end{array}
\right]
$$

For the circular RYB/BYR phasing, and for a six-phase transposition cycle, the balanced phase susceptance and corresponding sequence susceptance matrices are equal to

$$
\mathbf{B}_{\text{Phase}} =
\begin{array}{c}
1 \\ 2 \\ 3 \\ 4 \\ 5 \\ 6
\end{array}
\left[
\begin{array}{ccc|ccc}
3.875 & -0.604 & -0.604 & -0.306 & -0.306 & -0.170 \\
-0.604 & 3.875 & -0.604 & -0.306 & -0.170 & -0.306 \\
-0.604 & -0.604 & 3.875 & -0.170 & -0.306 & -0.306 \\
\hline
-0.306 & -0.306 & -0.170 & 3.875 & -0.604 & -0.604 \\
-0.306 & -0.170 & -0.306 & -0.604 & 3.875 & -0.604 \\
-0.170 & -0.306 & -0.306 & -0.604 & -0.604 & 3.875
\end{array}
\right] \mu\text{S/km}
$$

and

$$
\mathbf{B}^{\text{PNZ}} =
\begin{array}{c}
1 \\ 2 \\ 3 \\ 4 \\ 5 \\ 6
\end{array}
\left[
\begin{array}{ccc|ccc}
4.479 & 0 & 0 & 0.136 & 0 & 0 \\
0 & 4.479 & 0 & 0 & 0.136 & 0 \\
0 & 0 & 2.667 & 0 & 0 & -0.782 \\
\hline
0.136 & 0 & 0 & 4.479 & 0 & 0 \\
0 & 0.136 & 0 & 0 & 4.479 & 0 \\
0 & 0 & -0.782 & 0 & 0 & 2.667
\end{array}
\right] \mu\text{S/km}
$$

For the ideal nine transpositions, the balanced phase susceptance and corresponding sequence susceptance matrices are equal to

$$
\mathbf{B}_{\text{Phase}} = \begin{array}{c} 1 \\ 2 \\ 3 \\ 4 \\ 5 \\ 6 \end{array}
\left[\begin{array}{ccc|ccc}
3.875 & -0.604 & -0.604 & -0.261 & -0.261 & -0.261 \\
-0.604 & 3.875 & -0.604 & -0.261 & -0.261 & -0.261 \\
-0.604 & -0.604 & 3.875 & -0.261 & -0.261 & -0.261 \\
-0.261 & -0.261 & -0.261 & 3.875 & -0.604 & -0.604 \\
-0.261 & -0.261 & -0.261 & -0.604 & 3.875 & -0.604 \\
-0.261 & -0.261 & -0.261 & -0.604 & -0.604 & 3.875
\end{array} \right] \mu\text{S/km}
$$

and

$$
\mathbf{B}^{\text{PNZ}} = \begin{array}{c} 1 \\ 2 \\ 3 \\ 4 \\ 5 \\ 6 \end{array}
\left[\begin{array}{ccc|ccc}
4.479 & 0 & 0 & 0 & 0 & 0 \\
0 & 4.479 & 0 & 0 & 0 & 0 \\
0 & 0 & 2.667 & 0 & 0 & -0.782 \\
0 & 0 & 0 & 4.479 & 0 & 0 \\
0 & 0 & 0 & 0 & 4.479 & 0 \\
0 & 0 & -0.782 & 0 & 0 & 2.667
\end{array} \right] \mu\text{S/km}
$$

Phase and sequence impedance matrices
The GMR of a stranded phase conductor is $\text{GMR}_C = 0.80987 \times 14.31 = 11.589$ mm. We have four subconductors per phase so $\text{GMR}_{\text{Eq}} = \sqrt[4]{4 \times 11.589 \times (353.553)^3} = 212.75$ mm. The ac resistance per phase $= 0.0684/4 = 0.0171\ \Omega/\text{km}$. The depth of equivalent earth return conductor $D_{\text{erc}} = 658.87 \times \sqrt{20/50} = 416.7$ m.

The GMR of the stranded earth wire is $\text{GMR}_C = 0.80987 \times 9.765 = 7.91$ mm. The phase impedances with earth return are equal to

$$
Z_{11} = 0.0171 + \pi^2 10^{-4} \times 50 + j4\pi 10^{-4} \times 50 \left[\frac{1}{4 \times 4} + \log_e \left(\frac{416.7}{212.75 \times 10^{-3}} \right) \right]
$$

$$
= (0.0664 + j0.480)\Omega/\text{km} = Z_{22} = Z_{33} = Z_{44} = Z_{55} = Z_{66}
$$

$$
Z_{77} = 0.0643 + \pi^2 10^{-4} \times 50 + j4\pi 10^{-4} \times 50 \left[\frac{1}{4} + \log_e \left(\frac{416.7}{7.91 \times 10^{-3}} \right) \right]
$$

$$
= (0.1136 + j0.6988)\Omega/\text{km}
$$

$$
Z_{12} = \pi^2 10^{-4} \times 50 + j4\pi 10^{-4} \times 50 \log_e \left(\frac{416.7}{\sqrt{(32.26 - 21.79)^2 + (10.16 - 6.93)^2}} \right)
$$

$$
= (0.0493 + j0.2286)\Omega/\text{km} = Z_{21} = Z_{45} = Z_{54}
$$

The rest of the self and mutual phase impedances are calculated similarly. The full phase impedance matrix in Ω/km is equal to

$$
\mathbf{Z} = \begin{array}{c} 1 \\ 2 \\ 3 \\ 4 \\ 5 \\ 6 \\ 7 \end{array} \begin{bmatrix} 0.0664 + j0.480 & 0.0493 + j0.2286 & 0.0493 + j0.1928 & 0.0493 + j0.2138 & 0.0493 + j0.1906 & 0.0493 + j0.1777 & 0.0493 + j0.2147 \\ 0.0493 + j0.2286 & 0.0664 + j0.480 & 0.0493 + j0.2407 & 0.0493 + j0.1906 & 0.0493 + j0.1898 & 0.0493 + j0.1892 & 0.0493 + j0.1780 \\ 0.0493 + j0.1928 & 0.0493 + j0.2407 & 0.0664 + j0.480 & 0.0493 + j0.1777 & 0.0493 + j0.1892 & 0.0493 + j0.2023 & 0.0493 + j0.1609 \\ 0.0493 + j0.2138 & 0.0493 + j0.1906 & 0.0493 + j0.1777 & 0.0664 + j0.480 & 0.0493 + j0.2286 & 0.0493 + j0.1928 & 0.0493 + j0.2147 \\ 0.0493 + j0.1906 & 0.0493 + j0.1898 & 0.0493 + j0.1892 & 0.0493 + j0.2286 & 0.0664 + j0.480 & 0.0493 + j0.2407 & 0.0493 + j0.1780 \\ 0.0493 + j0.1777 & 0.0493 + j0.1892 & 0.0493 + j0.2023 & 0.0493 + j0.1928 & 0.0493 + j0.2407 & 0.0664 + j0.480 & 0.0493 + j0.1609 \\ 0.0493 + j0.2147 & 0.0493 + j0.1780 & 0.0493 + j0.1609 & 0.0493 + j0.2147 & 0.0493 + j0.1780 & 0.0493 + j0.1609 & 0.1136 + j0.6988 \end{bmatrix}
$$

Eliminating the earth wire, the phase impedance matrix of the untransposed double-circuit-line is equal to

$$
\mathbf{Z}_{\text{Phase}} = \begin{array}{c} 1 \\ 2 \\ 3 \\ 4 \\ 5 \\ 6 \end{array} \begin{bmatrix} 0.047 + j0.414 & 0.03 + j0.174 & 0.031 + j0.144 & 0.030 + j0.148 & 0.030 + j0.136 & 0.031 + j0.129 \\ 0.03 + j0.174 & 0.049 + j0.435 & 0.032 + j0.200 & 0.030 + j0.136 & 0.031 + j0.145 & 0.032 + j0.149 \\ 0.031 + j0.144 & 0.032 + j0.200 & 0.05 + j0.444 & 0.031 + j0.129 & 0.032 + j0.149 & 0.032 + j0.166 \\ 0.030 + j0.148 & 0.030 + j0.136 & 0.031 + j0.129 & 0.047 + j0.414 & 0.03 + j0.174 & 0.031 + j0.144 \\ 0.030 + j0.136 & 0.031 + j0.145 & 0.032 + j0.149 & 0.03 + j0.174 & 0.049 + j0.435 & 0.032 + j0.200 \\ 0.031 + j0.129 & 0.032 + j0.149 & 0.032 + j0.166 & 0.031 + j0.144 & 0.032 + j0.200 & 0.05 + j0.444 \end{bmatrix}
$$

The corresponding sequence impedance matrix is $\mathbf{Z}^{\text{PNZ}} = \mathbf{H}_{\text{dc}}^{-1} \mathbf{Z}_{\text{Phase}} \mathbf{H}_{\text{dc}}$. For similar phasing of the two circuits, i.e. RYB/RYB, as illustrated in Figure 3.18:

$$
\mathbf{Z}^{\text{PNZ}} = \begin{array}{c} 1 \\ 2 \\ 3 \\ 4 \\ 5 \\ 6 \end{array} \begin{bmatrix} 0.017 + j0.258 & -0.020 + j0.019 & -0.008 - j0.023 & 0.00 + j0.015 & -0.010 + j0.009 & 0.003 - j0.008 \\ 0.020 + j0.019 & 0.017 + j0.258 & 0.005 - j0.022 & 0.011 + j0.008 & 0.00 + j0.015 & -0.005 - j0.008 \\ 0.005 - j0.022 & -0.008 - j0.023 & 0.11 + j0.777 & -0.005 - j0.008 & 0.003 - j0.008 & 0.093 + j0.429 \\ 0.00 + j0.015 & -0.010 + j0.009 & 0.003 - j0.008 & 0.017 + j0.258 & -0.020 + j0.019 & -0.008 - j0.023 \\ 0.011 + j0.008 & 0.00 + j0.015 & -0.005 - j0.008 & 0.020 + j0.019 & 0.017 + j0.258 & 0.005 - j0.022 \\ -0.005 - j0.008 & 0.003 - j0.008 & 0.093 + j0.429 & 0.005 - j0.022 & -0.008 - j0.023 & 0.11 + j0.777 \end{bmatrix}
$$

For the circular phasing of the two circuits of RYB/BYR, as illustrated in Figure 3.18, we obtain

$$
\mathbf{Z}^{\text{PNZ}} = \begin{array}{c} 1 \\ 2 \\ 3 \\ 4 \\ 5 \\ 6 \end{array} \begin{bmatrix} 0.017 + j0.258 & -0.020 + j0.019 & -0.008 - j0.023 & -0.002 - j0.013 & 0.013 - j0.008 & 0.003 - j0.008 \\ 0.020 + j0.019 & 0.017 + j0.258 & 0.005 - j0.022 & -0.013 - j0.007 & 0.002 - j0.013 & -0.005 - j0.008 \\ 0.005 - j0.022 & -0.008 - j0.023 & 0.11 + j0.777 & 0.006 + j0.006 & -0.004 + j0.008 & 0.093 + j0.429 \\ 0.002 - j0.013 & 0.013 - j0.008 & -0.004 + j0.008 & 0.017 + j0.258 & -0.027 + j0.008 & -0.021 + j0.006 \\ -0.013 - j0.007 & -0.002 - j0.013 & 0.006 + j0.006 & 0.027 + j0.008 & 0.017 + j0.258 & 0.023 + j0.005 \\ -0.005 - j0.008 & 0.003 - j0.008 & 0.093 + j0.429 & 0.023 + j0.005 & -0.021 + j0.006 & 0.11 + j0.777 \end{bmatrix}
$$

For the RYB/BYR phasing, and for a six-phase transposition cycle, the balanced phase impedance and corresponding sequence impedance matrices are

$$
\mathbf{Z}_{\text{Phase}} = \begin{array}{c} 1 \\ 2 \\ 3 \\ 4 \\ 5 \\ 6 \end{array} \begin{bmatrix} 0.048 + j0.431 & 0.031 + j0.173 & 0.031 + j0.173 & 0.031 + j0.147 & 0.031 + j0.147 & 0.031 + j0.134 \\ 0.031 + j0.173 & 0.048 + j0.431 & 0.031 + j0.173 & 0.031 + j0.147 & 0.031 + j0.134 & 0.031 + j0.147 \\ 0.031 + j0.173 & 0.031 + j0.173 & 0.048 + j0.431 & 0.031 + j0.134 & 0.031 + j0.147 & 0.031 + j0.147 \\ 0.031 + j0.147 & 0.031 + j0.147 & 0.031 + j0.134 & 0.048 + j0.431 & 0.031 + j0.173 & 0.031 + j0.173 \\ 0.031 + j0.147 & 0.031 + j0.134 & 0.031 + j0.147 & 0.031 + j0.173 & 0.048 + j0.431 & 0.031 + j0.173 \\ 0.031 + j0.134 & 0.031 + j0.147 & 0.031 + j0.147 & 0.031 + j0.173 & 0.031 + j0.173 & 0.048 + j0.431 \end{bmatrix}
$$

and

$$
Z^{PNZ} =
\begin{array}{c} 1 \\ 2 \\ 3 \\ 4 \\ 5 \\ 6 \end{array}
\left[
\begin{array}{ccc|ccc}
0.017 + j0.258 & 0 & 0 & 0 - j0.013 & 0 & 0 \\
0 & 0.017 + j0.258 & 0 & 0 & 0 - j0.013 & 0 \\
0 & 0 & 0.110 + j0.777 & 0 & 0 & 0.093 + j0.429 \\
0 - j0.013 & 0 & 0 & 0.017 + j0.258 & 0 & 0 \\
0 & 0 - j0.013 & 0 & 0 & 0.017 + j0.258 & 0 \\
0 & 0 & 0.093 + j0.429 & 0 & 0 & 0.110 + j0.777
\end{array}
\right]
$$

For the ideal nine transpositions, the balanced phase impedance and corresponding sequence impedance matrices are

$$
Z_{Phase} =
\begin{array}{c} 1 \\ 2 \\ 3 \\ 4 \\ 5 \\ 6 \end{array}
\left[
\begin{array}{ccc|ccc}
0.048 + j0.431 & 0.031 + j0.173 & 0.031 + j0.173 & 0.031 + j0.143 & 0.031 + j0.143 & 0.031 + j0.143 \\
0.031 + j0.173 & 0.048 + j0.431 & 0.031 + j0.173 & 0.031 + j0.143 & 0.031 + j0.143 & 0.031 + j0.143 \\
0.031 + j0.173 & 0.031 + j0.173 & 0.048 + j0.431 & 0.031 + j0.143 & 0.031 + j0.143 & 0.031 + j0.143 \\
0.031 + j0.143 & 0.031 + j0.143 & 0.031 + j0.143 & 0.048 + j0.431 & 0.031 + j0.173 & 0.031 + j0.173 \\
0.031 + j0.143 & 0.031 + j0.143 & 0.031 + j0.143 & 0.031 + j0.173 & 0.048 + j0.431 & 0.031 + j0.173 \\
0.031 + j0.143 & 0.031 + j0.143 & 0.031 + j0.143 & 0.031 + j0.173 & 0.031 + j0.173 & 0.048 + j0.431
\end{array}
\right]
$$

$$
Z^{PNZ} =
\begin{array}{c} 1 \\ 2 \\ 3 \\ 4 \\ 5 \\ 6 \end{array}
\left[
\begin{array}{ccc|ccc}
0.017 + j0.258 & 0 & 0 & 0 & 0 & 0 \\
0 & 0.017 + j0.258 & 0 & 0 & 0 & 0 \\
0 & 0 & 0.110 + j0.777 & 0 & 0 & 0.093 + j0.429 \\
0 & 0 & 0 & 0.017 + j0.258 & 0 & 0 \\
0 & 0 & 0 & 0 & 0.017 + j0.258 & 0 \\
0 & 0 & 0.093 + j0.429 & 0 & 0 & 0.110 + j0.177
\end{array}
\right]
$$

The reader will benefit from carefully studying the results of the above examples which provide several insights of significant practical importance on natural overhead line unbalance and the effects of transpositions in the practice of electrical power systems engineering.

3.3 Phase and sequence modelling of three-phase cables

3.3.1 Background

Cables are generally classified as underground, submarine or aerial. Underground cables may be buried directly in the ground, laid in trenches, pipes or in underground tunnels. Cables are also classified depending on the type of core insulation, e.g. oil-impregnated paper, ethylene propylene rubber (EPR), or cross-linked polyethylene (XLPE), etc. Some cables may have a protective armour in addition to the metallic sheath or screen. Other classification is based on the number of cores, e.g. single core, two core or three core, etc. Single-core cables are coaxial cables with insulation that can either be extruded, e.g. XLPE or oil-impregnated paper. Self-contained fluid-filled cables consist of three single-core cables each having a hollow copper core which permits the flow of pressurised dielectric. High voltage cables are always designed with a metallic sheath conductor which may be copper, corrugated aluminium or lead. Three-core cables consist of three single-core cables contained within a common shell which may be an insulating shell or a steel pipe. There are a large number of different cable types, designs and layouts in use for the transmission and distribution of electrical energy and operating at voltages from 1 to 500 kV. It is therefore impossible to cover but a few cables focussing on those designs and layouts that are most commonly used. In some countries, e.g. the UK, distribution networks in large cities are almost entirely underground cable networks. Even at transmission voltage levels, increasing environmental and visual amenity pressures are resulting in more undergrounding of cables in addition to some of the usual requirements imposed by wide rivers and sea crossings. At the lower end of distribution voltages, belted-type cables have been used and these are three-core cables usually insulated with solid oil-impregnated paper with the three conductors covered in a single metallic sheath. At operating voltages between 10 and 36 kV, three-core cables are becoming mainly of the screened type with polymeric insulation and outer sheath applied to each core. Above 36 kV, cables have historically been fluid-filled paper-insulated but more recently, the use of such cables has mainly been above 132 kV. In the last few years, XLPE-insulated cables are increasingly being used at voltages up to 500 kV. Cables used at transmission voltages are mostly single-core cables that may be laid in a variety of configurations in the ground, tunnels or pipes.

The ac submarine cables are similar to underground cables except that they invariably have a steel armour for mechanical protection and this consists of a number of steel wires or tapes. Submarine cables may be three-core cables buried underneath the seabed at between 1 to 2 m depth to avoid damage by ships' anchors, fishing trawlers or possibly physical displacement in areas of high sea currents. Submarine cables may also be single core and buried under the seabed but where this is not the case, e.g. in deep water installations, the individual phases may be laid a significant distance away from each other to minimise the risk of anchors damaging more than one phase. Pipe-type cables consist of three single-core cables each usually having a stranded copper conductor, impregnated paper insulation and a metallic sheath. The three cables may be installed asymmetrically or in a trefoil formation inside a steel pipe which is filled with pressurised low-viscosity oil or gas. The three sheaths are in contact with the pipe inside wall and they may touch each other. The steel pipe is normally coated on the outside to prevent corrosion. In the USA, over 90% of underground cables from 69 kV up to 345 kV are of pipe-type design but many modern replacement and new installations are employing oil-free XLPE cables with forced water cooling inside the pipe. In England and Wales, almost all transmission cables are of the self-contained single-core design laid either directly in trenches underground or in tunnels. At the time of writing, single-core XLPE cables up to 500 kV and three-core XLPE cables up to 245 kV are in use.

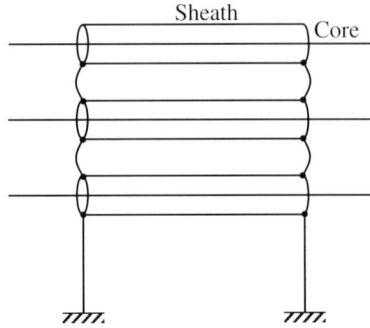

Figure 3.19   Solidly bonded three-phase cable

3.3.2 Cable sheath bonding and earthing arrangements

Cable sheaths are metallic conductors that perform a variety of important functions. The sheaths prevent moisture ingress into the insulation, contain cable pressure in fluid-filled cables, provide a continuous circuit for short-circuit fault current return and help prevent mechanical damage. Three cables form a three-phase single circuit and two such three-phase circuits may be laid in close proximity to each other. The ac current flowing in the core of one cable, as well as the currents flowing in adjacent cores, induce voltages on the metallic sheath(s) of the cable(s). In order to limit these voltages and prevent cable damage, the conducting sheaths may be bonded in a variety of methods. The three most common methods are briefly described below.

Solidly bonded cables

Solid bonding of cable sheaths is where both ends of the cable are bonded and connected to earth as shown in Figure 3.19.

Under PPS voltage conditions, voltages are induced in the sheaths and because these are solidly bonded and earthed at both ends, circulating currents flow such that the sheaths are at zero voltage along their entire length. However, these currents can cause significant sheath losses and heating which can adversely affect the thermal rating of the cable's core conductor. This arrangement is most suitable for three-core cables and is not usually used at voltages above 66 kV. Under ZPS conditions, the sheaths provide an excellent path for the earth fault return current.

Single-point bonded cables

Bonding the cable sheaths to earth at one end and using sheath voltage limiters at the other end is a good arrangement for short cable lengths. This limits the sheath voltages and prevents circulation of sheath currents during normal operation because of the absence of a closed sheath circuit. A similar arrangement that provides similar benefits and allows doubling the cable length is bonding and

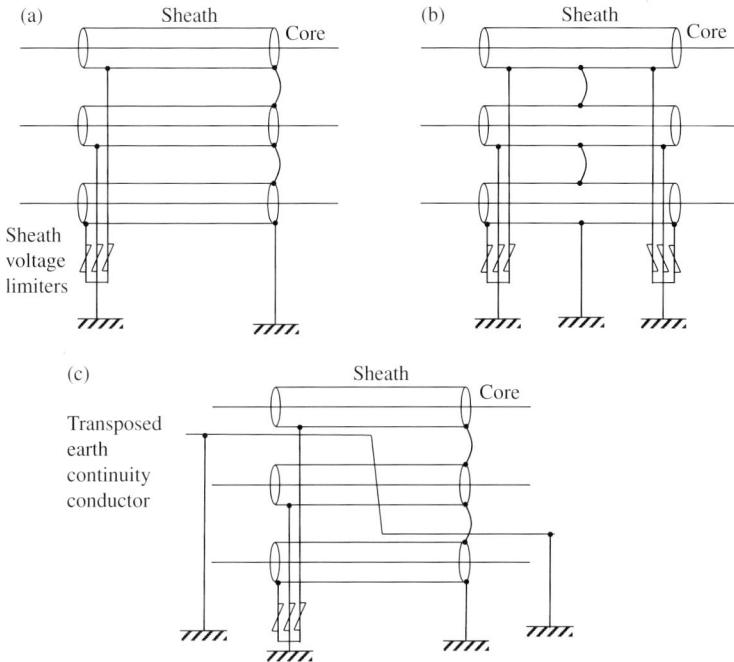

Figure 3.20 Single-point bonded three-phase cable: (a) end-point bonded cable; (b) mid-point bonded cable and (c) end-point bonded cable with a transposed earth continuity conductor

earthing the cable sheath at the cable mid-point and using sheath voltage limiters at the ends. Because no sheath current can flow, no sheath losses occur. However, in practical installations, a transposed earth continuity conductor earthed at both ends is also used in order to provide a return path for the current under ZPS earth fault conditions. Under PPS conditions, the transposed earth continuity conductor carries no current. Figure 3.20 illustrates these three earthing arrangements.

Cross-bonded cables

For long cable circuits, cross-bonding is commonly used. This method aims at minimising the total induced voltages in the sheath in order to minimise circulating currents and losses. The cable circuit is divided into major and minor sections so that each major section consists of three minor sections of equal length. Figure 3.21 illustrates a major section of a cross-bonded cable.

The best arrangement is where the cores of the three minor sections within each major section are perfectly transposed but the sheaths are not. The sheaths at both ends of the major section are solidly bonded and earthed but at the other two positions within the major section, they are bonded to sheath voltage limiters. The bonding and earthing at the ends of a major section eliminate the need for the earth continuity conductor required in single-point bonded and earthed arrangements.

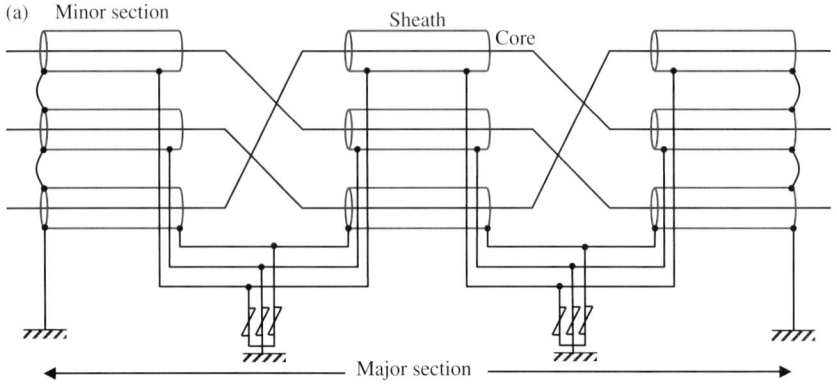

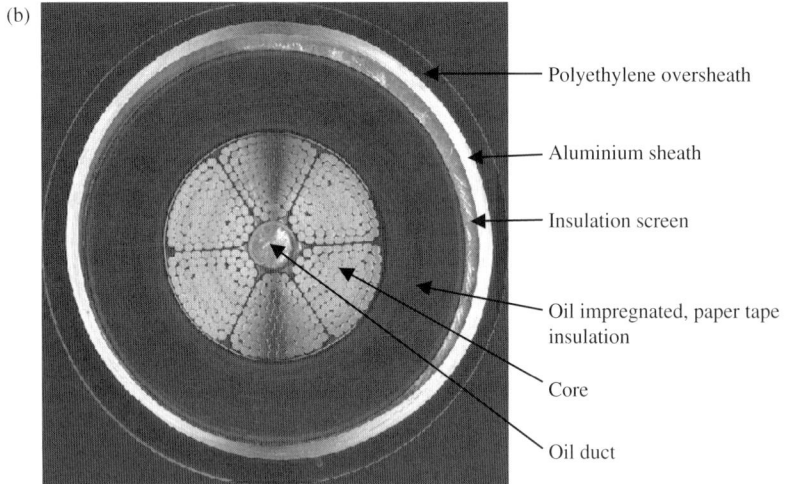

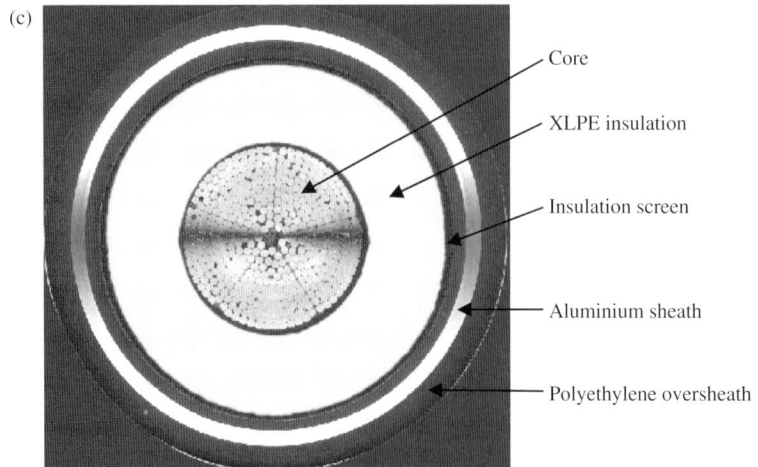

Figure 3.21 (a) Cross-bonded three-phase cable; (b) 275 kV oil-filled paper insulated cable and (c) 400 kV cross-linked polyethylene (XLPE) cable

Under PPS conditions, voltages are induced in the sheaths, but because the cores are perfectly transposed, the resulting voltages in each minor section are separated by 120° and thus sum to zero. Therefore, no sheath currents flow. Under ZPS conditions, e.g. earthed faults, the cross-bonding does not affect the excellent path which the sheaths provide for the return current. The earlier and rather less satisfactory cross-bonding method of transposing the sheaths but not the cores cannot achieve a good balance of induced sheath voltages unless the cables are laid in a trefoil formation. Therefore, the cores of cross-bonded cables laid in flat formations are generally transposed. Cross-bonding with core transposition is a general practice in the UK at 275 and 400 kV.

3.3.3 Overview of the calculation of cable parameters

General

Single-core and three-core cables are characterised by close coupling between the core conductor and the conducting sheath. Also, for buried self-contained single-core cables, there is a conducting earth path between the three adjacent conductors. In pipe-type cables, the steel pipe is also a conducting medium. Therefore, a single circuit self-contained three-phase cable is a multi-conductor system that consists of three cores and three sheaths (six conductors) together with the earth path. If the three cables are in a steel pipe, induced voltages in the pipe will cause currents to circulate. For single-core submarine cables where each has a conducting armour, a three further conductors are present. Therefore, for N three-phase cable circuits, the number of coupled conductors is $6N$ for core–sheath single-core or three-core cables, and $9N$ for core–sheath–armour single-core cables. For core–sheath–armour single-circuit three-core cables, the total number of coupled conductors is seven.

Similar to an overhead line, the basic electrical parameters of cables are the self and mutual impedances between conductors, and conductor shunt admittances. For power frequency steady state analysis, the small conductance of the insulation is usually neglected. Practical calculations of multi-conductor cable parameters with series impedance expressed in Ω per-unit length and shunt susceptance in μS unit length, are carried out using digital computer programs. These parameters are then used to form the cable series impedance and shunt admittance matrices in the phase frame of reference as will be described later. The cable capacitances or susceptances are calculated from the cable physical dimensions and geometry. The self and mutual impedances are calculated from the conductor material, construction, physical dimensions and earth resistivity. The significant variety of available cable designs and installations cannot be covered here. The equations of the parameters for some widely used cable designs are presented below.

Shunt capacitances and susceptances

For single-core screened cables where the phases are laid parallel to each other and to the earth's surface, the earth the screens are connected to acts as an electrostatic

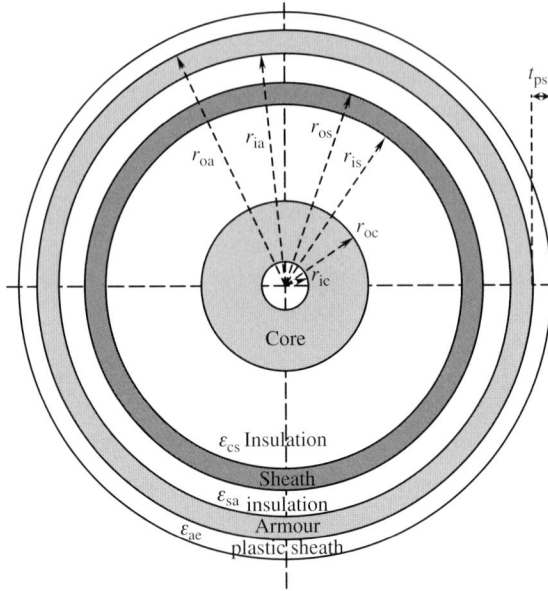

Figure 3.22   Cross-section of one phase of a self-contained armoured cable

shield so that there is no electrostatic coupling between the phases. This also applies to three-phase three-core screened cables where each core is individually screened. Therefore, there is no mutual capacitance among the three phases and the individual cable capacitance is independent of the spacings between the phases. Figure 3.22 is a cross-section of a general cable showing a hollow core conductor, core insulation, sheath/screen conductor, sheath insulation, armour conductor and a further insulation layer such as a plastic sheath.

Using Figure 3.22, the capacitance between the core and sheath of a screened single core or a screened three-core cable is given by

$$C_{cs} = \frac{0.0556325 \times \varepsilon_{cs}}{\log_e \left(\frac{r_{is}}{r_{oc}} \right)} \, \mu\text{F/km/phase} \tag{3.96a}$$

and the corresponding susceptance is given by

$$B_{cs} = \frac{0.34954933 \times f \times \varepsilon_{cs}}{\log_e \left(\frac{r_{is}}{r_{oc}} \right)} \, \mu\text{S/km/phase} \tag{3.96b}$$

The capacitance between the sheath and armour, where present, is given by

$$C_{sa} = \frac{0.0556325 \times \varepsilon_{sa}}{\log_e \left(\frac{r_{ia}}{r_{os}} \right)} \, \mu\text{F/km/phase} \tag{3.97a}$$

and the corresponding susceptance is given by

$$B_{sa} = \frac{0.34954933 \times f \times \varepsilon_{sa}}{\log_e\left(\frac{r_{ia}}{r_{os}}\right)} \ \mu\text{S/km/phase} \tag{3.97b}$$

Similarly, the capacitance between the armour and earth is given by

$$C_{ae} = \frac{0.0556325 \times \varepsilon_{ae}}{\log_e\left(\frac{r_{oa}+t_{ps}}{r_{oa}}\right)} \ \mu\text{F/km/phase} \tag{3.98a}$$

and the corresponding susceptance is given by

$$C_{ae} = \frac{0.34954933 \times f \times \varepsilon_{ae}}{\log_e\left(\frac{r_{oa}+t_{ps}}{r_{oa}}\right)} \ \mu\text{S/km/phase} \tag{3.98b}$$

where
r_{ic} = inner radius of core tubular conductor
r_{oc} = outer radius of core conductor = inner radius of core conductor insulation
   assuming no core semiconductor tape
r_{is} = outer radius of core conductor insulation = inner radius of metallic sheath
   assuming no tape over insulation
r_{os} = outer radius of metallic sheath = inner radius of sheath insulation
r_{ia} = inner radius of metallic armour = outer radius of sheath insulation
r_{oa} = outer radius of metallic armour = inner radius of metallic armour's plastic
   sheath
t_{ps} = thickness of metallic armour's plastic sheath
ε_{cs}, ε_{sa} and ε_{ae} are the relative permittivities of core-to-sheath, sheath-to-armour
   and armour-to-earth insulation.

The capacitance between the armour and earth would be redundant if there were no armour insulation or where the armour is directly earthed. Similarly, this capacitance would not exist for cables that have no armour and in this case, the cable would generally be characterised by two capacitances only. The relative permittivity of insulation is in general a complex number with frequency dependent real and imaginary parts, i.e. $\varepsilon_r = \varepsilon'(f) - j\varepsilon''(f)$. However, for power frequency steady state analysis, only the real part of ε_r is of significance and this is equivalent to assuming zero shunt conductance. Typical values for the real part of ε_r are: 4 for solid-impregnated paper, 3.5 for oil-filled-impregnated paper, 3.7 for oil-pressure pipe-type, 3.5 for gas-pressure, 4 for butyl rubber (BR), 3 for EPR, 2.5 for XLPE, 2.3 for polyethylene (PE) and 8 for polyvinyl chloride (PVC).

For three-phase pipe-type cables, the cables inside the pipe may be laid in a touching triangular or a cradle formation as illustrated in Figure 3.23. The core-to-sheath capacitance of each cable phase can be calculated using Equation (3.96a).

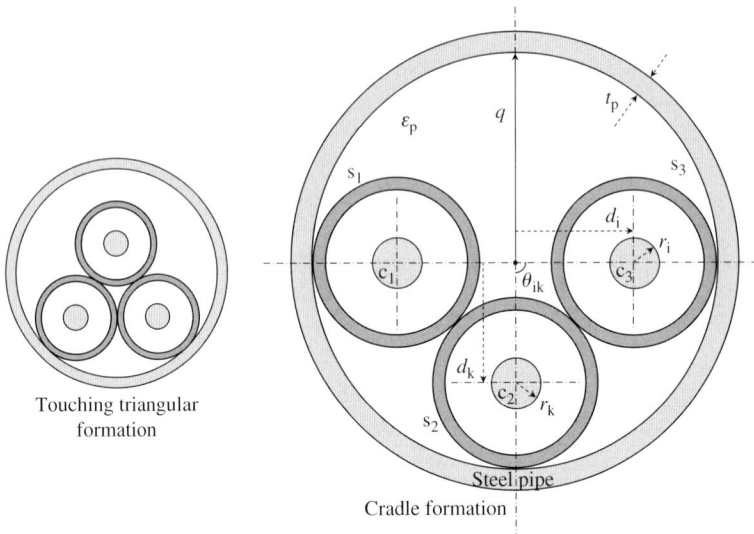

Figure 3.23   Pipe-type three-phase cables

Mutual capacitances between the sheaths of the three phases and between the sheath of each phase and the pipe can be calculated using

$$C_{ss(ik)} = \frac{0.0556325 \times \varepsilon_p}{\log_e\left[\frac{q}{\sqrt{d_i^2+d_k^2-2d_id_k\cos\theta_{ik}}}\right] - \sum_{n=1}^{\infty}\left(\frac{d_id_k}{q^2}\right)^n \times \frac{\cos(n\theta_{ik})}{n}} \, \mu F/km/phase$$

(3.99a)

and

$$C_{sp(ii)} = \frac{0.0556325 \times \varepsilon_p}{\log_e\left(\frac{q^2-d_i^2}{q\times r_i}\right)} \, \mu F/km/phase$$

(3.99b)

where q is the inner radius of the steel pipe, ε_p is the permittivity of insulation inside the pipe, d_i is the distance between cable i core centre and pipe centre, r_i is the outer radius of cable i core conductor and θ_{ik} is the angle subtended between the pipe's centre and the centres of cables i and k. The infinite series term in Equation (3.99a) is usually very small and for most d_i and q distances encountered in practical installations, it may be neglected with insignificant loss of accuracy. It should be noted that the mutual sheath capacitance would be short-circuited where the sheaths are in contact with each other, and the pipe's inside wall. This applies for both cradle and triangular formations.

For three-core unscreened or belted cables, there is no simple reasonably accurate formula that can be used because of the effect of two insulation layers that have to be considered namely the core insulation and the belt insulation. For these cables, it is recommended that resort is made either to empirical design data from

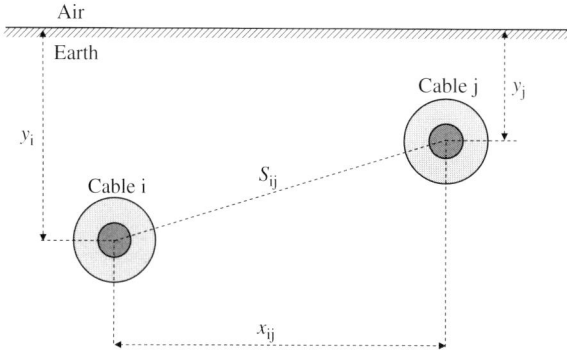

Figure 3.24 Illustration of spatial dimensions of two underground cable phases

the manufacturers or better still to field test measurements. This is dealt with in Section 3.6.3.

Series self and mutual impedances

For a three-phase screened cable buried underground and having armour protection, the core, sheath and armour will each have a self-impedance with earth return. Electromagnetic coupling, and hence mutual impedances, exist between the core, sheath and armour within each phase. Even though the three phases are usually laid parallel to each other and the earth's surface, the earth does not act as an electromagnetic shield between them. Therefore, mutual impedances exist between all the conductors of all the phases. Figures 3.22 and 3.24 are used in the calculation of self and mutual impedances of underground cables for use in steady state power frequency analysis.

The self-impedance of a core conductor with earth return is given by

$$Z_{cc} = R_{c(ac)} + \pi^2 10^{-4}f + j4\pi 10^{-4}f\left[\frac{\mu_c}{4} \times f(r_{oc}, r_{ic}) + \log_e\left(\frac{D_{erc}}{r_{oc}}\right)\right] \Omega/\text{km}$$

(3.100a)

where

$$f(r_{oc}, r_{ic}) = 1 - \frac{2r_{ic}^2}{(r_{oc}^2 - r_{ic}^2)} + \frac{4r_{ic}^4}{(r_{oc}^2 - r_{ic}^2)^2}\log_e\left(\frac{r_{oc}}{r_{ic}}\right)$$

(3.100b)

and D_{erc} is the depth of equivalent earth return conductor given in Equation (3.15) and μ_c is the relative permeability of the core conductor.

The self-impedance of a sheath with earth return is given by

$$Z_{ss} = R_{S(ac)} + \pi^2 10^{-4}f + j4\pi 10^{-4}f\left[\frac{\mu_s}{4} \times f(r_{os}, r_{is}) + \log_e\left(\frac{D_{erc}}{r_{os}}\right)\right] \Omega/\text{km}$$

(3.101a)

where

$$f(r_{os}, r_{is}) = 1 - \frac{2r_{is}^2}{(r_{os}^2 - r_{is}^2)} + \frac{4r_{is}^4}{(r_{os}^2 - r_{is}^2)^2} \log_e \left(\frac{r_{os}}{r_{is}} \right) \tag{3.101b}$$

and μ_s is the relative permeability of the sheath conductor.

The self-impedance of an armour with earth return is

$$Z_{aa} = R_{a(ac)} + \pi^2 10^{-4} f + j4\pi 10^{-4} f \left[\frac{\mu_a}{4} \times f(r_{oa}, r_{ia}) + \log_e \left(\frac{D_{erc}}{r_{oa}} \right) \right] \Omega/km \tag{3.102a}$$

where

$$f(r_{os}, r_{is}) = 1 - \frac{2r_{is}^2}{(r_{os}^2 - r_{is}^2)} + \frac{4r_{is}^4}{(r_{os}^2 - r_{is}^2)^2} \log_e \left(\frac{r_{os}}{r_{is}} \right) \tag{3.102b}$$

and μ_a is the relative permeability of the armour conductor.

The mutual impedance between core or sheath or armour i, and core or sheath or armour j, with earth return, is given by

$$Z_{ij} = \pi^2 10^{-4} f + j4\pi 10^{-4} f \times \log_e \left(\frac{D_{erc}}{S_{ij}} \right) \Omega/km \tag{3.103}$$

where S_{ij} is the distance between the centres of cables i and j if the conductors belong to different cables. If the conductors belong to the same cable, S_{ij} is the geometric mean distance between the two conductors, e.g. the GMD between the core and the sheath of cable j is given by $S_{jj} = (r_{os} + r_{is})/2$ which is sufficiently accurate for practical cable dimensions.

ac resistance

The ac resistance of the core, sheath or armour can be calculated from the dc resistance using the following formula:

$$R_{(ac)} = R_{(dc)}[1 + y(k_S + k_P)] \Omega/km \tag{3.104a}$$

where $y = 1$ for single-core, two-core and three-core cables but $y = 1.5$ for pipe-type cables. k_S and k_P are skin and proximity effect factors, respectively. Also,

$$R_{(dc)} = \frac{1000\rho}{A}[1 + \alpha_{20} \times (T - 20)] \Omega/km \tag{3.104b}$$

ρ is the conductor resistivity in Ωm, A is conductor nominal cross-sectional area in m^2, T is conductor temperature in °C and α_{20} in °C^{-1} is the constant mass temperature coefficient at 20°C. Table 3.1 illustrates typical values for α_{20} and resistivity at 20°C.

Table 3.1   Typical values of α_{20} and conductor resistivity at 20°C

Material	Temperature coefficient $\alpha_{20}(°C^{-1})$ at 20°C	Resistivity ρ_{20} (Ωm) at 20°C
Cores		
Copper	3.93×10^{-3}	1.7241×10^{-8}
Aluminium	4.03×10^{-3}	2.8264×10^{-8}
Sheaths or armour		
Lead	4×10^{-3}	21.4×10^{-8}
Bronze	3×10^{-3}	3.5×10^{-8}
Steel	4.5×10^{-3}	13.8×10^{-8}
Stainless steel	0	70×10^{-8}

The skin effect factor is the incremental resistance factor produced by ac current in an isolated conductor due to skin effect and is given by

$$k_S = \begin{cases} \dfrac{z^4}{0.8 \times z^4 + 192} & 0 < z \le 2.8 \\[2ex] 0.0563 \times z^2 - 0.0177 \times z - 0.136 & 2.8 < z \le 3.8 \\[2ex] 0.354 \times z - 0.733 & z > 3.8 \end{cases} \qquad (3.105)$$

where $z = \sqrt{8\pi f a_z/(10^4 R_{dc})}$. For copper conductors, $a_z = 1$ for normally stranded circular and sector-shaped conductors but $a_z = 0.43$ for segmental or Milliken shaped conductors. For stranded annular conductors, $a_z = [(r_o - r_i)/(r_o + r_i)][(r_o + 2r_i)/(r_o + r_i)]^2$, where r_i and r_o are the conductor's inner and outer radii, respectively. Generally, z is less than 2.8 for the majority of practical applications. Although based on a simplified approach, Equation (3.105) for the skin effect factor involves an error of less than 0.5% at power frequency.

The proximity effect factor is the incremental resistance factor due to the proximity of other ac current carrying conductors and is given by

$$k_P = \begin{cases} 2.9 \times F(p)\left(\dfrac{d_c}{S}\right)^2 & \text{for two-core and two single-core cables} \\[3ex] F(p)\left(\dfrac{d_c}{S}\right)^2 \left[0.312\left(\dfrac{d_c}{S}\right)^2 + \dfrac{1.18}{F(p) + 0.27} \right] & \text{for three-core and three single-core cables} \end{cases}$$

$$(3.106a)$$

where

$$F(p) = \frac{p^4}{0.8 \times p^4 + 192} \quad \text{and} \quad p = \sqrt{\frac{8\pi f a_p}{10^4 R_{dc}}} \qquad (3.106b)$$

d_c is conductor diameter and S is the axial spacing between conductors. For both copper and Aluminium, a_p is equal to 0.8 for round stranded, sector shaped and annular stranded conductors. a_p is equal to 0.37 for round segmental conductors.

The values of the various factors used in Equations (3.104), (3.105) and (3.106) are usually supplied by the cable manufacturer.

For a three-phase submarine cable, the power frequency equations given for underground cables can be used. However, the sea will now predominantly replace earth as the return path and is represented as follows:

$$R_{Sea} = 399.63 \times \sqrt{\frac{\rho_{Sea}}{f}} \text{ m} \qquad (3.107)$$

where R_{Sea} is the outer radius of sea return represented as an equivalent conductor and ρ_{Sea} is the resistivity of sea water. This concept is derived from the calculation of sea return impedance where the cable is assumed to be completely surrounded by an indefinite sea that acts as an equivalent return conductor having an outer radius of R_{Sea}. For example, for a typical value of sea water resistivity of $\rho_{Sea} = 0.5 \, \Omega\text{m}$, $R_{Sea} \cong 40 \, \text{m}$ at 50 Hz. It is interesting to note that in deep sea water, where the cable phases are laid 100 to 500 m apart, electromagnetic coupling between the phases would be very weak and can be normally ignored. The internal impedance of the sea return represented as an equivalent conductor of radius R_{Sea} is given by

$$Z_{Sea} = \pi^2 10^{-4} f \left[1 + \frac{4}{\pi} \text{kei}(\alpha) \right] + j4\pi 10^{-4} f \left[\log_e \left(\frac{R_{Sea}}{r} \right) - \text{ker}(\alpha) \right] \Omega/\text{km}$$

$$(3.108)$$

where $\alpha = 1.123 \times D/R_{Sea}$, D is mean spacing between the cable's phases in m, r is conductor radius in m, and $\text{ker}(\alpha)$ and $\text{kei}(\alpha)$ are Kelvin functions with a real argument α.

For pipe-type cables, the calculation of the self and mutual impedances is more complex than for buried underground cables. The calculation of the flux linkages within the wall of the steel pipe and outside the pipe is further complicated by the non-linear permeability of the steel pipe which itself varies with the magnitude of ZPS current flowing through the pipe under earth fault conditions due to pipe saturation. The effect of saturation is to cause a reduction in the effective ZPS impedance of the cable; the larger the ZPS current, the larger the reduction in ZPS impedance. Cable manufacturers are usually required to provide such data to network utilities. For power frequency analysis, a usual assumption is that the pipe's thickness is greater than the depth of penetration into the pipe wall and that this assumption remains approximately true under increased pipe ZPS current. This means that besides the sheaths of the three cables, the pipe is the only current return path and that no current returns through the earth. The cables inside the pipe can then be considered as three self-contained single-core cables but with the pipe replacing the earth as the current return path. The depth of penetration into the pipe can be calculated using the skin depth formula of Equation (3.7b)

$\delta = 503.292 \times \sqrt{\rho_p / f \mu_p}$ where ρ_p and μ_p are the pipe's resistivity and relative permeability respectively. To illustrate the assumption of infinite pipe thickness, consider a steel pipe of a 132 kV pipe-type cable having a thickness of 6.3 mm, a resistivity of $\rho_p = 3.8 \times 10^{-8}$ Ωm and a relative permeability $\mu_p = 400$. The depth of penetration into the pipe at 50 Hz is equal to $\delta = 1.32$ mm. This is smaller than the pipe's thickness and illustrates that the return current will flow towards the inner wall of the pipe and that the earth return can be ignored. In Figure 3.23, we assume that each cable phase consists of a solid core conductor of radius r_c, core insulation and a sheath conductor of inner and outer radii r_{is} and r_{os}, respectively. From Figure 3.23, the self-impedance matrix of cable phase k inside the pipe is given by

$$Z_k = \begin{bmatrix} Z_{cc-k} & Z_{cs-k} \\ Z_{cs-k} & Z_{ss-k} \end{bmatrix} = \begin{bmatrix} Z_1 + Z_2 + Z_3 + Z_4 + Z_5 - 2Z_6 & Z_4 + Z_5 - Z_6 \\ Z_4 + Z_5 - Z_6 & Z_4 + Z_5 \end{bmatrix}$$

(3.109)

where Z_1 is as given in Equation (3.7), and

$$Z_2 = j4\pi f 10^{-4} \log_e(r_{is}/r_c)$$

$$Z_3 = \frac{1000\rho m}{2\pi r_{is} D} [I_0(mr_{is})K_1(mr_{os}) + K_0(mr_{is})I_1(mr_{os})] \qquad (3.110a)$$

$$Z_4 = \frac{1000\rho m}{2\pi r_{os} D} [I_0(mr_{os})K_1(mr_{is}) + K_0(mr_{os})I_1(mr_{is})]$$

$$Z_5 = j4\pi f 10^{-4} \log_e \left(\frac{q^2 - d_k^2}{qr_{os}} \right) \qquad (3.110b)$$

$$Z_6 = \frac{1000\rho}{2\pi r_{is} r_{os} D} \qquad m = \sqrt{2}\delta^{-1} e^{j\pi/4}$$

$$D = I_1(mr_{os})K_1(mr_{is}) - K_1(mr_{os})I_1(mr_{is}) \qquad (3.110c)$$

where Z_1 to Z_6 are in Ω/km. I_i and K_i are modified Bessel functions of the first and second kind of order i, respectively. The pipe's internal impedance with the return path being the inside wall of the pipe is given by

$$Z_{p-int} = j4\pi f \mu_p 10^{-4} \left\{ \frac{K_0(mq)}{mqK_1(mq)} + 2\sum_{n=1}^{\infty} \left[\left(\frac{d_i}{q} \right)^{2n} \frac{K_n(mq)}{n\mu_p K_n(mq) - K_n'(mq)} \right] \right\} \Omega/km$$

(3.111)

The mutual impedance in Ω/km between the ith and kth conductor with respect to the pipe's inner wall is given by

$$Z_{i-k} = j4\pi f 10^{-4} \left\{ \frac{\mu_p K_0(mq)}{mqK_1(mq)} + \log_e \left[\frac{q}{\sqrt{d_i^2 + d_k^2 - 2d_i d_k \cos \theta_{ik}}} \right] \right.$$

$$\left. + \sum_{n=1}^{\infty} \left(\frac{d_i d_k}{q^2} \right)^n \cos(n\theta_{ik}) \left[2\mu_p \frac{K_n(mq)}{n\mu_p K_n(mq) - mqK_n'(mq)} - \frac{1}{n} \right] \right\}$$

$$(3.112)$$

where K_n' is the derivative of K_n.

3.3.4 Series phase and sequence impedance matrices of single-circuit cables

General

Having calculated the basic phase parameters in Section 3.3.3, we will now derive the cable's phase and sequence impedance matrices. We use the suffixes C, S and A to denote core, sheath and armour conductors respectively for cable phases 1, 2 and 3. Figure 3.25 shows six practical cable layouts for three-core and single-core cables. Pipe-type cables were shown in Figure 3.23.

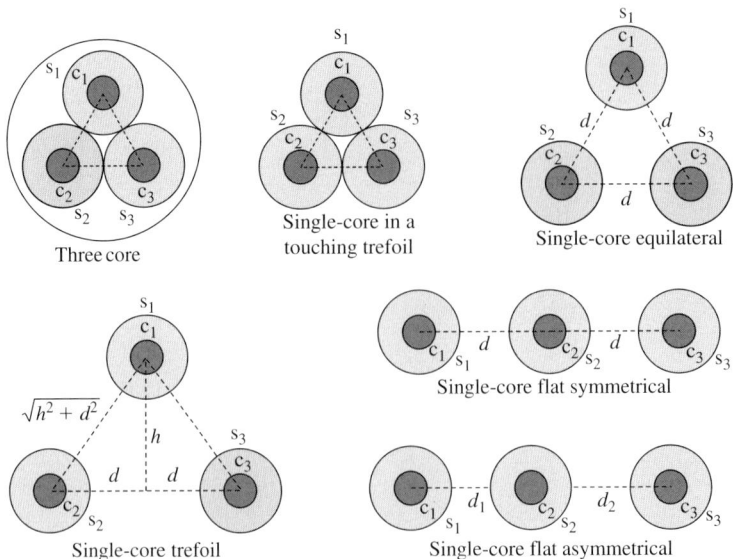

Figure 3.25 Typical practical single- and three-core cable layouts

Cables with no armour

A three-phase single-core or shielded three-core cable has a total of six conductors; three cores and three sheaths as shown in Figure 3.26a.

The series voltage drop per-unit length across the cores and sheaths is calculated from the cable's full impedance matrix, core and sheath conductor currents and is given by

$$\begin{bmatrix} V_{C1} \\ V_{C2} \\ V_{C3} \\ \hline V_{S1} \\ V_{S2} \\ V_{S3} \end{bmatrix} = \begin{array}{r} \\ \\ \text{C1} \\ \text{Cores C2} \\ \text{C3} \\ \text{S1} \\ \text{Sheaths S2} \\ \text{S3} \end{array} \overset{\begin{array}{cccccc} \multicolumn{3}{c}{\text{Cores}} & \multicolumn{3}{c}{\text{Sheaths}} \\ \text{C1} & \text{C2} & \text{C3} & \text{S1} & \text{S2} & \text{S3} \end{array}}{\left[\begin{array}{ccc|ccc} Z_{C1C1} & Z_{C1C2} & Z_{C1C3} & Z_{C1S1} & Z_{C1S2} & Z_{C1S3} \\ Z_{C2C1} & Z_{C2C2} & Z_{C2C3} & Z_{C2S1} & Z_{C2S2} & Z_{C2S3} \\ Z_{C3C1} & Z_{C3C2} & Z_{C3C3} & Z_{C3S1} & Z_{C3S2} & Z_{C3S3} \\ \hline Z_{S1C1} & Z_{S1C2} & Z_{S1C3} & Z_{S1S1} & Z_{S1S2} & Z_{S1S3} \\ Z_{S2C1} & Z_{S2C2} & Z_{S2C3} & Z_{S2S1} & Z_{S2S2} & Z_{S2S3} \\ Z_{S3C1} & Z_{S3C2} & Z_{S3C3} & Z_{S3S1} & Z_{S3S2} & Z_{S3S3} \end{array}\right]} \begin{bmatrix} I_{C1} \\ I_{C2} \\ I_{C3} \\ \hline I_{S1} \\ I_{S2} \\ I_{S3} \end{bmatrix} \tag{3.113}$$

where $\mathbf{Z}_{CC}$ and $\mathbf{Z}_{SS}$ are the core and sheath self-impedances with earth return, respectively, and $\mathbf{Z}_{CS}$ is the mutual impedance between core and sheath with earth

Figure 3.26 Multi-conductor cables with sheaths and armours: (a) cable conductors, no armour and (b) cable conductors, with armour

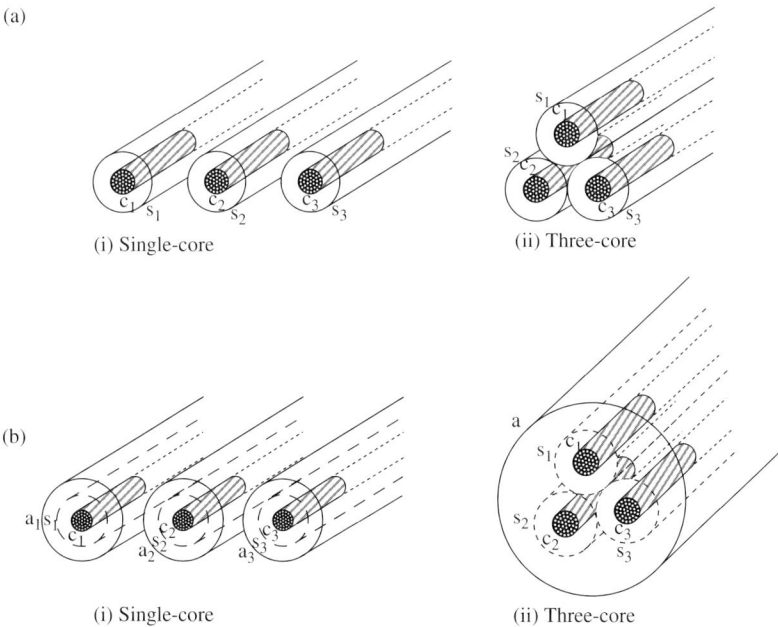

return. Equation (3.113) can be simplified depending on the cable layout and spacings between the phases as shown in Figure 3.25.

Three-core cables, three single-core cables in touching trefoil layout and three-single-core cables in equilateral layout

These three cable layouts, shown in Figure 3.25 are fully symmetrical. From the configuration of these cables, a number of practical equalities can be deduced as follows:

$$a = Z_{C1S1} = Z_{C2S2} = Z_{C3S3} = Z_{S1C1} = Z_{S2C2} = Z_{S3C3}$$

$$b = Z_{C1C2} = Z_{C2C1} = Z_{C1C3} = Z_{C3C1} = Z_{C2C3} = Z_{C3C2} = Z_{C1S2} = Z_{S2C1}$$
$$= Z_{C1S3} = Z_{S3C1} = Z_{C2S1} = Z_{S1C2} = Z_{C2S1} = Z_{S1C2} = Z_{C2S3} = Z_{S3C2}$$
$$= Z_{C3S1} = Z_{S1C3} = Z_{C3S2} = Z_{S2C3} = Z_{S1S2} = Z_{S2S1} = Z_{S1S3} = Z_{S3S1}$$
$$= Z_{S2S3} = Z_{S3S2}$$

$$e = Z_{C1C1} = Z_{C2C2} = Z_{C3C3} \quad f = Z_{S1S1} = Z_{S2S2} = Z_{S3S3} \tag{3.114}$$

Therefore, the cable full impedance matrix of Equation (3.113) can be written as

$$
\begin{array}{c}
 \text{C1 C2 C3 S1 S2 S3} \\
\begin{bmatrix} V_{C1} \\ V_{C2} \\ V_{C3} \\ V_{S1} \\ V_{S2} \\ V_{S3} \end{bmatrix}
=
\begin{array}{c}
\text{C1} \\ \text{C2} \\ \text{C3} \\ \text{S1} \\ \text{S2} \\ \text{S3}
\end{array}
\left[\begin{array}{ccc|ccc}
e & b & b & a & b & b \\
b & e & b & b & a & b \\
b & b & e & b & b & a \\
\hline
a & b & b & f & b & b \\
b & a & b & b & f & b \\
b & b & a & b & b & f
\end{array}\right]
\begin{bmatrix} I_{C1} \\ I_{C2} \\ I_{C3} \\ I_{S1} \\ I_{S2} \\ I_{S3} \end{bmatrix}
\end{array}
\tag{3.115a}
$$

or

$$
\begin{bmatrix} V_C \\ V_S \end{bmatrix} = \begin{bmatrix} Z_{CC} & Z_{CS} \\ Z_{CS} & Z_{SS} \end{bmatrix} \begin{bmatrix} I_C \\ I_S \end{bmatrix}
\tag{3.115b}
$$

where

$$
V_C = \begin{bmatrix} V_{C1} \\ V_{C2} \\ V_{C3} \end{bmatrix} \quad V_S = \begin{bmatrix} V_{S1} \\ V_{S2} \\ V_{S3} \end{bmatrix} \quad I_C = \begin{bmatrix} I_{C1} \\ I_{C2} \\ I_{C3} \end{bmatrix} \quad \text{and} \quad I_S = \begin{bmatrix} I_{S1} \\ I_{S2} \\ I_{S3} \end{bmatrix}
\tag{3.115c}
$$

$$
Z_{CC} = \begin{bmatrix} e & b & b \\ b & e & b \\ b & b & e \end{bmatrix} \quad Z_{CS} = \begin{bmatrix} a & b & b \\ b & a & b \\ b & b & a \end{bmatrix} \quad Z_{SS} = \begin{bmatrix} f & b & b \\ b & f & b \\ b & b & f \end{bmatrix}
$$

It is interesting to observe that the matrices Z_{CC}, Z_{CS} and Z_{SS} are all balanced matrices since all diagonal terms are equal and all off-diagonal terms are equal. This is due to the symmetrical cable layouts. Where the cable sheaths are solidly bonded and earthed at both ends, the sheaths can be eliminated. Therefore, the phase impedance matrix involving the cores only can be calculated by setting $\mathbf{V}_S = 0$ in Equation (3.115b). The resultant core or phase impedance matrix is given by

$$\mathbf{Z}_{\text{Phase}} = \mathbf{Z}_{\text{CC}} - \mathbf{Z}_{\text{CS}} \mathbf{Z}_{\text{SS}}^{-1} \mathbf{Z}_{\text{CS}} = \begin{bmatrix} Z_{C(\text{Self})} & Z_{C(\text{Mut})} & Z_{C(\text{Mut})} \\ Z_{C(\text{Mut})} & Z_{C(\text{Self})} & Z_{C(\text{Mut})} \\ Z_{C(\text{Mut})} & Z_{C(\text{Mut})} & Z_{C(\text{Self})} \end{bmatrix} \qquad (3.116)$$

Because the individual impedance matrices $\mathbf{Z}_{\text{CC}}$, $\mathbf{Z}_{\text{CS}}$ and $\mathbf{Z}_{\text{SS}}$ are all balanced, the resultant phase impedance matrix $\mathbf{Z}_{\text{Phase}}$, that includes the effect of the eliminated sheaths, will also be balanced. The reader is encouraged to prove this statement. Therefore, the sequence impedance matrix is given by $\mathbf{Z}^{\text{PNZ}} = \mathbf{H}^{-1}\mathbf{Z}_{\text{Phase}}\mathbf{H}$ or

$$\mathbf{Z}^{\text{PNZ}} = \begin{bmatrix} Z^P & 0 & 0 \\ 0 & Z^N & 0 \\ 0 & 0 & Z^Z \end{bmatrix} \qquad (3.117a)$$

where

$$Z^P = Z^N = Z_{C(\text{Self})} - Z_{C(\text{Mut})} \quad \text{and} \quad Z^Z = Z_{C(\text{Self})} + 2Z_{C(\text{Mut})} \qquad (3.117b)$$

Single-core cables in a trefoil layout

Using Figure 3.25 and following similar steps, it can be shown that the cable full impedance matrix of Equation (3.113) can be written as

$$
\begin{array}{c}
 \\
\begin{bmatrix} V_{C1} \\ V_{C2} \\ V_{C3} \\ V_{S1} \\ V_{S2} \\ V_{S3} \end{bmatrix}
=
\begin{array}{c} \\ \\ \\ \\ \\ \\ \end{array}
\begin{array}{c}
\begin{array}{cccccc} C1 & C2 & C3 & S1 & S2 & S3 \end{array} \\
\begin{array}{c} C1 \\ C2 \\ C3 \\ S1 \\ S2 \\ S3 \end{array}
\left[\begin{array}{ccc|ccc}
e & b & b & a & b & b \\
b & e & c & b & a & c \\
b & c & e & b & c & a \\
\hline
a & b & b & f & b & b \\
b & a & c & b & f & c \\
b & c & a & b & c & f
\end{array}\right]
\end{array}
\begin{bmatrix} I_{C1} \\ I_{C2} \\ I_{C3} \\ I_{S1} \\ I_{S2} \\ I_{S3} \end{bmatrix}
\end{array}
\qquad (3.118a)
$$

where

$$e = Z_{C1C1} = Z_{C2C2} = Z_{C3C3} \quad f = Z_{S1S1} = Z_{S2S2} = Z_{S3S3}$$

$$a = Z_{C1S1} = Z_{C2S2} = Z_{C3S3} = Z_{S1C1} = Z_{S2C2} = Z_{S3C3}$$

$$c = Z_{C2C3} = Z_{C3C2} = Z_{C2S3} = Z_{S3C2} = Z_{S2C3} = Z_{S3C2} = Z_{S2S3} = Z_{S3S2}$$

$$b = Z_{C1C2} = Z_{C2C1} = Z_{C1C3} = Z_{C3C1} = Z_{C1S2} = Z_{S2C1} = Z_{C1S3} = Z_{C2S1}$$
$$= Z_{S1C2} = Z_{C3S1} = Z_{S1C3} = Z_{S1S2} = Z_{S2S1} = Z_{S1S3} = Z_{S3S1} \qquad (3.118b)$$

The calculation of the phase impedance matrix involving the cores only and the corresponding sequence impedance matrix follows the same steps as above. However, in this case, the phase impedance matrix will not be balanced and as a result, the sequence impedance matrix will contain off-diagonal intersequence mutual terms.

Single-core cables in a flat layout

The general asymmetrical flat layout of single-core three-phase cables is shown in Figure 3.25. From the configuration of the cable, we can write the following equalities

$$a = Z_{C1S1} = Z_{C2S2} = Z_{C3S3} = Z_{S1C1} = Z_{S2C2} = Z_{S3C3}$$

$$b = Z_{C1C2} = Z_{C2C1} = Z_{C1S2} = Z_{S2C1} = Z_{C2S1} = Z_{S1C2} = Z_{S1S2} = Z_{S2S1}$$

$$c = Z_{C2C3} = Z_{C3C2} = Z_{C2S3} = Z_{S3C2} = Z_{C3S2} = Z_{S2C3} = Z_{S2S3} = Z_{S3S2}$$

$$d = Z_{C1C3} = Z_{C3C1} = Z_{C1S3} = Z_{S3C1} = Z_{C3S1} = Z_{S1C3} = Z_{S1S3} = Z_{S3S1}$$

$$e = Z_{C1C1} = Z_{C2C2} = Z_{C3C3} \quad f = Z_{S1S1} = Z_{S2S2} = Z_{S3S3} \qquad (3.119)$$

Therefore, the cable full impedance matrix of Equation (3.113) can be written as

$$
\mathbf{Z} =
\begin{array}{c}
\\
C1 \\
C2 \\
C3 \\
S1 \\
S2 \\
S3
\end{array}
\overset{\begin{array}{cccccc} C1 & C2 & C3 & S1 & S2 & S3 \end{array}}{
\left[
\begin{array}{ccc|ccc}
e & b & d & a & b & d \\
b & e & c & b & a & c \\
d & c & e & d & c & a \\
\hline
a & b & d & f & b & d \\
b & a & c & b & f & c \\
d & c & a & d & c & f
\end{array}
\right]}
\qquad (3.120a)
$$

For cross-bonded cables shown in Figure 3.21, the cores of the minor sections are transposed so that each occupies the three positions, but the sheaths are not. This is illustrated in Figure 3.27.

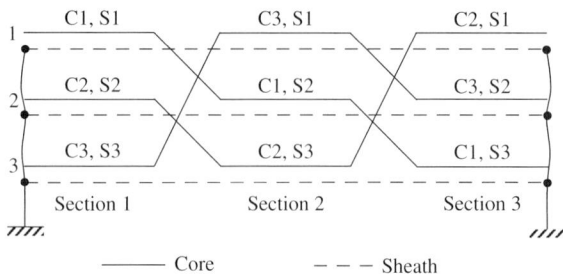

Figure 3.27 Core transposition of cross-bonded cables

Equation (3.120a) is the full impedance matrix of the first minor section. The full impedance matrices of the second and third minor sections are given by

$$\mathbf{Z}_{\text{Section-2}} = \begin{bmatrix} e & c & b & b & a & c \\ c & e & d & d & c & a \\ b & d & e & a & b & d \\ b & d & a & f & b & d \\ a & c & b & b & f & c \\ c & a & d & d & c & f \end{bmatrix} \quad \mathbf{Z}_{\text{Section-3}} = \begin{bmatrix} e & d & c & d & c & a \\ d & e & b & a & b & d \\ c & b & e & b & a & c \\ d & a & b & f & b & d \\ c & b & a & b & f & c \\ a & d & c & d & c & f \end{bmatrix}$$

$$(3.120b)$$

Using Equation (3.120) and all impedances are in per unit length i.e. Ω/km, the full impedance matrix of a cross-bonded cable with transposed cores in each major section is given by

$$\mathbf{Z} = \frac{1}{3} \sum_{i=1}^{3} \mathbf{Z}_{\text{Section-}i}$$

or

$$\mathbf{Z} = \frac{1}{3} \begin{bmatrix} 3e & b+c+d & b+c+d & a+b+d & a+b+c & a+c+d \\ b+c+d & 3e & b+c+d & a+b+d & a+b+c & a+c+d \\ b+c+d & b+c+d & 3e & a+b+d & a+b+c & a+c+d \\ a+b+d & a+b+d & a+b+d & 3f & 3b & 3d \\ a+b+c & a+b+c & a+b+c & 3b & 3f & 3c \\ a+c+d & a+c+d & a+c+d & 3d & 3c & 3f \end{bmatrix}$$

$$(3.121a)$$

or including core and sheath voltage vectors

$$\begin{bmatrix} \mathbf{V}_C \\ \mathbf{V}_S \end{bmatrix} = \begin{bmatrix} \mathbf{Z}_{CC} & \mathbf{Z}_{CS} \\ \mathbf{Z}_{CS}^t & \mathbf{Z}_{SS} \end{bmatrix} \begin{bmatrix} \mathbf{I}_C \\ \mathbf{I}_S \end{bmatrix}$$

$$(3.121b)$$

As expected, the core matrix $\mathbf{Z}_{CC}$ is balanced, due to the perfect core transpositions assumed but the sheath matrix $\mathbf{Z}_{SS}$ is not because the sheaths are untransposed. The mutual impedance matrix between the cores and the sheaths $\mathbf{Z}_{CS}$ is also unbalanced.

In both cases of a solidly bonded cable where the sheaths are bonded and earthed at the cable ends, and the case of a cross-bonded cable where the sheaths are bonded and earthed at the ends of major sections, the phase impedance matrix involving the cores only can be calculated by setting $\mathbf{V}_S = 0$ in Equation (3.121b). Therefore, the core or phase impedance matrix takes the following general form

$$\mathbf{Z}_{\text{Phase}} = \mathbf{Z}_{CC} - \mathbf{Z}_{CS} \mathbf{Z}_{SS}^{-1} \mathbf{Z}_{CS}^t = \begin{bmatrix} Z_{C1C1-\text{Sheath}} & Z_{C1C2-\text{Sheath}} & Z_{C1C3-\text{Sheath}} \\ Z_{C2C1-\text{Sheath}} & Z_{C2C2-\text{Sheath}} & Z_{C2C3-\text{Sheath}} \\ Z_{C3C1-\text{Sheath}} & Z_{C3C2-\text{Sheath}} & Z_{C3C3-\text{Sheath}} \end{bmatrix}$$

$$(3.122)$$

The elements of the matrix $\mathbf{Z}_{Phase}$ are the self-impedances of each phase (core) and the mutual impedances between phases (cores) with the sheaths and earth return included. Generally, for solidly bonded cables, the self or diagonal terms of $\mathbf{Z}_{Phase}$ are not equal among each other nor are the off-diagonal mutual terms. In other words, the matrix is generally not balanced. This might also be expected to be the case for cross-bonded cables but in practical cable spacings and layouts, the core transposition eliminates the unbalance. The general sequence impedance matrix equation that corresponds to Equation (3.122), calculated using $\mathbf{Z}^{PNZ} = \mathbf{H}^{-1}\mathbf{Z}_{Phase}\mathbf{H}$, is given by

$$\mathbf{Z}^{PNZ} = \begin{bmatrix} Z^{PP} & Z^{PN} & Z^{PZ} \\ Z^{NP} & Z^{NN} & Z^{NZ} \\ Z^{ZP} & Z^{ZN} & Z^{ZZ} \end{bmatrix} \tag{3.123}$$

where Z^{PP}, Z^{NN} and Z^{ZZ} are the PPS, NPS and ZPS impedances of the cable and $Z^{PP} = Z^{NN}$. The off-diagonal terms are intersequence mutual terms which are normally small in comparison with the diagonal terms for solidly bonded cables and practically negligible for cross-bonded cables.

For a cable that is solidly bonded and earthed at its centre and a transposed earth continuity conductor is used, two full impedance matrices would need to be built for each half of the cable including the earth continuity conductor then combined into a single matrix. The calculation of the phase impedance matrix of such a cable makes use of the fact that no current can flow in the sheath that is $\mathbf{I}_S = 0$. The derivation of the phase impedance matrix for such a cable is left for the reader.

Cables with armour

Some land based cables and almost all submarine cables have metallic armour. The armour acts as a third conductor in addition to the core and the sheath.

Single-core armoured cables
Using Figure 3.26(b)(i) for single-core armoured cables, the series voltage drop per-unit length is given by

$$\begin{bmatrix} V_{C1} \\ V_{C2} \\ V_{C3} \\ V_{S1} \\ V_{S2} \\ V_{S3} \\ V_{A1} \\ V_{A2} \\ V_{A3} \end{bmatrix} = \begin{array}{l} \text{Cores} \\ \\ \text{Sheaths} \\ \\ \text{Armours} \end{array} \begin{bmatrix} Z_{C1C1} & Z_{C1C2} & Z_{C1C3} & Z_{C1S1} & Z_{C1S2} & Z_{C1S3} & Z_{C1A1} & Z_{C1A2} & Z_{C1A3} \\ Z_{C2C1} & Z_{C2C2} & Z_{C2C3} & Z_{C2S1} & Z_{C2S2} & Z_{C2S3} & Z_{C2A1} & Z_{C2A2} & Z_{C2A3} \\ Z_{C3C1} & Z_{C3C2} & Z_{C3C3} & Z_{C3S1} & Z_{C3S2} & Z_{C3S3} & Z_{C3A1} & Z_{C3A2} & Z_{C3A3} \\ Z_{S1C1} & Z_{S1C2} & Z_{S1C3} & Z_{S1S1} & Z_{S1S2} & Z_{S1S3} & Z_{S1A1} & Z_{S1A2} & Z_{S1A3} \\ Z_{S2C1} & Z_{S2C2} & Z_{S2C3} & Z_{S2S1} & Z_{S2S2} & Z_{S2S3} & Z_{S2A1} & Z_{S2A2} & Z_{S2A3} \\ Z_{S3C1} & Z_{S3C2} & Z_{S3C3} & Z_{S3S1} & Z_{S3S2} & Z_{S3S3} & Z_{S3A1} & Z_{S3A2} & Z_{S3A3} \\ Z_{A1C1} & Z_{A1C2} & Z_{A1C3} & Z_{A1S1} & Z_{A1S2} & Z_{A1S3} & Z_{A1A1} & Z_{A1A2} & Z_{A1A3} \\ Z_{A2C1} & Z_{A2C2} & Z_{A2C3} & Z_{A2S1} & Z_{A2S2} & Z_{A2S3} & Z_{A2A1} & Z_{A2A2} & Z_{A2A3} \\ Z_{A3C1} & Z_{A3C2} & Z_{A3C3} & Z_{A3S1} & Z_{A3S2} & Z_{A3S3} & Z_{A3A1} & Z_{A3A2} & Z_{A3A3} \end{bmatrix} \begin{bmatrix} I_{C1} \\ I_{C2} \\ I_{C3} \\ I_{S1} \\ I_{S1} \\ I_{S1} \\ I_{A1} \\ I_{A1} \\ I_{A1} \end{bmatrix}$$

with column group headers: Cores | Sheaths | Armours

$$(3.124)$$

In order to control the voltages between the sheath and the armour, the sheath is usually bonded to the armour at a number of points along the route. The armour covering is not normally an electric insulation so that the armour, and the sheath bonded to it, are effectively earthed. Therefore, the calculation of the cable's phase (core) impedance matrix requires the elimination of both sheath and armour. Writing Equation (3.124) in partitioned matrix form, we have

$$
\begin{bmatrix} \mathbf{V_C} \\ \mathbf{V_S} \\ \mathbf{V_A} \end{bmatrix} = \begin{bmatrix} \mathbf{Z_{CC}} & \mathbf{Z_{CS}} & \mathbf{Z_{CA}} \\ \mathbf{Z_{CS}^t} & \mathbf{Z_{SS}} & \mathbf{Z_{SA}} \\ \mathbf{Z_{CA}^t} & \mathbf{Z_{SA}^t} & \mathbf{Z_{AA}} \end{bmatrix} \begin{bmatrix} \mathbf{I_C} \\ \mathbf{I_S} \\ \mathbf{I_A} \end{bmatrix}
\tag{3.125}
$$

The sheaths and armours are eliminated by setting $\mathbf{V_S} = \mathbf{V_A} = 0$ in Equation (3.125) and combining their matrices as follows:

$$
\begin{bmatrix} \mathbf{V}_{C(3\times1)} \\ \mathbf{0}_{S,A(6\times1)} \end{bmatrix} = \begin{bmatrix} \mathbf{Z}_{CC(3\times3)} & \mathbf{Z}_{CS,CA(3\times6)} \\ (\mathbf{Z}_{CS,CA(3\times6)})^t & \mathbf{Z}_{SS,AA(6\times6)} \end{bmatrix} \begin{bmatrix} \mathbf{I}_{C(3\times1)} \\ \mathbf{I}_{S,A(6\times1)} \end{bmatrix}
\tag{3.126a}
$$

Therefore, the phase (core) impedance matrix of the cable with sheaths, armours and earth, or sea water in the case of a submarine cable, constituting the current return path, is obtained as follows:

$$
\mathbf{Z}_{Phase(3\times3)} = \mathbf{Z}_{CC(3\times3)} - \mathbf{Z}_{CS,CA(3\times6)} \mathbf{Z}_{SS,AA(6\times6)}^{-1} (\mathbf{Z}_{CS,CA(3\times6)})^t
\tag{3.126b}
$$

For single-core submarine cables where the individual phases are laid physically far apart from each other, the mutual electromagnetic coupling between the phases may be so weak that it can be neglected. Therefore, from the configuration of such three-phase submarine cables, a number of practical equalities can be deduced as follows:

$$a = Z_{C1S1} = Z_{C2S2} = Z_{C3S3} = Z_{S1C1} = Z_{S2C2} = Z_{S3C3} = Z_{CS}$$
$$b = Z_{C1A1} = Z_{C2A2} = Z_{C3A3} = Z_{A1C1} = Z_{A2C2} = Z_{A3C3} = Z_{CA}$$
$$c = Z_{S1A1} = Z_{S2A2} = Z_{S3A3} = Z_{A1S1} = Z_{A2S2} = Z_{A3S3} = Z_{SA}$$
$$d = Z_{C1C1} = Z_{C2C2} = Z_{C3C3} = Z_{CC} \quad e = Z_{S1S1} = Z_{S2S2} = Z_{S3S3} = Z_{SS}$$
$$f = Z_{A1A1} = Z_{A2A2} = Z_{A3A3} = Z_{AA}
\tag{3.127}$$

and all mutual terms between the three phases of the cables are zero. Therefore, Equation (3.124) simplifies to

$$
\mathbf{Z} = \begin{array}{c} \\ \text{Cores} \\ \\ \text{Sheaths} \\ \\ \text{Armours} \end{array}
\begin{bmatrix}
d & 0 & 0 & a & 0 & 0 & b & 0 & 0 \\
0 & d & 0 & 0 & a & 0 & 0 & b & 0 \\
0 & 0 & d & 0 & 0 & a & 0 & 0 & b \\
a & 0 & 0 & e & 0 & 0 & c & 0 & 0 \\
0 & a & 0 & 0 & e & 0 & 0 & c & 0 \\
0 & 0 & a & 0 & 0 & e & 0 & 0 & c \\
b & 0 & 0 & c & 0 & 0 & f & 0 & 0 \\
0 & b & 0 & 0 & c & 0 & 0 & f & 0 \\
0 & 0 & b & 0 & 0 & c & 0 & 0 & f
\end{bmatrix}
\begin{array}{c} \text{Cores} \quad \text{Sheaths} \quad \text{Armours} \end{array}
\tag{3.128}
$$

Using Equation (3.126b), the phase (cores) impedance matrix is given by

$$\mathbf{Z}_{Phase(3\times3)} = \begin{bmatrix} Z_\alpha & 0 & 0 \\ 0 & Z_\alpha & 0 \\ 0 & 0 & Z_\alpha \end{bmatrix} \qquad (3.129a)$$

where $Z_\alpha = d - (a^2f + b^2e - 2abc)$ or using Equation (3.127),

$$Z_\alpha = Z_{CC} - (Z_{CS}^2 Z_{AA} + Z_{CA}^2 Z_{SS} - 2Z_{CS}Z_{CA}Z_{SA}) \qquad (3.129b)$$

Z_α is the impedance of each core or phase including the effect of the combined sheath, armour and sea water return. The corresponding PPS/NPS and ZPS sequence impedances for the cable are all equal to Z_α which is a direct result of the absence of mutual coupling between the three phases.

In the case where the distances between the phases are not so large that the mutual coupling between them cannot be ignored, three further impedance parameters need to be defined to account for the interphase coupling. Let the mutual impedance parameter between phases 1 and 2 be g such that

$$g = Z_{C1C2} = Z_{C1S2} = Z_{C1A2} = Z_{S1C2} = Z_{S1S2} = Z_{S1A2} = Z_{A1C2} = Z_{A1S2} = Z_{A1A2}$$
$$= Z_{C2C1} = Z_{S2C1} = Z_{A2C1} = Z_{C2S1} = Z_{S2S1} = Z_{A2S1} = Z_{C2A1} = Z_{S2A1} = Z_{A2A1}$$

The mutual impedances between phases 1 and 3, and phase 2 and 3 are similarly defined. The analysis for calculating the cable phase impedance matrix proceeds as above.

Three-core armoured cables

Using Figure 3.26(ii) for three-core armoured cables, there are three cores and three sheaths but only one armour. Thus, the series voltage drop per-unit length is given by

		Cores			Sheaths			Armour	
V_{C1}		Z_{C1C1}	Z_{C1C2}	Z_{C1C3}	Z_{C1S1}	Z_{C1S2}	Z_{C1S3}	Z_{C1A}	I_{C1}
V_{C2}	Cores	Z_{C2C1}	Z_{C2C2}	Z_{C2C3}	Z_{C2S1}	Z_{C2S2}	Z_{C2S3}	Z_{C2A}	I_{C2}
V_{C3}		Z_{C3C1}	Z_{C3C2}	Z_{C3C3}	Z_{C3S1}	Z_{C3S2}	Z_{C3S3}	Z_{C3A}	I_{C3}
V_{S1} =		Z_{S1C1}	Z_{S1C2}	Z_{S1C3}	Z_{S1S1}	Z_{S1S2}	Z_{S1S3}	Z_{S1A}	I_{S1}
V_{S2}	Sheaths	Z_{S2C1}	Z_{S2C2}	Z_{S2C3}	Z_{S2S1}	Z_{S2S2}	Z_{S2S3}	Z_{S2A}	I_{S2}
V_{S3}		Z_{S3C1}	Z_{S3C2}	Z_{S3C3}	Z_{S3S1}	Z_{S3S2}	Z_{S3S3}	Z_{S3A}	I_{S3}
V_A	Armour	Z_{AC1}	Z_{AC2}	Z_{AC3}	Z_{AS1}	Z_{AS2}	Z_{AS3}	Z_{AA}	I_A

$$(3.130)$$

Using the equalities of Equation (3.114) for the case with no armour, and the new equalities

$$c = Z_{C1A} = Z_{C2A} = Z_{C3A} = Z_{AC1} = Z_{AC2} = Z_{AC3} = Z_{S1A} = Z_{S2A} = Z_{S3A}$$
$$= Z_{AS1} = Z_{AS2} = Z_{AS3}, \quad \text{and} \quad d = Z_{AA}$$

we can write

$$
\begin{bmatrix} V_{C1} \\ V_{C2} \\ V_{C3} \\ \hline V_{S1} \\ V_{S2} \\ V_{S3} \\ \hline V_A \end{bmatrix}
=
\begin{array}{c}
\text{Cores} \\ \\ \\ \\ \text{Sheaths} \\ \\ \\ \text{Armour}
\end{array}
\overset{\displaystyle \text{Cores} \quad\quad \text{Sheaths} \quad \text{Armour}}{
\left[\begin{array}{ccc|ccc|c}
e & b & b & a & b & b & c \\
b & e & b & b & a & b & c \\
b & b & e & b & b & a & c \\
\hline
a & b & b & f & b & b & c \\
b & a & b & b & f & b & c \\
b & b & a & b & b & f & c \\
\hline
c & c & c & c & c & c & d
\end{array}\right]}
\begin{bmatrix} I_{C1} \\ I_{C2} \\ I_{C3} \\ \hline I_{S1} \\ I_{S2} \\ I_{S3} \\ \hline I_A \end{bmatrix}
\qquad (3.131)
$$

As for the single-core cable, where the sheaths and armours are bonded and earthed, we can now eliminate the sheaths and armour by combining the matrices that correspond to them by setting $\mathbf{V}_S = \mathbf{V}_A = 0$. Writing Equation (3.131) in partitioned form, we have

$$
\begin{bmatrix} \mathbf{V}_{C(3\times1)} \\ \mathbf{0}_{S,A(4\times1)} \end{bmatrix}
=
\begin{bmatrix} \mathbf{Z}_{CC(3\times3)} & \mathbf{Z}_{CS,CA(3\times4)} \\ (\mathbf{Z}_{CS,CA(3\times4)})^t & \mathbf{Z}_{SS,AA(4\times4)} \end{bmatrix}
\begin{bmatrix} \mathbf{I}_{C(3\times1)} \\ \mathbf{I}_{S,A(4\times1)} \end{bmatrix}
\qquad (3.132a)
$$

The phase (core) impedance matrix of the cable with sheath, armour and earth, or sea water in the case of a submarine cable, constituting the current return path, is given by

$$
\mathbf{Z}_{\text{Phase}(3\times3)} = \mathbf{Z}_{CC(3\times3)} - \mathbf{Z}_{CS,CA(3\times4)} \mathbf{Z}_{SS,AA(4\times4)}^{-1} (\mathbf{Z}_{CS,CA(3\times4)})^t
\qquad (3.132b)
$$

The phase (core) impedance matrix will virtually be balanced. The corresponding sequence impedance matrix calculated using $\mathbf{Z}^{\text{PNZ}} = \mathbf{H}^{-1} \mathbf{Z}_{\text{Phase}} \mathbf{H}$ will be a diagonal matrix, i.e. having virtually no intersequence terms. The diagonal terms are the required PPS/NPS and ZPS impedances.

Pipe-type cables

Using Figure 3.23 for a three-phase pipe-type cable, there are three cores, three sheaths and one pipe which represents the return path, i.e. a total of seven conductors. Therefore, the full cable impedance matrix takes the form of Equation (3.130) with armour 'A' replaced with pipe 'P'. The elimination of the pipe only or pipe and sheaths can be carried out as described above.

3.3.5 Shunt phase and sequence susceptance matrices of single-circuit cables

In this section, we derive the cable phase and sequence susceptance matrices using the basic phase parameters calculated in Section 3.3.3.

Screened cables with no armour

For screened three-core and screened single-core cables, we have already established that there is no capacitance between one cable phase and another. From Figure 3.22, we note that there are two capacitances, or susceptances, corresponding to the core insulation and the sheath insulation layers and these are illustrated in Figure 3.28(a).

For each phase, the core and sheath current/voltage equations can be written as $I_C = jB_{CS}(V_C - V_S)$ and $I_C + I_S = jB_{SE}V_S$ where B_{CS} and B_{SE} are the core-to-sheath and sheath-to-earth susceptances, respectively. Therefore, writing similar equations for the three cores and three sheaths, the full shunt susceptance matrix is given by

$$\begin{bmatrix} I_{C1} \\ I_{C2} \\ I_{C3} \\ \hline I_{S1} \\ I_{S2} \\ I_{S3} \end{bmatrix} = \begin{matrix} \text{Cores} \\ \\ \text{Sheaths} \end{matrix} \overset{\text{Cores} \qquad\qquad\qquad\qquad \text{Sheaths}}{\left[\begin{array}{ccc|ccc} jB_{CS} & 0 & 0 & -jB_{CS} & 0 & 0 \\ 0 & jB_{CS} & 0 & 0 & -jB_{CS} & 0 \\ 0 & 0 & jB_{CS} & 0 & 0 & -jB_{CS} \\ \hline -jB_{CS} & 0 & 0 & j(B_{CS}+B_{SE}) & 0 & 0 \\ 0 & -jB_{CS} & 0 & 0 & j(B_{CS}+B_{SE}) & 0 \\ 0 & 0 & -jB_{CS} & 0 & 0 & j(B_{CS}+B_{SE}) \end{array}\right]} \begin{bmatrix} V_{C1} \\ V_{C2} \\ V_{C3} \\ \hline V_{S1} \\ V_{S2} \\ V_{S3} \end{bmatrix} \tag{3.133}$$

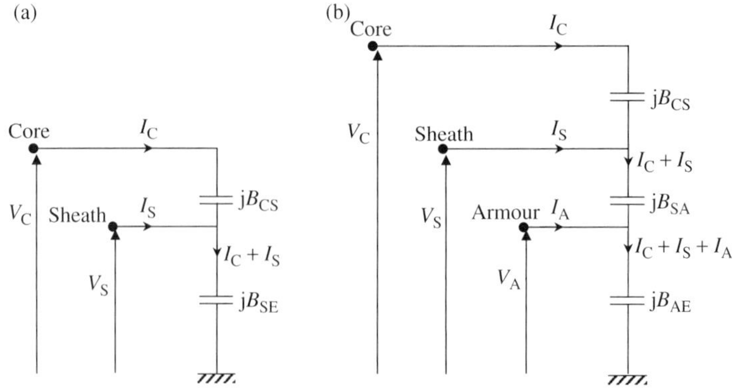

Figure 3.28 Equivalent capacitances of screened cables: (a) no armour and (b) with armour

For solidly bonded and cross-bonded cables, the sheaths can be eliminated using $V_{S1} = V_{S2} = V_{S3} = 0$. The resultant shunt phase (cores) susceptance matrix is given by

$$\mathbf{B}_{\text{Phase}} = \begin{bmatrix} B_{CS} & 0 & 0 \\ 0 & B_{CS} & 0 \\ 0 & 0 & B_{CS} \end{bmatrix} \quad (3.134)$$

that is the PPS, NPS and ZPS susceptances are all equal, and equal to B_{CS}.

For cables that are bonded and earthed at one end only or at the cable centre, there will also be a capacitance or susceptance between the earth continuity conductor and earth with $I_{ECC} = jB_{ECC-E}V_{ECC}$. This increases the dimension of the full admittance matrix given in Equation (3.133) by one row and one column, i.e. from 6×6 to 7×7. The calculation of the 3×3 phase susceptance matrix for such cables, the constraint that no sheath current can flow, i.e. $I_{S1} = I_{S2} = I_{S3} = 0$ is applied. The resultant phase (cores) susceptance matrix is given by

$$\mathbf{B}_{\text{Phase}} = \begin{bmatrix} \dfrac{B_{CS}B_{SE}}{B_{CS} + B_{SE}} & 0 & 0 \\ 0 & \dfrac{B_{CS}B_{SE}}{B_{CS} + B_{SE}} & 0 \\ 0 & 0 & \dfrac{B_{CS}B_{SE}}{B_{CS} + B_{SE}} \end{bmatrix} \quad (3.135a)$$

This result shows that the cable-to-sheath capacitance and the sheath-to-earth capacitance of each core are effectively in series as can be seen from Figure 3.28(a).

The corresponding shunt sequence susceptance matrix of Equations (3.134) and (3.135a) is given by

$$\mathbf{B}^{PNZ} = \mathbf{B}_{\text{Phase}} \quad (3.135b)$$

that is the PPS, NPS and ZPS susceptances are all equal to the relevant phase susceptances.

Screened cables with armour

From Figure 3.22, we note that there are three capacitances, or susceptances, corresponding to the core insulation layer, the sheath insulation layer and the armour coating where this is made of insulation material. These are illustrated in Figure 3.28(b). For each phase, the core, sheath and armour current/voltage equations can be written as

$$I_C = jB_{CS}(V_C - V_S) \quad I_C + I_S = jB_{SA}(V_S - V_A)$$

and

$$I_C + I_S + I_A = jB_{AE}V_A$$

Therefore, writing similar equations for the three cores, three sheaths and three armours, the full shunt susceptance matrix $\mathbf{B}$ is given by

	Cores			Sheaths			Armours		
	C1	C2	C3	S1	S2	S3	A1	A2	A3
C1	B_{CS}	0	0	$-B_{CS}$	0	0	0	0	0
C2	0	B_{CS}	0	0	$-B_{CS}$	0	0	0	0
C3	0	0	B_{CS}	0	0	$-B_{CS}$	0	0	0
S1	$-B_{CS}$	0	0	$B_{CS}+B_{SA}$	0	0	$-B_{SA}$	0	0
S2	0	$-B_{CS}$	0	0	$B_{CS}+B_{SA}$	0	0	$-B_{SA}$	0
S3	0	0	$-B_{CS}$	0	0	$B_{CS}+B_{SA}$	0	0	$-B_{SA}$
A1	0	0	0	$-B_{SA}$	0	0	$B_{SA}+B_{AE}$	0	0
A2	0	0	0	0	$-B_{SA}$	0	0	$B_{SA}+B_{AE}$	0
A3	0	0	0	0	0	$-B_{SA}$	0	0	$B_{SA}+B_{AE}$

$$(3.136)$$

If the sheath and armour are bonded together and earthed at both ends, then circulating currents can flow through the sheath and armour in parallel with earth return, or sea water return in the case of a submarine cable. The phase (core) susceptance matrix is easily obtained by deleting the last six rows and columns that correspond to the sheath and armours. The earthing of the sheath and armour, in effect, short circuits the sheath-to-armour and armour-to-earth capacitances as can be seen in Figure 3.28(b). The PPS, NPS and ZPS susceptances in equal to the phase susceptance B_{CS}. If the sheath and armour are unearthed or earthed at one point only, then no sheath or armour current would flow and the equivalent core or phase capacitance is equal to the three capacitances of Figure 3.28(b) in series that is

$$B_{Phase} = \frac{1}{\frac{1}{B_{CS}} + \frac{1}{B_{SA}} + \frac{1}{B_{AE}}}$$

The PPS, NPS and ZPS susceptances are equal to B_{Phase}.

Unscreened or belted cables

Belted cables are three-core cables where each core has an insulation layer but no screen or metallic sheath. There is also a belt insulation layer around all three cores and a single metallic sheath cover as illustrated in Figure 3.29(a).

There are capacitances among the three cores, between each core and sheath and between the sheath and earth. Figure 3.29(b) shows the equivalent capacitance circuit of the cable where C_{CC} is the core-to-core capacitance, C_{CS} is the core-to-sheath capacitance and C_{SE} is the sheath-to-earth capacitance. Where the sheath is solidly earthed, as is usually the case, the equivalent capacitance circuit reduces to that shown in Figure 3.29(c). The injected currents into the three cores can be expressed as follows:

$$I_{C1} = jB_{CS}V_{C1} + jB_{CC}(V_{C1} - V_{C2}) + jB_{CC}(V_{C1} - V_{C3})$$

(a)

→ Belt insulation
→ Sheath
→ Insulation
→ Filler
→ Core

(b)

(c)

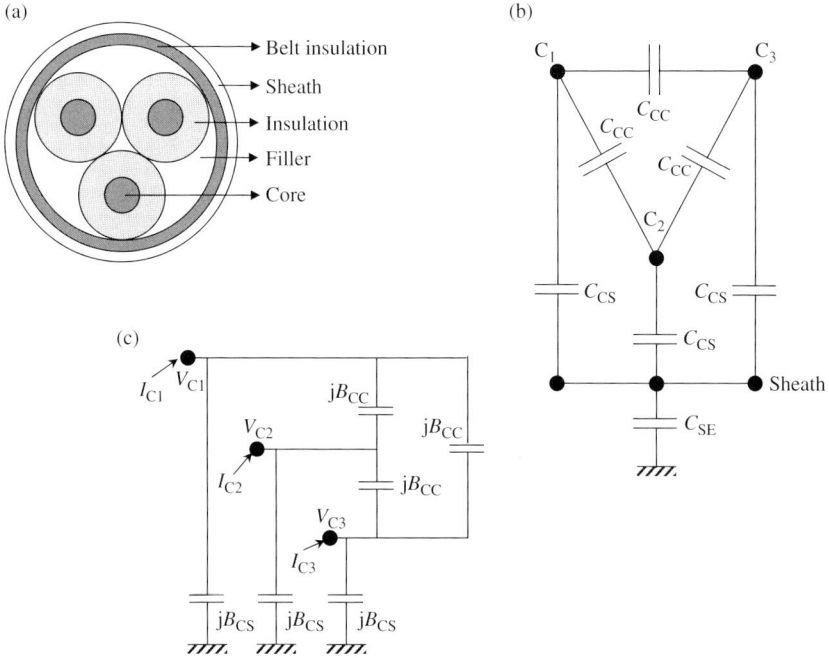

Figure 3.29 Three-core belted cable and capacitance equivalent circuits: (a) cross-section of a three-phase belted cable; (b) capacitance equivalent circuit and (c) capacitance equivalent circuit with earthed sheath

or

$$I_{C1} = j(B_{CS} + 2B_{CC})V_{C1} - jB_{CC}V_{C2} - jB_{CC}V_{C3} \qquad (3.137)$$

and similarly for the two remaining cores. The nodal phase susceptance matrix is

$$\begin{bmatrix} I_{C1} \\ I_{C2} \\ I_{C3} \end{bmatrix} = j \begin{bmatrix} B_{CS} + 2B_{CC} & -B_{CC} & -B_{CC} \\ -B_{CC} & B_{CS} + 2B_{CC} & -B_{CC} \\ -B_{CC} & -B_{CC} & B_{CS} + 2B_{CC} \end{bmatrix} \begin{bmatrix} V_{C1} \\ V_{C2} \\ V_{C3} \end{bmatrix} \qquad (3.138a)$$

and the corresponding sequence susceptance matrix is given by

$$\mathbf{B}^{PNZ} = \begin{bmatrix} B^P & 0 & 0 \\ 0 & B^N & 0 \\ 0 & 0 & B^Z \end{bmatrix} \qquad (3.138b)$$

where

$$B^P = B^N = B_{CS} + 3B_{CC} \quad \text{and} \quad B^Z = B_{CS} \qquad (3.138c)$$

It is interesting to note that the ZPS susceptance of a belted cable is smaller than its PPS/NPS susceptance. This contrasts with the equal PPS/NPS and ZPS susceptances of screened three-core and single-core cables. It is useful to calculate the core-to-core cable susceptance in terms of the sequence susceptances giving $B_{CC} = (B^P - B^Z)/3$.

3.3.6 Three-phase double-circuit cables

In some installations, double-circuit cables may be laid close to each other. These may be three-core circuits or single-core circuits laid in flat or trefoil arrangements. Near vertical arrangements may be used in tunnels. We will describe the case of single-core unarmoured cables. As each circuit consists of six conductors, three cores and three sheaths, twelve conductors form this complex mutually coupled multi-conductor arrangement. Let the two circuits be designated A and B and let C1, C2, C3, S1, S2 and S3 be the cores and sheaths of circuit A, C4, C5, C6, S4, S5 and S6 be the cores and sheaths of circuit B. The full 12×12 impedance matrix is given by

		Circuit A Cores			Circuit B Cores			Circuit A Sheaths			Circuit B Sheaths		
		C1	C2	C3	C4	C5	C6	S1	S2	S3	S4	S5	S6
A	C1	Z_{C1C1}	Z_{C1C2}	Z_{C1C3}	Z_{C1C4}	Z_{C1C5}	Z_{C1C6}	Z_{C1S1}	Z_{C1S2}	Z_{C1S3}	Z_{C1S4}	Z_{C1S5}	Z_{C1S6}
	C2	Z_{C2C1}	Z_{C2C2}	Z_{C2C3}	Z_{C2C4}	Z_{C2C5}	Z_{C2C6}	Z_{C2S1}	Z_{C2S2}	Z_{C2S3}	Z_{C2S4}	Z_{C2S5}	Z_{C2S6}
	C3	Z_{C3C1}	Z_{C3C2}	Z_{C3C3}	Z_{C3C4}	Z_{C3C5}	Z_{C3C6}	Z_{C3S1}	Z_{C3S2}	Z_{C3S3}	Z_{C3S4}	Z_{C3S5}	Z_{C3S6}
B	C4	Z_{C4C1}	Z_{C4C2}	Z_{C4C3}	Z_{C4C4}	Z_{C4C5}	Z_{C4C6}	Z_{C4S1}	Z_{C4S2}	Z_{C4S3}	Z_{C4S4}	Z_{C4S5}	Z_{C4S6}
	C5	Z_{C5C1}	Z_{C5C2}	Z_{C5C3}	Z_{C5C4}	Z_{C5C5}	Z_{C5C6}	Z_{C5S1}	Z_{C5S2}	Z_{C5S3}	Z_{C5S4}	Z_{C5S5}	Z_{C5S6}
	C6	Z_{C6C1}	Z_{C6C2}	Z_{C6C3}	Z_{C6C4}	Z_{C6C5}	Z_{C6C6}	Z_{C6S1}	Z_{C6S2}	Z_{C6S3}	Z_{C6S4}	Z_{C6S5}	Z_{C6S6}
A	S1	Z_{S1C1}	Z_{S1C2}	Z_{S1C3}	Z_{S1C4}	Z_{S1C5}	Z_{S1C6}	Z_{S1S1}	Z_{S1S2}	Z_{S1S3}	Z_{S1S4}	Z_{S1S5}	Z_{S1S6}
	S2	Z_{S2C1}	Z_{S2C2}	Z_{S2C3}	Z_{S2C4}	Z_{S2C5}	Z_{S2C6}	Z_{S2S1}	Z_{S2S2}	Z_{S2S3}	Z_{S2S4}	Z_{S2S5}	Z_{S2S6}
	S3	Z_{S3C1}	Z_{S3C2}	Z_{S3C3}	Z_{S3C4}	Z_{S3C5}	Z_{S3C6}	Z_{S3S1}	Z_{S3S2}	Z_{S3S3}	Z_{S3S4}	Z_{S3S5}	Z_{S3S6}
	S4	Z_{S4C1}	Z_{S4C2}	Z_{S4C3}	Z_{S4C4}	Z_{S4C5}	Z_{S4C6}	Z_{S4S1}	Z_{S4S2}	Z_{S4S3}	Z_{S4S4}	Z_{S4S5}	Z_{S4S6}
B	S5	Z_{S5C1}	Z_{S5C2}	Z_{S5C3}	Z_{S5C4}	Z_{S5C5}	Z_{S5C6}	Z_{S5S1}	Z_{S5S2}	Z_{S5S3}	Z_{S5S4}	Z_{S5S5}	Z_{S5S6}
	S6	Z_{S6C1}	Z_{S6C2}	Z_{S6C3}	Z_{S6C4}	Z_{S6C5}	Z_{S6C6}	Z_{S6S1}	Z_{S6S2}	Z_{S6S3}	Z_{S6S4}	Z_{S6S5}	Z_{S6S6}

where $Z =$ the above matrix.

$$(3.139a)$$

or in concise matrix form, including voltage drop and current vectors

$$
\begin{bmatrix} \mathbf{V}_{C(A)} \\ \mathbf{V}_{C(B)} \\ \hline \mathbf{V}_{S(A)} \\ \mathbf{V}_{S(B)} \end{bmatrix}
=
\begin{bmatrix} \mathbf{Z}_{CC(A)} & \mathbf{Z}_{C(A)C(B)} & \mathbf{Z}_{C(A)S(A)} & \mathbf{Z}_{C(A)S(B)} \\ \mathbf{Z}^t_{C(A)C(B)} & \mathbf{Z}_{CC(B)} & \mathbf{Z}^t_{C(A)S(B)} & \mathbf{Z}_{C(B)S(B)} \\ \hline \mathbf{Z}^t_{C(A)S(A)} & \mathbf{Z}^t_{C(B)S(A)} & \mathbf{Z}_{SS(A)} & \mathbf{Z}_{S(A)S(B)} \\ \mathbf{Z}^t_{C(A)S(B)} & \mathbf{Z}^t_{C(B)S(B)} & \mathbf{Z}^t_{S(A)S(B)} & \mathbf{Z}_{SS(B)} \end{bmatrix}
\begin{bmatrix} \mathbf{I}_{C(A)} \\ \mathbf{I}_{C(B)} \\ \hline \mathbf{I}_{S(A)} \\ \mathbf{I}_{S(B)} \end{bmatrix}
$$

$$(3.139b)$$

In practical installations, various double-circuit cable earthing arrangements may be used. Some of these are: circuit A and circuit B are solidly bonded, circuit A and circuit B are cross-bonded, circuit A and circuit B are single-point bonded, circuit A is solidly bonded in a trefoil formation and circuit B is single-point bonded in a

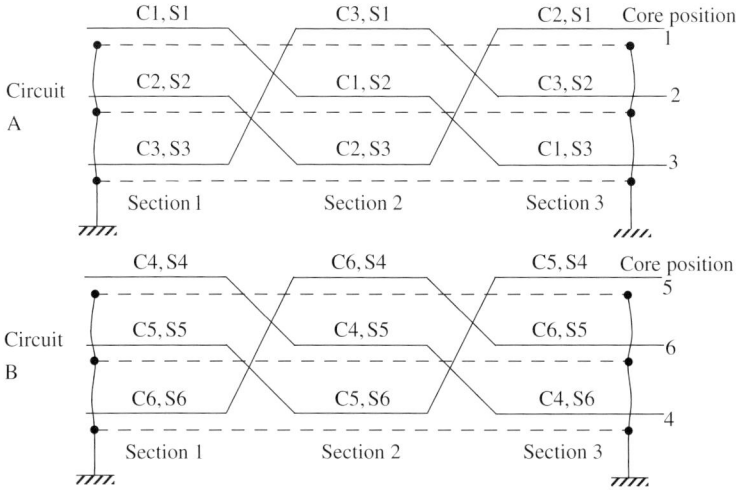

Figure 3.30 Double-circuit cross-bonded cables

flat formation, etc. The matrix of Equation (3.139) can be modified for any of these arrangements to calculate the reduced 6×6 phase impedance matrix for the two circuits that include their cores only. Figure 3.30 shows one arrangement where circuit A and circuit B are both cross-bonded with identical core transposition arrangements.

It is instructive for the reader to use the methodology presented in the single-circuit case to calculate the average 12×12 impedance matrix for the transposed double-circuit cable from the 12×12 matrix of each minor section. Using Equation (3.139b), this average 12×12 matrix can be written as follows:

$$
\begin{bmatrix} \mathbf{V}_{C(AB)6 \times 1} \\ \mathbf{V}_{S(AB)6 \times 1} \end{bmatrix} = \begin{bmatrix} \mathbf{Z}_{CC(AB)6 \times 6} & \mathbf{Z}_{CS(AB)6 \times 6} \\ \mathbf{Z}^t_{CS(AB)6 \times 6} & \mathbf{Z}_{SS(AB)6 \times 6} \end{bmatrix} \begin{bmatrix} \mathbf{I}_{C(AB)6 \times 1} \\ \mathbf{I}_{S(AB)6 \times 1} \end{bmatrix} \tag{3.140a}
$$

The calculation of the 6×6 phase impedance matrix of the cores only, with both circuits loaded, makes use of $\mathbf{V}_{S(A)} = \mathbf{V}_{S(B)} = 0$ or $\mathbf{V}_{S(AB)} = 0$ in Equation (3.140a) giving

$$
\mathbf{V}_{C(AB)6 \times 1} = \mathbf{Z}_{Phase6 \times 6} \mathbf{I}_{C(AB)6 \times 1} \tag{3.140b}
$$

where

$$
\mathbf{Z}_{Phase6 \times 6} = \mathbf{Z}_{CC(AB)6 \times 6} - \mathbf{Z}_{CS(AB)6 \times 6} \mathbf{Z}^{-1}_{SS(AB)6 \times 6} \mathbf{Z}^t_{CS(AB)6 \times 6} \tag{3.140c}
$$

Equation (3.140b) can be rewritten as follows:

$$
\begin{bmatrix} \mathbf{V}_{C(A)3 \times 1} \\ \mathbf{V}_{C(B)3 \times 1} \end{bmatrix} = \begin{bmatrix} \mathbf{Z}_{Phase(A)3 \times 3} & \mathbf{Z}_{Phase(AB)3 \times 3} \\ \mathbf{Z}^t_{Phase(AB)3 \times 3} & \mathbf{Z}_{Phase(B)3 \times 3} \end{bmatrix} \begin{bmatrix} \mathbf{I}_{C(A)3 \times 1} \\ \mathbf{I}_{C(B)3 \times 1} \end{bmatrix} \tag{3.141a}
$$

If one circuit, e.g. circuit A, is open-circuited, i.e. the cores carry no currents, $\mathbf{I}_{C(A)} = 0$ and the resultant 3×3 matrix for circuit B is directly obtained as $\mathbf{Z}_{\text{Phase(B)}3\times3}$. If circuit A is open and earthed, then the resultant 3×3 impedance matrix of circuit B is derived using $\mathbf{V}_{C(A)} = 0$ in Equation (3.141a) giving

$$\mathbf{Z}_{\text{Eq(B)}3\times3} = \mathbf{Z}_{\text{Phase(B)}3\times3} - \mathbf{Z}^t_{\text{Phase(AB)}3\times3}\mathbf{Z}^{-1}_{\text{Phase(A)}3\times3}\mathbf{Z}_{\text{Phase(AB)}3\times3} \quad (3.141b)$$

These are practical cases that can arise in normal power system design and operation for which the cable impedance parameters would be required.

3.3.7 Examples

Example 3.5   A 275 kV underground self-contained three-phase cable laid out in a symmetrical flat arrangement as shown in Figure 3.31.
The cable's geometrical and physical data are as follows:

Copper core: inner radius = oil duct radius = 6.8 mm,   outer radius = 21.9 mm, relative permeability = 1.0, ac resistance = 0.01665 Ω/km.
Core insulation: paper, relative permeability = 1.0, relative permittivity = 3.8.
Lead sheath: inner radius = 37.9 mm,   outer radius = 40.9 mm,   relative permeability = 1.0, ac resistance = 0.28865 Ω/km.
PVC over-sheath: thickness = 3mm,   relative permittivity = 3.5,   relative permeability = 1.0.

Earth resistivity = 20 Ωm. Nominal frequency $f = 50$ Hz.
Calculate the phase and sequence susceptance and impedance parameters for a solidly bonded cable and a cross-bonded cable whose cores are perfectly transposed in each major section.

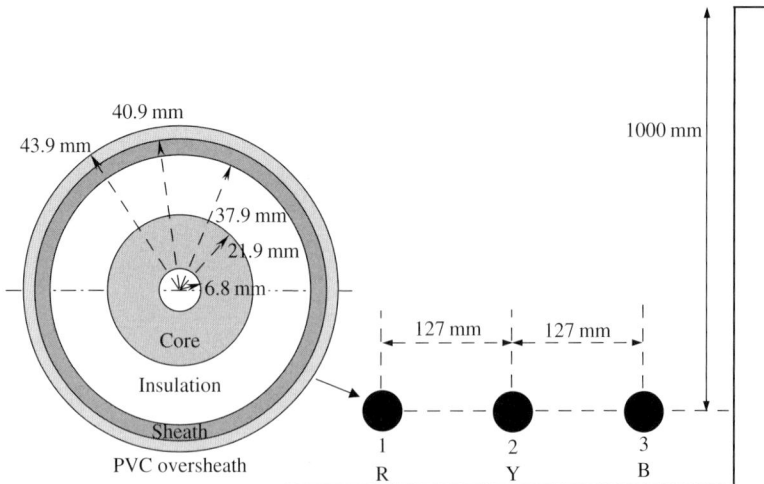

Figure 3.31   Self-contained single-core cable layout used in Example 3.5

Cable's susceptances
The core insulation susceptance is equal to

$$B_{CS} = \frac{0.34954933 \times 50 \times 3.8}{\log_e \left(\frac{37.9}{21.9}\right)} = 121.091 \, \mu S/km$$

and the PVC over-sheath susceptance is

$$B_{SE} = \frac{0.34954933 \times 50 \times 3.5}{\log_e \left(\frac{43.9}{40.9}\right)} = 864.191 \, \mu S/km$$

The full susceptance matrix of the cable is equal to

$$\mathbf{B} = \begin{array}{c} C1 \\ C2 \\ C3 \\ S1 \\ S2 \\ S3 \end{array} \left[\begin{array}{ccc|ccc} 121.091 & 0 & 0 & -121.091 & 0 & 0 \\ 0 & 121.091 & 0 & 0 & -121.091 & 0 \\ 0 & 0 & 121.091 & 0 & 0 & -121.091 \\ \hline -121.091 & 0 & 0 & 985.282 & 0 & 0 \\ 0 & -121.091 & 0 & 0 & 985.282 & 0 \\ 0 & 0 & -121.091 & 0 & 0 & 985.282 \end{array} \right]$$

The phase and sequence susceptance matrix for a solidly bonded or cross-bonded cable is given by

$$\mathbf{B}_{Phase} = \mathbf{B}^{PNZ} = \begin{bmatrix} 121.091 & 0 & 0 \\ 0 & 121.091 & 0 \\ 0 & 0 & 121.091 \end{bmatrix}$$

Cable's impedances
Depth of earth return conductor $= 658.87\sqrt{20/50} = 416.7$ m. For the tubular core conductor

$$f(r_o, r_i) = 1 - \frac{2 \times (6.8^2)}{21.9^2 - 6.8^2} + \frac{4 \times (6.8^4)}{(21.9^2 - 6.8^2)^2} \log_e \left(\frac{21.9}{6.8}\right) = 0.839863$$

thus

$$Z_{C1C1} = 0.01665 + \pi^2 10^{-4} \times 50 + j4\pi 10^{-4}$$
$$\times 50 \left[\frac{1}{4} \times 0.839863 + \log_e \left(\frac{416.7}{21.9 \times 10^{-3}}\right) \right]$$
$$= (0.066 + j0.632)\Omega/km = Z_{C2C2} = Z_{C3C3}$$

For the sheath,

$$f(r_0, r_i) = 1 - \frac{2 \times (37.9^2)}{40.9^2 - 37.9^2} + \frac{4 \times (37.9^4)}{(40.9^2 - 37.9^2)^2} \log_e\left(\frac{40.9}{37.9}\right) = 0.09774$$

$$Z_{S1S1} = 0.28865 + \pi^2 10^{-4} \times 50 + j4\pi 10^{-4}$$
$$\times 50\left[\frac{1}{4} \times 0.09774 + \log_e\left(\frac{416.7}{40.9 \times 10^{-3}}\right)\right]$$
$$= (0.338 + j0.5814)\Omega/\text{km} = Z_{S2S2} = Z_{S3S3}$$

$$Z_{C1S1} = \pi^2 10^{-4} \times 50 + j4\pi 10^{-4} \times 50 \log_e\left(\frac{416.7}{\frac{(40.9+37.9)}{2} \times 10^{-3}}\right)$$
$$= (0.0493 + j0.5822)\Omega/\text{km} = Z_{S1C1} = Z_{C2S2} = Z_{S2C2}$$
$$= Z_{C3S3} = Z_{S3C3}$$

$$Z_{C1C2} = \pi^2 10^{-4} \times 50 + j4\pi 10^{-4} \times 50 \log_e\left(\frac{416.7}{127 \times 10^{-3}}\right)$$
$$= (0.0493 + j0.5086)\Omega/\text{km}$$

$$Z_{C1C3} = \pi^2 10^{-4} \times 50 + j4\pi 10^{-4} \times 50 \log_e\left(\frac{416.7}{2 \times 127 \times 10^{-3}}\right)$$
$$= (0.0493 + j0.4651)\Omega/\text{km}$$

All mutual impedances between cable phase 1 core and sheath conductors, and cable phase 2 core and sheath conductors are equal, and similarly for those impedances between cable phases 2 and 3. The full impedance matrix is equal to

$$
\mathbf{z} = \begin{array}{c} \text{C1} \\ \text{C2} \\ \text{C3} \\ \text{S1} \\ \text{S2} \\ \text{S3} \end{array}
\left[\begin{array}{ccc|ccc}
0.066 + j0.632 & 0.0493 + j0.508 & 0.0493 + j0.465 & 0.0493 + j0.582 & 0.0493 + j0.508 & 0.0493 + j0.465 \\
0.0493 + j0.508 & 0.066 + j0.632 & 0.0493 + j0.508 & 0.0493 + j0.508 & 0.0493 + j0.582 & 0.0493 + j0.508 \\
0.0493 + j0.465 & 0.0493 + j0.508 & 0.066 + j0.632 & 0.0493 + j0.465 & 0.0493 + j0.508 & 0.0493 + j0.582 \\
0.0493 + j0.582 & 0.0493 + j0.508 & 0.0493 + j0.465 & 0.338 + j0.5814 & 0.0493 + j0.508 & 0.0493 + j0.465 \\
0.0493 + j0.508 & 0.0493 + j0.582 & 0.0493 + j0.508 & 0.0493 + j0.508 & 0.338 + j0.5814 & 0.0493 + j0.508 \\
0.0493 + j0.465 & 0.0493 + j0.508 & 0.0493 + j0.582 & 0.0493 + j0.465 & 0.0493 + j0.508 & 0.338 + j0.5814
\end{array}\right]
$$

For a solidly bonded cable, the phase impedance matrix is equal to

$$
\mathbf{Z}_{\text{Phase}} = \mathbf{Z}_{CC} - \mathbf{Z}_{CS}\mathbf{Z}_{SS}^{-1}\mathbf{Z}_{CS} = \begin{array}{c} \text{C1} \\ \text{C2} \\ \text{C3} \end{array}
\left[\begin{array}{ccc}
0.129 + j0.126 & 0.089 - j0.003 & 0.072 - j0.025 \\
0.089 - j0.003 & 0.12 + j0.104 & 0.089 - j0.003 \\
0.072 - j0.025 & 0.089 - j0.003 & 0.129 + j0.126
\end{array}\right]
$$

and the corresponding sequence impedance matrix is equal to

$$
\mathbf{Z}^{PNZ} = \mathbf{H}^{-1}\mathbf{Z}_{Phase}\mathbf{H} = \begin{array}{c} P \\ N \\ Z \end{array}
\begin{bmatrix}
0.043 + j0.129 & -0.012 + j0.023 & -0.001 + j0.002 \\
0.026 - j0.002 & 0.043 + j0.129 & -0.001 - j0.002 \\
-0.001 - j0.002 & -0.001 + j0.002 & 0.292 + j0.1
\end{bmatrix}
$$

$\mathbf{Z}^P = \mathbf{Z}^N = (0.043 + j0.129)\Omega/\text{km}$ and $\mathbf{Z}^Z = (0.292 + j0.1)\Omega/\text{km}$.

For a cross-bonded cable, the full impedance matrix for a major cable section that includes three perfectly transposed minor sections is equal to

$$
\mathbf{Z} = \begin{array}{c} C1 \\ C2 \\ C3 \\ S1 \\ S2 \\ S3 \end{array}
\left[\begin{array}{ccc|ccc}
0.066 + j0.632 & 0.0493 + j0.494 & 0.0493 + j0.494 & 0.0493 + j0.518 & 0.0493 + j0.533 & 0.0493 + j0.518 \\
0.0493 + j0.494 & 0.066 + j0.632 & 0.0493 + j0.494 & 0.0493 + j0.518 & 0.0493 + j0.533 & 0.0493 + j0.518 \\
0.0493 + j0.494 & 0.0493 + j0.494 & 0.066 + j0.632 & 0.0493 + j0.518 & 0.0493 + j0.533 & 0.0493 + j0.518 \\
0.0493 + j0.518 & 0.0493 + j0.518 & 0.0493 + j0.518 & 0.338 + j0.5814 & 0.0493 + j0.508 & 0.0493 + j0.465 \\
0.0493 + j0.533 & 0.0493 + j0.533 & 0.0493 + j0.533 & 0.0493 + j0.508 & 0.338 + j0.5814 & 0.0493 + j0.508 \\
0.0493 + j0.518 & 0.0493 + j0.518 & 0.0493 + j0.518 & 0.0493 + j0.465 & 0.0493 + j0.508 & 0.338 + j0.5814
\end{array}\right]
$$

The phase impedance matrix is equal to

$$
\mathbf{Z}_{Phase} = \mathbf{Z}_{CC} - \mathbf{Z}_{CS}\mathbf{Z}_{SS}^{-1}\mathbf{Z}_{SC} = \begin{array}{c} C1 \\ C2 \\ C3 \end{array}
\begin{bmatrix}
0.108 + j0.125 & 0.092 - j0.013 & 0.092 - j0.013 \\
0.092 - j0.013 & 0.108 + j0.125 & 0.092 - j0.013 \\
0.092 - j0.013 & 0.092 - j0.013 & 0.108 + j0.125
\end{bmatrix}
$$

and the corresponding sequence impedance matrix is equal to

$$
\mathbf{Z}^{PNZ} = \mathbf{H}^{-1}\mathbf{Z}_{Phase}\mathbf{H} = \begin{array}{c} P \\ N \\ Z \end{array}
\begin{bmatrix}
0.017 + j0.138 & 0 & 0 \\
0 & 0.017 + j0.138 & 0 \\
0 & 0 & 0.292 + j0.100
\end{bmatrix}
$$

$\mathbf{Z}^P = \mathbf{Z}^N = (0.017 + j0.138)\Omega/\text{km}$ and $\mathbf{Z}^Z = (0.292 + j0.1)\Omega/\text{km}$.

3.4 Sequence π models of single-circuit and double-circuit overhead lines and cables

3.4.1 Background

In Section 3.3, we presented calculations of the PPS/NPS and ZPS impedances and susceptances of overhead lines and cables per km of circuit length. The electrical parameters of lines and cables are distributed over their length. The derivation of T and π equivalent circuits using lumped parameters for the entire length of a line or a cable is covered in most basic power system textbooks and will not be repeated here. The π equivalent circuit is most extensively used in practical applications. Figure 3.32(a) illustrates the distributed nature of the parameters of a line or cable.

(a)

Self series impedance $= Z' = R + jX\ \Omega/km$
Self shunt admittance $= Y' = jB$ S/km, $\quad G \approx 0$ at 50/60 Hz
$\ell =$ line length (km)

(b)

$$Z = Z'\,\ell\left(\frac{\sinh\sqrt{Z'Y'}\ell}{\sqrt{Z'Y'}\ell}\right)$$

$$Y = \frac{Y'\ell}{2}\left(\frac{\tanh\dfrac{\sqrt{Z'Y'}\ell}{2}}{\dfrac{\sqrt{Z'Y'}\ell}{2}}\right)$$

(c)

$$Z = Z'\ell$$

$$Y = \frac{Y'\ell}{2}$$

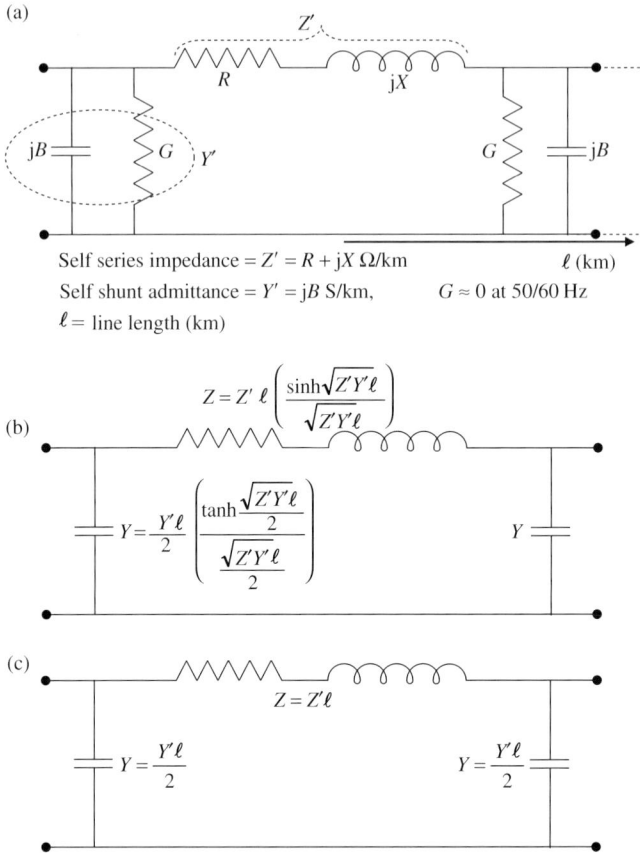

Figure 3.32 Representation of lines and cables: (a) distributed parameter circuit; (b) accurate π lumped parameter equivalent circuit and (c) nominal π lumped parameter equivalent circuit

Figure 3.32(c) is the nominal π lumped parameter equivalent circuit of the line or cable where the total impedance and susceptance are calculate by multiplying the per km parameters by the line/cable length. In Figure 3.32b, the terms in brackets represent correction terms that allow for the distributed nature of the circuit parameters over the entire circuit length. These terms can be represented as infinite series as follows:

$$\frac{\sinh\sqrt{Z'Y'}\ell}{\sqrt{Z'Y'}\ell} = 1 + \frac{Z'\ell Y'\ell}{6} + \frac{(Z'\ell Y'\ell)^2}{120} + \frac{(Z'\ell Y'\ell)^3}{5040} + \cdots \qquad (3.142a)$$

$$\frac{\tanh\dfrac{\sqrt{Z'Y'}\ell}{2}}{\dfrac{\sqrt{Z'Y'}\ell}{2}} = 1 - \frac{Z'\ell Y'\ell}{12} + \frac{(Z'\ell Y'\ell)^2}{120} - \frac{17(Z'\ell Y'\ell)^3}{20\,160} + \cdots \qquad (3.142b)$$

The hyperbolic sine and tangent of a complex argument are equal to the complex argument itself when this has small values. For a lossless line or cable, the complex argument is equal to $(2\pi/\lambda)\ell$ where λ is the wavelength in km and is given by $\lambda \cong 3 \times 10^5/(f \times \sqrt{\varepsilon_r})$ where ε_r is the relative permittivity of the dielectric. For overhead lines ($\varepsilon_r = 1$ for air), the wavelength is approximately 6000 km at 50 Hz and 5000 km at 60 Hz. The definition of an 'electrically short' line or cable for which a nominal π circuit can be used depends on the acceptable error magnitude in the total series impedance and total shunt susceptance, and this determines the physical length in km for such a short line or cable. For an error of less than 1% in these parameters, the physical length should be less than about 3% of the wavelength. For overhead lines, this corresponds to 180 km for 50 Hz and 150 km for 60 Hz systems. For underground cables, with ε_r being typically between 2.4 and 4.2, the corresponding cable lengths using $\varepsilon_r = 4.2$ are 88 km for 50 Hz and 73 km for 60 Hz systems. In practice, these cable lengths are not reached above voltages of around 220 kV due to the adverse impact of the cable charging current on the cable thermal rating. For such lines or cables, the first term in the infinite series expansion of Equation (3.142), which is equal to unity, is taken and all other terms are neglected. Therefore, the exact equivalent π circuit of Figure 3.32(b) reduces to that shown in Figure 3.32(c).

3.4.2 Sequence π models of single-circuit overhead lines and cables

Overhead lines

The PPS, NPS and ZPS impedances and susceptances of perfectly transposed single-circuit overhead lines were derived in Section 3.2.4. Recalling the sequence impedance and susceptance matrices of Equations (3.50a) and (3.54a), denoting 1 and 2 as the sending and receiving ends of the line or cable circuit and connecting ½ the susceptance at each end, as shown in Figure 3.33, we can write

$$
\begin{bmatrix} V_1^P - V_2^P \\ V_1^N - V_2^N \\ V_1^Z - V_2^Z \end{bmatrix} = \begin{bmatrix} Z^P & 0 & 0 \\ 0 & Z^N & 0 \\ 0 & 0 & Z^Z \end{bmatrix} \begin{bmatrix} I_S^P \\ I_S^N \\ I_S^Z \end{bmatrix} \quad \begin{bmatrix} I_1^P \\ I_1^N \\ I_1^Z \end{bmatrix} = \begin{bmatrix} j\dfrac{B^P}{2} & 0 & 0 \\ 0 & j\dfrac{B^N}{2} & 0 \\ 0 & 0 & j\dfrac{B^Z}{2} \end{bmatrix} \begin{bmatrix} V_1^P \\ V_1^N \\ V_1^Z \end{bmatrix}
$$

and

$$
\begin{bmatrix} I_2^P \\ I_2^N \\ I_2^Z \end{bmatrix} = \begin{bmatrix} j\dfrac{B^P}{2} & 0 & 0 \\ 0 & j\dfrac{B^N}{2} & 0 \\ 0 & 0 & j\dfrac{B^Z}{2} \end{bmatrix} \begin{bmatrix} V_2^P \\ V_2^N \\ V_2^Z \end{bmatrix} \tag{3.143a}
$$

For the PPS, NPS and ZPS circuits, the total injected current at ends 1 and 2 are given by

$$I_{t1} = I_S + I_1 \quad \text{and} \quad I_{t2} = -I_S + I_2 \qquad (3.143b)$$

Therefore, using Equation (3.143), we obtain three separate PPS, NPS and ZPS sequence π models for the line or cable as follows:

$$\begin{bmatrix} I_{t1}^x \\ I_{t2}^x \end{bmatrix} = \begin{bmatrix} \dfrac{1}{Z^x} + j\dfrac{B^x}{2} & -\dfrac{1}{Z^x} \\ -\dfrac{1}{Z^x} & \dfrac{1}{Z^x} + j\dfrac{B^x}{2} \end{bmatrix} \begin{bmatrix} V_1^x \\ V_2^x \end{bmatrix} \quad x = P, N \text{ or } Z \qquad (3.144)$$

As the PPS and NPS impedances and susceptances are equal, only two sequence π models are needed as shown in Figure 3.32. For convenience, we will only refer to the PPS and ZPS sequence parameters and models in subsequent sections.

Cables

The sequence impedance matrices of single-circuit cables of various sheath earthing arrangements were derived in Section 3.3.4. The sequence impedance matrix of Equation (3.123), combined with the circuit sequence currents and voltages, gives

$$\begin{bmatrix} V_1^P - V_2^P \\ V_1^N - V_2^N \\ V_1^Z - V_2^Z \end{bmatrix} = \begin{bmatrix} Z^{PP} & Z^{PN} & Z^{PZ} \\ Z^{NP} & Z^{NN} & Z^{NZ} \\ Z^{ZP} & Z^{ZN} & Z^{ZZ} \end{bmatrix} \begin{bmatrix} I^P \\ I^N \\ I^Z \end{bmatrix} \qquad (3.145)$$

The mutual intersequence terms are normally much smaller than the self-terms for solidly bonded cables and practically zero for cross-bonded cables. For the former, it is usual practice to ignore the mutual terms and use the self-terms only for the PPS, NPS and ZPS impedances of the cable. In the case of submarine cables where the three phases are laid a significant distance apart, the sequence impedance matrix is diagonal with no mutual intersequence terms. This is also the case for three-core cables, and single-core cables in touching trefoil or equilateral arrangements. The sequence susceptance matrix of all shielded and belted cables is diagonal and hence the PPS, NPS and ZPS susceptances are equal to the phase terms unaffected by any intersequence terms. Therefore, as in the case of single-circuit overhead lines, PPS and ZPS sequence π models can be derived as shown in Figure 3.33.

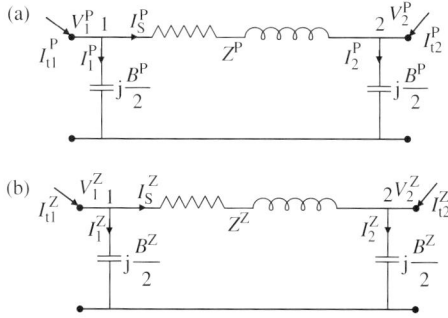

Figure 3.33   (a) PPS/NPS and (b) ZPS π models for single-circuit overhead line or cable

3.4.3 Sequence π models of double-circuit overhead lines

The PPS, NPS and ZPS impedances and susceptances of double-circuit overhead lines were derived in Section 3.2.6. We recall the sequence impedance and susceptance matrices of Equations (3.76a) and (3.78a) and illustrate their physical meaning as shown in Figure 3.34(a). Equation (3.76a) suggests that each circuit has a self PPS, NPS and ZPS impedance. In addition, PPS, NPS and ZPS inter-circuit mutual coupling exists between the PPS, NPS and ZPS impedances of each circuit. Importantly, we note that the three PPS, NPS and ZPS circuits are separate and independent. Equation (3.78a) contains similar information for susceptances. Allocating ½ the susceptances to each circuit end, Equations (3.76a) and (3.78b) of circuit 1 and circuit 2 are represented in Figure 3.34 as a π model.

Using Figure 3.34 and Equations (3.76a) and (3.78a), we can write

$$
\begin{bmatrix} V_1'^P - V_1''^P \\ V_1'^N - V_1''^N \\ V_1'^Z - V_1''^Z \\ \hline V_2'^P - V_2''^P \\ V_2'^N - V_2''^N \\ V_2'^Z - V_2''^Z \end{bmatrix} = \begin{bmatrix} Z_1^P & 0 & 0 & Z_{12}^P & 0 & 0 \\ 0 & Z_1^P & 0 & 0 & Z_{12}^P & 0 \\ 0 & 0 & Z_1^Z & 0 & 0 & Z_{12}^Z \\ \hline Z_{12}^P & 0 & 0 & Z_2^P & 0 & 0 \\ 0 & Z_{12}^P & 0 & 0 & Z_2^P & 0 \\ 0 & 0 & Z_{12}^Z & 0 & 0 & Z_2^Z \end{bmatrix} \begin{bmatrix} I_1^P \\ I_1^N \\ I_1^Z \\ \hline I_2^P \\ I_2^N \\ I_2^Z \end{bmatrix} = \begin{bmatrix} \mathbf{Z}_1^{PNZ} & \mathbf{Z}_{12}^{PNZ} \\ \mathbf{Z}_{12}^{PNZ} & \mathbf{Z}_2^{PNZ} \end{bmatrix} \begin{bmatrix} \mathbf{I}_1^{PNZ} \\ \mathbf{I}_2^{PNZ} \end{bmatrix}
$$

$$(3.146a)$$

$$
\begin{bmatrix} I_{1,SH}'^P \\ I_{1,SH}'^N \\ I_{1,SH}'^Z \\ \hline I_{2,SH}'^P \\ I_{2,SH}'^N \\ I_{2,SH}'^Z \end{bmatrix} = j\frac{1}{2} \begin{bmatrix} B_1^P & 0 & 0 & B_{12}^P & 0 & 0 \\ 0 & B_1^P & 0 & 0 & B_{12}^P & 0 \\ 0 & 0 & B_1^Z & 0 & 0 & B_{12}^Z \\ \hline B_{12}^P & 0 & 0 & B_2^P & 0 & 0 \\ 0 & B_{12}^P & 0 & 0 & B_2^P & 0 \\ 0 & 0 & B_{12}^Z & 0 & 0 & B_2^Z \end{bmatrix} \begin{bmatrix} V_1'^P \\ V_1'^N \\ V_1'^Z \\ \hline V_2'^P \\ V_2'^N \\ V_2'^Z \end{bmatrix} = j\frac{1}{2} \begin{bmatrix} \mathbf{B}_1^{PNZ} & \mathbf{B}_{12}^{PNZ} \\ \mathbf{B}_{12}^{PNZ} & \mathbf{B}_2^{PNZ} \end{bmatrix} \begin{bmatrix} \mathbf{V}_1'^{PNZ} \\ \mathbf{V}_2'^{PNZ} \end{bmatrix}
$$

$$(3.146b)$$

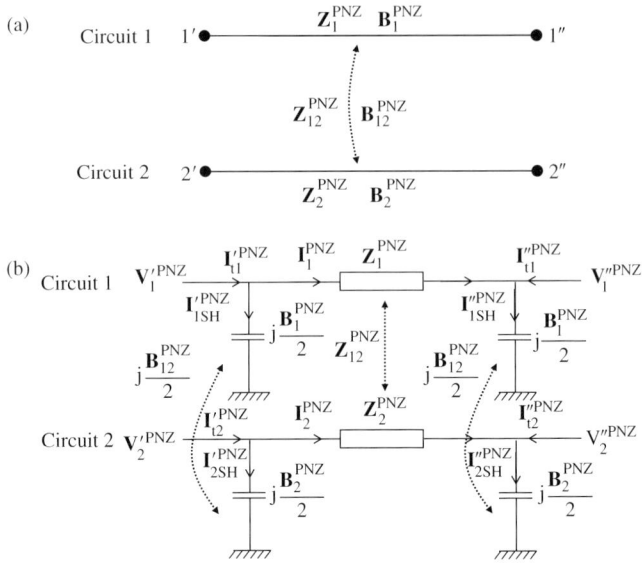

Figure 3.34 Double-circuit overhead line model in the sequence reference frame using sequence matrices: (a) schematic and (b) sequence π model

$$
\begin{bmatrix} I''^{P}_{1,SH} \\ I''^{N}_{1,SH} \\ I''^{Z}_{1,SH} \\ I''^{P}_{2,SH} \\ I''^{N}_{2,SH} \\ I''^{Z}_{2,SH} \end{bmatrix} = j\frac{1}{2} \left[\begin{array}{ccc|ccc} B^{P}_{1} & 0 & 0 & B^{P}_{12} & 0 & 0 \\ 0 & B^{P}_{1} & 0 & 0 & B^{P}_{12} & 0 \\ 0 & 0 & B^{Z}_{1} & 0 & 0 & B^{Z}_{12} \\ \hline B^{P}_{12} & 0 & 0 & B^{P}_{2} & 0 & 0 \\ 0 & B^{P}_{12} & 0 & 0 & B^{P}_{2} & 0 \\ 0 & 0 & B^{Z}_{12} & 0 & 0 & B^{Z}_{2} \end{array} \right] \begin{bmatrix} V''^{P}_{1} \\ V''^{N}_{1} \\ V''^{Z}_{1} \\ V''^{P}_{2} \\ V''^{N}_{2} \\ V''^{Z}_{2} \end{bmatrix} = j\frac{1}{2} \begin{bmatrix} \mathbf{B}^{PNZ}_{1} & \mathbf{B}^{PNZ}_{12} \\ \mathbf{B}^{PNZ}_{12} & \mathbf{B}^{PNZ}_{2} \end{bmatrix} \begin{bmatrix} \mathbf{V}''^{PNZ}_{1} \\ \mathbf{V}''^{PNZ}_{2} \end{bmatrix}
$$

$$(3.146c)$$

The equivalent admittance matrix can be derived for each sequence circuit independently because the PPS, NPS and ZPS circuits are separate. The form of the admittance matrix is the same for the PPS, NPS and ZPS circuits so the derivation of one would suffice. Collecting the PPS terms from Equation (3.146a), we have

$$
\begin{bmatrix} V'^{P}_{1} - V''^{P}_{1} \\ V'^{P}_{2} - V''^{P}_{2} \end{bmatrix} = \begin{bmatrix} Z^{P}_{1} & Z^{P}_{12} \\ Z^{P}_{12} & Z^{P}_{2} \end{bmatrix} \begin{bmatrix} I^{P}_{1} \\ I^{P}_{2} \end{bmatrix}
$$

or

$$
\begin{bmatrix} I^{P}_{1} \\ I^{P}_{2} \end{bmatrix} = \frac{1}{D} \begin{bmatrix} Z^{P}_{2} & -Z^{P}_{12} \\ -Z^{P}_{12} & Z^{P}_{1} \end{bmatrix} \begin{bmatrix} V'^{P}_{1} - V''^{P}_{1} \\ V'^{P}_{2} - V''^{P}_{2} \end{bmatrix}
$$

$$(3.147a)$$

where $D = Z_1^P Z_2^P - (Z_{12}^P)^2$. Similarly, collecting the shunt PPS terms from Equations (3.146b) and (3.146c), we have

$$
\begin{bmatrix} I'^P_{1,\text{SH}} \\ I'^P_{2,\text{SH}} \end{bmatrix} = j\frac{1}{2} \begin{bmatrix} B_1^P & B_{12}^P \\ B_{12}^P & B_2^P \end{bmatrix} \begin{bmatrix} V'^P_1 \\ V'^P_2 \end{bmatrix} \quad \text{and} \quad \begin{bmatrix} I''^P_{1,\text{SH}} \\ I''^P_{2,\text{SH}} \end{bmatrix} = j\frac{1}{2} \begin{bmatrix} B_1^P & B_{12}^P \\ B_{12}^P & B_2^P \end{bmatrix} \begin{bmatrix} V''^P_1 \\ V''^P_2 \end{bmatrix}
$$

$$(3.147b)$$

The total injected currents into circuit 1 and circuit 2 at both ends are given by

$$
\begin{bmatrix} I'^P_{1,t} \\ I''^P_{1,t} \\ I'^P_{2,t} \\ I''^P_{2,t} \end{bmatrix} = \begin{bmatrix} I'^P_{1,\text{SH}} \\ I''^P_{1,\text{SH}} \\ I'^P_{2,\text{SH}} \\ I''^P_{2,\text{SH}} \end{bmatrix} + \begin{bmatrix} I_1^P \\ -I_1^P \\ I_2^P \\ -I_2^P \end{bmatrix}
$$

$$(3.148a)$$

Substituting Equations (3.147a) and (3.147b) into Equation (3.148a), we obtain

$$
\begin{bmatrix} I'^P_{1,t} \\ I''^P_{1,t} \\ I'^P_{2,t} \\ I''^P_{2,t} \end{bmatrix} = \begin{bmatrix} \left(j\frac{B_1^P}{2} + \frac{Z_2^P}{D} \right) & \frac{-Z_2^P}{D} & \left(j\frac{B_{12}^P}{2} - \frac{Z_{12}^P}{D} \right) & \frac{Z_{12}^P}{D} \\[2mm] \frac{-Z_2^P}{D} & \left(j\frac{B_1^P}{2} + \frac{Z_2^P}{D} \right) & \frac{Z_{12}^P}{D} & \left(j\frac{B_{12}^P}{2} - \frac{Z_{12}^P}{D} \right) \\[2mm] \left(j\frac{B_{12}^P}{2} - \frac{Z_{12}^P}{D} \right) & \frac{Z_{12}^P}{D} & \left(j\frac{B_2^P}{2} + \frac{Z_1^P}{D} \right) & \frac{-Z_1^P}{D} \\[2mm] \frac{Z_{12}^P}{D} & \left(j\frac{B_{12}^P}{2} - \frac{Z_{12}^P}{D} \right) & \frac{-Z_1^P}{D} & \left(j\frac{B_2^P}{2} + \frac{Z_1^P}{D} \right) \end{bmatrix} \begin{bmatrix} V'^P_1 \\ V''^P_1 \\ V'^P_2 \\ V''^P_2 \end{bmatrix}
$$

$$(3.148b)$$

The NPS and ZPS admittance matrices of the double-circuit line with NPS and ZPS impedance and susceptance coupling between the two circuits are identical to that given in Equation (3.148b) except that the ZPS parameters are different.

The sequence model for a double-circuit overhead line derived above represents the case where the line is assumed to have had six transpositions. However, as we showed in Equation (3.89), when the line is assumed to be ideally transposed with nine transpositions, there would only be ZPS coupling between the two circuits. In this case, the ZPS admittance matrix would be identical to that of Equation (3.148b) but the PPS/NPS admittance matrices would have zero inter-circuit mutual impedances and susceptances as follows:

$$
\begin{bmatrix} I'^{P}_{1,t} \\ I''^{P}_{1,t} \\ I'^{P}_{2,t} \\ I''^{P}_{2,t} \end{bmatrix} =
\begin{bmatrix}
\left(j\dfrac{B^{P}_1}{2} + \dfrac{1}{Z^{P}_1} \right) & \dfrac{-1}{Z^{P}_1} & 0 & 0 \\[2ex]
\dfrac{-1}{Z^{P}_1} & \left(j\dfrac{B^{P}_1}{2} + \dfrac{1}{Z^{P}_1} \right) & 0 & 0 \\[2ex]
0 & 0 & \left(j\dfrac{B^{P}_2}{2} + \dfrac{1}{Z^{P}_2} \right) & \dfrac{-1}{Z^{P}_2} \\[2ex]
0 & 0 & \dfrac{-1}{Z^{P}_2} & \left(j\dfrac{B^{P}_2}{2} + \dfrac{1}{Z^{P}_2} \right)
\end{bmatrix}
\begin{bmatrix} V'^{P}_1 \\ V''^{P}_1 \\ V'^{P}_2 \\ V''^{P}_2 \end{bmatrix}
$$

$$(3.149)$$

3.4.4 Sequence π models of double-circuit cables

As presented in Section 3.3.3, the series sequence impedance matrix of each cable circuit in a double-circuit cable will effectively be diagonal. There will also be inter-circuit mutual inductive coupling between the two circuits but the inter-circuit phase impedance mutual matrix is in general not balanced. For cross-bonded single-core double circuit cables, there would be very small intersequence mutual coupling between the two circuits in the PPS and NPS but a non-negligible coupling in the ZPS depending on spacing between the two circuits. The same applies for a three-core double-circuit cable. As presented in Section 3.3.4, the shunt sequence susceptance matrix of a single-circuit cable is diagonal, i.e. there is no intersequence mutual coupling. The presence of a second cable circuit has no effect other than introducing a second diagonal shunt sequence susceptance matrix for this circuit. Therefore, the PPS and ZPS sequence π models shown in Figure 3.33 for a single-circuit cable may also be used for double-circuit cables. Where ZPS impedance coupling between the two cable circuits is significant and needs to be represented, then the approach described in Section 3.4.3 for double-circuit overhead lines can be used.

3.5 Sequence π models of three-circuit overhead lines

Using a similar approach to that of Section 3.4.3, Figure 3.35 illustrates three mutually coupled circuits and their self and mutual sequence impedance and susceptance matrices.

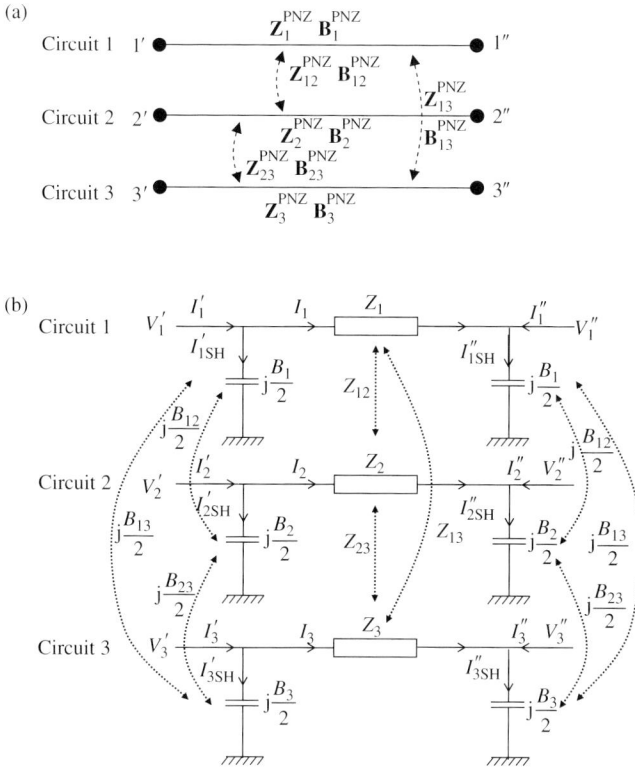

Figure 3.35 Three-circuit overhead line model in the sequence reference frame: (a) schematic and (b) ZPS π model

The sequence π model represents either the PPS, NPS or ZPS equivalent circuits. Where ideal transposition is used with only ZPS inter-circuit mutual coupling, the PPS and NPS π model of each circuit is completely independent from the other circuits. We will now derive the admittance matrix for the general case where inter-circuit mutual coupling exists and we will use the ZPS π model parameters but drop the superscript Z to simplify notation.

From Figure 3.35(b), we can write

$$\begin{bmatrix} V_1' - V_1'' \\ V_2' - V_2'' \\ V_3' - V_3'' \end{bmatrix} = \begin{bmatrix} Z_1 & Z_{12} & Z_{13} \\ Z_{12} & Z_2 & Z_{23} \\ Z_{13} & Z_{23} & Z_3 \end{bmatrix} \begin{bmatrix} I_1 \\ I_2 \\ I_3 \end{bmatrix} \quad \text{or} \quad \mathbf{I}_{1,2,3} = \mathbf{Z}_{1,2,3}^{-1} \begin{bmatrix} V_1' - V_1'' \\ V_2' - V_2'' \\ V_3' - V_3'' \end{bmatrix}$$

or

$$\mathbf{I}_{1,2,3} = \begin{bmatrix} \mathbf{Z}_{1,2,3}^{-1} & -\mathbf{Z}_{1,2,3}^{-1} \end{bmatrix} \begin{bmatrix} \mathbf{V}_{1,2,3}' \\ \mathbf{V}_{1,2,3}'' \end{bmatrix} \qquad (3.150a)$$

where

$$\mathbf{Z}_{1,2,3} = \begin{bmatrix} Z_1 & Z_{12} & Z_{13} \\ Z_{12} & Z_2 & Z_{23} \\ Z_{13} & Z_{23} & Z_3 \end{bmatrix} \qquad (3.150b)$$

The total injected currents into each circuit at both ends are given by

$$\begin{bmatrix} I'_1 \\ I'_2 \\ I'_3 \\ I''_1 \\ I''_2 \\ I''_3 \end{bmatrix} = \begin{bmatrix} \dfrac{jB_1}{2} & \dfrac{jB_{12}}{2} & \dfrac{jB_{13}}{2} & 0 & 0 & 0 \\ \dfrac{jB_{12}}{2} & \dfrac{jB_2}{2} & \dfrac{jB_{23}}{2} & 0 & 0 & 0 \\ \dfrac{jB_{13}}{2} & \dfrac{jB_{23}}{2} & \dfrac{jB_3}{2} & 0 & 0 & 0 \\ 0 & 0 & 0 & \dfrac{jB_1}{2} & \dfrac{jB_{12}}{2} & \dfrac{jB_{13}}{2} \\ 0 & 0 & 0 & \dfrac{jB_{12}}{2} & \dfrac{jB_2}{2} & \dfrac{jB_{23}}{2} \\ 0 & 0 & 0 & \dfrac{jB_{13}}{2} & \dfrac{jB_{23}}{2} & \dfrac{jB_3}{2} \end{bmatrix} \begin{bmatrix} V'_1 \\ V'_2 \\ V'_3 \\ V''_1 \\ V''_2 \\ V''_3 \end{bmatrix} + \begin{bmatrix} I_1 \\ I_2 \\ I_3 \\ -I_1 \\ -I_2 \\ -I_3 \end{bmatrix} \qquad (3.151)$$

or written in concise matrix form using Equation (3.150a), we have

$$\begin{bmatrix} \mathbf{I}'_{1,2,3} \\ \mathbf{I}''_{1,2,3} \end{bmatrix} = \begin{bmatrix} \dfrac{j\mathbf{B}_{1,2,3}}{2} & 0 \\ 0 & \dfrac{j\mathbf{B}_{1,2,3}}{2} \end{bmatrix} \begin{bmatrix} \mathbf{V}'_{1,2,3} \\ \mathbf{V}''_{1,2,3} \end{bmatrix} + \begin{bmatrix} \mathbf{Z}^{-1}_{1,2,3} & -\mathbf{Z}^{-1}_{1,2,3} \\ -\mathbf{Z}^{-1}_{1,2,3} & \mathbf{Z}^{-1}_{1,2,3} \end{bmatrix} \begin{bmatrix} \mathbf{V}'_{1,2,3} \\ \mathbf{V}''_{1,2,3} \end{bmatrix}$$

or

$$\begin{bmatrix} \mathbf{I}'_{1,2,3} \\ \mathbf{I}''_{1,2,3} \end{bmatrix} = \begin{bmatrix} \dfrac{j\mathbf{B}_{1,2,3}}{2} + \mathbf{Z}^{-1}_{1,2,3} & -\mathbf{Z}^{-1}_{1,2,3} \\ -\mathbf{Z}^{-1}_{1,2,3} & \dfrac{j\mathbf{B}_{1,2,3}}{2} + \mathbf{Z}^{-1}_{1,2,3} \end{bmatrix} \begin{bmatrix} \mathbf{V}'_{1,2,3} \\ \mathbf{V}''_{1,2,3} \end{bmatrix} \qquad (3.152a)$$

where

$$\mathbf{B}_{1,2,3} = \begin{bmatrix} B_1 & B_{12} & B_{13} \\ B_{12} & B_2 & B_{23} \\ B_{13} & B_{23} & B_3 \end{bmatrix} \qquad (3.152b)$$

3.6 Three-phase modelling of overhead lines and cables (phase frame of reference)

3.6.1 Background

In advanced studies such as multiphase loadflow and short-circuit fault analysis where the natural unbalance of untransposed lines and cables is to be taken into account, the modelling and analysis may be carried out in the phase frame of reference. Such analysis allows the calculation of the current distribution on overhead line earth wires and cable sheaths under unbalanced power flow and short-circuit faults.

3.6.2 Single-circuit overhead lines and cables

For single-circuit overhead lines, the series phase impedance matrices with earth wires present and eliminated were given in Equations (3.29) and (3.32c), respectively. The corresponding shunt phase susceptance matrices are obtained from the inverse of Equation (3.36) and from Equation (3.39). Considering the case of an overhead line with one earth wire, the line can be represented by a three-phase equivalent nominal π model as shown in Figure 3.36(a).

In Figure 3.36, the line is represented by its series phase impedance matrix and ½ its shunt phase susceptance matrix connected at each end. We denote as 1 and 2 the two ends of the line, and we use 'ph' for 'phase' quantities, which also include the earth wires if these have not been eliminated. From Figure 3.36(b), we have

$$\mathbf{V}_1^{ph} - \mathbf{V}_2^{ph} = \mathbf{Z}^{ph}\mathbf{I}_S^{ph} \quad \mathbf{I}_{SH(1)}^{ph} = j\frac{\mathbf{B}^{ph}}{2}\mathbf{V}_1^{ph} \quad \mathbf{I}_{SH(2)}^{ph} = j\frac{\mathbf{B}^{ph}}{2}\mathbf{V}_2^{ph} \qquad (3.153a)$$

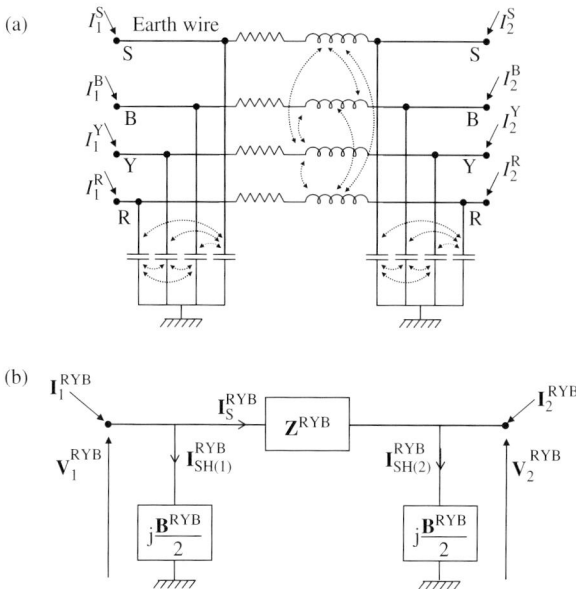

Figure 3.36 Three-phase model of a single-circuit line including earth wire(s): (a) π equivalent circuit and (b) π model (matrices)

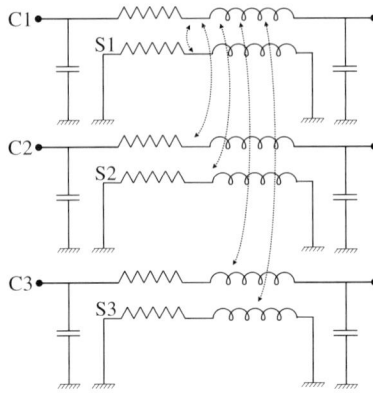

Figure 3.37 Three-phase π equivalent of a single-circuit solidly bonded underground cable

Therefore, the nodal phase admittance matrix is given by

$$
\begin{bmatrix} \mathbf{I}_1^{ph} \\ \mathbf{I}_2^{ph} \end{bmatrix} = \begin{bmatrix} j\dfrac{\mathbf{B}^{ph}}{2} + (\mathbf{Z}^{ph})^{-1} & -(\mathbf{Z}^{ph})^{-1} \\ -(\mathbf{Z}^{ph})^{-1} & j\dfrac{\mathbf{B}^{ph}}{2} + (\mathbf{Z}^{ph})^{-1} \end{bmatrix} \begin{bmatrix} \mathbf{V}_1^{ph} \\ \mathbf{V}_2^{ph} \end{bmatrix} \qquad (3.153b)
$$

If the earth wire has already been eliminated, then it is simply removed from Figure 3.36(a) and the dimension of the series phase impedance and shunt phase susceptance matrices is reduced accordingly. The series phase impedance and shunt phase admittance matrices for cables, with their sheaths present or eliminated, can be represented using a similar approach taking into account the particular sheath earthing arrangement. Figure 3.37 shows a three-phase model for a single-circuit underground cable with a solidly bonded sheath.

3.6.3 Double-circuit overhead lines and cables

For double-circuit overhead lines, the series phase impedance matrices with earth wires present and eliminated are given in Equations (3.57a) and (3.58b), respectively. The corresponding shunt phase susceptance matrices are obtained from the inverse of Equation (3.60a) and (3.62c), respectively. Using the π matrix model shown in Figure 3.36(b) together with the inter-circuit mutual impedance and susceptance matrices, we obtain the π model representation of a double-circuit overhead line in the phase frame of reference as shown in Figure 3.38.

We denote 1 and $1'$, and 2 and $2'$ as the two ends of circuit 1 and circuit 2, respectively. We assume that the inter-circuit mutual coupling is bilateral, i.e. $\mathbf{B}_{21} = \mathbf{B}_{12}^t$. From Figure 3.38, the series voltage drop across the two circuits is

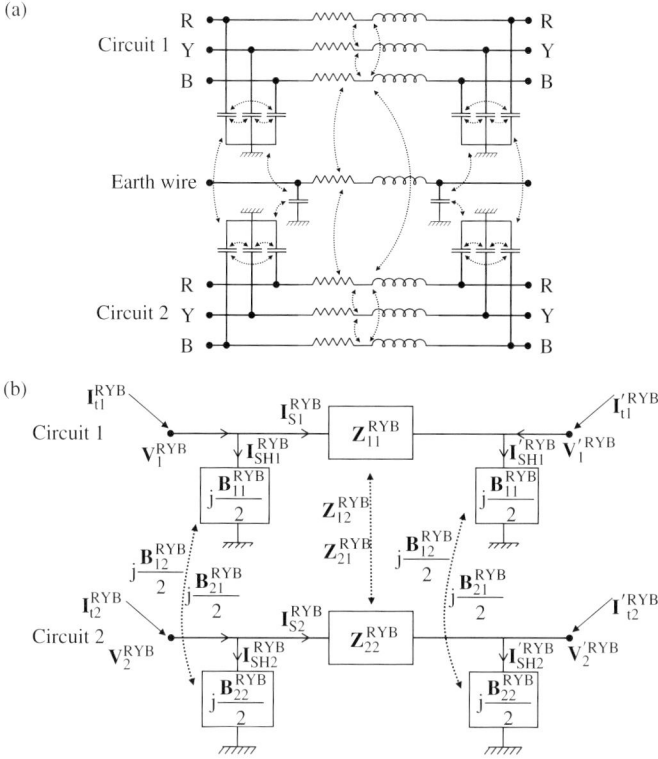

Figure 3.38 Three-phase model of a double-circuit line including earth wire(s): (a) π equivalent circuit and (b) π model (matrices)

given by

$$
\begin{bmatrix} \mathbf{V}_1^{ph} - \mathbf{V}_{1'}^{ph} \\ \mathbf{V}_2^{ph} - \mathbf{V}_{2'}^{ph} \end{bmatrix} = \begin{bmatrix} \mathbf{Z}_{11}^{ph} & \mathbf{Z}_{12}^{ph} \\ \mathbf{Z}_{21}^{ph} & \mathbf{Z}_{22}^{ph} \end{bmatrix} \begin{bmatrix} \mathbf{I}_{S(1)}^{ph} \\ \mathbf{I}_{S(2)}^{ph} \end{bmatrix}
$$

or

$$
\begin{bmatrix} \mathbf{I}_{S(1)}^{ph} \\ \mathbf{I}_{S(2)}^{ph} \end{bmatrix} = \begin{bmatrix} \mathbf{Y}_{11}^{ph} & \mathbf{Y}_{12}^{ph} \\ \mathbf{Y}_{21}^{ph} & \mathbf{Y}_{22}^{ph} \end{bmatrix} \begin{bmatrix} \mathbf{V}_1^{ph} - \mathbf{V}_{1'}^{ph} \\ \mathbf{V}_2^{ph} - \mathbf{V}_{2'}^{ph} \end{bmatrix}
\tag{3.154}
$$

The shunt susceptance currents are given by

$$
\begin{bmatrix} \mathbf{I}_{SH(1)}^{ph} \\ \mathbf{I}_{SH(2)}^{ph} \end{bmatrix} = \frac{j}{2} \begin{bmatrix} \mathbf{B}_{11}^{ph} & \mathbf{B}_{12}^{ph} \\ \mathbf{B}_{21}^{ph} & \mathbf{B}_{22}^{ph} \end{bmatrix} \begin{bmatrix} \mathbf{V}_1^{ph} \\ \mathbf{V}_2^{ph} \end{bmatrix} \quad \text{and} \quad \begin{bmatrix} \mathbf{I}_{SH(1')}^{ph} \\ \mathbf{I}_{SH(2')}^{ph} \end{bmatrix} = \frac{j}{2} \begin{bmatrix} \mathbf{B}_{11}^{ph} & \mathbf{B}_{12}^{ph} \\ \mathbf{B}_{21}^{ph} & \mathbf{B}_{22}^{ph} \end{bmatrix} \begin{bmatrix} \mathbf{V}_{1'}^{ph} \\ \mathbf{V}_{2'}^{ph} \end{bmatrix}
\tag{3.155}
$$

Since the total current injected into each circuit end is the sum of the series and shunt currents, it can be shown that the nodal phase admittance matrix equation of the mutually coupled double-circuit line is given by

$$
\begin{bmatrix} \mathbf{I}_{t(1)}^{ph} \\ \mathbf{I}_{t(1')}^{ph} \\ \mathbf{I}_{t(2)}^{ph} \\ \mathbf{I}_{t(2')}^{ph} \end{bmatrix} = \begin{bmatrix} \mathbf{Y}_{11}^{ph} + j\frac{\mathbf{B}_{11}^{ph}}{2} & -\mathbf{Y}_{11}^{ph} & \mathbf{Y}_{12}^{ph} + j\frac{\mathbf{B}_{12}^{ph}}{2} & -\mathbf{Y}_{12}^{ph} \\ -\mathbf{Y}_{11}^{ph} & \mathbf{Y}_{11}^{ph} + j\frac{\mathbf{B}_{11}^{ph}}{2} & -\mathbf{Y}_{12}^{ph} & \mathbf{Y}_{12}^{ph} + j\frac{\mathbf{B}_{12}^{ph}}{2} \\ \mathbf{Y}_{21}^{ph} + j\frac{(\mathbf{B}_{12}^{ph})^t}{2} & -\mathbf{Y}_{21}^{ph} & \mathbf{Y}_{22}^{ph} + j\frac{\mathbf{B}_{22}^{ph}}{2} & -\mathbf{Y}_{22}^{ph} \\ -\mathbf{Y}_{21}^{ph} & \mathbf{Y}_{21}^{ph} + j\frac{(\mathbf{B}_{12}^{ph})^t}{2} & -\mathbf{Y}_{22}^{ph} & \mathbf{Y}_{22}^{ph} + j\frac{\mathbf{B}_{22}^{ph}}{2} \end{bmatrix} \begin{bmatrix} \mathbf{V}_1^{ph} \\ \mathbf{V}_{1'}^{ph} \\ \mathbf{V}_2^{ph} \\ \mathbf{V}_{2'}^{ph} \end{bmatrix}
$$

$$(3.156)$$

All quantities in Equation (3.156) are matrices. Where the earth wires are eliminated, the dimensions of the impedances and susceptances are 3×3 and those of the current and voltage vectors are 3×1. Therefore, the dimension of the nodal current and voltage vectors are 12×1 and the dimension of the admittance matrix of the double-circuit line is 12×12. The same approach presented above can be followed in the case of double-circuit cables although the dimensions of the matrices will be increased to a greater extent by the number of sheaths and armours as appropriate.

3.7 Computer calculations and measurements of overhead line and cable parameters

3.7.1 Computer calculations of overhead line and cable parameters

The days of hand or analytical calculations of overhead line and cable phase and sequence impedances and susceptances have long gone together with, unfortunately, some of the great insight such methods provided. The calculations of overhead line and cable electrical parameters at power frequency, i.e. 50 or 60 Hz, as well as higher frequencies, are nowadays efficiently carried out using digital computer programs. For overhead lines, the input data are phase conductor and earth wire material and resistivity, physical dimensions, dc or ac resistance, conductor bundle data, tower geometry, i.e. spacing between conductors and conductor height above ground, average conductor sag and earth resistivity. The output information usually comprises the line's ac resistance matrix, inductance matrix, series phase reactance matrix, series phase impedance matrix, potential coefficient matrix, shunt capacitance matrix and shunt susceptance matrix, all including and excluding the earth wires. Assumptions of phase transpositions can be easily implemented using efficient matrix analysis techniques. The sequence data calculated consists of the series sequence impedance and shunt sequence susceptance matrices. The dimension of these sequence matrices is $3N$ where N is the number of three-phase circuits. The output information is usually calculated in physical units, e.g. μF/km/phase or Ω/km/phase and in pu or % on some defined voltage and MVA base.

For cables, the input data are core, sheath and, where applicable, armour conductor material and resistivity, their physical dimensions, dc or ac resistance, the cable layout or spacing between phases, relative permittivity of core, sheath and, where applicable, armour insulation, as well as earth resistivity. The output information usually comprises the cable's full and reduced series phase impedance shunt susceptance matrices, i.e. including and excluding the sheaths and armour(s). The sequence data calculated consists of the sequence series impedance matrix and sequence shunt susceptance matrix.

3.7.2 Measurement of overhead line parameters

General

Measurements of overhead lines power frequency electrical parameters or constants can be made at the time of commissioning when the line construction has been completed so that the tests replicate the line in normal service operation. The parameters that may be measured are phase conductor self and mutual impedances with earth return, and self and mutual susceptances. The series phase impedance and shunt phase susceptance matrices are then constructed and the sequence impedances and susceptances calculated. There are numerous designs of overhead lines used in practical installations but we will illustrate the techniques that can be used for single-circuit line with one earth wire and double-circuit line with one earth wire. The technique can be extended for application to any other line construction.

Single-circuit overhead lines with one earth wire

Figures 3.39(a) illustrates a technique for phase impedance measurements where the three-phase conductors are earthed at the far end of the line but kept free from earth at the near measurement end.

A test voltage is applied to one-phase conductor and the current flowing through it is measured. The induced voltages to earth on the remaining conductors are also measured. In addition to the calculation of the impedances from the measured voltages and currents, a wattmeter or phase angle meter can be used to find the phase angle of these voltages with respect to the injected current. This allows the calculation of the resistance and reactance components of the impedance using $R = P/I^2$ (P is measured phase power in Watts) and $X = \sqrt{Z^2 - R^2}$ in the case of a wattmeter, but $R = Z \times \cos \phi$ and $X = Z \times \sin \phi$ in the case of a phase angle meter measurement. The process is repeated for all other conductors. Where earth wire(s) are present, the measured impedances should represent values with the earth wire in parallel with earth return path to represent normal service operation.

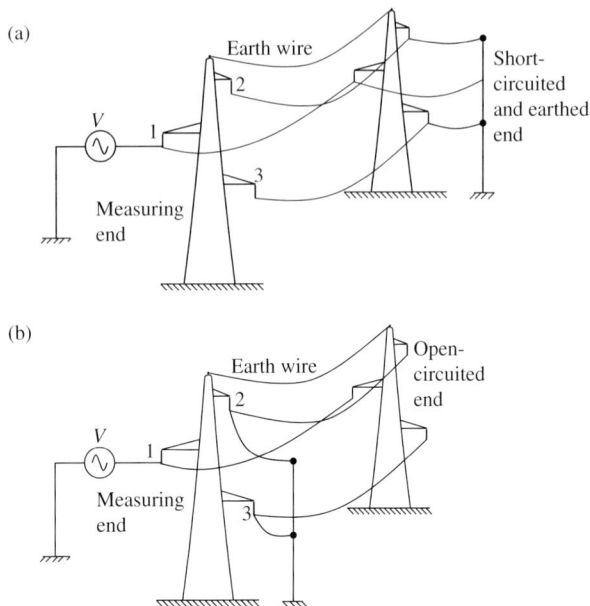

Figure 3.39 Measurement of the electrical parameters of a single-circuit overhead line: measurement of self and mutual (a) series phase impedances with earth return and (b) shunt phase susceptances

Using the single-circuit phase impedance matrix including the effect of earth wire that is with the earth wire mathematically eliminated, we have

$$V_1 = Z_{11}I_1 + Z_{12}I_2 + Z_{13}I_3 \qquad (3.157a)$$

$$V_2 = Z_{21}I_1 + Z_{22}I_2 + Z_{23}I_3 \qquad (3.157b)$$

$$V_3 = Z_{31}I_1 + Z_{32}I_2 + Z_{33}I_3 \qquad (3.157c)$$

With the voltage source applied to conductor 1, V_1 is known and the measured values are I_1, V_2 and V_3 with $I_2 = I_3 = 0$. Using the measured values in Equations (3.157), we obtain

$$Z_{11} = \frac{V_1}{I_1} \quad Z_{21} = \frac{V_2}{I_1} \quad Z_{31} = \frac{V_3}{I_1} \qquad (3.158a)$$

Applying the voltage source to conductor 2 then conductor 3, we obtain

$$Z_{12} = \frac{V_1}{I_2} \quad Z_{22} = \frac{V_2}{I_2} \quad Z_{32} = \frac{V_3}{I_2} \qquad (3.158b)$$

and

$$Z_{13} = \frac{V_1}{I_3} \quad Z_{23} = \frac{V_2}{I_3} \quad Z_{33} = \frac{V_3}{I_3} \tag{3.158c}$$

For sequence impedance calculations, and with the line assumed to be perfectly transposed, the self and mutual impedances per phase are calculated as the average of the measured self and mutual impedances as follows:

$$Z_S = \frac{Z_{11} + Z_{22} + Z_{33}}{3} \tag{3.159a}$$

and

$$Z_M = \frac{Z_{12} + Z_{21} + Z_{13} + Z_{31} + Z_{23} + Z_{32}}{6} \tag{3.159b}$$

Therefore, the PPS and ZPS impedances are given by

$$Z^P = \frac{2(Z_{11} + Z_{22} + Z_{33}) - (Z_{12} + Z_{21} + Z_{13} + Z_{31} + Z_{23} + Z_{32})}{6} \tag{3.160a}$$

$$Z^Z = \frac{(Z_{12} + Z_{22} + Z_{33}) + (Z_{12} + Z_{21} + Z_{13} + Z_{31} + Z_{23} + Z_{32})}{3} \tag{3.160b}$$

Figure 3.39(b) illustrates a technique for phase susceptance measurements where the three-phase conductors are earthed through milliammeters at the near measurement end of the line but kept free from earth, i.e. open circuit at the far end. A test voltage is applied in series with one-phase conductor and the currents flowing through all conductors are measured. The process is repeated for all other conductors. As in the case of impedance measurements the measured susceptances should include the effect of the earth wire. Using the single-circuit phase susceptance matrix including the effect of earth wire that is with the earth wire mathematically eliminated, we have

$$I_1 = jB_{11}V_1 - jB_{12}V_2 - jB_{13}V_3 \tag{3.161a}$$

$$I_2 = -jB_{21}V_1 + jB_{22}V_2 - jB_{23}V_3 \tag{3.161b}$$

$$I_3 = -jB_{31}V_1 - jB_{32}V_2 + jB_{33}V_3 \tag{3.161c}$$

With the test voltage source applied to conductor 1, V_1 is known and the measured values are I_1, I_2 and I_3 with $V_2 = V_3 = 0$. Using the measured values in Equations (3.161), we obtain

$$jB_{11} = \frac{I_1}{V_1} \quad jB_{21} = \frac{-I_2}{V_1} \quad jB_{31} = \frac{-I_3}{V_1} \tag{3.162a}$$

Applying the test voltage source to conductor 2 then conductor 3, we obtain

$$jB_{12} = \frac{-I_1}{V_2} \quad jB_{22} = \frac{I_2}{V_2} \quad jB_{32} = \frac{-I_3}{V_2} \tag{3.162b}$$

and

$$jB_{13} = \frac{-I_1}{V_3} \quad jB_{23} = \frac{-I_2}{V_3} \quad jB_{33} = \frac{I_3}{V_3} \tag{3.162c}$$

For sequence susceptance calculations and with the line assumed to be perfectly transposed, the self and mutual susceptances per phase are calculated as the average of the measured self and mutual susceptances as follows:

$$B_S = \frac{B_{11} + B_{22} + B_{33}}{3} \tag{3.163a}$$

and

$$B_M = \frac{B_{12} + B_{21} + B_{13} + B_{31} + B_{23} + B_{32}}{6} \tag{3.163b}$$

Therefore, the PPS and ZPS impedances are given by

$$B^P = \frac{2(B_{11} + B_{22} + B_{33}) + (B_{12} + B_{21} + B_{13} + B_{31} + B_{23} + B_{32})}{6} \tag{3.164a}$$

$$B^Z = \frac{(B_{12} + B_{22} + B_{33}) - (B_{12} + B_{21} + B_{13} + B_{31} + B_{23} + B_{32})}{3} \tag{3.164b}$$

The measured voltages and currents inherently take into account the distributed nature of the line's impedances and susceptances. However, we have used lumped linear impedance and susceptance matrix models for the line which will introduce small errors in the calculated impedances and susceptances depending on line length. The calculated impedances and susceptances can be corrected for the line length effect if the equivalent π circuit of a distributed multi-conductor line derived in Appendix 1 is used. The impedance and susceptance constants now become impedance and susceptance matrices and the general nodal admittance equations for a multi-conductor line are given by

$$\mathbf{I}_S = \mathbf{A} \times \mathbf{V}_S - \mathbf{B} \times \mathbf{V}_R \tag{3.165a}$$

$$\mathbf{I}_R = -\mathbf{B} \times \mathbf{V}_S + \mathbf{A} \times \mathbf{V}_R \tag{3.165b}$$

where

$$A = Y(\sqrt{ZY})^{-1} \coth(\sqrt{ZY}\ell) \quad \text{and} \quad B = Y(\sqrt{ZY})^{-1} \operatorname{cosech}(\sqrt{ZY}\ell)$$

$$(3.165c)$$

and I_S and I_R are the line's sending and receiving end current vectors, respectively; and V_S and V_R are the line's sending and receiving end voltage vectors, respectively.

In the impedance measurement technique, the voltage and current vectors V_S and I_S are known from the measurements, $V_R = 0$ but I_R is unknown. Therefore, from Equation (3.165), we can write $V_S = A^{-1}I_S = [Y^{-1}\sqrt{YZ}\tanh(\sqrt{YZ}\ell)]I_S$. Expanding the hyperbolic tangent into an infinite series, we obtain

$$V_S = \left[Z - \frac{1}{3}ZYZ\ell^2 + \frac{2}{15}ZYZYZ\ell^4 - \cdots \right] \times \ell \times I_S \qquad (3.166)$$

In the susceptance measurement technique, the voltage and current vectors V_S and I_S are known from the measurements, $I_R = 0$ but V_R is unknown. Therefore, from Equation (3.165), we can write $I_S = (A - B \times A^{-1} \times B) \times V_S$. Using the values of A and B from Equation (3.165) and expanding the hyperbolic tangent into an infinite series, we obtain

$$I_S = \left[Y - \frac{1}{3}YZY\ell^2 + \frac{2}{15}YZYZY\ell^4 - \cdots \right] \times \ell \times V_S \qquad (3.167)$$

We can use the voltages and currents obtained from the test measurements to calculate to a first approximation of the elements of the impedance and susceptance matrices using the approximate lumped parameter equations as follows:

$$V_S \approx Z_1 \times \ell \times I_S \quad \text{and} \quad I_S \approx Y_1 \times \ell \times V_S \qquad (3.168)$$

Z_1 and Y_1 can now be equated to the series impedance and susceptance values of Equations (3.166) and (3.167) and used to obtain a second approximation as follows:

$$Z_2 \approx Z_1 + \frac{1}{3}Z_1Y_1Z_1\ell^2 - \frac{2}{15}Z_1Y_1Z_1Y_1Z_1\ell^4 - \cdots \qquad (3.169a)$$

$$Y_2 \approx Y_1 + \frac{1}{3}Y_1Z_1Y_1\ell^2 - \frac{2}{15}Y_1Z_1Y_1Z_1Y_1\ell^4 - \cdots \qquad (3.169b)$$

Similarly, the new values of the impedance and susceptance matrices can again be used to obtain an improved approximation and the process can be repeated until the desired level of accuracy is obtained. In general, we can write

$$\mathbf{Z}_{N+1} \approx \mathbf{Z}_N + \frac{1}{3}\mathbf{Z}_N\mathbf{Y}_N\mathbf{Z}_N\ell^2 - \frac{2}{15}\mathbf{Z}_N\mathbf{Y}_N\mathbf{Z}_N\mathbf{Y}_N\mathbf{Z}_N\ell^4 - \cdots \qquad (3.170a)$$

$$\mathbf{Y}_{N+1} \approx \mathbf{Y}_N + \frac{1}{3}\mathbf{Y}_N\mathbf{Z}_N\mathbf{Y}_N\ell^2 - \frac{2}{15}\mathbf{Y}_N\mathbf{Z}_N\mathbf{Y}_N\mathbf{Z}_N\mathbf{Y}_N\ell^4 - \cdots \qquad (3.170b)$$

Double-circuit overhead lines

The technique presented above for a single-circuit overhead line with one earth wire can be extended to double-circuit overhead lines with any number of earth wires. Figure 3.40 shows the impedance and susceptance test arrangements.

The measured phase impedance and susceptance matrices are given as

$$
\begin{bmatrix} V_1 \\ V_2 \\ V_3 \\ V_4 \\ V_5 \\ V_6 \end{bmatrix}
=
\begin{array}{c}
\text{Circuit 1} \\ \\ \\ \text{Circuit 2}
\end{array}
\overset{\overset{\text{Circuit 1}\qquad\qquad\text{Circuit 2}}{}}{
\begin{bmatrix}
Z_{11} & Z_{12} & Z_{13} & Z_{14} & Z_{15} & Z_{16} \\
Z_{21} & Z_{22} & Z_{23} & Z_{24} & Z_{25} & Z_{26} \\
Z_{31} & Z_{32} & Z_{33} & Z_{34} & Z_{35} & Z_{36} \\
Z_{41} & Z_{42} & Z_{43} & Z_{44} & Z_{45} & Z_{46} \\
Z_{51} & Z_{52} & Z_{53} & Z_{54} & Z_{55} & Z_{56} \\
Z_{61} & Z_{62} & Z_{63} & Z_{64} & Z_{65} & Z_{66}
\end{bmatrix}}
\begin{bmatrix} I_1 \\ I_2 \\ I_3 \\ I_4 \\ I_5 \\ I_6 \end{bmatrix}
\qquad (3.171a)
$$

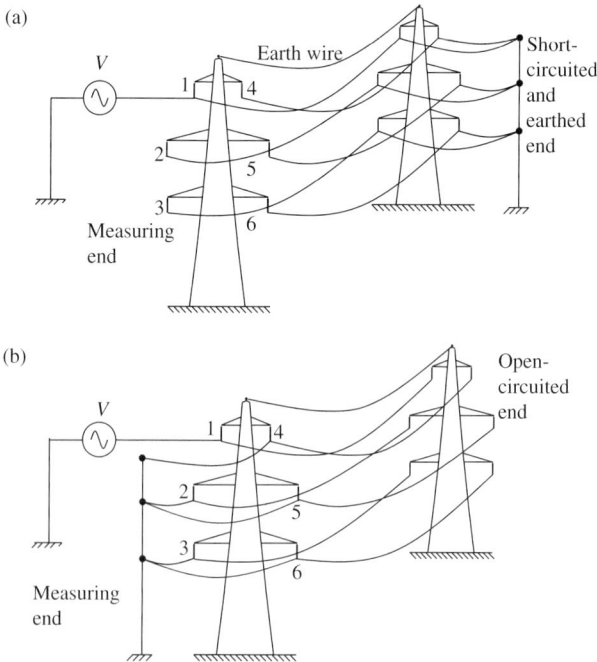

Figure 3.40   Measurement of electrical parameters of a double-circuit overhead line: measurement of self and mutual (a) series phase impedances with earth return and (b) shunt phase susceptances

and

$$
\begin{bmatrix} I_1 \\ I_2 \\ I_3 \\ I_4 \\ I_5 \\ I_6 \end{bmatrix} =
\begin{array}{c} \\ \text{Circuit 1} \\ \\ \\ \text{Circuit 2} \\ \\ \end{array}
\begin{bmatrix}
jB_{11} & -jB_{12} & -jB_{13} & -jB_{14} & -jB_{15} & -jB_{16} \\
-jB_{21} & jB_{22} & -jB_{23} & -jB_{24} & -jB_{25} & -jB_{26} \\
-jB_{31} & -jB_{32} & jB_{33} & -jB_{34} & -jB_{35} & -jB_{36} \\
-jB_{41} & -jB_{42} & -jB_{43} & jB_{44} & -jB_{45} & -jB_{46} \\
-jB_{51} & -jB_{52} & -jB_{53} & -jB_{54} & jB_{55} & -jB_{56} \\
-jB_{61} & -jB_{62} & -jB_{63} & -jB_{64} & -jB_{65} & jB_{66}
\end{bmatrix}
\begin{bmatrix} V_1 \\ V_2 \\ V_3 \\ V_4 \\ V_5 \\ V_6 \end{bmatrix}
$$

(3.171b)

In Figure 3.40(a), we apply a voltage source V_1 to conductor 1 and measure I_1, V_2, V_3, V_4, V_5 and V_6. In Figure 3.40(b), we apply a voltage source V_1 to conductor 1 and measure I_1, I_2, I_3, I_4, I_5 and I_6. From these tests, and using Equation (3.171), the self and mutual phase impedances and susceptances are calculated as follows:

$$
Z_{11} = \frac{V_1}{I_1} \quad Z_{21} = \frac{V_2}{I_1} \quad Z_{31} = \frac{V_3}{I_1} \quad Z_{41} = \frac{V_4}{I_1} \quad Z_{51} = \frac{V_5}{I_1} \quad Z_{61} = \frac{V_6}{I_1}
$$

(3.172a)

$$
jB_{11} = \frac{I_1}{V} \quad jB_{21} = \frac{-I_2}{V} \quad jB_{31} = \frac{-I_3}{V}
$$

$$
jB_{41} = \frac{-I_4}{V} \quad jB_{51} = \frac{-I_5}{V} \quad jB_{61} = \frac{-I_6}{V}
$$

(3.172b)

In both impedance and susceptance tests, the voltage source is applied to each phase conductor in turn in order to measure all the self and mutual impedances and susceptances of Equation (3.171). The balanced phase and resultant sequence impedances can be calculated assuming an appropriate number of phase transpositions such as three, six or ideal nine transpositions. The reader is encouraged to derive the balanced phase and sequence parameters. The sequence susceptance parameters can be similarly calculated.

3.7.3 Measurement of cable parameters

General

Measurements of power frequency electrical parameters or constants of cables can be made at the time of commissioning when the cable installation has been completed so that the tests replicate the cable in normal service operation. The relevant parameters that are usually measured are the conductor dc resistances

at prevailing ambient temperature, the PPS/NPS and ZPS impedances and sometimes the PPS/NPS/ZPS susceptances. As there are numerous designs and layouts
of cables in use in practical installations, we will illustrate the techniques that
can be used for a typical single-circuit single-core three-phase coaxial cable with
core and sheath conductors. An illustration of the PPS/NPS susceptance and ZPS
susceptance measurement of a three-core belted cable is also presented.

Measurement of dc resistances

The dc resistances of the cores and sheaths (screens) are measured using a dc
voltage source as illustrated in Figure 3.41(a) and (b). In the first case, the dc
voltage source V is applied between cores 1 and 2 with the third core floating and
remote ends of the three cores short-circuited. The dc source V is next applied
between cores 1 and 3, then 2 and 3. The measured core loop resistance in each
case is calculated from $R_{Cc(loop)} = V/I$ and the three measured loop resistances are
given by

$$R_{C1C2} = R_{C1} + R_{C2} \quad R_{C1C3} = R_{C1} + R_{C3} \quad R_{C2C3} = R_{C2} + R_{C3} \quad (3.173a)$$

Therefore, the individual core resistances are given by

$$R_{C1} = \frac{R_{C1C2} + R_{C1C3} - R_{C2C3}}{2} \quad R_{C2} = \frac{R_{C1C2} + R_{C2C3} - R_{C1C3}}{2}$$

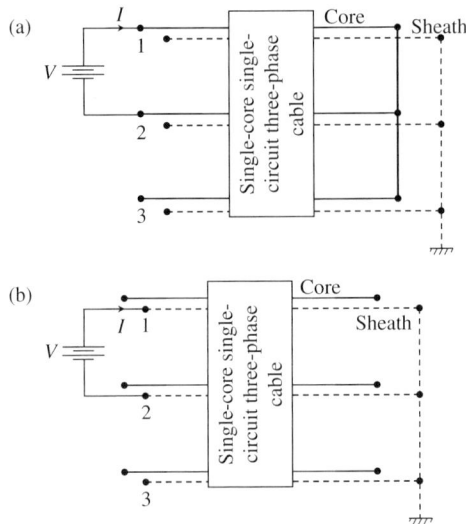

Figure 3.41   Measurement of dc resistances of (a) core and (b) sheath/screen of a three-phase cable

and

$$R_{C3} = \frac{R_{C1C3} + R_{C2C3} - R_{C1C2}}{2} \qquad (3.173b)$$

The test procedure for the sheath or screen dc resistances is similar to that for the core dc resistances but the dc voltage source is now applied between two sheaths at a time as shown in Figure 3.41(b).

Measurement of PPS/NPS impedance

The measurement of the cable's PPS/NPS impedance per phase requires a three-phase power frequency source. On each of the single cores, measurements of voltage, current and phase power or phase angle are made. The sheath's bonding and earthing must correspond to that in actual circuit operation. Figure 3.42(a) illustrates the PPS/NPS impedance measurement for a cable with a sheath that is solidly bonded at both ends. From the measurements, the following calculations for each phase are made.

If a wattmeter is used to measure input power, we have

$$Z^P = \frac{V^P}{I^P} \qquad R^P = \frac{P^P}{(I^P)^2} \qquad X^P = \sqrt{(Z^P)^2 - (R^P)^2} \qquad (3.174)$$

If a phase angle meter is used to measure the angle between the input voltage and current

$$Z^P = \frac{V^P}{I^P} \qquad R^P = Z^P \times \cos\phi \qquad X^P = Z^P \times \sin\phi \qquad (3.175)$$

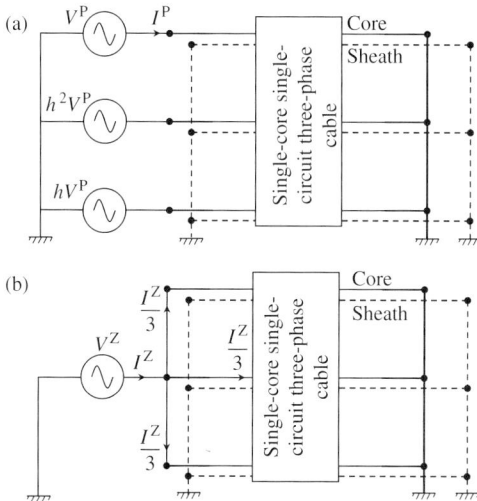

Figure 3.42   Measurement of (a) PPS/NPS and (b) ZPS impedances of a three-phase cable

where P^P is the measured phase power in Watts. The three values obtained for each phase are then averaged and this is particularly important in the case of untransposed cable systems. The final PPS/NPS cable parameters are given as

$$Z^P = \frac{\sum_{i=1}^{3} Z_i^P}{3} \; \Omega/\text{phase} \quad R^P = \frac{\sum_{i=1}^{3} R_i^P}{3} \; \Omega/\text{phase} \quad X^P = \frac{\sum_{i=1}^{3} X_i^P}{3} \; \Omega/\text{phase}$$

$$(3.176)$$

Measurement of ZPS impedance

The measurement of the cable's ZPS impedance per phase requires a single-phase power frequency source. The three cores at the test end are bonded together and at the remote end are bonded together and to the sheaths. Single-phase currents are injected into the cable cores to return via the sheath and earth. Figure 3.42(b) illustrates the test circuit. The sheaths are solidly bonded at both ends as in actual cable operation. The calculations of the ZPS impedance, resistance and reactance use similar equations to those used for the PPS/NPS impedance except that the current should now be divided by 3, i.e. the impedance equation becomes

$$Z^Z = \frac{3V^Z}{I^Z} \tag{3.177}$$

Measurement of earth loop impedances

As in the case of an overhead line, and if required, the earth loop impedances for each core and sheath may also be measured using a single-phase power frequency source. The technique is similar to that presented in Section 3.7.2.

Measurement of PPS/NPS/ZPS susceptance

The measurement of a screened cable core-to-sheath, and if required, sheath-to-earth susceptances uses a single-phase power frequency source. The three cores at the remote end are left floating, i.e. open circuit. The applied voltage V, charging current I and phase angle ϕ between applied voltage and current, the latter measured using a low power factor meter, are measured. This is usually called the power factor and capacitance test of the dielectric. The calculations made are

$$Y = \frac{I}{V} \quad G = Y \times \cos\phi \quad B = Y \times \sin\phi \tag{3.178}$$

B is the PPS/NPS/ZPS susceptance per phase. G is very small and is usually neglected in power frequency steady state analysis.

For three-core belted cables, the core-to-core and core-to-sheath capacitances can be calculated from two measurements. The core to sheath, i.e. ZPS susceptance can be measured by short-circuiting the three cores and applying a voltage source

V between them and the earthed sheath, see Figure 3.29(c). From the measured injected current I, we can write

$$B^Z = B_{CS} = \frac{I}{3V} \tag{3.179}$$

The core-to-core capacitance can be obtained from a measurement where a voltage source V is applied between two cores with the third core short-circuited to the sheath. Let the measured susceptance be B_{Meas}. Therefore, using Figure 3.29(c), we can write

$$B_{Meas} = \frac{I}{V} = B_{CC} + \frac{B_{CC} + B_{CS}}{2} = \frac{3B_{CC} + B_{CS}}{2}$$

Therefore, the core-to-core susceptance is calculated as

$$B_{CC} = \frac{2B_{Meas} - B_{CS}}{3} \tag{3.180}$$

Using Equation (3.138c), the PPS susceptance of the cable is given by

$$B^P = 2B_{Meas} \tag{3.181}$$

3.8 Practical aspects of phase and sequence parameters of overhead lines and cables

3.8.1 Overhead lines

During the installation and commissioning of new overhead line designs, it was common practice to carry out field test measurements of the phase self and mutual impedances and susceptances. From these tests, the sequence impedances and susceptances are obtained. Measurements of sequence parameters could also be made. Measurements served as benchmark values for the validation of line's power frequency models and calculations carried out on digital computers. The range of errors in the calculations of the sequence impedance parameters is usually very small. The assumption of conductor temperature used in the calculation is important primarily for resistance calculation and to a small extent reactance and susceptance calculations due to changes in conductor sag. Also, a reasonable value of uniform earth resistivity may be used and this may be an average value to represent the route over which the line runs. In England and Wales and in America, average uniform earth resistivities of 20 and 100 Ωm are generally used.

3.8.2 Cables

For cables, available models and computer calculations of the PPS/NPS impedances and sequence susceptances usually provide results with good accuracy in comparison with field test measurements despite the large variety of cable designs and installations. However, this is not normally the case for the ZPS impedance particularly for high voltage cables. In addition to the cable design including its sheath or screen and cable layout, the ZPS impedance strongly depends on any other local and nearby metal objects that may form a return path in parallel with the cable sheath itself. These might be other parallel cables, nearby water pipes and pipelines, gas mains, subway structures, railway tracks and lines, steel or concrete tunnels and any earth electrodes that may be used, etc. The information about most of these additional factors is practically unknown and impossible to obtain so they cannot be appropriately included in the calculation. The only practical solution to obtain accurate results is to carry out field test measurement of the ZPS impedance with the cable installed as in actual service operation. In addition, such a measurement will only be accurate at the time it is taken but it may be different a few years later. This may be caused by underground construction or modifications of conducting material e.g. water and gas mains.

Further reading

Books

[1] Bickford, J.P., *et al.*, *Computation of Power-System Transients*, Peter Peregrinus Ltd., 1980, ISBN 0-906048-35-4.
[2] McAllister, D., *Electric Cables Handbook*, Granada, 1982, ISBN 0-246-11467-3.
[3] Abramowitz, M. and Stegun, I., *et al.*, *Handbook of Mathematical Functions*, Dover publications, Inc., 1964, Library of Congress catalog card number: 65-122253.
[4] IEC 60287-1-3, *Electric Cables – Calculation of Current Rating*, 1st Edn 2002–2005.

Papers

[5] Carson, J.R., Wave propagation in overhead wires with ground return, *Bell System Technical Journal*, Vol. 5, 1926, 539–554.
[6] Pollaczek, F., Ueber das feld einer unendlich langen wechsel stromdurchflossenen einfachleitung, *Elektrische Nachrichlen Technik*, Vol. 3, No. 5, 1926, 339–359.
[7] Schelkunoff, S.A., The electromagnetic theory of coaxial transmission lines and cylindrical shells, *Bell System Technical Journal* Vol. XIII, 1934, 532–578.
[8] Haberland, G., Theorie der Leitung von Wechselstrom durch die Erde, *Zeitschrifi fir Angewand te Mathematic and Mechanik*, Vol. 5, No. 6, 1926, 366.
[9] Wang, Y.J., A review of methods for calculation of frequency-dependent impedance of overhead power transmission lines, *Proceedings of the National Science Council Republic of China(A)*, Vol. 25, No. 6, 2001, 329–338.
[10] Galloway, R.H., *et al.*, Calculation of electrical parameters for short and long polyphase transmission lines, *Proceedings IEE*, Vol. 111, December 1964, 2051–2059.
[11] Dommel, H.W., Overhead line parameters from handbook formulas and computer programs, *IEEE Transactions on PAS*, Vol. PAS-104, No. 2, February 1985, 366–372.

[12] Nielsen, H., *et al.*, *Underground Cables and Overhead Lines Earth Return Path Impedance Calculations with Reference to Single Line to Ground Faults*, Aalborg University, Denmark, 2004, 1–11.

[13] Ametani, A., A general formulation of impedance and admittance of cables, *IEEE Transactions on PAS*, Vol. PAS-99, No. 3, May/June 1980, 902–910.

[14] Brown, G.W., *et al.*, Surge propagation in three phase pipe-type cables, Part 1_ unsaturated pipe, *IEEE Transactions on PAS*, Vol. PAS-95, No. 1, January/February 1976, 89–95.

[15] Bianchi, G., *et al.*, Induced currents and losses in single-core submarine cables, *IEEE Transactions on PAS*, Vol. PAS-95, No. 1, January/February 1976, 49–58.

[16] Wedepohl, L.M., *et al.*, Transient analysis of underground power-transmission systems, *IEE Proceedings*, Vol. 120, No. 2, February 1973, 253–260.

4

Modelling of transformers, static power plant and static load

4.1 General

In this chapter, we describe modelling techniques of various types of transformers, quadrature boosters (QBs), phase shifters (PSs), series and shunt reactors, series and shunt capacitors, static variable compensators and static power system load. Sequence, i.e. positive phase sequaence (PPS), negative phase sequence (NPS) and zero phase sequence (ZPS) models, as well as three-phase models suitable for analysis in the phase frame reference are presented. Sequence impedance measurement techniques of transformers, QBs and PSs are also described.

4.2 Sequence modelling of transformers

4.2.1 Background

The invention of the transformer machine towards the end of the nineteenth century having the basic function of stepping up or down the voltages (and currents but in opposite ratios) created the advantage of transmitting alternating current (ac) electric power over larger distances. This was not possible with direct current (dc) electric power since the locations of power generation and demand centres could not be too far apart. This is to ensure that power loss and voltage drops are kept to a reasonably low level. The technology of transformers has undergone significant developments during the twentieth century and many types of transformers are nowadays used in ac power systems.

Three banks of single-phase transformers are generally used in power stations to connect very large generators to transmission networks with the low voltage (LV) and high voltage (HV) windings connected as required, e.g. in delta or star. The main reasons for using separate banks are to reduce transport weight, improve overall transformer availability and reliability, and reduce the cost of spare holding. Single-phase transformers are also used in remote rural distribution networks to supply small single-phase demand consumers where three-phase supplies are uneconomic to use. There are also special applications where single-phase transformers are used such as in the case of high capacity traction transformers connected to three-phase power networks, usually connected to two phases of the three-phase transmission or distribution network. Other applications of single-phase transformers are those used to provide neutral earthing for generators.

Three-phase power transformers are invariably used in transmission, subtransmission and distribution substations for essentially voltage transformation. In addition, they are also used for voltage or reactive power flow control and active power flow control. For the latter application, they are called PS or QB transformers. Other applications of three-phase power transformers are for connections to converters in high voltage direct current (HVDC) links, for earthing of isolated systems, as well as in large industrial applications such as arc furnaces.

Transformers used for voltage control are equipped with off-load or on-load tap-changers that vary the number of turns on the associated winding and hence the turns ratio of the transformer. In some countries, e.g. the UK, generator transformers are generally equipped with on-load tap-changers on the HV winding with the HV side voltage being the controlled voltage. Typical voltage transformations are 11–142 kV, 15–300 kV and 23.5–432 kV. Two-winding transformers having large transformation ratios may be equipped with on-load tap-changers on the HV winding but with the LV winding voltage being the controlled voltage. Typical transformations are 275–33 kV, 132–11 kV, 380–110 kV and 110–20 kV. Distribution transformers, e.g. 11–0.433 kV usually have off-load tap-changers.

Where the transformation ratio is generally less than about three, e.g. 400 kV/132 kV and 500 kV/230 kV, autotransformers are generally more economic to use. Nowadays with increased trends towards automation, autotransformers that supply power to distribution systems tend to be equipped with on-load tap-changers. These may be located either at the LV winding line-end or neutral-end with the LV winding voltage being the controlled voltage. Autotransformers may also have a third low MVA capacity delta-connected winding called 'tertiary' winding. These may be used for the connection of reactive compensation plant e.g. capacitors, reactors or synchronous compensators (condensers). They also provide a path for the circulation of triplen harmonic currents and hence prevent or reduce their flow on the power network as well as reduce network voltage unbalance.

The calculation of unbalanced currents and voltages due, for example, to short-circuit faults requires correct and practical modelling of power transformers. In this section, we present the theory of modelling single-phase transformers, and

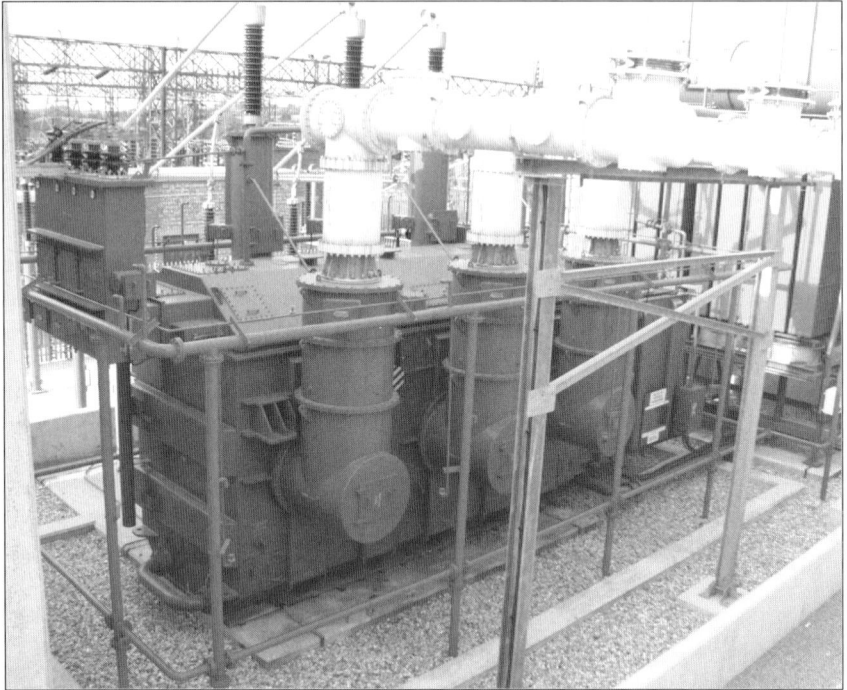

Figure 4.1   240 MVA 400 kV/132 kV autotransformer

three-phase power transformers having various numbers of windings and different types of winding connections. Knowledge of basic electromagnetic transformer theory is assumed. Figures 4.1 and 4.2 show two large autotransformers used in a 400 kV/275 kV/132 kV transmission system.

4.2.2 Single-phase two-winding transformers

The power transformer is a complex static electromagnetic machine with windings and a non-linear iron core. We will first present the transformer equivalent circuit in actual physical units then derive the per-unit equivalent circuit for the cases where the transformer has nominal and off-nominal turns ratios. The latter corresponds to the common case where the transformer is equipped with an on-load tap-changer.

Equivalent circuit in actual physical units

The conventional equivalent circuit of a single-phase two-winding transformer is shown in Figure 4.3(a).

Where all quantities shown are in physical units of volts, amps and ohms, and $Z_H = R_H + jX_H$, $R_H = $ HV winding resistance, $X_H = $ HV winding leakage reactance. $Z_L = R_L + jX_L$, $R_L = $ LV winding resistance, $X_L = $ LV winding leakage reactance. $R_I = $ resistance representing core iron losses which are assumed to vary

Figure 4.2 1000 MVA 400 kV/275 kV autotransformer

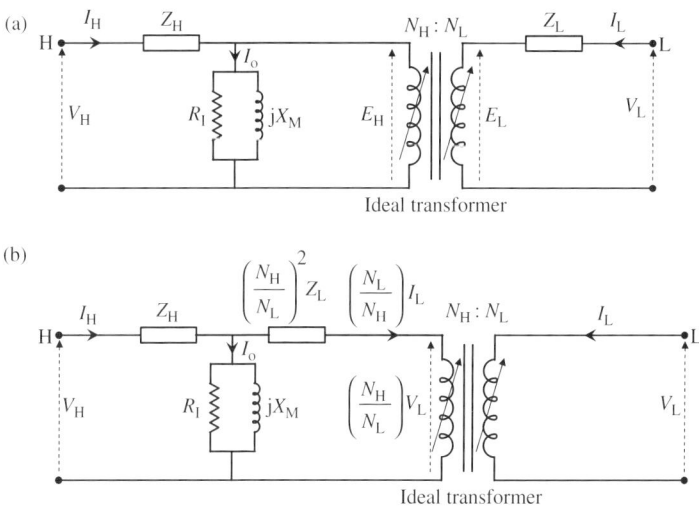

Figure 4.3 Single-phase transformer equivalent circuit in actual physical units: (a) basic equivalent circuit and (b) L winding impedance referred to H winding

with the square of applied HV winding voltage. X_M = core magnetising reactance referred to the HV side and representing the rms value of the magnetising current. N_H and N_L are the actual number of turns of HV and LV windings, respectively. $N_{H(nominal)}$ and $N_{L(nominal)}$ are HV and LV winding number of turns at nominal tap positions, respectively.

From basic transformer theory and since the same flux will thread both HV and LV windings, the induced voltage per turn on the HV and LV windings are equal, hence

$$\frac{E_H}{E_L} = \frac{N_H}{N_L} = N_{HL} \tag{4.1a}$$

In the derivation of the equivalent circuit, we will initially neglect the ampere-turns or MMF due to the exciting shunt branch that represents the very small core iron and magnetising losses. The HV winding and LV winding MMFs, for the current directions shown in Figure 4.3(a) are related by

$$N_H I_H + N_L I_L = 0 \quad \text{or} \quad \frac{I_H}{I_L} = -\frac{1}{N_{HL}} \tag{4.1b}$$

Also, the voltage drop equations for the transformer HV and LV windings can be written as follows:

$$V_H - E_H = Z_H I_H \tag{4.2a}$$

$$V_L - E_L = Z_L I_L \tag{4.2b}$$

Using Equations (4.1a) and (4.2b), Equation (4.2a) can be rewritten as

$$\frac{1}{N_{HL}}(V_H - Z_H I_H) = V_L - Z_L I_L \tag{4.3}$$

Substituting Equation (4.1b) into Equation (4.3), we have

$$V_H - (Z_H + N_{HL}^2 Z_L)I_H = N_{HL} V_L \tag{4.4a}$$

or

$$V_H - Z_{HL} I_H = N_{HL} V_L \tag{4.4b}$$

where

$$Z_{HL} = Z_H + N_{HL}^2 Z_L \tag{4.4c}$$

Z_{HL} is the equivalent transformer leakage impedance referred to the HV side and $N_{HL}^2 Z_L$ is the LV impedance referred to the HV side. Figure 4.3(b) is drawn using Equation (4.4a) ignoring the shunt exciting branch. The LV voltage V_L and LV current I_L are referred to the HV side as $N_{HL} V_L$ and I_L/N_{HL}, respectively. The exciting shunt branch can be reinserted between Z_H and $N_{HL}^2 Z_L$ on the input HV terminals. Alternatively, Z_H could have been referred to the LV side and the exciting branch inserted similarly. The exciting impedance is much bigger than the leakage impedances of either winding by almost a factor of 400–1. However, modern short-circuit, power flow and transient stability analyses practice includes the iron loss branch as well as the magnetising branch.

Equivalent circuit in per unit

In many types of steady state analysis, we are interested in the transformer equivalent circuit where all quantities are in per unit on some defined base quantities. Let the base quantities be defined as

$$S_{(B)} = \text{Transformer rating in VA}$$

$$V_{H(B)} = \text{HV network base voltage} \qquad V_{L(B)} = \text{LV network base voltage}$$

$$I_{H(B)} = \text{HV network base current} \qquad I_{L(B)} = \text{LV network base current}$$

$$V_{H(B)} = Z_{H(B)}I_{H(B)} \qquad\qquad V_{L(B)} = Z_{L(B)}I_{L(B)} \qquad (4.5a)$$

$$V_{H(B)}I_{H(B)} = V_{L(B)}I_{L(B)} = S_{(B)} \qquad \frac{V_{H(B)}}{V_{L(B)}} = \frac{N_{H(\text{nominal tap position})}}{N_{L(\text{nominal tap position})}} \qquad (4.5b)$$

In Equation (4.5b), the HV and LV windings' nominal tap positions are chosen to correspond to the HV and LV base voltages of the network sections to which the windings are connected.

Let us now define the following per-unit quantities:

$$Z_{H(pu)} = \frac{Z_H}{Z_{H(B)}} \quad Z_{L(pu)} = \frac{Z_L}{Z_{L(B)}} \qquad (4.6a)$$

$$V_{H(pu)} = \frac{V_H}{V_{H(B)}} \quad V_{L(pu)} = \frac{V_L}{V_{L(B)}} \qquad (4.6b)$$

$$I_{H(pu)} = \frac{I_H}{I_{H(B)}} \quad I_{L(pu)} = \frac{I_L}{I_{L(B)}} \qquad (4.6c)$$

Dividing Equation (4.3) by $S_{(B)}$ and using Equations (4.5) and (4.6), we obtain

$$\frac{1}{\frac{N_H}{N_{H(\text{nominal})}}} \left(\frac{V_H}{V_{H(B)}} - \frac{Z_H I_H}{Z_{H(B)}I_{H(B)}} \right) = \frac{1}{\frac{N_L}{N_{L(\text{nominal})}}} \left(\frac{V_L}{V_{L(B)}} - \frac{Z_L I_L}{Z_{L(B)}I_{L(B)}} \right)$$

which can be rewritten as

$$\frac{1}{t_{H(pu)}}[V_{H(pu)} - Z_{H(pu)}I_{H(pu)}] = \frac{1}{t_{L(pu)}}[V_{L(pu)} - Z_{L(pu)}I_{L(pu)}] \qquad (4.7a)$$

where the pu tap ratios of the HV and LV windings are given by

$$t_{H(pu)} = \frac{N_H}{N_{H(\text{nominal})}} = \frac{N_{H(\text{at a given HV tap position})}}{N_{H(\text{at nominal HV tap position})}} \qquad (4.7b)$$

and

$$t_{L(pu)} = \frac{N_L}{N_{L(\text{nominal})}} = \frac{N_{L(\text{at a given LV tap position})}}{N_{L(\text{at nominal LV tap position})}} \qquad (4.7c)$$

We will now assume that the impedance of each winding is proportional to the square of the number of turns of the winding. We know from basic transformer theory that this is generally valid for the winding inductive reactance but not for the winding resistance. However, because the resistance is generally much smaller than the reactance, this approximation is accepted. Thus, we define $Z_{H(nominal)}$ as the pu HV winding impedance at nominal HV winding tap position and $Z_{L(nominal)}$ as the pu LV winding impedance at nominal LV winding tap position. Therefore, for each winding, we have

$$Z_{H(pu)} = t_{H(pu)}^2 \times Z_{H(pu,nominal)} \tag{4.8a}$$

$$Z_{L(pu)} = t_{L(pu)}^2 \times Z_{L(pu,nominal)} \tag{4.8b}$$

Substituting Equation (4.8) into Equation (4.7a), we obtain

$$\frac{1}{t_{H(pu)}}[V_{H(pu)} - t_{H(pu)}^2 Z_{H(pu,nominal)} I_{H(pu)}]$$

$$= \frac{1}{t_{L(pu)}}[V_{L(pu)} - t_{L(pu)}^2 Z_{L(pu,nominal)} I_{L(pu)}] \tag{4.9}$$

The general pu equivalent circuit shown in Figure 4.4(a) represents Equation (4.9).

In practice, we are more interested in deriving an equivalent where both impedances appear on one side of the ideal transformer. This can be done once we have derived the relationship between the HV and LV pu currents from Equation (4.1b) as follows:

$$N_L I_L = -N_H I_H$$

which, using Equation (4.5b), can be written as

$$\frac{N_L I_L}{V_{L(B)} I_{L(B)}} = -\frac{N_H I_H}{V_{H(B)} I_{H(B)}}$$

or, using Equations (4.7b) and (4.7c), reduces to

$$t_{L(pu)} I_{L(pu)} = -t_{H(pu)} I_{H(pu)} \tag{4.10}$$

Substituting Equation (4.10) into Equation (4.9) and rearranging, we obtain the following two equations

$$\frac{V_{H(pu)}}{t_{HL(pu)}} - t_{HL(pu)} I_{H(pu)} \times t_{L(pu)}^2 Z_{LH(pu,nominal)} = V_{L(pu)} \tag{4.11a}$$

and

$$\frac{V_{L(pu)}}{t_{LH(pu)}} - t_{LH(pu)} I_{L(pu)} \times t_{H(pu)}^2 Z_{HL(pu,nominal)} = V_{H(pu)} \tag{4.11b}$$

where

$$Z_{LH(pu,nominal)} = Z_{HL(pu,nominal)} = Z_{H(pu,nominal)} + Z_{L(pu,nominal)} \tag{4.11c}$$

$$t_{HL(pu)} = \frac{t_{H(pu)}}{t_{L(pu)}} \quad \text{and} \quad t_{LH(pu)} = \frac{t_{L(pu)}}{t_{H(pu)}} \tag{4.11d}$$

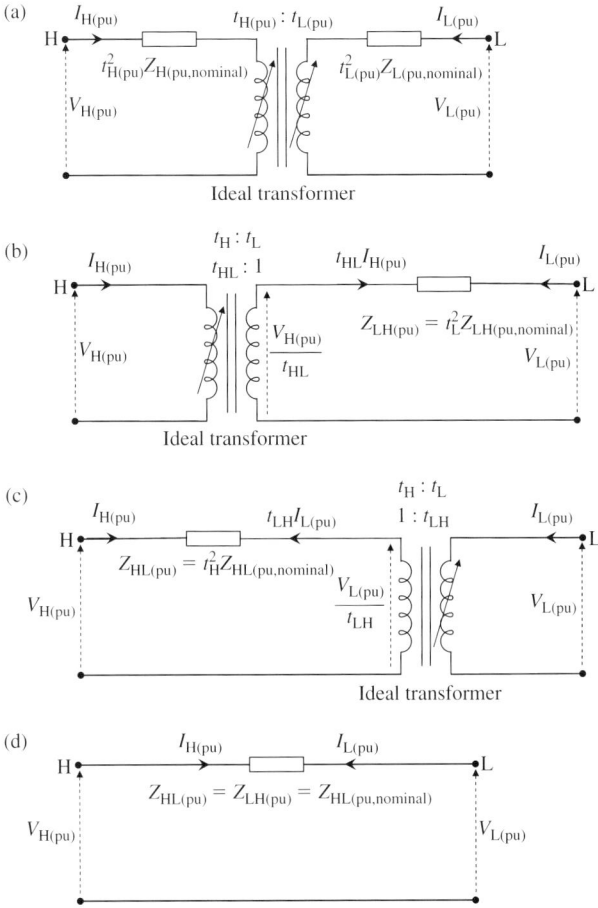

Figure 4.4 Transformer equivalent circuits in pu: (a) pu equivalent circuit of general off-nominal-ratio transformer; (b) pu equivalent circuit with H side off-nominal ratio; (c) pu equivalent circuit with L side off-nominal ratio and (d) pu equivalent circuit of nominal-ratio transformer

$$t_{H(pu)} = \frac{N_{H(at\ a\ given\ HV\ tap\ position)}}{N_{H(at\ nominal\ HV\ tap\ position)}}$$

$$= \frac{\text{Rated voltage of transformer HV winding for given tap position}}{\text{Rated voltage of transformer HV winding for nominal tap position}}$$

(4.11e)

$$t_{L(pu)} = \frac{N_{L(at\ a\ given\ LV\ tap\ position)}}{N_{L(at\ nominal\ LV\ tap\ position)}}$$

$$= \frac{\text{Rated voltage of transformer LV winding for given tap position}}{\text{Rated voltage of transformer LV winding for nominal tap position}}$$

(4.11f)

Figure 4.4(b) and (c) represent Equations (4.11a) and (4.11b), respectively. Both equivalents are identical in terms of representing the general equivalent circuit of an off-nominal ratio transformer. However, in practice, it is more convenient to use one or the other depending on the actual transformer characteristics. That is, Figure 4.4(b) would be used to represent a transformer with the following characteristics:

(a) The HV side is equipped with an on-load tap-changer and the LV side is fixed nominal or fixed off-nominal ratio.
(b) The HV side is fixed off-nominal ratio and the LV side is fixed nominal ratio or fixed off-nominal ratio.

Figure 4.4(c), however, would be used to represent a transformer with the following characteristics:

(a) The LV side is equipped with an on-load tap-changer and the HV side is fixed nominal or fixed off-nominal ratio.
(b) The LV side is fixed off-nominal ratio and the HV side is fixed nominal ratio or fixed off-nominal ratio.

We will make a number of important practical observations that apply to Figure 4.4(b) and (c). Considering Figure 4.4(b), and in the case of a tap-changer located on the HV winding with a fixed nominal or off-nominal ratio on the LV winding, then only t_H changes but Z_{LH} does not change with HV tap position. Similar observation applies to Figure 4.4(c). In both circuits, the variable tap ratio of the ideal transformer, t_{HL} or t_{LH}, is located on the side equipped with the on-load tap-changer. That is, it is located away from the transformer impedance which has a value that correspond to the fixed nominal or off-nominal ratio of the other side. It should be noted that the location of the variable tap ratio is a purely arbitrary choice and it could have been placed on the impedance side provided that Equations (4.11a) and (4.11b) continue to be satisfied.

A nominal-ratio-transformer is one where the chosen base voltages of the network sections to which the transformer is connected are the same as the transformer turns ratio or rated voltages. For such a transformer, Equation (4.7) reduces to $t_H = 1$ and $t_L = 1$, hence the ideal transformers shown in Figure 4.4(b) and (c) can be eliminated as shown in Figure 4.4(d).

However, for many transformers used in power systems, the transformer voltage ratio is not the same as the ratio of base or rated voltages of the associated network sections. Therefore, the choice of voltage base quantities for the network has to be made independent of the transformer actual voltage ratio or turns ratio. The most common reason for this is that a very large number of transformers used in power systems are equipped with tap-changers, on the HV or LV winding, whose function is to vary the number of turns of their associated winding. In addition, some distribution transformers, in particular, can have one winding with a fixed off-nominal ratio.

To ensure a proper understanding of the concepts introduced in deriving Figure 4.4, we will present a number of practical cases. In the first case, consider a single-phase 11 kV/0.25 kV distribution transformer interconnecting an 11 kV network and a 0.25 kV network. The transformation ratio or turns ratio of 11–0.25 kV corresponds exactly to the network base voltages. Therefore, Figure 4.4(d) represents such a transformer. We also note that with the transformer on open circuit, $V_{H(pu)} = V_{L(pu)}$ but with the transformer on-load, $I_{H(pu)} = -I_{L(pu)}$.

In the second case, we consider a 10 kV/0.25 kV single-phase transformer interconnecting an 11 kV network and a 0.25 kV network. The HV winding is clearly a fixed off-nominal-ratio winding. The pu tap ratio t_{HL} in Figure 4.4(b) is calculated as

$$t_{HL} = \frac{t_H}{t_L} = \frac{10/11}{0.25/0.25} = \frac{10}{11} = 0.9091$$

In the third case, we consider a transformer equipped with an on-load tap-changer on the HV side and we will also consider the effect on the transformer impedance. Consider a 239 MVA, 231 kV/21.5 kV single-phase transformer (this is actually one of a 717 MVA, 400 kV/21.5 kV three-phase bank) having a nominal-ratio leakage reactance of 15% on rating (the resistance is ignored for the purposes of this example but not in practical calculations). The transformer is connected to a network with 231 and 21.5 kV base voltages. The transformer, therefore, has a fixed LV nominal-ratio, i.e. $t_L = 1$. The tap-changer is located on the 231 kV winding hence t_H is variable. We will calculate the transformer reactance and redraw the equivalent circuit of Figure 4.4(b) to correspond to the transformation ratios 207.9 kV/21.5 kV, 231 kV/21.5 kV and 254.1 kV/21.5 kV. The three voltage transformation ratios correspond to t_H values of 0.9, 1.0 and 1.1, respectively. Clearly, the transformer impedance value does not change from nominal as illustrated in Figure 4.5(a). It is instructive for the reader to demonstrate that if the tap ratio is placed on the impedance side, then, by making appropriate use of Figure 4.4(c), the reactance values that correspond to $t_H = 0.9$, 1.0 and 1.1 would become 12.15%, 15% and 18.15%, respectively. This is illustrated in Figure 4.5(b). The reader should also demonstrate that the corresponding circuits of Figure 4.5(a) and (b) are equivalent to each other. This can be done by calculating, for the corresponding equivalent circuits, the short-circuit impedances, i.e. the impedance seen from one side, e.g. H side with the other side, L side, short-circuited and vice versa. The calculated impedances will be identical.

The variation of the tap position of a transformer equipped with an on-load tap-changer results in a value of t_{HL}, for most transformers used in electric power systems, to be generally within 0.8 and 1.2 pu.

π Equivalent circuit model

The ideal transformer in the general pu equivalent circuit of an off-nominal-ratio transformer shown in Figure 4.4(b) and (c) is, although suitable for use on network analysers, not satisfactory for modern calculations that are almost entirely based on digital computation. Therefore, the ideal transformer must be replaced with a suitable equivalent circuit and this will be achieved by deriving an equivalent

(a)

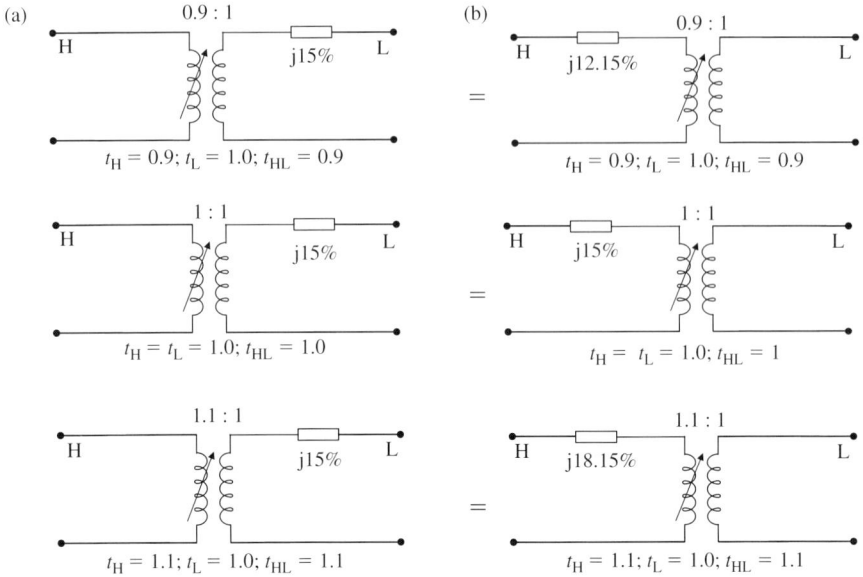

$$0.9 : 1$$

H · · · L

$$j15\%$$

$$t_H = 0.9; t_L = 1.0; t_{HL} = 0.9$$

$$1 : 1$$

H · · · L

$$j15\%$$

$$t_H = t_L = 1.0; t_{HL} = 1.0$$

$$1.1 : 1$$

H · · · L

$$j15\%$$

$$t_H = 1.1; t_L = 1.0; t_{HL} = 1.1$$

(b)

$$0.9 : 1$$

H · · · L

$$j12.15\%$$

$$t_H = 0.9; t_L = 1.0; t_{HL} = 0.9$$

$$1 : 1$$

H · · · L

$$j15\%$$

$$t_H = t_L = 1.0; t_{HL} = 1$$

$$1.1 : 1$$

H · · · L

$$j18.15\%$$

$$t_H = 1.1; t_L = 1.0; t_{HL} = 1.1$$

Figure 4.5 Example of variation of transformer impedance with tap position

π circuit model for Figure 4.4(b). The reader is encouraged to do so for Figure 4.4(c). For convenience, we will drop the explicit pu notation and recall that all quantities are in pu. From Figure 4.4(b), we can write

$$I_L = \frac{V_L - (V_H/t_{HL})}{Z_{LH}}$$

or

$$I_L = \frac{-1}{t_{HL}Z_{LH}} V_H + \frac{1}{Z_{LH}} V_L \qquad (4.12a)$$

also

$$-t_{HL}I_H = I_L$$

or

$$I_H = \frac{1}{t_{HL}^2 Z_{LH}} V_H - \frac{1}{t_{HL}Z_{LH}} V_L \qquad (4.12b)$$

or in matrix form

$$\begin{bmatrix} I_H \\ I_L \end{bmatrix} = \begin{bmatrix} \dfrac{1}{t_{HL}^2 Z_{LH}} & \dfrac{-1}{t_{HL}Z_{LH}} \\ \dfrac{-1}{t_{HL}Z_{LH}} & \dfrac{1}{Z_{LH}} \end{bmatrix} \begin{bmatrix} V_H \\ V_L \end{bmatrix} \qquad (4.13a)$$

The admittance matrix of Figure 4.4(c) can be similarly derived and the result is

$$
\begin{bmatrix} I_H \\ I_L \end{bmatrix} = \begin{bmatrix} \dfrac{1}{Z_{HL}} & \dfrac{-1}{t_{LH}Z_{HL}} \\ \dfrac{-1}{t_{LH}Z_{HL}} & \dfrac{1}{t_{LH}^2 Z_{HL}} \end{bmatrix} \begin{bmatrix} V_H \\ V_L \end{bmatrix}
\tag{4.13b}
$$

The equations of the general equivalent π circuit shown in Figure 4.6(a) are

$$
I_H = \left(Y_B + \frac{1}{Z_A} \right) V_H - \frac{1}{Z_A} V_L
\tag{4.14a}
$$

$$
I_L = -\frac{1}{Z_A} V_H + \left(Y_C + \frac{1}{Z_A} \right) V_L
\tag{4.14b}
$$

or in matrix form

$$
\begin{bmatrix} I_H \\ I_L \end{bmatrix} = \begin{bmatrix} Y_B + \dfrac{1}{Z_A} & \dfrac{-1}{Z_A} \\ \dfrac{-1}{Z_A} & Y_C + \dfrac{1}{Z_A} \end{bmatrix} \begin{bmatrix} V_H \\ V_L \end{bmatrix}
\tag{4.14c}
$$

Equating the admittance terms in Equations (4.13a) and (4.14c), we have

$$
Z_A = t_{HL} Z_{LH}
\tag{4.15a}
$$

and

$$
Y_B = \frac{t_{HL} - 1}{t_{HL} Z_{LH}} \left(\frac{-1}{t_{HL}} \right)
\tag{4.15b}
$$

and

$$
Y_C = \frac{t_{HL} - 1}{t_{HL} Z_{LH}}
\tag{4.15c}
$$

Similarly, equating the admittance terms in Equations (4.13b) and (4.14c), we have

$$
Z_A = t_{LH} Z_{HL}
\tag{4.16a}
$$

and

$$
Y_B = \frac{t_{LH} - 1}{t_{LH} Z_{HL}}
\tag{4.16b}
$$

and

$$
Y_C = \frac{t_{LH} - 1}{t_{LH} Z_{HL}} \left(\frac{-1}{t_{LH}} \right)
\tag{4.16c}
$$

The equivalent π circuit models of the off-nominal-ratio transformer of Figure 4.4(b) and (c) are shown in Figure 4.6(b) and (c), respectively. The equivalent π circuit model parameters for the three conditions shown in Figure 4.5(a) are calculated and shown in Figure 4.7. Also shown in Figure 4.7 are the short-circuit impedances calculated from side H with side L short-circuited, and vice versa.

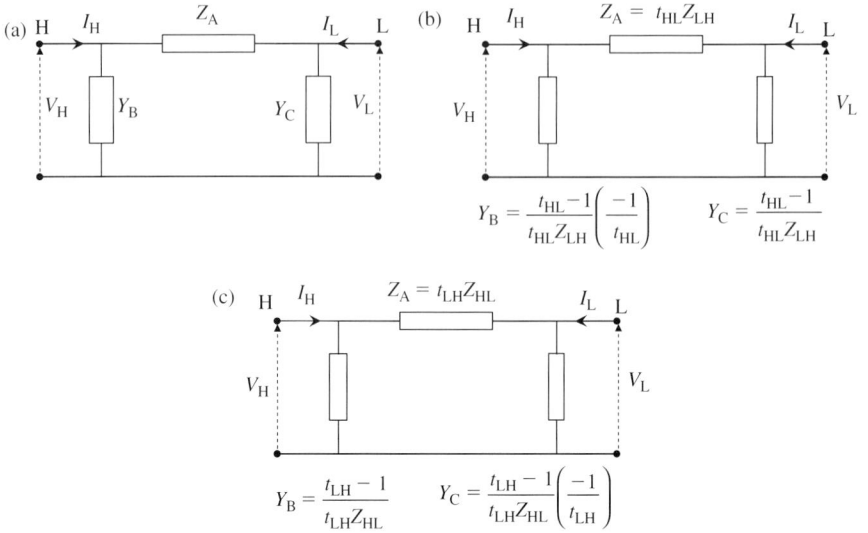

Figure 4.6 π equivalent circuit of an off-nominal-ratio transformer: (a) general π equivalent circuit, (b) π equivalent circuit of Figure 4.4(b) and (c) π equivalent circuit of Figure 4.4(c)

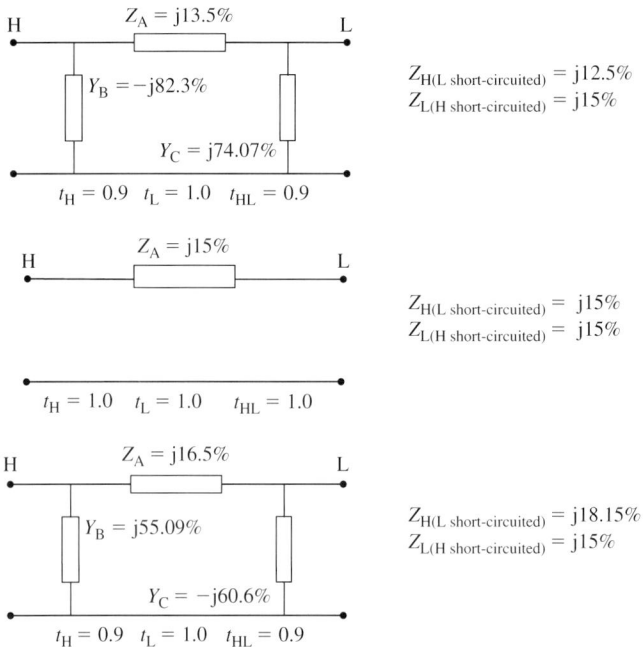

Figure 4.7 π equivalent circuit of example shown in Figure 4.5(a)

It should not come as a surprise to the reader that these impedances are identical to the ones that the reader may have calculated, as recommended previously, for the equivalent circuits shown in Figure 4.5(a) and (b).

As mentioned before, it is generally no longer the case in modern industry practice to neglect the shunt exciting branch representing the iron losses and magnetising losses. The exciting branch admittance can be calculated from the corresponding impedance and inserted in parallel with either Y_B or Y_C or it can also be inserted as a shunt admittance midway through the series impedance. There is also a usual approximation to split it in half and connect each half at either end of the equivalent circuit. The admittance of the exciting or magnetising branch is given by

$$Y_0 = G_I - jB_M \quad \text{where} \quad G_I = \frac{1}{R_I} \quad \text{and} \quad B_M = \frac{1}{X_M}$$

The general 2×2 admittance matrix is given by

$$\begin{bmatrix} I_H \\ I_L \end{bmatrix} = \begin{bmatrix} Y_{11} & Y_{12} \\ Y_{21} & Y_{22} \end{bmatrix} \begin{bmatrix} V_H \\ V_L \end{bmatrix} \tag{4.17a}$$

whose elements, using Equation (4.15), are given by

$$Y_{11} = \frac{1}{t_{HL}^2 Z_{LH}} \quad Y_{12} = Y_{21} = \frac{-1}{t_{HL} Z_{LH}} \quad Y_{22} = \frac{1}{Z_{LH}} \tag{4.17b}$$

whereas, using Equation (4.16), the matrix elements are given by

$$Y_{11} = \frac{1}{Z_{HL}} \quad Y_{12} = Y_{21} = \frac{-1}{t_{LH} Z_{HL}} \quad Y_{22} = \frac{1}{t_{LH}^2 Z_{HL}} \tag{4.17c}$$

4.2.3 Three-phase two-winding transformers

The majority of power transformers used in power systems are three-phase transformers since these have significant economic benefits compared to three banks of single-phase units that have the same total rating and perform similar duties. On the other hand, reliability considerations for very large three-phase transformers may result in the use of three banks of single-phase units. In double-wound transformers, there is a low voltage winding and an electrically separate high voltage winding.

Winding connections

In three-phase double-wound power transformers, the three-phases of each winding are usually connected as star or delta. There are thus four possible connections of the primary and secondary windings namely star–star, star–delta, delta–star and delta–delta connections. Interconnected star or zig-zag windings are dealt with in Section 4.2.6.

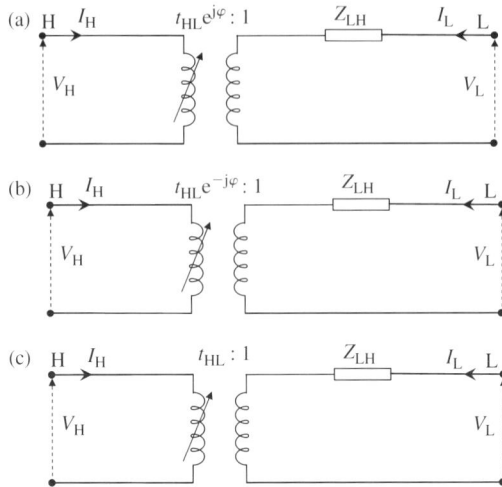

Figure 4.8 PPS and NPS equivalent circuits of Figure 4.4(b) for an off-nominal ratio two-winding three-phase transformer: (a) PPS equivalent circuit, (b) NPS equivalent circuit and (c) PPS and NPS equivalent circuit ignoring phase shift

PPS and NPS equivalent circuits

Like all static three-phase power system plant, the impedance such plant presents to the flow of PPS or NPS currents is independent of the phase rotation or sequence of the three-phase applied voltages R, Y, B or R, B, Y. Therefore, the plant PPS and NPS impedances are equal.

It will be shown later in this section that three-phase transformers can introduce a phase shift between the primary and secondary winding currents, and the same phase shift between the primary and secondary voltages, depending on the winding connections. In addition, where the phase shift of PPS currents and voltages is φ, it is $-\varphi$ for NPS currents and voltages. The effect of this phase shift is to change the ideal transformer off-nominal turns ratio from a real to a complex number.

Therefore, the general PPS and NPS equivalent circuits of a three-phase two-winding transformer that correspond to the single-phase equivalent circuit of Figure 4.4(b) are shown in Figure 4.8(a) and (b), respectively. Similarly, the general PPS and NPS equivalent circuits of a three-phase two-winding transformer that correspond to the single-phase equivalent circuit of Figure 4.4(c) are shown in Figure 4.9(a) and (b), respectively. As discussed later in this section, the phase shift that may be introduced by a three-phase transformer is not required to be represented initially in short-circuit analysis (or PPS based power flow or stability analysis). Due account of any phase shifts can be made from knowledge of the location of transformers that introduce phase shifts in the system once the sequence currents and voltages are calculated by initially ignoring the phase shifts. Therefore, the transformer PPS and NPS equivalent circuits, ignoring the phase shift, are shown in Figures 4.8(c) and 4.9(c) and the corresponding π equivalent circuit

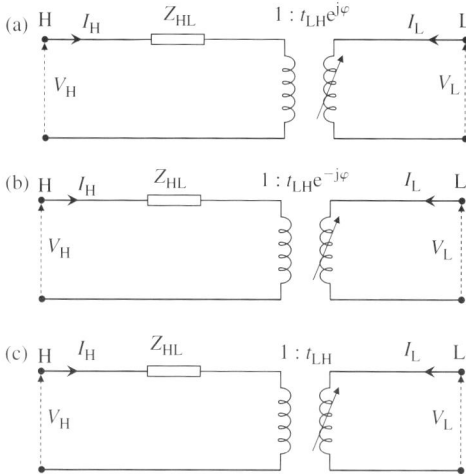

Figure 4.9 PPS and NPS equivalent circuits of Figure 4.4(c) for an off-nominal ratio two-winding three-phase transformer: (a) PPS equivalent circuit, (b) NPS equivalent circuit and (c) PPS and NPS equivalent circuit ignoring phase shift

model is the same as that in Figure 4.6(b) and (c). It should be noted that for a three-phase transformer, the transformer nominal turns ratio is equal to the ratio of the phase to phase base voltages on the HV and LV sides of the transformer irrespective of the primary and secondary winding connections. This means that the base voltage ratio and the nominal turns ratio are equal for star–star and delta–delta winding connections but also includes the factor $\sqrt{3}$ for star–delta winding connection.

ZPS equivalent circuits

The ZPS equivalent circuit of a two-winding transformer is primarily dependent on the method of connection of the primary and secondary windings because the ZPS currents in each phase are equal in magnitude and are in phase. Also, the ZPS equivalent circuit and currents are affected by the winding earthing arrangements. In addition, a practical factor that can influence the magnitude of the ZPS impedances and make them appreciably different from the PPS impedances is the type of construction of the transformer magnetic circuit. The effect of transformer core construction will be dealt with later in this section, but we will now focus on the effect of the winding connection and any neutral earthing that may be present.

Before considering three-phase two-winding transformers of various winding connections, we will restate a basic understanding regarding ZPS currents in simple star-connected or delta-connected three sets of impedances. For ZPS currents to flow in a star-connected set of impedances, the neutral point of the star should be earthed either directly or indirectly, e.g. through an impedance. The ZPS current that flows to earth through this path is the sum of the ZPS currents in the three phases of the star impedances or three times the phase R ZPS current. For a delta-connected three-phase set of impedances, no ZPS currents will flow in the output terminals and hence inside the delta connection. However, ZPS currents can circulate inside the delta by mutual coupling but no ZPS current can emerge out into the output terminals of the delta. Figure 4.10(a) shows the impedance connections and ZPS current flows for the above three cases and Figure 4.10(b) shows the three-phase circuits and their corresponding ZPS equivalent circuits. It should be noted that because the voltage drop across the earthing impedance Z_E is $3I^Z Z_E$, the effective earthing impedance appearing in the ZPS equivalent circuit is $3Z_E$. The reader should easily demonstrate this.

We will now derive approximate ZPS equivalent circuits for two-winding transformers of various winding connection arrangements. We emphasise that these are approximate because for now we ignore the effect of transformer core or magnetic circuit construction on the magnetising impedance. This is an important practical aspect that should not be ignored in practical short-circuit analysis. This is discussed in detail in Section 4.2.8.

The basic and fundamental principle that underpins the derivation of the approximate ZPS equivalent circuits is based on Equation (4.1b). This states that there is always a magnetic circuit MMF or ampere-turn balance for the transformer primary and secondary windings. That is, no current can flow in one winding unless a corresponding current, allowing for the winding turns ratio, flows in the other winding. In deriving the ZPS equivalent circuits of two-winding transformers of various winding connections, the winding terminal is connected to the external circuit if ZPS current can flow into and out of the winding. If ZPS current can circulate inside the winding without flowing in the external circuit, the winding terminal is connected directly to the reference zero voltage node. Figure 4.11 shows ZPS equivalent circuits for an off-nominal ratio transformer that correspond to

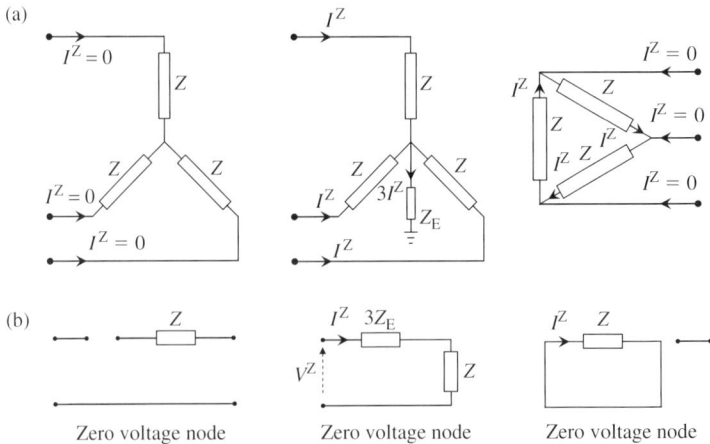

Figure 4.10 ZPS equivalent circuits for three-phase impedances connected in star or delta: (a) three-phase star- and delta-connected impedances and (b) ZPS equivalent circuits

Figures 4.8(c) and 4.9(c). These take account of any HV and LV winding connection arrangements, and any neutral impedance earthing that may be used.

By making use of Figure 4.11(a) and (b), and noting that an arrow on a winding signifies the presence of a tap-changer on this winding, Figure 4.12 can be derived. This shows approximate ZPS equivalent circuits for the most common three-phase

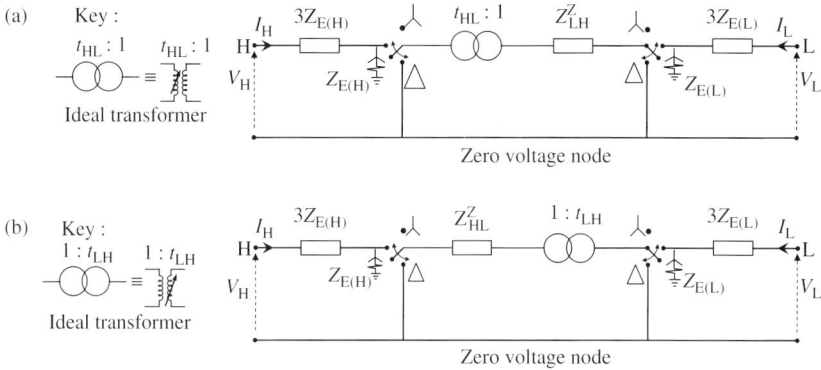

Figure 4.11 Generic ZPS equivalent circuits for two-winding transformers: (a) generic ZPS equivalent circuit of Figure 4.8(c) and (b) Generic ZPS equivalent circuit of Figure 4.9(c)

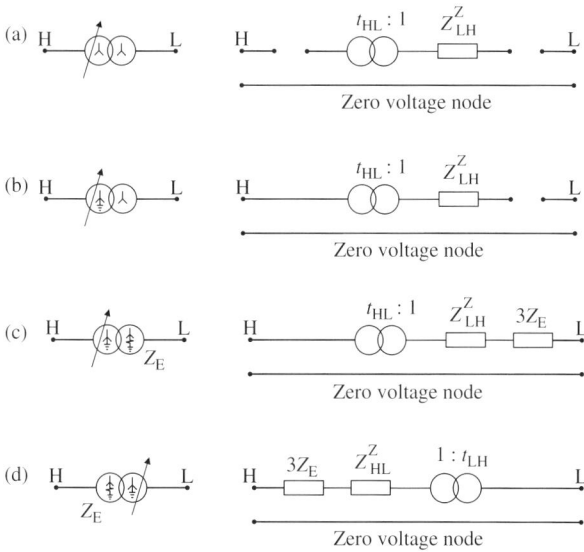

Figure 4.12 Approximate ZPS equivalent circuits for common three-phase two-winding transformers: (a) star–star with isolated neutrals; (b) star with solidly earthed neutral – star isolated neutral; (c) star solidly earthed – star neutral earthing impedance; (d) star neutral earthing impedance – star solidly earthed neutral; (e) star isolated neutral – delta; (f) delta–delta; (g) star neutral earthing impedance – delta and (h) delta–star neutral earthing impedance

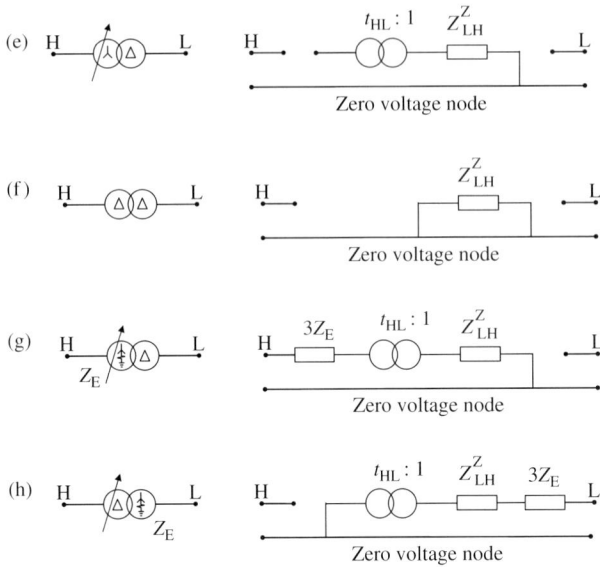

(e) H ⟨Y△⟩ L   H ⊶⊶ $t_{HL}:1$ Z_{LH}^Z ▭ L   Zero voltage node

(f) H ⟨△△⟩ L   H Z_{LH}^Z ▭ L   Zero voltage node

(g) H ⟨✦△⟩ L Z_E H $3Z_E$ ▭ $t_{HL}:1$ Z_{LH}^Z ▭ L   Zero voltage node

(h) H ⟨△✦⟩ L Z_E H $t_{HL}:1$ Z_{LH}^Z $3Z_E$ ▭ L   Zero voltage node

Figure 4.12 (*Continued*)

two-winding transformers used in power systems. It should be noted that Z_E is in per unit on the voltage base of the winding to which it is connected.

Returning to the example shown in Figure 4.5, and assuming the transformer is a three-phase two-winding star (HV winding)–delta (LV winding) transformer, the ZPS equivalent circuits that correspond to each tap ratio condition are shown in Figure 4.13. The reader should find it instructive to obtain the three equivalent circuits shown in Figure 4.13 from the corresponding equivalent π circuits shown in Figure 4.7 by applying a short-circuit at end L and calculating the equivalent impedance seen from the H side.

The elements of the admittance matrix of Equation (4.17a), for the transformer winding connections and their equivalent circuits shown in Figure 4.12, can be easily derived using the following equations:

$$Y_{11} = \frac{I_H}{V_H}\bigg|_{V_L=0} \quad Y_{22} = \frac{I_L}{V_L}\bigg|_{V_H=0} \quad Y_{12} = \frac{I_H}{V_L}\bigg|_{V_H=0} \quad Y_{21} = \frac{I_L}{V_H}\bigg|_{V_L=0} \quad Y_{12} = Y_{21}$$

The results are shown in Table 4.1. The reader is encouraged to derive these results.

Effect of winding connection phase shifts on sequence voltages and currents

The effect of three-phase transformer phase shifts on the sequence currents and voltages will now be considered. The presence of a phase shift between the transformer primary and secondary voltages, and currents, depends on the transformer

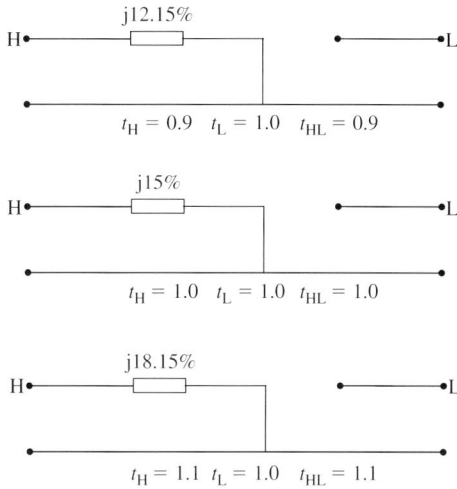

H•————[ ]————   •————————•L

j12.15%

$t_H = 0.9$ $t_L = 1.0$ $t_{HL} = 0.9$

H•————[ ]————   •————————•L

j15%

$t_H = 1.0$ $t_L = 1.0$ $t_{HL} = 1.0$

H•————[ ]————   •————————•L

j18.15%

$t_H = 1.1$ $t_L = 1.0$ $t_{HL} = 1.1$

Figure 4.13   ZPS equivalent circuits of two-winding transformer of Figure (4.5)

primary and secondary winding connections. For transformers with a star–star or a delta–delta winding connections, the primary and secondary currents, and voltages, in each of the three phases are either in phase or out of phase, i.e. the windings are connected so that the phase shifts are either 0° or ±180°. The former case is illustrated in Figure 4.14(a) and (b). British and IEC practices use 'vector group reference' number and symbol. In the symbol Yd1, the capital and small letters Y and d indicate HV winding star and LV winding delta connections, respectively, and the digit 1 indicates a phase shift of −30° using a 12 × 30° clock reference. For example, 0° indicates 12 o'clock, 180° indicates 6 o'clock, −30° indicates 1 o'clock and +30° indicates 11 o'clock.

In Figure 4.14, the 0° phase shift is achieved by ensuring that the parallel windings, i.e. same phase windings, are linked by the same magnetic flux. Figure 4.14 also shows that the absence of phase shifts in the phase currents and voltages also translates into the PPS, and NPS, currents and voltages. Consequently, the presence of such transformers in the three-phase network requires no special treatment in the formed PPS and NPS networks under balanced or unbalanced conditions. It should be noted that for the delta winding, although a physical neutral point does not exist, a voltage from each phase terminal to neutral does still exist because the network to which the delta winding is connected would in practice contain a neutral point.

In the case of transformers with windings connected in star–delta (or delta–star), voltages and currents on the star winding side will be phase shifted by a ±30° angle with respect to those on the delta side (or vice versa depending on the chosen reference). According to British practice, Yd11 results in the PPS phase-to-neutral voltages on the star side lagging by 30° the corresponding ones on the delta side. Also, Yd1 results in the PPS phase-to-neutral voltages on the star

Table 4.1   Elements of ZPS admittance matrix of two-winding three-phase transformers

Figure	Two-winding three-phase transformer winding connections	Elements of ZPS admittance matrix of Equation (4.19b)
4.12(a)	H — L	$Y_{11} = Y_{12} = Y_{21} = Y_{22} = 0$
4.12(b)	H — L	$Y_{11} = Y_{12} = Y_{21} = Y_{22} = 0$
4.12(c)	H — L, Z_E	$Y_{11} = \dfrac{1}{t_{HL}^2 (Z_{LH}^Z + 3Z_E)}$ $Y_{12} = Y_{21} = \dfrac{-1}{t_{HL}(Z_{LH}^Z + 3Z_E)}$ $Y_{22} = \dfrac{1}{Z_{LH}^Z + 3Z_E}$
4.12(d)	H — L, Z_E	$Y_{11} = \dfrac{1}{Z_{HL}^Z + 3Z_E}$ $Y_{12} = Y_{21} = \dfrac{-1}{t_{LH}(Z_{HL}^Z + 3Z_E)}$ $Y_{22} = \dfrac{1}{t_{LH}^2 (Z_{HL}^Z + 3Z_E)}$
4.12(e)	H — L	$Y_{11} = Y_{12} = Y_{21} = Y_{22} = 0$
4.12(f)	H — L	$Y_{11} = Y_{12} = Y_{21} = Y_{22} = 0$
4.12(g)	H — L, Z_E	$Y_{11} = \dfrac{1}{t_{HL}^2 Z_{LH}^Z + 3Z_E}$ $Y_{12} = Y_{21} = Y_{22} = 0$
4.12(h)	H — L, Z_E	$Y_{11} = Y_{12} = Y_{21} = 0$ $Y_{22} = \dfrac{1}{Z_{LH}^Z + 3Z_E}$

side leading by 30° the corresponding ones on the delta side. The example vector diagrams shown in Figure 4.15 for Yd1 and Yd11 illustrate this effect.

For RBY/rby NPS phase sequence or rotation, Figure 4.15 also shows the effect of the Yd1 and Yd11 on the NPS voltage phase shifts and show that these are now reversed. These phase shifts also apply to the PPS and NPS currents in these windings because the phase angles of the currents with respect to their associated voltages are determined by the balanced load impedances only. In summary, where

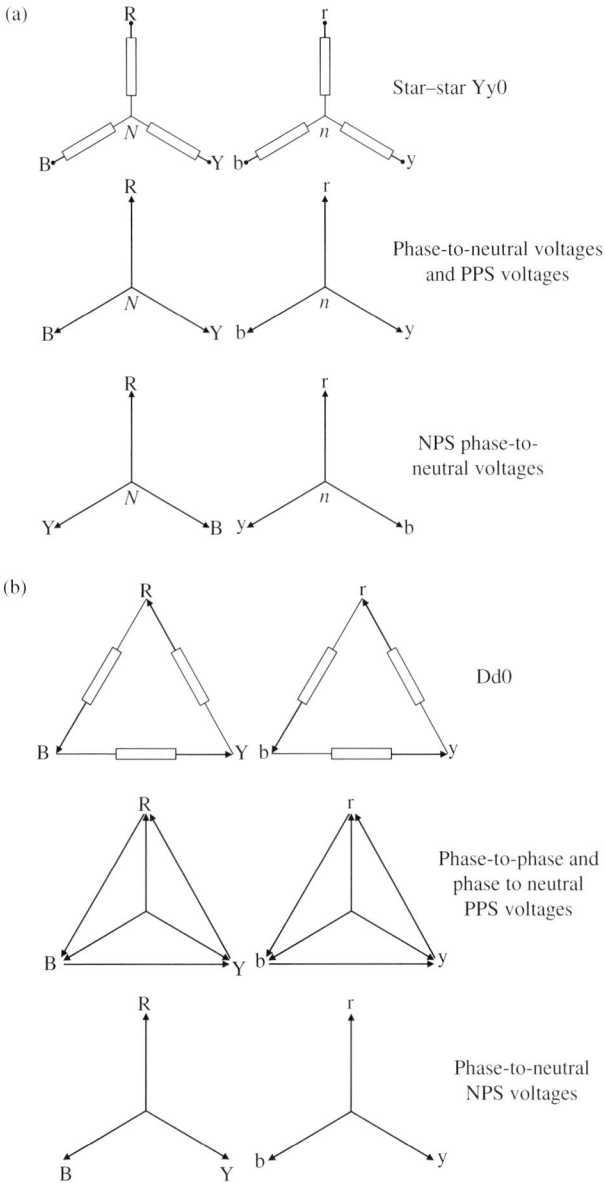

(a)

Star–star Yy0

Phase-to-neutral voltages
and PPS voltages

NPS phase-to-
neutral voltages

(b)

Dd0

Phase-to-phase and
phase to neutral
PPS voltages

Phase-to-neutral
NPS voltages

Figure 4.14   PPS and NPS voltage phase shifts for Yy0 and Dd0 connected transformers

the PPS voltages and currents are shifted by $+30°$, the corresponding NPS voltages and currents are shifted by $-30°$ and vice versa depending on the specified connection and phase shift, i.e. Yd1 or Yd11. Mathematically, this is derived for the Yd1 transformer shown in Figure 4.15 where n is the turns ratio as follows. The red phase

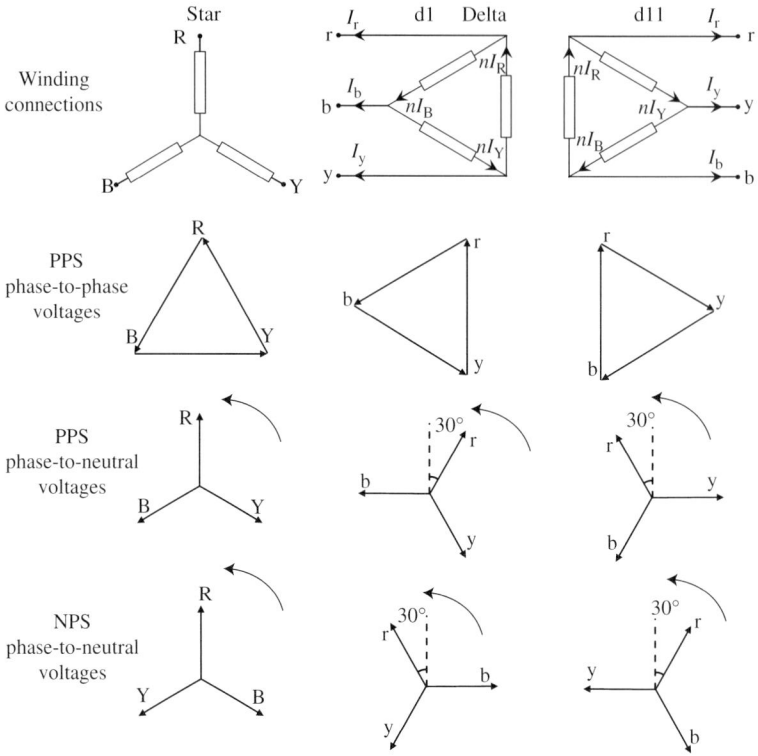

Figure 4.15   PPS and NPS voltage phase shifts for Yd1 and Yd11 transformers

current in amps flowing out of the d winding r phase is equal to $I_r = n(I_R - I_B)$. Using Equation (2.9) from Chapter 2 for phase currents and noting that $I_R^Z = 0$ because the in-phase ZPS currents cannot exit the d winding, we can write

$$I_r = n[(1 - h)I_R^P + (1 - h^2)I_R^N] = n\sqrt{3}I_R^P e^{-j30°} + n\sqrt{3}I_R^N e^{j30°}$$

or

$$I_r = I_r^P + I_r^N$$

where

$$I_r^P = n\sqrt{3}I_R^P e^{-j30°} \quad \text{and} \quad I_r^N = n\sqrt{3}I_R^N e^{j30°} \qquad (4.18a)$$

or

$$I_R^P = \frac{1}{n\sqrt{3}}I_r^P e^{j30°} \quad \text{and} \quad I_R^N = \frac{1}{n\sqrt{3}}I_r^N e^{-j30°} \qquad (4.18b)$$

or in per unit, where $n = \frac{1}{\sqrt{3}}$,

$$I_R^P = I_r^P e^{j30°} \quad \text{and} \quad I_R^N = I_r^N e^{-j30°} \qquad (4.18c)$$

Similarly, from Figure 4.15, the phase-to-neutral voltage in volts on the star winding R phase is

$$V_R = n(V_r - V_y)$$

and using Equation (2.9) for phase r and y voltages, we have

$$V_R = n[(1 - h^2)V_r^P + (1 - h)V_r^N] = n\sqrt{3}V_r^P e^{j30°} + n\sqrt{3}V_r^N e^{-j30°}$$

or

$$V_R = V_R^P + V_R^N$$

where

$$V_R^P = n\sqrt{3}V_r^P e^{j30°} \quad \text{and} \quad V_R^N = n\sqrt{3}V_r^N e^{-j30°} \tag{4.19a}$$

or

$$V_r^P = \frac{1}{n\sqrt{3}}V_R^P e^{-j30°} \quad \text{and} \quad V_r^N = \frac{1}{n\sqrt{3}}V_R^N e^{j30°} \tag{4.19b}$$

or in per unit, where $n = \frac{1}{\sqrt{3}}$,

$$V_R^P = V_r^P e^{j30°} \quad \text{and} \quad V_R^N = V_r^N e^{-j30°} \tag{4.19c}$$

The reader is encouraged to derive the equations for the Yd11 transformer.

The American standard for designating winding terminals on star–delta transformers requires that the PPS (NPS) phase-to-neutral voltages on the high voltage winding to lead (lag) the corresponding PPS (NPS) phase-to-neutral voltages on the low voltage winding. This is so regardless of whether the star or the delta winding is on the high voltage side. In terms of sequence analysis, this means that when stepping up from the low voltage to the high voltage side of a star–delta or delta–star transformer, the PPS voltages and currents should be advanced by 30° whereas the NPS voltages and currents should be retarded by 30°. It is interesting to note the following observation on the British and American standards. In American practice, when the star winding in a star–delta transformer is the high voltage winding, this would correspond, in terms of phase shifts, to the Yd1 in British practice. However, when, in American practice, the delta winding in a star–delta transformer is the high voltage winding, this would correspond, in terms of phase shifts, to the Yd11 in British practice.

In terms of fault analysis in power system networks using the PPS and NPS networks, it is common practice to initially 'ignore' the phase shifts introduced by all star–delta transformers by assuming them as equivalent star–star transformers and to calculate the sequence voltages and currents on this basis. Then, having noted the locations in the network of such star–delta transformers, the appropriate phase shifts can be easily applied using the above equations as appropriate for the specified Yd transformer.

4.2.4 Three-phase three-winding transformers

Three-phase three-winding transformers are widely used in power systems. When the VA rating of the third winding is appreciably lower than the primary or secondary winding ratings, the third winding is called a tertiary winding. These windings are usually used for the connection of reactive compensation plant such as shunt rectors, shunt capacitors, static variable compensators or synchronous compensators or synchronous condensers. Tertiary windings may also be used to supply auxiliary load in substations and to generators. Delta-connected tertiary windings may also be used (and left unloaded) in order to provide a low impedance path to ZPS triplen harmonic currents. The *B–H* curve of the transformer magnetic circuit is non-linear and, under normal conditions, transformers do operate on the non-linear knee part of the curve. Thus, for a sinusoidal primary voltage, the magnetising current will be non-linear and will contain harmonic components which are mainly third harmonics. Since in three-phase systems, the third-order harmonic currents in each phase are in phase, they can be considered as ZPS currents of three times the fundamental frequency. Consequently, as for the ZPS fundamental frequency currents, the tertiary winding allows the circulation of third-order harmonic currents. Other benefits of Delta-connected tertiary windings is an improvement, i.e. a reduction in the unbalance of three-phase voltages.

Sometimes, for economic benefits, a third winding is usually provided to form a transformer with double-secondary windings. These transformers tend to be used to supply high density load in cities and this also provide additional benefits of fewer HV switchgear and also limiting LV system short-circuit currents where the two secondary LV terminals are not connected to the same busbar. Other uses are for the connection of two generators to a power network or for the connection of networks operating at different voltage levels.

Winding connections

There are several winding connections for three-phase three-winding transformers which may be used in power systems. Some examples are YNdd, Ydd, YNynyn, YNynd, YNyd and Yyd.

PPS and NPS equivalent circuits

In a three-winding transformer, the three windings are generally mutually coupled although the degree of coupling between the two LV windings of a double secondary transformer can be chosen by design so that the LV windings are either closely or loosely coupled. Figure 4.16(a) shows a single-phase representation of a three-winding three-phase transformer ignoring, for now, the shunt exciting impedance.

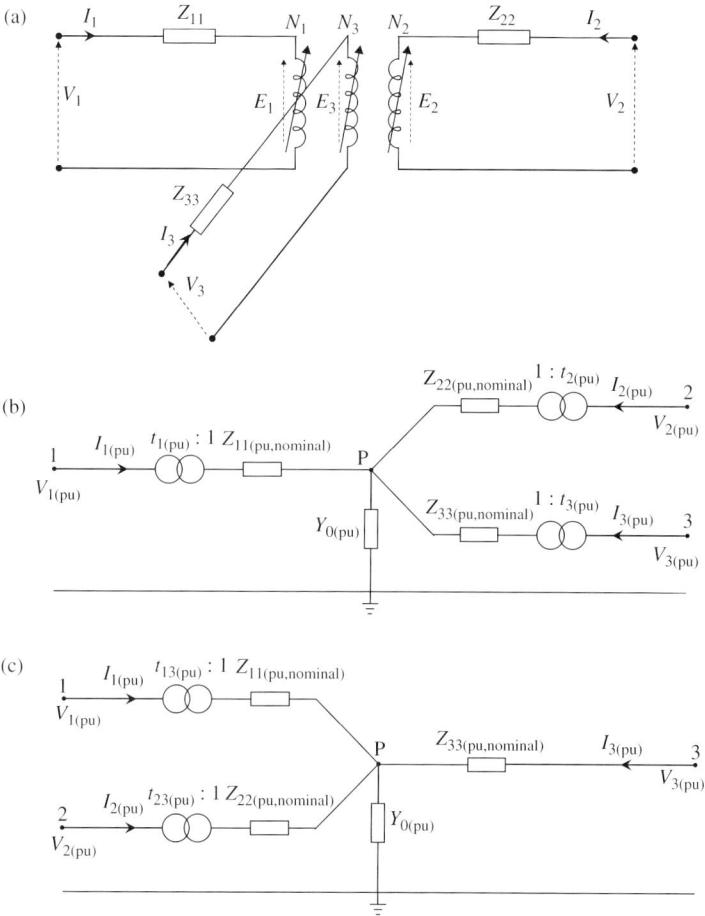

Figure 4.16 Equivalent circuits for three-winding three-phase transformers: (a) single-phase representation in actual physical units; (b) PPS and NPS equivalent circuit with three off-nominal turns ratios and (c) PPS and NPS equivalent circuit with two off-nominal turns ratios

From Figure 4.16(a), the following equations in actual physical units can be written

$$V_1 - E_1 = Z_{11}I_1 \tag{4.20a}$$

$$V_2 - E_2 = Z_{22}I_2 \tag{4.20b}$$

$$V_3 - E_3 = Z_{33}I_3 \tag{4.20c}$$

$$N_1I_1 + N_2I_2 + N_3I_3 = N_3I_0 \tag{4.20d}$$

Neglecting the no-load current, $I_0 = 0$, and dividing Equation (4.20d) by N_3, we obtain

$$N_{13}I_1 + N_{23}I_2 + I_3 = 0 \tag{4.21a}$$

where

$$N_{13} = \frac{N_1}{N_3} \quad \text{and} \quad N_{23} = \frac{N_2}{N_3} \tag{4.21b}$$

Also

$$\frac{E_1}{E_3} = \frac{N_1}{N_3} = N_{13} \quad \text{and} \quad \frac{E_2}{E_3} = \frac{N_2}{N_3} = N_{23} \tag{4.21c}$$

Using Equations (4.21b), (4.21c) and (4.20c), Equations (4.20a) and (4.20b) become

$$\frac{1}{N_1}(V_1 - Z_{11}I_1) = \frac{1}{N_3}(V_3 - Z_{33}I_3) \tag{4.22a}$$

$$\frac{1}{N_2}(V_2 - Z_{22}I_2) = \frac{1}{N_3}(V_3 - Z_{33}I_3) \tag{4.22b}$$

In most power system steady state analysis, we are more interested in deriving the three-winding transformer equivalent circuit in pu rather than in actual physical units. The following base and pu quantities are defined

$$V_{1(B)} = Z_{11(B)}I_{1(B)} \quad V_{2(B)} = Z_{22(B)}I_{2(B)} \quad \text{and} \quad V_{3(B)} = Z_{33(B)}I_{3(B)} \tag{4.23a}$$

$$V_{1(pu)} = \frac{V_1}{V_{1(B)}} \quad V_{2(pu)} = \frac{V_2}{V_{2(B)}} \quad \text{and} \quad V_{3(pu)} = \frac{V_3}{V_{3(B)}} \tag{4.23b}$$

$$I_{1(pu)} = \frac{I_1}{I_{1(B)}} \quad I_{2(pu)} = \frac{I_2}{I_{2(B)}} \quad \text{and} \quad I_{3(pu)} = \frac{I_3}{I_{3(B)}} \tag{4.23c}$$

$$Z_{11(pu)} = \frac{Z_{11}}{Z_{11(B)}} \quad Z_{22(pu)} = \frac{Z_{22}}{Z_{22(B)}} \quad \text{and} \quad Z_{33(pu)} = \frac{Z_{33}}{Z_{33(B)}} \tag{4.24a}$$

$$S_{1(B)} = S_{2(B)} = S_{3(B)} = V_{1(B)}I_{1(B)} = V_{2(B)}I_{2(B)} = V_{3(B)}I_{3(B)} \tag{4.24b}$$

$$\frac{V_{1(B)}}{V_{3(B)}} = \frac{I_{3(B)}}{I_{1(B)}} = \frac{N_{1(nominal)}}{N_{3(nominal)}} \quad \frac{V_{2(B)}}{V_{3(B)}} = \frac{I_{3(B)}}{I_{2(B)}} = \frac{N_{2(nominal)}}{N_{3(nominal)}} \tag{4.24c}$$

Using Equations (4.23) and (4.24), Equations (4.22a) and (4.22b) become

$$\frac{1}{\frac{N_1}{N_{1(nominal)}}}[V_{1(pu)} - I_{1(pu)}Z_{11(pu)}] = \frac{1}{\frac{N_3}{N_{3(nominal)}}}[V_{3(pu)} - Z_{33(pu)}I_{3(pu)}] \tag{4.25a}$$

$$\frac{1}{\frac{N_2}{N_{2(nominal)}}}[V_{2(pu)} - I_{2(pu)}Z_{22(pu)}] = \frac{1}{\frac{N_3}{N_{3(nominal)}}}[V_{3(pu)} - Z_{33(pu)}I_{3(pu)}] \tag{4.25b}$$

Equations (4.25a) and (4.25b) can be rewritten as follows:

$$\frac{1}{t_{1(pu)}}[V_{1(pu)} - I_{1(pu)}Z_{11(pu)}] = \frac{1}{t_{3(pu)}}[V_{3(pu)} - Z_{33(pu)}I_{3(pu)}] \tag{4.26a}$$

$$\frac{1}{t_{2(pu)}}[V_{2(pu)} - I_{2(pu)}Z_{22(pu)}] = \frac{1}{t_{3(pu)}}[V_{3(pu)} - Z_{33(pu)}I_{3(pu)}] \qquad (4.26b)$$

where, the following pu tap ratios are defined

$$t_{1(pu)} = \frac{N_1}{N_{1(nominal)}} = \frac{N_{1(at\ a\ given\ winding\ 1\ tap\ position)}}{N_{1(winding\ 1\ nominal\ tap\ position)}}$$

$$= \frac{\text{Rated voltage of winding 1 for given tap position}}{\text{Rated voltage of winding 1 for nominal tap position}}$$

$$(4.27a)$$

$$t_{2(pu)} = \frac{N_2}{N_{2(nominal)}} = \frac{N_{2(at\ a\ given\ winding\ 2\ tap\ position)}}{N_{2(winding\ 2\ nominal\ tap\ position)}}$$

$$= \frac{\text{Rated voltage of winding 2 for given tap position}}{\text{Rated voltage of winding 2 for nominal tap position}}$$

$$(4.27b)$$

$$t_{3(pu)} = \frac{N_3}{N_{3(nominal)}} = \frac{N_{3(at\ a\ given\ winding\ 3\ tap\ position)}}{N_{3(winding\ 3\ nominal\ tap\ position)}}$$

$$= \frac{\text{Rated voltage of winding 3 for given tap position}}{\text{Rated voltage of winding 3 for nominal tap position}}$$

$$(4.27c)$$

Now, as in the case of two-winding transformers, we define

$$Z_{11(pu)} = t_{1(pu)}^2 Z_{11(pu,nominal)} \qquad (4.28a)$$

$$Z_{22(pu)} = t_{2(pu)}^2 Z_{22(pu,nominal)} \qquad (4.28b)$$

$$Z_{33(pu)} = t_{3(pu)}^2 Z_{33(pu,nominal)} \qquad (4.28c)$$

Substituting Equation (4.28) into Equation (4.26), we obtain

$$\frac{V_{1(pu)}}{t_{1(pu)}} - t_{1(pu)}I_{1(pu)}Z_{11(pu,nominal)} = \frac{V_{3(pu)}}{t_{3(pu)}} - t_{3(pu)}I_{3(pu)}Z_{33(pu,nominal)} \qquad (4.29a)$$

$$\frac{V_{2(pu)}}{t_{2(pu)}} - t_{2(pu)}I_{2(pu)}Z_{22(pu,nominal)} = \frac{V_{3(pu)}}{t_{3(pu)}} - t_{3(pu)}I_{3(pu)}Z_{33(pu,nominal)} \qquad (4.29b)$$

Using Equations (4.24c) and (4.27) in Equation (4.21a), we obtain

$$t_{1(pu)}I_{1(pu)} + t_{2(pu)}I_{2(pu)} + t_{3(pu)}I_{3(pu)} = 0 \qquad (4.29c)$$

Equations (4.29) are represented by the equivalent circuit shown in Figure 4.16(b) where we have replaced the three-winding transformer with three two-winding transformers connected as three branches in a star configuration. Each branch corresponds to a winding and has its own general off-nominal tap ratio to represent a nominal, fixed off-nominal ratio or an on-load tap-changer, as required. The impedance of the magnetising branch Y_0 can be inserted at the fictitious point P in the star equivalent circuit.

For improved applications in practice, Figure 4.16(b) can be further simplified to include only two off-nominal tap ratios instead of three as follows. Let

$$t_{13(pu)} = \frac{t_{1(pu)}}{t_{3(pu)}} \tag{4.30a}$$

$$t_{23(pu)} = \frac{t_{2(pu)}}{t_{3(pu)}} \tag{4.30b}$$

$$Z_{11(pu)} = t_{13(pu)}^2 Z_{11(pu,nominal)} \tag{4.31a}$$

$$Z_{22(pu)} = t_{23(pu)}^2 Z_{22(pu,nominal)} \tag{4.31b}$$

and

$$Z_{33(pu)} = Z_{33(pu,nominal)} \tag{4.31c}$$

Substituting Equations (4.30) and (4.31) into Equation (4.26), we obtain

$$\frac{V_{1(pu)}}{t_{13(pu)}} - t_{13(pu)}I_{1(pu)}Z_{11(pu,nominal)} = V_{3(pu)} - Z_{33(pu,nominal)}I_{3(pu)} \tag{4.32a}$$

$$\frac{V_{2(pu)}}{t_{23(pu)}} - t_{23(pu)}I_{2(pu)}Z_{22(pu,nominal)} = V_{3(pu)} - Z_{33(pu,nominal)}I_{3(pu)} \tag{4.32b}$$

also, from Equation (4.29c), we have

$$t_{13(pu)}I_{1(pu)} + t_{23(pu)}I_{2(pu)} + I_{3(pu)} = 0 \tag{4.32c}$$

Equations (4.32) are represented by the equivalent circuit shown in Figure 4.16(c) where, again, the three-winding transformer is replaced by three two-winding transformers connected as three branches in a star configuration. However, branch 3 now has an effective tap ratio of unity, $t_{33(pu)} = t_{3(pu)}/t_{3(pu)} = 1$ as implied from Equations (4.30a) and (4.30b), although $t_{3(pu)}$ itself does not necessarily have to be equal to unity. The tap ratios included in the remaining two branches include the effect of branch 3 off-nominal tap ratio. This model would need to be used with care since, for example, the presence of a variable tap on one winding, say winding 3, must result in a coordinated and consistent change in the turns ratios between windings 1 and 3, and between windings 2 and 3.

Figure 4.16(b) or (c) represent the three-winding transformer PPS and NPS equivalent circuit if we ignore the windings phase shift. Otherwise, the only difference between the PPS and NPS equivalent circuits would be introduced by the windings phase shift. This was described in Section 4.2.3 and shown to result in a complex transformation ratio instead of a real one. In practice, the impedances of the three-winding transformer required in the equivalent circuit of Figure 4.16 are calculated from short-circuit test data supplied by the manufacturer. This will be covered in detail in Section 4.2.9.

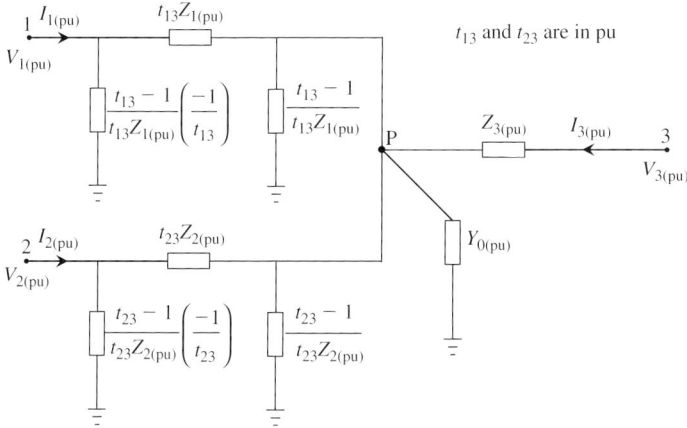

Figure 4.17 PPS and NPS π equivalent circuit for three-winding transformer

The PPS/NPS star equivalent circuit of the three-winding transformer shown in Figure 4.16(b) can be converted into a delta equivalent circuit. We encourage the reader to do so.

As before, we will convert Figure 4.16(c) into π equivalents suitable for digital computer computation. We obviously obtain three π equivalents, one for each transformer branch as shown in Figure 4.17. It should be remembered that the π equivalents were derived with the convention that the currents are injected into them at both ends. Therefore, the currents flowing into the point P from the three branches would need to be reversed in sign.

ZPS equivalent circuits

The approximate ZPS equivalent circuits of three-winding transformers using various windings arrangements will be derived with the aid of a generic ZPS equivalent circuit as already presented in the case of two-winding transformers. Remembering that in Figure 4.16(c), we have converted the three-winding transformer into three two-winding transformers connected in star. Therefore, the generic ZPS equivalent circuit, ignoring the shunt exciting branch for now is shown in Figure 4.18 and is based on the following assumptions:

(a) The secondary of each two-winding transformer is assumed to be star connected with solidly earthed neutral, i.e. directly connected to point P.
(b) The primary of each two-winding transformer is assumed to have the same winding connection as one of the windings of the three-winding transformer.

The equivalent ZPS circuits for three-winding transformers having YNdd, Ydd, YNyd, YNynd, YNynyn and Yyd winding arrangements are shown in Figure 4.19.

The elements of the admittance matrix of Equation (4.17a) can be derived for each equivalent ZPS circuit shown in Figure 4.19. This is similar to the derivation steps that resulted in Table 4.1. We will leave this exercise for the motivated reader!

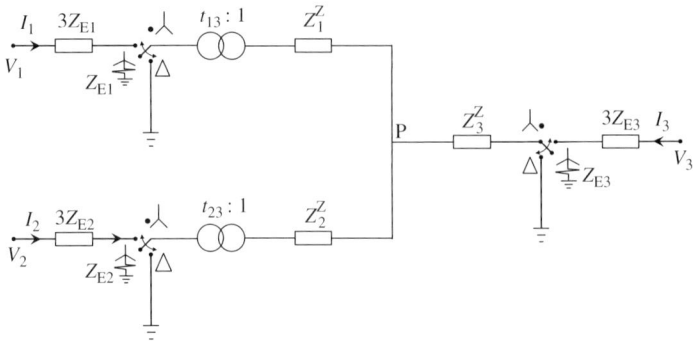

Figure 4.18 Generic ZPS equivalent circuit of three-phase three-winding transformers

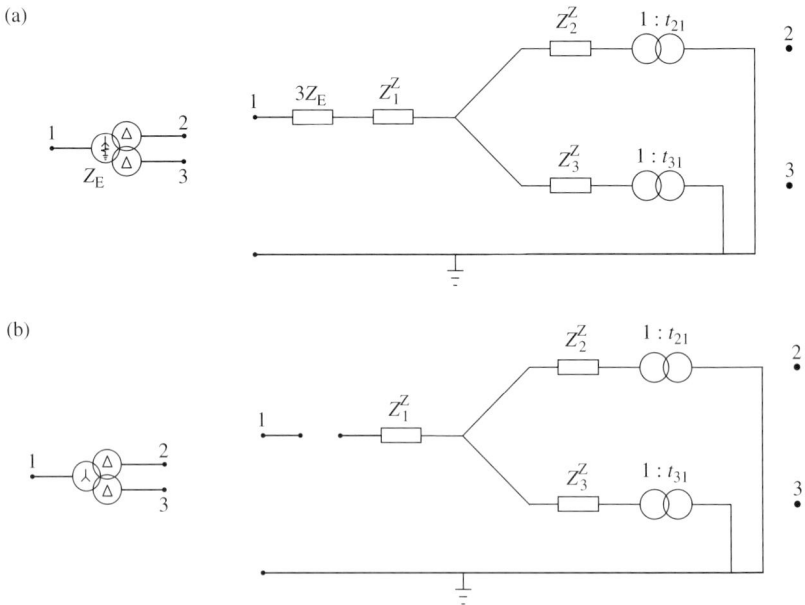

Figure 4.19 ZPS equivalent circuits of three-phase three-winding transformer: (a) star neutral earthing impedance – delta–delta; (b) star isolated neutral – delta–delta; (c) star neutral earthing impedance – delta–star isolated neutral; (d) star neutral earthing impedance – delta–star neutral earthing impedance; (e) all windings are star neutral earthing impedances and (f) star isolated neutral – star isolated neutral – delta

4.2.5 Three-phase autotransformers with and without tertiary windings

The autotransformer consists of a single continuous winding part of which is shared by the high and low voltage circuits. This is called the 'common' winding and is

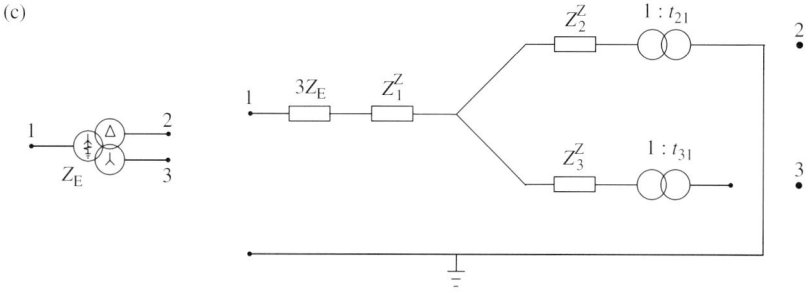

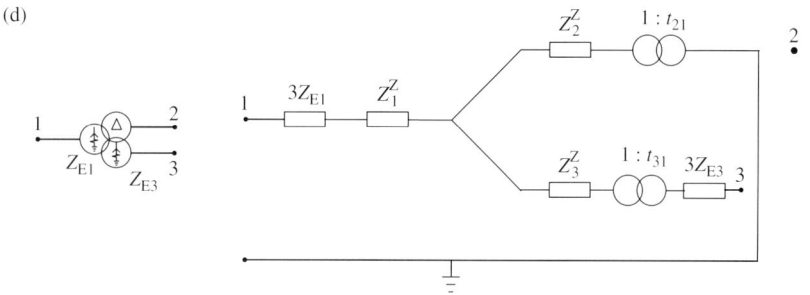

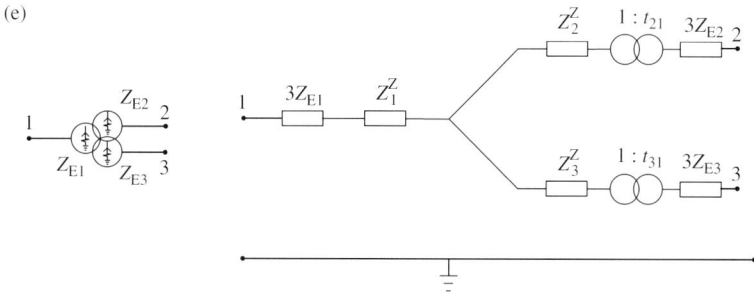

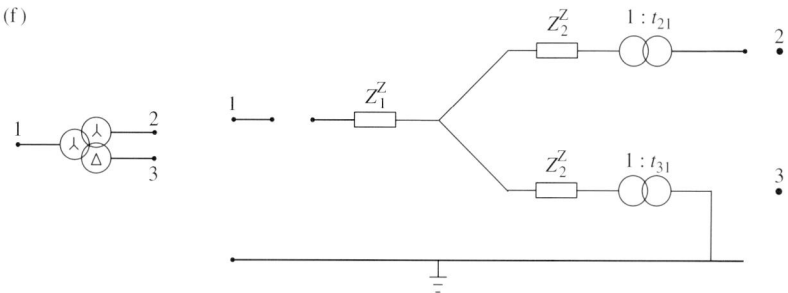

Figure 4.19   (*Continued*)

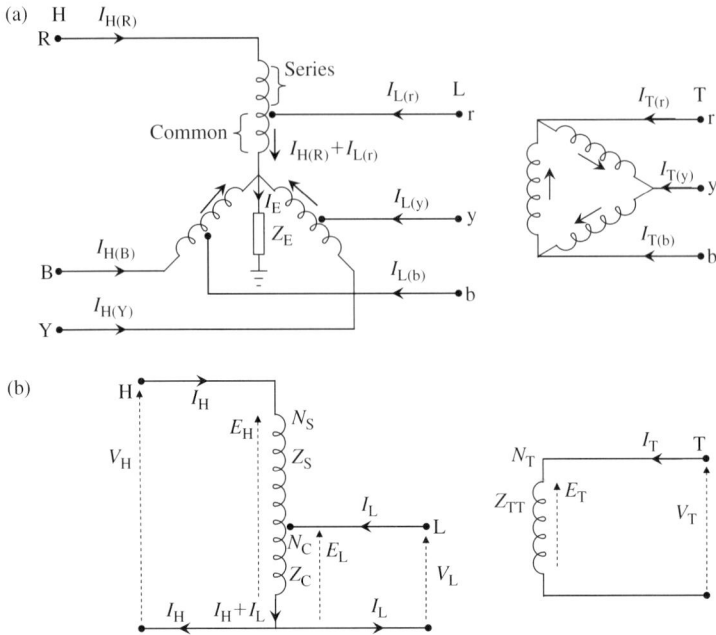

Figure 4.20 Star–star autotransformer with a delta tertiary winding: (a) three-phase representation and (b) single-phase representation

connected between the low voltage terminals. The rest of the total winding is called the 'series' winding and is the remaining part of the high voltage circuits. The combination of the series–common windings' produces the high voltage terminals. Figure 4.20 shows a three-phase representation of a star–star autotransformer with a delta tertiary winding. In the UK, autotransformers interconnect the 400 and 275 kV networks, the 400/275 and 132 kV networks, whereas in North America, they interconnect the 345 kV and 138 kV networks, the 500 kV and 230 kV networks, etc.

Winding connections

Three-phase autotransformers are most often star–star connected as depicted in Figure 4.20(a). The common neutral point of the star–star connection require the two networks they interconnect to have the same earthing arrangement, e.g. solid earthing in the UK although impedance earthing may infrequently be used. Most autotransformers have a low MVA rating tertiary winding connected in delta.

PPS and NPS equivalent circuits

Autotransformers that interconnect extra high voltage transmission systems are not generally equipped with tap-changers due to high costs. However, those that interconnect the transmission and subtransmission or distribution networks are usually equipped with on-load tap-changers in order to control or improve the quality of their LV output voltage under heavy or light load system conditions. Although some tap-changers are connected on the HV winding, most tend to be connected on the LV winding. Most of these are connected at the LV winding line-end and only a few are connected at the winding neutral-end.

A single-phase representation of the general case of an autotransformer with a tertiary winding is shown in Figure 4.20(b). Using S, C and T to denote the series, common and tertiary windings, we can write in actual physical units

$$V_H - E_H = Z_S I_H + Z_C(I_H + I_L) \tag{4.33a}$$

$$V_L - E_L = Z_C(I_H + I_L) \tag{4.33b}$$

$$V_T - E_T = Z_{TT} I_T \tag{4.33c}$$

Neglecting the no-load current, the MMF balance is expressed as

$$N_S I_H + N_C(I_H + I_L) + N_T I_T = 0$$

or

$$I_H + \frac{I_L}{N_{HL}} + \frac{I_T}{N_{HT}} = 0 \tag{4.34a}$$

where

$$N_{HL} = \frac{N_H}{N_L} = \frac{N_S + N_C}{N_C} \quad \text{and} \quad N_{HT} = \frac{N_H}{N_T} = \frac{N_S + N_C}{N_T} \tag{4.34b}$$

Also

$$\frac{E_H}{E_L} = N_{HL} = \frac{1}{N_{LH}} \quad \text{and} \quad \frac{E_H}{E_T} = N_{HT} = \frac{1}{N_{TH}} \tag{4.34c}$$

Using Equations (4.34b), (4.34c) and (4.33a), Equations (4.33b) and (4.33c) can be written as

$$\frac{1}{N_{LH}} \left[V_L - I_L \left(\frac{N_{HL} - 1}{N_{HL}} \right) Z_C \right] = V_H - I_H[Z_S - (N_{HL} - 1)Z_C] \tag{4.35a}$$

$$\frac{1}{N_{TH}} \left[V_T - I_T \left(Z_{TT} + \frac{N_{HL}}{N_{HT}^2} Z_C \right) \right] = V_H - I_H[Z_S - (N_{HL} - 1)Z_C] \tag{4.35b}$$

Equation (4.35) can be represented by the star equivalent circuit shown in Figure 4.21(a) containing two ideal transformers as for a three-winding transformer.

The measurement of the autotransformer PPS and ZPS impedances using short-circuit tests between two winding terminals is dealt with later in this section. However, it is instructive to use Equation (4.35) to demonstrate the results that can be obtained from such tests. Using Equations (4.34a) and (4.35), the PPS impedance measured from the H terminals with the L terminals short-circuited and T terminals open-circuited is

$$Z_{HL} = \left. \frac{V_H}{I_H} \right|_{V_L=0, I_T=0}$$

(a)

(b)

(c)

(d)

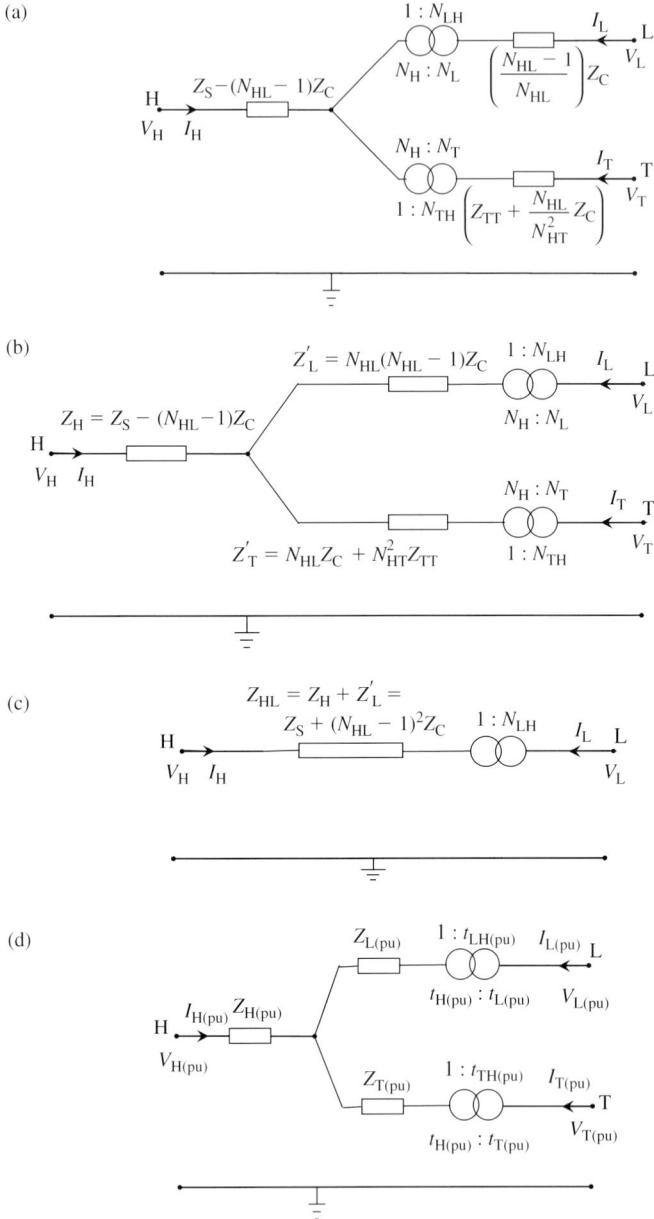

Figure 4.21 PPS/NPS equivalent circuit of an autotransformer with a tertiary winding: (a) equivalent circuit in actual physical units; (b) as (a) above but with L and T branch impedances referred to H voltage base; (c) as (b) above but an autotransformer without a tertiary winding and (d) as (b) above but all quantities are in pu

hence

$$Z_{HL} = Z_S + (N_{HL} - 1)^2 Z_C \qquad (4.36a)$$

Also, the impedance measured from the H terminals with the T terminals short-circuited and L terminals open-circuited is

$$Z_{HT} = \frac{V_H}{I_H}\bigg|_{V_T=0, I_L=0}$$

hence

$$Z_{HT} = Z_S + Z_C + N_{HT}^2 Z_{TT} \qquad (4.36b)$$

Similarly, the impedance measured from the L terminals with the T terminals short-circuited and H terminals open-circuited is

$$Z_{LT} = \frac{V_L}{I_L}\bigg|_{V_T=0, I_H=0}$$

hence

$$Z_{LT} = Z_C + \frac{N_{HT}^2}{N_{HL}^2} Z_{TT} \qquad (4.36c)$$

To calculate the impedance of each branch of the T equivalent circuit in ohms with all impedances referred to the H side voltage base, let us define the measured impedances as follows:

$$Z_{HL} = Z_H + Z_L' \qquad (4.36d)$$

$$Z_{HT} = Z_H + Z_T' \qquad (4.36e)$$

$$Z_{LT} = \frac{1}{N_{HL}^2}(Z_L' + Z_T') \qquad (4.36f)$$

where the prime indicates quantities referred to the H side.

Solving Equations (4.36d), (4.36e) and (4.36f) for each branch impedance, we obtain

$$Z_H = \frac{1}{2}(Z_{HL} + Z_{HT} - N_{HL}^2 Z_{LT}) \qquad (4.37a)$$

$$Z_L' = \frac{1}{2}(Z_{HL} + N_{HL}^2 Z_{LT} - Z_{HT}) \qquad (4.37b)$$

$$Z_T' = \frac{1}{2}(N_{HL}^2 Z_{LT} + Z_{HT} - Z_{HL}) \qquad (4.37c)$$

Now, substituting Equations (4.36a), (4.36b) and (4.36c) into Equations (4.37a), (4.37b) and (4.37c), we obtain

$$Z_H = Z_S - (N_{HL} - 1)Z_C \qquad (4.38a)$$

$$Z'_L = N_{HL}(N_{HL} - 1)Z_C \tag{4.38b}$$

$$Z'_T = N_{HL}Z_C + N^2_{HT}Z_{TT} \tag{4.38c}$$

Figure 4.21(b) shows the autotransformer PPS T equivalent circuit with all impedances in ohms referred to the H terminals voltage base. In the absence of a tertiary winding, Figure 4.21c shows the equivalent circuit of the autotransformer. Using Equations (4.38) in Equations (4.35), we obtain

$$\frac{1}{N_{LH}}\left[V_L - I_L\frac{Z'_L}{N^2_{HL}}\right] = V_H - I_H Z_H \tag{4.39a}$$

$$\frac{1}{N_{TH}}\left[V_T - I_T\frac{Z'_T}{N^2_{HT}}\right] = V_H - I_H Z_H \tag{4.39b}$$

Now, we will convert Equations (4.39) from actual units to pu values. To do so, we define the following pu quantities

$$V_{pu} = \frac{V}{V_{(B)}} \quad I_{pu} = \frac{I}{I_{(B)}} \quad Z_{H(pu)} = \frac{Z_H}{Z_{H(B)}} \quad Z_{L(pu)} = \frac{Z'_L}{Z_{H(B)}} \quad Z_{T(pu)} = \frac{Z'_T}{Z_{H(B)}}$$
$$\tag{4.40a}$$

$$V_{H(B)} = Z_{H(B)}I_{H(B)} \quad V_{L(B)} = Z_{L(B)}I_{L(B)} \tag{4.40b}$$

$$S_{H(B)} = S_{L(B)} = S_{T(B)} = V_{H(B)}I_{H(B)} = V_{L(B)}I_{L(B)} = V_{T(B)}I_{T(B)} \tag{4.40c}$$

$$\frac{V_{H(B)}}{V_{L(B)}} = \frac{N_{H(nominal)}}{N_{L(nominal)}} \quad \frac{V_{H(B)}}{V_{T(B)}} = \frac{N_{H(nominal)}}{N_{T(nominal)}} \tag{4.40d}$$

Using Equations (4.40) in Equations (4.39a) and (4.39b), we obtain

$$\frac{V_{L(pu)}}{\frac{N_{LH}V_{H(B)}}{V_{L(B)}}} - \frac{N_{LH}V_{H(B)}}{V_{L(B)}}I_{L(pu)}Z_{L(pu)} = V_{H(pu)} - Z_{H(pu)}I_{H(pu)} \tag{4.41a}$$

$$\frac{V_{T(pu)}}{\frac{N_{TH}V_{H(B)}}{V_{T(B)}}} - \frac{N_{TH}V_{H(B)}}{V_{T(B)}}I_{T(pu)}Z_{T(pu)} = V_{H(pu)} - Z_{H(pu)}I_{H(pu)} \tag{4.41b}$$

Equations (4.41a) and (4.41b) can be rewritten as

$$\frac{V_{L(pu)}}{t_{LH(pu)}} - t_{LH(pu)}I_{L(pu)}Z_{L(pu)} = V_{H(pu)} - Z_{H(pu)}I_{H(pu)} \tag{4.42a}$$

$$\frac{V_{T(pu)}}{t_{TH(pu)}} - t_{TH(pu)}I_{T(pu)}Z_{T(pu)} = V_{H(pu)} - Z_{H(pu)}I_{H(pu)} \tag{4.42b}$$

where the following pu tap ratios are defined

$$t_{LH(pu)} = \frac{N_{LH}V_{H(B)}}{V_{L(B)}} = \frac{N_L V_{H(B)}}{N_H V_{L(B)}} = \frac{N_L N_{H(nominal)}}{N_H N_{L(nominal)}} = \frac{\frac{N_{L(at\ a\ given\ tap\ position)}}{N_{L(nominal\ tap\ position)}}}{\frac{N_{H(at\ a\ given\ tap\ position)}}{N_{H(nominal\ tap\ position)}}} = \frac{t_{L(pu)}}{t_{H(pu)}}$$

(4.43a)

$$t_{TH(pu)} = \frac{N_{TH}V_{H(B)}}{V_{T(B)}} = \frac{N_T V_{H(B)}}{N_H V_{T(B)}} = \frac{N_T N_{H(nominal)}}{N_H N_{T(nominal)}} = \frac{\frac{N_{T(at\ a\ given\ tap\ position)}}{N_{T(nominal\ tap\ position)}}}{\frac{N_{H(at\ a\ given\ tap\ position)}}{N_{H(nominal\ tap\ position)}}} = \frac{t_{T(pu)}}{t_{H(pu)}}$$

(4.43b)

Equations (4.42) are represented by the pu equivalent circuit shown in Figure 4.21(d) which represents the autotransformer PPS/NPS equivalent circuit ignoring the delta tertiary phase shift. The autotransformer is clearly represented as three two-winding transformers that are star or T connected. Two of these transformers have off-nominal tap ratios that can represent any off-nominal tap ratios on any winding or a combination of tap-ratios. In some cases, the two variable ratios must be consistent and coordinated where an active tap-changer on only one winding can in effect change the effective turns ratio on another. For example, for a 400 kV/132 kV/13 kV autotransformer having a tap-changer acting on the neutral end of the common winding, the variation of $t_{LH(pu)}$ caused by HV to LV turns ratio changes will also cause corresponding changes in the HV to TV turns ratio and hence in $t_{TH(pu)}$. Therefore, $t_{TH(pu)}$ is a function of $t_{LH(pu)}$ which varies as a result of controlling the LV (132kV) terminal voltage to a specified target value around a deadband.

Where the autotransformer does not have a tertiary winding or where the tertiary winding is unloaded, the T terminal in Figure 4.21(d) would be unconnected to the power system network and its branch impedance has no effect on the network currents and voltages. Thus, this branch can be disregarded and the effective auto-transformer impedance would then be the sum of the H and L branch impedances given by $Z_{HL(pu)} = Z_{H(pu)} + Z_{L(pu)}$. In this case, the PPS/NPS equivalent circuit of an autotransformer is similar to that already derived for a two-winding trans-former and shown in Figures 4.8(c) or 4.9(c). These can be used to represent an autotransformer with 'series' winding tap-changer or 'common' winding tap-changer, respectively. The latter represents British practice irrespective of whether the tap-changer is connected to the line-end or neutral-end of the 'common' winding.

The impedances of the autotransformer required in the equivalent circuit of Figure 4.21(d) are calculated from short-circuit test data supplied by the manufacturer. This is covered in detail in Section 4.2.9.

ZPS equivalent circuits

In addition to the primary, secondary and, where present, tertiary winding connections, the presence of an impedance in the neutral point of the star-connected

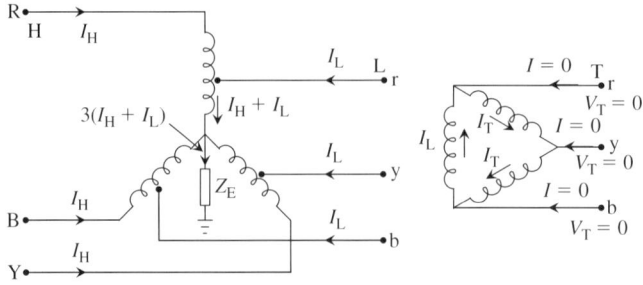

Figure 4.22 Derivation of ZPS equivalent circuit of an autotransformer

'common' winding must be taken into account in deriving the autotransformers ZPS equivalent circuit. We will ignore, as before, the shunt exciting impedance, and consider the general case of a star–star autotransformer with a tertiary winding with the common star neutral point earthed through an impedance Z_E. The ZPS equivalent circuit, taking into account the presence of Z_E in the neutral, can be derived in a similar way to the PPS/NPS circuit. However, we will leave this as a minor challenge for the interested reader but we have drawn Figure 4.22 to help the reader in such derivation.

For us, it suffices to say that in Equations (4.35a) and (4.35b), Z_C should be replaced by $Z_C + 3Z_E$. Therefore, the ZPS equations are given by

$$\frac{1}{N_{LH}}\left[V_L - I_L\left(\frac{N_{HL}-1}{N_{HL}}\right)(Z_C + 3Z_E)\right] = V_H - I_H[Z_S - (N_{HL}-1)(Z_C + 3Z_E)]$$
(4.44a)

$$\frac{1}{N_{TH}}\left[0 - I_T\left\{Z_{TT} + \frac{N_{HL}}{N_{HT}^2}(Z_C + 3Z_E)\right\}\right] = V_H - I_H[Z_S - (N_{HL}-1)(Z_C + 3Z_E)]$$
(4.44b)

Similar to the derivation of the PPS impedances, the measured ZPS impedances between two terminals and the corresponding T equivalent branch impedances are

$$Z_{HL} = Z_S + (N_{HL}-1)^2(Z_C + 3Z_E)$$
(4.45a)

$$Z_{HT} = Z_S + Z_C + 3Z_E + N_{HT}^2 Z_{TT}$$
(4.45b)

$$Z_{LT} = Z_C + 3Z_E + \frac{N_{HT}^2}{N_{HL}^2}Z_{TT}$$
(4.45c)

$$Z_H = Z_S - (N_{HL}-1)(Z_C + 3Z_E)$$
(4.45d)

$$Z_L' = N_{HL}(N_{HL}-1)(Z_C + 3Z_E)$$
(4.45e)

$$Z_T' = N_{HL}(Z_C + 3Z_E) + N_{HT}^2 Z_{TT}$$
(4.45f)

Substituting Equations (4.45d), (4.45e) and (4.45f) into Equations (4.44a) and (4.44b), we obtain

$$\frac{1}{N_{LH}}\left[V_L - I_L\left\{\frac{Z_L'}{N_{HL}^2} + \frac{3(N_{HL} - 1)}{N_{HL}}Z_E\right\}\right] = V_H - I_H[Z_H - 3(N_{HL} - 1)Z_E]$$

$$(4.46a)$$

$$\frac{1}{N_{TH}}\left[0 - I_T\left\{\frac{Z_T'}{N_{HT}^2} + \frac{3N_{HL}}{N_{HT}^2}Z_E\right\}\right] = V_H - I_H[Z_H - 3(N_{HL} - 1)Z_E] \quad (4.46b)$$

To convert Equations (4.46) to pu, we will use the base quantities defined in Equations (4.40) and define $Z_{E(pu)} = Z_E/Z_{L(B)}$, i.e. we choose to use the L terminal as the base voltage for converting Z_E to pu. It should be noted that the choice of the voltage base for Z_E is arbitrary. Therefore, Equations (4.46) become

$$\frac{V_{L(pu)}}{t_{LH(pu)}} - t_{LH(pu)}I_{L(pu)}\left[Z_{L(pu)} + \frac{3(N_{HL} - 1)}{N_{HL}}Z_{E(pu)}\right]$$

$$= V_{H(pu)} - I_{H(pu)}\left[Z_{H(pu)} - \frac{3(N_{HL} - 1)}{N_{HL}^2}Z_{E(pu)}\right] \quad (4.46c)$$

$$\frac{V_{T(pu)}}{t_{TH(pu)}} - t_{TH(pu)}I_{T(pu)}\left[Z_{T(pu)} + \frac{3}{N_{HL}}Z_{E(pu)}\right]$$

$$= V_{H(pu)} - I_{H(pu)}\left[Z_{H(pu)} - \frac{3(N_{HL} - 1)}{N_{HL}^2}Z_{E(pu)}\right] \quad (4.46d)$$

where $V_{T(pu)} = 0$. Equations (4.46c) and (4.46d) are represented by the ZPS T equivalent circuit of Figure 4.23 for a star–star autotransformer with a delta tertiary winding and with the common neutral point earthed through an impedance Z_E. It is informative to observe how the neutral earthing impedance appears in the three branches connected to the H, L and T terminals

$$Z_{E(pu)}(\text{appearing in H terminal}) = \frac{-3(N_{HL} - 1)}{N_{HL}^2}Z_{E(pu)} \quad (4.47a)$$

$$Z_{E(pu)}(\text{appearing in L terminal}) = \frac{3(N_{HL} - 1)}{N_{HL}}Z_{E(pu)} \quad (4.47b)$$

$$Z_{E(pu)}(\text{appearing in T terminal}) = \frac{3}{N_{HL}}Z_{E(pu)} \quad (4.47c)$$

where $Z_{E(pu)}$ is in pu on L terminal voltage base.

In the absence of a tertiary winding, the equivalent ZPS circuit cannot be obtained from Figure 4.23 by removing the branch that corresponds to the tertiary winding. A shunt branch to the zero voltage node may indeed still be required depending on the transformer core construction. This is covered in detail in Section 4.2.8.

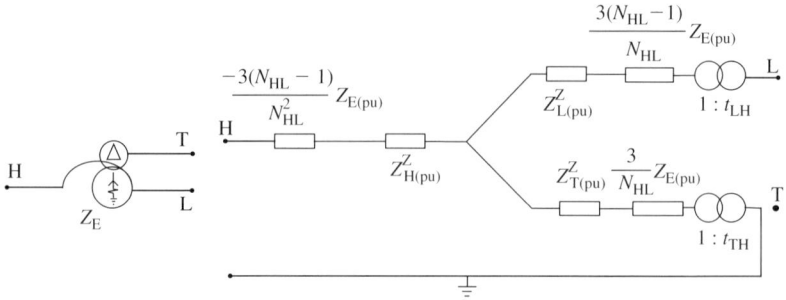

Figure 4.23 ZPS equivalent circuit of a star–star autotransformer with a neutral earthing impedance and a delta tertiary winding

The case of a solidly earthed neutral is represented in Figure 4.23 by setting $Z_E = 0$. However, the case of an isolated neutral, i.e. $Z_E \to \infty$ cannot be represented by Figure 4.23 because the branch impedances of the equivalent circuit become infinite. This indicates no apparent paths for ZPS currents between the windings although a physical circuit does exist. A mathematical solution is to convert the star circuit of Figure 4.23 to delta or π and then setting $Z_E \to \infty$. However, we prefer the physical engineering approach to derive the equivalent ZPS circuit of such a transformer. Therefore, using the actual circuit of an unearthed autotransformer with a delta tertiary winding, shown in Figure 4.24(a), we can write

$$V_H - V_L - E_H = Z_S I_H \qquad (4.48a)$$

$$0 - E_T = Z_{TT} I_T \qquad (4.48b)$$

$$N_S I_H + N_T I_T = 0 \qquad (4.48c)$$

$$\frac{E_H}{E_T} = \frac{N_S}{N_T} \qquad (4.48d)$$

Using Equations (4.48b), (4.48c) and (4.48d) in Equation (4.48a), we obtain

$$V_H - I_H \left[Z_S + \left(\frac{N_S}{N_T} \right)^2 Z_{TT} \right] = V_L \qquad (4.48e)$$

The measured ZPS leakage impedance in ohms

$$Z_{HL}^Z = \frac{V_H}{I_H} \bigg|_{V_L = 0} \quad \text{or} \quad Z_{LH}^Z = \frac{V_L}{I_L} \bigg|_{V_H = 0}$$

is given by

$$Z_{HL}^Z = Z_{LH}^Z = Z_{S-T} = Z_S + Z_{TT}' = Z_S + \left(\frac{N_S}{N_T} \right)^2 Z_{TT} \qquad (4.48f)$$

The ZPS leakage impedance measured from the H side or L side is the leakage impedance between the series and tertiary winding that is the sum of the series

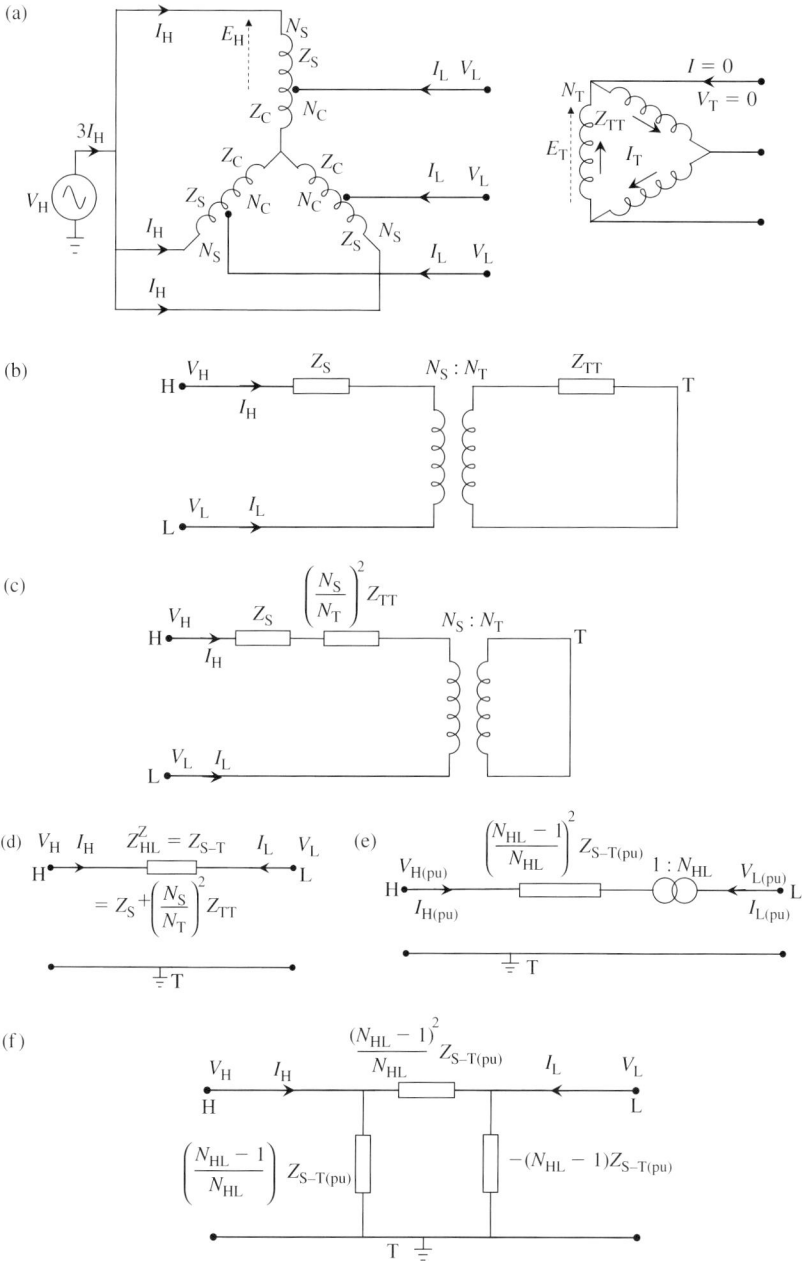

Figure 4.24 Derivation of ZPS equivalent circuit of an autotransformer with an isolated neutral and a delta tertiary winding: (a) autotranformer with isolated neutral and a delta tertiary winding; (b) single-phase representation; (c) tertiary impedance referred to series winding base turns; (d) equivalent ZPS circuit and impedance in physical units; (e) equivalent ZPS circuit in pu and (f) π ZPS equivalent circuit of (c)

winding leakage impedance and the tertiary winding leakage impedance referred to the series side base turns. Figure 4.24(b) shows a single-phase representation of the series and tertiary winding impedances and turns ratios. Figure 4.24(c) shows the tertiary winding impedance referred to the series winding side and Figure 4.24(d) represents the ZPS equivalent circuit and total ZPS equivalent impedance in actual physical units. The physical explanation for this result is as follows. The ZPS currents flow on the H side through the series windings only and, without transformation, flow out of the L side. This is possible because the delta tertiary winding circulates equal ampere-turns to balance that of the series winding but no ZPS currents flow in the common windings of the autotransformer.

To convert Equation (4.48e) to pu, we first calculate a pu value for Z_{S-T} and note that this is the same whether referred to the series winding or tertiary winding because the base voltages in the windings are directly proportional to the number of turns in the windings. Using Equation (4.40), we have

$$V_{H(pu)}V_{H(B)} - I_{H(pu)}I_{H(B)}Z_{S-T(pu)}Z_{S(B)} = V_{L(pu)}V_{L(B)}$$

Dividing by $V_{H(B)}$ and using

$$\frac{Z_{S(B)}}{Z_{H(B)}} = \left(\frac{N_S}{N_S + N_C}\right)^2 = \left(\frac{N_{HL} - 1}{N_{HL}}\right)^2$$

we obtain

$$V_{H(pu)} - I_{H(pu)}\left(\frac{N_{HL} - 1}{N_{HL}}\right)^2 Z_{S-T(pu)} = \frac{1}{N_{HL}}V_{L(pu)} \qquad (4.49a)$$

Equation (4.49a) is represented by the pu equivalent ZPS circuit shown in Figure 4.24(e) from which the ZPS leakage impedance measured from the H side with L side short-circuited is given by

$$Z^Z_{HL(pu\ referred\ to\ H\ side)} = \left(\frac{N_{HL} - 1}{N_{HL}}\right)^2 Z_{S-T(pu)} \qquad (4.49b)$$

Similarly, the ZPS leakage impedance measured from the L side with H side short-circuited is given by

$$Z^Z_{LH(pu\ referred\ to\ L\ side)} = (N_{HL} - 1)^2 Z_{S-T(pu)} \qquad (4.49c)$$

The delta or π equivalent circuit can be derived using Figure 4.6(c). The result is shown in Figure 4.24(f) using shunt impedances instead of admittances.

Another similar test may be made but with the delta tertiary open to obtain the shunt ZPS magnetising impedance which may be low enough to be included depending on the core construction. This is dealt with in detail in Section 4.2.8.

4.2.6 Three-phase earthing or zig-zag transformers

Three-phase earthing transformers are connected to unearthed networks in order to provide a neutral point for connection to earth. In the UK, this arises in delta–connected tertiary windings and 33 kV distribution networks supplied via star–delta transformers from 132 or 275 kV networks which are solidly earthed. The purpose of using these transformers is to provide a low ZPS impedance path to the flow of ZPS short-circuit fault currents under unbalanced earthed faults. Such an aim can be achieved by the use of:

(a) A star–delta transformer where the star winding terminals are connected to the network and the delta winding is left unconnected. The star neutral may be earthed through an impedance.
(b) An interconnected star or zig-zag transformer which may also have a secondary star auxiliary winding to provide a neutral and substation supplies. The zig-zag neutral may be earthed through an impedance.
(c) An interconnected star or zig-zag as (b) above but with a delta-connected auxiliary winding.

The star–delta transformer has already been covered. However, in this case, its PPS and NPS impedance is effectively the shunt exciting impedance which is very large and can be neglected. The ZPS impedance is the leakage impedance presented by the star–delta connection discussed previously. The interconnected star or zig-zag transformer has each phase winding split into two halves and interconnected as shown in Figure 4.25(a) and (b). It should be obvious from the vector diagram, Figure 4.25(c) that this winding connection produces a phase shift of 30° as if the zig-zag winding had a delta-connected characteristic. Again, the PPS/NPS impedance presented by this transformer if PPS/NPS voltages are applied is the very large exciting impedance. However, under ZPS excitation, MMF balance or cancellation occurs between the winding halves wound on the same core due to equal and opposite currents flowing in each winding half. The ZPS impedance per phase is therefore the leakage impedance between the two winding's halves. In practical installations, the impedance connected to the neutral is usually much greater than the transformer leakage impedance. The zig-zag transformer ZPS equivalent circuit is shown in Figure 4.25(d).

The zig-zag delta-connected transformer presents a similar ZPS equivalent circuit as the zig-zag transformer shown in Figure 4.25(d). If the delta winding is loaded, the transformer PPS/NPS equivalent circuit includes a 90° phase shift and is similar to that of a two-winding transformer with an impedance derived from the star equivalent of the three windings.

4.2.7 Single-phase traction transformers connected to three-phase systems

Single-phase traction transformers are used to provide electricity supply to trains overhead catenary systems and are connected on the HV side to two phases of the transmission or distribution network. In the UK, the secondary nominal voltage to earth of these single-phase transformers is 25 kV and their HV voltage may be 132, 275 or 400 kV. The 132 kV/25 kV are typically two-winding transformers

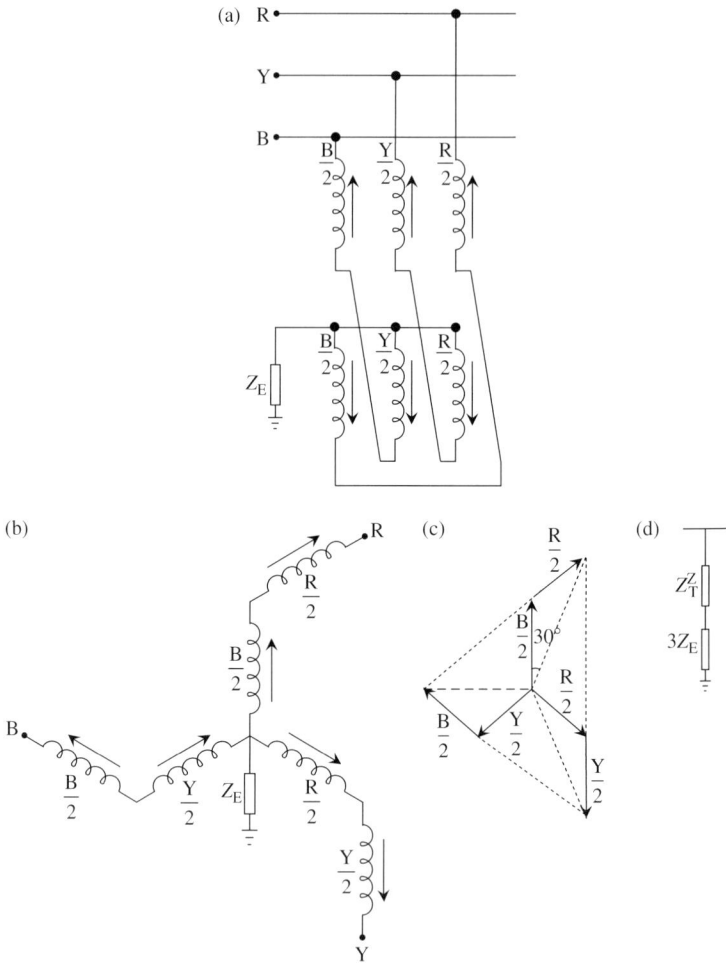

Figure 4.25 Interconnected-star or zig-zag transformer: (a) connection to a three-phase network, (b) winding connections, (c) vector diagram and (d) zps equivalent impedance

whereas modern large units derive their supplies from 400 kV and tend to be three-winding transformers, e.g. 400 kV/26.25 kV-0-26.25 kV. Figure 4.26 illustrate such connections.

The model of two-winding single-phase traction transformers is the same as the single-phase model presented at the beginning of this chapter. Three-winding single-phase traction transformers can also be represented as a star equivalent of three leakage impedances. Under normal operating conditions, the traction load impedance referred to the primary side of the traction transformer appears in series with the transformer leakage impedance and is very large. However, under a short-circuit fault on the LV side of the traction transformer, the transformer winding leakage impedance appears directly on the HV side as an impedance

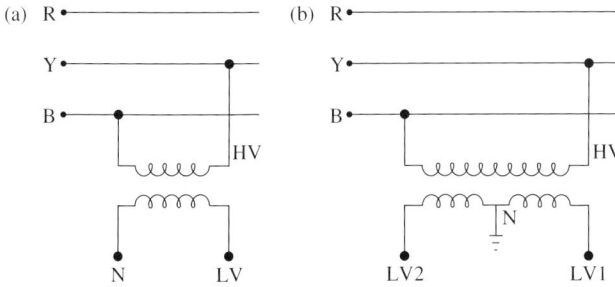

Figure 4.26 Single-phase traction transformer connected to a three-phase system: (a) two windings and (b) three windings

connected between two phases. The reader should establish that this can be modelled as a phase-to-phase short-circuit fault through a fault impedance equal to the transformer leakage impedance.

4.2.8 Variation of transformer's PPS leakage impedance with tap position

We have so far covered the modelling of the transformer with an off-nominal turns ratio that may be due to the use of a variable tap-changer or fixed off-nominal ratio. We have shown that, for some models, the leakage impedance, both PPS and ZPS, needs to be multiplied by the square of the off-nominal ratio. In addition, since the operation of the tap-changer involves the addition or cancellation of turns from a given winding, the variation of the number of turns causes variations in the leakage flux patterns and flux linkages.

The magnitude of the impedance variation and its direction in terms of whether it remains broadly unaffected, increases or decreases by the addition or removal of turns, depends on a number of factors. One factor is the tapped winding, e.g. whether in the case of a two-winding transformer, the HV or LV winding is the tapped winding. For an autotransformer, it is whether the series or common winding and in the latter case, whether the common winding is tapped at the line, i.e. output end or at the neutral-end of the winding. Other factors are the range of variation of the number of turns and the location of the taps, e.g. for a two-winding transformer, whether they are located in the body of the tapped winding itself, inside the LV winding or outside the HV winding. In summary, the addition of turns above nominal tap position or the removal of turns below nominal tap position may produce any of the following effects depending on specific transformer and tap-changer design factors:

(a) The leakage impedance remains fairly constant and unaffected.
(b) The leakage impedance may increase either side of nominal tap position.
(c) The leakage impedance may consistently decrease across the entire tap range as the number of turns is increased from minimum to maximum.

(d) The leakage impedance may consistently increase across the entire tap range as the number of turns is increased from minimum to maximum.

From a network analysis viewpoint, the leakage impedance variation across the tap range can be significant and this should be taken account of in the analysis. It is general practice in industry that the data of the impedance variation across the entire tap range is usually requested by transformer purchasers and supplied by manufacturers in short-circuit test certificates. The measurement of transformer impedances is covered in Section 4.2.10.

4.2.9 Practical aspects of ZPS impedances of transformers

In deriving the ZPS equivalent circuits for the various transformers we have presented so far, we only considered the primary effects of the winding connections and neutral earthing impedances in the case of star-connected windings. We have temporarily neglected the effect of the transformer core construction and hence the characteristics of the ZPS flux paths on the ZPS leakage impedances. Contrary to what is generally published in the literature, we recommend that the effect of transformer core construction on the magnitude of the ZPS leakage impedance is fully taken into account in setting up transformer ZPS equivalent circuits in network models for use in short-circuit analysis. We consider this as international best practice that in the author's experience, can have a material effect during the assessment of short-circuit duties on circuit-breakers.

We will now consider this aspect in some detail and in order to aid our discussion, we will recall some of the basics of magnetic and electromagnetic circuit theory. The relative permeability of transformer iron or steel core is hundreds of times greater than that of air. The reluctance of the magnetic core is its ability to oppose the flow of flux and is inversely proportional to its permeability. The reluctance and flux of a magnetic circuit are analogous to the resistance and current in an electric circuit. Therefore, transformer iron or steel cores present low reluctance paths to the flow of flux in the core. Also, the core magnetising reactance is inversely proportional to its reluctance. Therefore, where the flux flows within the transformer core, the transformer magnetising reactance will have a very large value and will therefore not have a material effect on the transformer leakage impedance. However, where the flux is forced to flow out of the transformer core, e.g. into the air and complete its circuit through the tank and/or oil, then the effect of this external very high reluctance path is to significantly lower the magnetising reactance. This will substantially lower the leakage impedance of the transformer. It should be remembered that under PPS/NPS excitation, nearly all flux is confined to the iron or steel magnetic circuit, the magnetising current is very low and hence the PPS/NPS magnetising reactance is very large and has no practical effect on the PPS/NPS leakage impedance. We will now discuss the effect of various transformer core constructions on the ZPS impedance.

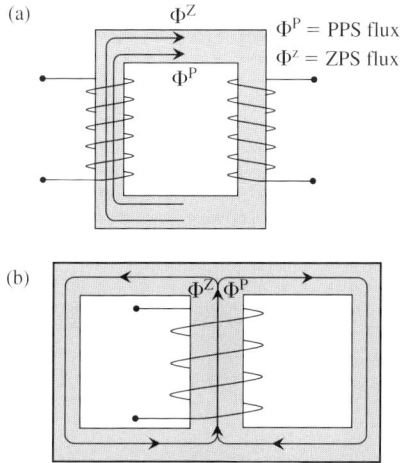

Figure 4.27 Three single-phase banks three-phase transformer shows ZPS flux remains in core: (a) both limbs are wound and (b) centre limb is wound

Three-phase transformers made up of three single-phase banks

Figure 4.27 illustrates two typical single-phase transformer cores. In both cases, the ZPS flux set up by ZPS voltage excitation can flow within the core in a similar way to PPS flux. Consequently, the ZPS magnetising impedance will be very large and the ZPS leakage impedance of such transformers will be substantially equal to the PPS leakage impedance.

Three-phase transformers of 5-limb core construction and shell-type construction

Figures 4.28 and 4.29 illustrate a 5-limb core-type and a shell-type three-phase transformers, respectively. The 5-limb design is widely used in the UK and Europe whereas the shell-type tends to be widely used in North America. In both designs, the ZPS flux set up by ZPS excitation can flow within the core and return in the outer limbs. Consequently, as in the case of three-banks of single-phase transformers, the ZPS magnetising impedance will be very large and the ZPS leakage impedance of such transformers will be substantially equal to the PPS leakage impedance. The measurement of sequence impedances will be dealt within the next section. However, it is appropriate to explain now that PPS leakage impedances are normally measured at nearly rated current whereas ZPS impedances are measured, where this is done, at quite low current values. Therefore, under actual earth fault conditions giving rise to sufficiently high ZPS voltages on nearby transformers with currents similar to those of the PPS tests, the transformer outer limbs may approach saturation and this lowers the ZPS magnetising impedance. Consequently, the measured ZPS impedance value may be somewhat higher than the actual value for core-type design and substantially higher for shell-type design that exhibits much

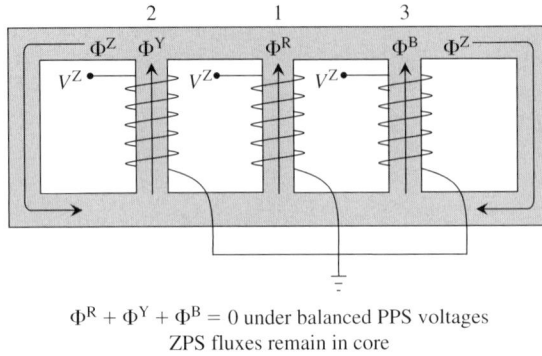

$\Phi^R + \Phi^Y + \Phi^B = 0$ under balanced PPS voltages
ZPS fluxes remain in core

Figure 4.28 Three-phase 5-limb core transformer

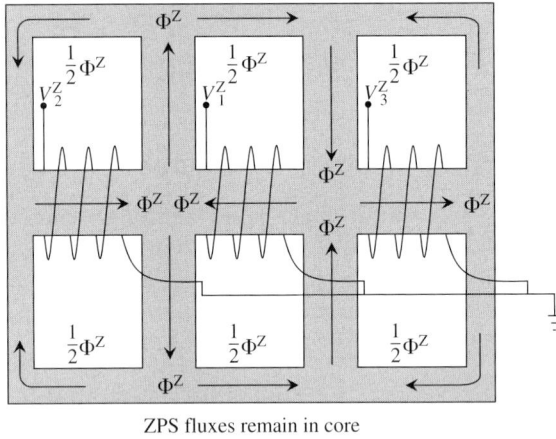

ZPS fluxes remain in core

Figure 4.29 Three-phase shell-type core transformer

more variable core saturation than the limb-type core. Nonetheless, these magnetising impedances remain relatively large in comparison with the leakage impedances and hence, it is usually assumed that the PPS and ZPS leakage impedances of such transformers are equal.

Three-phase transformers of 3-limb core construction

Figure 4.30 illustrates a three-phase transformer of 3-limb core-type construction which is extensively used in the UK and the rest of the world. Under ZPS voltage excitation, the ZPS flux must exit the core and its return path must be completed through the air with the dominant part being the tank then the oil and perhaps the core support framework. This ZPS flux will induce large ZPS currents through the central belt of the transformer tank and the overall effect of this very high reluctance path is to significantly lower the ZPS magnetising impedance. This

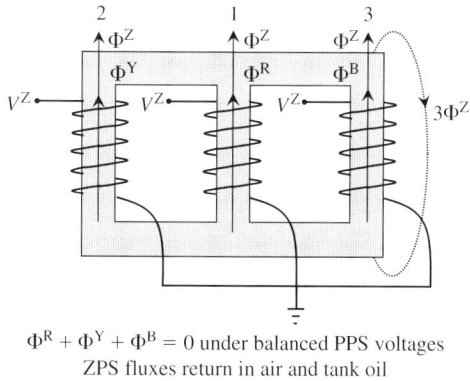

$\Phi^R + \Phi^Y + \Phi^B = 0$ under balanced PPS voltages
ZPS fluxes return in air and tank oil

Figure 4.30   Three-phase 3-limb core transformer

ZPS magnetising impedance will be comparable to other plant values and may be 60–120% on rating for two and three-winding transformers and perhaps 150–250% for autotransformers. These figures can be compared with 5000–20 000% for the corresponding PPS values (these correspond to 2–0.5% no-load currents). Therefore, the effect of the tank for two-winding and three-winding transformers with star earthed neutral windings, and for autotransformers without tertiary windings, can be treated as if the transformer has a virtual magnetic delta-connected winding. We will provide practical examples of this in the next section.

Similar to 5-limb and shell-type transformers, ZPS leakage impedance measurements may be made at low current values and hence may give slightly higher impedances than the values inferred from PPS measurements made at rated current value. In addition, the effect of the tank in the 3-limb construction is non-linear with the reluctance increasing with increasing current that is the magnetising reactance decreasing with increasing current. Therefore, the actual ZPS leakage impedance of three-limb transformers may be slightly lower than the measured values.

4.2.10 Measurement of sequence impedances of three-phase transformers

Transformers are subjected to a variety of tests by their manufacturers to establish correct design parameters, quality and suitability for 30 or 40 years service life. The tests from which the transformer impedances are calculated are iron loss test, no-load current test, copper loss test, short-circuit PPS and sometimes ZPS impedance tests.

The iron loss and no-load current tests are carried out simultaneously. The rated voltage at rated frequency is applied to the LV winding, or tertiary winding where present, with the HV winding open-circuited. The shunt exciting impedance or admittance is calculated from the applied voltage and no-load current, and its resistive part is calculated from the iron loss. The magnetising reactance is then calculated from the resistance or conductance and impedance or admittance. The no-load current is obtained from ammeter readings in each phase and usually an average value is taken. The iron loss would be the same if measured from the HV side but the current would obviously be different as it will be in inverse proportion to the turns ratio. Also, the application of rated voltage to the LV or tertiary is more easily obtainable and the current magnitude would be larger and more conveniently read. The measured no-load loss is usually considered equal to the iron loss by ignoring the dielectric loss and the copper loss due to the small exciting current.

The copper loss and short-circuit impedance tests are carried out simultaneously. A low voltage is applied to the HV winding with the LV winding being short-circuited. The applied voltage is gradually increased until the HV current is equal to the full load or rated current. The applied voltage, measured current and measured load loss are read. As the current is at rated value, the short-circuit impedance voltage in pu is given by the applied voltage in pu of rated voltage of the HV winding. The load-loss would be the same if measured from the LV side. It is usual for transformers having normal impedance values to ignore the iron loss component of the load loss and the measured loss to be taken as the copper loss. The reason being is that the iron loss is negligible in comparison with the copper loss at the reduced short-circuit test voltage. Where high impedance transformers are specified, however, the iron loss at the short-circuit test voltage may not be negligible. In this case, the copper-loss can be found after reading the load-loss from the short-circuit test by removing the short-circuit (i.e. convert the test to an open-circuit test at the short-circuit applied voltage) reading the iron-loss and subtracting it from the load-loss. The resistance is calculated from the measured copper-loss and the reactance is calculated from the impedance voltage and the resistance.

The measurements made in volts and amps allow the calculation of the impedances in ohms but for general power system analysis, these need to be converted to per unit on some defined base. The equations presented below are straightforward to derive and we recommend that the reader does so.

From the open-circuit test, the exciting admittance $Y_{(pu)} = G_{(pu)} - jB_{M(pu)}$ is defined as follows:

$$Y_{(pu \text{ on rated MVA})} = \frac{\text{No load MVA}}{S_{\text{Rated (MVA)}}} \quad \text{or} \quad Y_{(pu \text{ on rated MVA})} = \frac{I_{\text{No load (A)}}}{I_{\text{Rated (A)}}} \quad (4.50a)$$

where

$$\text{No load MVA} = \sqrt{3}V_{\text{LL rated (kV)}}I_{\text{No load (kA)}} \quad (4.50b)$$

or on a new MVA base S_{Base}

$$Y_{(pu \text{ on } S_{\text{Base}})} = \frac{\text{No load MVA}}{S_{\text{Base (MVA)}}} \quad \text{or} \quad Y_{(pu \text{ on } S_{\text{Base}})} = \frac{I_{\text{No load (A)}}}{I_{\text{Base (A)}}} \quad (4.50c)$$

and

$$Z_{(pu)} = \frac{1}{Y_{(pu)}}$$

The resistive (conductance) part of the exciting admittance is calculated as

$$G_{\text{(pu on rated MVA)}} = \frac{\text{Fe}_{[3\text{-Phase loss (MW)}]}}{S_{\text{Rated (MVA)}}} \quad \text{and} \quad R_{\text{(pu on rated MVA)}} = \frac{1}{G_{\text{(pu on rated MVA)}}}$$

$$(4.50\text{d})$$

where $\text{Fe}_{[3\text{-Phase loss (MW)}]}$ is the iron loss.

The inductive part is calculated as

$$B_{\text{M (pu on rated MVA)}} = \frac{1}{S_{\text{Rated (MVA)}}} \sqrt{(\text{No load MVA})^2 - (\text{Fe}_{3\text{-Phase loss (MW)}})^2}$$

and

$$X_{\text{M (pu on rated MVA)}} = \frac{1}{B_{\text{M (pu on rated MVA)}}} \quad (4.50\text{e})$$

or on a new MVA base S_{Base}

$$G_{\text{(pu on } S_{\text{Base}})} = \frac{\text{Fe}_{[3\text{-Phase loss (MW)}]}}{S_{\text{Base (MVA)}}}$$

and

$$B_{\text{M (pu on } S_{\text{Base}})} = \frac{1}{S_{\text{Base (MVA)}}} \sqrt{(\text{No load MVA})^2 - (\text{Fe}_{3\text{-Phase loss (MW)}})^2} \quad (4.50\text{f})$$

From the short-circuit test, the short-circuit leakage impedance $Z_{\text{(pu)}} = R_{\text{(pu)}} + jX_{\text{(pu)}}$ is defined as follows:

$$Z_{\text{(pu on rated MVA)}} = \frac{V_{\text{LL test (kV)}}}{V_{\text{LL rated (kV)}}} \quad (4.51\text{a})$$

This is equal to the applied test voltage in pu which is why the term short-circuit impedance voltage is used by transformer manufacturers. This applies at nominal and off-nominal tap position.

The resistive part of the leakage impedance is calculated as

$$R_{\text{(pu on rated MVA)}} = \frac{\text{Copper}_{[3\text{-Phase loss (MW)}]}}{S_{\text{Rated (MVA)}}}$$

or on a new MVA base S_{Base}

$$R_{\text{(pu on } S_{\text{Base}})} = \frac{\text{Copper}_{[3\text{-Phase loss (MW)}]} \times S_{\text{Base (MVA)}}}{(S_{\text{Rated (MVA)}})^2} \quad (4.51\text{b})$$

and the inductive part is calculated as

$$X_{\text{(pu on rated MVA)}}$$

$$= \frac{1}{S_{\text{Rated (MVA)}}} \sqrt{(\sqrt{3} V_{\text{LL test (kV)}} I_{\text{Rated (kA)}})^2 - (\text{Copper}_{[3\text{-Phase loss (MW)}]})^2}$$

or on a new MVA base S_{Base}

$$
\begin{aligned}
&X_{(\text{pu on } S_{Base})} \\
&= \frac{S_{\text{Base (MVA)}}}{(S_{\text{Rated (MVA)}})^2} \sqrt{(\sqrt{3} V_{\text{LL test (kV)}} I_{\text{rated (kA)}})^2 - (\text{Copper}_{[\text{3-Phase loss (MW)}]})^2}
\end{aligned}
$$

$$(4.51c)$$

Whereas PPS impedance measurement tests for a number of tap positions such as minimum, maximum, nominal and mean have always been the norm, it is only recently becoming standard industry practice to similarly specify ZPS impedance tests. In the case of ZPS impedance measurements, the three-phase terminals of the winding from which the measurement is made are joined together and a single-phase voltage source is applied between this point and neutral. Voltage, current and copper loss may be measured; the ZPS resistance is calculated from the copper loss and input ZPS current. The ZPS impedance is calculated from the applied voltage and 1/3rd of the source current because the source ZPS current divides equally between the three phases. The basic principle of the PPS and ZPS leakage impedance tests is illustrated in Figure 4.31 for a star with neutral solidly earthed-delta two-winding transformer. It should be noted that, in the ZPS test, if the LV winding is delta connected, it must be closed but not necessarily short-circuited.

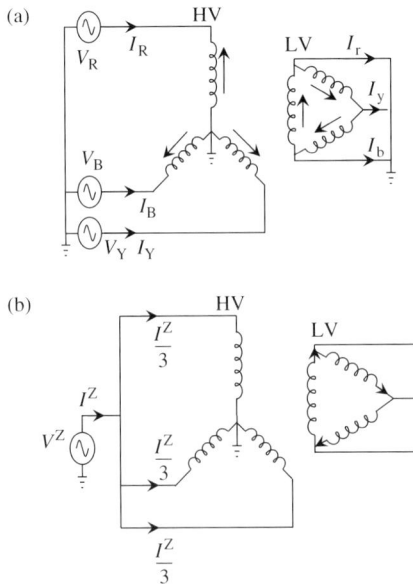

Figure 4.31 Illustration of PPS and ZPS test circuits on a star–delta two winding transformer: (a) PPS leakage impedance test, HV winding supplied, LV winding short circuited $Z_{HL}^P = V_R / I_R$ and (b) ZPS leakage impedance test, HV winding terminals joined and supplied, LV winding closed $Z_{HL}^Z = 3V^Z/I^Z$

PPS and ZPS impedance tests on two-winding transformers

The PPS impedance test is HV–LV//N, i.e. the HV winding is supplied with the LV winding phase terminals joined together, i.e. short-circuited and connected to neutral. The same test may be done for the ZPS impedance if the secondary winding is delta connected. However, for a star earthed–star earthed transformer, the ZPS equivalent circuit that correctly represents a 3-limb core transformer, is a star or T equivalent circuit since we represent the tank ZPS contribution as a magnetic delta tertiary winding. Therefore, at least three measurements are required. It is the author's practice that four ZPS tests are specified to enable a better estimate of the impedance to neutral branch that represents the tank contribution. The ZPS tests are HV–N with LV phase terminals open-circuited, LV–N with HV phase terminals open-circuited, HV–LV//N with LV phase terminals short-circuited and LV–HV//N with HV phase terminals short-circuited.

PPS and ZPS impedance tests on three-winding transformers

In order to derive the star or T equivalent PPS circuit for such transformers from measurements, let the three windings be denoted as 1, 2 and 3, where:

Z_{12} is the PPS leakage impedance measured from winding 1 with winding 2 short-circuited and winding 3 open-circuited. P_{12} is the measured copper loss. Z_{13} is the PPS leakage impedance measured from winding 1 with winding 3 short-circuited and winding 2 open-circuited. P_{13} is the measured copper loss. Z_{23} is the PPS leakage impedance measured from winding 2 with winding 3 short-circuited and winding 1 open-circuited. P_{23} is the measured copper loss.

In three-winding transformers, at least one winding will have a different MVA rating but all the pu impedances must be expressed on the same MVA base. From the above tests, and with the impedances in ohms referred to the same voltage base, and copper losses in kW, we have

$$Z_{12} = Z_1 + Z_2 \quad P_{12} = P_1 + P_2 \quad R_{12} = R_1 + R_2 \quad \text{(4.52a)}$$

$$Z_{13} = Z_1 + Z_3 \quad P_{13} = P_1 + P_3 \quad R_{13} = R_1 + R_3 \quad \text{(4.52b)}$$

$$Z_{23} = Z_2 + Z_3 \quad P_{23} = P_2 + P_3 \quad R_{23} = R_2 + R_3 \quad \text{(4.52c)}$$

Solving Equations (4.52), we obtain

$$Z_1 = \frac{1}{2}(Z_{12} + Z_{13} - Z_{23}) \quad P_1 = \frac{1}{2}(P_{12} + P_{13} - P_{23}) \quad R_1 = \frac{1}{2}(R_{12} + R_{13} - R_{23})$$
$$\text{(4.53a)}$$

$$Z_2 = \frac{1}{2}(Z_{12} + Z_{23} - Z_{13}) \quad P_2 = \frac{1}{2}(P_{12} + P_{23} - P_{13}) \quad R_2 = \frac{1}{2}(R_{12} + R_{23} - R_{13})$$
$$\text{(4.53b)}$$

$$Z_3 = \frac{1}{2}(Z_{13} + Z_{23} - Z_{12}) \quad P_3 = \frac{1}{2}(P_{13} + P_{23} - P_{12}) \quad R_3 = \frac{1}{2}(R_{13} + R_{23} - R_{12})$$
$$\text{(4.53c)}$$

The winding copper loss measurements of Equation (4.52) can be used to calculate the corresponding winding resistances in ohms or in pu on a common MVA base. Then individual resistances are calculated from Equation (4.53). Alternatively, individual winding copper losses could be calculated using Equation (4.53) and used to calculate the corresponding individual winding resistances. When converted into pu, all impedances must be based on one common MVA base.

Regarding the ZPS impedances, the ZPS equivalent circuit that correctly represents a 3-limb core three-winding transformer even when all windings are star-connected, is a star or T equivalent circuit since we represent the tank ZPS contribution as a magnetic delta tertiary winding. Therefore, as in the case of the two-winding transformer, the ZPS tests are HV–N with LV phase terminals open-circuited, LV–N with HV phase terminals open-circuited, HV–LV//N with LV phase terminals short-circuited and LV–HV//N with HV phase terminals short-circuited.

Autotransformers

Generally, two cases are of most practical interest for autotransformers. The first is an autotransformer with a delta-connected tertiary winding and the second is without a tertiary winding. Without a tertiary winding, the PPS test is similar to that of a two-winding transformer described above. With a tertiary winding, the PPS tests are similar to those of a three-winding transformer. However, the ZPS equivalent circuit that correctly represents a 3-limb core autotransformer, with or without a tertiary winding, is a star or T equivalent circuit since we represent the tank ZPS contribution as a magnetic delta tertiary winding. Therefore, the ZPS tests are:

With delta tertiary	Without tertiary	Comments
HV–Tertiary//N	HV–N	LV phase terminals open-circuited
LV–Tertiary//N	LV–N	HV phase terminals open-circuited
HV–LV//Tertiary//N	HV–LV//N	LV phase terminals short-circuited
LV–HV//Tertiary//N	LV–HV//N	HV phase terminals short-circuited

4.2.11 Examples

Example 4.1   The rated and measured test data of a three-phase two-winding transformer are:

Rated data: 120 MVA, 275 kV star with neutral solidly earthed/66 kV delta, $\pm15\%$ HV tapping range
Test data: No-load current measured from LV terminals = 7 A
Iron loss = 103 kW, Full load copper loss = 574 kW

Short-circuit PPS impedance measured from HV side with LV side shorted $= 122.6\,\Omega$

Short-circuit ZPS impedance measured from HV side with LV side shorted $= 106.7\,\Omega$

Calculate the PPS and ZPS equivalent circuit parameters in pu on 100 MVA base

$$Y_{\text{pu on 100 MVA}} = \frac{\sqrt{3} \times 66\,\text{kV} \times 0.007\,\text{kA}}{100\,\text{MVA}} = \frac{0.8}{100} = 0.008\,\text{pu} \quad \text{or} \quad 0.8\% \quad \text{on}$$
100 MVA base

$$G_{\text{pu on 100 MVA}} = \frac{0.103}{100} = 0.00103\,\text{pu or}\ 0.103\%\ \text{on 100 MVA}$$

$$B_{\text{M pu on 100 MVA}} = \frac{1}{100}\sqrt{(0.8)^2 - (0.103)^2} = 0.00793\,\text{pu} \quad \text{or} \quad 0.793\% \quad \text{on}$$
100 MVA

$$Z_{\text{HL PPS pu on 100 MVA}} = \frac{122.6}{(275)^2/100} = 0.162\,\text{pu or}\ 16.2\%\ \text{on 100 MVA}$$

$$R_{\text{HL PPS pu on 100 MVA}} = \frac{0.574 \times 100}{(120)^2} = 0.00398\,\text{pu or}\ 0.398\%\ \text{on 100 MVA}$$

$$X_{\text{HL PPS pu on 100 MVA}} = \sqrt{(0.162)^2 - (0.00398)^2} = 0.16195\,\text{pu} \quad \text{or} \quad 16.195\%$$
on 100 MVA.

It is worth noting that the PPS X/R ratio is $16.195/0.398 = 40.7$

$$Z_{\text{HL ZPS pu on 100 MVA}} = \frac{106.7}{(275)^2/100} = 0.141\,\text{pu or}\ 14.1\%\ \text{on 100 MVA}.$$

This is 87% of the PPS impedance. In the absence of ZPS copper loss measurement, the ZPS X/R ratio may be assumed equal to the PPS X/R ratio of 40.7. Therefore,

$$R_{\text{HL ZPS pu on 100 MVA}} = 0.346\%\ \text{on 100 MVA}$$

and

$$X_{\text{HL ZPS pu on 100 MVA}} = 14.095\%\ \text{on 100 MVA}$$

If the PPS impedance were measured from the LV side, its ohmic value would be $122.6 \times (66\,\text{kV}/275\,\text{kV})^2 = 7.06\,\Omega$ and its pu value would also be equal to 16.2%.

Example 4.2   Consider a two-winding three-phase transformer with the following rated data: 45 MVA, 132 kV/33 kV star solidly earthed neutral – delta, HV winding tapped with -20% to $+10\%$ tap range. PPS impedance tests were carried out at all 19 tap positions with the HV terminals supplied by a three-phase voltage source and the LV winding short-circuited. The applied test voltages and full load currents on the HV side at minimum, nominal and maximum tap positions are:

Tap position	Phase–phase voltage (V)	Current (A)
1 (+10%)	17 800	179
7 (nominal)	15 560	197
19 (−20%)	11 130	246

Calculate the measured impedance at each tap position both in ohms and in pu on MVA rating:

Tap position 1
Rated tap voltage $= 132 \times 1.1 = 145.2\,\text{kV}$.

$$\text{Impedance in pu on 45 MVA} = \frac{17.8}{145.2} \times 100 = 12.3\%.$$

$$\text{Impedance in ohm} = \frac{17\,800/\sqrt{3}}{179} = 57.4\,\Omega.$$

$$\text{Alternatively, impedance in pu on 45 MVA} = \frac{57.4}{(145.2)^2/45} \times 100 = 12.3\%.$$

Tap position 7
Rated tap voltage $= 132 \times 1 = 132\,\text{kV}$, the impedance is $45.6\,\Omega$ or 11.8% on 45 MVA.

Tap position 19
Rated tap voltage $= 132 \times 0.8 = 105.74\,\text{kV}$, the impedance is $26.1\,\Omega$ or 10.6% on 45 MVA.

The reader should notice that the apparent small variation in impedance in % terms masks the significant change in the impedance when actual units, i.e. ohms are used. Also, the base impedance in each case is different because it is dependent on the rated tap voltage which is a function of the tap position.

Example 4.3   The rated and measured test data for a three-phase three-winding transformer are:

Primary HV winding 60.6 MVA, 132 kV star solidly earthed.
Secondary LV1 winding 30.3 MVA, 11.11 kV star solidly earthed.
Secondary LV2 winding 30.3 MVA, 11.11 kV star solidly earthed.

The construction is a 3-limb core. The short-circuit impedance and load loss measurement test data at nominal tap position is:

Test	Volts	Amps	Full load loss (kW)
PPS			
HV/LV1 shorted, LV2 open	43 843	132.6	201.5 on 30.3 MVA
HV/LV2 shorted, LV1 open	43 750	132.6	202.2 on 30.3 MVA
LV1/LV2 shorted, HV open	5940	1517	413 on 30.3 MVA
ZPS			
HV shorted, LV1 and LV2 open	21 960	80	

Derive the PPS and ZPS impedances and corresponding transformer equivalent circuits. Calculate the effective pu tap ratios at -10% turns HV winding tap position. The network base voltages are 132 and 11 kV. Use 60.6 MVA as a common base.

$$Z_{HL1} = \frac{43\,843/\sqrt{3}}{132.6} = 190.9\,\Omega \quad \text{or} \quad \frac{190.9}{132^2/60.6} \times 100 = 66.4\%$$

$$Z_{HL2} = \frac{43\,750/\sqrt{3}}{132.6} = 190.5\,\Omega \quad \text{or} \quad \frac{190.5}{132^2/60.6} \times 100 = 66.2\%$$

$$Z_{L1L2} = \frac{5940/\sqrt{3}}{1517} = 2.26\,\Omega \quad \text{or} \quad \frac{2.26}{11.11^2/60.6} \times \frac{11.11^2}{11^2} \times 100 = 113\%$$

$$Z_H(\Omega) = \frac{1}{2}\left[190.9 + 190.5 - 2.26 \times \left(\frac{132}{11}\right)^2\right] = 27.98\,\Omega$$

It is important to remember that the impedances in Ω in the star or T equivalent of the three-winding transformer must all be referred to the same voltage base:

$$Z_{L1}(\Omega) = \frac{1}{2}\left[190.9 + 2.26 \times \left(\frac{132}{11}\right)^2 - 190.5\right] = 162.92\,\Omega$$

$$Z_{L2}(\Omega) = \frac{1}{2}\left[190.5 + 2.26 \times \left(\frac{132}{11}\right)^2 - 190.9\right] = 162.52\,\Omega$$

$$Z_H(\%) = \frac{27.98}{132^2/60.6} \times 100 = 9.73\%$$

$$Z_{L1}(\%) = \frac{162.92}{132^2/60.6} \times 100 = 56.66\%$$

$$Z_{L2}(\%) = \frac{162.52}{132^2/60.6} \times 100 = 56.52\%$$

$$Z_{HN}^Z(\Omega) = \frac{21\,960}{80/3} = 823.5\,\Omega \quad \text{or} \quad \frac{823.5}{(132)^2/60.6} \times 100 = 286.4\%$$

but

$$Z_{HN}^Z = Z_H + Z_M^Z$$

thus

$$Z_M^Z = 823.5 - 27.98 = 795.5\,\Omega$$

or

$$Z_M^Z = 286.4 - 9.73 = 276.67\%$$

$$R_{HL1} = \frac{201.5 \times 10^3/3}{132.6^2} = 3.82\,\Omega \quad R_{HL2} = \frac{202.2 \times 10^3/3}{132.6^2} = 3.83\,\Omega$$

$$R_{L1L2} = \frac{413 \times 10^3/3}{1517^2} = 0.0598\,\Omega$$

$$R_H(\Omega) = \frac{1}{2}\left[3.82 + 3.83 - 0.0598 \times \left(\frac{132}{11}\right)^2\right] = -0.48\,\Omega$$

This should not alarm the reader because the T equivalent is a fictitious mathematical model!

$$R_{L1}(\Omega) = \frac{1}{2}\left[3.82 + 0.0598 \times \left(\frac{132}{11}\right)^2 - 3.83\right] = 4.3\,\Omega$$

$$R_{L2}(\Omega) = \frac{1}{2}\left[3.83 + 0.0598 \times \left(\frac{132}{11}\right)^2 - 3.82\right] = 4.31\,\Omega$$

$$R_H(\%) = \frac{-0.48}{(132)^2/60.6} \times 100 = -0.167\%$$

$$R_{L1}(\%) = \frac{4.3}{(132)^2/60.6} \times 100 = 1.49\%$$

$$R_{L1}(\%) = \frac{4.31}{(132)^2/60.6} \times 100 = 1.5\%$$

The reactances of the T equivalent are given by $X_H = 27.97\,\Omega$ or 9.7%;

$$X_{L1} = 162.86\,\Omega \text{ or } 56.6\% \quad \text{and} \quad X_{L2} = 162.46\,\Omega \text{ or } 56.5\%.$$

Designating terminals LV1 as 1, LV2 as 2 and HV as 3, the tap effective pu ratios at -10% HV winding turns are

$$t_{13}(\text{pu}) = t_{23}(\text{pu}) = \frac{11.11/11}{0.9 \times 132/132} = 1.1223\,\text{pu}$$

The PPS and ZPS equivalent circuits are shown in Figure 4.32.

(a)

(b)

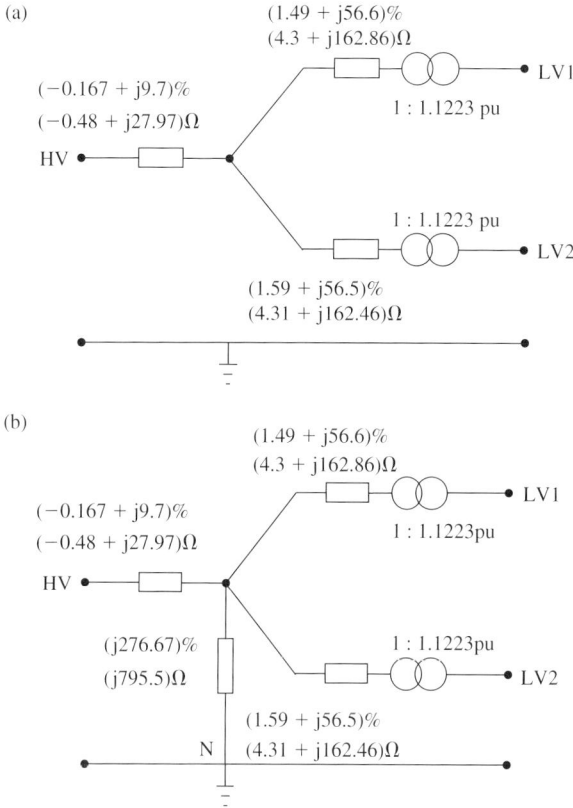

Figure 4.32 Example 4.3 three-winding transformer impedance tests: (a) PPS equivalent circuit and (b) ZPS equivalent circuit

Example 4.4 A 3 limb core autotransformer has the following rated data: 275 kV/132 kV, 240 MVA, star–star solidly earthed neutral and a 60 MVA/13 kV tertiary winding. The HV to LV PPS impedance is 18.95% on 240MVA rating. ZPS impedance tests were carried out with the delta winding closed and open to simulate the change in ZPS impedances if the transformer had no tertiary winding and to identify the effect of the core construction. The measured ZPS impedances at nominal tap position in both cases are:

ZPS test	Measured impedance (Ω/phase)	
	Delta closed	Delta open
Z^Z_{H-N}, H supplied, L open	107.3	741.4
Z^Z_{L-N}, L supplied, H open	14.3	200.5
$Z^Z_{H-L//N}$, H supplied, L short-circuited	50.2	50.0
$Z^Z_{L-H//N}$, L supplied, H short-circuited	6.8	13.5

Derive general equations for the ZPS T equivalent circuit impedances and calculate the impedances for the above two cases in % on 100 MVA:

$$Z_{H-N}^Z(\%) = \frac{Z_{H-N}^Z(\Omega)}{275^2/100} \times 100 \quad Z_{H-L//N}^Z(\%) = \frac{Z_{H-L//N}^Z(\Omega)}{275^2/100} \times 100$$

$$Z_{L-N}^Z(\%) = \frac{Z_{L-N}^Z(\Omega)}{132^2/100} \times 100 \quad Z_{L-H//N}^Z(\%) = \frac{Z_{L-H//N}^Z(\Omega)}{132^2/100} \times 100$$

Using these equations, we obtain

ZPS impedance (% on 100 MVA)	Delta closed	Delta open
Z_{H-N}^Z	14.2	98
Z_{L-N}^Z	8.2	115
$Z_{H-L//N}^Z$	6.6	6.6
$Z_{L-H//N}^Z$	3.9	7.8

The ZPS T equivalent circuit has three unknowns and therefore only three tests are required, e.g. the first three above. However, the fourth test, where available, can be used to improve the prediction of the shunt branch impedance. The four tests above are illustrated in Figure 4.33(a) from which we have the following:

$$Z_{H-N}^Z = Z_H^Z + Z_N^Z \quad \text{and} \quad Z_{L-N}^Z = Z_L^Z + Z_N^Z$$

$$Z_{H-L//N}^Z = Z_H^Z + \frac{Z_L^Z Z_N^Z}{Z_L^Z + Z_N^Z} \quad \text{and} \quad Z_{L-H//N}^Z = Z_L^Z + \frac{Z_H^Z Z_N^Z}{Z_H^Z + Z_N^Z}$$

Substituting Z_H^Z and Z_L^Z from the first two equations into the second two, we obtain

$$Z_{N(1)}^Z = \sqrt{Z_{H-N}^Z(Z_{L-N}^Z - Z_{L-H//N}^Z)} \quad \text{and} \quad Z_{N(2)}^Z = \sqrt{Z_{L-N}^Z(Z_{H-N}^Z - Z_{H-L//N}^Z)}$$

and a mean value is calculated as

$$Z_{N(mean)}^Z = \frac{Z_{N(1)}^Z + Z_{N(2)}^Z}{2}$$

The remaining H and L branch impedances are

$$Z_H^Z = Z_{H-N}^Z - Z_{N(mean)}^Z \quad \text{and} \quad Z_L^Z = Z_{L-N}^Z - Z_{N(mean)}^Z$$

The calculated Z_H^Z, Z_L^Z and Z_N^Z for the closed and open delta tertiary are shown in Figure 4.33(b).

Example 4.5   The autotransformer with a tertiary winding used in Example 4.3 has a neutral earthing reactor of 10 Ω connected to its neutral. Calculate the earthing impedance values in % on 100 MVA base that would appear in each branch of the autotransformer ZPS T equivalent circuit.

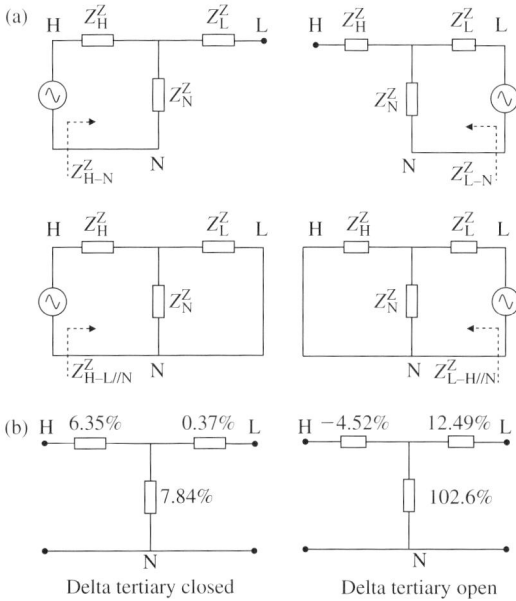

Figure 4.33 ZPS leakage impedance tests on an autotransformer: (a) ZPS leakage impedance tests; (b) ZPS equivalent circuits

The transformer turns ratio is $275/132 = 2.0834$.

The % impedance value of the earthing reactor is

$$Z_E^Z(\%) = \frac{10}{132^2/100} \times 100 = 5.739\%$$

The earthing impedances appearing in the three branches of the ZPS T equivalent circuit are

$$Z_E^Z \text{ (appearing in H terminal in %)} = \frac{-3 \times (2.0834 - 1)}{2.0834^2} \times 5.739 = -4.297\%$$

$$Z_E^Z \text{ (appearing in L terminal in %)} = \frac{3 \times (2.0834 - 1)}{2.0834} \times 5.739 = 8.953\%$$

$$Z_E^Z \text{ (appearing in T terminal in %)} = \frac{3}{2.0834} \times 5.739 = 8.264\%$$

4.3 Sequence modelling of QBs and PS transformers

4.3.1 Background

QBs and PS transformers are widely used in transmission and subtransmission power systems for controlling the magnitude and direction of active power flow

Figure 4.34 2000 MVA 400 kV ±11.3° QB consisting of a shunt and a series transformer

mainly over parallel circuits in order to increase the power transfer capability across boundaries or interfaces. A single-core design is characterised by its simplicity and economy but suffers from disadvantages. These include the overvoltages that may be imposed on the tap-changer due to its position in series with the main line as well as having a virtually zero impedance at nominal tap which may not be desirable if short-circuit current limitation is required. By far, the most commonly used design is the two-core design in either a single tank or two tanks where very large rating is required. Figure 4.34 shows a 400 kV QB rated at 2000 MVA and ±11.3° phase shift. Figures 4.35(a) and 4.36(a) show QB and PS one-line diagrams, respectively. A simplified representation of the three-phase connections of typical large QBs and PSs are shown in Figures 4.35(b) and 4.36(b), respectively. The designs shown are those where both QB and PS consist of two transformers; a shunt transformer connected in star–star and a series transformer connected in delta-series star. The primary winding of the shunt transformer is called the exciting winding and the secondary is called the regulating winding. The primary delta of the series transformer is called the booster winding and the secondary is called the series winding because it is in series with the line.

The principles of operation of the QB and PS are similar. The control of the magnitude and direction of active power flow on the line is achieved by varying the phase angle shift across the series winding. The phase shift and its variation are obtained by an on-load (under-load) tap-changer acting on the regulating winding and deriving a variable voltage component across two phases, e.g. Y and B. This voltage is in quadrature with the input voltage. It is then injected by the booster

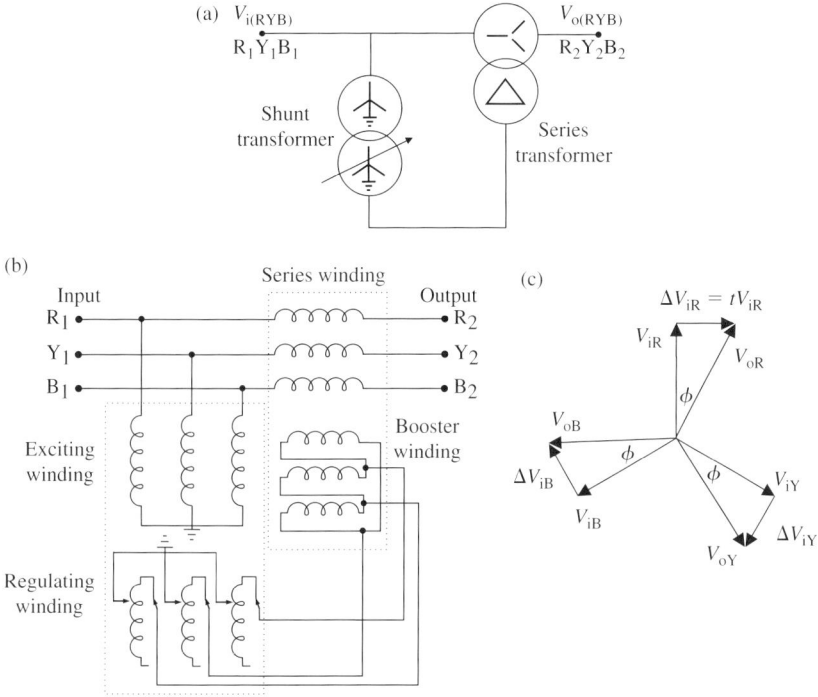

Figure 4.35 QB transformer: (a) one-line diagram, (b) winding connections and (c) open-circuit vector diagram

transformer delta winding across the series winding third phase, e.g. phase R, as shown in Figures 4.35(c) and 4.36(c). One important difference to notice between a QB and a PS is that for a QB, the input voltage of the shunt transformer is derived from one side of the series transformer so that the output voltage is slightly higher than the input voltage. However, for a PS, the shunt transformer input voltage is derived from the mid-point on the series winding of the series transformer so that the magnitudes of the input and output voltages remain equal.

4.3.2 PPS, NPS and ZPS modelling of QBs and PSs

PPS equivalent circuit model

The derivation of QB and PS detailed equivalent circuits needs to take into account both the shunt and series transformers, their winding connections, tap-changer operation, complex turns ratio and leakage impedance variations with tap position. However, we will instead present a simplified treatment but one that is still sufficient for use in practice in PPS power flow, stability and short-circuit analysis. The analysis below is based on the vector diagrams of Figures 4.35(c) and 4.36(c). These allow us to represent the QB or PS as an ideal transformer with a complex turns ratio in series with an appropriate impedance. This ratio is in series with a single effective leakage impedance representing both the series and shunt

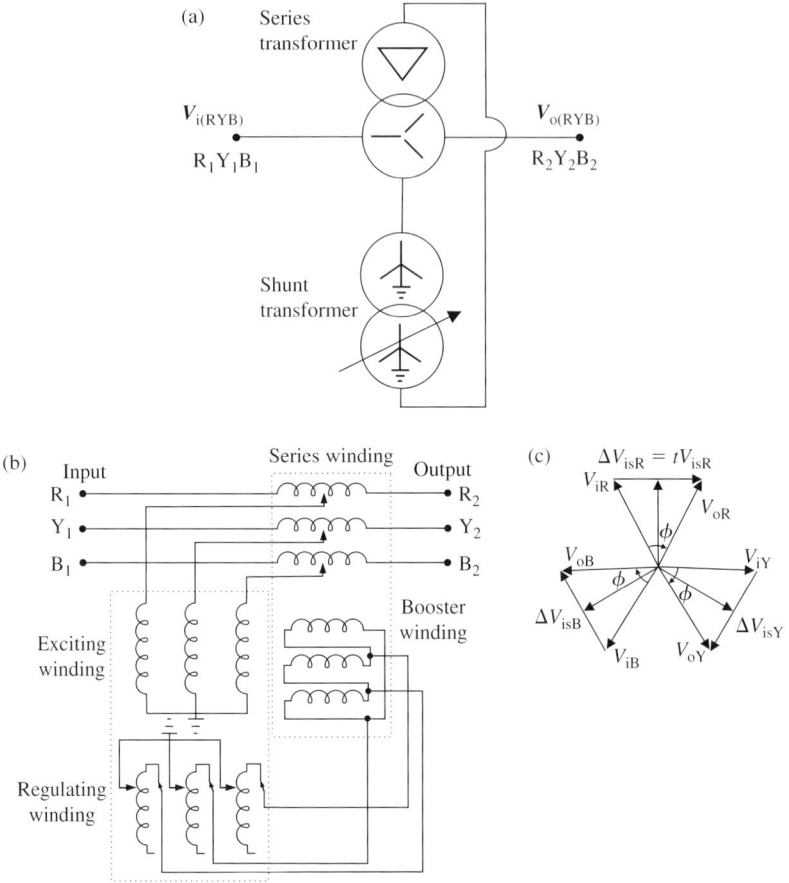

Figure 4.36 PS transformer: (a) one-line diagram, (b) winding connections and (c) open-circuit vector diagram

transformers' leakage impedances including the effect of the tap-changer on the shunt transformer's impedance. This representation is shown in Figure 4.37.

For both a QB and a PS, let $t = \Delta V_{i(R)}/V_{i(R)}$ be the magnitude of the injected quadrature voltage in pu of the input voltage. From the QB vector diagram shown in Figure 4.35(c), the phase R open-circuit output voltage phasor is given by

$$V_{o(R)} = V_{i(R)}(1 + jt) \tag{4.54a}$$

Let $\overline{N} = \frac{V_{o(R)}}{V_{i(R)}}$ be defined as the complex turns ratio, at no load, thus,

$$\overline{N} = 1 + jt = \sqrt{1 + t^2}e^{j\phi} \tag{4.54b}$$

where

$$\phi = \tan^{-1}(t) \tag{4.54c}$$

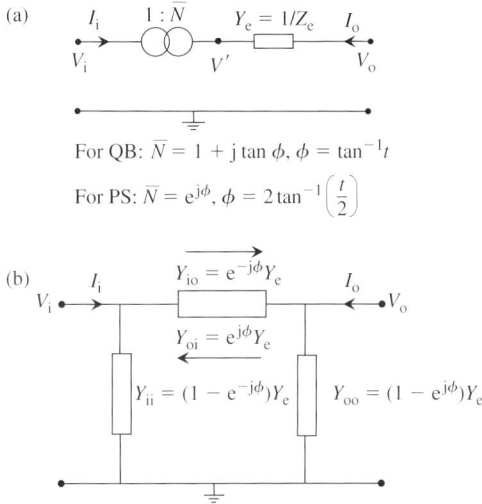

Figure 4.37 PPS equivalent circuits for a QB or a PS: (a) PPS equivalent circuit and (b) asymmetrical π equivalent circuit for a PS

is the phase angle shift that has a maximum or rated value for practical QBs used in power networks of typically less than or equal to $\pm 15°$. Similar equations apply for the Y and B phases.

Similarly, from the PS vector diagram shown in Figure 4.36(c), the phase R open-circuit output voltage is given by

$$V_{o(R)} = V_{iS(R)} \left(1 + j\frac{t}{2} \right) \qquad V_{i(R)} = V_{iS(R)} \left(1 - j\frac{t}{2} \right) \qquad (4.55a)$$

where $V_{iS(R)}$ is the voltage at the mid-point of the series winding.

The complex turns ratio of the PS, at no load, is given by

$$\overline{N} = \frac{V_{o(R)}}{V_{i(R)}} = \frac{\left(1 + j\dfrac{t}{2} \right)}{\left(1 - j\dfrac{t}{2} \right)} = e^{j\phi} \qquad (4.55b)$$

where

$$\phi = 2 \tan^{-1} \left(\frac{t}{2} \right) \qquad (4.55c)$$

is the phase angle shift that has a maximum or rated value for practical PSs used in power networks of typically less than or equal to $\pm 60°$.

Let $Y_e = 1/Z_e$ where Z_e is the effective QB or PS leakage impedance appearing in series with the line or the network where the device is connected. The QB and PS are both represented by the same PPS equivalent circuit shown in Figure 4.37. The PPS admittance matrix for this circuit can be easily derived as in the case of an off-nominal-ratio transformer but with an important difference now that the turns ratio is a complex number. From Figure 4.37, we can write

$$\frac{V'}{V_i} = \overline{N} \quad V_i I_i^* = V'(-I_o)^* \quad \text{hence} \quad \frac{I_i}{I_o} = -\overline{N}^* \quad \text{and} \quad V_o - \overline{N}V_i = \frac{I_o}{Y_e}$$

From these equations, the following admittance matrix is obtained

$$\begin{bmatrix} I_i \\ I_o \end{bmatrix} = \begin{bmatrix} |\overline{N}|^2 Y_e & -\overline{N}^* Y_e \\ -\overline{N} Y_e & Y_e \end{bmatrix} \begin{bmatrix} V_i \\ V_o \end{bmatrix} \qquad (4.56a)$$

which, using Equations (4.54b) and (4.54c), can be written as

$$\begin{bmatrix} I_i \\ I_o \end{bmatrix} = \begin{bmatrix} (1 + \tan^2\phi)Y_e & -(1 - j\tan\phi)Y_e \\ -(1 + j\tan\phi)Y_e & Y_e \end{bmatrix} \begin{bmatrix} V_i \\ V_o \end{bmatrix} \quad \text{for a QB} \qquad (4.56b)$$

or, using Equation (4.55b), can be written as

$$\begin{bmatrix} I_i \\ I_o \end{bmatrix} = \begin{bmatrix} Y_e & -e^{-j\phi} Y_e \\ -e^{j\phi} Y_e & Y_e \end{bmatrix} \begin{bmatrix} V_i \\ V_o \end{bmatrix} \quad \text{for a PS} \qquad (4.56c)$$

The above two admittance matrices for a QB and a PS cannot be represented by a physical π equivalent circuit because the complex turns ratio has resulted in matrices that are non-symmetric, i.e. the off-diagonal transfer terms are not equal. However, recognising that an asymmetrical non-physical π equivalent circuit is simply a mathematical tool, such a circuit is shown in Figure 4.37(b) for a PS. A similar π equivalent for a QB can be derived and the reader is encouraged to derive this equivalent.

NPS equivalent circuit model

Since the QB and PS are static devices, the NPS impedance should be the same as the PPS impedance. However, we have shown that the PPS model includes a complex turns ratio which is a mathematical operation that introduces a phase angle shift in the output voltage (and current) with respect to the input voltage (and current). The mathematical derivation of the phase shift in the QB or PS NPS model requires a detailed representation of all windings of the shunt and series transformers, which, as outlined above, is not presented here. However, we will use vector diagrams to show that the phase shift in the NPS circuit is equal to that in the PPS circuit but reversed in sign. Figures 4.38(i) show the PPS open-circuit vector diagrams of a QB and a PS and the resultant phase angle shift ϕ in each case. Figures 4.38(ii) show the NPS vector diagrams of a QB and a PS and the resultant phase angle shift in each case.

It is clear that with the reversed NPS phase rotation RBY, the injected series quadrature voltage vector is also reversed, for both QB and PS. Therefore, if the PPS phase angle shift is ϕ, then the NPS phase angle shift is $-\phi$. Figure 4.39 shows the QB or PS NPS equivalent circuit model where the complex ratio is given by

$$\overline{N} = \begin{cases} 1 - jt = \sqrt{1 + t^2}e^{-j\phi} & \text{where } \phi = \tan^{-1}(t) & \text{for a QB} \\ e^{-j\phi} & \text{where } \phi = 2\tan^{-1}\left(\frac{t}{2}\right) & \text{for a PS} \end{cases} \qquad (4.57a)$$

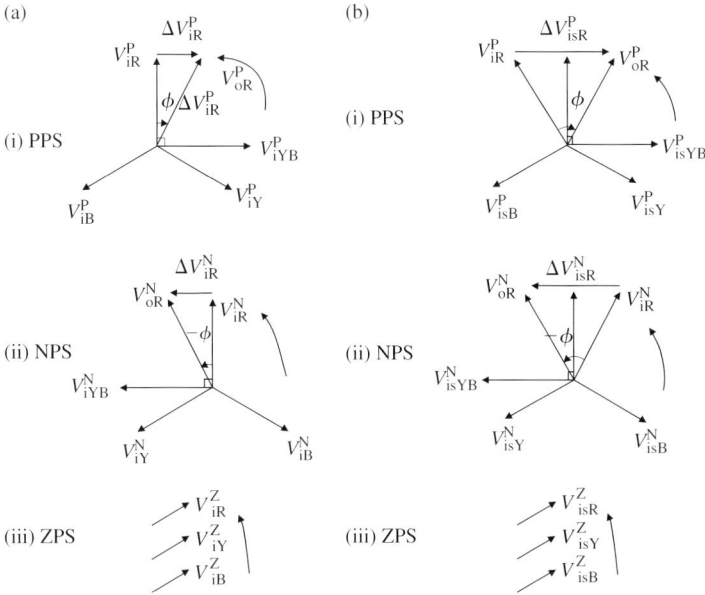

Figure 4.38 PPS, NPS and ZPS vector diagrams for a QB and a PS: (a) QB vector diagram and (b) PS vector diagram

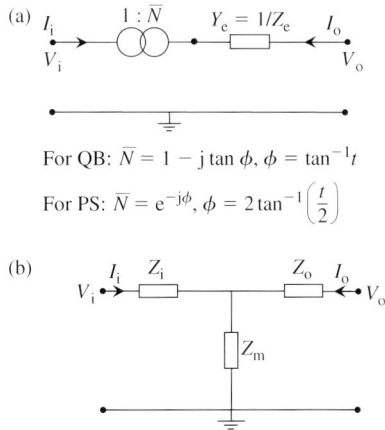

For QB: $\overline{N} = 1 - j \tan \phi$, $\phi = \tan^{-1} t$

For PS: $\overline{N} = e^{-j\phi}$, $\phi = 2 \tan^{-1} \left(\frac{t}{2} \right)$

Figure 4.39 NPS and ZPS equivalent circuits for a QB or a PS: (a) NPS equivalent circuit and (b) ZPS equivalent circuit

Using the PPS admittance matrices for a QB and a PS, the NPS admittance matrices are easily obtained by replacing ϕ with $-\phi$ in Equations (4.56b) and (4.56c).

ZPS equivalent circuit model

We have already stated that the basic principle of operation of the QB or PS is to inject a voltage across, say phase R, series transformer series winding that is in quadrature with the phase R input voltage. In the PPS and NPS circuits, this is achieved by deriving the injected voltage so as to be in phase or out of phase with the voltage difference between phases Y and B as shown in Figure 4.38(i) and (ii), respectively. However, since the three ZPS voltages are in phase with each other, a quadrature voltage component cannot be produced. Therefore, the phase shift between the input and open-circuit output voltages (and currents) in the ZPS circuit is zero. Therefore, the complex turns ratio in the ZPS equivalent circuit for both a QB and a PS becomes unity.

The QB or PS practical ZPS equivalent circuit, however, is different from the PPS and NPS equivalent circuits. The ZPS equivalent circuit is dependent on the winding connections of the series and shunt transformers, their core construction and whether the shunt transformer has any delta-connected tertiary winding. In a similar way to a three-winding transformer, or a 3-limb core autotransformer with or without a tertiary winding, the general ZPS equivalent circuit of a QB or a PS is a star or T equivalent as shown in Figure 4.39(b). The shunt branch in this equivalent represents either the impedance of a tertiary winding or the stray air path ZPS flux through the tank/oil. In addition, in some QBs and PSs, both series and shunt transformers are in the same tank so that the ZPS flux can link both windings either directly or through the tank. In practice, the QB or PS ZPS impedances are derived from a series of manufacturer works-tests and this is described in the next section. The ZPS admittance matrix model of the QB or PS is given by

$$\begin{bmatrix} I_i \\ I_o \end{bmatrix} = \begin{bmatrix} Y_i & -Y_m \\ -Y_m & Y_o \end{bmatrix} \begin{bmatrix} V_i \\ V_o \end{bmatrix} \tag{4.57b}$$

where

$$Y_i = \frac{1}{Z_i + \frac{Z_o Z_m}{Z_o + Z_m}} \quad Y_m = \frac{-1}{(Z_i + Z_o) + \frac{Z_i Z_o}{Z_m}} \quad Y_o = \frac{1}{Z_o + \frac{Z_i Z_m}{Z_i + Z_m}} \tag{4.57c}$$

4.3.3 Measurement of QB and PS sequence impedances

Like transformers, QBs and PSs are subjected to a variety of works-tests by their manufacturers. As most QB and PS are three-phase two-transformer devices, tests are usually carried out on each transformer alone and on the combined units. No-load loss, load loss, PPS impedance and ZPS impedance tests are carried out on the combined units. The PPS impedance and load loss tests are carried out simultaneously with the output end of the series transformer supplied by three-phase voltage sources and the shunt transformer input terminals short-circuited. The ohmic value of the impedance is the applied voltage divided by the measured current. The pu value of the impedance must be expressed on the voltage base of the winding energised or supplied under the tests. Under such a test, the PPS impedance measured is the series combination of the leakage impedances of the series and shunt transformers. This can be easily visualised by referring back to the QB and PS winding connections shown in Figures 4.35 and 4.36. Therefore, this impedance is the QB or PS effective PPS impedance Z_e that appears in series with the line as seen from the QB or PS series transformer output end.

The series transformer does not have a tap-changer so its leakage impedance, when measured from its output terminals, is constant. However, when measured from the same terminals, the shunt transformer leakage impedance varies with tap position as the tap-changer acts on the shunt transformer regulating winding. The variation of tap position varies the magnitude of the injected quadrature voltage and, the PPS/NPS leakage impedance will vary with the square of the injected quadrature voltage t. Therefore,

$$Z_e^P(t) = Z_e^N(t) = Z_{Series} + \left(\frac{t}{t_{max}} \right)^2 Z_{Shunt \text{ at end tap}} \tag{4.58}$$

which can be expressed in terms of the phase angle shift as follows:

$$Z_e^P(t) = Z_e^N(t) = \begin{cases} Z_{Series} + \left(\dfrac{\tan \phi}{\tan \phi_{Max}} \right)^2 \times Z_{Shunt \text{ at end tap}} & \text{for a QB} \\[4mm] Z_{Series} + \left(\dfrac{\tan \left(\frac{\phi}{2} \right)}{\tan \left(\frac{\phi_{Max}}{2} \right)} \right)^2 \times Z_{Shunt \text{ at end tap}} & \text{for a PS} \end{cases} \tag{4.59}$$

where ϕ_{Max} is the QB or PS rated or maximum phase shift design value, ϕ the actual phase shift in operation and $Z_{Shunt \text{ at end tap}}$ is the shunt transformer leakage impedance measured at end tap position that is at maximum quadrature voltage or maximum phase shift. It should be noted that $Z_e(\phi = 0)$ is equal to Z_{Series} because the booster delta winding of the series transformer is effectively short-circuited by the tap-changer when a zero quadrature voltage is injected.

Like transformers, the ZPS leakage impedance tests are carried out applying a single-phase voltage source to three-phase input or output terminals joined together. Usually three or four such tests may be carried out to derive the star ZPS equivalent circuit. Some tests that may be carried out are described below where T represents a tertiary winding, where present in the shunt transformer, and N represents neutral. Reference is made to Figures 4.35 and 4.36 to see the designated QB and PS terminals.

Terminals supplied	Designation	Comments
R1Y1B1 terminals joined together	Zo1–T//N	R2Y2B2 terminals open-circuit, tertiary closed, where present
R2Y2B2 terminals joined together	Zo2–T//N	R1Y1B1 terminals open-circuit, tertiary closed, where present
R1Y1B1 terminals joined together	Zo1–2//T//N	R2Y2B2 terminals short-circuited, tertiary closed, where present
R2Y2B2 terminals joined together	Zo2–1//T//N	R1Y1B1 terminals short-Circuited, tertiary closed, where present

Example 4.6   A 400 kV 2000 MVA throughput QB has a rated quadrature injected voltage of ±0.2 pu of input voltage. The following PPS and ZPS impedance tests were carried out. In the PPS tests, the output, i.e. series transformer terminals R2Y2B2 are supplied and the input of the shunt transformer terminals R1Y1B1 are short-circuited.

PPS Test	Phase–phase voltage (V)	Current (A)
Tap 1	56 800	2887
Tap 20 (nominal)	28 049	2886
Tap 39	56 920	2888

ZPS Test	Voltage (V)	Current (A)
Zo1–T//N		
Tap 1	20 817	260.1
Tap 20 (nominal)	20 823	260.4
Tap 39	20 818	260.4
Zo2–T//N		
Tap 1	21 395	260.8
Tap 20 (nominal)	21 360	260.4
Tap 39	21 377	260.5
Zo2–1//T//N		
Tap 1	506	260.0
Tap 20 (nominal)	507	260.7
Tap 39	506	260.4

Calculate the open-circuit rated phase shift and output voltage of the QB for a 1 pu input voltage. Also, calculate the effective PPS impedance of the QB and the ZPS star equivalent impedances at nominal, minimum and maximum tap positions.

The open-circuit rated phase shift is $\tan^{-1}(\pm 0.2) = \pm 11.3°$. The QB range of rated injected voltage or phase shift is from -0.2 pu corresponding to $-11.3°$ phase shift to $+0.2$ pu corresponding to $+11.3°$ phase shift. The open-circuit output voltage at rated phase shift is $\sqrt{1 + 0.2^2} \approx 1.02$ pu. The PPS impedance is calculated as follows:

Tap 1:

$$\text{Impedance in pu on 2000 MVA} = \frac{56.8}{400} \times 100 = 14.2\%.$$

$$\text{Alternatively, impedance in pu on 2000 MVA} = \frac{(56\,800/\sqrt{3})/2887}{(400)^2/2000} \times$$

$$100 = 14.2\%.$$

Tap 20:
Impedance is 7.01% on 2000 MVA.

Tap 39:
Impedance is 14.23% on 2000 MVA.

We note that the PPS impedance effectively doubles between nominal tap position (tap 20) and maximum/minimum tap positions (tap 39 and tap 1). The results of the calculated ZPS impedances are given as:

ZPS impedance	Zo1–T//N (Ω/phase)	Zo2–T//N (Ω/phase)	Zo2–1//T//N (Ω/phase)
Tap 1	240.1	246.1	5.84
Tap 20	239.9	246.1	5.83
Tap 39	239.8	246.2	5.83

We note that the ZPS impedances of the equivalent star are effectively unaffected by variation of tap position. The impedances of the individual T branch equivalent circuit at nominal tap position in % on 2000 MVA and 400 kV are:

$Z_1^Z = 7.582\%$ where 1 corresponds to the shunt transformer's input end R1Y1B1.

$Z_2^Z = -0.294\%$ where 2 corresponds to the series transformer's output end R2Y2B2.

$Z_N^Z = 300\%$ where N corresponds to neutral or zero voltage reference in this case.

Example 4.7   A 132 kV, 300 MVA throughput PS has a rated phase shift of $\pm 30°$ and the shunt transformer unit includes a tertiary winding. The PPS impedance measured at nominal tap position is 2% on 100 MVA at 132 kV. The ZPS impedances of the PS are largely unaffected by variation of tap position and the ZPS impedance test data are as follows:

Zo1–T//N (Ω/phase)	Zo2–T//N (Ω/phase)	Zo1–2//T//N (Ω/phase)	Zo2–1//T//N (Ω/phase)
14.38	14.9	3.16	3.16

Calculate the ZPS star equivalent circuit impedances in % on 100 MVA.

Like autotransformers, we have four tests but only three unknowns so we will calculate two values of the shunt impedance in the star equivalent and obtain a mean value as follows:

$$Z_{N(1)}^Z = \sqrt{14.38 \times (14.9 - 3.16)} = 13\,\Omega$$

and

$$Z_{N(2)}^Z = \sqrt{14.9 \times (14.38 - 3.16)} = 12.93\,\Omega$$

and the mean value is $Z_{N(mean)}^Z = 12.96\,\Omega$

$$Z_{(1)}^Z = 14.38 - 12.96 = 1.42\,\Omega \quad \text{and} \quad Z_{(2)}^Z = 14.9 - 12.96 = 1.94\,\Omega$$

The impedance values in % on 100 MVA are

$$Z_{(1)}^Z(\%) = \frac{1.42}{132^2/100} \times 100 = 0.81\%$$

$$Z_{(2)}^Z(\%) = \frac{1.94}{132^2/100} \times 100 = 1.11\%$$

$$Z_{N(\text{mean})}^Z(\%) = \frac{12.96}{132^2/100} \times 100 = 7.44\%$$

4.4 Sequence modelling of series and shunt reactors and capacitors

4.4.1 Background

Series reactors are widely used in power systems for power flow control or as fault current limiters. The latter application will be covered in Chapter 9. Series reactors are generally used in electric distribution, subtransmission and transmission systems as well as power stations and industrial power systems. Air core reactors are usually used at nominal system voltages up to and including 36 kV, but iron-cored reactors are generally used at higher voltages. The former is usually coreless and magnetically or electromagnetically shielded and the latter are usually oil filled and gapped. Three-phase gapped iron core reactors look in appearance similar to a three-phase three-limb core transformer. However, reactors have only one winding on each limb. The function of the gaps in the core is to lower the flux density so that under high current conditions, the core barely enters into saturation and the reactor's impedance remains substantially constant. In practice a decrease in the impedance from the rated current value by a few percent is typical. Magnetically shielded coreless reactors use shields to surround the windings in order to provide a return path for the winding flux. However, saturation and hence impedance reduction can still occur. In the electromagnetically shielded coreless reactor, the impedance remains virtually constant but with an increase in reactor cost.

Shunt reactors are widely used in power networks at various voltage levels for controlling and limiting transient and steady state voltages. Shunt reactors are cost-effective and robust reactive compensation devices used in absorbing surplus reactive power supply in cable systems where they may be switched in and out of service with the cable itself by the cable circuit-breakers. Many aspects of their design are similar to those of three-limb series reactors except that at higher voltages, many shunt reactors are star connected with their neutral point either solidly or impedance earthed.

Shunt capacitors are usually mechanically switched, i.e. by circuit-breakers and are very cost-effective reactive compensation devices that are extensively used in transmission and lower voltage power networks. They provide a source of reactive power supply and help to improve network voltage stability, voltage levels and power factors. Shunt capacitors can be star connected with the neutral isolated or solidly earthed, or delta connected, depending on the nominal system voltage.

Series capacitors are also widely used in transmission and distribution networks. In distribution, they are mainly used to improve the voltage profile of heavily loaded feeders. In transmission, they are mainly used to increase network power transfer capabilities by improving generator and network transient stability, damping of power oscillations, network voltage stability or load sharing on parallel circuits.

4.4.2 Modelling of series reactors

In a three-limb design of a three-phase series reactor, the winding around each limb represents one phase and the flux in any phase does not link that in the other two phases. Therefore, there is practically little or no mutual inductive coupling between the phases of the reactor. In addition, the phase windings are practically identical by design. A three-phase equivalent circuit is shown in Figure 4.40 where $Z = R + jX$ is the leakage impedance per phase. The series phase voltage drop is given by

$$\begin{bmatrix} V_R - V'_R \\ V_Y - V'_Y \\ V_B - V'_B \end{bmatrix} = \begin{bmatrix} \Delta V_R \\ \Delta V_Y \\ \Delta V_B \end{bmatrix} = \begin{bmatrix} Z & 0 & 0 \\ 0 & Z & 0 \\ 0 & 0 & Z \end{bmatrix} \begin{bmatrix} I_R \\ I_Y \\ I_B \end{bmatrix} \tag{4.60a}$$

and its inverse is given by

$$\begin{bmatrix} I_R \\ I_Y \\ I_B \end{bmatrix} = \begin{bmatrix} 1/Z & 0 & 0 \\ 0 & 1/Z & 0 \\ 0 & 0 & 1/Z \end{bmatrix} \begin{bmatrix} \Delta V_R \\ \Delta V_Y \\ \Delta V_B \end{bmatrix} \tag{4.60b}$$

or in the sequence component reference frame expressed in terms of phase R,

$$\begin{bmatrix} I_R^P \\ I_R^N \\ I_R^Z \end{bmatrix} = \begin{bmatrix} 1/Z & 0 & 0 \\ 0 & 1/Z & 0 \\ 0 & 0 & 1/Z \end{bmatrix} \begin{bmatrix} \Delta V_R^P \\ \Delta V_R^N \\ \Delta V_R^Z \end{bmatrix} \tag{4.60c}$$

Therefore, the PPS, NPS and ZPS admittances are all self-terms, as the phase admittances, and all are equal to $1/Z$.

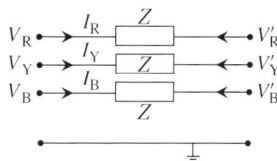

Figure 4.40   Equivalent circuit of a 3-limb core series reactor

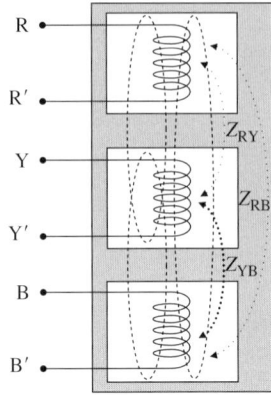

Figure 4.41   Series reactor with a vertically stacked three-phase windings

For some designs, the three-phase windings of the reactor are vertically stacked on top of each other as illustrated in Figure 4.41.

In such arrangement, the flux in one phase can link with the other phases and mutual inductive coupling between the three phases can therefore exist with the middle phase seeing more linked flux than the outer phases. This means that whilst the self-impedances are equal, the mutual impedance between the outer phases and that between the adjacent phases are not equal. Therefore, from Figure 4.41, we define the following:

$$Z_{RR} = Z_{YY} = Z_{BB} = Z_S \quad Z_{RY} = Z_{YR} = Z_{YB} = Z_{BY} = Z_{M1}$$
$$Z_{RB} = Z_{BR} = Z_{M2}$$

Therefore, the series phase impedance matrix is written as

$$\mathbf{Z}_{RYB} = \begin{bmatrix} Z_S & Z_{M1} & Z_{M2} \\ Z_{M1} & Z_S & Z_{M1} \\ Z_{M2} & Z_{M1} & Z_S \end{bmatrix} \tag{4.61a}$$

The series sequence impedance matrix can be easily calculated using $\mathbf{Z}^{PNZ} = \mathbf{H}^{-1}\mathbf{Z}_{RYB}\mathbf{H}$, where $\mathbf{H}$ is the transformation matrix.

Thus,

$$\mathbf{Z}^{PNZ} = \frac{1}{3}\begin{bmatrix} 3Z_S - (Z_{M2} + 2Z_{M1}) & 2h^2(Z_{M2} - Z_{M1}) & -h(Z_{M2} - Z_{M1}) \\ 2h(Z_{M2} - Z_{M1}) & 3Z_S - (Z_{M2} + 2Z_{M1}) & -h^2(Z_{M2} - Z_{M1}) \\ -h^2(Z_{M2} - Z_{M1}) & -h(Z_{M2} - Z_{M1}) & 3Z_S + 2(Z_{M2} + 2Z_{M1}) \end{bmatrix}$$
$$\tag{4.61b}$$

This sequence impedance matrix is full and is not symmetric, i.e. it shows unequal intersequence mutual coupling, which as discussed in Chapter 2, eliminates the

fundamental advantage of using the sequence component reference frame. However, in practice, for many vertical reactor designs, the vertical insulators between
the phases are deliberately chosen to be sufficiently long so that the mutual coupling between the adjacent phases is quite small. In such cases, two practical
options exist. The simplest, which is the one mostly used, is to ignore the small
mutual coupling and set $Z_{M1} = Z_{M2} = 0$ so that the sequence impedance matrix
reduces to

$$\mathbf{Z}^{PNZ} = \mathbf{Z}_{RYB} = \begin{bmatrix} Z_S & 0 & 0 \\ 0 & Z_S & 0 \\ 0 & 0 & Z_S \end{bmatrix} \tag{4.61c}$$

The second option, where the mutual impedances are known, is to set them to a
mean value $Z_{M(mean)} = (Z_{M1} + Z_{M2})/2$. In this case, the series phase impedance
matrix of Equation (4.16a) becomes

$$\mathbf{Z}_{RYB} = \begin{bmatrix} Z_S & Z_{M(mean)} & Z_{M(mean)} \\ Z_{M(mean)} & Z_S & Z_{M(mean)} \\ Z_{M(mean)} & Z_{M(mean)} & Z_S \end{bmatrix} \tag{4.61d}$$

The corresponding series sequence impedance matrix is given by

$$\mathbf{Z}^{PNZ} = \begin{bmatrix} Z_S - Z_{M(mean)} & 0 & 0 \\ 0 & Z_S - Z_{M(mean)} & 0 \\ 0 & 0 & Z_S + 2Z_{M(mean)} \end{bmatrix} \tag{4.61e}$$

The PPS and NPS impedances are equal to $Z_S - Z_{M(mean)}$ and the ZPS impedance
is equal to $Z_S + 2Z_{M(mean)}$. The series sequence admittance matrix is given by

$$\mathbf{Y}^{PNZ} = \begin{bmatrix} \dfrac{1}{Z_S - Z_{M(mean)}} & 0 & 0 \\ 0 & \dfrac{1}{Z_S - Z_{M(mean)}} & 0 \\ 0 & 0 & \dfrac{1}{Z_S + 2Z_{M(mean)}} \end{bmatrix} \tag{4.61f}$$

4.4.3 Modelling of shunt reactors and capacitors

In three-phase shunt reactors, there is practically negligible mutual coupling
between the phases. Three-phase shunt reactors and capacitors depicted in
Figure 4.42 are each represented as three identical shunt susceptances. The resistive
part of the shunt reactor is normally neglected since, by design, these are required to
have very high X/R ratios in the order of 100–200 to minimise active power losses.
The internal series resistance of shunt capacitors is negligibly small. The phase

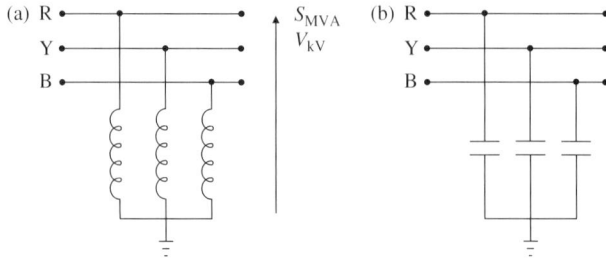

Figure 4.42   Star-connected three-phase shunt reactors and capacitors: (a) three-phase shunt reactor and (b) three-phase shunt capacitor

and sequence admittance matrix of a shunt reactor or shunt capacitor is given by

$$\mathbf{Y}_{RYB} = \mathbf{Y}^{PNZ} = \begin{bmatrix} Y & 0 & 0 \\ 0 & Y & 0 \\ 0 & 0 & Y \end{bmatrix} \qquad (4.62a)$$

where

$$Y = -jB \quad \text{for a shunt reactor}$$

and

$$Y = jB \quad \text{for a shunt capacitor}$$

The PPS, NPS and ZPS susceptances are equal. The magnitude of the susceptance is derived from the three-phase rated MVAr data of the reactor or capacitor as follows:

$$B_{(S)} = \frac{S_{Rating\ (MVAr)}}{V^2_{LL(kV)}} \qquad (4.62b)$$

The pu susceptance is given by

$$B_{(pu)} = \frac{\dfrac{S_{Rating\ (MVAr)}}{V^2_{LL(kV)}}}{\dfrac{S_{Base\ (MVA)}}{V^2_{LL(kV)}}}$$

or

$$B_{(\%)} = \frac{S_{Rating\ (MVAr)}}{S_{Base\ (MVA)}} \times 100 \qquad (4.62c)$$

For example, the susceptance of a 150 MVAr shunt reactor or capacitor in per cent on a 100 MVA base is equal to 150%.

Figure 4.43 shows two typical shunt capacitors used in 132 kV and 400 kV power systems.

(a)

(b)

Figure 4.43   (a) 60 MVAr 132 kV Mechanically Switched Capacitor (MSC) and (b) 400 kV 225 MVAr MSC (with an RLC damping network)

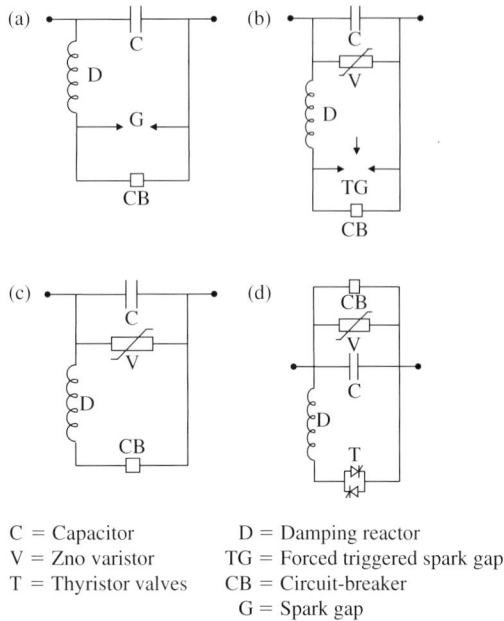

C = Capacitor　　　　　　D = Damping reactor
V = Zno varistor　　　　TG = Forced triggered spark gap
T = Thyristor valves　　CB = Circuit-breaker
　　　　　　　　　　　　G = Spark gap

Figure 4.44　Typical series capacitor schemes: (a) basic spark gap scheme, (b) modern scheme with varistor and triggered spark gap, (c) modern scheme with varistor but no spark gap and (d) modern thyristor controlled scheme with varistor

4.4.4 Modelling of series capacitors

Types of series capacitor schemes

There are several three-phase series capacitor scheme designs in use in power systems, four of which are depicted in Figure 4.44.

General modelling aspects of series capacitors

The modelling of series capacitor schemes in power flow analysis is straight-forward because the magnitude of current flowing through the capacitor would be within its appropriate rated design value. These rated values include not only the highest continuous current, but also the highest short-term overload value per-mitted and the highest permitted value under power oscillation conditions. The capacitors are therefore represented as three identical series reactances and the ohmic value of the reactance is known from the rated data. The resistive part of the series capacitor is negligibly small. Therefore, the sequence and phase admittance matrix of a three-phase series capacitor is given by

$$\mathbf{Y}^{\text{PNZ}} = \mathbf{Y}_{\text{RYB}} = \begin{bmatrix} Y & 0 & 0 \\ 0 & Y & 0 \\ 0 & 0 & Y \end{bmatrix} \qquad (4.63)$$

where

$$Y = j\frac{1}{X_\mathrm{C}}$$

The PPS, NPS and ZPS admittances are equal.

However, for short-circuit analysis, the modelling of series capacitors is not straightforward and requires knowledge of the capacitor protection scheme and capacitor design ratings. It is worth noting that IEC 60909-0:2001 short-circuit analysis standard states that the effect of series capacitors can be neglected in the calculation of short-circuit currents if they are equipped with voltage-limiting devices in parallel acting if a short-circuit occurs. However, as we will see later, this approach is generally inappropriate for varistor protected series capacitors and could result in significant underestimates in the magnitude of short-circuit currents.

Modelling of series capacitors for short-circuit analysis

Old series capacitor schemes were protected by spark gaps against a short-circuit fault that can cause a large capacitor current to flow and a substantial increase in the voltage across the capacitor. The flashover across the spark gap bypasses the series capacitor and the bypass circuit-breaker is immediately closed to extinguish the gap.

In modern series capacitor installations, non-linear resistors are used and these are called metal (usually zinc) oxide varistors (MOVs) known as ZnO. MOVs are used to provide the overvoltage protection mainly against external short-circuit faults, i.e. external to the series capacitor and its line zone. For internal faults, within the series capacitor protected line zone, a forced triggered spark gap can operate typically within 1 ms to limit the energy absorption duty on the varistors. This is then followed by closure of the circuit-breaker to extinguish the gap and hence the capacitor is completely bypassed and removed from the circuit.

The modelling and analysis of varistor protected series capacitors under external or remote short-circuit fault conditions is quite complex. However, we will briefly and qualitatively describe the operation of the modern varistor protected capacitor schemes under external through-fault conditions that result in varistor operation only. As the instantaneous short-circuit current flowing through the capacitor increases, so will the voltage across the capacitor or varistor until this voltage reaches a preset threshold where the varistor starts conducting current in order to keep the voltage across the capacitor constant. This applies under both positive and negative polarity current conditions so that for the first part of each half cycle, the current flows through the capacitor only but for the other part of the half cycle, the current switches to the varistor. The behaviour of the varistor and series capacitor combination is repeated every half cycle during the fault period whenever the voltage across the capacitor/varistor exceeds the preset capacitor protection level. Using Figure 4.45(a), the behaviour of the series capacitor/varistor parallel combination under a through fault condition is analysed using an electro-magnetic transient analysis program. The voltage across the capacitor/varistor and

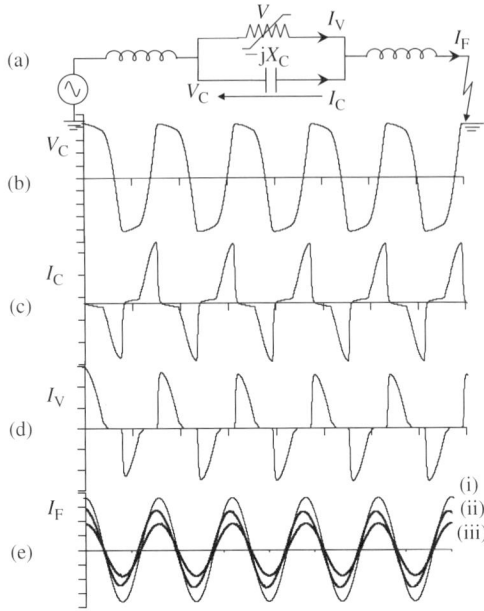

Figure 4.45   Behaviour of a series capacitor/varistor combination under external through-fault conditions: (a) generic circuit; (b) capacitor voltage; (c) capacitor current; (d) varistor current; and (e) fault current

the currents through the capacitor and varistor are shown in Figure 4.45(b)–(d), respectively. The fault current is shown in Figure 4.45(e) curve (ii).

Curve (i) of Figure 4.43(e) shows the fault current that would flow through the capacitor if there were no varistor to protect it. Clearly, the conduction of the MOV causes an effective reduction in the fault current magnitude. The extent of reduction is dependent on the electrical proximity of the fault location to the capacitor/varistor location. In addition, curve (iii) of Figure 4.45(e) shows the fault current if the capacitor is assumed to be permanently bypassed. This generic study illustrates that in general, both ignoring the presence of the varistor or assuming the capacitor is permanently bypassed during the fault period can result in a large error in the fault current magnitude. The former can result in an overestimate whereas the latter can result in an underestimate.

The modelling and analysis of short-circuit faults in large-scale power systems is based on the assumption that all power system plant are linear or can be approximated as piece-wise linear. The varistor is, however, a highly non-linear resistor that cannot be directly included in standard quasi-steady state short-circuit modelling and analysis simulations. Figure 4.46 shows a typical voltage/current characteristic of a varistor which presents an almost infinite resistance below the threshold conduction voltage then a very low resistance above it.

The impedance presented by the combined capacitor reactance and varistor resistance over a half cycle is dependent on the total current flowing through

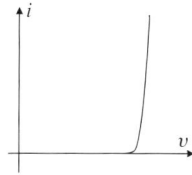

Figure 4.46 Typical varistor non-linear current–voltage characteristic

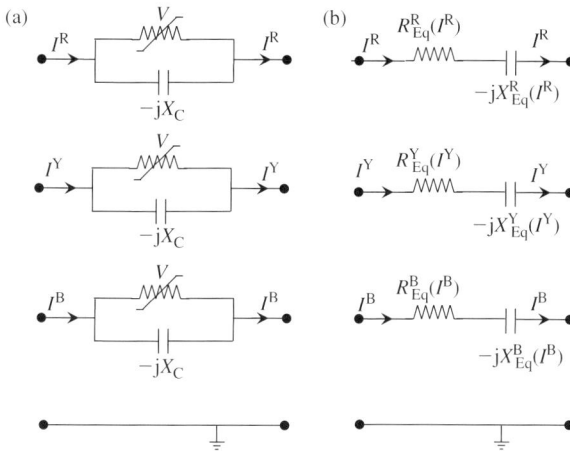

Figure 4.47 Varistor-protected three-phase series capacitor and equivalent circuit: (a) three-phase representation and (b) equivalent circuit model

their parallel combination. The capacitor current or voltage protection threshold is known because the capacitor reactance is a known design parameter. Therefore, below this threshold, the equivalent impedance of the parallel capacitor/varistor combination is the capacitor reactance in parallel with a very high resistance or alternatively, a very low resistance (practically zero) in series with the capacitor reactance. When a current higher than the threshold flows through the capacitor, the varistor will conduct current for most of the half cycle so that the combination will appear as a low resistance in parallel with the capacitor reactance or, alternatively, a higher resistance than zero in series with a much reduced capacitor reactance. The higher the current is, the more resistive the equivalent impedance becomes reflecting the increased short-circuiting of the capacitor by the varistor. Therefore, the capacitor/varistor parallel combination can be approximated as a resistance and a capacitance in series with both being functions of the total current flowing through them. Figure 4.47(a) shows a three-phase representation of a series capacitor protected by a varistor. Figure 4.47(b) shows the corresponding equivalent circuit where the equivalent series impedance is a function of the total current flowing through the parallel capacitor/varistor combination.

Denoting the capacitor current protective threshold as I_{thr}, the equivalent series impedance for each phase is given by

$$Z_{Eq}(I) = \begin{cases} -jX_C & \text{if } I < I_{thr} \\ R_{Eq}(I) - jX_{Eq}(I) & \text{if } I \geq I_{thr} \end{cases} \qquad (4.64)$$

where X_C is the capacitor reactance when all current flows through it, i.e. its nominal design value, I is the total current flowing through the capacitor/varistor parallel combination and I_{thr} is a known design threshold parameter. The mathematical functions that describe $R_{Eq}(I)$ and $X_{Eq}(I)$ as functions of I when $I \geq I_{thr}$ can be calculated from the known design data or parameters of the series capacitor and varistor. Figure 4.48 illustrates typical equivalent impedance characteristics.

The series phase admittance matrix of the three-phase series capacitor/varistor device is given by

$$\mathbf{Z}^{RYB} = \begin{bmatrix} Z^R_{Eq}(I) & 0 & 0 \\ 0 & Z^Y_{Eq}(I) & 0 \\ 0 & 0 & Z^B_{Eq}(I) \end{bmatrix} \qquad (4.65)$$

Prior to the short-circuit fault or when the current is less than I_{thr}, the equivalent phase admittances are balanced and equal. This will also be the case under balanced three-phase short-circuit faults. However, under unbalanced short-circuit faults, they are generally unequal depending on the fault type. Therefore, the sequence admittance matrix of Equation (4.65) is given by

$$\mathbf{Z}^{PNZ} = \begin{matrix} \\ P \\ N \\ Z \end{matrix} \begin{matrix} P & N & Z \\ \begin{bmatrix} Z_S & Z_{M1} & Z_{M2} \\ Z_{M2} & Z_S & Z_{M1} \\ Z_{M1} & Z_{M2} & Z_S \end{bmatrix} \end{matrix} \qquad (4.66a)$$

where

$$Z_S = \frac{1}{3}[Z^R_{Eq}(I) + Z^Y_{Eq}(I) + Z^B_{Eq}(I)]$$

$$Z_{M1} = \frac{1}{3}[Z^R_{Eq}(I) + h^2 Z^Y_{Eq}(I) + h Z^B_{Eq}(I)]$$

$$Z_{M2} = \frac{1}{3}[Z^R_{Eq}(I) + h Z^Y_{Eq}(I) + h^2 Z^B_{Eq}(I)] \qquad (4.66b)$$

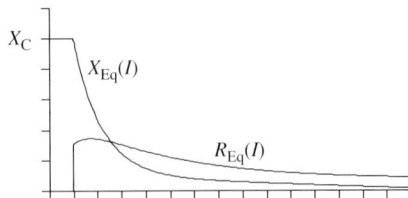

Figure 4.48 Typical equivalent impedance of a series capacitor/varistor combination under external through-fault conditions

Equation (4.66a) shows the presence of unequal intersequence mutual coupling which should be taken into account if significant errors are to be avoided.

Because fixed impedance short-circuit analysis techniques are linear, Equation (4.64) may be implemented in an iterative linear calculation process. The fault currents flowing in the network and the series capacitors are initially calculated without varistor action, the capacitor current is then used to calculate a new value for the series capacitor equivalent impedance using Equation (4.64). The new impedance is then used to calculate new fault currents in the network and the series capacitor. The final solution would be arrived at when there is a little change in the last two equivalent impedances or capacitor currents calculated.

In power system networks with modern varistor protected series capacitor installations, the operation of the varistor when the capacitor current exceeds I_{thr} cannot be ignored if very large and unacceptable errors (overestimates) in the calculated short-circuit currents in the network are to be avoided. Also, if these series capacitors are not included in the network model at all, i.e. assumed to be completely bypassed under short-circuit faults, as per IEC 60909, then this could result in significant underestimates in the magnitude of short-circuit currents.

4.5 Sequence modelling of static variable compensators

4.5.1 Background

Static variable compensators, with one typical arrangement shown in Figure 4.49(a), are shunt connected devices used in transmission, subtransmission, distribution and industrial power systems to provide fast acting dynamic reactive

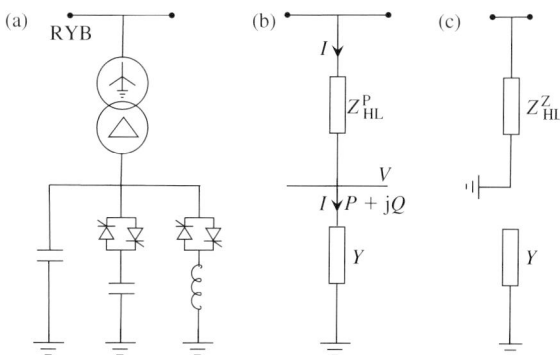

Figure 4.49 Typical static variable compensator (one-line diagram): (a) one-line diagram, (b) PPS/NPS equivalent circuit and (c) ZPS equivalent circuit

power and voltage control. The reactive power is usually provided by various combinations of fixed capacitors, thyristor switched capacitors and thyristor controlled reactors. These are usually connected at low voltage levels of typically 5–15 kV. Therefore, normally, a star–delta transformer is used to connect these reactive elements to the high voltage power network. Figure 4.50 shows a 400 kV Static Variable Compensator (SVC) rates at −75 to 150 MVAr.

4.5.2 PPS, NPS and ZPS modelling

The PPS and NPS equivalent circuits of a static variable compensator are identical and consist of the compensator transformer in series with the inductive and/or capacitive susceptance of the reactive elements. The admittance is calculated from Figure 4.49(b) using $S = P + jQ = VI^*$, $I = YV$ and $P = 0$. Thus, the PPS and NPS admittance is given by

$$Y^P = Y^N = \frac{-jQ}{|V|^2} \tag{4.67}$$

where Q is the reactive power output and V is the voltage at the LV side of the compensator transformer. Both Q and V are known from an initial load flow study.

In the ZPS network, the compensator is represented by its transformer leakage impedance only as shown in Figure 4.49(c).

Figure 4.50 400 kV −75/150 MVAr Static Variable Compensator (SVC)

4.6 Sequence modelling of static power system load

4.6.1 Background

The term power system load does not have an agreed and clear meaning in academia or industry and in practice, it can be interpreted to mean many different things by practising engineers. Much research and development has been carried out to define the term load as well as its parameters and characteristics for use in short-term power flow and transient stability studies. However, to my knowledge, the load has received little attention in large-scale network short-circuit analysis. A primary reason for this is that it has been a general practice to ignore the presence of load and calculate the short-circuit current changes in the network due to the short-circuit faults only. This is the case of IEC 60909 and American IEEE C 37.010 standards as will be discussed in Chapter 7. However, as we will see later, the pressures in recent years for increased accuracy and precision in short-circuit calculations resulted in the introduction of new techniques that require the inclusion of the load in the analysis.

It is straightforward to include a model of load equipment in the analysis where this equipment is clearly identifiable because its electrical characteristics would generally be known. For example, a heating load, lighting load or induction motor load, the latter will be covered in detail in Chapter 5, can easily be modelled if one knew how much there is of each component. Nearly all the load components in an industrial power system or a power station auxiliary system would be known. However, in distribution, subtransmission and transmission system short-circuit analysis, the load in MW and MVAr at any given time is that supplied at a major grid supply substation, e.g. 132 kV substation, or a major bulk supply substation, e.g. 33 kV. This load therefore consists of thousands or tens of thousands of components of different types and characteristics and such composition is generally unknown to the network company. The load will contain static components, i.e. those that do not provide a short-circuit contribution to a fault in the host network such as a fixed impedance load. Importantly, from a short-circuit analysis viewpoint, the load will contain a numerous number of single-phase induction motors, three-phase induction motors and embedded large or small scale three-phase synchronous generators all of which will feed a short-circuit current to a fault in the host network. In addition, these load components will be separated from the grid and bulk supply substations by distribution networks. Therefore, the general loads seen at these substations for modelling in large-scale analysis will consist of distribution networks as well as passive and active, i.e. machine, components. Better precision short-circuit analysis should model the contribution of rotating plant, i.e. motors and generators, forming part of the general load at each substation in the network. The sequence modelling of rotating machines is covered in detail in Chapter 5 but we will now restrict our attention to static or passive load as seen from major load supply substations.

4.6.2 PPS, NPS and ZPS modelling

The PPS and NPS model of a passive load supplied radially at a substation that includes the load components and the intervening distribution network can be generally represented as a shunt admittance to earth as follows:

$$Y^P = Y^N = \frac{P - jQ}{|V|^2} \tag{4.68}$$

where P and Q is the load supplied in pu MW and pu MVAr and V is the voltage at the substation just before the occurrence of the short-circuit fault, in pu.

The ZPS model of a passive load can be substantially different from the PPS/NPS model. Generally, this model is determined by the presence of low ZPS impedance paths for the flow of ZPS currents such as those provided by star–delta transformers. The ZPS model is therefore the minimum ZPS impedance seen from the supplied substation looking down into the distribution network topology and equipment characteristics including transformer winding connections.

4.7 Three-phase modelling of static power plant and load in the phase frame of reference

4.7.1 Background

In Chapter 6, we describe a short-circuit analysis technique in large-scale power systems in the phase frame of reference. This type of analysis requires the use of three-phase models for power plant and load rather than PPS, NPS and ZPS models. Three-phase models are presented in this section.

4.7.2 Three-phase modelling of reactors and capacitors

The three-phase series impedance and admittance matrix models for series reactors are given in Equations (4.60a) and (4.60b), respectively. Where interphase mutual coupling exists, the series phase impedance matrices given in Equations (4.61a) and (4.61d) apply for unequal and equal interphase couplings, respectively. The corresponding phase admittance matrices are given by

$$\mathbf{Y}^{RYB} = \frac{1}{Z_S(Z_S + Z_{M2}) - 2Z_{M1}^2} \begin{bmatrix} \dfrac{Z_S^2 - Z_{M1}^2}{Z_S - Z_{M2}} & -Z_{M1} & \dfrac{-(Z_S Z_{M2} - Z_{M1}^2)}{Z_S - Z_{M2}} \\ -Z_{M1} & Z_S + Z_{M2} & -Z_{M1} \\ \dfrac{-(Z_S Z_{M2} - Z_{M1}^2)}{Z_S - Z_{M2}} & -Z_{M1} & \dfrac{Z_S^2 - Z_{M1}^2}{Z_S - Z_{M2}} \end{bmatrix}$$

$$\tag{4.69a}$$

in the case of unequal interphase mutual coupling, and

$$\mathbf{Y}^{RYB} = \frac{1}{Z_S(Z_S + Z_M) - 2Z_M^2} \begin{bmatrix} Z_S + Z_M & -Z_M & -Z_M \\ -Z_M & Z_S + Z_M & -Z_M \\ -Z_M & -Z_M & Z_S + Z_M \end{bmatrix} \quad (4.69b)$$

in the case of equal interphase mutual coupling, i.e. $Z_{M1} = Z_{M2} = Z_M$.

The three-phase shunt admittance matrix of shunt reactors and capacitors was given in Equation (4.62a). For modern varistor protected series capacitors, the current-dependent three-phase series impedance matrix was given in Equation (4.65). The corresponding series phase admittance matrix is given by

$$\mathbf{Y}^{RYB} = \begin{bmatrix} 1/Z_{Eq}^R(I) & 0 & 0 \\ 0 & 1/Z_{Eq}^Y(I) & 0 \\ 0 & 0 & 1/Z_{Eq}^B(I) \end{bmatrix} \quad (4.70)$$

4.7.3 Three-phase modelling of transformers

General

The sequence modelling of single-phase and three-phase transformers of various winding connections was extensively covered in Section 4.2. In three-phase steady state analysis in the phase frame of reference, three-phase models are required. For the purpose of steady state e.g. short-circuit analysis, it is sufficient to model the transformer as a set of mutually coupled windings with a linear magnetising reactance. We will not provide a similar extensive coverage to that in Section 4.2 but will instead present the basic modelling technique of a single-phase two-winding transformer, three-phase transformer banks and a three-limb core three-phase transformer. For the three-phase transformer, only a star–delta winding connection is considered. The principles provide the reader with the information and methodology required to develop a three-phase model for a transformer of any winding connection.

Single-phase two-winding transformers

Figure 4.51(a) illustrates a single-phase two-winding transformer represented as a set of two mutually coupled windings where: $Z_{11}(Y_{11})$ is winding 1 self-impedance (admittance) and $Y_{11} = 1/Z_{11}$; $Z_{22}(Y_{22})$ is winding 2 self-impedance (admittance) and $Y_{22} = 1/Z_{22}$; $Z_{12} = Z_{21}(Y_{12} = Y_{21})$ is mutual impedance (admittance) between windings 1 and 2 and $Y_{12} = 1/Z_{12}$; $Z_{HL}(Y_{HL})$ is leakage impedance (admittance) of windings 1 and 2 and $Y_{HL} = 1/Z_{HL}$.

From Figure 4.51(a), we can write

$$\begin{bmatrix} V_1 \\ V_2 \end{bmatrix} = \begin{bmatrix} Z_{11} & Z_{12} \\ Z_{12} & Z_{22} \end{bmatrix} \begin{bmatrix} I_1 \\ I_2 \end{bmatrix} \quad (4.71)$$

From Equation (4.71), the self-impedance Z_{11} of winding 1 and the mutual impedance Z_{12} between the two windings are obtained from the results of an

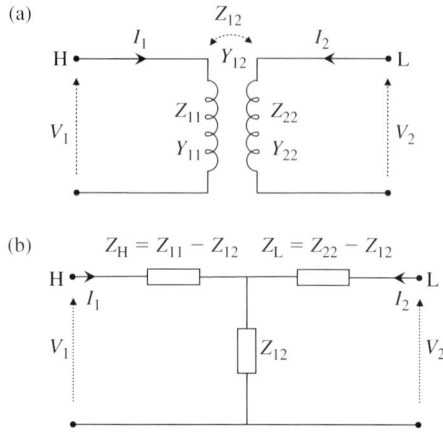

Figure 4.51   Single-phase two-winding transformer: (a) two mutually coupled windings and (b) equivalent circuit

open-circuit test with winding 1 supplied and winding 2 open-circuited and are given by

$$Z_{11} = \frac{V_1}{I_1}\bigg|_{I_2=0} \quad \text{and} \quad Z_{12} = \frac{V_2}{I_1}\bigg|_{I_2=0} \tag{4.72a}$$

From Equation (4.72a), the mutual impedance can be expressed as

$$Z_{12} = Z_{11}\frac{V_2}{V_1} \tag{4.72b}$$

where V_1/V_2 is the transformer no-load voltage ratio.

Similarly, the self-impedance Z_{22} of winding 2 and the mutual impedance Z_{12} between the two windings can be obtained from the results of an open-circuit test with winding 2 supplied and winding 1 open-circuited. Thus

$$Z_{22} = \frac{V_2}{I_2}\bigg|_{I_1=0} \quad \text{and} \quad Z_{12} = \frac{V_1}{I_2}\bigg|_{I_1=0} \tag{4.73a}$$

From Equation (4.73a), the mutual impedance can be expressed as

$$Z_{12} = Z_{22}\frac{V_1}{V_2} \tag{4.73b}$$

From Equations (4.72b) and (4.73b), we have

$$Z_{11} = Z_{22}\frac{V_1^2}{V_2^2} \tag{4.74}$$

The leakage impedance Z_{HL} between the two windings is obtained from the results of a short-circuit test with winding 1 supplied and winding 2 short-circuited. Thus

$$Z_{HL} = \frac{V_1}{I_1}\bigg|_{V_2=0} \tag{4.75a}$$

Using Equation (4.75a) in Equation (4.71), the leakage impedance Z_{HL} can be expressed in terms of the windings self and mutual impedances as follows:

$$Z_{HL} = Z_{11} - \frac{Z_{12}^2}{Z_{22}} \qquad (4.75b)$$

Having obtained all four elements of the impedance matrix of Equation (4.71), it is more convenient to use the corresponding admittance matrix given by

$$\begin{bmatrix} I_1 \\ I_2 \end{bmatrix} = \begin{bmatrix} Y_{11} & Y_{12} \\ Y_{12} & Y_{22} \end{bmatrix} \begin{bmatrix} V_1 \\ V_2 \end{bmatrix} \qquad (4.76a)$$

where

$$Y_{11} = \frac{Z_{22}}{Z_{11}Z_{22} - Z_{12}^2} \qquad Y_{12} = \frac{-Z_{12}}{Z_{11}Z_{22} - Z_{12}^2} \qquad Y_{22} = \frac{Z_{11}}{Z_{11}Z_{22} - Z_{12}^2} \qquad (4.76b)$$

Using Equations (4.73b), (4.74) and (4.75b) in Equations (4.76b), we obtain

$$Y_{11} = Y_{HL} \qquad Y_{12} = -\frac{V_1}{V_2}Y_{HL} \qquad Y_{22} = \frac{V_1^2}{V_2^2}Y_{HL} \qquad Y_{HL} = \frac{1}{Z_{HL}} \qquad (4.77a)$$

and

$$\begin{bmatrix} I_1 \\ I_2 \end{bmatrix} = \begin{bmatrix} Y_{HL} & -\dfrac{V_1}{V_2}Y_{HL} \\ -\dfrac{V_1}{V_2}Y_{HL} & \dfrac{V_1^2}{V_2^2}Y_{HL} \end{bmatrix} \begin{bmatrix} V_1 \\ V_2 \end{bmatrix} \qquad (4.77b)$$

In the derivation of the above impedance and admittance matrices, we note that, in practice, the only data most likely available are the measured short-circuit leakage impedance and no-load current. The derivation is illustrated using an example.

Example 4.8   Consider a 239 MVA 231 kV/21.5 kV single-phase two-winding transformer having a leakage reactance of 15% and a no-load current of 0.5% whether measured from the high or low voltage sides. Thus, $Z_{11} = Z_{22} = 1/0.005 = 200$ pu. Equation (4.71) can be represented by the equivalent circuit of Figure 4.51(b) where the self-impedance of a winding is defined as the sum of its leakage impedance and the mutual impedance, i.e. in per-unit, $Z_{11} = Z_{22} = Z_H + Z_{12} = Z_L + Z_{12}$ giving $Z_H = Z_L$. Since $Z_H = Z_L \ll Z_{12}$, the short-circuit leakage impedance is given by $Z_{HL} = Z_H + Z_L$ giving $Z_H = Z_L = 0.075$ pu. Ignoring the magnetising losses and the resistive part of the leakage impedance, the mutual reactance is equal to $jX_{12} = j200 - j0.075 = j199.925$ pu. Therefore, the impedance and admittance matrices, in pu, are given by

$$Z_{pu} = j\begin{bmatrix} 200 & 199.925 \\ 199.925 & 200 \end{bmatrix} \quad \text{and} \quad Y_{pu} = j\begin{bmatrix} -6.6679169 & 6.6654164 \\ 6.6654164 & -6.6679169 \end{bmatrix}$$

The difference between the self and mutual reactances is the very small value of $1/2$ the leakage reactance. This results in very small differences between the self and mutual susceptances and necessitates calculations to at least six or seven decimal places if the important leakage reactance is not to be lost. The corresponding reactance and susceptance matrices in Ω are given by

$$\mathbf{Z}_\Omega = j \begin{bmatrix} 200 \times \dfrac{231^2}{239} & 199.925 \times \dfrac{231 \times 21.5}{239} \\ 199.925 \times \dfrac{231 \times 21.5}{239} & 200 \times \dfrac{21.5^2}{239} \end{bmatrix}$$

$$= j \begin{bmatrix} 44653.5565 & 4154.5084 \\ 4154.5084 & 386.82 \end{bmatrix} \Omega$$

$$\mathbf{Y}_S = j \begin{bmatrix} -6.6679169 \times \dfrac{239}{231^2} & 6.6654164 \times \dfrac{239}{231 \times 21.5} \\ 6.6654164 \times \dfrac{239}{231 \times 21.5} & -6.6679169 \times \dfrac{239}{21.5^2} \end{bmatrix}$$

$$= j \begin{bmatrix} -0.029865 & 0.320756 \\ 0.320756 & -3.44755 \end{bmatrix} S$$

The admittance matrix could also be calculated using Equation (4.77b). The leakage impedance referred to the HV side is equal to $Z_{HL} = j0.15 \times \frac{231^2}{239} = j33.490\,\Omega$ giving $Y_{HL} = -j0.02986\,S$. The elements of the admittance matrix of Equation (4.77b) are calculated using Equation (4.77a).

The above approach can easily be extended to single-phase three-winding transformers where the dimension of the admittance matrix of Equation (4.76a) becomes 3×3. Obviously, we now have three self admittances, three mutual admittances and three short-circuit leakage admittances. This is a simple exercise in algebra left for the reader to derive the admittance equations following a similar method to the two-winding case.

Three-phase banks two-winding transformers (no interphase mutual couplings)

Consider a large two-winding star–delta transformer constructed as three single-phase banks as shown in Figure 4.52.

In such an arrangement, there is negligible or no interphase coupling between the three phases. Using the admittance matrix given in Equation (4.76a) for each phase and Figure 4.52, the admittance matrix equation that relates the winding

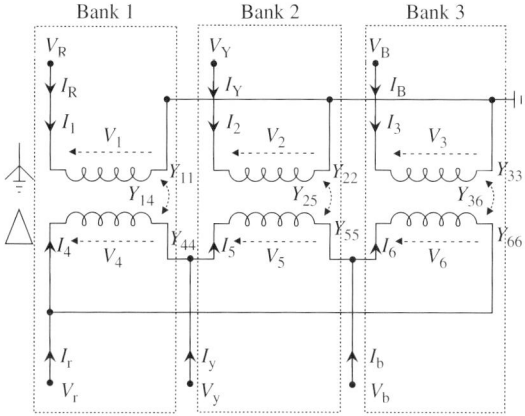

Figure 4.52   Three single-phase banks two-winding star–delta transformer

currents and voltages for the three-phases is given by

$$
\begin{bmatrix} I_1 \\ I_2 \\ I_3 \\ I_4 \\ I_5 \\ I_6 \end{bmatrix} = \begin{bmatrix} Y_{11} & 0 & 0 & Y_{14} & 0 & 0 \\ 0 & Y_{22} & 0 & 0 & Y_{25} & 0 \\ 0 & 0 & Y_{33} & 0 & 0 & Y_{36} \\ Y_{14} & 0 & 0 & Y_{44} & 0 & 0 \\ 0 & Y_{25} & 0 & 0 & Y_{55} & 0 \\ 0 & 0 & Y_{36} & 0 & 0 & Y_{66} \end{bmatrix} \begin{bmatrix} V_1 \\ V_2 \\ V_3 \\ V_4 \\ V_5 \\ V_6 \end{bmatrix}
\tag{4.78a}
$$

or

$$
\mathbf{I}_{\text{Winding}} = \mathbf{Y}_{\text{Winding}} \mathbf{V}_{\text{Winding}}
\tag{4.78b}
$$

Equation (4.78b) applies to a two set of three-phase windings irrespective of the type of winding connection. Thus, for our star–delta transformer, the three-phase nodal admittance matrix that relates the nodal currents and voltages can be derived from the relationship between the winding currents and voltages, and the nodal currents and voltages. With all nodal voltages being with respect to the reference earth, the relationship between the winding and nodal voltages is found by inspection from Figure 4.52 as follows:

$$
\begin{bmatrix} V_1 \\ V_2 \\ V_3 \\ V_4 \\ V_5 \\ V_6 \end{bmatrix} = \begin{bmatrix} 1 & 0 & 0 & 0 & 0 & 0 \\ 0 & 1 & 0 & 0 & 0 & 0 \\ 0 & 0 & 1 & 0 & 0 & 0 \\ 0 & 0 & 0 & 1 & -1 & 0 \\ 0 & 0 & 0 & 0 & 1 & -1 \\ 0 & 0 & 0 & -1 & 0 & 1 \end{bmatrix} \begin{bmatrix} V_R \\ V_Y \\ V_B \\ V_r \\ V_y \\ V_b \end{bmatrix} \quad \text{or} \quad \mathbf{V}_{\text{Winding}} = \mathbf{C} \mathbf{V}_{\text{Node}}
$$

$$
\tag{4.79a}
$$

where $\mathbf{C}$ is defined as the connection matrix. Similarly, the relationship between the nodal and winding currents is found by inspection from Figure 4.52 and is given by

$$
\begin{bmatrix} I_R \\ I_Y \\ I_B \\ \hline I_r \\ I_y \\ I_b \end{bmatrix} =
\begin{bmatrix} 1 & 0 & 0 & 0 & 0 & 0 \\ 0 & 1 & 0 & 0 & 0 & 0 \\ 0 & 0 & 1 & 0 & 0 & 0 \\ \hline 0 & 0 & 0 & 1 & 0 & -1 \\ 0 & 0 & 0 & -1 & 1 & 0 \\ 0 & 0 & 0 & 0 & -1 & 1 \end{bmatrix}
\begin{bmatrix} I_1 \\ I_2 \\ I_3 \\ I_4 \\ I_5 \\ I_6 \end{bmatrix}
\quad \text{or} \quad \mathbf{I}_{\text{Node}} = \mathbf{C}^t \mathbf{I}_{\text{Winding}} \quad (4.79b)
$$

Using Equations (4.79a) and (4.79b) in Equation (4.78b), we obtain

$$
\mathbf{I}_{\text{Node}} = \mathbf{Y}_{\text{Node}} \mathbf{V}_{\text{Node}} \tag{4.80a}
$$

where

$$
\mathbf{Y}_{\text{Node}} = \mathbf{C}^t \mathbf{Y}_{\text{Winding}} \mathbf{C} \tag{4.80b}
$$

Denoting the star-connected winding as H and the delta-connected winding as L, Equation (4.80a) can be written in partitioned form as follows:

$$
\begin{bmatrix} \mathbf{I}_{H(\text{node})} \\ \mathbf{I}_{L(\text{node})} \end{bmatrix} =
\begin{bmatrix} \mathbf{Y}_{HH} & \mathbf{Y}_{HL} \\ \mathbf{Y}_{HL}^t & \mathbf{Y}_{LL} \end{bmatrix}
\begin{bmatrix} \mathbf{V}_{H(\text{node})} \\ \mathbf{V}_{L(\text{node})} \end{bmatrix} \tag{4.81a}
$$

Using Equation (4.77a), it can be shown that the self and mutual admittance matrices of Equation (4.81a) are given by

$$
\mathbf{Y}_{HH} = \begin{bmatrix} Y_{HL14} & 0 & 0 \\ 0 & Y_{HL25} & 0 \\ 0 & 0 & Y_{HL36} \end{bmatrix}
\quad
\mathbf{Y}_{HL} = \begin{bmatrix} -\dfrac{V_1}{V_4}Y_{HL14} & \dfrac{V_1}{V_4}Y_{HL14} & 0 \\ 0 & -\dfrac{V_2}{V_5}Y_{HL25} & \dfrac{V_2}{V_5}Y_{HL25} \\ \dfrac{V_3}{V_6}Y_{HL36} & 0 & -\dfrac{V_3}{V_6}Y_{HL36} \end{bmatrix}
$$

$$
\mathbf{Y}_{LL} = \begin{bmatrix}
\dfrac{V_1^2}{V_4^2}Y_{HL14} + \dfrac{V_3^2}{V_6^2}Y_{HL36} & -\dfrac{V_1^2}{V_4^2}Y_{HL14} & -\dfrac{V_3^2}{V_6^2}Y_{HL36} \\
-\dfrac{V_1^2}{V_4^2}Y_{HL14} & \dfrac{V_1^2}{V_4^2}Y_{HL14} + \dfrac{V_2^2}{V_5^2}Y_{HL25} & \dfrac{V_2^2}{V_5^2}Y_{HL25} \\
-\dfrac{V_3^2}{V_6^2}Y_{HL36} & -\dfrac{V_2^2}{V_5^2}Y_{HL25} & \dfrac{V_2^2}{V_5^2}Y_{HL25} + \dfrac{V_3^2}{V_6^2}Y_{HL36}
\end{bmatrix}
\tag{4.81b}
$$

where $\mathbf{Y}_{HL14} = 1/Z_{HL14}$ and so on.

In the per-unit system with both the star and delta voltages equal to one per unit, the effective turns ratio on the delta side is $\frac{V_4}{V_1} = \frac{V_5}{V_2} = \frac{V_6}{V_3} = \sqrt{3}$. If the three phases of the transformer are considered identical, i.e. having identical leakage admittances that is $Y_{HL14} = Y_{HL25} = Y_{HL36} = Y_{HL} = \frac{1}{Z_{HL}}$, then using Equation (4.81), Equation (4.80b) becomes

$$\mathbf{Y}_{Node} = \begin{bmatrix} Y_{HL} & 0 & 0 & -\dfrac{Y_{HL}}{\sqrt{3}} & \dfrac{Y_{HL}}{\sqrt{3}} & 0 \\[2mm] 0 & Y_{HL} & 0 & 0 & -\dfrac{Y_{HL}}{\sqrt{3}} & \dfrac{Y_{HL}}{\sqrt{3}} \\[2mm] 0 & 0 & Y_{HL} & \dfrac{Y_{HL}}{\sqrt{3}} & 0 & -\dfrac{Y_{HL}}{\sqrt{3}} \\[2mm] -\dfrac{Y_{HL}}{\sqrt{3}} & 0 & \dfrac{Y_{HL}}{\sqrt{3}} & \dfrac{2Y_{HL}}{3} & -\dfrac{Y_{HL}}{3} & -\dfrac{Y_{HL}}{3} \\[2mm] \dfrac{Y_{HL}}{\sqrt{3}} & -\dfrac{Y_{HL}}{\sqrt{3}} & 0 & -\dfrac{Y_{HL}}{3} & \dfrac{2Y_{HL}}{3} & -\dfrac{Y_{HL}}{3} \\[2mm] 0 & \dfrac{Y_{HL}}{\sqrt{3}} & -\dfrac{Y_{HL}}{\sqrt{3}} & -\dfrac{Y_{HL}}{3} & -\dfrac{Y_{HL}}{3} & \dfrac{2Y_{HL}}{3} \end{bmatrix} \tag{4.82}$$

To allow for the effect of off-nominal tap ratio, e.g. on the star side, then according to Equation (4.13a), the terms of the self admittance $\mathbf{Y}_{HH}$ should be divided by t_{HL}^2 whereas the terms of the mutual admittances $\mathbf{Y}_{HL}$ and $\mathbf{Y}_{HL}^t$ should be divided by t_{HL}.

The reader is encouraged to show that the nodal admittance matrix of a two-winding star–star transformer with both neutrals solidly earthed and no interphase mutual couplings has the form as the winding admittance given in Equation (4.78a). Using these principles, the reader can easily obtain the nodal admittance matrix for any winding connections of a three-phase banks of transformers.

Three-phase common-core two-winding transformers (with interphase mutual couplings)

For three-phase 3-limb, 5-limb or shell-type core transformers, interphase mutual couplings exist and these are generally asymmetrical, as can be expected from Figures 4.28–4.30, between the outer-limb phases and the phases connected to adjacent limbs. However, in practice, little or no information is available on these asymmetrical coupling affects. Standard transformer test certificates include the normally measured parameters, i.e. the PPS and ZPS impedances and the magnetising current from which the magnetising impedance is calculated. Since the PPS and ZPS impedances are generally different, this implies that equal interphase mutual couplings exist in the three-phase admittance matrix. The asymmetrical interphase coupling effects may also be indirectly averaged out by their short-circuit measurement procedures as described in Section 4.2.10. Figure 4.53 illustrates a common-core star–delta transformer where each phase winding is coupled to all other phase windings and where windings 1 and 4, 2 and 5, and 3 and 6, are wound on the same limb. With the assumption of full flux symmetry, we

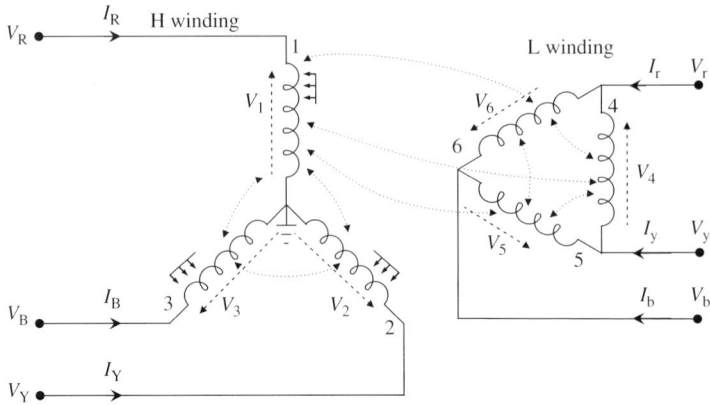

Figure 4.53 Common-core three-phase star–delta transformer with symmetrical interphase coupling

obtain a 6×6 winding admittance matrix that consists of four 3×3 symmetrical submatrices. The corresponding nodal admittance matrix can be obtained using Equation (4.80b).

However, since in nearly all practical situations only sequence and magnetising impedances are available, we will present a different technique for the formulation of the transformer three-phase model. The technique is based on using the known PPS and ZPS leakage impedances, ignoring the magnetising impedances for now. To illustrate the technique, we consider the star–delta transformer of Figure 4.53 with a tap-changer on the star side. In Figure 4.53, phase R of the star winding is assumed symmetrically coupled with phases Y and B of the star winding on different limbs, with phase 'ry' on the delta winding on the same limb, and with phases 'rb' and 'by' on the delta winding on different limbs. Similar symmetrical couplings are assumed for other phases. The PPS and NPS equivalent circuits are shown in Figure 4.8(a) and (b). The ZPS equivalent circuit is shown in Figure 4.12(g) with $Z_E = 0$. Equation (4.13a) corresponds to Figure 4.8(c) where the transformer winding phase shift of 30° for Yd1 and −30° for Yd11 connections was neglected. In the derivation to follow, we denote the star side as H and the delta side as L. Therefore, denoting the phase shift as ϕ and letting all impedances be expressed in pu, it can be shown that the PPS and NPS nodal admittance matrices of Figures 4.8(a) and (b) are given by

$$
\begin{bmatrix} I_H^P \\ I_L^P \end{bmatrix} = \begin{bmatrix} \dfrac{1}{t_{HL}^2 Z_{LH}^P} & -\dfrac{e^{j\phi}}{t_{HL} Z_{LH}^P} \\ -\dfrac{e^{-j\phi}}{t_{HL} Z_{LH}^P} & \dfrac{1}{Z_{LH}^P} \end{bmatrix} \begin{bmatrix} V_H^P \\ V_L^P \end{bmatrix} \tag{4.83a}
$$

$$
\begin{bmatrix} I_H^N \\ I_L^N \end{bmatrix} = \begin{bmatrix} \dfrac{1}{t_{HL}^2 Z_{LH}^P} & -\dfrac{e^{-j\phi}}{t_{HL} Z_{LH}^P} \\ -\dfrac{e^{j\phi}}{t_{HL} Z_{LH}^P} & \dfrac{1}{Z_{LH}^P} \end{bmatrix} \begin{bmatrix} V_H^N \\ V_L^N \end{bmatrix}
\tag{4.83b}
$$

and the ZPS nodal admittance matrix of Figure 4.12(g) with $Z_E = 0$ is given by

$$
\begin{bmatrix} I_H^Z \\ I_L^Z \end{bmatrix} = \begin{bmatrix} \dfrac{1}{t_{HL}^2 Z_{LH}^Z} & 0 \\ 0 & 0 \end{bmatrix} \begin{bmatrix} V_H^Z \\ V_L^Z \end{bmatrix}
\tag{4.83c}
$$

Equations (4.83a), (4.83b) and (4.83c) can be assembled to obtain the transformer sequence admittance matrix, including sequence current and voltage vectors, as follows:

$$
\begin{bmatrix} I_H^P \\ I_H^N \\ I_H^Z \\ I_L^P \\ I_L^N \\ I_L^Z \end{bmatrix} = \begin{bmatrix} \dfrac{1}{t_{HL}^2 Z_{LH}^P} & 0 & 0 & -\dfrac{e^{j\phi}}{t_{HL} Z_{LH}^P} & 0 & 0 \\ 0 & \dfrac{1}{t_{HL}^2 Z_{LH}^P} & 0 & 0 & -\dfrac{e^{-j\phi}}{t_{HL} Z_{LH}^P} & 0 \\ 0 & 0 & \dfrac{1}{t_{HL}^2 Z_{LH}^Z} & 0 & 0 & 0 \\ -\dfrac{e^{-j\phi}}{t_{HL} Z_{LH}^P} & 0 & 0 & \dfrac{1}{Z_{LH}^P} & 0 & 0 \\ 0 & -\dfrac{e^{j\phi}}{t_{HL} Z_{LH}^P} & 0 & 0 & \dfrac{1}{Z_{LH}^P} & 0 \\ 0 & 0 & 0 & 0 & 0 & 0 \end{bmatrix} \begin{bmatrix} V_H^P \\ V_H^N \\ V_H^Z \\ V_L^P \\ V_L^N \\ V_L^Z \end{bmatrix}
\tag{4.84a}
$$

or

$$
\begin{bmatrix} \mathbf{I}_{H(node)}^{PNZ} \\ \mathbf{I}_{L(node)}^{PNZ} \end{bmatrix} = \begin{bmatrix} \mathbf{Y}_{HH}^{PNZ} & \mathbf{Y}_{HL}^{PNZ} \\ \mathbf{Y}_{LH}^{PNZ} & \mathbf{Y}_{LL}^{PNZ} \end{bmatrix} \begin{bmatrix} \mathbf{V}_{H(node)}^{PNZ} \\ \mathbf{V}_{L(node)}^{PNZ} \end{bmatrix}
\tag{4.84b}
$$

It should be noted that $\mathbf{Y}_{LH}^{PNZ} = (\mathbf{Y}_{HL}^{PNZ})^*$. The phase admittance matrix is obtained using the sequence to phase transformation matrix $\mathbf{H}$ noting that in pu, the effective delta winding turns ratio is $\sqrt{3}$. Thus

$$
\begin{bmatrix} \mathbf{I}_{H(node)}^{RYB} \\ \mathbf{I}_{L(node)}^{ryb} \end{bmatrix} = \begin{bmatrix} \mathbf{Y}_{HH}^{RYB} & \mathbf{Y}_{HL}^{phase} \\ \mathbf{Y}_{LH}^{phase} & \mathbf{Y}_{LL}^{ryb} \end{bmatrix} \begin{bmatrix} \mathbf{V}_{H(node)}^{RYB} \\ \mathbf{V}_{L(node)}^{ryb} \end{bmatrix}
\tag{4.85a}
$$

where

$$\mathbf{I}_{H(node)}^{RYB} = [\mathbf{I}_H^R \quad \mathbf{I}_H^Y \quad \mathbf{I}_H^B]^t \qquad \mathbf{V}_{H(node)}^{RYB} = [\mathbf{V}_H^R \quad \mathbf{V}_H^Y \quad \mathbf{V}_H^B]^t$$

$$\mathbf{I}_{L(node)}^{ryb} = [\mathbf{I}_L^r \quad \mathbf{I}_L^y \quad \mathbf{I}_L^b]^t \qquad \mathbf{V}_{L(node)}^{ryb} = [\mathbf{V}_L^r \quad \mathbf{V}_L^y \quad \mathbf{V}_L^b]^t$$

$$\mathbf{Y}_{HH}^{RYB} = \mathbf{H}\mathbf{Y}_{HH}^{PNZ}\mathbf{H}^{-1} \qquad \mathbf{Y}_{LL}^{ryb} = 3\mathbf{H}\mathbf{Y}_{LL}^{PNZ}\mathbf{H}^{-1} \qquad (4.85b)$$

$$\mathbf{Y}_{HL}^{phase} = \sqrt{3}\mathbf{H}\mathbf{Y}_{HL}^{PNZ}\mathbf{H}^{-1} \qquad \mathbf{Y}_{LH}^{phase} = \sqrt{3}\mathbf{H}\mathbf{Y}_{LH}^{PNZ}\mathbf{H}^{-1} \qquad (4.85c)$$

Therefore, using Equations (4.84) in Equations (4.85b) and (4.85c), the self (H and L windings) and mutual (H–L and L–H windings) phase admittance matrices are given by

$$\mathbf{Y}_{HH}^{RYB} = \frac{1}{3t_{HL}^2} \begin{bmatrix} \dfrac{1}{Z_{LH}^Z} + \dfrac{2}{Z_{LH}^P} & \dfrac{1}{Z_{LH}^Z} - \dfrac{1}{Z_{LH}^P} & \dfrac{1}{Z_{LH}^Z} - \dfrac{1}{Z_{LH}^P} \\ \dfrac{1}{Z_{LH}^Z} - \dfrac{1}{Z_{LH}^P} & \dfrac{1}{Z_{LH}^Z} + \dfrac{2}{Z_{LH}^P} & \dfrac{1}{Z_{LH}^Z} - \dfrac{1}{Z_{LH}^P} \\ \dfrac{1}{Z_{LH}^Z} - \dfrac{1}{Z_{LH}^P} & \dfrac{1}{Z_{LH}^Z} - \dfrac{1}{Z_{LH}^P} & \dfrac{1}{Z_{LH}^Z} + \dfrac{2}{Z_{LH}^P} \end{bmatrix} \qquad (4.86a)$$

$$\mathbf{Y}_{LL}^{ryb} = 3\mathbf{H}\mathbf{Y}_{LL}^{PNZ}\mathbf{H}^{-1} = \begin{bmatrix} \dfrac{2}{Z_{LH}^P} & \dfrac{-1}{Z_{LH}^P} & \dfrac{-1}{Z_{LH}^P} \\ \dfrac{-1}{Z_{LH}^P} & \dfrac{2}{Z_{LH}^P} & \dfrac{-1}{Z_{LH}^P} \\ \dfrac{-1}{Z_{LH}^P} & \dfrac{-1}{Z_{LH}^P} & \dfrac{2}{Z_{LH}^P} \end{bmatrix} \qquad (4.86b)$$

$$\mathbf{Y}_{HL}^{phase} = \sqrt{3}\frac{2}{3t_{HL}Z_{LH}^P} \begin{bmatrix} -\cos\phi & -\cos(\phi + 2\pi/3) & -\cos(\phi - 2\pi/3) \\ -\cos(\phi - 2\pi/3) & -\cos\phi & -\cos(\phi + 2\pi/3) \\ -\cos(\phi + 2\pi/3) & -\cos(\phi - 2\pi/3) & -\cos\phi \end{bmatrix} \qquad (4.87a)$$

and

$$\mathbf{Y}_{LH}^{phase} = (\mathbf{Y}_{HL}^{phase})^t \qquad (4.87b)$$

It should be noted that for star–star and delta–delta connected two-winding transformers, $\phi = 0$ or $\phi = 180°$. However, for the Yd1 transformer of Figure (4.53), $\phi = 30°$ and the mutual phase admittance matrices become

$$\mathbf{Y}_{HL}^{phase} = \frac{1}{t_{HL}Z_{LH}^P} \begin{bmatrix} -1 & 1 & 0 \\ 0 & -1 & 1 \\ 1 & 0 & -1 \end{bmatrix} \quad \text{and} \quad \mathbf{Y}_{LH}^{phase} = \frac{1}{t_{HL}Z_{LH}^P} \begin{bmatrix} -1 & 0 & 1 \\ 1 & -1 & 0 \\ 0 & 1 & -1 \end{bmatrix}$$

$$(4.87c)$$

The transformer PPS magnetising impedance is generally available from the open-circuit test. The ZPS magnetising impedance may also be available from a direct test for some transformers. Alternatively, where it is not available, it can be estimated from the measured PPS and ZPS leakage impedances as discussed in Section 4.2.9. Where no measured ZPS leakage impedance is available, the magnetising impedance and the ZPS leakage impedance may be estimated as discussed in Section 4.2.9. The PPS magnetising impedance is so high in comparison with the PPS leakage impedance that it may be connected on the high voltage side of the transformer with practically negligible error. The ZPS magnetising impedance is also better and more correctly connected nearer the high voltage side of the transformer because the LV winding is wound nearer the core but the HV winding is wound on the outside over the LV winding. Therefore, the effect of this connection is to modify the sequence admittance matrix $\mathbf{Y}_{HH}^{PNZ}$ and the corresponding phase admittance matrix $\mathbf{Y}_{HH}^{RYB}$ to

$$\mathbf{Y}_{HH}^{PNZ} = \begin{bmatrix} \dfrac{1}{t_{HL}^2 Z_{LH}^P} + \dfrac{1}{Z_M^P} & 0 & 0 \\ 0 & \dfrac{1}{t_{HL}^2 Z_{LH}^P} + \dfrac{1}{Z_M^P} & 0 \\ 0 & 0 & \dfrac{1}{t_{HL}^2 Z_{LH}^Z} + \dfrac{1}{Z_M^Z} \end{bmatrix} \tag{4.88a}$$

$$\mathbf{Y}_{HH}^{RYB} = \frac{1}{3}\begin{bmatrix} Y^Z + 2Y^P & Y^Z - Y^P & Y^Z - Y^P \\ Y^Z - Y^P & Y^Z + 2Y^P & Y^Z - Y^P \\ Y^Z - Y^P & Y^Z - Y^P & Y^Z + 2Y^P \end{bmatrix} \tag{4.88b}$$

where

$$Y^P = \frac{1}{t_{HL}^2 Z_{LH}^P} + \frac{1}{Z_M^P}, \qquad Y^Z = \frac{1}{t_{HL}^2 Z_{LH}^Z} + \frac{1}{Z_M^Z} \tag{4.88c}$$

and Z_M^P and Z_M^Z are the PPS and ZPS magnetising impedances, respectively.

Three-phase models for other transformer winding connections including three-winding transformers that result in a 9×9 nodal phase admittance matrix can be derived using a similar technique.

4.7.4 Three-phase modelling of QBs and PSs

In some special analysis, both the shunt and series transformers of QBs and PSs need to be explicitly modelled as three-phase devices. However, for most steady state e.g. unbalanced short-circuit analysis on the external network to which these equipment are connected, the two transformers may be combined into the same single model. Similar to a three-phase transformer with symmetrical interphase mutual couplings, we will make use of the PPS and ZPS impedances and equivalent circuits of the combined transformers to derive the three-phase admittance matrix that relates the input and output phase currents and voltages. Therefore, using Figures 4.37(a), 4.39(a) and (b) as well as the sequence admittance Equations (4.56a), (4.57a) and (4.57b), the nodal sequence admittance matrix including sequence current and voltage vectors is given by

$$
\begin{bmatrix} I_i^P \\ I_i^N \\ I_i^Z \\ \hline I_o^P \\ I_o^N \\ I_o^Z \end{bmatrix} =
\begin{bmatrix}
\left|\overline{N}^P\right|^2 Y_e^P & 0 & 0 & -(\overline{N}^P)^* Y_e^P & 0 & 0 \\
0 & \left|\overline{N}^N\right|^2 Y_e^P & 0 & 0 & -(\overline{N}^N)^* Y_e^P & 0 \\
0 & 0 & Y_i^Z & 0 & 0 & -Y_m^Z \\
\hline
-\overline{N}^P Y_e^P & 0 & 0 & Y_e^P & 0 & 0 \\
0 & -\overline{N}^N Y_e^P & 0 & 0 & Y_e^P & 0 \\
0 & 0 & -Y_m^Z & 0 & 0 & Y_o^Z
\end{bmatrix}
\begin{bmatrix} V_i^P \\ V_i^N \\ V_i^Z \\ \hline V_o^P \\ V_o^N \\ V_o^Z \end{bmatrix}
$$

(4.89a)

or

$$
\begin{bmatrix} \mathbf{I}_i^{PNZ} \\ \mathbf{I}_o^{PNZ} \end{bmatrix} =
\begin{bmatrix} \mathbf{Y}_{ii}^{PNZ} & \mathbf{Y}_{io}^{PNZ} \\ \mathbf{Y}_{oi}^{PNZ} & \mathbf{Y}_{oo}^{PNZ} \end{bmatrix}
\begin{bmatrix} \mathbf{V}_i^{PNZ} \\ \mathbf{V}_o^{PNZ} \end{bmatrix} =
\mathbf{Y}^{PNZ} \begin{bmatrix} \mathbf{V}_i^{PNZ} \\ \mathbf{V}_o^{PNZ} \end{bmatrix}
$$

(4.89b)

where $\overline{N}^P$ and $\overline{N}^N$ are the PPS and NPS complex turns ratios and $(\overline{N}^P)^*$ and $(\overline{N}^N)^*$ are their conjugates. $\overline{N}^P$ is given in Equations (4.54b) and (4.55b) for a QB and a PS, respectively and $\overline{N}^N$ is given in Equation (4.57a). Therefore, using $\mathbf{Y}^{RYB} = \mathbf{H}\mathbf{Y}^{PNZ}\mathbf{H}^{-1}$ it can be shown that the nodal phase admittance matrix is given by

$$
\begin{bmatrix} \mathbf{I}_i^{RYB} \\ \mathbf{I}_o^{RYB} \end{bmatrix} =
\begin{bmatrix} \mathbf{Y}_{ii}^{RYB} & \mathbf{Y}_{io}^{RYB} \\ \mathbf{Y}_{oi}^{RYB} & \mathbf{Y}_{oo}^{RYB} \end{bmatrix}
\begin{bmatrix} \mathbf{V}_i^{RYB} \\ \mathbf{V}_o^{RYB} \end{bmatrix}
$$

(4.90a)

where

$$
\mathbf{Y}_{ii}^{RYB} = \frac{1}{3}
\begin{bmatrix}
a & b & b \\
b & a & b \\
b & b & a
\end{bmatrix}
$$

(4.90b)

$$
a = \begin{cases} Y_i^Z + 2(1 + \tan^2\phi)Y_e^P & \text{for a QB} \\ Y_i^Z + 2Y_e^P & \text{for a PS} \end{cases}
$$

(4.90c)

$$
b = \begin{cases} Y_i^Z - (1 + \tan^2\phi)Y_e^P & \text{for a QB} \\ Y_i^Z - Y_e^P & \text{for a PS} \end{cases}
$$

(4.90d)

and

$$
\mathbf{Y}_{oo}^{RYB} = \frac{1}{3}
\begin{bmatrix}
Y_o^Z + 2Y_e^P & Y_o^Z - Y_e^P & Y_o^Z - Y_e^P \\
Y_o^Z - Y_e^P & Y_o^Z + 2Y_e^P & Y_o^Z - Y_e^P \\
Y_o^Z - Y_e^P & Y_o^Z - Y_e^P & Y_o^Z + 2Y_e^P
\end{bmatrix}
$$

(4.91)

for both QBs and PSs. And

$$\mathbf{Y}_{io}^{RYB} = \frac{1}{3}\begin{bmatrix} x & y & z \\ z & x & y \\ y & z & x \end{bmatrix} \tag{4.92a}$$

where

$$x = \begin{cases} -(Y_m^Z + 2Y_e^P) & \text{for a QB} \\ -(Y_m^Z + 2\cos\phi Y_e^P) & \text{for a PS} \end{cases} \tag{4.92b}$$

$$y = \begin{cases} -[Y_m^Z + (\sqrt{3}\tan\phi - 1)Y_e^P] & \text{for a QB} \\ -[Y_m^Z + 2\cos(\phi - 2\pi/3)Y_e^P] & \text{for a PS} \end{cases} \tag{4.92c}$$

$$z = \begin{cases} -[Y_m^Z - (1 + \sqrt{3}\tan\phi)Y_e^P] & \text{for a QB} \\ -[Y_m^Z + 2\cos(\phi + 2\pi/3)Y_e^P] & \text{for a PS} \end{cases} \tag{4.92d}$$

The reader should easily find that $\mathbf{Y}_{oi}^{RYB} = (\mathbf{Y}_{io}^{RYB})^t$.

4.7.5 Three-phase modelling of static load

Static three-phase load was represented by equal PPS and NPS impedances or admittances as given by Equation (4.68) and a ZPS impedance or admittance as described in Section 4.6.2. The sequence admittance matrix of the load is given by

$$\mathbf{Y}_L^{PNZ} = \begin{matrix} & \begin{matrix} P & N & Z \end{matrix} \\ \begin{matrix} P \\ N \\ Z \end{matrix} & \begin{bmatrix} Y_L^P & 0 & 0 \\ 0 & Y_L^P & 0 \\ 0 & 0 & Y_L^Z \end{bmatrix} \end{matrix} \tag{4.93a}$$

The corresponding three-phase admittance matrix is calculated using $\mathbf{Y}_L^{RYB} = \mathbf{H}\mathbf{Y}_L^{PNZ}\mathbf{H}^{-1}$ giving the following balanced three-phase admittance matrix

$$\mathbf{Y}_L^{RYB} = \frac{1}{3}\begin{bmatrix} Y_L^Z + 2Y_L^P & Y_L^Z - Y_L^P & Y_L^Z - Y_L^P \\ Y_L^Z - Y_L^P & Y_L^Z + 2Y_L^P & Y_L^Z - Y_L^P \\ Y_L^Z - Y_L^P & Y_L^Z - Y_L^P & Y_L^Z + 2Y_L^P \end{bmatrix} \tag{4.93b}$$

In the special case where the ZPS admittance Y_L^Z is equal to the PPS admittance Y_L^P, then the load can be represented as three shunt admittances as follows:

$$\mathbf{Y}_L^{RYB} = \begin{bmatrix} Y_L^P & 0 & 0 \\ 0 & Y_L^P & 0 \\ 0 & 0 & Y_L^P \end{bmatrix} \tag{4.93c}$$

Further reading

Books

[1] The Electricity Council, 'Power system protection', Vol. 1, '*Principles and Components*', 1981, ISBN 0-906048-47-8.
[2] Arrilaga, J., *et al.*, '*Computer Modeling of Electrical Power Systems*', John Wiley & Sons, 1983, ISBN 0-471-10406-X.
[3] Grainger, J.J, '*Power System Analysis*', McGraw-Hill Int. Ed., 1994, ISBN 0-07-061293-5.
[4] Weedy, B.M., '*Electric Power Systems*', John Wiley & Sons, 1967, ISBN 0-471-92445-8.

Papers

[5] Bratton, E., *et al.*, 'Tertiary windings in star/star connected power transformers', *Electrical Power Engineer*, June 1969, pp. 6–9.
[6] Riedel, P., 'Modeling of zigzag-transformers in the three-phase systems', Institute for Electrical Engineering D-3000 Hanover 1, FRG, pp. T3/1-T3/9.
[7] Chen, Mo-Shing, *et al.*, 'Power system modeling, IEE Proceedings', Vol. 62, No. 7, July 1974, 901–915.
[8] Laugton, M.A., 'Analysis of unbalanced polyphase networks by the method of phase coordinates', IEE Proceedings, Vol. 115, No. 8, August 1968, 1163–1171.
[9] Bowman, W.I., *et al.*, 'Development of equivalent Pi and T matrix circuits for long untransposed transmission lines', *IEEE Summer General Meeting and Nuclear Radiation Effects Conference*, Toronto, Ont., June 1964, pp. 625–631.
[10] Jimenez, R., *et al.*, 'Protecting a 138 KV phase shifting transformer: EMTP modelling and model power system testing', 2002, 1–15.
[11] Mahseredjian, J., *et al.*, 'Superposition technique for MOV-protected series capacitors in short-circuit calculations', *IEEE Transactions On Power Delivery*, Vol. 10, No. 3, July 1995, 1394–1400.
[12] Goldworthy, D.L., 'A linearised model for MOV-protected series capacitors', IEEE PES Transactions Paper 86 SM 357–8. Mexico City, 1986.

5

Modelling of ac rotating machines

5.1 General

The short-circuit performance of ac rotating machines in power systems is of fundamental importance in the analysis of short-circuit currents and hence the safety of people and reliability of power systems. The modelling of rotating machines presents, in general, one of the most complex problems in the analysis of power systems. Fortunately, however, the required modelling detail of rotating machines depends on the type of study to be undertaken. In practice, different models are developed and used for load flow, short circuit, dynamic or electromechanical analysis and three-phase electromagnetic transient analysis, etc. The eventual aim of this chapter is to present the modelling of rotating machines in the sequence frame of reference for use in short-circuit analysis of power systems. However, in order to obtain an insight into the behaviour of rotating machines during transient fault conditions, the well-known machine model in the $dq0$ axis reference frame will be briefly presented and the machine models in the sequence frame of reference will then be derived. The presentation of the machine model in the $dq0$ reference is important since it provides an understanding of the meaning and origin of machine parameters, such as the machine d and q axes reactances and time constants, that can affect the short-circuit fault current.

Until the end of the twentieth century, almost all of the world's electric power supply was virtually produced by synchronous generators. With emerging concerns over climate change and the need for reduced CO_2 emissions, renewable energy, particularly using energy, is now becoming a main stream source of electric energy. Wind turbines utilising a range of novel non-synchronous generators are being used as a major new energy source in Western Europe (Denmark, Germany, Spain, UK, Holland, Ireland, etc.), North America (the US and Canada) and Asia (India and Japan). Wind farms that consist of tens to hundreds of wind turbine generators with individual turbine sizes of up to 5 MW are being built and turbine sizes in excess of 6 MW are under development and field testing. In addition to traditional synchronous and induction machines, short-circuit modelling of these new and emerging machines including those that utilise power electronics converters will also be described. The reader is expected to have a basic understanding of the theory of ac rotating machines.

5.2 Overview of synchronous machine modelling in the phase frame of reference

A brief overview of synchronous machine modelling in the phase frame of reference is presented. This model is a prerequisite to the derivation of the $dq0$ machine model as well as to the process of transformations from one reference to the other. The basic structure of a salient two-pole three-phase synchronous machine is shown in Figure 5.1(a) and a schematic is shown in Figure 5.1(b) illustrating stator and rotor circuits. The machine comprises three windings, r, y, b on the stator displaced by 120°; a field winding f that carries the direct current (dc) excitation and a short-circuited damper winding k on the rotor. The rotor can be cylindrical or salient pole in construction with the former mainly used in steam power plant and confined to two or four poles turbo-machines. Salient-pole machines are mainly used in hydro plant and low speed plant in general, and can have more than 100 poles. From the basic theory of magnetic coupling between circuits, the self and mutual inductances between stator and rotor circuits are defined as shown in Figure 5.1(b).

The damper winding is represented as two short-circuited windings; one in the same magnetic axis as the field winding f on the d-axis and termed kd and the other is a winding in an axis that is 90° ahead of the field axis and termed kq. We include one damper winding on the q-axis in our model. Figure 5.1(b) illustrates the field and damper windings on the rotor with respect to the d and q axes. The flux linkages in all six windings namely stator phases r, y and b, field winding f, damper windings k on d and q axes, using the generator convention, i.e. stator currents flowing out of the machine, are written in terms of self and mutual inductances as follows:

$$
\begin{bmatrix} \psi_{\text{ryb}} \\ \psi_{(\text{f}d)(\text{k}d)(\text{k}q)} \end{bmatrix} = \begin{bmatrix} \mathbf{L}_{\text{ss}}(\theta) & \mathbf{L}_{\text{sr}}(\theta) \\ \mathbf{L}_{\text{rs}}(\theta) & \mathbf{L}_{\text{rr}} \end{bmatrix} \begin{bmatrix} -\mathbf{i}_{\text{ryb}} \\ \mathbf{i}_{(\text{f}d)(\text{k}d)(\text{k}q)} \end{bmatrix} \tag{5.1}
$$

where

$$
\mathbf{L}_{\text{ss}}(\theta)
$$

$$
= \begin{array}{c} \\ \text{r} \\ \text{y} \\ \text{b} \end{array} \begin{array}{cccc} \quad\text{r} & \quad\quad\text{y} & \quad\quad\text{b} \\ \begin{bmatrix} L_{\text{ss}} + L_{\text{M}} \cos 2\theta & \dfrac{-L_{\text{ms}}}{2} + L_{\text{M}} \cos(2\theta - 2\pi/3) & \dfrac{-L_{\text{ms}}}{2} + L_{\text{M}} \cos(2\theta + 2\pi/3) \\ \dfrac{-L_{\text{ms}}}{2} + L_{\text{M}} \cos(2\theta - 2\pi/3) & L_{\text{ss}} + L_{\text{M}} \cos(2\theta + 2\pi/3) & \dfrac{-L_{\text{ms}}}{2} + L_{\text{M}} \cos 2\theta \\ \dfrac{-L_{\text{ms}}}{2} + L_{\text{M}} \cos(2\theta + 2\pi/3) & \dfrac{-L_{\text{ms}}}{2} + L_{\text{M}} \cos 2\theta & L_{\text{ss}} + L_{\text{M}} \cos(2\theta - 2\pi/3) \end{bmatrix} \end{array}
$$

$$
\tag{5.2a}
$$

(a)

(b)

ℓ_{rfd} – Mutual inductance
stator (r phase) to field

ℓ_{rkd} – Mutual inductance
stator (r phase) to d-axis
damper winding kd

ℓ_{rkq} – Mutual inductance
stator (r phase) to q-axis
damper winding kq

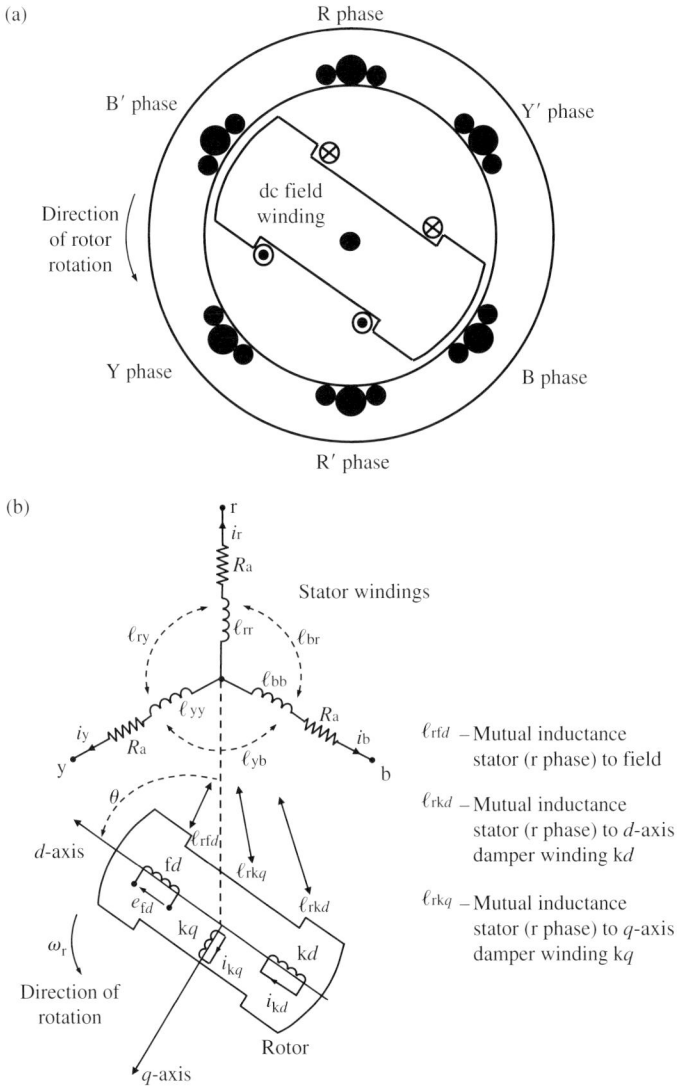

Figure 5.1 Three-phase salient-pole synchronous machine: (a) cross section and (b) illustration of stator and rotor circuits including mutual coupling

$$\mathbf{L}_{sr}(\theta) = \begin{array}{c} \text{r} \\ \text{y} \\ \text{b} \end{array} \begin{array}{ccc} \text{f} & \text{k}d & \text{k}q \\ \begin{bmatrix} L_{rfd}\cos\theta & L_{rkd}\cos\theta & L_{rkq}\sin\theta \\ L_{rfd}\cos(\theta - 2\pi/3) & L_{rkd}\cos(\theta - 2\pi/3) & L_{rkq}\sin(\theta - 2\pi/3) \\ L_{rfd}\cos(\theta + 2\pi/3) & L_{rkd}\cos(\theta + 2\pi/3) & L_{rkq}\cos(\theta + 2\pi/3) \end{bmatrix} \end{array}$$

(5.2b)

$$\begin{matrix} & \text{f} & \text{k}d & \text{k}q \end{matrix}$$

$$\mathbf{L}_{rr} = \begin{matrix} \text{f} \\ \text{k}d \\ \text{k}q \end{matrix} \begin{bmatrix} L_{ffd} & L_{fkd} & 0 \\ L_{kdf} & L_{kkd} & 0 \\ 0 & 0 & L_{kkq} \end{bmatrix} \qquad \mathbf{L}_{rs}(\theta) = (\mathbf{L}_{sr}(\theta))^{T} \quad L_{ss} = L_{\sigma} + L_{ms} \qquad (5.2c)$$

L_{σ} is stator leakage inductance and L_{ms} is stator magnetising inductance. The expressions $L_M \cos[f(\theta)]$ in Equation (5.2a) represent the fluctuating part of the stator self and mutual inductances caused by the magnetically unsymmetrical structure of the rotor of a salient-pole machine. For a symmetrical round rotor machine, $L_M = 0$. The subscripts rf$d$, rk$d$ and rk$q$ in Equation (5.2b) denote mutual inductances between stator winding phase r, and rotor field, d-axis damper and q-axis damper windings, respectively.

The zeros in the rotor inductance matrix $\mathbf{L}_{rr}$ represent the fact that there is no magnetic coupling between the d and q axes because they are orthogonal, i.e. displaced by $90°$. If the machine were a static device such as a transformer, then all inductances in Equation (5.1) would be constant. However, as can be seen in Figure 5.1(b), because of the rotation of the rotor, the self-inductances of the stator winding and the mutual inductances between the stator and the rotor windings vary with rotor angular position. This is defined as $\theta = \omega_r t$, where ω_r is rotor angular velocity, and is the angle by which the d-axis leads the magnetic axis of phase r winding in the direction of rotation.

The stator and rotor voltage relations, i.e. for all six windings are given by

$$\begin{bmatrix} e_r \\ e_y \\ e_b \\ e_{fd} \\ 0 \\ 0 \end{bmatrix} = \frac{d}{dt} \begin{bmatrix} \psi_r \\ \psi_y \\ \psi_b \\ \psi_{fd} \\ \psi_{kd} \\ \psi_{kq} \end{bmatrix} + \begin{bmatrix} -R_a i_r \\ -R_a i_y \\ -R_a i_b \\ R_{fd} i_{fd} \\ R_{kd} i_{kd} \\ R_{kq} i_{kq} \end{bmatrix} \qquad (5.3)$$

The zeros on the left-hand side of Equation (5.3) represent the short-circuited damper windings. The inductances of Equation (5.2) show that the flux linkages of Equation (5.1), and also the voltages of Equation (5.3) are non-linear functions of rotor angle positions. Generally, with the exception of studies of three-phase electromagnetic transients and sub-synchronous resonance analysis, this modelling is rather unwieldy and impractical for use in large-scale multi-machine short circuit, load flow or electromechanical stability analysis.

5.3 Synchronous machine modelling in the *dq*0 frame of reference

5.3.1 Transformation from phase ryb to *dq*0 frame of reference

The inductances of Equations (5.2a) and (5.2b) are time varying through their dependence on rotor angle position. Thus, the flux linkages are also functions of rotor angle position as given by Equation (5.1). Using Equation (5.2) in Equation (5.1), the rotor d and q axes damper winding flux linkages are given by

$$\psi_{kd} = L_{fkd}i_{fd} + L_{kkd}i_{kd} - L_{rkd}[i_r \cos\theta + i_y \cos(\theta - 2\pi/3) + i_b \cos(\theta + 2\pi/3)]$$

(5.4a)

$$\psi_{kq} = L_{kkq}i_{kq} + L_{rkq}[i_r \sin\theta + i_y \sin(\theta - 2\pi/3) + i_b \sin(\theta + 2\pi/3)]$$ (5.4b)

Equations (5.4a) and (5.4b) show a pattern involving the stator currents and rotor position. The outcome is new currents which can be expressed as follows:

$$i_d = \frac{2}{3}[i_r \cos\theta + i_y \cos(\theta - 2\pi/3) + i_b \cos(\theta + 2\pi/3)]$$ (5.5a)

$$i_q = -\frac{2}{3}[i_r \sin\theta + i_y \sin(\theta - 2\pi/3) + i_b \sin(\theta + 2\pi/3)]$$ (5.5b)

It can be shown that the constant multiplier in Equation (5.5) is arbitrary and the choice of 2/3 results in the peak value of i_d being equal to the peak value of the stator current. A third variable can be conveniently defined as the zero sequence current which is $i_0 = \frac{1}{3}(i_r + i_y + i_b)$. Therefore, the transformation from the stator ryb reference frame to the dq0 reference frame, written using stator currents, is given as

$$\mathbf{i}_{ryb} = \mathbf{M}(\theta)\mathbf{i}_{dq0}$$ (5.6a)

and

$$\mathbf{i}_{dq0} = \mathbf{M}^{-1}(\theta)\mathbf{i}_{ryb}$$ (5.6b)

where

$$\mathbf{M}(\theta) = \begin{bmatrix} \cos\theta & -\sin\theta & 1 \\ \cos(\theta - 2\pi/3) & -\sin(\theta - 2\pi/3) & 1 \\ \cos(\theta + 2\pi/3) & -\sin(\theta + 2\pi/3) & 1 \end{bmatrix}$$ (5.7a)

and

$$\mathbf{M}^{-1}(\theta) = \frac{2}{3}\begin{bmatrix} \cos\theta & \cos(\theta - 2\pi/3) & \cos(\theta + 2\pi/3) \\ -\sin\theta & -\sin(\theta - 2\pi/3) & -\sin(\theta + 2\pi/3) \\ \frac{1}{2} & \frac{1}{2} & \frac{1}{2} \end{bmatrix}$$ (5.7b)

$$\mathbf{i}_{ryb} = \begin{bmatrix} i_r \\ i_y \\ i_b \end{bmatrix} \quad \mathbf{i}_{dq0} = \begin{bmatrix} i_d \\ i_q \\ i_0 \end{bmatrix}$$ (5.7c)

The transformation matrix $\mathbf{M}^{-1}(\theta)$ allows us to transform quantities from the ryb reference frame to the dq0 reference frame, and vice versa using matrix $\mathbf{M}(\theta)$. This transformation method with the frame of reference fixed on the rotor applies to the stator fluxes, voltages and currents, and is generally known as Park's transformation.

This transformation process produces a new set of variables associated with two fictitious d and q stator windings that rotate with the rotor such that the stator inductances become constants as seen from the rotor during steady state operation.

Therefore, applying the transformation matrices of Equations (5.6a) and (5.6b) to Equation (5.1), and after much matrix and trigonometric analysis, we obtain

$$
\begin{bmatrix} \psi_d \\ \psi_q \\ \psi_0 \\ \psi_{fd} \\ \psi_{kd} \\ \psi_{kq} \end{bmatrix}
=
\begin{bmatrix}
L_d & 0 & 0 & L_{rfd} & L_{rkd} & 0 \\
0 & L_q & 0 & 0 & 0 & L_{rkq} \\
0 & 0 & L_0 & 0 & 0 & 0 \\
1.5L_{rfd} & 0 & 0 & L_{ffd} & L_{fkd} & 0 \\
1.5L_{rkd} & 0 & 0 & L_{fkd} & L_{kkd} & 0 \\
0 & 1.5L_{rkq} & 0 & 0 & 0 & L_{kkq}
\end{bmatrix}
\begin{bmatrix} -i_d \\ -i_q \\ -i_0 \\ i_{fd} \\ i_{kd} \\ i_{kq} \end{bmatrix}
\tag{5.8a}
$$

where

$$
L_d = L_\sigma + \frac{3}{2}(L_{ms} + L_M) \quad L_q = L_\sigma + \frac{3}{2}(L_{ms} - L_M) \quad L_0 = L_\sigma \tag{5.8b}
$$

Similarly, we apply the transformation matrices of Equations (5.6a) and (5.6b) to the stator voltages of Equation (5.3) and note that the matrix $\mathbf{M}(\theta)$ is derivable because its elements are functions of time through the rotor angular position. Therefore, after much trigonometric analysis, we obtain

$$
\begin{bmatrix} e_d \\ e_q \\ e_0 \end{bmatrix}
=
\frac{d}{dt}
\begin{bmatrix} \psi_d \\ \psi_q \\ \psi_0 \end{bmatrix}
+
\begin{bmatrix} -\psi_q \dfrac{d\theta}{dt} \\ \psi_d \dfrac{d\theta}{dt} \\ 0 \end{bmatrix}
-
\begin{bmatrix}
R_a & 0 & 0 \\
0 & R_a & 0 \\
0 & 0 & R_a
\end{bmatrix}
\begin{bmatrix} i_d \\ i_q \\ i_0 \end{bmatrix}
\tag{5.9}
$$

5.3.2 Machine $dq0$ equations in per unit

The equations derived so far are in physical units. However, the analysis is greatly simplified by if they are converted into a per-unit form. Several per-unit systems have been proposed in the literature but we will use the equations that correspond to the system known as the L_{ad} base reciprocal per-unit system. In this system, the pu mutual inductances among the three-stator field and damper windings are reciprocal and all mutual inductances between the stator and, field and damper windings, in the d and q axes are equal thus

$$
L_{rfd\,(pu)} = L_{fdr\,(pu)} = L_{rkd\,(pu)} = L_{kdr\,(pu)} = L_{ad\,(pu)}
$$

$$
L_{rkq\,(pu)} = L_{kqr\,(pu)} = L_{aq\,(pu)} \quad \text{and} \quad L_{fkd\,(pu)} = L_{kdf\,(pu)}
$$

where

$$
L_{d\,(pu)} = L_{\sigma\,(pu)} + L_{ad\,(pu)}
$$

$$
L_{q\,(pu)} = L_{\sigma\,(pu)} + L_{aq\,(pu)} \tag{5.10}
$$

Dropping the explicit pu notation for convenience, it can be shown that Equations (5.8a) and (5.9) can be written in pu form as follows:

$$
\begin{bmatrix} \psi_d \\ \psi_q \\ \psi_0 \\ \hline \psi_{fd} \\ \psi_{kd} \\ \psi_{kq} \end{bmatrix} =
\begin{bmatrix} L_d & 0 & 0 & L_{ad} & L_{ad} & 0 \\ 0 & L_q & 0 & 0 & 0 & L_{aq} \\ 0 & 0 & L_0 & 0 & 0 & 0 \\ \hline L_{ad} & 0 & 0 & L_{ffd} & L_{fkd} & 0 \\ L_{ad} & 0 & 0 & L_{fkd} & L_{kkd} & 0 \\ 0 & L_{aq} & 0 & 0 & 0 & L_{kkq} \end{bmatrix}
\begin{bmatrix} -i_d \\ -i_q \\ -i_0 \\ \hline i_{fd} \\ i_{kd} \\ i_{kq} \end{bmatrix}
\tag{5.11a}
$$

and

$$
\begin{bmatrix} e_d \\ e_q \\ e_0 \\ e_{fd} \\ 0 \\ 0 \end{bmatrix} =
\frac{d}{dt}
\begin{bmatrix} \psi_d \\ \psi_q \\ \psi_0 \\ \psi_{fd} \\ \psi_{kd} \\ \psi_{kq} \end{bmatrix} +
\begin{bmatrix} -\omega_r \psi_q \\ \omega_r \psi_d \\ 0 \\ 0 \\ 0 \\ 0 \end{bmatrix} +
\begin{bmatrix} -R_a i_d \\ -R_a i_q \\ -R_a i_0 \\ R_{fd} i_{fd} \\ R_{kd} i_{kd} \\ R_{kq} i_{kq} \end{bmatrix}
\tag{5.11b}
$$

In this pu system, we note that time is also in pu where $t_{Base} = 1/\omega_{Base} = 1/(2\pi f_{Base})$ with $\omega_{Base} = 314.159$ rad/s for a 50 Hz system. Also, since any reactance, e.g. $X_d(\Omega) = \omega L_d = 2\pi f L_d$ and $X_{Base}(\Omega) = \omega_{Base} L_{Base} = 2\pi f_{Base} L_{Base}$, it follows that if the stator frequency is equal to the base frequency, i.e. $f = f_{Base}$, then

$$
X_d(pu) = \frac{X_d(\Omega)}{X_{Base}(\Omega)} = \frac{2\pi f L_d}{2\pi f_{Base} L_{Base}} = L_d(pu)
$$

that is the pu values of X_d and L_d are equal. It is interesting to note that pu Equations (5.11a) and (5.11b) retain the same form as the physical ones, but the factor 1.5 has been eliminated from the rotor flux equations. Equations (5.11a) can be substituted in Equation (5.11b) and the result can be visualised using the simplified d and q axes equivalent circuits shown in Figure 5.2.

where

$$
X_{fd} = X_{ffd} - X_{ad} \quad X_{kd} = X_{kkd} - X_{ad} \quad X_{kq} = X_{kkq} - X_{aq}
$$
$$
X_{fkd} = X_{ad} \qquad X_d = X_\sigma + X_{ad} \qquad X_q = X_\sigma + X_{aq}
\tag{5.12}
$$

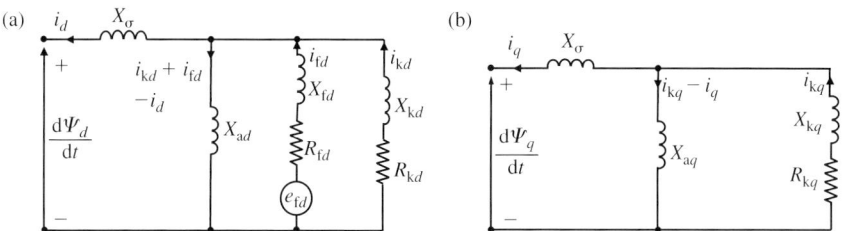

Figure 5.2 Synchronous machine *d* and *q* axes equivalent circuits: (a) *d*-axis and (b) *q*-axis

Figure 5.2(a) shows that the machine's d-axis equivalent circuit, ignoring stator resistance, consists of a field winding resistance and reactance in series with a voltage source e_{fd}, a damper winding resistance and reactance, a mutual reactance and a stator leakage reactance. Figure 5.2(b) shows the q-axis machine equivalent circuit. This is similar to that of the d-axis except that there is no field winding on the q-axis.

5.3.3 Machine operator reactance analysis

In order to gain an understanding of the origins of the various machine parameters, e.g. subtransient and transient reactances and time constants, we will use the method of operator d and q axes reactances for its simplicity and the clear insight it provides.

q-axis operator reactance

We will take the Laplace transform of the stator and rotor flux linkages of Equation (5.11a) remembering that Laplace $[\frac{d}{dt}f(t)] = sf(s) - f(0)$ and $s = j\omega$. Also, we replace the inductance symbol L by the symbol X for reactance since pu inductance and pu reactance are equal. It is also convenient to express the variables in terms of changes about the initial operating point in order to remove the terms that correspond to the initial condition that is $\Delta f(s) = f(s) - f(0)/s$. Thus

$$\Delta \psi_q(s) = -X_q \Delta i_q(s) + X_{aq} \Delta i_{kq}(s) \tag{5.13a}$$

and

$$\Delta \psi_{kq}(s) = -X_{aq} \Delta i_q(s) + X_{kkq} \Delta i_{kq}(s) \tag{5.13b}$$

Taking the Laplace transform of the q-axis damper voltage equation, last row of Equation (5.11b), and rearranging, we obtain

$$\Delta \psi_{kq}(s) = \frac{-R_{kq}}{s} \Delta i_{kq}(s) \tag{5.14a}$$

Substituting Equation (5.14a) into Equation (5.13b) and rearranging, we obtain

$$\Delta i_{kq}(s) = \frac{sX_{aq}}{R_{kq} + sX_{kkq}} \Delta i_q(s) \tag{5.14b}$$

Substituting Equation (5.14b) into Equation (5.13a) and rearranging, we obtain

$$\Delta \psi_q(s) = -\left(X_q - \frac{sX_{aq}^2}{R_{kq} + sX_{kkq}}\right) \Delta i_q(s)$$

or

$$\Delta \psi_q(s) = -X_q(s) \Delta i_q(s) \tag{5.15a}$$

where the q-axis operator reactance is defined as $X_q(s)$ and is given by

$$X_q(s) = X_q - \frac{sX_{aq}^2}{R_{kq} + sX_{kkq}} \tag{5.15b}$$

d-axis operator reactance

Taking the Laplace transform of the first, fourth and fifth rows of Equation (5.11a),

$$\begin{bmatrix} \Delta\psi_d(s) \\ \Delta\psi_{fd}(s) \\ \Delta\psi_{kd}(s) \end{bmatrix} = \begin{bmatrix} X_d & X_{ad} & X_{ad} \\ X_{ad} & X_{ffd} & X_{ad} \\ X_{ad} & X_{ad} & X_{kkd} \end{bmatrix} \begin{bmatrix} -\Delta i_d(s) \\ \Delta i_{fd}(s) \\ \Delta i_{kd}(s) \end{bmatrix} \tag{5.16a}$$

Also, taking the Laplace transform of the fourth and fifth rows of Equation (5.11b),

$$\begin{bmatrix} \Delta e_{fd}(s) \\ 0 \end{bmatrix} = \begin{bmatrix} s\Delta\psi_{fd}(s) \\ s\Delta\psi_{kd}(s) \end{bmatrix} + \begin{bmatrix} R_{fd} & 0 \\ 0 & R_{kd} \end{bmatrix} \begin{bmatrix} \Delta i_{fd}(s) \\ \Delta i_{kd}(s) \end{bmatrix} \tag{5.16b}$$

By substituting $\Delta\psi_{fd}(s)$ and $\Delta\psi_{kd}(s)$ from Equation (5.16a) into Equation (5.16b), we solve for $\Delta i_{fd}(s)$ and $\Delta i_{kd}(s)$ in terms of $\Delta e_{fd}(s)$ and $\Delta i_d(s)$. We then substitute this result back into Equation (5.16a), and use Equation (5.12), we obtain, after some algebra

$$\Delta\psi_d(s) = -X_d(s)\Delta i_d(s) + G_d(s)\Delta e_{fd}(s) \tag{5.17a}$$

$$\Delta i_{fd}(s) = -sG_d(s)\Delta i_d(s) + F_d(s)\Delta e_{fd}(s) \tag{5.17b}$$

where the d-axis operator reactance $X_d(s)$ is given by

$$X_d(s) = X_d - \frac{X_{ad}^2(R_{fd} + R_{kd})s + X_{ad}^2(X_{fd} + X_{kd})s^2}{A(s)} \tag{5.18a}$$

also

$$G_d(s) = \frac{X_{ad}(R_{kd} + sX_{kd})}{A(s)} \tag{5.18b}$$

$$F_d(s) = \frac{R_{kd} + sX_{kd}}{A(s)} \tag{5.19a}$$

and

$$A(s) = R_{fd}R_{kd} + (R_{fd}X_{kkd} + R_{kd}X_{ffd})s + (X_{kkd}X_{ffd} - X_{ad}^2)s^2 \tag{5.19b}$$

5.3.4 Machine parameters: subtransient and transient reactances and time constants

q-axis parameters

Using Equation (5.12), we can rewrite the q-axis operator reactance of Equation (5.15b) as follows:

$$X_q(s) = X_q - \frac{sX_{aq}^2}{R_{kq} + sX_{kkq}} = X_q - \frac{s\frac{X_{aq}^2}{R_{kq}}}{1 + s\frac{X_{kkq}}{R_{kq}}} = \frac{X_q + \left(X_q - \frac{X_{aq}^2}{X_{kkq}}\right)\frac{X_{kkq}}{R_{kq}}s}{1 + s\frac{X_{kkq}}{R_{kq}}}$$

or

$$X_q(s) = X_q \frac{1 + sT_q''}{1 + sT_{qo}''} \tag{5.20a}$$

where

$$T_{qo}'' = \frac{X_{kq} + X_{aq}}{\omega_s R_{kq}} s \tag{5.20b}$$

and

$$T_q'' = \frac{1}{\omega_s R_{kq}}\left(X_{kq} + \frac{1}{\frac{1}{X_{aq}} + \frac{1}{X_\sigma}}\right) s \tag{5.20c}$$

are the q-axis open-circuit and short-circuit subtransient time constants. We note that we divided these time constants by ω_s to convert from pu to seconds.

These time constants could also be derived by inspection from Figure 5.3(a). In the case of the open-circuit time constant, the equivalent reactance seen looking from the q-axis damper winding with the machine terminals open-circuited is X_{kq} in series with X_{aq}. This is then divided by the q-axis damper winding resistance R_{kq} to obtain the time constant. In the case of the short-circuit time constant, the equivalent reactance seen looking from the q-axis damper winding with the machine terminals short-circuited is X_{kq} in series with the parallel combination of X_{aq} and X_σ. This is then divided by the q-axis damper winding resistance R_{kq} to obtain the time constant.

The effective machine q-axis reactance at the instant of an external disturbance is defined as the subtransient reactance X_q''. Using Equation (5.20a), this is given by

$$X_q'' = \lim_{s \to \infty} X_q(s) = X_q \frac{T_q''}{T_{qo}''} \quad \text{and} \quad T_q'' = \frac{X_q''}{X_q} T_{qo}'' \tag{5.21a}$$

Substituting Equations (5.20b) and (5.20c) into Equation (5.21a) and using Equation (5.12), we obtain after some algebra

$$X_q'' = X_\sigma + \frac{1}{\frac{1}{X_{aq}} + \frac{1}{X_{kq}}} \tag{5.21b}$$

Again, the above q-axis subtransient reactance could be derived by inspection from Figure 5.3(a) with $R_{kq} = 0$ because the rotor flux linkages cannot change instantly

(a)

Subtransient time constants

Subtransient reactance   Steady state reactance

(b)

Subtransient reactance   Transient reactance   Steady state reactance

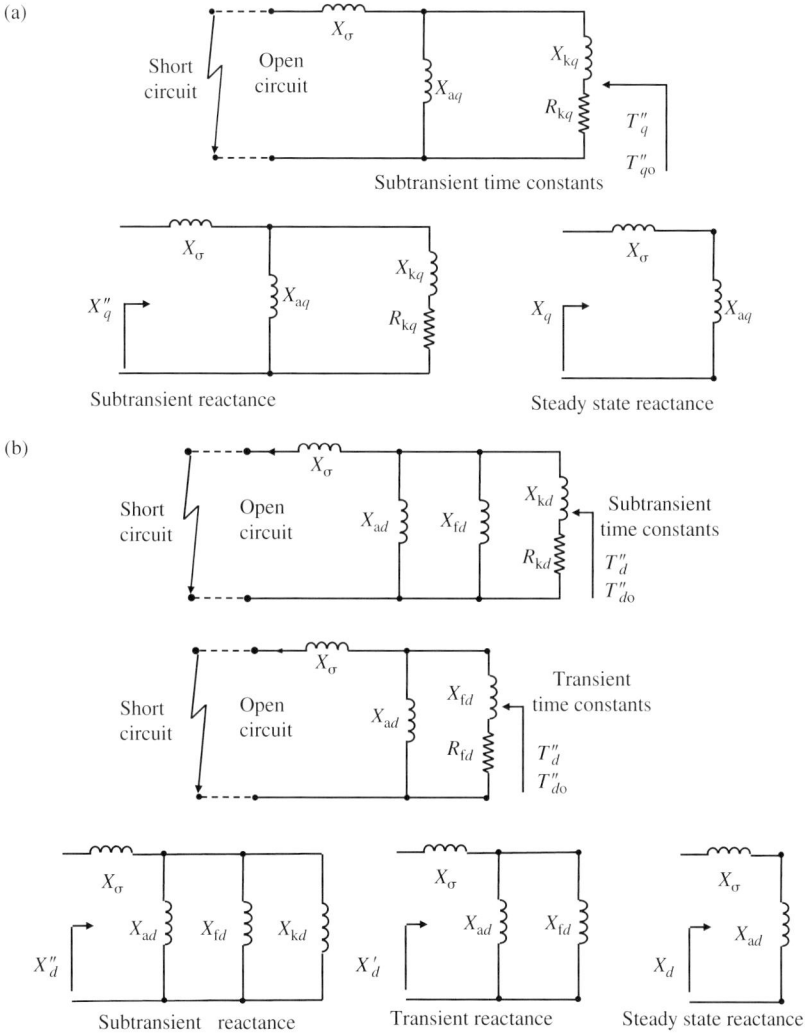

Figure 5.3   Machine parameters from its equivalent circuits: (a) *q*-axis: subtransient and steady state equivalent circuits and (b) *d*-axis: subtransient, transient and steady state equivalent circuits

following a disturbance. This is equivalent to looking into the machine from its *q*-axis terminals. This shows X_σ is in series with the parallel combination of X_{aq} and X_{kq}.

In the absence of a disturbance or under steady state conditions, the effective machine *q*-axis reactance is defined as the synchronous reactance X_q. Using Equation (5.20a)

$$X_q(0) = \lim_{s \to 0} X_q(s) = X_q = X_\sigma + X_{aq} \tag{5.21c}$$

It should be noted that using the q-axis operator reactance of Equation (5.20a), no q-axis transient time constant or transient reactance are defined and the latter can be assumed equal to the steady state value. This is due to our use of one q-axis damper winding on the q-axis as shown in Figure 5.2(b). This representation is accurate for laminated salient-pole machines and still reasonably accurate for a round rotor machine is sometimes that represented as two parallel damper winding circuits on the q-axis. In the latter case, the second may sometimes be used to represent the body of the solid rotor.

The q-axis operator reactance of Equation (5.20a) can be expressed in terms of partial fractions as follows:

$$\frac{1}{X_q(s)} = \frac{1 + sT''_{qo}}{X_q(1 + sT''_q)} = \frac{1}{X_q} + \left(\frac{1}{X''_q} - \frac{1}{X_q}\right)\frac{sT''_q}{(1 + sT''_q)} \tag{5.21d}$$

d-axis parameters

Following extensive algebraic manipulations, we can show that Equation (5.18a) can be written in the following form:

$$X_d(s) = X_d \frac{1 + (T_4 + T_5)s + (T_4 T_6)s^2}{1 + (T_1 + T_2)s + (T_1 T_3)s^2} \tag{5.22a}$$

where

$$T_1 = \frac{X_{ad} + X_{fd}}{\omega_s R_{fd}} \qquad\qquad T_2 = \frac{X_{ad} + X_{kd}}{\omega_s R_{kd}}$$

$$T_3 = \frac{1}{\omega_s R_{kd}}\left(X_{kd} + \frac{1}{\frac{1}{X_{ad}} + \frac{1}{X_{fd}}}\right) \qquad T_4 = \frac{1}{\omega_s R_{fd}}\left(X_{fd} + \frac{1}{\frac{1}{X_{ad}} + \frac{1}{X_\sigma}}\right)$$

$$T_5 = \frac{1}{\omega_s R_{kd}}\left(X_{kd} + \frac{1}{\frac{1}{X_{ad}} + \frac{1}{X_\sigma}}\right) \qquad T_6 = \frac{1}{\omega_s R_{kd}}\left(X_{kd} + \frac{1}{\frac{1}{X_\sigma} + \frac{1}{X_{ad}} + \frac{1}{X_{fd}}}\right)$$

$$\tag{5.22b}$$

The numerator and denominator of Equation (5.22a) can be expressed in terms of factors as follows:

$$X_d(s) = X_d \frac{(1 + sT'_d)(1 + sT''_d)}{(1 + sT'_{do})(1 + sT''_{do})} \tag{5.23a}$$

where

T'_{do} and T''_{do} are the transient and subtransient open-circuit time constants
T'_d and T''_d are the transient and subtransient short-circuit time constants

Accurate expressions for the machine open-circuit and short-circuit time constants can be derived by equating the numerator of Equation (5.22a) with that of Equation (5.23a), and similarly for the denominators. However, such a procedure would involve the solutions of quadratic equations and the values of the open-circuit and short-circuit time constants would result in rather involved expressions in terms of the time constants of Equation (5.22b). Reasonable approximations can be made on the basis that the field winding resistance is much smaller than the damper winding resistance. This means that T_2 (and T_3) $\ll T_1$ and T_5 (and T_6) $\ll T_4$. Therefore, Equation (5.22a) can be written as

$$X_d(s) \cong X_d \frac{(1 + sT_4)(1 + sT_6)}{(1 + sT_1)(1 + sT_3)} \tag{5.23b}$$

Therefore, from Equations (5.23a) and (5.23b), we have

$$T'_{do} = T_1 \quad T''_{do} = T_3 \quad T'_d = T_4 \quad T''_d = T_6 \tag{5.23c}$$

The effective machine *d*-axis reactance at the instant of an external disturbance is defined as the subtransient reactance X''_d. Using Equation (5.23a), this is given by

$$X''_d = \lim_{s \to \infty} X_d(s) = X_d \frac{T'_d}{T'_{do}} \frac{T''_d}{T''_{do}} \tag{5.24a}$$

This can also be expressed in terms of the internal machine *d*-axis parameters using Equations (5.22b) and (5.23c), hence

$$X''_d = X_\sigma + \frac{1}{\frac{1}{X_{ad}} + \frac{1}{X_{fd}} + \frac{1}{X_{kd}}} \tag{5.24b}$$

The *d*-axis operator reactance after the decay of the currents in the damper winding following an external disturbance can be represented by setting $X_{kd} = 0$ and $R_{kd} \to \infty$ in Equation (5.23b) and making use of Equation (5.23c). The result is given by

$$X_d(s) = X_d \frac{1 + sT'_d}{1 + sT'_{do}} \tag{5.25a}$$

Equation (5.25a) in fact also represents the case of a machine with only a field winding on the rotor but no damper winding. The effective machine *d*-axis reactance at the beginning of a disturbance is now defined as the transient reactance X'_d. Using Equation (5.25a), this is given by

$$X'_d = \lim_{s \to \infty} X_d(s) = X_d \frac{T'_d}{T'_{do}} \tag{5.25b}$$

This can also be expressed in terms of the internal machine *d*-axis parameters using Equations (5.22b) and (5.23c), hence

$$X'_d = X_\sigma + \frac{1}{\frac{1}{X_{ad}} + \frac{1}{X_{fd}}} \tag{5.25c}$$

From Equations (5.24a) and (5.25b), the following useful expression can be obtained

$$X''_d = X'_d \frac{T''_d}{T''_{do}} \quad \text{also} \quad T''_d = \frac{X''_d}{X'_d} T''_{do} \tag{5.26a}$$

In the absence of a disturbance or under steady state conditions, the effective machine d-axis reactance is defined as the synchronous reactance X_d. Using Equations (5.23a) or (5.25a)

$$X_d(0) = \lim_{s \to 0} X_d(s) = X_d = X_\sigma + X_{ad} \qquad (5.26b)$$

Again, the d-axis subtransient and transient time constants and reactances as well as the steady state reactance can be calculated by inspection from Figure 5.3(b).

The d-axis operator reactance of Equation (5.23a) can be rewritten as follows:

$$\frac{1}{X_d(s)} = \frac{1}{X_d}\frac{(1 + sT'_{do})(1 + sT''_{do})}{(1 + sT'_d)(1 + sT''_d)} = \frac{T'_{do}T''_{do}}{X_d T'_d T''_d}\frac{(s + 1/T'_{do})(s + 1/T''_{do})}{(s + 1/T'_d)(s + 1/T''_d)}$$

This expression can be approximated to a sum of three partial fractions assuming that T''_d and T''_{do} are small compared with T'_d and T'_{do}. The result is given by

$$\frac{1}{X_d(s)} = \frac{1}{X_d} + \left(\frac{1}{X'_d} - \frac{1}{X_d}\right)\frac{sT'_d}{(1 + sT'_d)} + \left(\frac{1}{X''_d} - \frac{1}{X'_d}\right)\frac{sT''_d}{(1 + sT''_d)} \qquad (5.27)$$

Equation (5.27) describes the machine reactance variation in the s domain as given in both IEC and IEEE standards for synchronous machines.

5.4 Synchronous machine behaviour under short-circuit faults and modelling in the sequence reference frame

Having established the machine d and q axes operator reactances, we are now able to proceed with analysing the behaviour of the machine under sudden network changes such as the occurrence of balanced and unbalanced short-circuit faults. Our objective is to understand the nature of short-circuit currents of synchronous machines and in order to do so, it is important to understand the meaning of positive phase sequence (PPS), negative phase sequence (NPS) and zero phase sequence (ZPS) machine impedances.

5.4.1 Synchronous machine sequence equivalent circuits

In Section 5.3, we defined the machine reactances and time constants in the $dq0$ reference frame. In short-circuit analysis, the machine PPS, NPS and ZPS machines models or equivalent circuits need to be defined as well as the sequence reactances. The three-phase synchronous generator is designed to produce a set of balanced three-phase voltages having the same magnitude and displaced by 120°. As we have already seen in Chapter 2, the transformation of these balanced three-phase voltages into the sequence reference frame produces a PPS voltage source in the PPS network but zero voltages in the NPS and ZPS networks. The stator windings

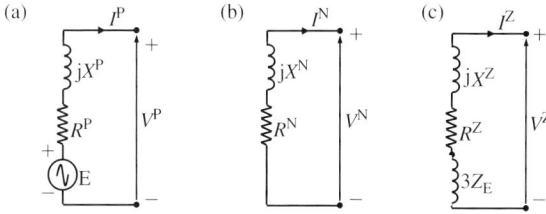

Figure 5.4   Synchronous machine: (a) PPS, (b) NPS and (c) ZPS equivalent circuits

of the machine are usually star connected and the neutral is usually earthed through an appropriate impedance. The earthing is usually used to limit the machine current under a single-phase short-circuit terminal fault to the rated current of the machine. We designate the machine PPS, NPS and ZPS reactances as X^P, X^N and X^Z, respectively. The machine PPS, NPS and ZPS equivalent circuits are shown in Figure 5.4 with the sequence quantities being those that correspond to phase r. As expected, the neutral earthing impedance appears only in the ZPS equivalent circuit and is multiplied by a factor of 3. The values of the sequence reactances and resistances is discussed in the next sections.

5.4.2 Three-phase short-circuit faults

Short-circuit currents

Before the occurrence of the short circuit, the machine is assumed in an open-circuit steady state condition and the rotor speed is the synchronous speed ω_s. The field voltage and current are constant, the damper currents are zero and the armature phase voltages are balanced three-phase quantities. Let $t = 0$ be the instant of short circuit that occurs at the machine terminals and let θ_o be the angle between the axis of phase r and direct axis at $t = 0$. Thus, θ_o defines the point in the voltage waveform at which the short-circuit occurs. The following initial conditions just before the short circuit, using Equations (5.11a) and (5.11b), are obtained

$$e_r(t) = \sqrt{2}E_o \cos(\omega_s t + \theta_o)$$

$$i_{do} = i_{qo} = i_{kdo} = i_{kqo} = \psi_{qo} = e_{do} = 0$$

$$\psi_{do} = L_{ad}i_{fdo} \quad e_{qo} = \psi_{do} = \sqrt{2}E_o \tag{5.28a}$$

We assume that, during the short-circuit period, the machine rotor speed and field voltage remain constant. The latter assumes no automatic voltage regulator (AVR) action. Therefore, immediately at the instant of the short circuit, we have

$$\Delta e_{fd} = 0 \quad e_{fd} = e_{fdo} = R_{fd}i_{fdo} \quad e_d = e_q = 0 \tag{5.28b}$$

Remembering that $\Delta f(s) = f(s) - f(0)/s$, the stator d- and q-axis voltages in Equation (5.11b), transformed into the Laplace domain and using Equations (5.28), can be written as follows:

$$0 = -R_a i_d(s) + s\left(\psi_d(s) - \frac{\psi_{do}}{s}\right) - \omega_s \psi_q(s) \tag{5.29a}$$

$$0 = -R_a i_q(s) + s\psi_q(s) + \omega_s \psi_d(s) \tag{5.29b}$$

Using Equation (5.28a), Equation (5.15a) becomes

$$\psi_q(s) = -X_q(s) i_q(s) \tag{5.30a}$$

Using Equations (5.28), Equation (5.17a) becomes

$$\psi_d(s) - \frac{\psi_{do}}{s} = -X_d(s) i_d(s) \tag{5.30b}$$

Substituting Equations (5.30a) and (5.30b) into Equations (5.29a) and (5.29b), and rearranging, we obtain

$$[R_a + sX_d(s)] i_d(s) - \omega_s X_q(s) i_q(s) = 0 \tag{5.31a}$$

$$[R_a + sX_q(s)] i_q(s) + \omega_s X_d(s) i_d(s) = \frac{\omega_s \sqrt{2} E_o}{s} \tag{5.31b}$$

Solving Equations (5.31a) and (5.31b) for d and q axes stator currents, we obtain

$$i_d(s) = \sqrt{2} E_o \omega_s^2 \frac{X_q(s)}{sB(s)} \quad \text{and} \quad i_q(s) = \sqrt{2} E_o \omega_s \frac{R_a + sX_d(s)}{sB(s)} \tag{5.32a}$$

where

$$B(s) = [R_a + sX_d(s)][R_a + sX_q(s)] + \omega_s^2 X_d(s) X_q(s) \tag{5.32b}$$

We will make a number of simplifications that help us to obtain the time domain solution of the d and q axes stator currents of Equation (5.32). From Equation (5.32b)

$$B(s) = X_d(s) X_q(s) \left\{ \omega_s^2 + \frac{[R_a + sX_d(s)][R_a + sX_q(s)]}{X_d(s) X_q(s)} \right\}$$

$$= X_d(s) X_q(s) \left\{ \omega_s^2 + \frac{R_a^2}{X_d(s) X_q(s)} + sR_a \left(\frac{1}{X_d(s)} + \frac{1}{X_q(s)} \right) + s^2 \right\}$$

$$\approx X_d(s) X_q(s) \left\{ s^2 + sR_a \left(\frac{1}{X_d''} + \frac{1}{X_q''} \right) + \omega_s^2 \right\}$$

$$\approx X_d(s) X_q(s) \{ (s + 1/T_a)^2 + \omega_s^2 \} \tag{5.33a}$$

where

$$T_a = \frac{2}{\omega_s R_a \left(\frac{1}{X_d''} + \frac{1}{X_q''} \right)} = \frac{2 X_d'' X_q''}{\omega_s R_a (X_d'' + X_q'')} s \tag{5.33b}$$

T_a is the stator or armature short-circuit time constant and is divided by ω_s to convent it from pu to seconds and R_a is the stator dc resistance. In arriving at

Equation (5.33a), we made two approximations which, in practice, would have a negligible effect. First, we ignored $R_a^2/[X_d(s)X_q(s)]$ because it is too small in comparison with ω_s^2. We then added $1/(T_a)^2$ where $1/(T_a)^2 \ll \omega_s^2$. We also replaced $X_d(s)$ with X_d'', and $X_q(s)$ with X_q'' in the term that multiplies sR_a because we assumed that the rotor field and damper winding resistances are very small. This means that the factors $(1+sT)$ in Equations (5.20a) and (5.23a) can be replaced with sT. Therefore, substituting Equation (5.33a) into Equations (5.32a), we obtain

$$i_d(s) = \frac{\sqrt{2}E_o\omega_s^2}{s[(s+1/T_a)^2 + \omega_s^2]X_d(s)} \tag{5.34a}$$

and

$$i_q(s) = \frac{\sqrt{2}E_o\omega_s}{[(s+1/T_a)^2 + \omega_s^2]X_q(s)} \tag{5.34b}$$

Substituting Equations (5.21d) and (5.27) into Equations (5.34) and taking the inverse Laplace transform, it can be shown that the time domain d and q axes currents are given by

$$i_d(t) = \sqrt{2}E_o\left[\frac{1}{X_d} + \left(\frac{1}{X_d'} - \frac{1}{X_d}\right)e^{-t/T_d'} + \left(\frac{1}{X_d''} - \frac{1}{X_d'}\right)e^{-t/T_d''}\right.$$
$$\left. - \frac{1}{X_d''}e^{-t/T_a}\cos\omega_s t\right] \tag{5.35a}$$

$$i_q(t) = \sqrt{2}E_o\frac{1}{X_q''}e^{-t/T_a}\sin\omega_s t \tag{5.35b}$$

The stator or armature phase r current can be calculated by transforming the d and q axes currents into the phase frame of reference using Equations (5.6a) and (5.6b) and noting that under balanced conditions $i_o = 0$. Therefore, using $\theta = \omega_s t + \theta_o$, $i_r(t) = i_d(t)\cos\theta - i_q(t)\sin\theta$, and after much trigonometric analysis, we obtain

$$i_r(t) \approx \underbrace{\left[\frac{1}{X_d} + \left(\frac{1}{X_d'} - \frac{1}{X_d}\right)e^{-t/T_d'} + \left(\frac{1}{X_d''} - \frac{1}{X_d'}\right)e^{-t/T_d''}\right]\sqrt{2}E_o\cos(\omega_s t + \theta_o - \pi/2)}_{\text{Power frequency (ac) component}}$$

$$\underbrace{-\frac{1}{2}\left(\frac{1}{X_d''} + \frac{1}{X_q''}\right)e^{-t/T_a}\sqrt{2}E_o\cos(\theta_o - \pi/2)}_{\text{Unidirectional (dc) component}}$$

$$\underbrace{-\frac{1}{2}\left(\frac{1}{X_d''} - \frac{1}{X_q''}\right)e^{-t/T_a}\sqrt{2}E_o\cos(2\omega_s t + \theta_o - \pi/2)}_{\text{Double frequency component}} \tag{5.36}$$

The currents in the two stator phases $i_y(t)$ and $i_b(t)$ are calculated with θ_o replaced by $\theta_o - 2\pi/3$ and $\theta_o + 2\pi/3$, respectively.

Equation (5.36) represents the total phase r current following a sudden three-phase short circuit fault at the machine terminals with the machine initially unloaded but running at rated speed. The short-circuit current contains three main components; a power frequency (e.g. 50 or 60 Hz) component, a transient unidirectional or dc component and a transient double-frequency (e.g. 100 or 120 Hz) component. Both the dc component and the double-frequency component decay to zero with a time constant equal to T_a which is typically between 0.1 and 0.4 s. These current components are illustrated in Figure 5.5.

The initial magnitude of the dc component is dependent on the instant of time at which the short circuit occurs, i.e. θ_0 and therefore the initial magnitudes of the dc components in the three phases are different but they all decay at the same rate given by the time constant T_a. The dc component in each phase appears in order to satisfy the physical condition that the current cannot change instantaneously at the instant of fault. The initial value of the transient double-frequency component is small as it is due to the difference between the d and q subtransient reactances $(X_q'' - X_d'')$; an effect termed as subtransient saliency. This current component disappears if the d and q subtransient reactances are equal. This component is neglected in network short-circuit analysis.

The power frequency component itself consists of three subcomponents as shown in Equation (5.36) and illustrated in Figure 5.5(b). These are termed the subtransient, transient and steady state components. The subtransient component decays to zero with a time constant T_d'' and typically lasts for up to 0.15 s. The transient component decays much more slowly to zero with a time constant T_d' and typically lasts for up to 5 s. The steady state or sustained component is constant. The parameters of the machine that determine the magnitude and rate of decay of each current component are the various reactances and short-circuit time constants shown in Equation (5.36). The sum of the dc and power frequency ac components produces an asymmetrical short-circuit current waveform. The phase r asymmetrical current is illustrated in Figure 5.5(c).

PPS reactance and resistance

The three-phase short-circuit is a balanced condition resulting in balanced ac currents in the three phases of the machine. From Equation (5.36), we can express the instantaneous power frequency component of the short-circuit current as $i_r(t) = \text{Real}[I_r(t)]$ where $I_r(t)$ is phase r complex instantaneous current given by

$$I_r(t) = \frac{1}{X_{(3\phi)}^{\text{p}}(t)} \sqrt{2} E_0 e^{j(\omega_s t + \theta_0 - \pi/2)} = \frac{E_r(t)}{jX_{(3\phi)}^{\text{p}}(t)} \qquad (5.37a)$$

where $E_r(t) = \sqrt{2} E_0 e^{j(\omega_s t + \theta_0)}$ is phase r complex instantaneous voltage and

$$\frac{1}{X_{(3\phi)}^{\text{p}}(t)} = \frac{1}{X_d} + \left(\frac{1}{X_d'} - \frac{1}{X_d}\right) e^{-t/T_d'} + \left(\frac{1}{X_d''} - \frac{1}{X_d'}\right) e^{-t/T_d''} \qquad (5.37b)$$

is an equivalent time-dependent machine PPS reactance as shown in Figure 5.6.

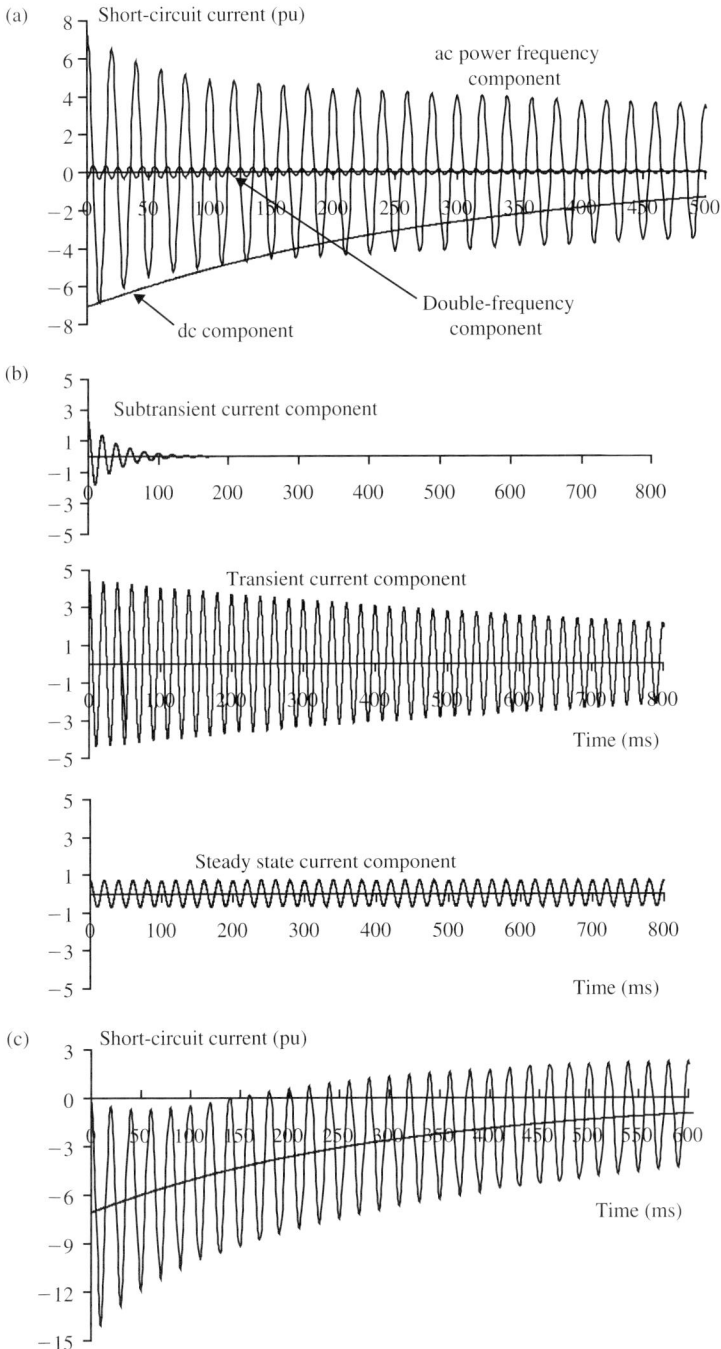

Figure 5.5   Three-phase short-circuit fault at a synchronous machine terminals: (a) the three components of the short-circuit current, (b) the three subcomponents of the ac power frequency component and (c) phase r short-circuit asymmetrical current

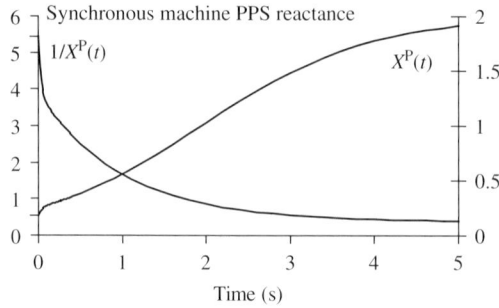

Figure 5.6 Synchronous machine time-dependent PPS reactance under a three-phase short-circuit at machine terminals

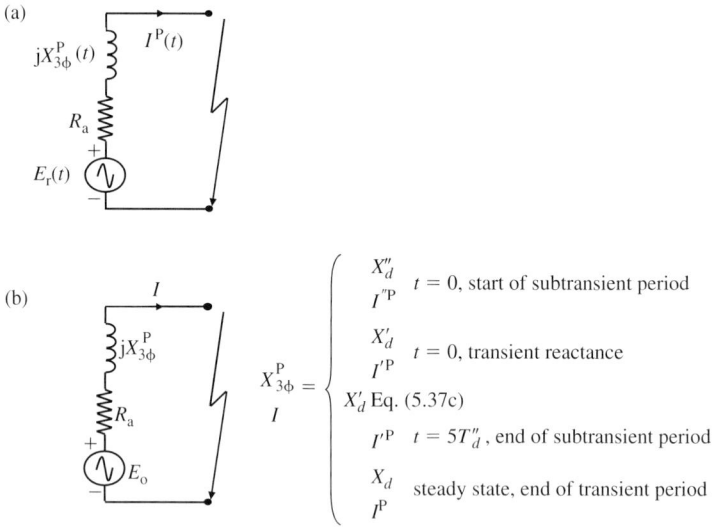

Figure 5.7 Synchronous machine PPS equivalent circuits for a three-phase short-circuit at machine terminals: (a) transient PPS time-dependent equivalent circuit and (b) fixed impedance equivalent circuits at various time instants

As presented in Chapter 2, the phase r complex voltage can also be written as $E_r(t) = E_r e^{j\omega_s t}$ where E_r is complex phasor given by $E_r = \sqrt{2}E_o e^{j\theta_o}$. As in the case of instantaneous currents, the complex phase y and phase b currents $I_y(t)$ and $I_b(t)$ are obtained by replacing θ_o of $I_r(t)$ with $\theta_o - 2\pi/3$ and $\theta_o + 2\pi/3$, respectively. The complex instantaneous PPS current is given by $I^P(t) = [I_r(t) + hI_y(t) + h^2 I_b(t)]/3 = I_r(t)$. Figure 5.7(a) shows the machine PPS equivalent circuit with a time-dependent equivalent reactance. We define such an equivalent circuit as a transient PPS symmetrical component equivalent circuit. Figure 5.7(b) shows the fixed impedance approach of the machine PPS reactance that is conventionally considered to consist of three components; subtransient,

transient and steady state components. If a time domain short-circuit analysis technique is used, then Equation (5.37b) or Equation (5.36) can be used directly to calculate the machine reactance or current at any time instant following the occurrence of the fault. However, the vast majority of large-scale short-circuit analysis computer programs used in practice use fixed impedance analysis techniques and this will be discussed in Chapter 6. Essentially, different values of reactance are used to calculate short-circuit currents at different times following the instant of fault. At the instant of short-circuit fault, $t = 0$, i.e. the start of the subtransient period, Equation (5.37b) gives $X^P = X_d''$. Also, neglecting the subtransient current component, the value of the transient reactance at $t = 0$ is obtained by putting $t = 0$ in Equation (5.37b) giving $X^P = X_d'$. The steady state reactance $X^P = X_d$ applies from the end of the transient period when the transient current component has vanished, i.e. $t \geq 5T_d'$. Another machine PPS reactance, found useful in practice, is the reactance that applies at the end of the subtransient period, i.e. at $t = 5T_d''$. Thus, using Equation (5.37b), this is given by

$$X^P = \frac{X_d'}{\frac{X_d'}{X_d}(1 - e^{-5T_d''/T_d'}) + e^{-5T_d''/T_d'}} \tag{5.37c}$$

In summary, the four PPS reactances are given by

$$X^P = \begin{cases} X_d'' & t = 0, \text{ start of subtransient period} \\ \text{Equation (5.37c)} & t = 5T_d'', \text{ end of subtransient period} \\ X_d' & t = 0, \text{ neglecting the subtransient current component} \\ X_d & t \geq 5T_d', \text{ steady state, end of the transient period} \end{cases} \tag{5.37d}$$

For typical salient-pole and round rotor synchronous machines, Equation (5.37c) shows that the PPS reactance at the end of the subtransient period is typically equal to $1.1X_d'$ to $1.5X_d'$.

The stator or armature dc resistance R_a is very small. The PPS power frequency stator resistance includes, in addition to stator losses, hysteresis and eddy current losses and may be 1.5–2 times the stator dc resistance. The machine PPS equivalent circuit, depending on the calculation time period of interest, are shown in Figure 5.7(b).

Effect of short-circuit fault through an external impedance

In many practical situations, the location of short-circuit fault will be on the network to which the machine is connected either directly or through a dedicated transformer. Therefore, an equivalent impedance will be present between the machine and the fault location. The effect of such an external machine impedance, denoted $(R_e + sX_e)$, and shown in Figure 5.8, is now considered.

The effect of the external impedance $(R_e + sX_e)$ can be considered to be equivalent to modifying the stator leakage reactance and stator resistance as shown in Figure 5.8. The analysis is in fact identical to the previous case without an external impedance except that the operator reactances will need to be modified to

(a)

Machine terminals

(b)

Machine terminals

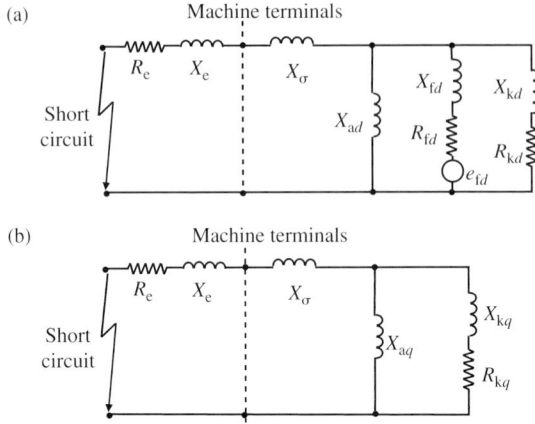

Figure 5.8 Three-phase short-circuit seen by a synchronous machine through an external impedance: (a) d-axis and (b) q-axis

include the external impedance $(R_e + sX_e)$. Using Equation (5.32a) and (5.32b), it can be shown that

$$i_d(s) = \sqrt{2}E_o \frac{\omega_s^2[X_q(s) + X_e]}{sD(s)} \quad i_q(s) = \sqrt{2}E_o \frac{\omega_s[R_a + R_e] + s[X_d(s) + X_e]}{sD(s)}$$

where

$$D(s) = \{(R_a + R_e) + s[X_d(s) + X_e]\}\{(R_a + R_e) + s[X_q(s) + X_e]\}$$
$$+ \omega_s^2[X_d(s) + X_e][X_q(s) + X_e]$$

The direct analytical solution of these equations is very tedious. An alternative approach is to calculate the short-circuit current components individually by substituting the relevant reactances. For example, to calculate the subtransient, transient and steady state power frequency current components, we substitute X_d'', X_d' and X_d in place of $\omega_s X_d(s)$, and X_q'' and X_q in place of $\omega_s X_q(s)$. Without going through the long mathematical analysis, it can be shown that the phase r short-circuit current, ignoring the double-frequency component, is given by

$$i_r(t) = \left[\frac{1}{X_d + X_e} + \left(\frac{1}{X_d' + X_e} - \frac{1}{X_d + X_e} \right) e^{-t/T_{de}'} \right.$$
$$+ \left(\frac{1}{X_d'' + X_e} - \frac{1}{X_d' + X_e} \right) e^{-t/T_{de}''} \left. \right] \sqrt{2}E_o \cos(\omega_s t + \theta_o - \pi/2)$$
$$- \frac{1}{2} \left(\frac{1}{X_d'' + X_e} + \frac{1}{X_q'' + X_e} \right) \sqrt{2}E_o e^{-t/T_{ae}} \cos(\theta_o - \pi/2) \quad (5.38a)$$

As before, the other two stator phases $i_y(t)$ and $i_b(t)$ are obtained with θ_o replaced by $\theta_o - 2\pi/3$ and $\theta_o + 2\pi/3$, respectively. Similar to the case with no external impedance, it can be shown that the effective short-circuit time constants in Equation (5.38a) can be expressed in terms of the open-circuit time constants as follows:

$$T'_{de} = \frac{(R_a + R_e)^2 + (X'_d + X_e)^2}{(R_a + R_e)^2 + (X'_d + X_e)(X_d + X_e)} T'_{do}$$

$$= \frac{(R_a + R_e)^2 + (X'_d + X_e)^2}{(R_a + R_e)^2 + (X'_d + X_e)(X_d + X_e)} \times \frac{X_d}{X'_d} T'_d \qquad (5.38b)$$

$$T''_{de} = \frac{(R_a + R_e)^2 + (X''_d + X_e)^2}{(R_a + R_e)^2 + (X''_d + X_e)(X'_d + X_e)} T''_{do}$$

$$= \frac{(R_a + R_e)^2 + (X''_d + X_e)^2}{(R_a + R_e)^2 + (X''_d + X_e)(X'_d + X_e)} \times \frac{X'_d}{X''_d} T''_d \qquad (5.38c)$$

$$T_{ae} = \frac{2}{\omega_s(R_a + R_e)\left[\frac{1}{X''_d + X_e} + \frac{1}{X''_q + X_e}\right]} \qquad (5.38d)$$

Equation (5.38d) can be approximated, with insignificant loss of accuracy, to

$$T_{ae} = \frac{\frac{1}{2}(X''_d + X''_q) + X_e}{\omega_s(R_a + R_e)} \approx \frac{X''_d + X_e}{\omega_s(R_a + R_e)} \quad \text{assuming } X''_d = X''_q \qquad (5.38e)$$

Usually the stator and external resistances are much smaller than the reactances. Thus, the transient and subtransient time constants reduce to the following:

$$T'_{de} = \frac{X'_d + X_e}{X_d + X_e} T'_{do} = \frac{X'_d + X_e}{X_d + X_e} \times \frac{X_d}{X'_d} T'_d \qquad (5.39a)$$

$$T''_{de} = \frac{X''_d + X_e}{X'_d + X_e} T''_{do} = \frac{X''_d + X_e}{X'_d + X_e} \times \frac{X'_d}{X''_d} T''_d \qquad (5.39b)$$

The effect of the external resistance R_e should not be neglected in calculating the armature or dc time constant given in Equation (5.38e).

The PPS equivalent circuits of the machine, 'seeing' a short circuit through an external impedance, depending on the calculation time period of interest, are shown in Figure 5.9.

Simplified machine short-circuit current equations

For most practical short-circuit calculations, simplified short-circuit equations can be used assuming $X''_d = X''_q$. Assuming maximum dc current offset and ignoring the double-frequency component, the peak current envelope of Equation (5.36), at any time instant, is given by

$$\hat{i}_r(t) = \sqrt{2}E_o\left[\frac{1}{X_d} + \left(\frac{1}{X'_d} - \frac{1}{X_d}\right)e^{-t/T'_d} + \left(\frac{1}{X''_d} - \frac{1}{X'_d}\right)e^{-t/T''_d} + \frac{1}{X''_d}e^{-t/T_a}\right]$$

$$(5.40a)$$

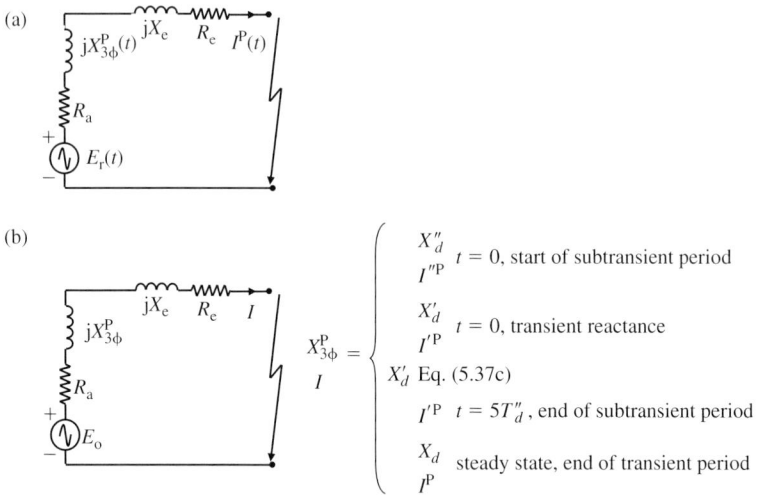

Figure 5.9 Synchronous machine PPS equivalent circuits for a three-phase short-circuit through an external impedance: (a) transient PPS time-dependent equivalent circuit and (b) fixed impedance equivalent circuits of various time instants

where

$$T_a = \frac{X_d''}{\omega_s R_a}$$

In the case of a fault through a predominantly inductive external impedance, the peak current envelope of Equation (5.38a) at any time instant is given by

$$
\hat{i}_r(t) = \sqrt{2} E_0 \left[\frac{1}{X_d + X_e} + \left(\frac{1}{X_d' + X_e} - \frac{1}{X_d + X_e} \right) e^{-t/T_{de}'} \right.
$$
$$
\left. + \left(\frac{1}{X_d'' + X_e} - \frac{1}{X_d' + X_e} \right) e^{-t/T_{de}''} + \frac{1}{X_d'' + X_e} e^{-t/T_{ae}} \right] \quad (5.40b)
$$

5.4.3 Unbalanced two-phase (phase-to-phase) short-circuit faults

Consider a two-phase or phase-to-phase short-circuit fault at the machine terminals involving phases y and b with the machine initially running at synchronous speed, rated terminal voltage and on open circuit. The initial conditions prior to the fault are given in Equations (5.28a) and (5.28b). The constraint equations that define the two-phase short-circuit fault are given by

$$e_y = e_b \quad \text{or} \quad e_y - e_b = 0 \quad\quad (5.41a)$$

$$i_r = 0 \quad \text{and} \quad i_y + i_b = 0 \quad \text{or} \quad i_y = -i_b \quad\quad (5.41b)$$

Transforming Equation (5.41a) and (5.41b) to the $dq0$ reference using Equation (5.7b), we obtain

$$e_d \sin(\omega_s t + \theta_o) + e_q \cos(\omega_s t + \theta_o) = 0 \qquad (5.41c)$$

$$i_d \cos(\omega_s t + \theta_o) = i_q \sin(\omega_s t + \theta_o) \qquad (5.41d)$$

$$i_o = 0 \qquad (5.41e)$$

Equations (5.41c) to (5.41e) and the machine Equations (5.11) that relate currents to fluxes and rate of change of fluxes to voltages provide a set of equations that are sufficient to obtain the solution of currents under a phase-to-phase short-circuit fault at the machine terminals. However, the resulting current equations are complex and non-linear and a closed form solution is extremely tedious. Instead, a step-by-step simplification process, similar to that described for a three-phase fault, can be used. The short-circuit current can be shown to contain a power frequency component, a dc component and both even and odd harmonic orders of the power frequency, the latter due to subtransient saliency. However, for practical calculations of machine phase-to-phase short-circuit currents involving the power frequency and dc components only and neglecting the harmonic components, it can be shown that the phase y and b short-circuit currents are given by

$$i_y(t) = -i_b(t) \approx -\left[\frac{1}{X_d + X^N} + \left(\frac{1}{X'_d + X^N} - \frac{1}{X_d + X^N} \right) e^{-t/T'_{d(2\phi)}} \right.$$
$$\left. + \left(\frac{1}{X''_d + X^N} - \frac{1}{X'_d + X^N} \right) e^{-t/T''_{d(2\phi)}} \right]$$
$$\times \sqrt{3}\sqrt{2}E_o \cos(\omega_s t + \theta_o) - \frac{1}{X''_d + X^N} e^{-t/T_{a(2\phi)}} \sqrt{3}\sqrt{2}E_o \cos\theta_o$$

$$(5.42a)$$

where

$$T'_{d(2\phi)} = \frac{X'_d + X^N}{X_d + X^N} T'_{do} \qquad (5.42b)$$

$$T''_{d(2\phi)} = \frac{X''_d + X^N}{X'_d + X^N} T''_{do} \qquad (5.42c)$$

$$T_{a(2\phi)} = \frac{X^N}{\omega_s R_a} \qquad (5.43a)$$

$$X^N = \sqrt{X''_d X''_q} \qquad (5.43b)$$

X^N is the machine NPS reactance.

The maximum dc current component in Equation (5.42a) occurs when $\theta_o = 0$.

From Equation (5.42a), we can express the instantaneous power frequency component of the short-circuit current as $i_y(t) = \text{Real}[I_y(t)]$ where $I_y(t)$ is phase y complex instantaneous current and is given by

$$I_y(t) = -\frac{\sqrt{3}\sqrt{2}E_o}{X^P_{(2\phi)}(t)}e^{j(\omega_s t + \theta_o)} = -\frac{\sqrt{3}E_r(t)}{X^P_{(2\phi)}(t)} = -I_b(t) \tag{5.44a}$$

and

$$E_r(t) = \sqrt{2}E_o e^{j(\omega_s t + \theta_o)} = E_r e^{j\omega_s t}$$

is phase r complex instantaneous voltage, E_r is phase r complex phasor given by

$$E_r = \sqrt{2}E_o e^{j\theta_o}$$

and

$$\frac{1}{X^P_{(2\phi)}(t)} = \frac{1}{X_d + X^N} + \left(\frac{1}{X'_d + X^N} - \frac{1}{X_d + X^N}\right)e^{-t/T'_{d(2\phi)}}$$

$$+ \left(\frac{1}{X''_d + X^N} - \frac{1}{X'_d + X^N}\right)e^{-t/T''_{d(2\phi)}} \tag{5.44b}$$

is an equivalent time-dependent machine PPS reactance. At the fault instant $t = 0$, i.e. start of subtransient period, $X^P_{(2\phi)} = X''_d + X^N$. At $t = 0$ and neglecting the subtransient component, $X^P_{(2\phi)} = X'_d + X^N$. At $t \geq 5T'_{d(2\phi)}$, i.e. end of transient period, $X^P_{(2\phi)} = X_d + X^N$. The machine reactance at the end of the subtransient period, i.e. at $t \approx 5T''_{d(2\phi)}$ is given by

$$X^P_{(2\phi)} = \frac{X'_d + X^N}{\frac{X'_d + X^N}{X_d + X^N}[1 - \exp(-5T''_{d(2\phi)}/T'_{d(2\phi)})] + \exp(-5T''_{d(2\phi)}/T'_{d(2\phi)})} \tag{5.44c}$$

In summary, the four fixed machine reactances are given by

$$X^P_{(2\phi)} = \begin{cases} X''_d + X^N & t = 0, \text{ start of subtransient period} \\ \text{Equation}(5.44c) & t = 5T''_{d(2\phi)}, \text{ end of subtransient period} \\ X'_d + X^N & t = 0, \text{ neglecting subtransient component} \\ X_d + X^N & t \geq 5T'_{d(2\phi)}, \text{ steady state} \end{cases} \tag{5.44d}$$

As we presented in Chapter 2, the transient PPS and NPS currents of the power frequency components of the complex instantaneous phase currents $I_y(t)$ and $I_b(t)$ of Equation (5.44a) can be calculated using

$$I^P(t) = \frac{1}{3}[I_r(t) + hI_y(t) + h^2 I_b(t)] \quad \text{and} \quad I^N(t) = \frac{1}{3}[I_r(t) + h^2 I_y(t) + hI_b(t)] \tag{5.45a}$$

Substituting Equation (5.44a) into Equation (5.45a), we obtain

$$I^P(t) = \frac{E_r(t)}{jX^P_{(2\phi)}(t)} \quad \text{and} \quad I^N(t) = -I^P(t) \tag{5.45b}$$

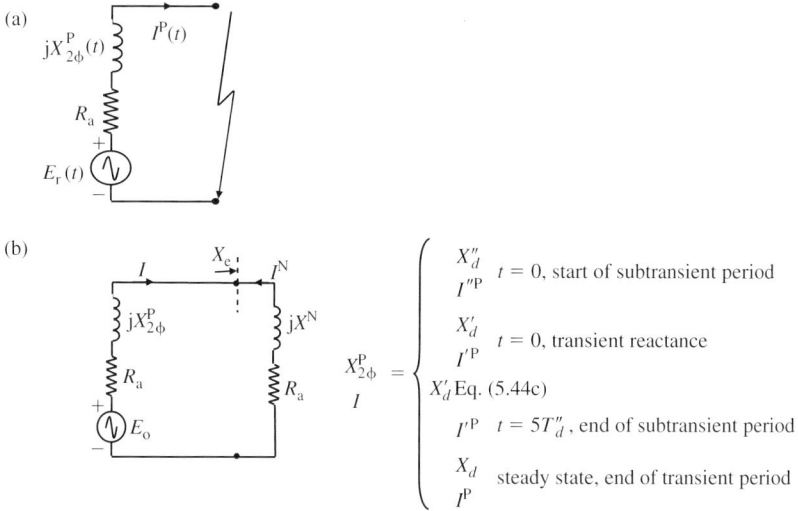

Figure 5.10 Synchronous machine equivalent circuits for a two-phase short-circuit fault at machine terminals: (a) transient PPS time-dependent equivalent circuit and (b) fixed impedance equivalent circuits of various time instants

Figure 5.10(a) shows the resultant transient symmetrical component equivalent circuit that satisfies Equation (5.45b) with the reinstated machine stator resistance for completeness. Figure 5.10(b) shows equivalent circuits using the fixed reactances of Equation (5.44d). The time constants of Equation (5.42b) and (5.42c), and the time-dependent machine reactance of Equation (5.44b) show, from a PPS current view point, that under a two-phase short-circuit fault the machine behaves as if it has 'seen' a balanced three-phase short-circuit through an external reactance X^N.

NPS reactance and resistance

We return to the NPS reactance X^N given in Equation (5.43b). This is the machine reactance that results from the flow of NPS stator currents. Let us first recall that during a balanced three-phase short-circuit, only PPS currents flow into the short circuit. These currents set up a MMF wave that rotates at synchronous speed in the same direction of rotation of the rotor. Because of the very high X/R ratio of the machine, the MMF wave lines up almost exactly with the d-axis hence d-axis currents only meet d-axis reactances. However, when NPS currents flow in the stator of the machine, these currents produce a MMF wave that rotates at synchronous speed in an opposite direction to the rotation of the rotor. It therefore rotates backward at twice synchronous speed with respect to the rotor. Thus, the currents induced in the rotor field and damper windings are double-frequency currents. At such a high frequency, the machine reactances are effectively the subtransient reactances and as the MMF wave sweeps rapidly over the d and q axes of the rotor, the equivalent NPS reactance alternates to d and q axes subtransient reactances. Therefore, the NPS reactance is defined, by both IEC and IEEE standards, as the arithmetic mean of the d and q axes subtransient reactances or

$$X^N = \frac{1}{2}(X_d'' + X_q'') \tag{5.46}$$

We have stated previously that when subtransient saliency is present, the short-circuit current will contain both even and odd harmonics. In this case, we have a different machine NPS reactance which is given by the geometric mean of the d and q axes subtransient reactances as shown in Equation (5.43b). However, because X_d'' and X_q'' are nearly equal, the difference between the arithmetic mean and the geometric mean is very small. The difference can be illustrated for both a round rotor machine and a salient-pole machine as follows. For a typical round rotor synchronous generator with $X_d'' = 0.2$ pu and $X_q'' = 0.24$ pu, their arithmetic and geometric means are 0.22 and 0.2191 pu, respectively. For a salient-pole machine with $X_d'' = 0.18$ pu and $X_q'' = 0.275$ pu, their arithmetic and geometric means are 0.2275 and 0.22248 pu, respectively. Given the uncertainty associated with the measurements of these quantities which can be up to 10%, differences in X^N of less than 2.5% are generally acceptable.

X^N takes the same value under subtransient, transient and steady state conditions as illustrated in Figure 5.10(b).

The stator NPS resistance is significantly higher (some 10–20 times) than the PPS resistance because the second harmonic backward rotating MMF causes significant additional $I^2 R$ losses in the rotor circuits due to double frequency currents in the rotor field and damper windings. For short-circuit analysis, the use of R_a instead of the NPS resistance is conservative.

Simplified machine short-circuit current equations

For maximum dc current offset, the peak current envelope of Equation (5.42a) at any instant of time is given by

$$\hat{i}_y(t) = \sqrt{3}\sqrt{2}E_o \left[\frac{1}{X_d + X^N} + \left(\frac{1}{X_d' + X^N} - \frac{1}{X_d + X^N} \right) e^{-t/T_{d(2\phi)}'} \right.$$
$$\left. + \left(\frac{1}{X_d'' + X^N} - \frac{1}{X_d' + X^N} \right) e^{-t/T_{d(2\phi)}''} + \frac{1}{X_d'' + X^N} e^{-t/T_{a(2\phi)}} \right]$$

$$\tag{5.47}$$

5.4.4 Unbalanced single-phase to earth short-circuit faults

Consider a single-phase to earth short-circuit fault at the machine terminals involving phase r with the machine initially running at synchronous speed, rated terminal voltage and on open circuit. The machine neutral is assumed solidly earthed and the initial conditions prior to the fault are given in Equations (5.28a) and (5.28b).

The constraint equations that define the single-phase short-circuit fault are given by

$$e_r = 0 \tag{5.48a}$$

$$i_y = i_b = 0 \tag{5.48b}$$

Transforming Equation (5.48) to the $dq0$ reference using Equation (5.7b), we obtain

$$e_d \cos(\omega_s t + \theta_o) - e_q \sin(\omega_s t + \theta_o) + e_o = 0 \tag{5.48c}$$

$$i_d \sin(\omega_s t + \theta_o) = -i_q \cos(\omega_s t + \theta_o) \tag{5.48d}$$

$$i_o = \frac{i_d}{2 \cos(\omega_s t + \theta_o)} \tag{5.48e}$$

As in the case of a two-phase short-circuit, Equations (5.48) and (5.11) are sufficient to obtain the solution of current under a single-phase short-circuit fault at the machine terminals. However, the resulting current equations are complex and non-linear and a closed form solution is extremely tedious. Instead, a step-by-step simplification process, similar to that described for a three-phase fault can be used. The single-phase short-circuit current can be shown to contain, in general, a power frequency component, a dc component and both even and odd harmonic orders of the power frequency, the latter due to subtransient saliency. However, for practical calculations of machine single-phase short-circuit currents involving the power frequency and dc component only, neglecting the harmonic components, it can be shown that the phase r short-circuit current is given by

$$i_r(t) \approx \left[\frac{1}{X_d + X^N + X^Z} + \left(\frac{1}{X'_d + X^N + X^Z} - \frac{1}{X_d + X^N + X^Z} \right) e^{-t/T'_{d(1\phi)}} \right. $$
$$\left. + \left(\frac{1}{X''_d + X^N + X^Z} - \frac{1}{X'_d + X^N + X^Z} \right) e^{-t/T''_{d(1\phi)}} \right]$$
$$\times 3\sqrt{2} E_o \cos(\omega_s t + \theta_o - \pi/2)$$
$$- \frac{1}{X''_d + X^N + X^Z} e^{-t/T_{a(1\phi)}} 3\sqrt{2} E_o \cos(\theta_o - \pi/2) \tag{5.49a}$$

where

$$T'_{d(1\phi)} = \frac{X'_d + X^N + X^Z}{X_d + X^N + X^Z} T'_{do} \tag{5.49b}$$

$$T''_{d(1\phi)} = \frac{X''_d + X^N + X^Z}{X'_d + X^N + X^Z} T''_{do} \tag{5.49c}$$

$$T_{a(1\phi)} = \frac{X''_d + X^N + X^Z}{\omega_s (3R_a)} \tag{5.49d}$$

Equation (5.49a) shows that the magnitude of the dc component of short-circuit current is maximum when $\theta_o = \pi/2$. In the time constant equations, X^Z is the machine ZPS reactance. X^N is the machine NPS reactance under a single-phase short-circuit fault condition and is given by

$$X^N = \sqrt{\left(X_d'' + \frac{X^Z}{2}\right)\left(X_q'' + \frac{X^Z}{2}\right)} - \frac{X^Z}{2} \tag{5.49e}$$

Since, X_d'' and X_q'' are nearly equal, then if we set $X_d'' = X_q''$ in Equation (5.49e), we obtain $X^N = X_d''$. Although the expression of Equation (5.49e) is different from the expressions given in Equations (5.43b) and (5.46), the difference in numerical values is practically insignificant.

From Equation (5.49a), we can express the instantaneous power frequency component of the short-circuit current as $i_r(t) = \text{Real}[I_r(t)]$ where $I_r(t)$ is phase r complex instantaneous current given by

$$I_r(t) = \frac{3\sqrt{2}E_o}{X_{(1\phi)}^P(t)} e^{j(\omega_s t + \theta_o - \pi/2)} = \frac{3E_r(t)}{jX_{(1\phi)}^P(t)} \tag{5.50a}$$

where $E_r(t) = \sqrt{2}E_o e^{j(\omega_s t + \theta_o)} = E_r e^{j\omega_s t}$ is phase r complex instantaneous voltage, E_r is phase r complex phasor given by $E_r = \sqrt{2}E_o e^{j\theta_o}$ and

$$\frac{1}{X_{(1\phi)}^P(t)} = \frac{1}{X_d + X^N + X^Z} + \left(\frac{1}{X_d' + X^N + X^Z} - \frac{1}{X_d + X^N + X^Z}\right) e^{-t/T_{d(1\phi)}'}$$

$$+ \left(\frac{1}{X_d'' + X^N + X^Z} - \frac{1}{X_d' + X^N + X^Z}\right) e^{-t/T_{d(1\phi)}''} \tag{5.50b}$$

is the machine equivalent time-dependent machine PPS reactance under a single-phase short-circuit fault condition. At the fault instant $t = 0$, i.e. start of subtransient period, $X_{(1\phi)}^P = X_d'' + X^N + X^Z$. At $t = 0$ and neglecting the subtransient component, $X_{(1\phi)}^P = X_d' + X^N + X^Z$. At $t \geq 5T_{d(1\phi)}'$, $X_{(1\phi)}^P = X_d + X^N + X^Z$. The machine reactance at the end of the subtransient period is given by

$$X_{(1\phi)}^P = \frac{X_d' + X^N + X^Z}{\frac{X_d' + X^N + X^Z}{X_d + X^N + X^Z}[1 - \exp(-5T_{d(1\phi)}''/T_{d(1\phi)}')] + \exp(-5T_{d(1\phi)}''/T_{d(1\phi)}')} \tag{5.50c}$$

In summary, the four fixed reactances are given by

$$X_{(1\phi)}^P = \begin{cases} X_d'' + X^N + X^Z & t = 0, \text{ start of subtransient period} \\ \text{Equation (5.50c)} & t = 5T_{d(1\phi)}'', \text{ end of subtransient period} \\ X_d' + X^N + X^Z & t = 0, \text{ neglecting subtransient component} \\ X_d + X^N + X^Z & t \geq 5T_{d(1\phi)}', \text{ steady state} \end{cases} \tag{5.50d}$$

The transient PPS, NPS and ZPS currents of the complex phase current $I_r(t)$ of Equation (5.50a) are calculated using

$$I^P(t) = I^N(t) = I^Z(t) = \frac{1}{3}I_r(t) \tag{5.51a}$$

(a) $I^P(t) = I^N(t) = I^Z(t)$

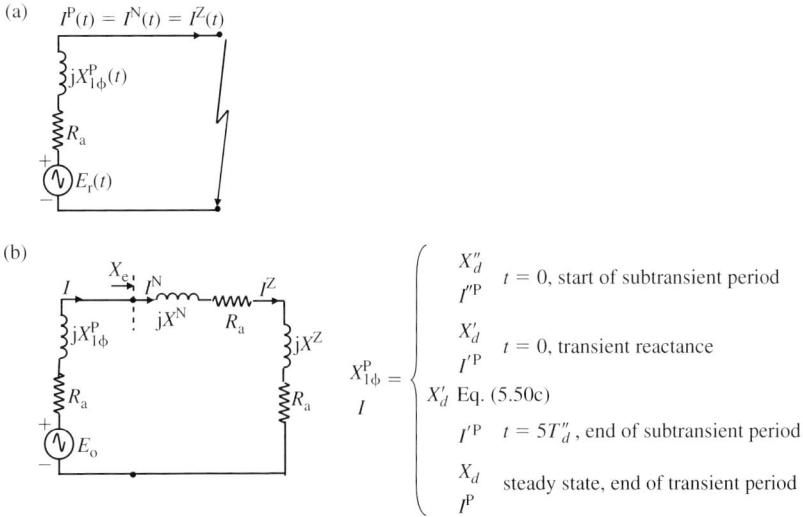

(b)

$$X^P_{1\phi} = \begin{cases} \dfrac{X''_d}{I''^P} & t = 0, \text{ start of subtransient period} \\[2ex] \dfrac{X'_d}{I'^P} & t = 0, \text{ transient reactance} \\[2ex] X'_d \text{ Eq. (5.50c)} \\[1ex] I'^P & t = 5T''_d, \text{ end of subtransient period} \\[2ex] \dfrac{X_d}{I^P} & \text{steady state, end of transient period} \end{cases}$$

Figure 5.11 Synchronous machine equivalent circuits for a one-phase to earth short-circuit fault at machine terminals: (a) transient PPS time-dependent equivalent circuit and (b) fixed impedance equivalent circuits of various time instants

Substituting Equation (5.50a) into Equation (5.51a), we obtain

$$I^P(t) = I^N(t) = I^Z(t) = \frac{E_r(t)}{jX^P_{(1\phi)}(t)} \tag{5.51b}$$

Figure 5.11(a) shows the transient PPS equivalent circuit that satisfies Equation (5.51b) with the reinstated machine stator resistance for completeness. Figure 5.11(b) shows equivalent circuits using the fixed machine reactances of Equation (5.50d). The time constants of Equation (5.49) and the time-dependent machine reactance of Equation (5.50b) show, from a PPS current view point that, under a single-phase to earth short-circuit fault, the machine behaves as if it has 'seen' a balanced three-phase short-circuit through an external reactance equal to $X_e = X^N + X^Z$.

Figure 5.11 shows the resultant symmetrical component equivalent circuits with the machine PPS reactance connected in series with the machine NPS and ZPS reactances to satisfy the single-phase short-circuit fault condition.

ZPS reactance and resistance

The machine ZPS impedance to the flow of ZPS armature currents applies where the stator winding is star connected and the neutral is not isolated. Recalling the spacial distribution of the stator phases and that ZPS currents are equal in magnitude and phase, these currents produce no net power frequency MMF wave across the air gap. Therefore, the ZPS impedance to the flow of ZPS currents is only due to some stator winding slot leakage flux which is greatly dependent on the machine coil pitch design. The ZPS reactance is always smaller than the stator leakage reactance X_σ and may be less than half of it depending on coil pitch.

Since there is no power frequency stator MMF wave across the air gap, the ZPS reactance is independent of the rotor's rotation. It is therefore constant and has the same value under subtransient, transient and steady state conditions as illustrated in Figure 5.11(b). The ZPS resistance can be assumed equal to the stator or armature ac resistance.

Simplified machine short-circuit current equations

For maximum dc current offset, the peak current envelope of Equation (5.49a) at any instant of time is given by

$$
\hat{i}_r(t) = 3\sqrt{2}E_o \left[\frac{1}{X_d + X^N + X^Z} + \left(\frac{1}{X_d' + X^N + X^Z} - \frac{1}{X_d + X^N + X^Z} \right) e^{-t/T_{d(1\phi)}'} \right.
$$

$$
+ \left(\frac{1}{X_d'' + X^N + X^Z} - \frac{1}{X_d' + X^N + X^Z} \right) e^{-t/T_{d(1\phi)}''}
$$

$$
\left. + \frac{1}{X_d'' + X^N + X^Z} e^{-t/T_{d(1\phi)}} \right]
\tag{5.52}
$$

5.4.5 Unbalanced two-phase to earth short-circuit faults

Consider a two-phase to earth short-circuit fault at the machine terminals involving phases y and b with the machine initially running at synchronous speed, rated terminal voltage and on open circuit. The machine neutral is assumed solidly earthed and the constraint equations that define the two-phase-to-earth short-circuit are given by

$$
e_y = e_b = 0
\tag{5.53a}
$$

$$
i_r = 0
\tag{5.53b}
$$

Transforming Equation (5.53) to the $dq0$ reference using Equation (5.7b), we obtain

$$
i_d \cos(\omega_s t + \theta_o) - i_q \sin(\omega_s t + \theta_o) + i_o = 0
\tag{5.54a}
$$

$$
e_d \sin(\omega_s t + \theta_o) = -e_q \cos(\omega_s t + \theta_o)
\tag{5.54b}
$$

$$
e_o = \frac{e_d}{2 \cos(\omega_s t + \theta_o)}
\tag{5.54c}
$$

As in the previous short-circuit cases, Equations (5.54) and (5.11) are theoretically sufficient to obtain the solution of currents under a two-phase to earth short-circuit fault at the machine terminals. However, the resulting equations are very complex and non-linear and a closed form solution is extremely unwieldy. Instead, a step-by-step simplification process can be used to derive the various components of the short-circuit currents. For practical calculations of machine short-circuit currents involving the power frequency and dc components only assuming that $X_d'' = X_q''$, it can be shown that the phase y and phase b short-circuit currents are given by

$$
\begin{aligned}
i_y(t) \approx \frac{3\sqrt{2}E_o}{2} &\left[\frac{-1}{(X_d'' + 2X^Z)} \frac{(X_d'' + X_e)}{X_{(2\phi-E)}^P(t)} \cos(\omega_s t + \theta_o) \right. \\
&\left. + \frac{1}{(X_d'' + 2X^Z)} \cos\theta_o e^{-t/T_{a1(2\phi-E)}} \right] \\
+ \frac{\sqrt{3}\sqrt{2}E_o}{2} &\left[\frac{1}{X_d''} \frac{(X_d'' + X_e)}{X_{(2\phi-E)}^P(t)} \sin(\omega_s t + \theta_o) - \frac{1}{X_d''} \sin\theta_o e^{-t/T_{a2(2\phi-E)}} \right]
\end{aligned}
$$

$$(5.55a)$$

$$
\begin{aligned}
i_b(t) \approx \frac{3\sqrt{2}E_o}{2} &\left[\frac{-1}{(X_d'' + 2X^Z)} \frac{(X_d'' + X_e)}{X_{(2\phi-E)}^P(t)} \cos(\omega_s t + \theta_o) \right. \\
&\left. + \frac{1}{(X_d'' + 2X^Z)} \cos\theta_o e^{-t/T_{a1(2\phi-E)}} \right] \\
+ \frac{\sqrt{3}\sqrt{2}E_o}{2} &\left[\frac{-1}{X_d''} \frac{(X_d'' + X_e)}{X_{(2\phi-E)}^P(t)} \sin(\omega_s t + \theta_o) + \frac{1}{X_d''} \sin\theta_o e^{-t/T_{a2(2\phi-E)}} \right]
\end{aligned}
$$

$$(5.55b)$$

The earth fault current is the sum of phase y and phase b currents, hence

$$
i_E(t) = \frac{3\sqrt{2}E_o}{(X_d'' + 2X^Z)} \left[-\frac{(X_d'' + X_e)}{X_{(2\phi-E)}(t)} \cos(\omega_s t + \theta_o) + \cos\theta_o e^{-t/T_{a1(2\phi-E)}} \right]
$$

$$(5.55c)$$

where

$$
\begin{aligned}
\frac{1}{X_{(2\phi-E)}(t)} = \frac{1}{X_d + X_e} &+ \left(\frac{1}{X_d' + X_e} - \frac{1}{X_d + X_e} \right) e^{-t/T_{d(2\phi-E)}'} \\
&+ \left(\frac{1}{X_d'' + X_e} - \frac{1}{X_d' + X_e} \right) e^{-t/T_{d(2\phi-E)}''}
\end{aligned}
$$

$$(5.56a)$$

$$
X_e = \frac{X^N X^Z}{X^N + X^Z}
$$

$$(5.56b)$$

$$
T_{d(2\phi-E)}' = \frac{X_d' + X_e}{X_d + X_e} T_{do}' \qquad T_{d(2\phi-E)}'' = \frac{X_d'' + X_e}{X_d' + X_e} T_{do}''
$$

$$(5.57a)$$

$$
T_{a1(2\phi-E)} = \frac{X^N + 2X^Z}{\omega_s(3R_a)} \qquad T_{a2(2\phi-E)} = \frac{X^N}{\omega_s R_a}
$$

$$(5.57b)$$

X^N is the machine NPS reactance under a two-phase to earth short-circuit fault condition which, in the general case when $X''_d \neq X''_q$, is given by

$$X^N = \frac{X''_d X''_q + \sqrt{X''_d X''_q (X''_d + 2X^Z)(X''_q + 2X^Z)}}{X''_d + X''_q + 2X^Z} \tag{5.58a}$$

The limiting values, i.e. the lower and upper bounds of Equation (5.58a) can be found by letting $X^Z \to 0$ and $X^Z \to \infty$, respectively. Thus, X^N is bound as follows

$$\frac{2X''_d X''_q}{X''_d + X''_q} < X^N < \sqrt{X''_d X''_q} \tag{5.58b}$$

The lower limit is the same as that found in the case of a three-phase fault given in Equation (5.33b). Also, the upper limit is the same as that found in the case of a two-phase fault given in Equation (5.43b). In practice, the differences between the various expressions for X^N, that depend on the short-circuit fault type, are negligible. The value of X^N calculated from the arithmetic average of X''_d and X''_q given in Equation (5.46) is sufficiently accurate for practical short-circuit analysis purposes. It should be noted that if we let $X''_d = X''_q$ in Equation (5.58a), we obtain, as we should expect, $X^N = X''_d$. Equations (5.55a) and (5.55b) of phase y and phase b short-circuit currents are quite different from those obtained for all short-circuit fault types considered previously. Each phase current now consists of two asymmetrical current components and each one of these consists of a power frequency ac component and a dc component. As expected, at $t = 0$, the sum of the ac and dc components of each asymmetrical current is zero. The earth fault current, on the other hand, consists of one asymmetrical current with only one power frequency ac component and one dc component as the other terms cancel out when the phase currents are added together. It is important to note that the two dc components have different initial magnitudes and different armature time constants.

The maximum dc offset occurs at $\theta_o = 0$ for the first dc component but at $\theta_o = \pi/2$ for the second dc current. It can be shown the value of θ_o that results in maximum asymmetrical phase y and phase b currents is given by

$$\theta_{o(y)} = -\theta_{o(b)} = -\tan^{-1} \left[\frac{1}{\sqrt{3}} \left(1 + \frac{2X^Z}{X''_d} \right) \exp \left\{ -t \left(\frac{1}{T_{a2(2\phi-E)}} - \frac{1}{T_{a1(2\phi-E)}} \right) \right\} \right] \tag{5.59a}$$

Typically $X''_d \approx 2X^Z$ hence at $t = 0$, $\theta_{o(y)} = -49.1°$ and $\theta_{o(b)} = 49.1°$. Also, for a typical 50 Hz synchronous machine with $X''_d = 0.2$ pu, $X^Z = 0.1$ pu and $R_a = 0.002$ pu, the armature dc time constants calculated using Equation (5.57b) are $T_{a1(2\phi-E)} = 0.2122$ s and $T_{a2(2\phi-E)} = 0.318$ s. Thus, using Equation (5.59a), we obtain $\theta_{o(y)} = -49.87°$, $(\theta_{o(b)} = 49.87°)$ at $t = 10$ ms and $\theta_{o(y)} = -53.1°$ $(\theta_{o(b)} = 53.1°)$ at $t = 50$ ms.

From Equation (5.56a) of the machine equivalent reactance under a two-phase to earth short-circuit, we distinguish four machine fixed impedance values. At

$t = 0$, i.e the start of the subtransient period, $X^{\mathrm{p}}_{(2\phi-\mathrm{E})} = X''_d + X_{\mathrm{e}}$. At $t = 0$, and neglecting the subtransient component, $X^{\mathrm{p}}_{(2\phi-\mathrm{E})} = X'_d + X_{\mathrm{e}}$. At $t \geq 5T'_{d(2\phi-\mathrm{E})}$, i.e. end of transient period, $X^{\mathrm{p}}_{(2\phi-\mathrm{E})} = X_d + X_{\mathrm{e}}$. The machine reactance at the end of the subtransient period, i.e. at $t \approx 5T''_{d(2\phi-\mathrm{E})}$, is given by

$$X^{\mathrm{p}}_{(2\phi-\mathrm{E})} = \frac{X'_d + X_{\mathrm{e}}}{\frac{X'_d + X_{\mathrm{e}}}{X_d + X_{\mathrm{e}}}[1 - \exp(-5T''_{d(2\phi-\mathrm{E})}/T'_{d(2\phi-\mathrm{E})})] + \exp(-5T''_{d(2\phi-\mathrm{E})}/T'_{d(2\phi-\mathrm{E})})}$$

$$(5.59\mathrm{b})$$

In summary, the four fixed reactances are given by

$$X^{\mathrm{p}}_{(2\phi-\mathrm{E})} = \begin{cases} X''_d + X_{\mathrm{e}} & t = 0, \text{ start of subtransient period} \\ \text{Equation (5.59b)} & t = 5T''_{d(2\phi-\mathrm{E})}, \text{ end of subtransient period} \\ X'_d + X_{\mathrm{e}} & t = 0, \text{ neglecting subtransient component} \\ X_d + X_{\mathrm{e}} & t \geq 5T'_{d(2\phi-\mathrm{E})}, \text{ steady state} \end{cases}$$

$$(5.59\mathrm{c})$$

From Equations (5.55a) and (5.55b), we can express the instantaneous power frequency component of the short-circuit currents as $i_y(t) = \mathrm{Real}[I_y(t)]$ and $i_b(t) = \mathrm{Real}[I_b(t)]$ where $I_y(t)$ and $I_b(t)$ are the phase y and phase b complex instantaneous currents given by

$$I_y(t) = \frac{(X''_d + X_{\mathrm{e}})}{2X^{\mathrm{p}}_{(2\phi-\mathrm{E})}(t)}\left[\frac{-3\sqrt{2}E_{\mathrm{o}}}{(X''_d + 2X^Z)}e^{\mathrm{j}(\omega_s t + \theta_{\mathrm{o}})} + \frac{\sqrt{3}\sqrt{2}E_{\mathrm{o}}}{X''_d}e^{\mathrm{j}(\omega_s t + \theta_{\mathrm{o}} - \pi/2)}\right]$$

or

$$I_y(t) = \frac{(X''_d + X_{\mathrm{e}})}{2X^{\mathrm{p}}_{(2\phi-\mathrm{E})}(t)}\left[\frac{-3E_{\mathrm{r}}(t)}{\mathrm{j}(X''_d + 2X^Z)} - \frac{\sqrt{3}E_{\mathrm{r}}(t)}{X''_d}\right] \qquad (5.60\mathrm{a})$$

$$I_b(t) = \frac{(X''_d + X_{\mathrm{e}})}{2X^{\mathrm{p}}_{(2\phi-\mathrm{E})}(t)}\left[\frac{-3\sqrt{2}E_{\mathrm{o}}}{(X''_d + 2X^Z)}e^{\mathrm{j}(\omega_s t + \theta_{\mathrm{o}})} - \frac{\sqrt{3}\sqrt{2}E_{\mathrm{o}}}{X''_d}e^{\mathrm{j}(\omega_s t + \theta_{\mathrm{o}} - \pi/2)}\right]$$

or

$$I_b(t) = \frac{(X''_d + X_{\mathrm{e}})}{2X^{\mathrm{p}}_{(2\phi-\mathrm{E})}(t)}\left[\frac{-3E_{\mathrm{r}}(t)}{\mathrm{j}(X''_d + 2X^Z)} + \frac{\sqrt{3}E_{\mathrm{r}}(t)}{X''_d}\right] \qquad (5.60\mathrm{b})$$

where $E_{\mathrm{r}}(t) = \sqrt{2}E_{\mathrm{o}}\,e^{\mathrm{j}(\omega_s t + \theta_{\mathrm{o}} + \pi/2)} = E_{\mathrm{r}}e^{\mathrm{j}(\omega_s t)}$ and $E_{\mathrm{r}} = \mathrm{j}\sqrt{2}E_{\mathrm{o}}\,e^{\mathrm{j}\theta_{\mathrm{o}}}$. The transient PPS, NPS and ZPS currents of the complex phase currents $I_y(t)$ and $I_b(t)$ of Equations (5.60) are calculated using

$$I^{\mathrm{P}}(t) = \frac{1}{3}[hI_y(t) + h^2I_b(t)] \quad I^{\mathrm{N}}(t) = \frac{1}{3}[h^2I_y(t) + hI_b(t)] \quad I^Z(t) = \frac{1}{3}[I_y(t) + I_b(t)]$$

$$(5.60\mathrm{c})$$

Substituting Equations (5.60a) and (5.60b) in Equation (5.60c), and using Equation (5.56b) for X_e with $X^N = X''_d$, we obtain, after some algebraic manipulations, which the reader is encouraged to prove, the following:

$$I^P(t) = \frac{E_r(t)}{jX^P_{(2\phi-E)}(t)} \tag{5.61a}$$

$$I^N(t) = \frac{-X^Z}{(X^N + X^Z)} \times \frac{E_r(t)}{jX^P_{(2\phi-E)}(t)} \tag{5.61b}$$

$$I^Z(t) = \frac{-X^N}{(X^N + X^Z)} \times \frac{E_r(t)}{jX^P_{(2\phi-E)}(t)} \tag{5.61c}$$

Figure 5.12(a) shows the transient PPS equivalent circuit that satisfies Equation (5.61a) with the reinstated stator resistance for completeness. Figure 5.12(b) shows equivalent circuits using the fixed machine reactances of Equation (5.59c). From Figure 5.12(b) and using complex phasor notation, the PPS phasor current is given by

$$I^P = \frac{E_o}{X^P + \frac{X^N X^Z}{X^N + X^Z}} \tag{5.62a}$$

where the value of the PPS reactance X^P is as shown in Figure 5.12(b) and depends on the calculation time instant. The NPS and ZPS phasor currents are given by

$$I^N = \frac{-X^Z}{X^N + X^Z} \times I^P \tag{5.62b}$$

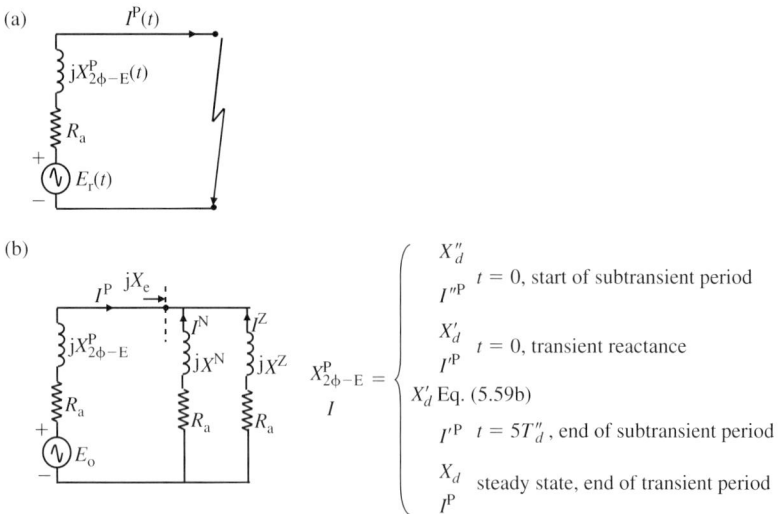

Figure 5.12 Synchronous machine equivalent circuits for a two-phase to earth short-circuit fault at machine terminals: (a) transient PPS time-dependent equivalent circuit and (b) fixed impedance equivalent circuits of various time instants

and

$$I^Z = \frac{-X^N}{X^N + X^Z} \times I^P \tag{5.62c}$$

The machine equivalent reactance of Equation (5.56a) and the time constants of Equation (5.57a) show that from a PPS current view point under a two-phase to earth short-circuit fault, the machine behaves practically as if it has 'seen' a balanced three-phase short circuit through an external impedance X_e. This impedance is given by the parallel combination of the NPS and ZPS machine reactances given in Equation (5.56b).

5.4.6 Modelling the effect of initial machine loading

Machine internal voltages

The short-circuit current equations already presented cover all cases of short-circuit currents at the machine terminals assuming the machine is initially unloaded that is the internal machine rms voltage is equal to E_o. However, the equations can be slightly modified to represent the situation where the machine is initially loaded by accounting for its prefault active and reactive power outputs as well as its terminal voltage. Because the machine is initially only producing a balanced set of three-phase voltages into a balanced network, the NPS and ZPS terminal voltages and output currents are all zero. Similarly, the initial machine active and reactive power outputs are PPS MW and PPS MVAr quantities. Assuming that the initial machine terminal voltage, active and reactive power outputs are V_t, P and Q, the machine stator current I can be calculated using the apparent power equation $P + jQ = V_t I^*$. We have already established that the machine PPS reactance X^P can be equated to one of four reactances depending on the calculation time instant after the occurrence of the fault. Therefore, using Figure 5.13, the internal machine voltages that correspond to these four reactances, for a given terminal voltage V_t, are E'', E'(at $t = 0$), E'(at $t \approx 5T''_d$) and E, respectively, and are given by

$$E'' = V_t + (R_a + jX''_d) \left(\frac{P - jQ}{V_t^*} \right) \tag{5.63a}$$

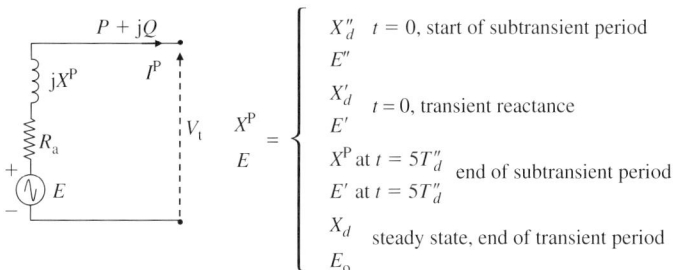

Figure 5.13   PPS equivalent circuits for an initially loaded synchronous machine

$$E' = V_t + (R_a + jX'_d)\left(\frac{P - jQ}{V_t^*}\right) \tag{5.63b}$$

$$E'(t \approx 5T''_d) = V_t + [R_a + jX^P(t \approx 5T''_d)]\left(\frac{P - jQ}{V_t^*}\right) \tag{5.63c}$$

$$E = V_t + (R_a + jX_d)\left(\frac{P - jQ}{V_t^*}\right) \tag{5.63d}$$

Clearly, when the machine is initially unloaded, $I = 0$ and $E'' = E' = E = E_o = V_t$.

Machine short-circuit currents

Using Equations (5.63) for E'', E'(at $t = 0$) or E'(at $t \approx 5T''_d$) and E, the equations that describe the machine peak short-circuit current envelope for each fault type are given below.

Three-phase short circuit

Using Equation (5.40a) and neglecting the double-frequency component, we have

$$\hat{i}_r(t) = \sqrt{2}\left[\frac{E}{X_d} + \left(\frac{E'}{X'_d} - \frac{E}{X_d}\right)e^{-t/T'_{d(3\phi)}} + \left(\frac{E''}{X''_d} - \frac{E'}{X'_d}\right)e^{-t/T''_{d(3\phi)}} + \frac{E''}{X''_d}e^{-t/T_{a(3\phi)}}\right] \tag{5.64}$$

Two-phase short circuit

Using Equation (5.47), we have

$$\hat{i}_y(t) = \sqrt{3}\sqrt{2}\left[\frac{E}{X_d + X^N} + \left(\frac{E'}{X'_d + X^N} - \frac{E}{X_d + X^N}\right)e^{-t/T'_{d(2\phi)}} \right.$$
$$\left. + \left(\frac{E''}{X''_d + X^N} - \frac{E'}{X'_d + X^N}\right)e^{-t/T''_{d(2\phi)}} + \frac{E''}{X''_d + X^N}e^{-t/T_{a(2\phi)}}\right] \tag{5.65}$$

One-phase to earth short circuit

Using Equation (5.52), we have

$$\hat{i}_r(t) = 3\sqrt{2}\left[\frac{E}{X_d + X^N + X^Z} + \left(\frac{E'}{X'_d + X^N + X^Z} - \frac{E}{X_d + X^N + X^Z}\right)e^{-t/T'_{d(1\phi)}} \right.$$
$$+ \left(\frac{E''}{X''_d + X^N + X^Z} - \frac{E'}{X'_d + X^N + X^Z}\right)e^{-t/T''_{d(1\phi)}}$$
$$\left. + \frac{E''}{X''_d + X^N + X^Z}e^{-t/T_{a(1\phi)}}\right] \tag{5.66}$$

Two-phase to earth short circuit

In the ac components of Equations (5.55a) and (5.55b), we simply replace the term $E_o/X_{(2\phi-E)}(t)$, using Equation (5.56a), with the following term

$$\frac{E_o}{X_d + X_e} + \left(\frac{E'}{X'_d + X_e} - \frac{E}{X_d + X_e} \right) e^{-t/T'_{d(2\phi-E)}}$$

$$+ \left(\frac{E''}{X''_d + X_e} - \frac{E'}{X'_d + X_e} \right) e^{-t/T''_{d(2\phi-E)}} \qquad (5.67)$$

In the dc components of Equations (5.55a) and (5.55), the term E_o should be replaced with E''.

5.4.7 Effect of AVRs on short-circuit currents

Discussion of the function, design, analysis or tuning of AVRs is the subject of power system dynamics, stability and control. In our book, we simply mention that the AVR generally attempts to control the machine terminal voltage by sensing its variations from a given set point or target and causing an increase or a decrease in the excitation or field voltage. In the case of short-circuit faults on the host network to which a synchronous machine is connected, the machine terminal voltage can see a significant drop during the fault period depending on the electrical distance, i.e. impedance to the fault point. The question we are interested in is what effect can the AVR have, if any, on the machine short-circuit current for a three-phase fault at the machine terminals from an unloaded initial condition? We recall that we have already determined the machine short-circuit current assuming no AVR action with $\Delta e_{fd} = 0$. To determine the machine current due to a change in field voltage Δe_{fd}, representing automatic AVR action, we recall Equation (5.17a), $\Delta \psi_d(s) = -X_d(s)\Delta i_d(s) + G_d(s)\Delta e_{fd}(s)$ from which we obtain

$$\Delta i_d(s)\Big|_{\text{caused by}\,\Delta efd} = \frac{G_d(s)}{X_d(s)}\Delta e_{fd}(s) \qquad (5.68a)$$

and from Equation (5.15a)

$$\Delta i_q(s) = 0 \qquad (5.68b)$$

Equations (5.18a) for $X_d(s)$ and (5.18b) for $G_d(s)$ were derived for the general case of a machine with a damper winding. For a machine without a damper winding, a similar method can be used to derive these expressions and the reader is encouraged to do so. However, for us, we will directly derive the expressions for $X_d(s)$ and $G_d(s)$ from the general ones of Equations (5.18a), (5.18b) and (5.19a). To do so, we represent the absence of the damper winding by substituting $X_{kd} = 0$ and letting $R_{kd} \to \infty$. After a little algebra, the results can be easily shown as

$$X_d(s) = X_d - \frac{X_{ad}^2 s}{R_{fd} + sX_{ffd}} \qquad (5.69a)$$

and

$$G_d(s) = \frac{X_{ad}}{R_{fd} + sX_{ffd}}$$
(5.69b)

Substituting Equations (5.69a) and (5.69b) into Equation (5.68a) and using $X_{ffd} = X_{fd} + X_{ad}$ and $X_d = X_{ad} + X_\sigma$ from Equation (5.12), we obtain

$$\Delta i_d(s) = \frac{X_{ad}\Delta e_{fd}(s)}{R_{fd}X_d} \cdot \frac{1}{1 + \frac{1}{R_{fd}}\left(X_{fd} + \frac{1}{\frac{1}{X_\sigma} + \frac{1}{X_{ad}}}\right)s}$$

or using

$$\Delta E_{fd}(s) = \frac{X_{ad}}{R_{fd}}\Delta e_{fd}(s)$$

and Equations (5.22b) for T_4 and (5.23c) for T_d', we obtain

$$\Delta i_d(s)\Big|_{\text{due to }\Delta E_{fd}} = \frac{1}{X_d} \times \frac{\Delta E_{fd}(s)}{1 + sT_{d(3\phi)}'}$$
(5.70a)

For simplicity, we consider an instantaneous or a step change in field voltage ΔE_{fd} where $\Delta E_{fd}(s) = \Delta E_{fd}/s$. Therefore, Equation (5.70a) can be written as

$$\Delta i_d(s)\Big|_{\text{due to }\Delta E_{fd}} = \frac{\Delta E_{fd}}{X_d}\frac{1}{s(1 + sT_{d(3\phi)}')} = \frac{\Delta E_{fd}}{X_d}\left(\frac{1}{s} - \frac{1}{s + 1/T_{d(3\phi)}'}\right)$$

and taking the inverse Laplace transform, we obtain

$$\Delta i_d(t)\Big|_{\text{due to }\Delta E_{fd}} = \frac{\Delta E_{fd}}{X_d}(1 - e^{-t/T_{d(3\phi)}'})$$
(5.70b)

Also, from Equation (5.68b), we have

$$\Delta i_q(t) = 0$$
(5.70c)

Equation (5.70b) shows that an instantaneous increase in E_{fd} causes an exponential increase in short-circuit current with a time constant equal to $T_{d(3\phi)}'$. That is, despite the very fast change in E_{fd}, the resultant change in machine current is slowed down or delayed by the d-axis short-circuit transient time constant $T_{d(3\phi)}'$ which falls in the range of 0.5–2 s. For example, for a typical value of $T_{d(3\phi)}' = 1$ s, the factor $[1 - e^{-t/T_{d(3\phi)}'}]$ is equal to 0.11 at 120 ms. This indicates some change in current towards the end of the subtransient period, i.e. at around $t = 120$ ms. However, in practice, even for modern fast excitation control systems such as brushless ac rotating exciters or static exciters, the change in field voltage will itself occur after a definite time delay. Thus, the effective change in short-circuit current will only begin to occur after the subtransient period. In practice, a static excitation control system will not be able to continue to operate by maintaining free thyristor firing if the terminal voltage, where it derives its supply, drops below around 0.2–0.3 pu. In other words, the subtransient component of short-circuit current will have decayed to zero by the time the effect of the AVR is felt through the machine. Therefore, the effect of the AVR is to cause a possible increase in the transient and steady state components of short-circuit current. It should be noted that the change in field voltage has no effect on the dc component of short-circuit current as shown in Equations (5.70b) and (5.70c).

We have now obtained the machine current changes due to both the short-circuit itself and the change in field voltage. Therefore, using the superposition theorem, the total machine short-circuit current change is obtained by the addition of the two values. Thus, examining the effect on the ac component of short-circuit current after the subtransient component has decayed to zero, Equation (5.35a) can be rewritten as

$$
\Delta i_d(t)\Big|_{\text{due to short circuit}} = \sqrt{2}E_o\left[\frac{1}{X_d} + \left(\frac{1}{X_d'} - \frac{1}{X_d}\right)e^{-t/T_{d(3\phi)}'}\right] \qquad (5.71a)
$$

Therefore, using Equations (5.70b) and (5.71a), the total change in machine current is given by

$$
\Delta i_d(t)\Big|_{\text{total}} = \frac{\Delta E_{fd}}{X_d}\left(1 - e^{-t/T_{d(3\phi)}'}\right) + \sqrt{2}E_o\left[\frac{1}{X_d} + \left(\frac{1}{X_d'} - \frac{1}{X_d}\right)e^{-t/T_{d(3\phi)}'}\right] \qquad (5.71b)
$$

A simplification can be made by using $\Delta E_{fd} = E_{fd} - E_{fdo}$ and since the machine is initially unloaded, we have $E_{fdo} = \sqrt{2}E_o$. Therefore, Equation (5.71b) simplifies to

$$
\Delta i_d(t)\Big|_{\text{total}} = \frac{E_{fd}}{X_d}\left(1 - e^{-t/T_{d(3\phi)}'}\right) + \frac{\sqrt{2}E_o}{X_d'}e^{-t/T_{d(3\phi)}'} \qquad (5.72)
$$

Equation (5.72) shows two transient components; one is a rising exponential and one is decaying exponential and both have the same time constant $T_{d(3\phi)}'$. The decaying exponential component represents, as expected, the decay in flux due to induced currents in the field winding whereas the rising exponential component is due to field voltage change by AVR action. It should be noted that time t represents the beginning of the transient period. Clearly, whether the short-circuit current continues to decrease, remains constant or start to increase is dependent on the magnitude of E_{fd} caused by AVR action. This depends on the extent of voltage drop at the machine terminals 'seen' by the AVR and this in turn depends on the electrical distance to the fault point. If $E_{fd}/X_d = \sqrt{2}E_o/X_d'$, the transient current component due to the AVR effect cancels out the decaying exponential and the short-circuit current remains constant at $\sqrt{2}E_o/X_d'$. If $E_{fd}/X_d > \sqrt{2}E_o/X_d'$, the current component due to effect of the AVR is greater than the decaying exponential component and the short-circuit current will increase. The converse is true for $E_{fd}/X_d < \sqrt{2}E_o/X_d'$. For a typical turbo-generator running at no-load and rated

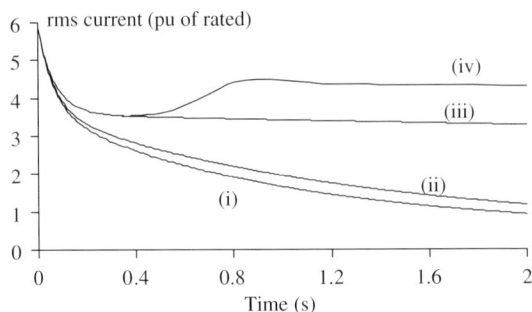

(i) no AVR,   (ii) with AVR and $\dfrac{E_{fd}}{X_d} < \dfrac{\sqrt{2}E_o}{X'_d}$

(iii) with AVR and $\dfrac{E_{fd}}{X_d} = \dfrac{\sqrt{2}E_o}{X'_d}$,   (iv) with AVR and $\dfrac{E_{fd}}{X_d} > \dfrac{\sqrt{2}E_o}{X'_d}$

Figure 5.14   Effect of synchronous machine AVR on short-circuit current

voltage, $E_o = 1$ pu, $X_d = 1.5$ pu and $X'_d = 0.25$ pu, $\sqrt{2}E_o/X'_d = 5.65$ pu. For high gain static excitation control systems, a small drop in terminal voltage of only a few percent is sufficient to cause a field voltage change to maximum ceiling, typically 7–8 pu where 1 pu field voltage is that which produces a 1 pu terminal voltage on open circuit, giving $E_{fd}/X_d = 5.34$ pu. The possible effects of the AVR on the machine short-circuit current are illustrated in Figure 5.14. It is worth noting that the phase r current after the subtransient contribution has decayed to zero is calculated by substituting Equations (5.70c) and (5.72) into

$$\Delta i_r(t) = \Delta i_d(t)\cos(\omega t + \theta_o) - \Delta i_q(t)\sin(\omega t + \theta_o)$$

giving

$$\Delta i_r(t) = \left[\frac{E_{fd}}{X_d} \left(\frac{\sqrt{2}E_o}{X'_d} - \frac{E_{fd}}{X_d} \right) e^{-t/T'_{d(3\Phi)}} \right] \cos(\omega_s t + \theta_o) \tag{5.73}$$

5.4.8 Modelling of synchronous motors/compensators/condensers

Although in the preceding analysis we used the general term of synchronous machine, we nonetheless deliberately biased our attention towards the synchronous generator. However, synchronous motors are also used in industry essentially in two general applications; as motors driving large mechanical loads, e.g. compressors or as reactive power compensators and these are generally known as synchronous condensers in North America or synchronous compensators in the UK. The preceding analysis for synchronous generators applies equally to synchronous motors. The only difference being the initial loading conditions of the motor.

5.4.9 Examples

Example 5.1   A 50 Hz three-phase synchronous generator has the following data: rated apparent power $= 165$ MVA, rated power $= 132$ MW, rated voltage $= 15$ kV, $X_d = 2.04$ pu, $X_d' = 0.275$ pu, $X_d'' = 0.19$ pu, $X_q'' = 0.2$ pu, $X^Z = 0.095$ pu, $T_0' = 8.16$ s, $T_0'' = 0.058$ s, $R_a = 0.002$ pu. All parameters are in pu on 165 MVA rating. Calculate the rms and dc short-circuit currents for a three-phase, two-phase and one-phase to earth short-circuit faults at the machine terminals at $t = 0$, 10 and 90 ms. The machine is initially operating on open circuit and has a terminal voltage of 1 pu. Although in practice, the neutral point of the star-connected stator winding of such a generator would be earthed through a high resistance, assume in this example that the neutral is solidly earthed.

Three-phase short-circuit fault

$$T_d' = \frac{0.275}{2.04} \times 8.16 = 1.1 \text{ s} \quad T_d'' = \frac{0.19}{0.275} \times 0.058 = 0.04 \text{ s}$$

$$T_a = \frac{0.19}{2\pi 50 \times 0.002} = 0.302 \text{ s}$$

From Equation (5.40a), the rms fault current is given by

$$i_{rms}(t) = \sqrt{3}\sqrt{2}e^{-t/0.31}/(0.19 + 0.195)$$

and the dc fault current is given by

$$i_{dc}(t) = \frac{\sqrt{2}}{0.19}e^{-t/0.302}$$

The rms fault current at $t = 0$, 10 and 100 ms is equal to 5.26, 4.87 and 3.5 pu, respectively. The dc fault current at $t = 0$, 10 and 100 ms is equal to 7.44, 7.2 and 5.34 pu, respectively.

Two-phase short-circuit fault
From either Equation (5.43b) (or (5.46)), we have $X^N = \sqrt{0.19 \times 0.2} = 0.195$ pu.

$$T_d' = \frac{0.275 + 0.195}{2.04 + 0.195} \times 8.16 = 1.716 \text{ s}$$

$$T_d'' = \frac{0.19 + 0.195}{0.275 + 0.195} \times 0.058 = 0.0475 \text{ s} \quad T_a = \frac{0.195}{2\pi 50 \times 0.002} = 0.31 \text{ s}$$

From Equation (5.47), the rms and dc fault currents are

$$i_{rms}(t) = \sqrt{3}\left[\frac{1}{2.04+0.195} + \left(\frac{1}{0.275+0.195} - \frac{1}{2.04+0.195}\right)e^{-t/1.716}\right.$$
$$\left. + \left(\frac{1}{0.19+0.195} - \frac{1}{0.275+0.195}\right)e^{-t/0.0475}\right]$$

$$i_{dc}(t) = \sqrt{3}\sqrt{2}e^{-t/0.31}/(0.19+0.195)$$

The rms fault current at $t=0$, 10 and 100 ms is equal to 4.5, 4.33 and 3.62 pu, respectively. The dc fault current at $t=0$, 10 and 100 ms is equal to 6.36, 6.16 and 4.6 pu, respectively.

One-phase short-circuit fault
From either Equation (5.49e), where $X^N = \sqrt{(0.19+0.095/2)(0.2,+0.095/2)}$
$-0.095/2 = 0.1949$ pu or a value of 0.195 pu as calculated above is acceptable from a practical viewpoint:

$$T_d' = \frac{0.275+0.195+0.095}{2.04+0.195+0.095} \times 8.16 = 1.978\,s$$

$$T_d'' = \frac{0.19+0.195+0.095}{0.275+0.195+0.095} \times 0.058 = 0.0493\,s$$

$$T_a = \frac{0.19+0.195+0.095}{2\pi50 \times (3 \times 0.002)} = 0.2546\,s$$

From Equation (5.52), the rms and dc fault currents are

$$i_{rms}(t) = 3\left[\frac{1}{2.04+0.195+0.095}\right.$$
$$+ \left(\frac{1}{0.275+0.195+0.095} - \frac{1}{2.04+0.195+0.095}\right)e^{-t/1.978}$$
$$\left. + \left(\frac{1}{0.19+0.195+0.095} - \frac{1}{0.275+0.195+0.095}\right)e^{-t/0.0493}\right]$$

$$i_{dc}(t) = \frac{3\sqrt{2}}{0.19+0.195+0.095}e^{-t/0.2546}$$

The rms fault current at $t=0$, 10 and 100 ms is equal to 6.25, 6.06 and 5.24 pu, respectively. The dc fault current at $t=0$, 10 and 100 ms is equal to 8.84, 8.5 and 5.97 pu, respectively.

Example 5.2　Repeat the calculations for the three-phase and one-phase short-circuit faults in Example 5.1 assuming the machine is initially loaded and operating at rated power and power factor and a terminal voltage of 1 pu.

Using Equation (5.63), let $V_t = 1$ pu. The machine rated lagging power factor is equal to $132/165 = 0.8$ and the rated lagging reactive power output is equal to $132\,\text{MW} \times \tan(\cos^{-1}0.8) = 99\,\text{MVAr}$. The machine's real and reactive power output in pu on machine MVA rating are equal to $132/165 = 0.8$ pu and $99/165 = 0.6$ pu. Thus,

$$E'' = 1 + (0.002 + \text{j}0.19)\left(\frac{0.8 - \text{j}0.6}{1}\right) = 1.1257\,\text{pu}\angle 7.7°$$

$$E'_{t=0} = 1 + (0.002 + \text{j}0.275)\left(\frac{0.8 - \text{j}0.6}{1}\right) = 1.187\,\text{pu}\angle 10.7°$$

$$E = 1 + (0.002 + \text{j}2.04)\left(\frac{0.8 - \text{j}0.6}{1}\right) = 2.759\,\text{pu}\angle 36.2°$$

Using Equations (5.64) for a three-phase fault, we have

$$i_{\text{rms}}(t) = \frac{2.759}{2.04} + \left(\frac{1.187}{0.275} - \frac{2.759}{2.04}\right)\text{e}^{-t/1.1} + \left(\frac{1.1257}{0.19} - \frac{1.187}{0.275}\right)\text{e}^{-t/0.04}$$

and the dc fault current is given by

$$i_{\text{dc}}(t) = \frac{\sqrt{2} \times 1.1257}{0.19}\text{e}^{-t/0.302}$$

The rms fault current at $t = 0$, 10 and 100 ms are equal to 5.93, 5.54 and 4.2 pu, respectively. The dc fault current at $t = 0$, 10 and 100 ms is equal to 8.38, 8.1 and 6.0 pu.

Using Equations (5.66) for a one-phase to earth fault, we have

$$i_{\text{rms}}(t) = 3\left[\frac{2.759}{2.04 + 0.195 + 0.095}\right.$$

$$+ \left(\frac{1.187}{0.275 + 0.195 + 0.095} - \frac{2.759}{2.04 + 0.195 + 0.095}\right)\text{e}^{-t/1.978}$$

$$+ \left.\left(\frac{1.1257}{0.19 + 0.195 + 0.095} - \frac{1.187}{0.275 + 0.195 + 0.095}\right)\text{e}^{-t/0.0493}\right]$$

$$i_{\text{dc}}(t) = \frac{3\sqrt{2} \times 1.1257}{0.19 + 0.195 + 0.095}\text{e}^{-t/0.2546}$$

The rms fault current at $t = 0$, 10 and 100 ms is equal to 7.17, 7.02 and 6.39 pu, respectively. The dc fault current at $t = 0$, 10 and 100 ms is equal to 8.84, 8.5 and 5.97 pu, respectively.

Example 5.3   The generator of Example 5.1 is now assumed to be connected to a high voltage 132 kV busbar through a star–delta transformer that has a PPS leakage impedance equal to $(0.00334 + j0.2)$pu on 165 MVA. The star winding is the high voltage winding. The transformer's ZPS reactance is equal to 95% of the PPS reactance. Calculate the rms and dc short-circuit currents for three-phase and one-phase short-circuit faults on the 132 kV busbar assuming that the machine is initially operating on open circuit and the initial voltages at its terminals and on the 132 kV busbar are both 1 pu.

Three-phase short-circuit fault
The new short-circuit time constants can be calculated using Equations (5.38b) and (5.38c). The external impedance 'seen' by the machine up to the fault point on the transformer high voltage side is simply that of the transformer. However, in this example, Equations (5.39a) and (5.39b) can be used because the armature resistance of the machine and the transformer resistance are very small in comparison with the machine and transformer reactances. Thus

$$T'_{de} = \frac{0.275 + 0.2}{2.04 + 0.2} \times 8.16 = 1.73\,\text{s} \quad T''_{de} = \frac{0.19 + 0.2}{0.275 + 0.2} \times 0.058 = 0.0476\,\text{s}$$

From Equation (5.38e), we have

$$T_{ae} = \frac{(0.19 + 0.2)/2 + 0.2}{2\pi 50 \times (0.002 + 0.0034)} = 0.2328\,\text{s}$$

The rms and dc fault currents are calculated using Equation (5.40b),

$$i_{\text{rms}}(t) = \frac{1}{2.04 + 0.2} + \left(\frac{1}{0.275 + 0.2} - \frac{1}{2.04 + 0.2} \right) e^{-t/1.73}$$

$$+ \left(\frac{1}{0.19 + 0.2} - \frac{1}{0.275 + 0.2} \right) e^{-t/0.047}$$

and

$$i_{\text{dc}}(t) = \frac{\sqrt{2}}{0.19 + 0.2} e^{-t/0.2328}$$

The rms fault current at $t = 0$, 10 and 100 ms is equal to 2.56, 2.46 and 2.06 pu, respectively. The dc fault current at $t = 0$, 10 and 100 ms is equal to 3.626, 3.47 and 2.36 pu, respectively.

One-phase to earth short-circuit fault
The first step is to calculate the external impedance 'seen' by the machine up to the fault point on the transformer high voltage side. The PPS, NPS and ZPS networks for this fault condition are connected in series as shown in figure in Example 5.3(a). The equivalent external impedance seen by the machine is shown in figure in Example 5.3(b).

(a) Connection of sequence networks   (b) Equivalent

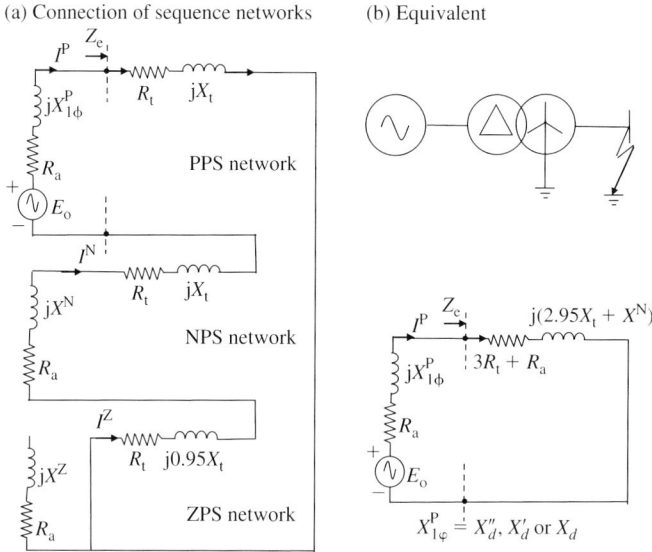

Thus, the external impedance is given by $Z_e = (3R_t + R_a) + j(2.95X_t + X^N)$. Thus, $R_e = 3 \times 0.00334 + 0.002 = 0.01202$ pu and $X_e = 2.95 \times 0.2 + 0.195 = 0.785$ pu. The short-circuit time constants can be calculated using Equations (5.38b) and (5.38c). However, we will use Equations (5.39a) and (5.39b) because the armature resistance and external resistance are very small in comparison with the machine's reactances and external reactance. Thus

$$T'_{de} = \frac{0.275 + 0.785}{2.04 + 0.785} \times 8.16 = 3.06\,\text{s} \quad T''_{de} = \frac{0.19 + 0.785}{0.275 + 0.785} \times 0.058 = 0.0533\,\text{s}$$

From Equation (5.38e), we have

$$T_{ae} = \frac{(0.19 + 0.2)/2 + 0.785}{2\pi 50 \times (0.002 + 0.01202)} = 0.2225\,\text{s}$$

The rms and dc fault currents are calculated using Equation (5.40b),

$$i_{rms}(t) = 3\left[\frac{1}{2.04 + 0.785} + \left(\frac{1}{0.275 + 0.785} - \frac{1}{2.04 + 0.785}\right)e^{-t/3.06}\right.$$
$$\left. + \left(\frac{1}{0.19 + 0.785} - \frac{1}{0.275 + 0.785}\right)e^{-t/0.0533}\right]$$

and

$$i_{dc}(t) = \frac{3\sqrt{2}}{0.19 + 0.785}e^{-t/0.2225}$$

The rms fault current at $t = 0$, 10 and 100 ms is equal to 3.07, 3.02 and 2.81 pu, respectively. The dc fault current at $t = 0$, 10 and 100 ms is equal to 4.35, 4.16 and 2.77 pu, respectively.

5.5 Determination of synchronous machines parameters from measurements

Although calculations of machine parameters are made by machine manufacturers at the design stage, factory or field tests are generally carried out on the built machine. These are to identify the machine parameters and confirm that they are within the guaranteed or declared design values which are typically $\pm 10\%$ or as agreed between the manufacturer and the customer. The reactances and time constants of the machine are determined from measurements as defined in IEC and IEEE standards.

5.5.1 Measurement of positive sequence reactances, positive sequence resistance and d-axis short-circuit time constants

Measurement and separation of ac and dc current components

Several parameters can be calculated from measurements of the stator short-circuit currents during a sudden three-phase short circuit at the machine terminals. These are the positive sequence or the d-axis subtransient, transient and steady state reactances, and the d-axis short-circuit subtransient, transient and armature (dc) time constants. The machine is on open circuit and running at rated speed just before the application of the simultaneous three-phase short-circuit fault. The unsaturated reactances and time constants are determined by performing tests at a few low values of prefault stator voltage, e.g. 0.1–0.4 pu. The saturated reactances are determined from tests at rated 1 pu prefault stator voltage. Oscillograms of the three-phase short-circuit currents are taken. Recalling Equation (5.40a).

$$\hat{i}_r(t) = \sqrt{2}\left[\frac{E_o}{X_d} + \left(\frac{E_o}{X'_d} - \frac{E_o}{X_d}\right)e^{-t/T'_{d(3\phi)}} + \left(\frac{E_o}{X''_d} - \frac{E_o}{X'_d}\right)e^{-t/T''_{d(3\phi)}} + \frac{E_o}{X''_d}e^{-t/T_{a(3\phi)}}\right]$$
(5.74a)

or

$$\hat{i}_r(t) = \hat{I}_{ac}(t) + i_{dc}(t)$$
(5.74b)

$$\hat{I}_{ac}(t) = \sqrt{2}I_d + \sqrt{2}\left(I'_d - I_d\right)e^{-t/T'_{d(3\phi)}} + \sqrt{2}\left(I''_d - I'_d\right)e^{-t/T''_{d(3\phi)}}$$
(5.75a)

and

$$i_{dc}(t) = \sqrt{2}I''_d e^{-t/T_{a(3\phi)}}$$
(5.75b)

Equation (5.75a) represents the envelope of the ac rms component of the short-circuit current where

$$I''_d = \frac{E_o}{X''_d} \quad I'_d = \frac{E_o}{X'_d} \quad I_d = \frac{E_o}{X_d}$$
(5.75c)

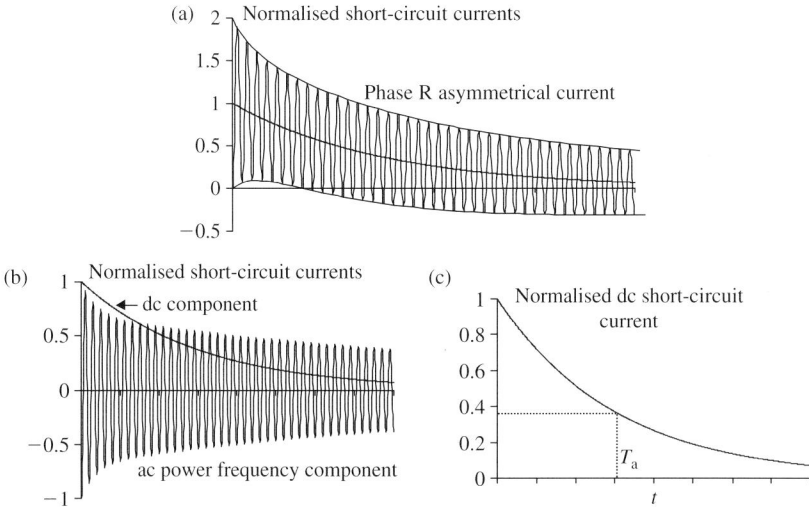

Figure 5.15 Measurement of short-circuit current and separation of ac and dc current components

To illustrate the process of determining the machine d-axis parameters, Figure 5.15(a) illustrates the measured phase r asymmetrical current. The first step is to separate the dc and ac components of the measured current. The dc component can be calculated as the algebraic half sum of the ordinates of the upper and lower envelopes of the current whereas the ac components can be similarly determined from the algebraic half difference of the upper and lower envelopes. In practice, experience shows that this manual process is dependent on the engineer doing the analysis and can be subject to some error whereas the use of numerical filters is both more accurate and consistent. The outcome of separating the ac and dc components of the short-circuit current is illustrated in Figure 5.15(b) and the dc component alone is shown in Figure 5.15(c).

The dc (armature) time constant

As a decaying exponential, the magnitude of the dc component drops from its initial value at $t = 0$ to $1/e = 0.36788$ of the initial value at $t = T_{a(3\phi)}$ where $T_{a(3\phi)}$ is the dc or stator or armature time constant. $T_{a(3\phi)}$ is indicated in Figure 5.15(c).

Steady state d-axis reactance

The peak envelope of the ac short-circuit current is shown in Figure 5.16(a). One method of determining the steady state reactance is from the oscillogram's current value after a sufficient time so that the transient current component has completely decayed or vanished. Therefore, with $\sqrt{2}I_d$ read from Figure 5.16(a), we have

$$X_d = \frac{E_o}{I_d} \qquad (5.76)$$

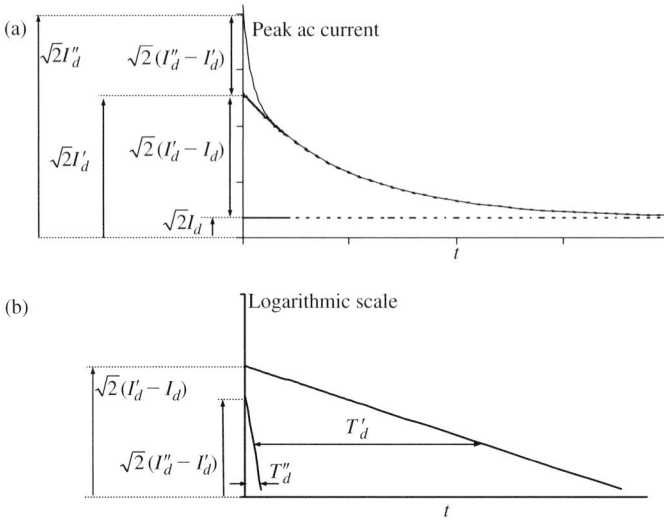

Figure 5.16 Measured ac current component plotted on linear and logarithmic scales: (a) peak ac current envelope and components and (b) time constants from current envelopes

An alternative method that is usually used by machine manufacturers is to calculate X_d from the ratio of the field current at rated short-circuit stator current to the air-gap field current at rated open-circuit stator voltage.

Transient reactance and transient short-circuit time constant

Using Figure 5.16(a), the machine current attributable to the transient reactance is determined from the transient envelope of the ac current after subtracting the steady state current value then the transient current result is plotted on linear-logarithmic coordinates as shown in Figure 5.16(b). The envelope of the transient current component is given by

$$\Delta I'_d(t) = \sqrt{2}(I'_d - I_d)e^{-t/T'_{d(3\phi)}} \tag{5.77a}$$

When this component is extrapolated back to zero time, it cuts the ordinate axis at $\Delta I'_d(t=0) = \sqrt{2}(I'_d - I_d)$ and this is read from the ordinate. Therefore, the transient current is calculated as follows:

$$I'_d = \frac{\Delta I'_d(t = 0)}{\sqrt{2}} + I_d \tag{5.77b}$$

The transient reactance can now be calculated as

$$X'_d = \frac{E_o}{I'_d} \tag{5.77c}$$

The transient time constant is the time required for the transient current component to drop to $1/e = 0.36788$ of its initial value as illustrated in Figure 5.16(b).

Subtransient reactance and subtransient short-circuit time constant

The machine current attributable to the subtransient reactance is determined from the upper envelope of the ac current after subtracting the transient current value. The subtransient current component is then plotted on linear-logarithmic coordinates as shown in Figure 5.16(b). The envelope of the subtransient current component is given by

$$\Delta I_d''(t) = \sqrt{2}(I_d'' - I_d')e^{-t/T_{d(3\phi)}''} \tag{5.78a}$$

When this component is extrapolated back to zero time, it cuts the ordinate axis at $\Delta I_d''(t=0) = \sqrt{2}(I_d'' - I_d')$ and this is read from the ordinate. Therefore, the subtransient current is calculated as follows:

$$I_d'' = \frac{\Delta I_d''(t=0)}{\sqrt{2}} + I_d' \tag{5.78b}$$

The subtransient reactance can now be calculated as

$$X_d'' = \frac{E_o}{I_d''} \tag{5.78c}$$

The subtransient time constant is the time required for the subtransient current component to drop to $1/e = 0.36788$ of its initial value as illustrated in Figure 5.16(b).

Transient reactance at the end of the subtransient period $t = 5T_d''$

The transient current component at the end of the subtransient period at $t = 5T_d''$ is given by $\Delta I_d'(t = 5T_d'') = \sqrt{2}[I_d'(t = 5T_d'') - I_d]$ and this is read from the ordinate of Figure 5.16(b). Therefore, the transient current at $t = 5T_d''$ is given by

$$I_d'(t = 5T_d'') = \frac{\Delta I_d'(t = 5T_d'')}{\sqrt{2}} + I_d \quad \text{and} \quad X_d'(t = 5T_d'') = \frac{E_o}{I_d'(t = 5T_d'')} \tag{5.78d}$$

Alternatively, Equation (5.37b) can be used to calculate $X_d'(t = 5T_d'')$.

The reactances and short-circuit time constants measured are usually the unsaturated values. Also, since the currents in three phases are measured, the above calculations are made for each phase and an average value of each parameter is taken.

PPS resistance

The total three-phase power during the three-phase short-circuit test is measured when the short-circuit current is equal to the rated machine current. The rated

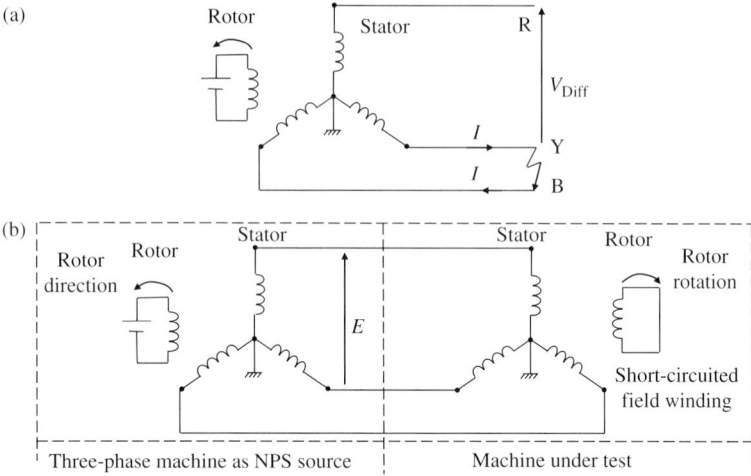

Figure 5.17　Measurement of synchronous machine NPS resistance and reactance: (a) sustained solid two-phase short-circuit and (b) three-phase NPS voltage source test

machine current is calculated from $I_{Rated} = MVA_{Rated}/(\sqrt{3} \times V_{Rated})$. The PPS resistance is given by

$$R^P = \frac{\text{Total losses under a three-phase short circuit}}{3 \times I_{Rated}^2} \tag{5.79}$$

5.5.2 Measurement of NPS impedance

The NPS reactance and resistance can be determined using one of two test methods. The first is by applying a solid two-phase sustained short circuit on any two phases when the machine is running at rated synchronous speed as illustrated in Figure 5.17(a). The short-circuit current $I(A)$, the voltage between the short-circuited phases and the healthy phase, $V_{Diff}(V)$, and the electric input power $P_{3\phi}(W)$ are measured. Using Equations (2.48a) to (2.48d) with $Z_F = 0$, it can be shown that the NPS impedance, reactance and resistance are given by

$$Z^N = \frac{V_{Diff}}{\sqrt{3} \times I} \, \Omega \quad X^N = \frac{P_{3\phi}}{\sqrt{3} \times I^2} \, \Omega \quad R^N = \sqrt{(Z^N)^2 - (X^N)^2} \, \Omega \tag{5.80}$$

The second test method is illustrated in Figure 5.17(b). It involves applying a three-phase voltage source having a NPS phase rotation RBY to the machine being tested with the machine running at rated speed in a PPS rotation RYB and its field winding short circuited, i.e. zero internal EMF. The reason for short-circuiting the field winding is that X^N, like X'_d and X''_d, is due to induced currents in the rotor and these currents must be able to flow unhindered. The slip of the machine being tested is therefore 200% and large double-frequency currents are induced in the damper windings. Because the NPS reactance is quite low, the applied test voltage must be very low, e.g. 0.02–0.2 pu, so as to avoid overheating. The current $I(A)$,

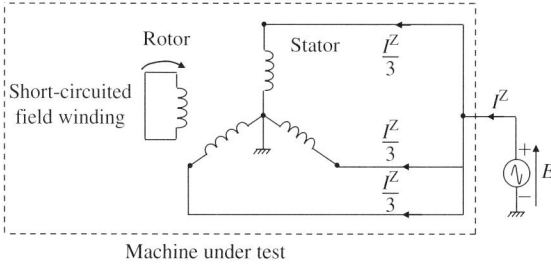

Figure 5.18 Measurement of synchronous machine ZPS resistance and reactance

the applied phase voltage $E(\text{V})$, and the electric input power $P(\text{W})$ are measured. The NPS reactance and resistance are given by

$$Z^{\text{N}} = \frac{E}{\sqrt{3} \times I} \,\Omega \quad R^{\text{N}} = \frac{P}{3 \times I^2} \,\Omega \quad X^{\text{N}} = \sqrt{(Z^{\text{N}})^2 - (R^{\text{N}})^2} \,\Omega \qquad (5.81)$$

The results obtained from calculations from the three phases are averaged.

5.5.3 Measurement of zero sequence impedance

The zero sequence reactance and resistance can be determined using one of two test methods. In the first, a solid two-phase to earth sustained short-circuit fault is applied with the machine initially running at rated speed and a very low stator voltage to avoid rotor overheating or vibration. The voltage to neutral (V) on the open healthy phase and the earth (neutral) current (A) are measured. It can be shown that, using Equations (2.57a) and (2.57b) with $Z_{\text{F}} = 0$, that the ZPS reactance can be calculated by dividing the measured voltage by the current.

The second test method is illustrated in Figure 5.18. It involves applying a single-phase voltage source to the machine being tested at the point where the three stator windings terminals are joined together. As for the NPS impedance test, the machine is running at rated speed and its field winding is short-circuited. The source current $I^Z(\text{A})$, the applied source voltage $E(\text{V})$, and the total electric input power $P(\text{W})$ are measured. The ZPS reactance and resistance are given by

$$Z^Z = \frac{3E}{I^Z} \,\Omega \quad R^Z = \frac{P}{3 \times (I^Z)^2} \,\Omega \quad X^Z = \sqrt{(Z^Z)^2 - (R^Z)^2} \,\Omega \qquad (5.82)$$

5.5.4 Example

Example 5.4 Consider a 50 Hz three-phase synchronous generator having rated apparent power of 165 MVA, rated power of 132 MW and rated terminal voltage of 15 kV. The results below are obtained from various tests and measurements on the generator. The armature dc resistance is 0.2% on rated voltage and rated MVA:

1. *Direct axis synchronous reactance*
   Unsaturated or air-gap field current at rated open-circuit terminal voltage is 682 A. The field current at rated armature (stator) current on three-phase terminal short-circuit is 1418 A.
2. *Sudden three-phase short-circuit test at rated synchronous speed*
   Prefault stator phase-to-phase voltage is 3 kV
   Steady state rms short-circuit current $I_d = 623$ A
   Three-phase load losses at rated current $= 477$ kW

   Figure 5.19 shows the results of the phase R current envelopes (peaks) plotted on a semi-logarithmic scale.
3. *NPS impedance test*
   A phase-to-phase sustained short-circuit test at the generator terminals gave the following results:

V_{Diff} (V)	453	626	806
rms current I(A)	1006	1407	1802
Three-phase power P (kW)	455	879	1451

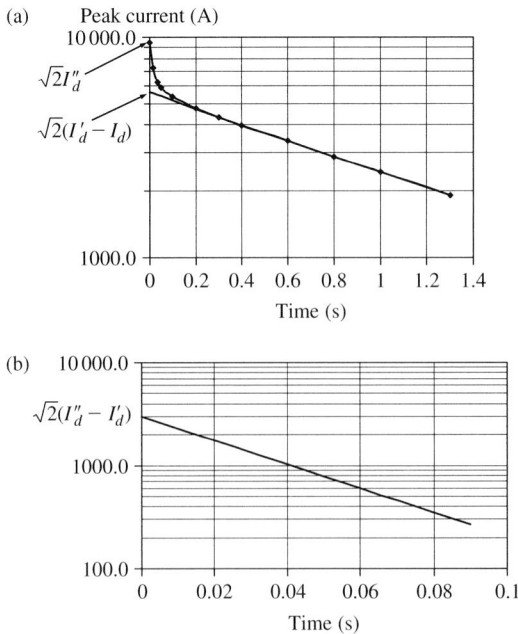

Figure 5.19   Measured short-circuit current envelopes plotted on semi-logarithmic scale: (a) subtransient and transient components and (b) subtransient component

4. *ZPS reactance test*

A two-phase to earth sustained short-circuit test at the generator terminals gave the following results:

Voltage to neutral, healthy phase (V)	149	194	232
rms current I(A)	1146	1490	1780

Solution

Direct axis synchronous reactance

$$\text{Unsaturated } X_d = \frac{1418 \text{ A}}{682 \text{ A}} \times 100 = 208\%$$

Alternatively, using $I_d = 623$ A from the sudden three-phase short-circuit test, we have

$$X_d = \frac{(3000/\sqrt{3})\text{V}}{623 \text{ A}} = 2.78 \, \Omega$$

$$\text{Base impedance} = \frac{(15 \text{ kV})^2}{165 \text{ MVA}} = 1.3636 \, \Omega$$

Thus

$$X_d = \frac{2.78 \, \Omega}{1.3636 \, \Omega} \times 100 = 204\%$$

Direct axis transient reactance and time constant
From Figure 5.19(a), we have

$$\sqrt{2}(I_d' - I_d) = 5655.4 \text{ A} \quad I_d' = \frac{5655.4}{\sqrt{2}} + 623 = 4622 \text{ A}$$

Thus

$$X_d' = \frac{3000/\sqrt{3} \text{ V}}{4622 \text{ A}} = 0.375 \, \Omega \quad \text{or} \quad X_d' = \frac{0.375 \, \Omega}{1.3636 \, \Omega} \times 100 = 27.5\%.$$

At $t = T_d'$, the transient current component is $1/e$ times the initial value of 5655.4 A, i.e. 2080.5 A. Thus, the transient time constant is found from Figure 5.19 as $T_d' = 1.1$ s.

Direct axis subtransient reactance and time constant
From Figure 5.19b, we have

$$\sqrt{2}(I_d'' - I_d') = 3010.9 \text{ A} \quad I_d'' = \frac{3010.9}{\sqrt{2}} + 4622 = 6751 \text{ A}$$

Thus

$$X_d'' = \frac{3000/\sqrt{3} \text{ V}}{6751 \text{ A}} = 0.25656 \, \Omega \quad \text{or} \quad X_d'' = \frac{0.25656 \, \Omega}{1.3636 \, \Omega} \times 100 = 18.8\%.$$

At $t = T_d''$, the subtransient current component is $1/e$ times the initial value of 3010.9 A, i.e. 1107.6 A. Thus, the subtransient time constant is found from Figure 5.19(b) as $T_d'' = 39\,\text{ms}$.

Transient reactance at the end of the subtransient period
The transient current component at the end of the subtransient period, i.e. at $t = 5T_d'' = 195\,\text{ms}$ is given by $\sqrt{2}[I_d'(t = 195\,\text{ms}) - I_d] = 4709\,\text{A}$ giving $I_d'(t = 200\,\text{ms}) = 3953\,\text{A}$. Therefore

$$X_d'(t = 200\,\text{ms}) = \frac{3000/\sqrt{3}\,\text{V}}{3953\,\text{A}} = 0.438\,\Omega$$

or

$$X_d'(t = 200\,\text{ms}) = \frac{0.438\,\Omega}{1.3636\,\Omega} \times 100 = 32.1\%$$

Alternatively, using Equation (5.37b), we obtain

$$X_d'(t = 200\,\text{ms}) = \frac{27.5\%}{\frac{27.5\%}{204\%}[1 - \exp(-5 \times 39/1100)] + \exp(-5 \times 39/1100)} = 32.0\%$$

Usually, similar analysis is carried out for the other two phases and an average is taken. The above test results were made at 20% rated voltage. The same tests may also be carried out at other voltages.

PPS resistance

$$R^P = \frac{\text{Load losses during short} - \text{circuit at rated current}}{3 \times I_{\text{Rated}}^2}$$

$$\text{Rated armature (stator) current} = \frac{165\,\text{MVA}}{\sqrt{3} \times 15\,\text{kV}} \times 1000 = 6351\,\text{A}$$

$$R^P = \frac{477 \times 10^3\,\text{W}}{3 \times (6351\,\text{A})^2} = 3.942 \times 10^{-3}\,\Omega$$

or

$$R^P = \frac{3.942 \times 10^{-3}\,\Omega}{1.3636\,\Omega} \times 100 = 0.289\%$$

NPS impedance
Using the first set of results, the NPS impedance, reactance and resistance are calculated as follows:

$$Z^N = \frac{453\,\text{V}}{\sqrt{3} \times 1006\,\text{A}} \times \frac{1}{1.3636\,\Omega} \times 100 = 19.06\%$$

$$X^N = \frac{455 \times 10^3\,\text{W}}{\sqrt{3} \times (1006\,\text{A})^2} \times \frac{1}{1.3636} \times 100 = 19.03\%$$

$$R^N = \sqrt{(19.06)^2 - (19.03)^2} = 1.07\%$$

Similar calculations are carried out using the other test results and these will give broadly similar results. The NPS impedance/reactance is then plotted against the short-circuit current and extrapolated to $\sqrt{3}I_{\text{Rated}} = 10\,827$ A. The NPS reactance value should be 19%.

ZPS reactance
Using the first set of results, the ZPS reactance is calculated as follows:

$$X^Z = \frac{149\text{ V}}{1146\text{ A}} \times \frac{1}{1.3636\,\Omega} \times 100 = 9.5\%$$

Similar calculations are carried out using the other test results and these will give broadly similar results. The ZPS reactance is then plotted against the short-circuit current and extrapolated to $3 \times I_{\text{Rated}} = 19\,053$ A. This value should be 9.5%.

Armature (dc) time constant
Using the measured PPS, NPS and ZPS reactances, and the known armature dc resistance, the armature dc time constants for various short-circuit faults are calculated as follows:

(a) Three-phase short circuit

$$T_{\text{a}} = \frac{18.8\%}{2\pi \times 50\text{ Hz} \times 0.2\%} = 0.30\text{ s}$$

(b) Two-phase short circuit

$$T_{\text{a}} = \frac{19.03\%}{2\pi \times 50\text{ Hz} \times 0.2\%} = 0.302\text{ s}$$

(c) One-phase to earth short circuit

$$T_{\text{a}} = \frac{(18.8 + 19.03 + 9.5)\%}{2\pi \times 50\text{ Hz} \times (3 \times 0.2\%)} = 0.251\text{ s}$$

(d) Two-phase to earth short circuit

$$T_{\text{a1}} = \frac{(19.03 + 2 \times 9.5)\%}{2\pi \times 50\text{ Hz} \times (3 \times 0.2)\%} = 0.20\text{ s},$$

$$T_{\text{a2}} = \frac{19.03\%}{2\pi \times 50\text{ Hz} \times 0.2\%} = 0.302\text{ s}$$

5.6 Modelling of induction motors in the phase frame of reference

5.6.1 General

For over a century, induction motors have been the workhorses of the electric energy industry. These motors are used to drive a variety of mechanical loads, e.g. fans, blowers, centrifugal pumps, hoists, conveyors, boring mills and textile machinery. The modelling of induction motors and the analysis of their short-circuit current contribution is recognised as critically important and is fully taken into account in industrial power systems, e.g. petrochemical plant, oil refineries, offshore oil platforms and power station auxiliary systems. The important characteristic in such installations is the concentrated large number of induction motors used which are clearly identifiable in terms of size, rating, location in the industrial network and, generally, electrical parameters.

Three-phase and single-phase motors are also used in some commercial installations with the latter being utilised in some domestic appliances. These motors can form an important part of the general power system load 'seen' at distribution network substations. The importance of the short-circuit contribution from such a cumulative motor load has only recently been recognised but, unfortunately, there is no international consensus regarding its treatment. This aspect is discussed in Chapter 7.

5.6.2 Overview of induction motor modelling in the phase frame of reference

Similar to a synchronous machine, the stator of an induction machine consists of a set of three-phase stator windings which may be star or delta connected. The rotor may be of a wound rotor or a squirrel-cage rotor construction with conducting bars placed in slots on the rotor surface. In squirrel-cage motors, the bars are short-circuited at both ends by end rings.

Due to the absence of slip rings or brushes, squirrel-cage induction motors are most rugged and are virtually maintenance-free machines. For some applications where high starting torques are required, e.g. boring mills, conveyor equipment, textile machinery, wood working equipment, etc., double squirrel-cage rotors are used. As the name implies two layers of bars are used; a high resistance one nearer the rotor surface and a low resistance one located below the first. The high resistance bars are effective at starting in order to give higher starting torque whereas the low resistance bars are effective under normal running operation.

The connection of a balanced three-phase power supply to the stator produces a MMF that rotates at synchronous speed $N_s = (120 \times f_s)/N_p$ where N_s is in rpm, f_s is supply frequency and N_P is the number of poles. This MMF induces voltages in the rotor windings and because these are short-circuited, rotor currents flow. The interaction of the rotor currents and the stator MMF produces a torque that acts to accelerate the rotor in the same direction of rotation of the stator MMF. The frequency of the rotor currents is determined by the relative motion of the stator MMF and rotor speed and is equal to $N_s - N_r$ where N_r is the rotor speed. The relative speed, of the rotor, called slip speed, in per unit of the synchronous speed, is given by

$$s = \frac{N_s - N_r}{N_s} \tag{5.83a}$$

Clearly, when the rotor is stationary, e.g. at starting, $N_r = 0$ and the slip $s = 1$.

Equation (5.83a) can be written in terms of stator MMF and rotor angular speeds ω_s and ω_r, in electrical radians per second, as follows:

$$s = \frac{\omega_s - \omega_r}{\omega_s} \tag{5.83b}$$

or

$$\omega_r = (1 - s)\omega_s \tag{5.83c}$$

As for synchronous machines, one of our objectives is to gain an insight into the meaning of the induction motor parameters required for use in short-circuit analysis. Prior to the derivation of the motor equations in the ryb phase frame of reference, we note that even for a squirrel-cage rotor, the rotor can be represented as a set of three-phase windings. This is because induced currents in them produce a MMF with the same number of poles as that produced by the stator windings. Also, the rotor is symmetrical as the air gap is uniform. This means that only the mutual inductances between stator and rotor windings are dependent on rotor angle position. Figure 5.20 illustrates the stator and rotor circuits of an induction motor.

Assuming that the axis of phase r rotor winding leads that of phase r stator winding by θ_r where

$$\theta_r = \omega_r t \tag{5.84a}$$

Using Equation (5.83c), we can write

$$\theta_r = (1 - s)\omega_s t \tag{5.84b}$$

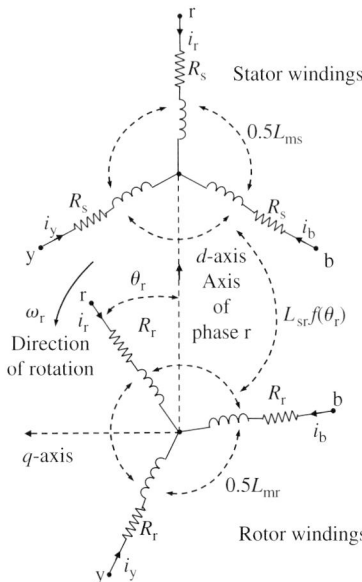

Figure 5.20   Illustration of stator and rotor circuits of an induction machine

Initially, we consider the rotor to have a single rotor winding to correspond to that of wound rotor or a single squirrel-cage rotor. The time domain stator and rotor voltage equations, with (s) and (r) denoting stator and rotor, respectively, can be written in matrix form as

$$\mathbf{e}_{ryb(s)} = \mathbf{R}_s \mathbf{i}_{ryb(s)} + \frac{d}{dt}\boldsymbol{\psi}_{ryb(s)} \tag{5.85a}$$

$$\mathbf{e}_{ryb(r)} = \mathbf{R}_r \mathbf{i}_{ryb(r)} + \frac{d}{dt}\boldsymbol{\psi}_{ryb(r)} \tag{5.85b}$$

where $\mathbf{e}$, $\mathbf{i}$ and $\boldsymbol{\psi}$ are 3×1 column matrices and $\mathbf{R}_s$ and $\mathbf{R}_r$ are 3×3 diagonal matrices of R_s and R_r elements, respectively. R_s and R_r are stator and rotor resistances, respectively.

The time domain stator and rotor flux linkage equations for a magnetically linear system are given by

$$\begin{bmatrix} \boldsymbol{\psi}_{ryb(s)} \\ \boldsymbol{\psi}_{ryb(r)} \end{bmatrix} = \begin{bmatrix} \mathbf{L}_{ss} & \mathbf{L}_{sr}(\theta_r) \\ \mathbf{L}_{rs}(\theta_r) & \mathbf{L}_{rr} \end{bmatrix} \begin{bmatrix} \mathbf{i}_{ryb(s)} \\ \mathbf{i}_{ryb(r)} \end{bmatrix} \tag{5.86a}$$

where

$$\mathbf{L}_{ss} = \begin{array}{c} \\ r(s) \\ y(s) \\ b(s) \end{array} \begin{array}{ccc} r(s) & y(s) & b(s) \\ \begin{bmatrix} L_{\sigma s} + L_{ms} & -\dfrac{1}{2}L_{ms} & -\dfrac{1}{2}L_{ms} \\ -\dfrac{1}{2}L_{ms} & L_{\sigma s} + L_{ms} & -\dfrac{1}{2}L_{ms} \\ -\dfrac{1}{2}L_{ms} & -\dfrac{1}{2}L_{ms} & L_{\sigma s} + L_{ms} \end{bmatrix} \end{array}$$

$$\mathbf{L}_{rr} = \begin{array}{c} \\ r(r) \\ y(r) \\ b(r) \end{array} \begin{array}{ccc} r(r) & y(r) & b(r) \\ \begin{bmatrix} L_{\sigma r} + L_{mr} & -\dfrac{1}{2}L_{mr} & -\dfrac{1}{2}L_{mr} \\ -\dfrac{1}{2}L_{mr} & L_{\sigma r} + L_{mr} & -\dfrac{1}{2}L_{mr} \\ -\dfrac{1}{2}L_{mr} & -\dfrac{1}{2}L_{mr} & L_{\sigma r} + L_{mr} \end{bmatrix} \end{array} \tag{5.86b}$$

$$\mathbf{L}_{sr}(\theta_r) = L_{sr} \begin{array}{ccc} r(r) & y(r) & b(r) \\ \begin{bmatrix} \cos\theta_r & \cos(\theta_r + 2\pi/3) & \cos(\theta_r - 2\pi/3) \\ \cos(\theta_r - 2\pi/3) & \cos\theta_r & \cos(\theta_r + 2\pi/3) \\ \cos(\theta_r + 2\pi/3) & \cos(\theta_r - 2\pi/3) & \cos\theta_r \end{bmatrix} \end{array} \begin{array}{c} r(s) \\ y(s) \\ b(s) \end{array} \tag{5.86c}$$

and

$$\mathbf{L}_{rs}(\theta_r) = \mathbf{L}_{sr}^{T}(\theta_r)$$

$L_{\sigma s}$ and L_{ms} are the leakage and magnetising inductances of the stator windings, $L_{\sigma r}$ and L_{mr} are the leakage and magnetising inductances of the rotor windings and L_{sr} is the amplitude of the mutual inductance between the stator and rotor windings. The magnetising and mutual inductances are associated with the same magnetic flux paths, hence L_{ms}, L_{mr} and L_{sr} are related by the stator/rotor turns ratio as follows:

$$L_{ms} = \frac{N_s}{N_r}L_{sr} \quad \text{and} \quad L_{mr} = \left(\frac{N_r}{N_s}\right)^2 L_{ms} \tag{5.87a}$$

Defining

$$L'_{sr} = \frac{N_s}{N_r}L_{sr} \quad \text{and} \quad L'_{\sigma r} = \left(\frac{N_s}{N_r}\right)^2 L_{\sigma r}$$

where the prime denotes rotor quantities that are referred to the stator. Thus

$$L'_{sr}(\theta_r) = L_{ms}\begin{bmatrix} \cos\theta_r & \cos(\theta_r + 2\pi/3) & \cos(\theta_r - 2\pi/3) \\ \cos(\theta_r - 2\pi/3) & \cos\theta_r & \cos(\theta_r + 2\pi/3) \\ \cos(\theta_r + 2\pi/3) & \cos(\theta_r - 2\pi/3) & \cos\theta_r \end{bmatrix} \tag{5.87b}$$

and

$$L'_{rr} = \begin{bmatrix} L'_{\sigma r} + L_{ms} & -\dfrac{1}{2}L_{ms} & -\dfrac{1}{2}L_{ms} \\ -\dfrac{1}{2}L_{ms} & L'_{\sigma r} + L_{ms} & -\dfrac{1}{2}L_{ms} \\ -\dfrac{1}{2}L_{ms} & -\dfrac{1}{2}L_{ms} & L'_{\sigma r} + L_{ms} \end{bmatrix} \tag{5.87c}$$

thus

$$\begin{bmatrix} \psi_{ryb(s)} \\ \psi_{ryb(r)} \end{bmatrix} = \begin{bmatrix} \mathbf{L}_{ss} & \mathbf{L}'_{sr}(\theta_r) \\ \mathbf{L}'_{rs}(\theta_r) & \mathbf{L}'_{rr} \end{bmatrix}\begin{bmatrix} i_{ryb(s)} \\ i'_{ryb(r)} \end{bmatrix} \tag{5.87d}$$

Substituting Equation (5.87d) into Equation (5.85), we obtain

$$\begin{bmatrix} e_{ryb(s)} \\ e_{ryb(r)} \end{bmatrix} = \begin{bmatrix} \mathbf{R}_s + p\mathbf{L}_{ss} & p\mathbf{L}'_{sr}(\theta_r) \\ p(\mathbf{L}'_{sr}(\theta_r))^{\mathrm{T}} & \mathbf{R}'_r + p\mathbf{L}'_{rr} \end{bmatrix}\begin{bmatrix} i_{ryb(s)} \\ i'_{ryb(r)} \end{bmatrix} \tag{5.87e}$$

where

$$\mathbf{R}'_r = \left(\frac{N_s}{N_r}\right)^2 \mathbf{R}_r \quad \text{and} \quad p = \frac{d}{dt}.$$

We have used p instead of s since the latter is used to denote slip speed.

5.7 Modelling of induction motors in the *dq* frame of reference

5.7.1 Transformation to *dq* axes

The next step is to transform the stator and rotor quantities to the d and q axes reference frame. However, because we are concerned in this book with short circuit rather than electromechanical analysis, we will fix our reference frame to the rotor as in the case of synchronous machines. This introduces an advantage that will become apparent later. The d-axis is chosen to coincide with stator phase r axis at $t = 0$ and the q-axis leads the d-axis by $90°$ in the direction of rotation. Therefore, the stator and rotor quantities can be transformed using the transformation matrix given in Equation (5.7b). It can be shown that the transformation of Equations (5.87d) and (5.87e) gives

$$
\begin{bmatrix} \psi_{ds} \\ \psi_{qs} \\ \psi_{dr} \\ \psi_{qr} \end{bmatrix} = \begin{bmatrix} L_{\sigma s} + L_{\mathrm{m}} & 0 & L_{\mathrm{m}} & 0 \\ 0 & L_{\sigma s} + L_{\mathrm{m}} & 0 & L_{\mathrm{m}} \\ L_{\mathrm{m}} & 0 & L'_{\sigma r} + L_{\mathrm{m}} & 0 \\ 0 & L_{\mathrm{m}} & 0 & L'_{\sigma r} + L_{\mathrm{m}} \end{bmatrix} \begin{bmatrix} i_{ds} \\ i_{qs} \\ i_{dr} \\ i_{qr} \end{bmatrix} \tag{5.88a}
$$

where

$$
L_{\mathrm{m}} = 1.5 L_{\mathrm{ms}}
$$

and

$$
\begin{bmatrix} e_{ds} \\ e_{qs} \\ e_{dr} \\ e_{qr} \end{bmatrix} = \begin{bmatrix} R_s i_{ds} \\ R_s i_{qs} \\ R'_r i_{dr} \\ R'_r i_{qr} \end{bmatrix} + \begin{bmatrix} \omega_{\mathrm{r}} \psi_{qs} \\ -\omega_{\mathrm{r}} \psi_{ds} \\ 0 \\ 0 \end{bmatrix} + \frac{\mathrm{d}}{\mathrm{d}t} \begin{bmatrix} \psi_{ds} \\ \psi_{qs} \\ \psi_{dr} \\ \psi_{qr} \end{bmatrix} \tag{5.88b}
$$

Because the rotor windings are short-circuited,

$$
e_{dr} = e_{qr} = 0 \tag{5.88c}
$$

We note that the speed voltage terms in the rotor voltages of Equation (5.88b) are zeros because of our choice of reference frame for the rotor quantities to coincide with the rotor rather than with the synchronously rotating MMF.

The above equations are in physical units. However, it can be easily shown that they take the same form in per unit because the stator rated quantities are taken as base quantities. The d and q axes transient equivalent circuits are obtained by substituting Equation (5.88a) into Equation (5.88b). The result is the operational equivalent circuit shown in Figure 5.21(a) that is identical in both the d and q axes because of the symmetrical structure of the rotor. The steady state equivalent circuit, shown in Figure 5.21(b), is derived by converting the time variables into phasors and setting all derivatives to zero. We note that in order to simplify the notation, we have dropped the prime from the rotor resistance and inductance parameters.

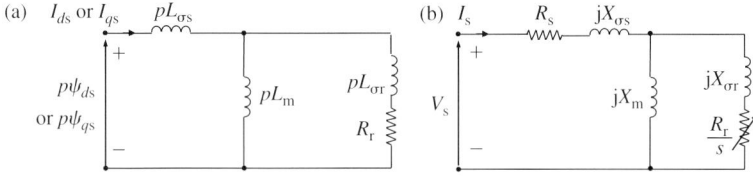

Figure 5.21 Equivalent circuits of an induction motor with one rotor winding: (a) operational equivalent circuit and (b) steady state equivalent circuit

5.7.2 Complex form of induction motor equations

To simplify the mathematics, we can replace all the voltage and flux linkage equations for the d and q axes of Equations (5.88a) and (5.88b) by half the number of equations if we define new complex variables V_s, I_s, I_r, ψ_s, ψ_r such that

$$V_s = e_{ds} + je_{qs} \quad I_s = i_{ds} + ji_{qs} \quad \psi_s = \psi_{ds} + j\psi_{qs} \quad \psi_r = \psi_{dr} + j\psi_{qr} \quad (5.89)$$

Therefore, combining the fluxes and voltages from Equations (5.88a) and (5.88b) according to Equation (5.89), we obtain

$$V_s = p\psi_s + j\omega_r\psi_s + R_s I_s \tag{5.90a}$$

$$0 = R_r I_r + p\psi_r \tag{5.90b}$$

$$\psi_s = (L_{\sigma s} + L_m)I_s + L_m I_r \tag{5.91a}$$

$$\psi_r = L_m I_s + (L_m + L_{\sigma r})I_r \tag{5.91b}$$

5.7.3 Operator reactance and parameters of a single-winding rotor

In order to gain an understanding of the origin and meaning of the various motor parameters, e.g. transient reactance and time constant, we will use the method of operator axis reactance analysis for the simplicity and clear insight it provides. Since the motor operational equivalent circuit of Figure 5.21(a) is identical for both the d and q axes, the d and q axes operator reactances will be identical. By substituting Equation (5.91b) for ψ_r into Equation (5.90b), we obtain

$$I_r = \frac{-L_m p}{R_r + (L_m + L_{\sigma r})p} I_s$$

which when substituted into Equation (5.91a) for ψ_s and simplifying, we obtain

$$\psi_s = \left[(L_{\sigma s} + L_m) - \frac{L_m^2 p}{R_r + (L_{\sigma r} + L_m)p}\right] I_s \tag{5.92a}$$

It can be shown that after further manipulation, Equation (5.92a) can be written as

$$\psi_s = (L_{\sigma s} + L_m)\frac{1 + pT'}{1 + pT'_0} I_s \tag{5.92b}$$

where

$$T'_0 = \frac{L_{\sigma r} + L_m}{R_r}$$ (5.93a)

and

$$T' = \frac{1}{R_r}\left(L_{\sigma r} + \frac{1}{\frac{1}{L_{\sigma s}} + \frac{1}{L_m}}\right)$$ (5.93b)

T'_0 and T' are the motor's open-circuit and short-circuit transient time constants, respectively. As for the synchronous machine, to obtain the time constants in seconds, the per-unit time constants of Equations (5.93a) and (5.93b) should be divided by $\omega_s = 2\pi f_s$ where f_s is the system power frequency. The motor operational reactance is defined as $X(p) = \omega_s \psi_s / I_s$, hence, from Equation (5.92b), we obtain

$$X(p) = (X_{\sigma s} + X_m)\frac{1 + pT'}{1 + pT'_0}$$ (5.94a)

The effective motor reactance at the instant of an external disturbance is defined as the transient reactance X'. Using Equation (5.94a), this is given by

$$X' = \lim_{p \to \infty} X(p) = (X_{\sigma s} + X_m)\frac{T'}{T'_0}$$ (5.94b)

This can also be expressed in terms of the electrical parameters of the stator and rotor circuits using Equations (5.93a) and (5.93b) as follows:

$$X' = X_{\sigma s} + \frac{1}{\frac{1}{X_m} + \frac{1}{X_{\sigma r}}}$$ (5.94c)

5.7.4 Operator reactance and parameters of double-cage or deep-bar rotor

Many induction motors are designed with a double squirrel-cage rotor. Also, many large MW size motors have deep-bar cage windings on the rotor designed to limit starting currents. For both double-cage and deep-bar rotors, we assume that the end-rings resistance and the part of the leakage flux which links the two rotor windings, but not the stator, are neglected. Therefore, we represent both rotor types by two parallel rotor circuits as shown in the operational equivalent circuit of Figure 5.22(a). The corresponding steady state equivalent circuit is shown in Figure 5.22(b). Figure 5.22(c) shows an alternative steady state equivalent circuit where the two rotor branches are converted into a single equivalent branch whose

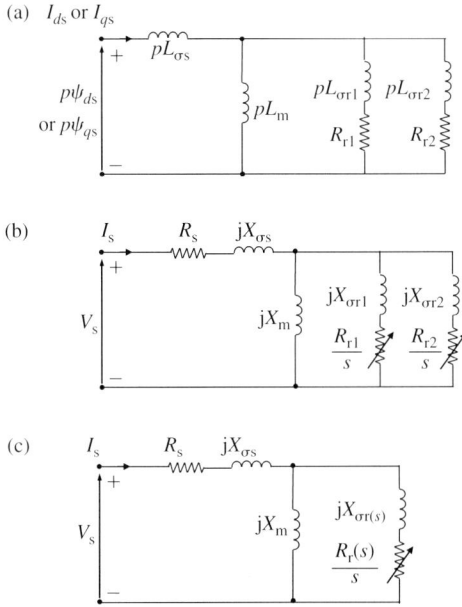

Figure 5.22   Equivalent circuits of an induction motor with a double squirrel-cage or deep-bar rotor: (a) operational equivalent circuit, (b) steady state equivalent circuit and (c) circuit (b) with a single equivalent rotor branch

resistance and reactance are functions of motor slip s as follows:

$$R_r(s) = \frac{R_{r1}R_{r2}(R_{r1} + R_{r2}) + s^2(R_{r2}X_{\sigma r1}^2 + R_{r1}X_{\sigma r2}^2)}{(R_{r1} + R_{r2})^2 + s^2(X_{\sigma r1} + X_{\sigma r2})^2} \qquad (5.95a)$$

$$X_r(s) = \frac{X_{\sigma r2}R_{r1}^2 + X_{\sigma r1}R_{r2}^2 + s^2 X_{\sigma r1}X_{\sigma r2}(X_{\sigma r1} + X_{\sigma r2})}{(R_{r1} + R_{r2})^2 + s^2(X_{\sigma r1} + X_{\sigma r2})^2} \qquad (5.95b)$$

From the operational circuit of Figure 5.22(a), and as in the case of a single rotor winding, the complex form of the motor voltage and flux linkage equations can be written by inspection as follows:

$$V_s = p\psi_s + j\omega_r\psi_s + R_sI_s \qquad (5.96a)$$

$$0 = R_{r1}I_{r1} + p\psi_{r1} \qquad (5.96b)$$

$$0 = R_{r2}I_{r2} + p\psi_{r2} \qquad (5.96c)$$

$$\psi_s = (L_{\sigma s} + L_m)I_s + L_mI_{r1} + L_mI_{r2} \qquad (5.97a)$$

$$\psi_{r1} = L_mI_s + (L_{\sigma r1} + L_m)I_{r1} + L_mI_{r2} \qquad (5.97b)$$

$$\psi_{r2} = L_mI_s + L_mI_{r1} + (L_{\sigma r2} + L_m)I_{r2} \qquad (5.97c)$$

The derivation of the operational reactance is similar to that in the case of a single rotor winding except that more algebra is involved. The steps are summarised as follows: substitute Equation (5.97b) in Equation (5.96b), Equation (5.97c) in Equation (5.96c), eliminate I_s and express I_{r1} in terms of I_{r2}, express I_{r1} and I_{r2} in terms of I_s and substitute these results back in Equation (5.97a). It can be shown that the result is given by

$$\frac{\psi_s}{I_s} = (L_{\sigma s} + L_m)$$

$$-\frac{L_m^2(R_{r1} + R_{r2})p + L_m^2(L_{\sigma r1} + L_{\sigma r2})p^2}{R_{r1}R_{r2} + [R_{r1}(L_{\sigma r2} + L_m) + R_{r2}(L_{\sigma r1} + L_m)]p + [L_{\sigma r1}L_m + L_{\sigma r2}(L_{\sigma r1} + L_m)]p^2}$$

(5.98a)

Using $X(p) = \omega_s \psi_s / I_s$, the motor operational reactance can be written as

$$X(p) = (X_{\sigma s} + X_m)$$

$$-\frac{X_m^2(R_{r1} + R_{r2})p + X_m^2(X_{\sigma r1} + X_{\sigma r2})p^2}{R_{r1}R_{r2} + [R_{r1}(X_{\sigma r2} + X_m) + R_{r2}(X_{\sigma r1} + X_m)]p + [X_{\sigma r1}X_m + X_{\sigma r2}(X_{\sigma r1} + X_m)]p^2}$$

(5.98b)

As in the case of synchronous machines, in order to derive the electrical parameters of the motor, it can be shown that Equation (5.98b) can be rewritten as

$$X(p) = (X_{\sigma s} + X_m)\frac{1 + (T_4 + T_5)p + (T_4 T_6)p^2}{1 + (T_1 + T_2)p + (T_1 T_3)p^2}$$

(5.99)

where

$$T_1 = \frac{X_{\sigma r1} + X_m}{R_{r1}} \quad T_2 = \frac{X_{\sigma r2} + X_m}{R_{r2}}$$

(5.100a)

$$T_3 = \frac{1}{R_{r2}}\left(X_{\sigma r2} + \frac{1}{\frac{1}{X_{\sigma r1}} + \frac{1}{X_m}}\right) \quad T_4 = \frac{1}{R_{r1}}\left(X_{\sigma r1} + \frac{1}{\frac{1}{X_{\sigma s}} + \frac{1}{X_m}}\right)$$

(5.100b)

$$T_5 = \frac{1}{R_{r2}}\left(X_{\sigma r2} + \frac{1}{\frac{1}{X_{\sigma s}} + \frac{1}{X_m}}\right) \quad T_6 = \frac{1}{R_{r2}}\left(X_{\sigma r2} + \frac{1}{\frac{1}{X_{\sigma s}} + \frac{1}{X_m} + \frac{1}{X_{\sigma r1}}}\right)$$

(5.100c)

Again, as in the case of synchronous machines, we make use of an approximation that reflects practical double squirrel-cage rotor or deep-bar rotor designs. The winding resistance of the lower cage winding R_{r1} is much smaller than R_{r2} of the second cage winding nearer the rotor surface. This means that T_2 (and T_3) is much smaller than T_1, and T_5 (and T_6) is much smaller than T_4. Therefore, Equation (5.99) can be expressed in terms of factors as follows:

$$X(p) = (X_{\sigma s} + X_m)\frac{(1 + pT')(1 + pT'')}{(1 + pT'_0)(1 + pT''_0)}$$

(5.101)

where

$$T'_0 = T_1 \quad T''_0 = T_3$$

(5.102a)

are the open-circuit transient and subtransient time constants, and

$$T' = T_4 \quad T'' = T_6 \tag{5.102b}$$

are the short-circuit transient and subtransient time constants.

The effective motor reactance at the instant of an external disturbance is defined as the subtransient reactance X''. Using Equation (5.101), this is given by

$$X'' = \lim_{p \to \infty} X(p) = (X_{\sigma s} + X_m) \frac{T'T''}{T_o'T_o''} \tag{5.103a}$$

which, using Equations (5.100) and (5.102), can be written in terms of the motor reactance parameters as follows:

$$X'' = X_{\sigma s} + \frac{1}{\frac{1}{X_m} + \frac{1}{X_{\sigma r 1}} + \frac{1}{X_{\sigma r 2}}} \tag{5.103b}$$

As in the case of synchronous machines, the second rotor cage is no longer effective at the end of the subtransient period and the beginning of the transient period. Thus, using Equation (5.95) previously derived for the single rotor cage, we have

$$X' = (X_{\sigma s} + X_m) \frac{T'}{T_o'} = X_{\sigma s} + \frac{1}{\frac{1}{X_m} + \frac{1}{X_{\sigma r 1}}} \tag{5.104}$$

Using Equation (5.104) for X' in Equation (5.103a) for X'', we obtain the following useful relationship between the subtransient and transient reactances and time constants

$$\frac{X''}{X'} = \frac{T''}{T_o''} \tag{5.105}$$

In order to determine the motor reactance under steady state conditions, we recall that the rotor slip is equal to s and thus the angular frequency of the dq axes stator currents is $s\omega_s$. Therefore, the steady state motor impedance can be determined by putting $p = js\omega_s$ in $X(p)$ expression given in Equation (5.101).

We can express the induction machine operational reactance given in Equation (5.101) as a sum of partial fractions as follows:

$$\frac{1}{X(p)} = \frac{1}{(X_{\sigma s} + X_m)} \frac{(1 + pT_o')(1 + pT_o'')}{(1 + pT')(1 + pT'')}$$

$$= \frac{T_o'T_o''}{(X_{\sigma s} + X_m)T'T''} \frac{(p + 1/T_o')(p + 1/T_o'')}{(p + 1/T')(p + 1/T'')}$$

which, using Equations (5.103a) for X'' and (5.104) for X', can be written as

$$\frac{1}{X(p)} = \frac{1}{X_{\sigma s} + X_m} + \left(\frac{1}{X'} - \frac{1}{X_{\sigma s} + X_m} \right) \frac{p}{(p + 1/T')} + \left(\frac{1}{X''} - \frac{1}{X'} \right) \frac{p}{(p + 1/T'')} \tag{5.106}$$

5.8 Induction motor behaviour under short-circuit faults and modelling in the sequence reference frame

The most common types of short-circuit faults studied in power systems are the balanced three-phase and unbalanced single-phase to earth faults. Having derived the motor reactances and time constants, we are now able to proceed with analysing the behaviour of the motor under such fault conditions.

5.8.1 Three-phase short-circuit faults

Armature (stator) short-circuit time constant

Using $X(p) = \omega_s \psi_s / I_s$ in Equation (5.96a), the stator current is given by

$$I_s = \frac{\omega_s V_s}{X(p)\left[(p + j\omega_r) + \frac{\omega_s R_s}{X(p)}\right]} \tag{5.107}$$

As for synchronous machines, we can make an assumption in the term $\omega_s R_s / X(p)$ to neglect the rotor resistances in the factors of $X(p)$ given in Equation (5.101). This is equivalent to setting infinite rotor time constants or neglecting the decay in the corresponding short-circuit currents. Mathematically speaking, this is equivalent to setting $p \to \infty$ in Equation (5.94a) and Equation (5.101). Thus, using Equation (5.94a) for a single rotor winding, we obtain $X(p) = X'$ giving

$$\frac{\omega_s R_s}{X(p)} = \frac{\omega_s R_s}{X'}$$

Also, using Equation (5.101) for a double squirrel-cage or deep-bar rotor, we obtain $X(p) = X''$ giving

$$\frac{\omega_s R_s}{X(p)} = \frac{\omega_s R_s}{X''}$$

Therefore, the armature or stator time constant T_a is defined as follows:

$$T_a = \begin{cases} \frac{X'}{\omega_s R_s} & \text{for a single rotor winding} \\ \frac{X''}{\omega_s R_s} & \text{for a double squirrel-cage or deep-bar rotor} \end{cases} \tag{5.108}$$

Total short-circuit current contribution from a motor on no load

Unlike a synchronous machine, since the induction motor has no rotor excitation, it draws its magnetising current from the supply network whatever the motor loading is, i.e. from no load to full load operation. Therefore, the motor cannot be operated open-circuited in a steady state condition! Prior to the short-circuit fault, the motor is assumed connected to a balanced three-phase supply and operating at rated terminal voltage. To simplify the mathematics, we assume that the motor is running at no load and hence the slip is very small and nearly equal to zero. Let the real instantaneous voltage, complex instantaneous voltage and complex phasor voltage values of the phase r stator voltage, just before the short-circuit, be given by

$$v_s(t) = \sqrt{2}V_{rms}\cos(\omega_s t + \theta_o) = \text{Real}[V_s(t)] \qquad (5.109a)$$

$$V_s(t) = \sqrt{2}V_{rms}e^{j(\omega_s t + \theta_o)} = V_s e^{j\omega_s t} \qquad (5.109b)$$

$$V_s = \sqrt{2}V_{rms}e^{j\theta_o} \qquad (5.109c)$$

where θ_o is the initial phase angle that defines the instant of fault on the phase r voltage waveform at $t = 0$. Neglecting stator resistance, i.e. letting $R_s = 0$, using Equation (5.83c) and putting $p = 0$ in Equation (5.107), we obtain

$$I_s = \frac{V_s}{jX(p = 0)(1 - s)}$$

Using Equation (5.101), $X(p = 0) = X_{\sigma s} + X_m$ and using Equation (5.109c), the stator steady state current phasor is given by

$$I_s = \sqrt{2}I_{rms}e^{j(\theta_o - \pi/2)} \qquad (5.110a)$$

where

$$I_{rms} = \frac{V_{rms}}{(X_{\sigma s} + X_m)(1 - s)} \qquad (5.110b)$$

Equation (5.110a) shows that the current lags the voltage by 90°, as expected.

Using the superposition principle, the total motor current after the application of the short circuit is the sum of the initial steady state current and the change in motor current due to the short circuit. The latter is obtained by applying a voltage source at the point of fault equal to but out of phase with the prefault voltage, i.e. $\Delta V_s = -\sqrt{2}V_{rms}e^{j\theta_o}$. Since the Laplace transform of ΔV_s is equal to

$$\frac{-\sqrt{2}V_{rms}e^{j\theta_o}}{p}$$

and

$$\frac{\omega_s R_s}{X(p)} = \frac{1}{T_a}$$

we obtain from Equation (5.107)

$$\Delta I_s(p) = \frac{-\omega_s\sqrt{2}V_{rms}e^{j\theta_o}}{p[(p + 1/T_a) + j\omega_r]X(p)} \qquad (5.111a)$$

Substituting the partial fractions Equation (5.106) for $X(p)$, we obtain

$$\Delta I_s(p) = \frac{-\omega_s \sqrt{2}V_{rms}e^{j\theta_o}}{p(p + 1/T_a + j\omega_r)} \left[\frac{1}{X_{\sigma s} + X_m} + \left(\frac{1}{X'} - \frac{1}{X_{\sigma s} + X_m} \right) \frac{p}{(p + 1/T')} \right.$$
$$\left. + \left(\frac{1}{X''} - \frac{1}{X'} \right) \frac{p}{(p + 1/T'')} \right] \tag{5.111b}$$

The complex instantaneous current $\Delta I_s(t)$ is obtained by taking the inverse Laplace transform of Equation (5.111b). The total short-circuit current $I_{s(total)}(t)$ is the sum of the prefault current given in Equation (5.110a) and $\Delta I_s(t)$. The real instantaneous phase r stator short-circuit current is given by $i_r(t) = \text{Real}[I_{s(total)}e^{j\omega_r t}]$. For all practical purposes, $1/T_a$, $1/T'$, and $1/T''$ are negligible in comparison with ω_s and ω_r. Therefore, and after much mathematical operations, it can be shown that the total phase r short-circuit current is given by

$$i_r(t) \approx \frac{\sqrt{2}V_{rms}}{(1 - s)} \left[\left(\frac{1}{X'} - \frac{1}{X_{\sigma s} + X_m} \right) e^{-t/T'} + \left(\frac{1}{X''} - \frac{1}{X'} \right) e^{-t/T''} \right]$$
$$\times \cos[(1 - s)\omega_s t + \theta_o - \pi/2]$$
$$- \frac{\sqrt{2}V_{rms}}{(1 - s)} \frac{1}{X''} e^{-t/T_a} \cos(\theta_o - \pi/2) \tag{5.112a}$$

The first term is the ac component that consists of a subtransient and a transient component which decay with time constants T' and T'', respectively. The second term is the dc component that has an initial magnitude that depends on θ_o and it decays with a time constant T_a. Unlike a synchronous machine, there is no steady state ac component term and hence the short-circuit current decays to zero. The frequency of the short-circuit current is slightly lower than the power frequency by the factor $(1 - s)$.

For a single cage/winding rotor, it can be shown that the instantaneous current is given by

$$i_r(t) \approx \frac{\sqrt{2}V_{rms}}{(1 - s)} \left[\left(\frac{1}{X'} - \frac{1}{X_{\sigma s} + X_m} \right) e^{-t/T'} \cos\{(1 - s)\omega_s t + \theta_o - \pi/2\} \right.$$
$$\left. - \frac{1}{X'} e^{-t/T_a} \cos(\theta_o - \pi/2) \right] \tag{5.112b}$$

In this case, the current consists of only one ac component and a dc component.

The phases y and b currents are obtained by replacing θ_o with $\theta_o - 2\pi/3$ and $\theta_o + 2\pi/3$ respectively.

Figure 5.23 shows typical short-circuit current waveforms of a large 6.6 kV 4700 hp motor with a deep-bar rotor having the following parameters: $X'' = 0.16$ pu, $X' = 0.2$ pu, $X = X_{\sigma s} + X_m = 3.729$ pu and $R_s = 0.0074$ pu.

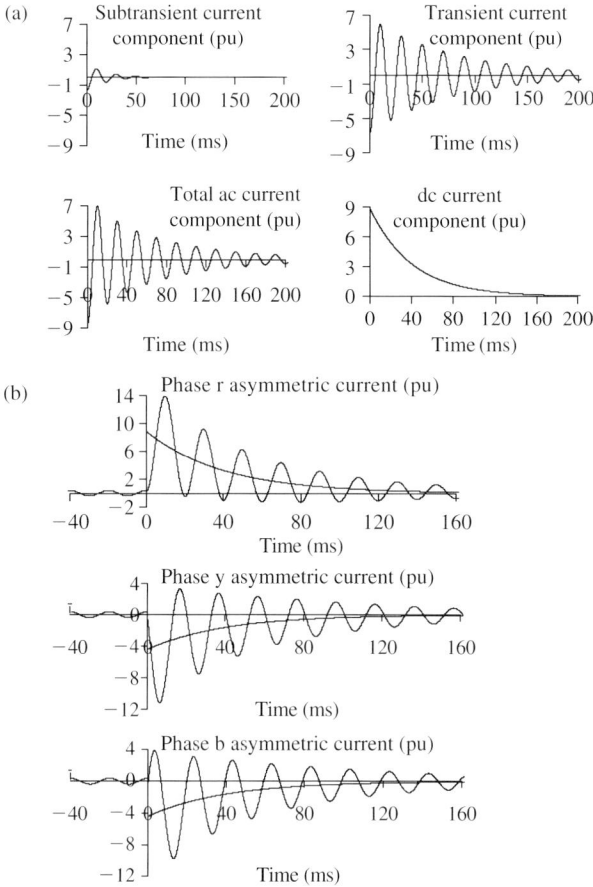

Figure 5.23 Three-phase short-circuit fault at the terminals of a double-cage induction motor from no load: (a) components of phase r short-circuit current and (b) asymmetric currents

Simplified motor short-circuit current equations

For practical applications, further simplifications in the short-circuits current equations can be made. The motor slip s is typically less than 1% for large motors and less than 4–5% for smaller motors. Therefore, if $(1 - s)$ is replaced by 1, then the peak current envelope at any time instant, for a double-cage or deep-bar rotor, assuming maximum dc offset, is given by

$$\hat{i}_r(t) = \sqrt{2}V_{rms}\left[\left(\frac{1}{X'} - \frac{1}{X_{\sigma s} + X_m}\right)e^{-t/T'} + \left(\frac{1}{X''} - \frac{1}{X'}\right)e^{-t/T''} + \frac{1}{X''}e^{-t/T_a}\right] \tag{5.113a}$$

or

$$\hat{i}_r(t) = \sqrt{2}[(I' - I)e^{-t/T'} + (I'' - I')e^{-t/T''}] + \sqrt{2}I''e^{-t/T_a} \tag{5.113b}$$

or

$$\hat{i}_r(t) = \sqrt{2}I_{ac}(t) + I_{dc}(t) \tag{5.113c}$$

where

$$I = \frac{V_{rms}}{X_{\sigma s} + X_m} \quad I' = \frac{V_{rms}}{X'} \quad I'' = \frac{V_{rms}}{X''} \tag{5.114a}$$

and

$$I_{dc}(t) = \sqrt{2}I''e^{-t/T_a} \tag{5.114b}$$

$$I_{ac}(t) = (I' - I)e^{-t/T'} + (I'' - I')e^{-t/T''} \tag{5.114c}$$

For a single cage/winding rotor, the peak current envelope is given by

$$\hat{i}_r(t) = \sqrt{2}V_{rms}\left[\left(\frac{1}{X'} - \frac{1}{X_{\sigma s} + X_m}\right)e^{-t/T'} + \frac{1}{X'}e^{-t/T_a}\right] \tag{5.115a}$$

or

$$\hat{i}_r(t) = \sqrt{2}(I' - I)e^{-t/T'} + \sqrt{2}I'e^{-t/T_a} \tag{5.115b}$$

or

$$\hat{i}_r(t) = \sqrt{2}I_{ac}(t) + I_{dc}(t) \tag{5.115c}$$

where

$$I_{dc}(t) = \sqrt{2}I'e^{-t/T_a} \tag{5.116a}$$

and

$$I_{ac}(t) = (I' - I)e^{-t/T'} \tag{5.116b}$$

Motor current change due to the short-circuit fault

In the above analysis, we determined the total motor current under a three-phase short-circuit at the motor terminals when the motor is running on no load. This current consisted of the steady state current just before the fault and the current change due to the short circuit. For some practical analysis, it is useful to calculate the current change only and account differently for the motor initial loading conditions. From Equations (5.113a) and (5.115a), we can show that the equations that give the current change for a double-cage and a single-cage rotor are given by

$$\hat{i}_r(t) = \sqrt{2}V_{rms}\left[\frac{1}{X'}e^{-t/T'} + \left(\frac{1}{X''} - \frac{1}{X'}\right)e^{-t/T''} + \frac{1}{X''}e^{-t/T_a}\right] \tag{5.117a}$$

or

$$\hat{i}_r(t) = \sqrt{2}[I'e^{-t/T'} + (I'' - I')e^{-t/T''}] + \sqrt{2}I''e^{-t/T_a} \tag{5.117b}$$

and

$$\hat{i}_r(t) = \frac{\sqrt{2}V_{rms}}{X'}(e^{-t/T'} + e^{-t/T_a}) \tag{5.118a}$$

or

$$\hat{i}_r(t) = \sqrt{2}I'(e^{-t/T'} + e^{-t/T_a}) \tag{5.118b}$$

Positive sequence reactance and resistance

From Equation (5.112a) for a double-cage or deep-bar rotor, the instantaneous current change due to the short-circuit fault is obtained by ignoring the prefault current $I_{rms} = \sqrt{2}V_{rms}/(X_{\sigma s} + X_m)$. Thus with $1 - s \approx 1$, we can express the instantaneous power frequency component of the short-circuit current as $i_r(t) = \text{Real}[I_r(t)]$ where $I_r(t)$ is phase r complex instantaneous current given by

$$I_r(t) = \frac{1}{X^P(t)}\sqrt{2}V_{rms}e^{j(\omega_s t + \theta_o - \pi/2)} = \frac{V_r(t)}{jX^P(t)} \tag{5.119a}$$

where

$$V_r(t) = \sqrt{2}V_{rms}e^{j(\omega_s t + \theta_o)} = V_r e^{j\omega_s t} \quad V_r = \sqrt{2}V_{rms}e^{j\theta_o}$$

and

$$\frac{1}{X^P(t)} = \frac{1}{X'}e^{-t/T'} + \left(\frac{1}{X''} - \frac{1}{X'}\right)e^{-t/T''} \tag{5.119b}$$

is the equivalent time-dependent motor PPS reactance. In the case of a single-winding rotor, using Equation (5.112b), the time-dependent motor PPS reactance is given by

$$\frac{1}{X^P(t)} = \frac{1}{X'}e^{-t/T'} \tag{5.119c}$$

From Equation (5.119b) for a motor with a double-cage or deep-bar rotor, $X^P = X''$ at $t = 0$ or the instant of short-circuit, and $X^P = X'$ at $t = 0$ with the subtransient current component neglected. At the end of the subtransient period, i.e. at $t \approx 5T''$, then from Equation (5.119b), we obtain

$$X^P(t) = X'e^{5T''/T'} \tag{5.119d}$$

The effective motor steady state PPS reactance is infinite because the motor does not supply a steady state short-circuit current. From Equation (5.119c) for a single winding rotor, $X^P = X'$ at the instant of short-circuit and is infinite in the steady state.

The complex instantaneous PPS current is given by

$$I^P(t) = \frac{I_r(t) + hI_y(t) + h^2 I_b(t)}{3} = I_r(t)$$

Figure 5.24 shows the PPS motor equivalent circuits for a three-phase short-circuit at the motor terminals.

(a)

(b)

$$X^P = \begin{cases} X'' & t = 0, \text{ start of subtransient period} \\ X' & t = 0, \text{ transient reactance} \\ X^P \text{ Eq. (5.119d)} & t = 5T'' \end{cases}$$

Steady state,
end of transient period

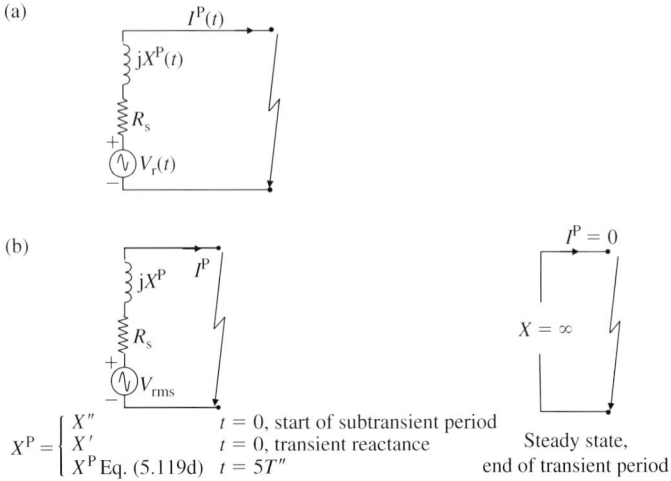

Figure 5.24 Induction motor PPS equivalent circuits for a three-phase short circuit at its terminals: (a) transient PPS time-dependent equivalent circuit and (b) PPS fixed impedance equivalent circuits at various time instants

The PPS impedance, $Z^P = R^P + jX^P$, at the instant of short-circuit fault can also be derived from the motor steady state equivalent circuit. For example, for a single winding rotor shown in Figure 5.21(b), Z^P is the impedance that would be seen at the motor terminals at starting (i.e. when $s = 1$). Thus, it can be shown that the PPS resistance and PPS reactance are given by

$$R^P = R_s + \frac{X_m^2 R_r}{R_r^2 + (X_m + X_{\sigma r})^2} \qquad X^P = X_{\sigma s} + \frac{X_m[R_r^2 + X_{\sigma r}(X_m + X_{\sigma r})]}{R_r^2 + (X_m + X_{\sigma r})^2} \quad (5.120a)$$

The term R_r^2 is negligible in comparison with either of the two terms $X_r(X_m + X_{\sigma r})$ or $(X_m + X_{\sigma r})^2$, thus, X^P and R^P reduce to

$$X^P = X_{\sigma s} + \frac{X_m X_{\sigma r}}{X_m + X_{\sigma r}} = X' \qquad (5.120b)$$

and

$$R^P = R_s + \frac{X_m R_r}{X_m + 2X_{\sigma r}} \qquad (5.120c)$$

Although unnecessary, further simplification in R^P and X^P can be made if we consider that $2X_{\sigma r}/X_m \ll 1$ to obtain $R^P = R_s + R_r$ and $X^P = X_s + X_{\sigma r}$.

Effect of short-circuit fault through an external impedance

The effect of external impedance $(R_e + jX_e)$ on the current change can be considered in the same manner as for synchronous machines. For the double-cage or deep-bar rotor, the peak current envelop and time constant equations are given by

$$\hat{i}_r(t) = \sqrt{2}V_{rms}\left[\frac{1}{X'+X_e}\exp(-t/T'_e) + \left(\frac{1}{X''+X_e} - \frac{1}{X'+X_e}\right)\exp(-t/T''_e)\right.$$

$$\left. + \frac{1}{X''+X_e}\exp(-t/T_{ae})\right] \tag{5.121a}$$

where

$$T''_e = \frac{(R_s+R_e)^2 + (X''+X_e)^2}{(R_s+R_e)^2 + (X''+X_e)(X'+X_e)}T''_o$$

$$= \frac{(R_s+R_e)^2 + (X''+X_e)^2}{(R_s+R_e)^2 + (X''+X_e)(X'+X_e)} \times \frac{X'}{X''}T'' \tag{5.121b}$$

and

$$T'_e = \frac{(R_s+R_e)^2 + (X'+X_e)^2}{(R_s+R_e)^2 + (X'+X_e)(X+X_e)}T'_o$$

$$= \frac{(R_s+R_e)^2 + (X'+X_e)^2}{(R_s+R_e)^2 + (X'+X_e)(X+X_e)} \times \frac{X}{X'}T' \tag{5.121c}$$

$$T_{ae} = \frac{X''+X_e}{\omega_s(R_s+R_e)} \qquad X = X_{\sigma s} + X_m \tag{5.121d}$$

Where the stator and external resistances are much smaller than the reactances, the transient and subtransient time constants reduce to

$$T'_e = \frac{X'+X_e}{X+X_e} \times \frac{X}{X'}T' \qquad T''_e = \frac{X''+X_e}{X'+X_e} \times \frac{X'}{X''}T'' \tag{5.121e}$$

For a single-cage rotor, the corresponding equations are given by

$$\hat{i}_r(t) = \frac{\sqrt{2}V_{rms}}{X'+X_e}[\exp(-t/T'_e) + \exp(-t/T_{ae})] \tag{5.121f}$$

where T'_e is as given in Equation (5.121c) or Equation (5.121e), and

$$T_{ae} = \frac{X'+X_e}{\omega_s(R_s+R_e)} \tag{5.121g}$$

5.8.2 Unbalanced single-phase to earth short-circuit faults

Short-circuit current equations

Consider an induction motor with a star-connected solidly earthed stator winding. When a loaded induction motor is subjected to an unbalanced single-phase short circuit at its terminals, the faulted phase will initially supply a short-circuit current.

However, under such a fault condition, there is an interaction between the motor and other short-circuit current sources in the network, e.g. synchronous machines through the ZPS network. The current contribution will therefore be affected by other sources in the network. Further, the phase current contribution would not decay to zero as in the case of a three-phase fault because the other two healthy phases remain supplied from the network and continue to provide some excitation which maintain some flux. Although the PPS current contribution will decay to zero, the NPS current contribution will not so long as the single-phase fault condition continues to exist. In Section 5.8.4, we describe a sudden short-circuit test immediately after motor disconnection from the supply network. It can be shown that the change in motor current supplied under a single-phase fault at the motor terminals, in the case of a double-cage or a deep-bar rotor, expressed in terms of the peak current envelope, is given by

$$\hat{i}_r(t) = \sqrt{2}V_{rms}\left[\frac{1}{X' + X^N + X^Z}\exp(-t/T'_{(1\phi)})\right.$$

$$+ \left(\frac{1}{X'' + X^N + X^Z} - \frac{1}{X' + X^N + X^Z}\right)\exp(-t/T''_{(1\phi)})$$

$$\left. + \frac{1}{X'' + X^N + X^Z}\exp(-t/T_{a(1\phi)})\right] \tag{5.122a}$$

where

$$T''_{(1\phi)} = \frac{X'' + X^N + X^Z}{X + X^N + X^Z}T''_o \quad T'_{(1\phi)} = \frac{X' + X^N + X^Z}{X + X^N + X^Z}T'_o$$

$$T_{a(1\phi)} = \frac{X'' + X^N + X^Z}{\omega_s(R_s + R^N + R^Z)} \tag{5.122b}$$

and for a single-cage rotor

$$\hat{i}_r(t) = \frac{\sqrt{2}V_{rms}}{X' + X^N + X^Z}[\exp(-t/T'_{(1\phi)}) + \exp(-t/T_{a(1\phi)})] \tag{5.123a}$$

where $T'_{(1\phi)}$ is as given in Equation (5.122b) and

$$T_{a(1\phi)} = \frac{X' + X^N + X^Z}{\omega_s(R_s + R^N + R^Z)} \tag{5.123b}$$

Negative sequence reactance and resistance

As for a synchronous machine, the NPS reactance of the induction motor X^N does not vary with time. $X^N = X''$ for a double-cage or deep-bar rotor and $X^N = X'$ for a single-cage rotor. In addition, the NPS impedance, $Z^N = R^N + jX^N$, can be derived from the motor steady state equivalent circuit shown in Figure 5.21(b) for a single-cage rotor and Figure 5.22(b) for a double-cage or deep-bar rotor. We recall that under a three-phase PPS excitation, both the stator MMF and the rotor rotate in the same direction. However, under three-phase NPS excitation, the NPS stator currents will produce a MMF that rotates in the opposite direction to the rotor. Using Equation (5.83a), the motor's NPS slip can be calculated as follows:

$$\text{NPS } s = \frac{N_s - (-N_r)}{N_s} = 2 - \left(\frac{N_s - N_r}{N_s}\right) = 2 - s \qquad (5.124a)$$

i.e. in the motor NPS equivalent circuit, s should be replaced by $(2-s)$. For large motors, s is typically less than around 1% and even for small motors, s is less than around 4% or 5%. Therefore, if we replace s with $(2-s)=2$ in Figure 5.21(b), we can calculate the motor NPS resistance and NPS reactance, as follows:

$$R^N = R_s + \frac{2X_m^2 R_r}{R_r^2 + 4(X_m + X_{\sigma r})^2} \qquad X^N = X_{\sigma s} + \frac{X_m[R_r^2 + 4X_{\sigma r}(X_m + X_{\sigma r})]}{R_r^2 + 4(X_m + X_{\sigma r})^2}$$

$$(5.124b)$$

As in the case of PPS impedance, the term R_r^2 is negligible and can be neglected. Therefore, Equation (5.124b) reduces to

$$X^N = X_{\sigma s} + \frac{X_m X_{\sigma r}}{X_m + X_{\sigma r}} = X' = X^P \qquad (5.124c)$$

and

$$R^N = R_s + \frac{X_m R_r}{2(X_m + 2X_{\sigma r})} \qquad (5.124d)$$

Further approximation in R^N can be made if we assume that $2X_{\sigma r}/X_m \ll 1$ giving

$$R^N = R_s + (R_r/2) \qquad (5.124e)$$

This produces an error in R^N of typically a couple of per cent only.

Zero sequence reactance and resistance

The ZPS impedance of an induction motor is infinite if the stator winding is either delta connected, or star connected with the neutral isolated. For a star-connected winding with an earthed neutral, then as for a synchronous machine, the ZPS reactance X^Z is finite, being smaller than the motor starting reactance (subtransient or transient as appropriate to the rotor construction) and does not vary with time. The ZPS resistance can be assumed equal to the stator ac resistance.

5.8.3 Modelling the effect of initial motor loading

When the motor is loaded, it draws active and reactive power from the supply network. Using the motor transient and subtransient reactances for a double-cage rotor or transient reactance for a single-cage rotor, the motor internal voltages

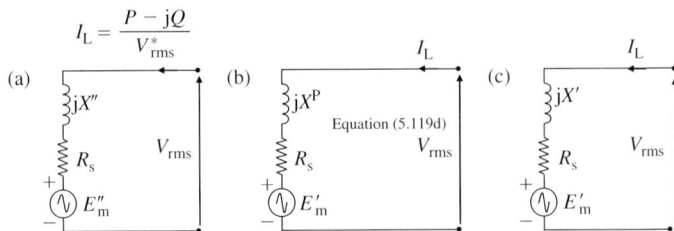

Figure 5.25 Representation of initial motor loading before the short-circuit fault: (a) motor internal voltage behind subtransient reactance, (b) motor internal voltage behind transient reactance at $t = 0$ and (c) motor internal voltage behind transient reactance at $t = 5T''$

behind subtransient and transient reactances, E'_m and E''_m, are given by

$$E''_m = V_{rms} - (R_s + jX'') \left(\frac{P - jQ}{V^*_{rms}} \right) \tag{5.125a}$$

$$E'_m = V_{rms} - (R_s + jX') \left(\frac{P - jQ}{V^*_{rms}} \right) \tag{5.125b}$$

$$E'_m(t \approx 5T'') = V_{rms} - [R_s + jX'(t \approx 5T'')] \left(\frac{P - jQ}{V^*_{rms}} \right) \tag{5.125c}$$

Equations (5.125) are illustrated in Figure 5.25.

The peak short-circuit current changes in the case of a three-phase fault given in Equations (5.117a) and (5.118a) are modified as follows:

$$\hat{\imath}_r(t) = \sqrt{2} \left[\frac{E'_m}{X'} e^{-t/T'} + \left(\frac{E''_m}{X''} - \frac{E'_m}{X'} \right) e^{-t/T''} + \frac{E''_m}{X''} e^{-t/T_a} \right] \tag{5.126a}$$

$$\hat{\imath}_r(t) = \sqrt{2} \frac{E'_m}{X'} (e^{-t/T'} + e^{-t/T_a}) \tag{5.126b}$$

From Equation (5.125) we observe that the motor internal voltage is equal to the terminal voltage less the voltage drop across the motor subtransient or transient impedance. For a 1 pu motor terminal voltage, typical values of motor subtransient and transient reactances, and various motor loading, the internal voltage magnitude can typically vary between 0.85 and 0.95 pu. Therefore, ignoring the initial motor loading and using 1 pu instead of the actual internal voltages represents a conservative assumption. This assumption is made by the American IEEE C37.010 Standard and UK ER G7/4 Guideline. This will be discussed in detail in Chapter 7.

5.8.4 Determination of motor's electrical parameters from tests

Steady state equivalent circuit: stator resistance test

The stator winding dc resistance of a three-phase induction motor is measured by connecting a dc voltage source across two stator terminals. No induced rotor

(a)

(b)

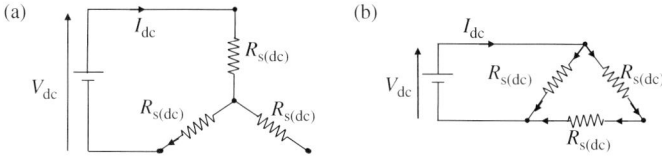

Figure 5.26 Measurement of stator dc resistance of an induction motor: (a) star-connected stator winding with isolated neutral and (b) delta-connected stator winding

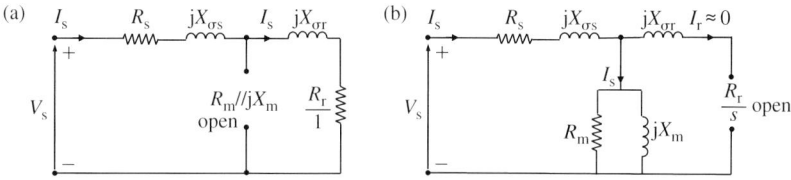

(a)

(b)

Figure 5.27 Measurement of electrical parameters of an induction motor: (a) locked rotor test and (b) no-load test

voltage can occur under dc stator excitation and hence the resistance 'seen' by the dc voltage source is the stator dc resistance. By applying a known dc voltage and measuring the dc current, the stator dc resistance can be calculated. The calculation depends on whether the stator winding is star or delta connected as shown in Figure 5.26(a) and (b) where the star is assumed to have an isolated neutral. In physical units, the stator dc resistance $R_{s(dc)}$ is given by

$$R_{s(dc)} = \begin{cases} \dfrac{1}{2}\dfrac{V_{dc}}{I_{dc}} & \text{for a star-connected stator winding} \\[2ex] \dfrac{3}{2}\dfrac{V_{dc}}{I_{dc}} & \text{for a delta-connected stator winding} \end{cases} \quad \Omega/\text{phase} \quad (5.127)$$

If the test is repeated for the other two sets of terminal connections, then an average is taken. $R_{s(dc)}$ has to be corrected for skin effect by multiplying it by the skin effect factor which is typically between 1.05 and 1.25.

Steady state equivalent circuit: locked rotor test

This test circuit is shown in Figure 5.27(a).

The stator is connected to a balanced three-phase ac voltage source and the rotor shaft is locked mechanically to prevent it from turning. Since the slip $s = 1$, the rotor impedance is several times smaller than that under rated running condition (because operating slip is typically between 0.001 and 0.05) and is typically 30–50 times smaller than the magnetising reactance. Thus, the locked rotor test is broadly analogous to a transformer short-circuit test. The test voltage applied should initially be very small and then gradually increased until rated or full load stator current is drawn. To avoid overheating, the applied voltage is significantly smaller than the rated voltage so that the motor current is limited to the rated value. If rated voltage is applied, the locked rotor current drawn is very large and is typically four to seven times the rated current value. The combination of a low-test voltage and a low rotor impedance results in a very small exciting current and hence the iron loss and magnetising branches can be neglected with an insignificant loss of accuracy. As shown in Figure 5.27(a), the equivalent motor impedance 'seen' from the test terminals is given by

$$Z_{Eq} = R_{Eq} + jX_{Eq} \approx R_s + R_r + j(X_{\sigma s} + X_{\sigma r}) \tag{5.128}$$

The measured stator quantities are line-to-line voltage V_{LL}, stator current I_s, three-phase input power $P_{3\phi}$ and three-phase input reactive power $Q_{3\phi}$. For either a star or a delta-connected stator winding, the equivalent resistance and reactance of Equation (5.128) are calculated as follows:

$$R_{Eq} = R_s + R_r = \frac{P_{3\phi}}{3I_s^2} \; \Omega/\text{phase} \tag{5.129a}$$

$$X_{Eq} = X_{\sigma s} + X_{\sigma r} = \frac{Q_{3\phi}}{3I_s^2} \; \Omega/\text{phase} \tag{5.129b}$$

Alternatively, using the line-to-line stator voltage instead of the reactive power

$$Z_{Eq} = \begin{cases} \dfrac{V_{LL}}{\sqrt{3}I_s} & \text{for a star-connected stator winding} \\[2ex] \dfrac{\sqrt{3}V_{LL}}{I_s} & \text{for a delta-connected stator winding} \end{cases} \; \Omega/\text{phase} \tag{5.130a}$$

and the equivalent reactance is calculated as follows:

$$X_{Eq} = X_{\sigma s} + X_{\sigma r} = \sqrt{Z_{Eq}^2 - R_{Eq}^2} \; \Omega/\text{phase} \tag{5.130b}$$

Using the measured stator dc resistance or better still its corrected skin effect value, the rotor resistance can be calculated from Equation (5.129a) as

$$R_r = R_{Eq} - R_s \; \Omega/\text{phase} \tag{5.130c}$$

In Equation (5.130b), the sum of the stator and rotor leakage reactances is calculated. In the absence of manufacturer design data that provides the ratio of these reactances, the division shown in Table 5.1 for different types of motors as classified by the National Electrical Manufacturers Association (NEMA) may be used.

Table 5.1   Estimation of induction motor stator and rotor leakage reactances using NEMA motor class in the absence of design data

NEMA Motor Design Class	$X_{\sigma s}$ in % of X_{Eq}	$X_{\sigma r}$ in % of X_{Eq}
Wound rotor	50	50
Class A	50	50
Class B	40	60
Class C	30	70
Class D	50	50

If the motor design class is not known, then $X_{\sigma s}$ and $X_{\sigma r}$ may be assumed to be equal. The locked rotor test may also be carried out at reduced frequency to account for skin depth and a more accurate rotor resistance under rated operating conditions where the rotor current frequency may be between 0.5 and 2.5 Hz. IEEE Standard 112 recommends a frequency of $1/4$ of the rated power frequency. In general, this method tends to provide better results for large motors above 20 hp but for small motors, the results obtained from the power frequency test are generally satisfactory.

Steady state equivalent circuit: no-load test

This test circuit is shown in Figure 5.27(b). The stator winding is connected to a three-phase source supplying rated voltage and frequency and the motor is run without shaft load. As a result, the rotor will rotate at almost synchronous speed, i.e. slip speed $s \approx 0$. Thus, $R_r/s \to \infty$ and the motor equivalent circuit is broadly similar to that of a transformer on open-circuit. Therefore, the rotor impedance branch is effectively open-circuit and the current drawn by the stator is fully attributed to the exciting current. The stator quantities measured are; line-to-line voltage V_{LL}, current I_s, three-phase input power $P_{3\phi}$ and three-phase input reactive power $Q_{3\phi}$. The iron loss resistance associated with the magnetising impedance can be calculated, if required, by connecting this resistance in parallel with the magnetising reactance. Using the inequality $R_m/X_m \gg 1$, the motor equivalent impedance as seen from the test terminals is approximately given by

$$Z_e = R_s + \frac{X_m^2}{R_m} + j(X_{\sigma s} + X_m) \tag{5.131a}$$

The measured input power includes rotational losses (friction and windage, iron and stray load losses) as well as stator copper losses and is given by

$$P_{3\phi} = 3I_s^2 R_s + 3I_s^2 \frac{X_m^2}{R_m} + \text{Losses}_{(\text{friction/windage}+\text{stray})} \tag{5.131b}$$

Using the measured input reactive power, we have

$$X_{\sigma s} + X_m = \frac{Q_{3\phi}}{3I_s^2} \tag{5.131c}$$

Using the measured input reactive power and stator current, and the calculated value of $X_{\sigma s}$, X_m is calculated using Equation (5.131c). Also, R_m is calculated from Equation (5.131b) using the measured input power and stator current, as well as the calculated quantities R_s and X_m. If the friction/windage and stray losses are not known, then they may be neglected with little error. If required, the very small stator copper losses in Equation (5.131b) may also be neglected and the calculation of R_m and X_m proceeds similarly.

Transient parameters from a sudden three-phase short-circuit immediately after motor disconnection

This method consists of two steps; the first is to disconnect the motor from the supply network and the second is to quickly apply a simultaneous three-phase short-circuit fault at the motor terminals. As for synchronous machines, this is usually a suitable test method for the determination of the motor subtransient and transient parameters.

Immediately upon supply disconnection, the flux linking the closed rotor inductive circuits prevents the rotor current from changing instantaneously to zero. This flux will decay in a manner determined by the rotor open-circuit subtransient and transient time constants T_0'' and T_0', respectively. It can be shown that the phase r open-circuit motor terminal voltage following motor disconnection, where the motor has a double squirrel-cage or deep-bar rotor, is given by

$$v_r(t) = \sqrt{2}I_{rms}[[(X_{\sigma s} + X_m) - X']e^{-t/T_0'} + (X' - X'')e^{-t/T_0''}]\cos(\omega_s t + \delta) \quad (5.132a)$$

where I_{rms} is the initial stator current before motor disconnection and δ is its phase angle.

The three-phase short circuit is applied a few cycles after the removal of the motor supply. The range of values of T_0'', even for large induction motors, tends to be between a few ms to about 30 ms so that the voltage component associated with T_0'' will be negligible by the time the short circuit is applied at time t_s as illustrated in Figure 5.28. The only relevant component is therefore the one associated with T_0'. It should be noted that this is the only component applicable in the case of a rotor with a single rotor winding, e.g. a single squirrel-cage rotor or a wound rotor. Therefore

$$v_r(t) = \sqrt{2}I_{rms}[(X_{\sigma s} + X_m) - X']e^{-t/T_0'}\cos(\omega_s t + \delta) \quad t \geq 5T_0'' \quad (5.132b)$$

Figure 5.28 illustrates the peak envelope of the motor phase r open-circuit voltage after supply disconnection. At the instant of short-circuit $t = t_s$, following the decay of the subtransient voltage component, the rms value of the motor open-circuit voltage is given by

$$V_{oc} = I_{rms}[(X_{\sigma s} + X_m) - X']e^{-t_s/T_0'} = V'e^{-t_s/T_0'} \quad (5.133)$$

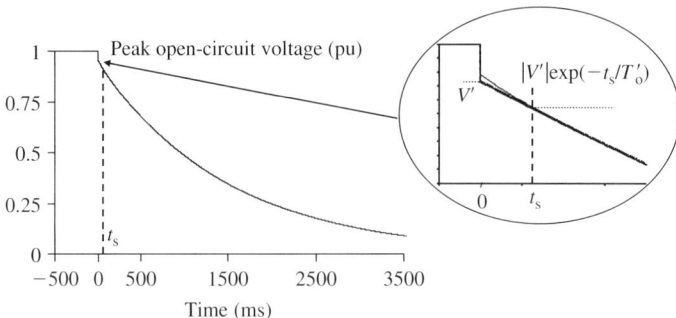

Figure 5.28   Upper envelope of open-circuit voltage of an induction motor upon supply disconnection

This voltage can be measured from the recorded voltage envelope. Since the motor is on open circuit, the equations that describe the motor short-circuit currents will be the same as those derived in Equations (5.117a) and (5.118a) or (5.122a) and (5.123a) but with V_{rms} replaced with the magnitude of V_{oc} of Equation (5.133). The envelope of the current can be used to determine the motor subtransient and transient reactances and time constants as described for synchronous machines.

5.8.5 Examples

Example 5.5   A 50 Hz three-phase induction motor of a single squirrel-cage rotor construction drives a condensate pump in a power generating station and has the following data: rated phase-to-phase voltage $= 3.3$ kV, rated three-phase apparent power $= 900$ kVA, $R_s = 0.0968$ Ω, $X_{\sigma s} = 1.331$ Ω, $X_m = 38.7$ Ω, and $R_r = 0.0726$ Ω, $X_{\sigma r} = 0.847$ Ω at rated slip. The rotor parameters referred to the stator voltage base. Calculate the motor parameters in pu on kVA rating, the motor transient reactance, open circuit, short circuit and armature time constants.

Base voltage and base apparent power are 3.3 kV and 0.9 MVA, respectively. Base impedance is equal to $(3.3)^2/0.9 = 12.1$ Ω. Therefore, $R_s = 0.0968/12.1 = 0.008$ pu, $X_{\sigma s} = 1.331/12.1 = 0.11$ pu, $X_m = 38.7/12.1 = 3.2$ pu, $R_r = 0.0726/12.1 = 0.006$ pu, $X_{\sigma r} = 0.847/12.1 = 0.07$ pu.

From Equations (5.108)

$$X' = 0.11 + \frac{3.2 \times 0.07}{3.2 + 0.07} = 0.1785 \text{ pu}$$

and

$$T_a = \frac{0.1785}{2\pi 50 \times 0.008} \times 1000 = 71 \text{ ms}$$

From Equations (5.93) and (5.95b)

$$T_o' = \frac{3.2 + 0.07}{2\pi \times 50 \times 0.006} = 1.73 \text{ s}$$

and

$$T' = \frac{0.07 + (3.2 \times 0.11)/(3.2 + 0.11)}{2\pi \times 50 \times 0.006} \times 1000 = 93.5 \text{ ms}$$

In many situations, the internal motor parameters may not be known but T' and X' may be known from measurements. In this case, the following empirical formula for T' that provides sufficient accuracy may be used $T' = X'/\omega_s R_r$. This is equivalent to replacing the numerator of Equation (5.93b) by X'. The numerator of Equation (5.93b) is equal to 0.1763 pu and hence the use of $T' = X'/\omega_s R_r$ to calculate the transient time constant gives a slightly higher and thus conservative time constant by about 1.2%.

Example 5.6   In the system shown in figure in Example 5.6, the two induction motors are identical and have the data given in Example 5.5.

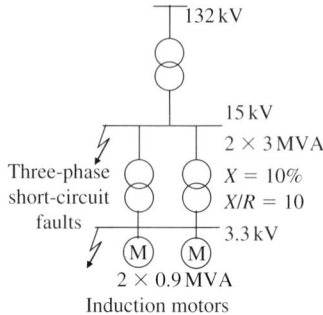

Calculate the peak short-circuit current contribution of the motors at $t = 0$, 10 and 100 ms for a three-phase short-circuit fault at (a) the 3.3 kV terminals of the motors, and (b) the 15 kV medium voltage busbar. Assume a 1 pu initial voltage at the 3.3 and 15 kV busbars and that the motors are initially unloaded. Consider only the change in motor current due to the fault.

(a) Three-phase short-circuit fault at the motor 3.3 kV busbar terminals
Using Equation (5.118a), the rms and dc short-circuit currents from each motor are equal to

$$i_{rms}(t) = \frac{1}{0.1785}e^{-t/93.5}$$

and

$$i_{dc}(t) = \frac{\sqrt{2} \times 1}{0.1785}e^{-t/71}$$

The rms current at $t = 0$, 10 and 100 ms is equal to 5.6, 5.03 and 1.92 pu, respectively. The dc current at $t = 0$, 10 and 100 ms is equal to 7.92, 6.88 and 1.94 pu, respectively. The peak current contribution at $t = 0$, 10 and 100 ms is equal to 15.84, 13.76 and 3.88 pu respectively. The total current contribution from both motors at $t = 0$, 10 and 100 ms is equal to 31.68, 27.52 and 7.76 pu, respectively.

(b) Three-phase short-circuit fault at the 15 kV medium voltage busbar
The equivalent transient reactance of the two motors is $0.1785/2 = 0.08925$ pu. The external impedance is equal to the transformer's impedance and is calculated on 0.9 MVA base as

$$Z_e = (0.01 + j0.1) \times \frac{0.9}{3} = (0.003 + j0.03)\text{pu}$$

The time constants of the equivalent motor representing the two parallel motors do not change but the stator resistance and transient reactance are halved so that $R_s = 0.004$ pu and $X' = 0.08925$ pu. The external impedance modifies the motor's short-circuit time constant according to Equation (5.121e), thus

$$T'_e = \frac{0.08925 + 0.03}{1.655 + 0.03} \times 1.73 = 122.4 \, \text{ms}$$

where $X = (3.2 + 0.11)/2 = 1.655$ pu.

The armature time constant is modified using Equation (5.121g) to

$$T_{ae} = \frac{0.08925 + 0.03}{2\pi 50 \times (0.004 + 0.003)} = 54.2 \, \text{ms}$$

Using Equation (5.121f), the rms and dc short-circuit currents are given by

$$i_{rms}(t) = \frac{1}{0.08925 + 0.03} e^{-t/122.4}$$

and

$$i_{dc}(t) = \frac{\sqrt{2}}{0.08925 + 0.03} e^{-t/54.2}$$

The rms current at $t = 0$, 10 and 100 ms is equal to 8.386, 7.73 and 3.7 pu, respectively. The dc current at $t = 0$, 10 and 100 ms is equal to 11.86, 9.86 and 1.87 pu, respectively. The peak current contribution at $t = 0$, 10 and 100 ms is equal to 23.7, 19.72 and 3.74 pu, respectively.

5.9 Modelling of wind turbine generators in short-circuit analysis

5.9.1 Types of wind turbine generator technologies

The most important feature of wind turbine generators that are directly connected to the ac network is that induction, rather than synchronous, generators are used in nearly all commercial wind turbine installations throughout the world. Synchronous generators are also widely used but, currently, these are not directly connected to the ac network. Rather, they are connected through a back-to-back power electronics converter. The use of directly connected synchronous generators may become possible if the development of variable speed gearbox technology proves sufficiently reliable and commercially attractive. The variable speed gearbox has a variable input speed on the turbine side but produces constant output synchronous speed. This technology is currently under research, development and field testing and large-scale commercial use may be imminent. The four currently most used wind turbine generator technologies in large-scale wind farm installations world wide are briefly described below.

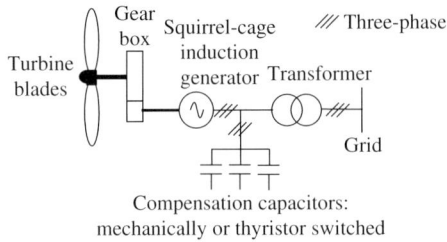

Figure 5.29   Fixed speed wind turbine induction generator

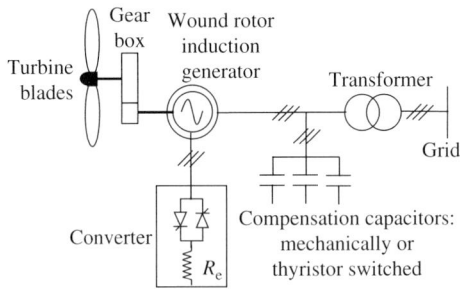

Figure 5.30   Small speed range wind turbine wound rotor induction generator

'Fixed' speed induction generators

These are essentially similar to squirrel-cage induction motors and are driven by a wind turbine prime mover at a speed just above synchronous speed, normally upto 1% rated slip for today's large wind turbines. Because the speed variation from no load to full load is very small, the term 'fixed' speed tends to be widely used. The generator is coupled to the wind turbine rotor via a gearbox as shown in Figure 5.29. The design and construction of the stator and rotor of an induction generator are similar to that of a large induction motor having a squirrel-cage rotor.

Small speed range wound rotor induction generators

This type of induction generator utilises a three-phase wound rotor winding that is accessible via slip rings. The rotor windings are usually connected to an external resistor circuit through a modern power electronics converter that modifies the rotor circuit resistance by injecting a variable external resistance on the rotor circuit. This method enables control of the magnitude of rotor currents and hence generator electromagnetic torque. This enables the speed of the generator to vary over a small range, typically upto 10%. Figure (5.30) illustrates this type of wind turbine generator.

Variable speed doubly fed induction generators

These are variable speed wound rotor induction generators that utilise two bidirectional back-to-back static power electronic converters of the voltage source type as shown in Figure 5.31. One converter is connected to the ac network either at the generator stator terminals or the tertiary winding of a three-winding generator

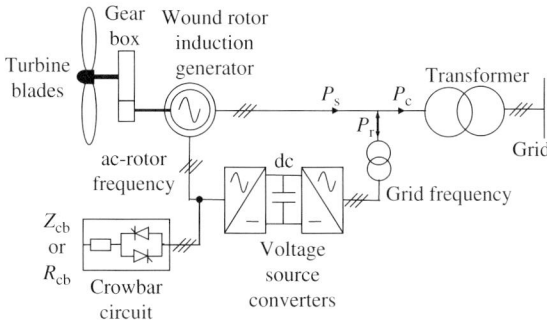

Figure 5.31   Wind turbine variable speed doubly fed induction generator

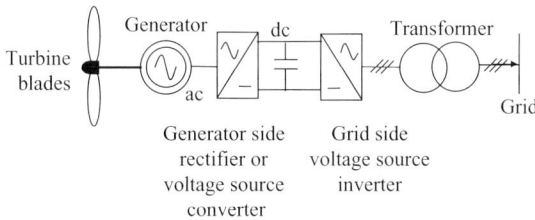

Figure 5.32   Variable speed series converter-connected wind turbine generator

step-up transformer. The other converter is connected to the rotor windings via a slip rings.

The purpose of the rotor-side converter is to inject a three-phase voltage at slip frequency onto the rotor circuit. The injected voltage can be varied in both magnitude and phase by the converter controller and this controls the rotor currents almost instantaneously. This control provides two important functions. The first is variation of generator electromagnetic torque and hence rotor speed. The second may be constant stator reactive power output control, stator power factor control or stator terminal voltage control. Typical speed control range for a modern MW class wind turbine doubly fed induction generator may be between 70% and 120% of nominal synchronous speed with the 120% usually referred to as the rated speed at which rated MW output is produced.

Variable speed series converter-connected generators

These are ac generators that are connected to the ac network through a ac/dc/ac series back-to-back converter with a rating that matches that of the generator. Many of such electrical generators presently in use are low speed multi-pole 'ring' type synchronous generators that are directly driven by the wind turbine rotor, i.e. there is no gearbox between the turbine rotor and generator. Other types of electrical generators are also used such as squirrel-cage induction generators but these are connected to the turbine rotor through a gearbox. In a series converter-connected generator, the full output of the generator is fed into a power electronics rectifier or a voltage sourced converter, then through a dc link into a static power electronics inverter as illustrated in Figure 5.32. In this type of technology, the electrical generator is isolated from the ac network. The generator side converter is usually used for controlling the generator speed and the network side inverter is usually used for voltage/reactive power control and dc link capacitor voltage control.

5.9.2 Modelling of fixed speed induction generators

We have already presented the modelling of synchronous generators and induction motors for short-circuit analysis. The PPS and NPS short-circuit contribution of a fixed speed induction generator can be represented in a similar way to that of an induction motor. The equations of subtransient and transient reactances and time constants derived for induction motors can also be used for induction generators. The stator windings of these generators are usually connected in delta or star with an isolated neutral. Thus, their ZPS impedance to the flow of ZPS currents is infinite.

5.9.3 Modelling of small speed range wound rotor induction generators

When a three-phase short-circuit fault occurs at the terminals of an induction generator, the generator will inherently supply a large stator short-circuit current. As we have already seen for synchronous generators and induction motors, due to the theorem of constant flux linkages, a corresponding increase in the generator's rotor currents occurs. For a small speed range wound rotor induction generator, the rotor converter used to control the rotor currents is suddenly subjected to large overcurrents. The junction temperature of the thyristor or transistor switches used in the converter cannot be exceeded or they will be damaged. The short-term overcurrent capability of the electronic switches used within the converter is extremely limited and almost practically non-existent if they are normally designed to operate at or close to their rated junction temperature. Therefore, when the instantaneous current limit of the switches is exceeded, they are immediately blocked and this effectively inserts the entire external resistance in series with the rotor circuit. The blocking of the switches is very fast and occurs typically in about 1 ms. This causes, the rotor winding of the generator to become effectively similar to that of a conventional wound rotor induction motor but with an additional external resistance. Therefore, the generator's initial short-circuit current contribution to a fault on the network can be calculated as for an induction motor with a single-winding rotor having an increased rotor resistance. Using Equation (5.100b) for T_4 and Equation (5.102b) for $T' = T_4$, and designating the external resistance as R_e, the transient time constant is given as

$$T'_e = \frac{1}{\omega_s(R_r + R_e)}\left(X_{\sigma r} + \frac{1}{\frac{1}{X_{\sigma s}} + \frac{1}{X_m}}\right) \approx \frac{X'}{\omega_s(R_r + R_e)}\text{s} \qquad (5.134)$$

The additional rotor resistance reduces the motor transient time constant and hence increases the rate of decay of the ac current component of the short-circuit current. Figure 5.33 illustrates the effect of external rotor resistance on the stator short-circuit rms current for a typical 2 MW induction generator.

As the stator short-circuit current decays, and the rotor current drops below the protection or the converter-controlled reference value, some generator manufacturers design their converters to quickly unblock the switches and regain control of the rotor currents back to some specified value. This means that the generator starts to supply a constant low value stator short-circuit current.

5.9.4 Modelling of doubly fed induction generators

The *dq* model of a doubly fed induction generator with a wound rotor is similar to that of an induction motor presented in Equation (5.88). However, there is an important exception; the *dq* rotor voltages e_{dr} and e_{qr} given in Equation (5.88b) are not equal to zero. They are equal to the voltages injected by the rotor-side converter. The steady state equivalent circuit of the doubly fed induction generator is shown in Figure 5.34 with all rotor quantities referred to the stator.

A three-phase short-circuit fault in the network will cause a symmetrical voltage dip at the generator terminals and large oscillatory currents in the rotor winding connected to the rotor-side converter. Controlling large rotor currents requires a large and uneconomic rotor voltage rating. Therefore, the large rotor currents can flow through and damage the converter switches. However, to protect these switches from damage, one method is to use a converter protective circuit called a 'crowbar' circuit that is connected to the rotor winding through anti-parallel thyristors as illustrated in Figure 5.31. When a large instantaneous rotor current in any phase in excess of the allowable converter limit is detected, the converter switches are immediately blocked and the thyristors of the crowbar circuit are

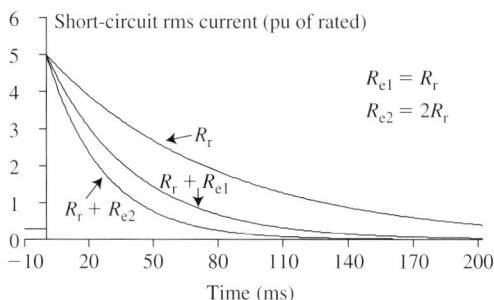

Figure 5.33 Effect of external rotor resistance on the short-circuit current of a small speed range wound rotor induction generator

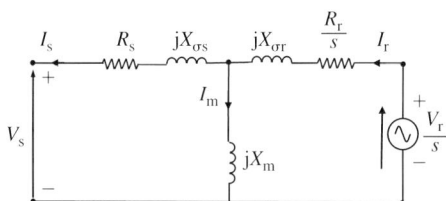

Figure 5.34 Steady state equivalent circuit of a doubly fed induction generator

fired to prevent a large overvoltage on the dc link. This crowbar action effectively short circuits and bypasses the converter and causes the rotor currents to flow into the crowbar circuit. Since the generator may be operating at a speed significantly above synchronous speed, some crowbar circuits may also insert an impedance, usually a resistance only, in series with the rotor winding in order to reduce the reactive power consumption of the generator and improve its electric torque/speed performance. The blocking of the converter switches and operation of the crowbar protective circuit are very fast and may occur well within 2 ms. Once the converter is bypassed, the rotor winding of the generator appears effectively similar to that of a conventional wound rotor induction generator with an external rotor resistance R_{cb} as shown in Figure 5.35a.

What happens next depends on the requirement of the power system to which the generator is connected. The whole wind turbine generator may be disconnected from the ac grid. Alternatively, only the rotor converter may be disconnected but not the wind turbine generator that continues to operate as a wound rotor induction machine with a higher rotor resistance. A more grid friendly converter control strategy is to keep the turbine generator connected to the grid and rotor converter connected to the rotor but quickly regain rotor current control once the rotor currents have decayed below a sufficiently low value.

Returning to the short-circuit current contribution of the generator, once the converter is bypassed and the rotor is short-circuited through a resistance R_{cb}, the transient short-circuit time constant of the doubly fed induction generator is modified by R_{cb} and is calculated using Equation (5.134) as follows:

$$T'_{cb} = \frac{1}{\omega_s(R_r + R_{cb})}\left(X_{\sigma r} + \frac{1}{\frac{1}{X_{\sigma s}} + \frac{1}{X_m}}\right) \approx \frac{X'}{\omega_s(R_r + R_{cb})}s \qquad (5.135)$$

However, unlike a conventional fixed speed induction machine whose slip is always close to zero, the slip for a typical doubly fed induction generator may vary between $s = 0.3$ pu sub-synchronous and $s = -0.2$ pu super-synchronous speed depending

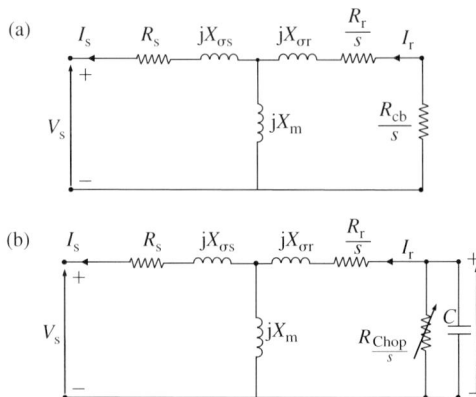

Figure 5.35 Doubly fed induction generator with operation of rotor side converter protection: (a) converter short-circuited by a crowbar through resistance R_{cb} and (b) converter protected by a dc chopper through resistance R_{Chop} in parallel with dc link capacitor

on machine rating and design. Therefore, using Equation (5.112b), and assuming that R_{cb} is not much smaller than X', the current change due to a three-phase short-circuit fault is approximately given by

$$i_r(t) \approx \frac{\sqrt{2}V_{rms}}{(1-s)\sqrt{X'^2 + R_{cb}^2}}[e^{-t/T'_{cb}}\cos\{(1-s)\omega_s t + \theta_o - \pi/2\} - e^{-t/T_a}\cos(\theta_o - \pi/2)]$$

(5.136a)

where X' and T_a are as given in Equations (5.94c) and (5.108) for a single rotor winding, respectively.

The envelope of the maximum short-circuit current is given by

$$\hat{i}_r(t) \approx \frac{\sqrt{2}V_{rms}}{(1-s)\sqrt{X'^2 + R_{cb}^2}}[e^{-t/T'_{cb}} + e^{-t/T_a}]$$

(5.136b)

Clearly, in the general case of a doubly fed induction generator, the factor $(1-s)$ in Equations (5.136a) and (5.136b) cannot be equated to unity unless when the generator's initial speed is close to synchronous speed. From Equation (5.136a), the frequency f_{sc} of the stator short-circuit current is given by

$$f_{sc} = (1-s)f_s \text{ Hz}$$

(5.136c)

where s is generator slip in per unit and f_s is the nominal system frequency in Hz. The difference between the frequency of the stator short-circuit current and nominal frequency affects the time instant that corresponds to the first and subsequent peaks of the short-circuit current of Equation (5.136a). Whereas this time instant is nearly equal to a half cycle of the nominal frequency for synchronous machines and fixed speed induction machines, this will only be the case for a doubly fed generator if its rotor speed is initially close to synchronous speed. If the doubly fed generator is initially operating at a high active power output, e.g. at a slip $s = -0.15$ pu, its speed will be super-synchronous and the frequency of the short-circuit current, assuming a 50 Hz system, is $f_{sc} = 1.15 \times 50 \text{ Hz} = 57.5 \text{ Hz}$. The effect of this is that the instant of the first peak of the current will occur before the usual half cycle of power frequency. Conversely, if the machine is initially operating at a sub-synchronous speed, e.g. at a slip $s = 0.2$ pu, we have $f_{sc} = 0.8 \times 50 \text{ Hz} = 40 \text{ Hz}$ and the instant of the first peak of the current will occur after the half cycle of nominal frequency. Equation (5.136a) also suggests that the magnitude of the current is inversely proportional to $(1-s)$ that is the initial current magnitude is higher if the generator is initially operating at minimum sub-synchronous speed.

In the grid friendly doubly fed generator design referred to above, the crowbar circuit is quickly switched out and the converter switches unblocked after some time delay from the instant of short-circuit current. The delay is to allow the rotor instantaneous current particularly its dc component to decay to a sufficiently low value. The switching out of the crowbar and unblocking of the converter switches allows the converter to regain control of the rotor currents. This enables the generator to supply a predefined and constant value of stator short-circuit current with a magnitude that is dependent on the retained stator voltage. It is usual that this constant current is the maximum reactive current that can be constantly supplied by the generator under reduced terminal voltage conditions without exceeding its rating.

If the fault location is on the network and is sufficiently remote from the generator, the rotor-side converter current limit may not be exceeded and hence the protective crowbar circuit would not operate. Thus, the magnitude of the stator current supplied may be affected and modified by the converter's voltage/reactive power control strategy. For example, consider a rotor side converter operating in stator terminal voltage control mode using an AVR acting through the converter to deliver the required change in rotor voltage. The extent to which the stator short-circuit current supplied by the generator is affected by the AVR is influenced by the converter control system parameters, e.g. gain and time constants as well as the generator short-circuit transient time constant T'. Similar analysis to that presented in Section 5.4.7 for a synchronous generator with no damper winding can also be used here for the doubly fed wound rotor induction generator. Thus, if an instantaneous or a step change in rotor voltage is assumed, the stator current begins to increase with a time constant equal to the induction generator short-circuit time constant T'. This change will be superimposed on the current caused by the short-circuit which decays with the same time constant T'. Therefore, the decay of the short-circuit current supplied by the generator may be arrested and at some point the current may even begin to increase. This behaviour is similar qualitatively to that of a synchronous generator shown in Figure 5.14 although the time when the current may begin to increase may be much shorter and in the order of only few cycles of the power frequency. If this occurs within circuit-breaker current interruption times, then this increase would need to be taken into consideration when assessing the short-circuit break duties of circuit-breakers.

The short-circuit current behaviour of the doubly fed generator depends on whether the fault is balanced or unbalanced. During unbalanced faults on the network that produce a large NPS voltage at the generator terminals and a corresponding large NPS rotor current, permanent crowbar operation may result until the fault is cleared.

Converter protection strategies used in doubly fed generator technology are still developing. An alternative to the use of a crowbar circuit is a strategy that blocks the converter switches but controls the dc link capacitor voltage using a dc chopper circuit in parallel with the dc link capacitor. The dc chopper circuit is a power electronics' controlled resistor, i.e. a variable resistor. The blocking of the converter switches and operation of the dc chopper circuit in parallel with the dc link capacitor is represented as shown in Figure 5.35(b). The effect of the dominant dc chopper resistance R_{Chop} on the machine's stator short-circuit current behaviour is similar to the effect of R_{cb} of the crowbar circuit shown in Figure 5.35(a).

The inclusion of short-circuit current contribution in steady state fixed impedance analysis programs, requires knowledge of the variation with time of the short-circuit current during the fault period. This variation may need to be established for near and remote faults, i.e. with and without crowbar, or other converter protective action, using detailed time domain simulation programs and taking into account the voltage/reactive current control strategy of the converter. Manufacturers are best placed to submit such information to power network utilities.

5.9.5 Modelling of series converter-connected generators

The static converters effectively isolate the electrical generator and its specific electrical performance characteristics from the electrical network in the event of short-circuit faults on the host network. Many modern static inverters tend to use insulated gate bipolar transistor (IGBT) switches. The performance of series converter-connected generators during short-circuit faults on the converter output terminals or elsewhere on the ac network is dictated by the control strategy of the inverter. Before a short circuit occurs, the inverter output current is controlled both in magnitude and phase. The phase is usually set within minimum and maximum limits with respect to the zero crossing of the output voltage in order to ensure that the inverter reactive power output is kept within corresponding minimum and maximum limits. When a three phase short-circuit occurs on the ac network, the inverter's inherent constant current control strategy ensures that the inverter continues to supply the same current as that supplied in the power frequency cycle just before the occurrence of the short-circuit fault. The magnitude of the inverter short circuit current contribution due to the inherent constant current control strategy in the event of a three-phase short circuit at bus i can be illustrated using Figure 5.36.

Figure 5.36(a) shows a series converter-connected generator, which may be induction or synchronous, with the grid inverter connected through a transformer to a distribution voltage level of typically 20–36 kV. This is then connected to a higher voltage system through another transformer. The largest pre short-circuit inverter current is that supplied at rated power and reactive power output of the inverter. Assume that at bus i the inverter rated power output is P, rated lagging reactive power output is Q and the rated lagging and (leading) power factor is $\cos \phi$. Thus, the rated inverter current delivered at bus i just before the short circuit is given by

$$I_{i(rated)} = \frac{P - jQ}{V_i^*} = |I_{i(rated)}| e^{-j\phi} \qquad (5.137a)$$

The relationship between bus i voltage and system voltage V_s at bus s, ignoring the transformer resistance R_t, because it is much smaller than its reactance X_t, is given by

$$V_s = V_i - jI_{i(rated)}X_t = |V_s| e^{-j\delta} \qquad (5.137b)$$

At bus s, the grid system is represented by its three-phase short-circuit infeed and X_s/R_s ratio. Since the grid inverter continues to supply a constant, balanced three-phase current during the short-circuit fault and the resultant voltage dip at its terminals, the inverter and its transformer can be represented as a constant PPS current source as shown in Figure 5.36(b). For a three-phase short-circuit fault at bus i, the PPS current supplied to the fault from the high voltage

(a)   Series converter-connected
      wind turbine generator,
      synchronous or induction

(b)

(c)

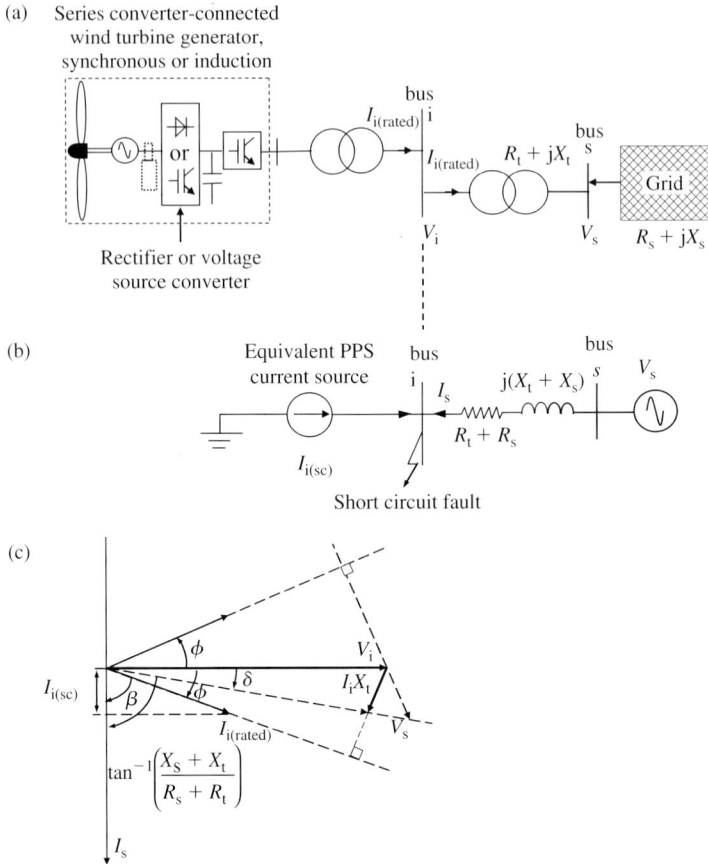

Figure 5.36 Modelling the short-circuit contribution of series converter-connected wind turbine generator: (a) initial prefault condition, (b) short-circuit fault condition and (c) vector diagram under fault conditions

grid system is given by

$$I_s = \frac{V_s}{Z_s + Z_t} = \frac{V_s}{\sqrt{(R_s + R_t)^2 + (X_s + X_t)^2}} \exp[-j\tan^{-1}[(X_s + X_t)/(R_s + R_t)]$$

(5.138)

In practice, $(X_s + X_t)/(R_s + R_t)$ will be dominated by the transformer X_t/R_t ratio which is typically between 14 and 30. These correspond to impedance phase angles $\tan^{-1}(X_t/R_t)$ of 86° and 88°, respectively. Figure 5.36(c) shows the vector diagram of the above voltages and currents including the system short-circuit current. The inverter PPS short-circuit current component that is in phase with I_s is given by

$$I_{i(sc)} = I_{i(rated)} \cos \beta = I_{i(rated)} \cos\left[\tan^{-1}\left(\frac{X_t}{R_t}\right) - (\phi - \delta)\right]$$

(5.139)

For example, for $X_t/R_t = 14.3$ or $\tan^{-1}(X_t/R_t) = 86°$, rated inverter lagging power factor on the high voltage side of its transformer is 0.95 or $\phi = \cos^{-1}(0.95) = 18.2°$ and load angle $\delta = 6°$, we have $I_{i(sc)} = I_{i(rated)} \cos[86 - (18.2 - 6)] \approx 0.28 I_{i(rated)}$.

The total short-circuit current is the sum of the high voltage system and inverter short-circuit current contributions, i.e. $I_{Total} = I_s + I_{i(sc)}$.

If the inverter were operating at rated leading power factor, it can be shown that

$$I_{i(sc)} = I_{i(rated)} \cos\left[\tan^{-1}\left(\frac{X_t}{R_t}\right) - (\phi + \delta)\right] \tag{5.140}$$

However, in this case, although this current has a larger magnitude, it is in anti-phase with the dominant current supplied from the higher voltage system and will thus act to reduce the total short-circuit current. Equation (5.139) shows that the inherent inverter short-circuit current contribution is a small fraction of the inverter rated current.

Unlike a doubly fed induction generator, series converter-connected generators with a voltage sourced grid-side converter can provide a controlled short-circuit current response during the entire fault duration. Therefore, network utilities usually require a larger short-circuit current contribution from such generators to assist network protection in the detection of short-circuit currents. Therefore, modern grid-side inverter controls are usually designed to supply a larger and constant value of three-phase short-circuit current that is related to the rated current by

$$I_{i(sc)} = \alpha I_{i(rated)} \tag{5.141}$$

where $\alpha \leq 3$ for most current inverter designs.

As we have already explained, the inverter can be considered to act as a PPS constant current source during the entire short-circuit period as given by Equation (5.141). In addition, the inverter current control strategy ensures that the PPS short-circuit current supplied does not contain a dc current component.

Under unbalanced short-circuit faults in the network, e.g. a one-phase to earth short-circuit fault, the control systems of most modern inverters are designed to continue to supply balanced three-phase currents irrespective of the degree of their voltage unbalance. Therefore, under unbalanced short-circuit faults, the inverter will only supply a PPS current with the NPS and ZPS currents being equal to zero.

In summary, series converter-connected generators with grid-side inverters that deliver an increased constant current can be represented in short-circuit studies by a PPS constant current source, given by Equation (5.141), and zero NPS and ZPS current sources. It is generally incorrect to assume that the inverter PPS short-circuit current contribution can be represented as a voltage source behind a PPS reactance in pu equal to $1/\alpha$. This representation will produce the correct inverter current contribution for one fault location only namely at the inverter transformer output terminals. At other fault locations on the grid network, the voltage source representation will produce a lower and incorrect current than that given by Equation (5.141). The inverter constant current control strategy is normally very fast so that the ac short-circuit current contribution does not change with time during the short circuit. This means that the initial peak current and break current are equal.

Further reading

Books

[1] Kundur, P., *Power System Stability and Control*, McGraw-Hill, Inc., 1994, ISBN 0-07-035958-X.
[2] Krause, P.C., *Analysis of Electric Machinery*, McGraw-Hill, 1986, New York.
[3] Adkins, B., *et al.*, *The General Theory of Alternating Current Machines*, Chapman and Hall, 1975, ISBN 0-412-15560-5.
[4] Concordia, C., *Synchronous Machines – Theory and Performance*, 1951, John Wiley, New York.

Papers

[5] Kai, T., *et al.*, A simplified fault currents analysis method considering transient of synchronous machine, *IEEE Transactions on Energy Conversion*, Vol. 12, No. 3, September 1997, 225–231.
[6] Hwang, H.H., Transient Analysis of unbalanced short circuits of synchronous machines, *IEEE Transactions on Power Apparatus and Systems*, Vol. PAS-88, No.1, January 1969, 67–71.
[7] Hwang, H.H., Mathematical analysis of double line to-ground short circuit of an alternator, *IEEE Transactions on Power Apparatus and Systems*, Vol. PAS-86, No. 10, October 1967, 1254–1257.
[8] Kamwa, I., *et al.*, Phenomenological models synchronous machines from short-circuit test during commissioning – a classical/modern approach, *IEEE Transactions on Energy Conversion*, Vol. 9, No. 1, March 1994, 85–97.
[9] Kamwa, I., *et al.*, Experience with computer-aided graphical analysis of sudden-short-circuit oscillograms of large synchronous machines, *IEEE Transactions on Energy Conversion*, Vol. 10, No. 3, September 1995, 407–414.
[10] Canay, I.M., Determination of model parameters of synchronous machines, *Proceedings IEE*, Vol. 130, No. 2, March 1983, 86–94.
[11] Luke, Y.Y., *et al.*, Motor contribution during three-phase short circuits, *IEEE Transactions On Industry Applications*, Vol. IA-18, No. 6, November/December 1982, 593–599.
[12] Kalsi, S.S., *et al.*, Calculating of system-fault currents due to induction motors, *Proceedings IEE*, Vol. 118, No. 1, January 1971, 201–215.
[13] Cooper, C.B., *et al.*, Application of test results to the calculation of short-circuit levels in large industrial systems with concentrated induction-motor loads, *Proceedings IEE*, Vol. 116, No. 11, November 1969, 1900–1906.
[14] Huening, W.C., Calculating short-circuit currents with contributions from induction motors, *IEEE Transactions on Industry Applications*, Vol. IA-18, No. 2, March/April 1982, 85–92.
[15] Niiranen, J., Voltage dip ride through of a doubly fed generator equipped with an active crowbar, *Nordic Wind Power Conference*, Chalmers University of Technology, Sweden, March 2004.
[16] Gertmar, L., *et al.*, New method and hardware for grid-connection of wind turbines and parks, *Nordic Wind Power Conference*, ESPOO, Finland, May 2006.

6

Short-circuit analysis techniques in ac power systems

6.1 General

In modern power systems, there are various types of active sources that can contribute short-circuit currents in the event of short-circuit faults on the power system. These include synchronous and induction generators, synchronous and induction motors and modern power electronics converter connected generators. Power system elements such as lines, cables, transformers, series reactors, etc. between the short-circuit fault location and the various current sources will affect the magnitude of the short-circuit currents infeed into the fault. The effect is generally to reduce the magnitude of the short-circuit currents and to increase the rate at which their ac and dc components will decay.

In Chapter 1, we illustrated in Figure 1.4a the make and break duties of a circuit-breaker and explained their meanings. The calculation of these duties requires the calculation of the ac and dc components of the short-circuit current at the fault location in systems that generally contain many different short-circuit sources. Generally, published literature appears to cover the inherent current decay rate of a single source, generally that of synchronous machine, but not the decay rate in a real system that includes several or many sources. In addition, it is surprising that the analysis of the magnitude or rate of decay of the dc component of short-circuit current in a system that includes several or many sources has received virtually no attention in published power system analysis books. In this chapter, we will briefly analyse the effect of various network configurations on the time constants of the ac and dc components of short-circuit currents and hence the effect on the variation of their magnitudes with time. We will illustrate the effects of various estimation methods of the ac and dc time constants. We will then present some of

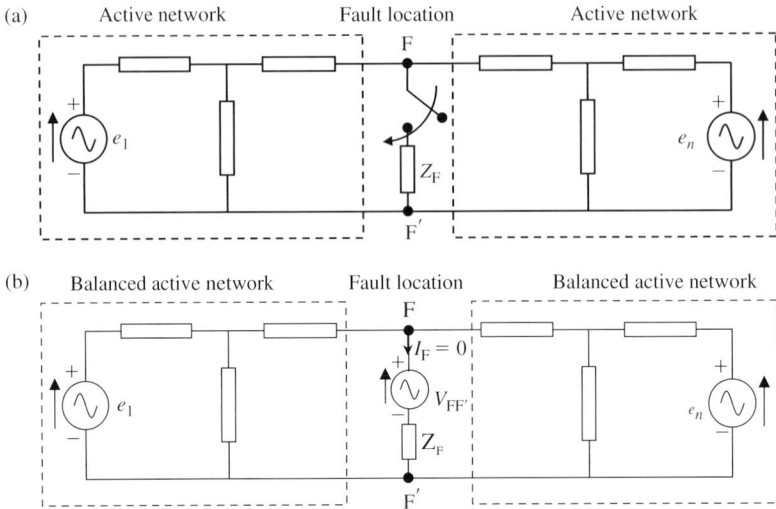

Figure 6.1 Representation of short-circuit faults and prefault network: (a) short-circuit fault through a fault impedance by switch closure and (b) initial network condition before the short-circuit fault

the practical techniques that can be used in the analysis of the ac and dc components of short-circuit currents in interconnected power systems.

6.2 Application of Thévenin's and superposition's theorems to the simulation of short-circuit and open-circuit faults

6.2.1 Simulation of short-circuit faults

Thévenin's and superposition's theorems are very useful in the calculation of currents and voltages due to short-circuit faults in power system networks. The application of a short-circuit fault through a fault impedance Z_F is represented by the closure of the switch between points F and F′ as shown in Figure 6.1(a).

The connection of a voltage source across the fault points FF′ equal in magnitude and phase to the prefault voltage, as shown in Figure 6.1(b), has no effect on the initial current and voltage conditions in the network. When the fault is applied, the voltage across the fault points FF′ drops to that across the fault impedance Z_F i.e. to $Z_F I_F$ or to zero for a solid fault where $Z_F = 0$. This is represented by the connection of a second voltage source in series with Z_F and the prefault voltage source as shown in Figure 6.2. The second voltage source is equal in magnitude but is 180° out of phase with the initial prefault voltage.

According to the superposition principle, the resultant currents and voltages in the network are the sum of those existed before the fault and the changes due to

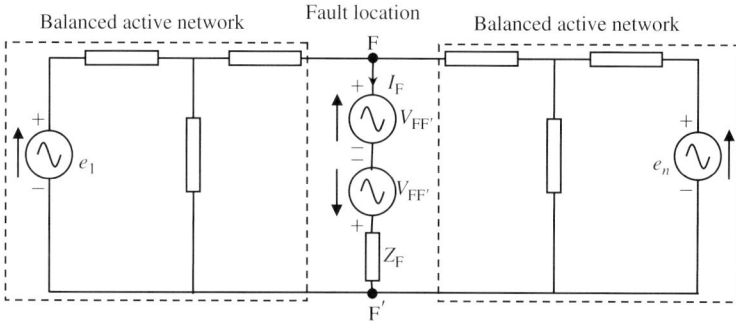

Figure 6.2   Simulation of a short-circuit fault across FF'

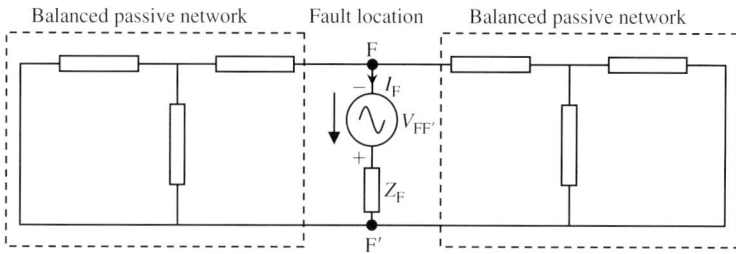

Figure 6.3   Calculation of current and voltage changes due to a short-circuit fault using Thévenin's theorem

the fault itself. Therefore, using the superposition principle, the simulation of the short-circuit fault shown in Figure 6.2 is achieved by the sum of the voltages and currents from two separate networks namely those in Figures 6.1(b) and 6.3.

The first network, shown in Figure 6.1(b), is an active network representing the initial voltage and current conditions before the fault. The second network, shown in Figure 6.3, is a passive network apart from a single voltage source injected across the fault points FF' with all other voltage sources in the rest of the network short-circuited. This is known as Thévenin's network in which the calculated currents and voltages represent the current and voltage changes due to the fault. The final or actual currents and voltages are obtained by the superposition of the current and voltage results obtained from the two networks. The fault current flowing out of the network shown in Figure 6.3 into the fault at the fault location is given by

$$I_F = \frac{V_{FF'}}{Z_{Thév.} + Z_F} = \frac{|V_{FF'}|e^{j\alpha}}{|Z_{Thév.} + Z_F|e^{j\delta}} = |I_F|e^{j\beta} \qquad (6.1a)$$

where $Z_{Thév.} = |Z_{Thév.}|e^{j\phi}$ is the Thévenin's equivalent impedance of the entire network as 'seen' from between points FF', ϕ is the Thévenin's impedance angle, α is the initial prefault voltage angle, δ is the phase angle of $Z_{Thev.} + Z_F$ and $\beta = \alpha - \delta$ is the fault current phase angle. The connection fault current between FF' calculated

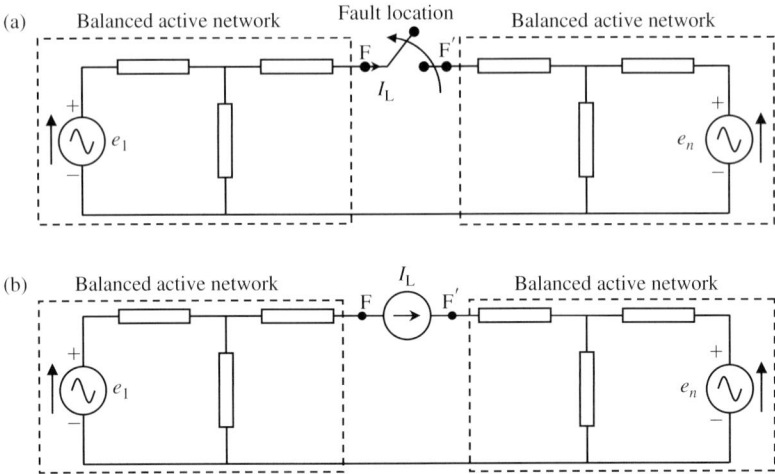

Figure 6.4 Representation of open-circuit faults and initial network: (a) open-circuit fault by switch opening and (b) initial network condition before the open circuit

from the passive network of Figure 6.3 is equal to the total fault current since the initial current is zero. The voltage change across the fault points FF' due to the fault is calculated from Figure (6.3) as $\Delta V_{FF'} = -V_{FF'} + Z_F I_F$ and the actual voltage during the fault is calculated using the superposition theorem from Figure 6.2 as

$$V_{FF'} = Z_F I_F \qquad (6.1b)$$

6.2.2 Simulation of open-circuit faults

The application of an open-circuit fault is represented by the opening of the switch between points F and F' as shown in Figure 6.4(a).

The connection of a current source across the fault points FF' equal in magnitude and phase to the prefault current, as shown in Figure 6.4(b), has no effect on the initial currents and voltages in the network. When the open-circuit fault is applied, the current across the fault points FF' drops to zero and this is represented by the connection of a second current source in parallel with the prefault current source. The second current source is equal in magnitude but is 180° out of phase with the initial prefault current as shown in Figure 6.5.

Using the superposition principle, the simulation of the open-circuit fault shown in Figure 6.5 is achieved by the sum of the voltages and currents from two separate networks as shown in Figures 6.4(b) and 6.6(a).

The first network shown in Figure 6.4(b) is an active network representing the initial voltage and current conditions before the fault. The second network, shown in Figure 6.6(a), is a passive network apart from a single current source injected across the fault points FF' with all other voltage sources short-circuited. The currents and voltages calculated as per Figure 6.6(a) represent the current and

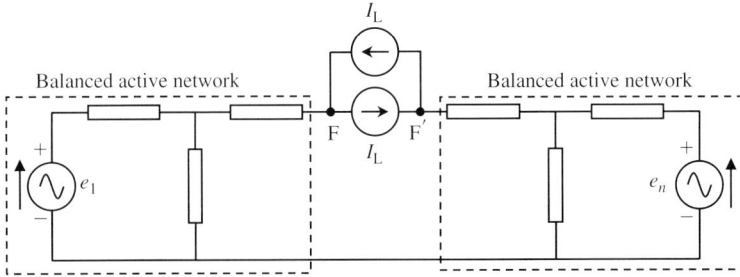

Figure 6.5 Simulation of an open-circuit fault across points FF′

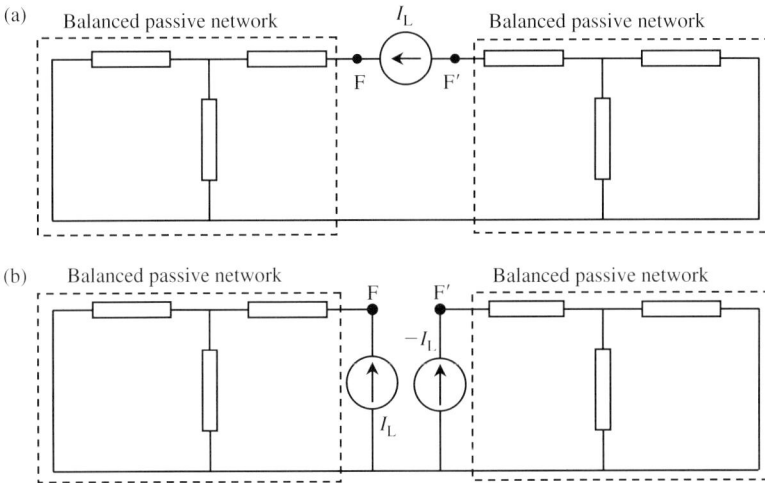

Figure 6.6 Calculation of current and voltage changes due to an open-circuit fault using Thévenin's theorem: (a) series injected current source and (b) series injected current source replaced by two equal shunt injected current sources

voltage changes due to the fault. The actual currents and voltages are obtained by the superposition of the currents and voltages obtained from the two networks. If the Thévenin's equivalent impedance of the entire network as 'seen' from between points FF′ is equal to $Z_{\text{Thév.}}$, the open-circuit voltage change across FF′ is given by

$$\Delta V_{FF'} = Z_{\text{Thév.}} I_L \tag{6.2}$$

where $I_{L'}$ is the known prefault current flowing from F to F′. The prefault voltage between FF′ is zero and hence the actual voltage across the open circuit is equal to that given in Equation (6.2).

An alternative representation to the series current source injected between points FF′ that is useful in practical large-scale analysis by matrix methods is shown in Figure 6.6(b). In this, the single injected series current source is replaced by two shunt injected current sources with that injected at F being equal to I_L and that injected at F' being equal to $-I_L$.

6.3 Fixed impedance short-circuit analysis techniques

6.3.1 Background

Mathematically speaking, the meaning of impedance is only valid when the power system network is in a sinusoidal steady state. Further, as their name implies, fixed impedance analysis techniques assume that the impedance of every element in the power system network is constant. Obviously, this is the case for nearly all passive linear elements in the network, ignoring saturation, but, as we have seen in Chapter 5, the positive phase sequence (PPS) reactances of synchronous and induction machines are time dependent. The fixed impedance analysis technique assumes a constant fixed impedance value for every machine in the system such as X_d'' for synchronous machines and X'' for induction machines. The short-circuit current calculated is therefore the root mean square (rms) subtransient current at $t = 0$ or the instant of fault. The calculation may also be repeated using a different value for each synchronous machine impedance, usually, X_d'. In this case, the short-circuit current calculated is the rms transient current and this may be considered in an empirical way to apply at a specific instant of time, such as 100 or 120 ms, in a meshed multi-machine network. Small induction machines forming part of the general power system load may or may not be represented depending on whether they will continue to provide a current contribution at this time. Some of the major international approaches to the representation of induction motor short-circuit current infeed will be covered in Chapter 7.

6.3.2 Passive short-circuit analysis techniques

The term passive short-circuit analysis means that the initial power system loading is ignored. The assumptions made are that the network voltage profile is uniform, transformer tap positions are nominal, and machine active and reactive power outputs are zero. No load currents flow in the network series branches, the shunt susceptances of network elements, e.g. lines, cables, capacitors, etc., as well as the shunt impedances of static load are not included in the network model. The calculated currents and voltages due to the application of the short-circuit fault are the changes that can be calculated as illustrated in Figure 6.3. The magnitude of the initial voltage at the fault location is usually taken as cU_n where U_n is the nominal system voltage and c is a voltage factor that is typically between 0.9 and 1.1 depending on whether maximum or minimum short-circuit currents are to be calculated. The voltage factor c is used in the international standard IEC 60909. This is discussed in Chapter 7.

6.3.3 The ac short-circuit analysis techniques

The term ac short-circuit analysis means that the initial power system load, i.e. magnitude and power factor, is taken into account prior to the application of the short-circuit fault. This is done using an ac loadflow study where the results are expressed in terms of machine active and reactive power outputs, transformer tap positions, voltage profile and load currents flowing in the network series branches. In the short-circuit study, the shunt susceptances of network elements, e.g. lines, cables, capacitors, etc., as well as the shunt impedances of static load may or may not be included in the network positive phase sequence (PPS), negative phase sequence (NPS) and zero phase sequence (ZPS) models depending on the approach adopted in a particular country. This is covered in Chapter 7. The calculated currents and voltages are obtained using the superposition theorem. The initial voltages and currents are taken from the initial loadflow study.

6.3.4 Estimation of dc short-circuit current component variation with time

Fixed impedance analysis enables the calculation of the ac component of short-circuit current at fixed time instants but the direct and explicit calculation of the dc component is obviously not possible. However, an estimate of the magnitude of the dc component at any time instant may be made in any type of network, i.e. radial or meshed by calculating the equivalent fixed impedance or Thévenin's impedance at the fault point looking back into the active sources of currents. The inherent assumption made in this method is that the mathematical function describing the dc current at the fault point is that of a single decaying exponential with a single time constant, i.e. $\exp(-t/T_{dc})$. The time constant T_{dc} is calculated from the equivalent system impedance 'seen' at the fault point by dividing its reactance to resistance ratio X/R by ω_s. The dc current magnitude can then be estimated at any required time instant. The accuracy of such a method is discussed in Sections 6.5.2 and 6.5.3.

In the case where passive short-circuit analysis techniques are used, as discussed in Section 6.3.2, the resistances of the network elements are included in the network model in order to calculate the equivalent system X/R ratio from the equivalent Thévenin's impedance at the fault point. Alternatively, the equivalent system X/R ratio can be calculated from the Thévenin's impedance phase angle ϕ as follows:

$$\frac{X}{R} = \tan \phi \tag{6.3a}$$

In some countries, e.g. in the American IEEE approach, the resistances of network elements may not be included in the network model used to calculate the current magnitude. Also, the equivalent X/R ratio at the fault point is calculated from separate reactance and resistance networks. The resistance network is obtained by setting all reactances to zero whereas the reactance network is obtained by setting all resistances to zero. An equivalent Thévenin's resistance and another equivalent

Thévenin's reactance are calculated and the equivalent X/R ratio at the fault point is calculated as

$$\frac{X}{R} = \frac{X_{\text{Thév.}}}{R_{\text{Thév.}}} \qquad (6.3b)$$

The reason for calculating the X/R ratio of the dc component of short-circuit current using separate networks is to improve the accuracy of calculation. This is discussed in Chapter 7.

6.3.5 Estimation of ac short-circuit current component variation with time

As stated in Section 6.3.1 fixed impedance analysis enables the calculation of the ac component of short-circuit current at fixed time instants. For example, the subtransient and transient rms currents are calculated using the subtransient and transient reactances of machines, respectively. However, since the PPS reactances of synchronous and induction machines are time dependent, the ac component of fault current will in general vary with time, the extent of variation being dependent on the proximity of the fault to machines. In practical calculations, particularly where an appreciable proportion of the short-circuit current is supplied from nearby machines, e.g. on transmission networks and industrial power systems, the decay of the ac current is appreciable and this needs to be considered in the calculation of the make and break currents, of circuit-breakers.

The variation of the ac current is directly calculated when time domain analysis is used. However, with fixed impedance analysis, there is no direct analytical or theoretical approach that can be used, particularly in meshed networks. This will be further discussed in Section 6.5.

6.4 Time domain short-circuit analysis techniques in large-scale power systems

In general, the most accurate short-circuit current solution method should be capable of explicitly calculating the ac and dc components of short-circuit current at any time instant. If such accurate calculation is required, then time-step analysis techniques may be used. Conventional transient stability rms time-step simulations with the ability to model network unbalance, i.e. the PPS, NPS and ZPS networks allow for accurate calculation of the time variation of the ac component of short-circuit current. The main advantage of this approach is that the various ac short-circuit time constants of synchronous and induction machines and their impact on the ac fault current decay are accurately accounted for. However, this approach is generally impractical in large-scale short-circuit analysis in large power systems where 1000s of short-circuit studies are routinely carried out. The difficulties are generally the large data volume to be managed (although this is becoming less demanding in modern power system analysis programs with advanced data management facilities) and the significant computer running time requirements and hence the very long turn around time. Furthermore, in conventional transient stability analysis simulations, the dc component of short-circuit current is generally not calculated because the dynamic models of machines generally ignore the stator flux transients, i.e. in Chapter 5 in Equation (5.9), $d\psi_d/dt$ and $d\psi_q/dt$ are normally set to zero. Therefore, some other method would still be needed to calculate the dc current component.

Where an appropriate time domain analysis utilising time-step calculation is used, allowing the explicit calculation of the dc current component, it may be possible, if required, to estimate the dc component X/R ratio from the dc current waveform and dc current rate of decay.

6.5 Analysis of the time variation of ac and dc short-circuit current components

We recall that for a single short-circuit current source, i.e. a single machine, the machine short-circuit ac and dc time constants enable us to calculate the magnitudes of the ac and dc current components at any instant of time following the initial short-circuit fault instant. In this section, we limit our analysis to the most common types of short-circuit calculation, used for the assessment of substation infrastructure and circuit-breaker duties. These are three-phase and single-phase short-circuit faults.

6.5.1 Single short-circuit source connected by a radial network

The effect of a single short-circuit current source supplying a short-circuit current contribution to the fault through an external impedance was presented in Chapter 5. Essentially, the ac and dc short-circuit time constants are affected by the external impedance between the source and the fault point. To simplify the analysis, we assume that the external network impedance is predominantly reactive for the purpose of ac time constant calculation.

Figure 6.7 shows a single short-circuit source taken here as an unloaded synchronous generator having a star-connected stator winding with a solidly earthed neutral feeding into the fault point F through a radial link. This link may be one or more power system elements connected in series and have an equivalent series impedance of $R_e + jX_e$. From Chapter 5, the machine's rms and dc short-circuit currents assuming a one per unit initial voltage are given by

$$i_{rms}(t) = \frac{1}{X_d + X_{Eq}} + \left(\frac{1}{X'_d + X_{Eq}} - \frac{1}{X_d + X_{Eq}} \right) e^{-t/T'_{ac}}$$

$$+ \left(\frac{1}{X''_d + X_{Eq}} - \frac{1}{X'_d + X_{Eq}} \right) e^{-t/T''_{ac}} \tag{6.4a}$$

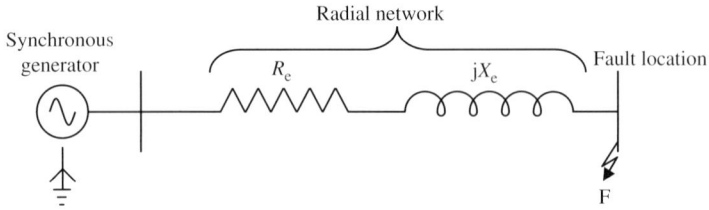

Figure 6.7 Synchronous generator connected through a radial network to the short-circuit fault location

and

$$i_{dc}(t) = \frac{\sqrt{2}}{X_d'' + X_{Eq}} e^{-t/T_{dc}} \tag{6.4b}$$

where X_{Eq}, the network ac and dc time constants for three-phase and single-phase short-circuit faults are given by

$$T_{ac}'' = \frac{X_d'' + X_{Eq}}{X_d' + X_{Eq}} T_{do}'' \quad T_{ac}' = \frac{X_d' + X_{Eq}}{X_d + X_{Eq}} T_{do}' \quad T_{dc} = \frac{X_d'' + X_{Eq}}{\omega_s (R_a + R_{Eq})} \tag{6.4c}$$

$$X_{Eq} = X_e \quad R_{Eq} = R_e \quad \text{for a three-phase short circuit} \tag{6.5a}$$

$$X_{Eq} = X^N + X^Z + 3X_e \quad R_{Eq} = 2R_a + 3R_e \quad \text{for a single-phase short circuit} \tag{6.5b}$$

From Equations (6.4) and (6.5), we can determine the effect of a generator external impedance on the equivalent network ac and dc time constants, i.e. those of the fault current at the fault location, and hence on the time variation of the magnitude of the short-circuit current.

Effect of faults near to and far from short-circuit sources

The effect of electrical distance between the short-circuit source and the fault point on the initial magnitude of the short-circuit current, i.e. at the instant of fault is easily understood. The closer the fault location to the short-circuit source is, the smaller is the impedance between the source and the fault location, and the higher is the initial magnitude of short-circuit current. However, what is the effect on the magnitudes of the ac and dc currents afterwards, e.g. a few cycles later? To answer this question we need to understand, besides the effect on the current magnitude, the effect of the external impedance on the ac and dc time constants that govern the rates of decay of the ac and dc currents. This is an important aspect in practical short-circuit analysis. From Equations (6.4c) and (6.5), we have the following.

Three-phase short-circuit fault

$$T''_{ac} = \frac{X''_d + X_e}{X'_d + X_e}T''_{do} = \frac{\frac{X''_d}{X_e} + 1}{\frac{X'_d}{X_e} + 1}\frac{X'_d}{X''_d}T''_d \tag{6.6}$$

The limiting values of the network subtransient time constant T''_{ac} are determined by setting $X_e = 0$ and $X_e \to \infty$ in Equation (6.6). Thus, the minimum value of T''_{ac} is equal to T''_d of the generator whereas the maximum value of T''_{ac} is equal to

$$T''_{do} = \frac{X'_d}{X''_d}T''_d$$

For most synchronous machines, $X'_d \approx 1.1X''_d$ to $1.5X''_d$ giving $T''_{ac} \approx T''_d$ to $1.5T''_d$. However, because the difference between X'_d and X''_d is much smaller than that between X'_d and X_d, the effect of external reactance on the initial magnitude of the subtransient component, as can be seen from Equation (6.4a), is much greater than the effect on the transient current component. In other words, the magnitude of the initial subtransient current component drops in a much greater proportion than the transient component. Following a similar approach for the network transient time constant T'_{ac} given in Equation (6.4c), we have

$$T'_{ac} = \frac{X'_d + X_e}{X_d + X_e}T'_{do} = \frac{\frac{X'_d}{X_e} + 1}{\frac{X_d}{X_e} + 1}\frac{X_d}{X'_d}T'_d \tag{6.7}$$

For most synchronous machines, X_d can be up to $8X'_d$ thus $T'_{ac} \approx T'_d \ldots 8T'_d$.

Similarly, the effect of the external reactance is to cause a proportionately much greater reduction in the transient current component than in the steady state current component. The network dc time constant T_{dc} given in Equation (6.4c) can be rewritten as

$$T_{dc} = \frac{X''_d + X_e}{\omega_s(R_a + R_e)} = \frac{X''_d}{\omega_s R_a} \times \frac{\left(1 + \frac{X_e}{X''_d}\right)}{\left(1 + \frac{R_e}{R_a}\right)} = T_{a(3\phi)} \times \frac{\left(1 + \frac{X_e}{X''_d}\right)}{\left(1 + \frac{R_e}{R_a}\right)} \tag{6.8}$$

where $T_{a(3\phi)}$ is the generator's armature time constant.

In practical power systems, generator X/R ratios X''_d/R_a are significantly greater than the X/R ratios of external network elements X_e/R_e of transformers, lines, cables, etc., or any combination of these. Therefore,

$$\frac{R_e}{R_a} > \frac{X_e}{X''_d}$$

giving

$$\frac{X''_d + X_e}{\omega_s(R_a + R_e)} < \frac{X''_d}{\omega_s R_a} \quad \text{or} \quad T_{dc} < T_{a(3\phi)}$$

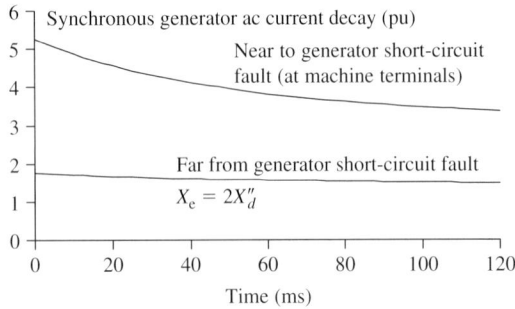

Figure 6.8　Illustration of near-to and far-from generator short-circuit faults

Thus, the effect of fault through an external impedance is to reduce the dc time constant and hence increase the rate of decay of the dc short-circuit current. The dc time constant is used to characterise the dc current component for circuit-breaker testing under IEC 62271-100:2001/2003 high voltage switchgear standard. In addition, it is usual power system practice to characterise the network dc current component by the network dc X/R ratio given by

$$\frac{X_d'' + X_e}{R_a + R_e} = \omega_s T_{dc} \tag{6.9}$$

Figure 6.8 shows the variation of the ac component of short-circuit current for a near to generator three-phase short-circuit fault and a far from generator similar fault. The latter is separated from the generator terminal by a large external reactance. The important observation is that the ac current decay is very pronounced in the near to generator case but is negligible in the far from generator case. Conversely, the dc current decay is now greater than that for a generator alone.

Single-phase short-circuit fault

Similar analysis to the case of a three-phase short circuit can be carried out using Equations (6.4c) and (6.5). The reader is encouraged to show that similar conclusions to those for a three-phase short-circuit fault can be obtained.

The analysis approach of Section 6.5.1 can easily be applied to other short current sources such as synchronous motors, induction generators and motors.

6.5.2 Parallel independent short-circuit sources connected by radial networks

In this section, we present the effect of parallel short-circuit sources, each connected by a radial network, on the total short-circuit current ac and dc time constants and the variation of the current magnitude with time. Figure 6.9 shows n parallel short-circuit sources each feeding through its own external impedance to the fault point F.

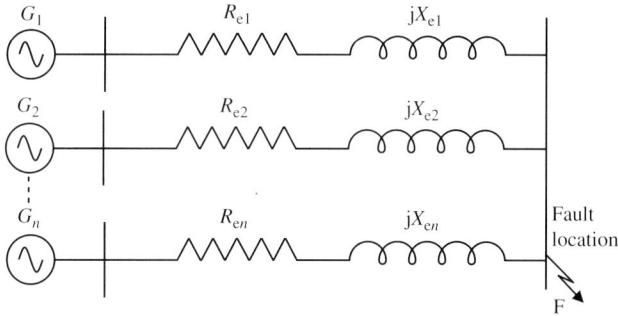

Figure 6.9 *n* parallel synchronous generators each connected by a radial network to the short-circuit fault location

Network ac time constant

We consider the network subtransient ac time constant only since similar analysis also applies to the transient time constant.

In the case of a three-phase short circuit at point F, the total subtransient current component at F, using Equation (6.4a) is the sum of the subtransient current components supplied by each individual generator. Thus,

$$
\begin{aligned}
i_{ac}(t) &= \left(\frac{1}{X_1'' + X_{e1}} - \frac{1}{X_1' + X_{e1}} \right) e^{-t/T_{ac(1)}''} \\
&\quad + \left(\frac{1}{X_2'' + X_{e2}} - \frac{1}{X_2' + X_{e2}} \right) e^{-t/T_{ac(2)}''} + \cdots \\
&\quad + \left(\frac{1}{X_n'' + X_{en}} - \frac{1}{X_n' + X_{en}} \right) e^{-t/T_{ac(n)}''} \\
&= \sum_{i=1}^{n} \left(\frac{1}{X_i'' + X_{ei}} - \frac{1}{X_i' + X_{ei}} \right) e^{-t/T_{ac(i)}''} \quad \text{for } t \geq 0 \\
&\neq \frac{1}{X_{Eq}} e^{-t/T_{ac(Eq)}''} \quad \text{for } t > 0
\end{aligned} \tag{6.10a}
$$

where

$$
T_{ac(i)}'' = \frac{X_{d(i)}'' + X_{e(i)}}{X_{d(i)}' + X_{e(i)}} T_{do(i)}'' \quad \text{for a synchronous generator/motor}
$$

or

$$
T_{ac(i)}'' = \frac{X_{(i)}'' + X_{e(i)}}{X_{(i)}' + X_{e(i)}} T_{o(i)}'' \quad \text{for an induction generator/motor}
$$

and

$$\frac{1}{X_{Eq}} = \sum_{i=1}^{n}\left(\frac{1}{X''_{(i)} + X_{e(i)}} - \frac{1}{X'_{(i)} + X_{e(i)}}\right) \tag{6.10b}$$

The expression for X_{Eq} in Equations (6.10a) and (6.10b) is a parallel equivalent and it ensures that the initial current magnitude, i.e. at $t=0$, of the single equivalent exponential in Equation (6.10a) is equal to that obtained from the sum of the individual exponentials.

In the general case, sources have different impedances and time constants and each current component decays at a different rate. Therefore, the magnitudes of the n subtransient current components for $t > 0$ will be different. The same observation would be made even if the machines were identical, i.e. have similar parameters, but their external impedances were different. Consequently, a single network ac time constant at the fault location for all the ac sources that enables the calculation of the total current magnitude for $t > 0$ does not exist. Mathematically speaking, the sum of several exponentials of different time constants cannot be made equal to a single equivalent exponential with a single equivalent time constant. However, in this radial network topology this is not required because each source is radially connected to the fault point and the accurate equation for the sum of n exponential terms can be used. It can be shown that similar conclusions apply for single-phase short-circuit faults.

Network dc time constant

In the case of a three-phase short circuit at point F in Figure 6.9, the total dc current component at point F, using Equation (6.4b), is the sum of the dc current components supplied by each individual machine. Thus

$$i_{dc}(t) = \frac{\sqrt{2}}{X''_1 + X_{e1}}e^{-t/T_{dc(1)}} + \frac{\sqrt{2}}{X''_2 + X_{e2}}e^{-t/T_{dc(2)}} + \cdots + \frac{\sqrt{2}}{X''_n + X_{en}}e^{-t/T_{dc(n)}}$$

$$= \sum_{i=1}^{n}\left(\frac{\sqrt{2}}{X''_i + X_{ei}}e^{-t/T_{dc(i)}}\right) \neq \frac{\sqrt{2}}{X_{Eq}}e^{-t/T_{dc(Eq)}} \quad \text{for } t > 0 \tag{6.11a}$$

where

$$T_{dc(i)} = \frac{X''_{di} + X_{ei}}{\omega_s(R_{ai} + R_{ei})} \quad \text{for a synchronous generator}$$

or

$$T_{dc(i)} = \frac{X''_i + X_{ei}}{\omega_s(R_{si} + R_{ei})} \quad \text{for an induction generator/motor}$$

and

$$\frac{1}{X_{Eq}} = \sum_{i=1}^{n}\frac{1}{X''_{di} + X_{ei}} \tag{6.11b}$$

Equation (6.11a) shows that at $t=0$, the magnitude of the total dc current is equal to the sum of the magnitudes of the individual dc current components. However, for $t > 0$, and because each current component decays at a different rate, a single equivalent exponential component that gives an equivalent rate of decay or time constant for the sum of the individual exponentials does not exist. This is best illustrated in the following example.

Example 6.1   Consider three parallel sources or $n = 3$ in Figure 6.9 having significantly different time constants as follows: $T_{dc(1)} = 50$ ms, $T_{dc(2)} = 150$ ms and $T_{dc(3)} = 15$ ms. Using Equation (6.11b), the corresponding initial dc current magnitudes supplied at $t = 0$ by the three sources are equal to $I_{dc(1)} = 0.6$ pu, $I_{dc(2)} = 0.3$ pu and $I_{dc(3)} = 0.1$ pu. These initial current magnitudes and their associated dc time constants can represent a variety of sources and external impedances such as synchronous machines, induction machines and equivalent infeeds from transmission or distribution networks. Figure 6.10 curve (a) shows the total dc fault current at point F supplied by the three parallel sources and calculated using Equation (6.11a), i.e. from the accurate sum of three exponential terms. Figure 6.10 also shows the total dc current derived from a single time constant as calculated by two different approximate methods. The first, curve (b), is calculated from the single X/R ratio of the equivalent impedance 'seen' at the fault point, $Z_{Thév.} = R + jX$, also known as IEC 60909 Standard Method B (this is discussed in Chapter 7). The second, curve (c), is calculated from the ratio of the equivalent reactance at the fault point, $X_{Thév.}$, (with all resistances set to zero) to the equivalent resistance at the fault point, $R_{Thév.}$, (with all reactances set to zero), as per IEEE C37.010 Standard (this is discussed in Chapter 7). It can be shown that the single equivalent time constant at the fault point using the IEC method is equal to 48.6 ms ($X/R = 15.2$ at 50 Hz). Using the IEEE method, the single equivalent time constant at the fault point is equal to 76.3 ms ($X/R = 24.0$ at 50 Hz). The equations for the dc currents calculated by the three methods are given by

$$i_{dc(accurate)}(t) = 0.6e^{-t/50} + 0.3e^{-t/150} + 0.1e^{-t/15}$$
$$i_{dc(IEC)}(t) = e^{-t/48.6} \quad i_{dc(IEEE)}(t) = e^{-t/76.3}$$

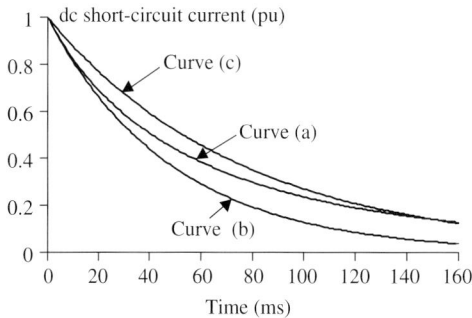

Figure 6.10   Illustration of IEC and IEEE approaches for the calculation of dc short-circuit current variation with time for Figure 6.9

where t is in ms. Figure 6.10 shows that the IEC Method B single equivalent time constant underestimates the actual dc current for $t \geq 10$ ms and this underestimate increases with time. Conversely, and because the IEEE $R_{\text{Thév.}}$ is significantly smaller than the equivalent resistance calculated using the IEC approach, the IEEE time constant is 1.57 times the IEC time constant. Figure 6.10 shows that the IEEE approach, in this example, significantly overestimates the actual dc current over the period up to 140 ms, then the overestimate disappears. It should be noted that the degree of underestimate of IEC Method B and overestimate of IEEE Method varies with the initial magnitude of each current source and the relative time constants between them. Larger and smaller underestimates and overestimates can be found in practice and the reader is encouraged to analyse the effect of different combinations.

6.5.3 Multiple short-circuit sources in interconnected networks

The topologies of transmission, subtransmission and many distribution power systems are such that multiple sources of short-circuit currents feed into the short-circuit fault via a meshed or interconnected network topology. For most practical networks, the effect of multiple short-circuit sources supplying currents through a meshed network configuration on the ac and dc time constants of the short-circuit current at the fault location is not amenable to analytical treatment. However, understanding this effect is of vital importance in practical short-circuit calculations if significant inaccuracies are to be avoided. The analysis presented next will explain the problem.

Network ac time constant

Figure 6.11(a) shows a simple meshed network that consists of three independent short-circuit current sources, each is connected by an external impedance to a common node N. The three sources are then connected to the short-circuit fault location F via a common impedance. To illustrate the problem, each source is assumed to be a synchronous machine whose time-dependent reactance is given by

$$\frac{1}{X_{\text{m}(i)}(t)} = \frac{1}{X_{d(i)}} + \left(\frac{1}{X'_{d(i)}} - \frac{1}{X_{d(i)}} \right) e^{-t/T'_{d(i)}} + \left(\frac{1}{X''_{d(i)}} - \frac{1}{X'_{d(i)}} \right) e^{-t/T''_{d(i)}} \quad (6.12)$$

where $i = 1$, 2 or 3.

The equivalent circuit is shown in Figure 6.11(b) where each machine is replaced by a constant voltage source E_{o} behind its time-dependent PPS reactance, given by Equation (6.12), and the armature resistance. The equivalent time-dependent

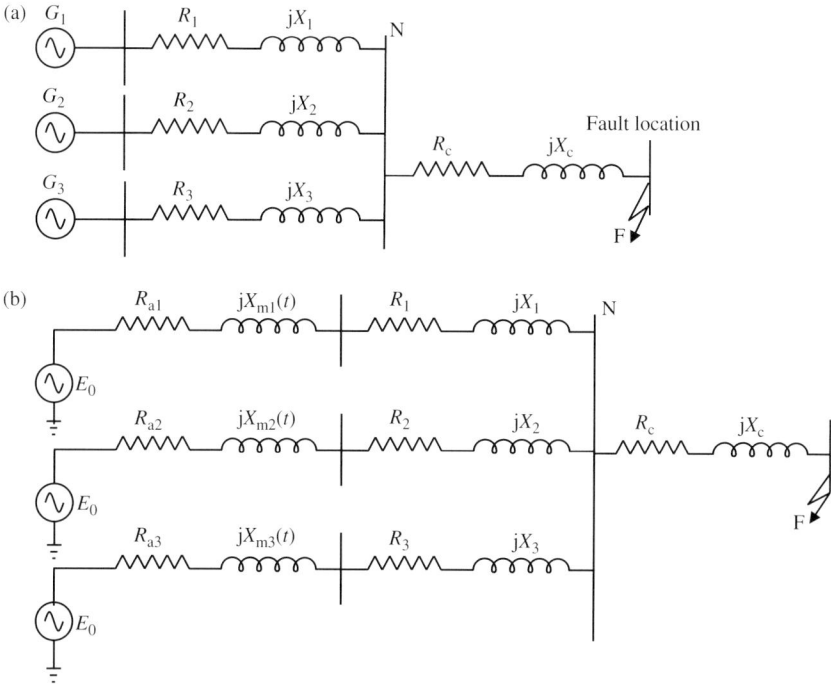

Figure 6.11 Simple meshed network for the analysis of ac short-circuit current decrement (variation with time): (a) network configuration and (b) equivalent circuit

PPS system impedance at the fault point F is given by

$$Z_F(t) = R_F(t) + jX_F(t) = R_c + jX_c + \cfrac{1}{\displaystyle\sum_{i=1}^{3} \cfrac{1}{(R_{a(i)} + R_{(i)}) + j[X_{m(i)}(t) + X_{(i)}]}} \tag{6.13}$$

From Equation (6.13), it is interesting to note that not only the equivalent reactance X_F at the fault point is time dependent but also the equivalent resistance R_F. In many practical calculations, we can assume that $X_{m(i)}(t) + X_{(i)}(t) \gg (R_{ai} + R_i)$ for $i = 1, 2$ and 3, and $X_c \gg R_c$. Therefore, from Equation (6.13), we have

$$X_F = X_c + \cfrac{1}{\cfrac{1}{X_{m(1)}(t) + X_1} + \cfrac{1}{X_{m(2)}(t) + X_2} + \cfrac{1}{X_{m(3)}(t) + X_3}} \tag{6.14}$$

In general, the magnitude of the time-dependent ac component of fault current at the fault point F is given by

$$|i_F(t)| = \frac{E_0}{|Z_F(t)|} \tag{6.15a}$$

or

$$i_F(t) = \frac{E_o}{X_F(t)} \tag{6.15b}$$

The substitution of Equation (6.12) in Equation (6.13) or (6.14) shows that the fault current magnitude given by Equation (6.15) would still exhibit subtransient and transient decay. However, the corresponding time constants are very complicated functions of the three individual machine time constants and no general analytical formula can be derived. The accuracy of various methods of calculating the time variation of the ac short-circuit current component is illustrated using an example.

Example 6.2   Consider Figure 6.11. The three machines are generators having the following parameters: $X_d = 2.04$ pu, $X_d' = 0.275$ pu, $X_d'' = 0.19$ pu, $T_d'' = 40$ ms, $T_d' = 1100$ ms, $X_1 = X_2 = X_3 = 0.19$ pu, $X_c = 0.2$ pu.

Curve (a) of Figure 6.12 shows the accurate time variation of the magnitude of the rms short-circuit current obtained from Equation (6.15b) with $E_o = 1$ pu. Curve (b) shows the current waveform obtained from the Thévenin's equivalent subtransient, transient and steady state impedances calculated at the fault point F as $X_F'' = 0.3267$ pu, $X_F' = 0.355$ pu and $X_F = 0.9434$ pu, respectively. The subtransient and transient time constants that include the effect of each machine's external impedance up to node N but exclude the effect of the common impedance Z_c are calculated as $T_{de(i)}'' = 47.3$ ms and $T_{de(i)}' = 1701.5$ ms. X_F'', X_F' and X_F are calculated using $X_{d(i)}''$, $X_{d(i)}'$ and $X_{d(i)}$, respectively, for each machine. The current obtained by using the Thévenin's equivalent subtransient,

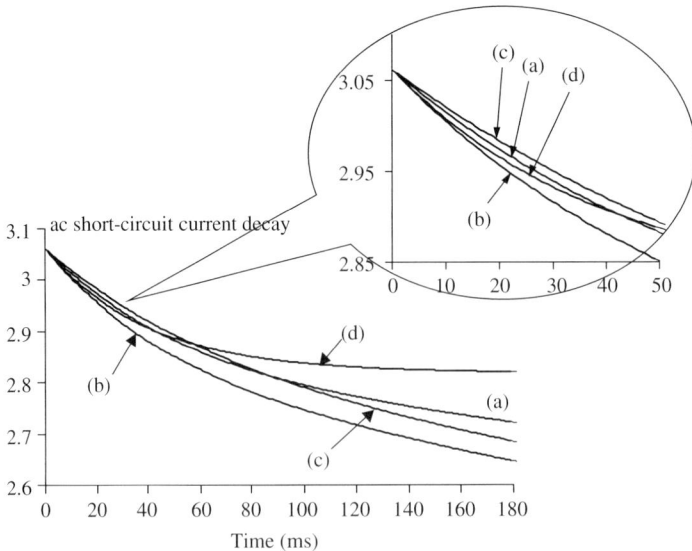

Figure 6.12   Illustration of different approaches for the calculation of ac short-circuit current variation with time for a meshed network

transient and steady state impedances is given by,

$$i_F(t) = \frac{1}{X_F} + \left(\frac{1}{X'_F} - \frac{1}{X_F}\right)e^{-t/T'_{de(i)}} + \left(\frac{1}{X''_F} - \frac{1}{X'_F}\right)e^{-t/T''_{de(i)}} \quad (6.16)$$

As expected, curve (b) Equation (6.16) underestimates the current magnitude for $t > 0$ and this underestimate increases with time. The reason is that the common impedance Z_c between node N and the fault point F will also cause some increase in the effective time constants of each source but Equation (6.16) fails to capture this effect. Curve (c) of Figure 6.12 shows the current variation if maximum known values of practical synchronous machine transient and subtransient time constants are used in Equation (6.16). These are 2000 and 70 ms, respectively. The result shows a small overestimate of up to around 90 ms, then an increasing underestimate thereafter. If the time range of interest is up to around 120 ms, as would be the case in high voltage and extra high voltage systems, then an approximation can be made in Equation (6.16) by setting $e^{-t/T'_{d(i)}} \approx 1$. Also, using an average value for synchronous machine subtransient time constants of around 40 ms, we can write

$$i_F(t) = \frac{1}{X'_F} + \left(\frac{1}{X''_F} - \frac{1}{X'_F}\right)e^{-t/40} \quad t \text{ in ms} \quad (6.17)$$

Equation (6.17) is plotted as curve (d) in Figure 6.12 where the current falls to the minimum transient value then remains constant. Equation (6.17) gives an increasing overestimate with time for $t \geq 50$ ms.

The above analysis can easily be extended to single-phase short-circuit faults using the machine time-dependent PPS reactance equations derived in Chapter 5.

There is no international consensus on how to calculate the time variation of the ac component of short-circuit current in practical meshed networks, when fixed impedance analysis techniques are used. Different empirical approaches are used, and this will be covered in Chapter 7.

Network dc time constant

We will continue to use the meshed network topology shown in Figure 6.11(a), but to simplify the analysis, we now use constant voltage sources instead of synchronous machines. The voltage source equation is given by $v(t) = \sqrt{2}V\sin(\omega t + \phi)$. When a short-circuit fault occurs at point F, the time domain solution of the short-circuit currents supplied by each source and the total fault current can be obtained by replacing all reactances with corresponding inductances and writing the differential equations relating the currents and voltages in this network. We encourage the reader to carry out this simple but time consuming exercise in Kirchoff's voltage and current laws to obtain the solutions of the four currents. It can be shown that the instantaneous short-circuit current supplied by each voltage source and the total short-circuit current at the fault point F are given by

$$i_{(i)}(t) = \sqrt{2}V\omega[A_i \cos(\omega t) + B_i \sin(\omega t) + C_i e^{-t/T_{dc(1)}} + D_i e^{-t/T_{dc(2)}} + E_i e^{-t/T_{dc(3)}}]$$

$$(6.18a)$$

$$i_F(t) = \sqrt{2}V\omega[A \cos(\omega t) + B \sin(\omega t) + C e^{-t/T_{dc(1)}} + D e^{-t/T_{dc(2)}} + E e^{-t/T_{dc(3)}}]$$

$$(6.18b)$$

where

$$A_i + C_i + D_i + E_i = 0 \quad \text{for } i = 1, 2 \text{ and } 3$$

and

$$A = \sum_{i=1}^{3} A_i \quad B = \sum_{i=1}^{3} B_i \quad C = \sum_{i=1}^{3} C_i \quad D = \sum_{i=1}^{3} D_i$$

$$E = \sum_{i=1}^{3} E_i \quad \text{and} \quad A + C + D + E = 0$$

All the constant multipliers and the three dc time constants are complicated functions of the network resistances and inductances.

Equation (6.18b) shows that the dc component of the total short-circuit current consists of three components having different time constants $T_{dc(1)}$, $T_{dc(2)}$ and $T_{dc(3)}$. The following example compares the accurate solution against the methods of characterising the dc current by a single decaying exponential having a time constant derived from the IEC 60909 Method B and the IEEE C37.010 approaches.

Example 6.3   In Figure 6.11(a), the X/R ratios of the four network branches are as follows: $X_1/R_1 = 15.7$, $X_2/R_2 = 47.0$, $X_3/R_3 = 4.7$, $X_c/R_c = 10$. Figure 6.13 shows that, as in the case of parallel independent sources, the equivalent X/R ratio at the fault point or single exponential method underestimates the

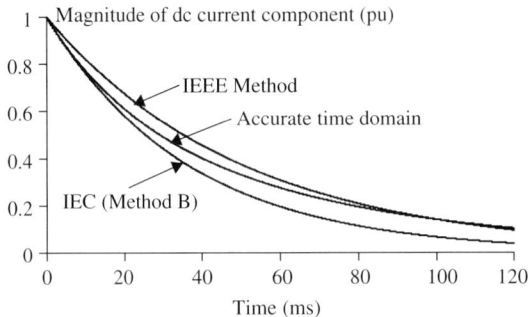

Figure 6.13   Illustration of IEC and IEEE approaches for the calculation of dc short-circuit current variation with time for Figure 6.11(a)

true magnitude of the dc current component. However, the underestimate produced using the IEC Method B is generally smaller in meshed network than that produced in a network having parallel independent sources, due to the common impedance branch that is shared by the three sources in the meshed network. The configuration of parallel independent short-circuit sources generally produces the worst case underestimate. An overestimate is observed using the IEEE C37.010 Method. The errors in the dc currents calculated using the IEC and IEEE approaches on one hand and the accurate solution reduce with increasing Z_c that is increasing distance between the machines and the fault location. In other words, as Z_c increases, the IEC and IEEE curves move towards the curve that represents the accurate solution. Although we have used simple parallel and meshed network configurations to present the problem of dc current estimation, the conclusions regarding the errors observed in such network configurations are generally applicable in other networks.

6.6 Fixed impedance short-circuit analysis of large-scale power systems

6.6.1 Background

Fixed impedance analysis allows the direct calculation of the rms value of the ac component of short-circuit current using phasor algebra. Hand calculations play a useful and important role in gaining an insight into problems. However, these can only be used in problems of low dimensionality. General analysis methods using matrix algebra suitable for digital computer calculations are almost universally used. Our presentation will be of practical nature that allows the inclusion, as required, of some or all of the following: resistive elements of series impedances, shunt capacitance of lines and cables, transformer magnetising reactances, transformer actual off-nominal tap positions determined from an initial loadflow study, shunt reactors and capacitors, etc. The sequence interconnection networks of various faults developed in Chapter 2 and the sequence models of various power plants developed in Chapters 3–5 are used to formulate a sequence model of the entire network. In this section, we will present a brief overview of the type of general analysis techniques that are directly applicable to large-scale N bus systems.

6.6.2 General analysis of balanced three-phase short-circuit faults

Thévenin's and Norton's PPS equivalent of models of machines

The topology of power system networks and the data of all network elements, as well as the initial network loading and generation operating conditions, allow

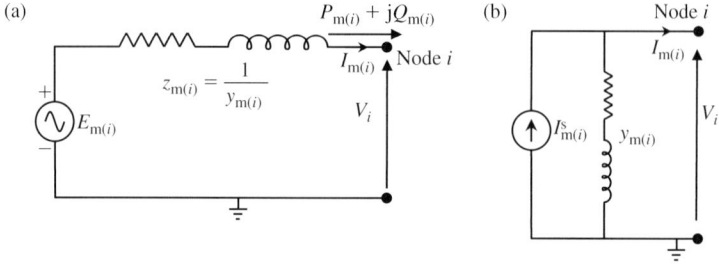

Figure 6.14 Machine equivalent PPS model: (a) Thévenin's equivalent and (b) Norton's equivalent

the formation of a nodal equivalent network of admittances and injected current sources. Short-circuit current sources such as generators and, where required, motors can be represented at the time instant of interest as constant voltage sources behind the appropriate machine fixed impedance, e.g. subtransient or transient impedance or any another value, as appropriate. Figure 6.14(a) and (b) shows equivalent models of a loaded synchronous machine. The latter shows a nodal or Norton's equivalent which consists of a current source in parallel with the machine admittance.

From Figure 6.14(a), the apparent power output of a machine connected to node (*i*) is given by

$$P_{m(i)} + jQ_{m(i)} = V_i I^*_{m(i)} \text{ or } I_{m(i)} = \frac{P_{m(i)} - jQ_{m(i)}}{V^*_i} \text{ and } E_{m(i)} = V_i + Z_{m(i)} I_{m(i)}$$

$$(6.19a)$$

The short-circuit current supplied by the machine for a fault at node *i* is given by

$$I^S_{m(i)} = y_{m(i)} E_{m(i)} \quad y_{m(i)} = \frac{1}{Z_{m(i)}} \tag{6.19b}$$

Using $E_{m(i)}$ from Equation (6.19a) in Equation (6.19b), we obtain

$$I^S_{m(i)} = y_{m(i)} V_i + I_{m(i)} \tag{6.19c}$$

PPS admittance and impedance matrix equations

Figure 6.15 shows a simple power system network that will be used to illustrate the process of formulating the sequences nodal matrices.

The formation of the admittance matrix starts by forming the positive sequence nodal model shown in Figure 6.16, where each network element is replaced by its PPS model.

Figure 6.17 shows the nodal equivalent model with all shunt elements at every node collected into an equivalent shunt branch. The result is only one shunt branch at each node and a series branch interconnecting two nodes. The nodal admittance

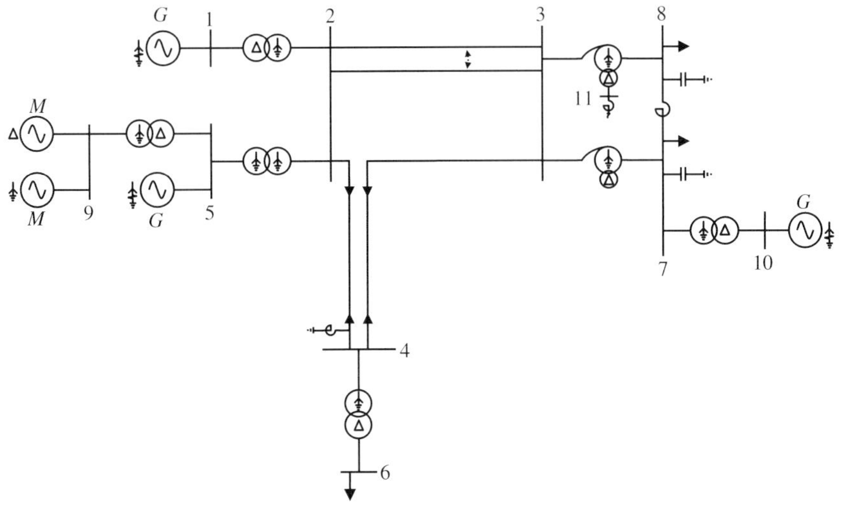

Figure 6.15 Simple power system network

matrix equation of the resultant nodal equivalent circuit is given by

$$
\begin{bmatrix} I_1 \\ I_2 \\ \cdot \\ I_N \end{bmatrix} = \begin{bmatrix} Y_{11} & Y_{12} & \cdot & Y_{1N} \\ Y_{21} & Y_{22} & \cdot & Y_{2N} \\ \cdot & \cdot & \cdot & \cdot \\ Y_{N1} & Y_{N2} & \cdot & Y_{NN} \end{bmatrix} \begin{bmatrix} V_1 \\ V_2 \\ \cdot \\ V_N \end{bmatrix}
\tag{6.20a}
$$

or in concise matrix form

$$
\mathbf{I} = \mathbf{YV}
\tag{6.20b}
$$

where

$$
Y_{ii} = \sum_j y_{ij} \quad \text{and} \quad Y_{ij} = -y_{ij} \ i \neq j
\tag{6.20c}
$$

The admittance matrix of Equation (6.20) is in general symmetric, and even for small power systems, it is quite sparse, i.e. it contains only a few non-zero elements, each representing an admittance element connecting two nodes. For example, for a medium size system of 4000 nodes and 3000 series branches, the number of non-zero elements is $4000 + 2 \times 3000 = 10\,000$. This is only 0.0625% of the total number of elements of the full matrix ($4000 \times 4000 = 16\,000\,000$).

The use of the nodal admittance matrix in calculating short-circuit currents is inefficient, as it requires a new solution for each fault location and fault type. Therefore, the nodal impedance matrix given in Equation (6.21) below is used. It is important to note that the current vector represents a set of injected nodal current sources. Thus, from Equation (6.20b),

$$
\mathbf{V} = \mathbf{Y}^{-1}\mathbf{I} = \mathbf{ZI}
\tag{6.21a}
$$

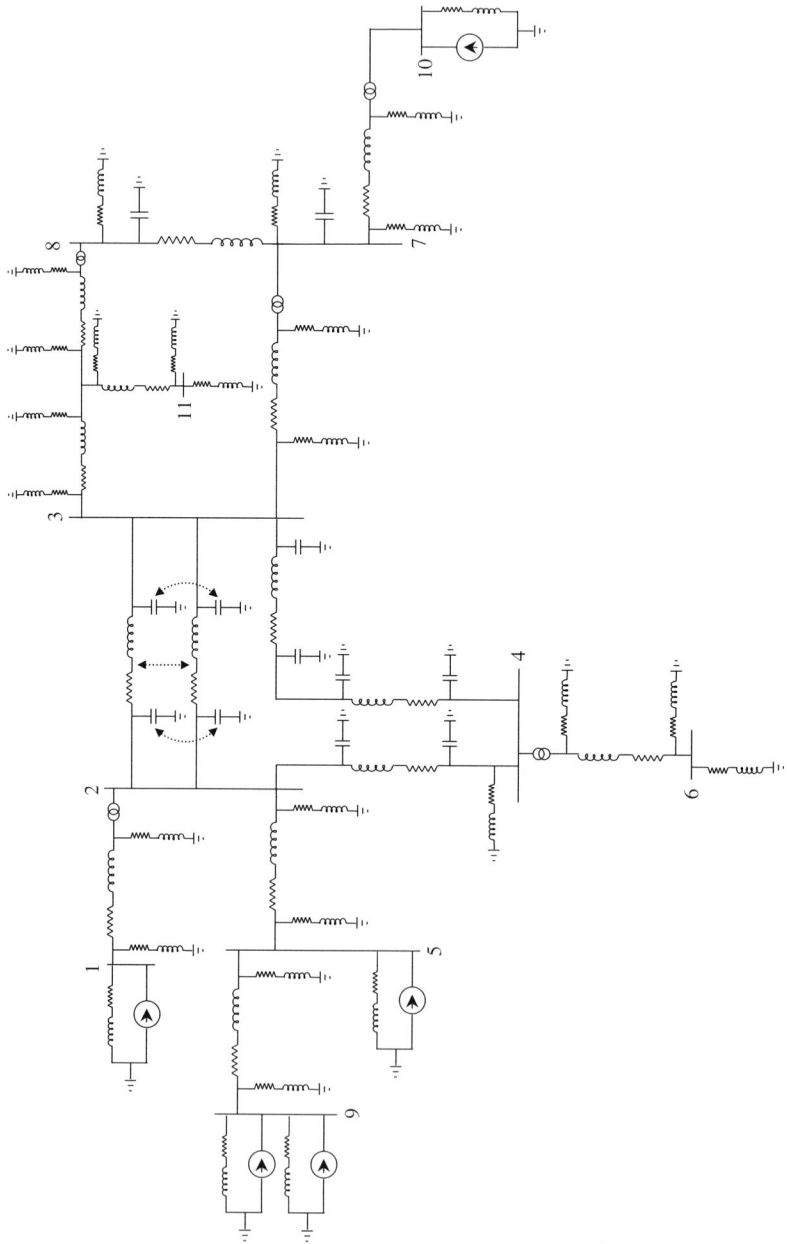

Figure 6.16 PPS network model of Figure 6.15

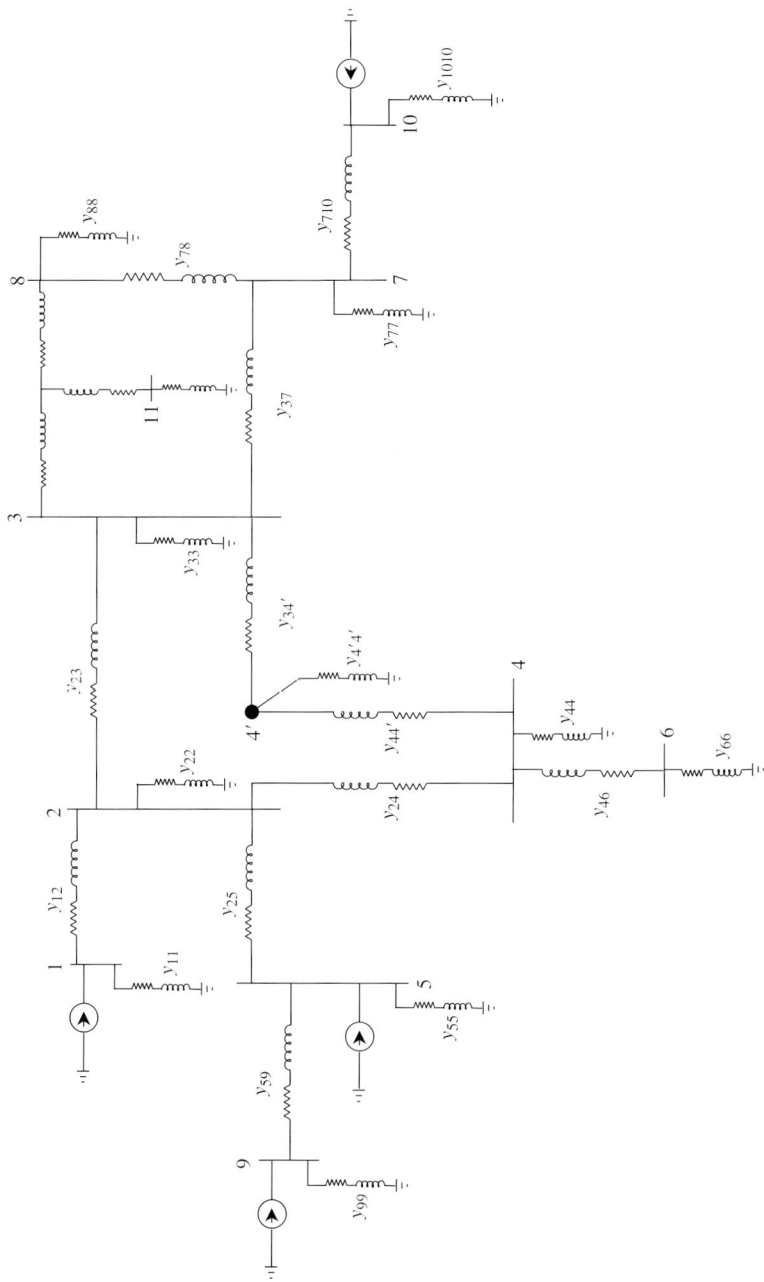

Figure 6.17 PPS network model of Figure 6.16 with lumped shunt branches at each node

or

$$
\begin{bmatrix} V_1 \\ \cdot \\ V_k \\ \cdot \\ V_N \end{bmatrix} = \begin{bmatrix} Z_{11} & \cdot & Z_{1k} & \cdot & Z_{1N} \\ \cdot & & \cdot & & \cdot \\ Z_{k1} & \cdot & Z_{kk} & \cdot & Z_{kN} \\ \cdot & & \cdot & & \cdot \\ Z_{N1} & \cdot & Z_{NK} & \cdot & Z_{NN} \end{bmatrix} \begin{bmatrix} I_1 \\ \cdot \\ I_k \\ \cdot \\ I_N \end{bmatrix} \tag{6.21b}
$$

The nodal impedance matrix is generally full, i.e. it contains elements in every position unless there are disconnected parts of the network, as would be the case in the ZPS network as we discuss later. However, in practice, it is inefficient to obtain this matrix by direct inversion of the admittance matrix for medium and large size systems. Direct inversion of large matrices is avoided by the use of various computer-based numerical calculation techniques, such as the method of successive forward elimination and backward substitution (also known as Gaussian elimination). Other similar methods are also used, such as Kron reduction, triangular factorisation, LU decomposition and others. Generally, the sparse admittance matrix can be retained and the impedance matrix is calculated or built directly from the factorised admittance matrix.

Referring back to Equation (6.21b), it is important to appreciate the meaning of the diagonal and off-diagonal elements of the nodal impedance matrix. The diagonal terms $Z_{kk} = \frac{V_k}{I_k}$, with all other current sources set to zero, is termed the driving point impedance or the Thévenin's impedance as 'seen' from node k. This impedance is equal to the voltage produced at node k divided by the current injected at node k. The off-diagonal terms are, for example, $Z_{1k} = \frac{V_1}{I_k}$ again with all other current sources set to zero is termed the transfer impedance between nodes 1 and k. It is equal to the voltage produced at node 1 divided by the current injected at node k.

General analysis of three-phase short-circuit faults

We will present a general analysis technique suitable for the calculation of not only the short-circuit current at the fault location but also the currents and voltages throughout the network. As explained in Section 6.3.2, the application of the fault will cause current and voltage changes in the network. The calculation of these quantities in a large-scale system of N nodes makes use of the efficiently calculated nodal bus impedance matrix of the system. Since the three-phase short circuit is a balanced fault, the system nodal admittance and nodal impedance matrices are in effect PPS admittance/impedance matrices since all the network elements are replaced by their PPS models. The actual currents and voltages in the network due to the fault are obtained by the superposition of the initial currents and voltages and the Thévenin's changes. Thus,

$$\mathbf{V} = \mathbf{ZI} = \mathbf{Z}(\mathbf{I}_o + \Delta\mathbf{I}) = \mathbf{ZI}_o + \mathbf{Z}\Delta\mathbf{I}$$

or

$$\mathbf{V} = \mathbf{V}_o + \Delta\mathbf{V} \quad \mathbf{I} = \mathbf{I}_o + \Delta\mathbf{I} \tag{6.22a}$$

$$\Delta\mathbf{V} = \mathbf{Z}\Delta\mathbf{I} \tag{6.22b}$$

Three-phase short-circuit fault at one location

Denote the faulted node as k, the fault impedance as Z_F, the connection fault current as I_F, the Thévenin's impedance at node k as Z_{kk} and the prefault voltage at node k as $V_{k(o)}$. The calculation of the current and voltage changes at the fault location and elsewhere is illustrated using Figure 6.18 and the nodal bus impedance matrix of Equation (6.21b).

Since we are considering a fault at one location only, namely node k, there is only one injected node current $I_k = -I_F$ and all other injected currents are zero. Therefore, from Equation (6.21b), the voltage changes are given by

$$
\begin{bmatrix} \Delta V_1 \\ \cdot \\ \Delta V_k \\ \cdot \\ \Delta V_N \end{bmatrix} = \begin{bmatrix} Z_{11} & \cdot & Z_{1k} & \cdot & Z_{1N} \\ \cdot & \cdot & \cdot & \cdot & \cdot \\ Z_{k1} & \cdot & Z_{kk} & \cdot & Z_{kN} \\ \cdot & \cdot & \cdot & \cdot & \cdot \\ Z_{N1} & \cdot & Z_{NK} & \cdot & Z_{NN} \end{bmatrix} \begin{bmatrix} 0 \\ \cdot \\ -I_F \\ \cdot \\ 0 \end{bmatrix} \quad (6.23a)
$$

From Equation (6.23a), the voltage changes caused by the fault are given by

$$
\begin{bmatrix} \Delta V_1 \\ \cdot \\ \Delta V_k \\ \cdot \\ \Delta V_N \end{bmatrix} = \begin{bmatrix} -Z_{1k}I_F \\ \cdot \\ -Z_{kk}I_F \\ \cdot \\ -Z_{Nk}I_F \end{bmatrix} \quad (6.23b)
$$

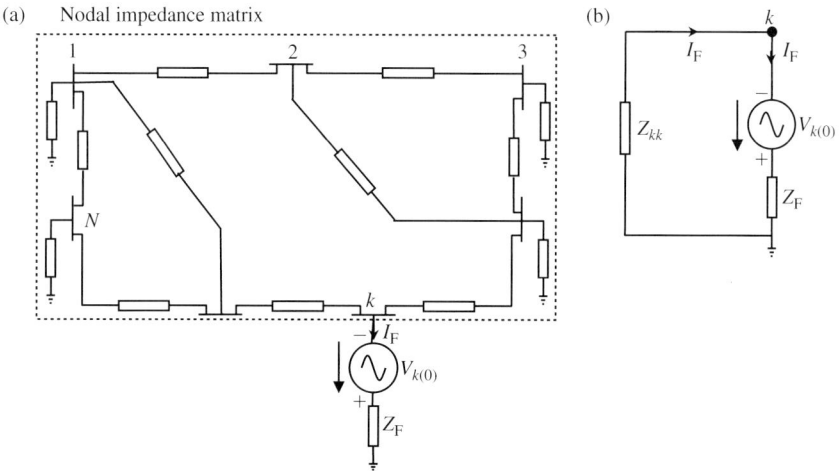

Figure 6.18 Calculation of current and voltage changes using Thévenin's equivalent circuit: (a) representation of the entire network and (b) equivalent at the faulted node k

The actual voltage during the fault at the faulted node k is given by

$$V_k = Z_F I_F \qquad (6.24)$$

From Equation (6.23b) row k, we can write $\Delta V_k = V_k - V_{k(o)} = -Z_{kk} I_F$ and using Equation (6.24), we obtain

$$I_F = \frac{V_{k(o)}}{Z_{kk} + Z_F} \qquad (6.25)$$

Having calculated the fault current, the nodal voltage changes throughout the network are calculated using Equation (6.23b). The changes in branch currents are calculated from the nodal voltage changes using

$$\Delta I_{ij} = y_{ij}(\Delta V_i - \Delta V_j) \qquad (6.26)$$

where y_{ij} is the retained admittance of branch ij.

Denoting the actual nodal voltages during the fault as $V_1, V_k \ldots V_N$, then from Equation (6.23b), we can write

$$\begin{bmatrix} V_1 \\ \cdot \\ V_k \\ \cdot \\ V_N \end{bmatrix} = \begin{bmatrix} V_{1(o)} - Z_{1k} I_F \\ \cdot \\ V_{k(o)} - Z_{kk} I_F \\ \cdot \\ V_{N(o)} - Z_{Nk} I_F \end{bmatrix} \qquad (6.27)$$

Substituting Equation (6.25) in Equations (6.24) or (6.27), the actual nodal voltages can be expressed as

$$V_k = \frac{Z_F}{Z_{kk} + Z_F} V_{k(o)} \qquad (6.28)$$

and

$$V_i = V_{i(o)} - \frac{Z_{ik}}{Z_{kk} + Z_F} V_{k(o)} \quad i \neq k \qquad (6.29)$$

For a solid short-circuit fault, $Z_F = 0$. The actual branch currents are calculated using

$$I_{ij} = I_{ij(o)} + \Delta I_{ij} = y_{ij}(V_i - V_j) \quad i \neq j \qquad (6.30)$$

The short-circuit current contribution of each machine is calculated from the actual nodal voltages where each machine is connected and using Equation (6.19c)

$$I_{m(i)} = I^s_{m(i)} - y_{m(i)} V_i \qquad (6.31)$$

or, alternatively, using Equation (6.19b) in Equation (6.31)

$$I_{m(i)} = y_{m(i)}[E_{m(i)} - V_i] \qquad (6.32)$$

Simultaneous three-phase short-circuit faults at two different locations

In Figure 6.19(a), denote the faulted nodes as j and k, the corresponding fault impedances as $Z_{j(F)}$ and $Z_{k(F)}$ and the connection fault currents as $I_{j(F)}$ and $I_{k(F)}$. The Thévenin's impedances at nodes j and k are Z_{jj} and Z_{kk}, the transfer impedances between nodes j and k are Z_{jk} and Z_{kj} and the prefault voltages at nodes j and k are $V_{j(o)}$ and $V_{k(o)}$.

From Equation (6.21b), we can write

$$
\begin{bmatrix} \Delta V_1 \\ \cdot \\ \Delta V_j \\ \Delta V_k \\ \cdot \\ \Delta V_N \end{bmatrix} = \begin{bmatrix} Z_{11} & \cdot & Z_{1j} & Z_{1k} & \cdot & Z_{1N} \\ \cdot & & \cdot & \cdot & \cdot & \cdot \\ Z_{j1} & \cdot & Z_{jj} & Z_{jk} & \cdot & Z_{jN} \\ Z_{k1} & \cdot & Z_{kj} & Z_{kk} & \cdot & Z_{kN} \\ \cdot & & \cdot & \cdot & \cdot & \cdot \\ Z_{N1} & \cdot & Z_{Nj} & Z_{Nk} & \cdot & Z_{NN} \end{bmatrix} \begin{bmatrix} 0 \\ \cdot \\ -I_{j(F)} \\ -I_{k(F)} \\ \cdot \\ 0 \end{bmatrix}
\tag{6.33}
$$

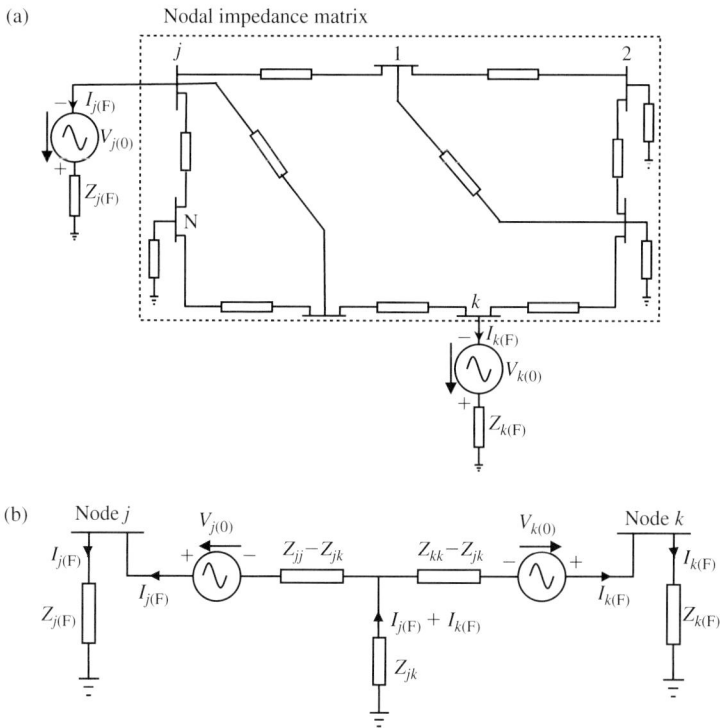

Figure 6.19 Calculation of current and voltage changes for two simultaneous short-circuit faults using Thévenin's equivalent circuits: (a) representation of the entire network for two simultaneous short-circuit faults and (b) equivalent between faulted nodes j and k

Since the nodal impedance matrix is symmetric, $Z_{jk} = Z_{kj}$. Therefore, the voltage changes in Equation (6.33) caused by the faults are given by

$$
\begin{bmatrix} \Delta V_1 \\ \cdot \\ \Delta V_j \\ \Delta V_k \\ \cdot \\ \Delta V_N \end{bmatrix} = \begin{bmatrix} -Z_{1j}I_{j(F)} - Z_{1k}I_{k(F)} \\ \cdot \\ -Z_{jj}I_{j(F)} - Z_{jk}I_{k(F)} \\ -Z_{jk}I_{j(F)} - Z_{kk}I_{k(F)} \\ \cdot \\ -Z_{Nj}I_{j(F)} - Z_{Nk}I_{k(F)} \end{bmatrix} \tag{6.34}
$$

The actual voltages during the faults at the faulted nodes j and k are given by

$$V_{j(F)} = Z_{j(F)}I_{j(F)} \tag{6.35a}$$

$$V_{k(F)} = Z_{k(F)}I_{k(F)} \tag{6.35b}$$

From Equation (6.34) rows j and k, we can write

$$\Delta V_j = V_{j(F)} - V_{j(o)} = -Z_{jj}I_{j(F)} - Z_{jk}I_{k(F)} \tag{6.36a}$$

and

$$\Delta V_k = V_{k(F)} - V_{k(o)} = -Z_{jk}I_{j(F)} - Z_{kk}I_{k(F)} \tag{6.36b}$$

Substituting Equations (6.35a) and (6.35b) in Equations (6.36a) and (6.36b), respectively, and solving for the fault currents, we obtain

$$
\begin{bmatrix} I_{j(F)} \\ I_{k(F)} \end{bmatrix} = \frac{1}{[Z_{j(F)} + Z_{jj}][Z_{k(F)} + Z_{kk}] - Z_{jk}^2} \begin{bmatrix} Z_{k(F)} + Z_{kk} & -Z_{jk} \\ -Z_{jk} & Z_{j(F)} + Z_{jj} \end{bmatrix} \begin{bmatrix} V_{j(o)} \\ V_{k(o)} \end{bmatrix} \tag{6.37}
$$

It is instructive to visualise the Thévenin's equivalent circuit seen from the two faulted nodes j and k. Adding and subtracting $Z_{jk}I_{j(F)}$ to Equation (6.36a) and $Z_{jk}I_{k(F)}$ to Equation (6.36b), we obtain

$$V_{j(F)} = V_{j(o)} - Z_{jk}[I_{j(F)} + I_{k(F)}] - [Z_{jj} - Z_{jk}]I_{j(F)} \tag{6.38a}$$

and

$$V_{k(F)} = V_{k(o)} - Z_{jk}[I_{j(F)} + I_{k(F)}] - [Z_{kk} - Z_{jk}]I_{k(F)} \tag{6.38b}$$

Equations (6.35) and (6.38) can be represented by the equivalent circuit shown in Figure 6.19(b). As expected, the Thévenin's impedance from nodes j and k to earth, or to the reference node, are Z_{jj} and Z_{kk}, respectively. The equivalent Thévenin's impedance between the faulted nodes j and k is given by

$$Z_{\text{Thév.}(jk)} = Z_{jj} + Z_{kk} - 2Z_{jk} \tag{6.39}$$

Having calculated the fault currents using Equation (6.37), the nodal voltage changes are calculated using Equation (6.34). The actual voltages are then calculated using Equation (6.22a). The actual branch and machine currents are calculated using Equations (6.30) and (6.32), respectively.

Three-phase short-circuit fault between two nodes

In Figure 6.20(a), consider a fault between nodes j and k. The fault impedance is Z_F, the connection fault currents are $I_{j(F)}$ and $I_{k(F)}$, the Thévenin's impedances at nodes j and k are Z_{jj} and Z_{kk}, the transfer impedance between nodes j and k is $Z_{jk} = Z_{kj}$, and the prefault voltages at nodes j and k are $V_{j(o)}$ and $V_{k(o)}$. Consider the fault current to flow out of node j and into node k hence the injected currents are $I_{j(F)} = -I_{(F)}$ and $I_{k(F)} = I_{(F)}$.

From Equation (6.21b), we can write

$$
\begin{bmatrix} \Delta V_1 \\ . \\ \Delta V_j \\ \Delta V_k \\ . \\ \Delta V_N \end{bmatrix} = \begin{bmatrix} Z_{11} & . & Z_{1j} & Z_{1k} & . & Z_{1N} \\ . & & . & . & & . \\ Z_{j1} & . & Z_{jj} & Z_{jk} & . & Z_{jN} \\ Z_{k1} & . & Z_{jk} & Z_{kk} & . & Z_{kN} \\ . & & . & . & & . \\ Z_{N1} & . & Z_{Nj} & Z_{Nk} & . & Z_{NN} \end{bmatrix} \begin{bmatrix} 0 \\ . \\ -I_{(F)} \\ I_{(F)} \\ . \\ 0 \end{bmatrix} \tag{6.40}
$$

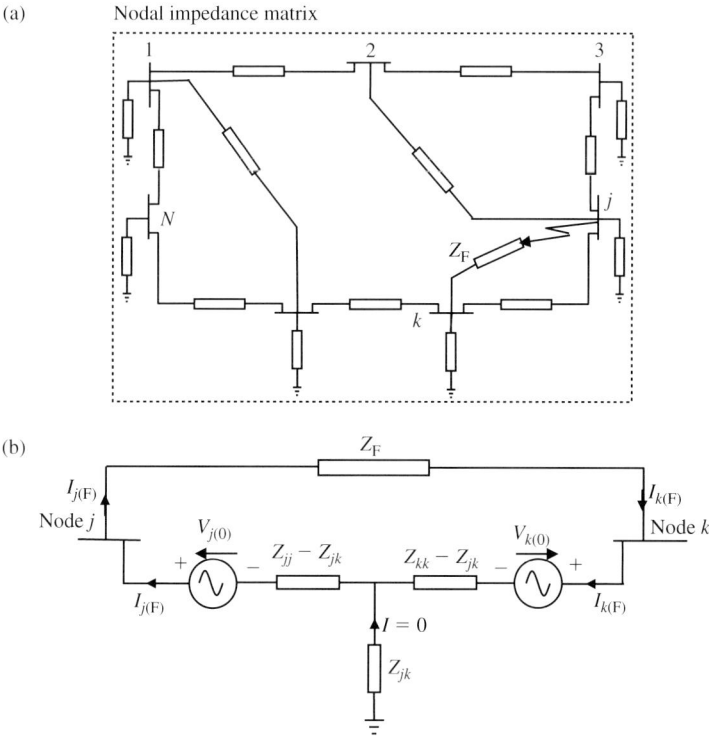

Figure 6.20 Representation of a three-phase short-circuit fault between two nodes: (a) Thévenin's representation of the entire network and (b) Thévenin's equivalent between nodes j and k for a short-circuit fault between them

From Equation (6.40), the voltage changes caused by the fault are given by

$$
\begin{bmatrix} \Delta V_1 \\ \cdot \\ \Delta V_j \\ \Delta V_k \\ \cdot \\ \Delta V_N \end{bmatrix} = \begin{bmatrix} (-Z_{1j} + Z_{1k})I_{(F)} \\ \cdot \\ (-Z_{jj} + Z_{jk})I_{(F)} \\ (-Z_{jk} + Z_{kk})I_{(F)} \\ \cdot \\ (-Z_{Nj} + Z_{Nk})I_{(F)} \end{bmatrix}
\tag{6.41}
$$

The actual voltage during the fault between the faulted nodes j and k is given by

$$
V_{jk(F)} = V_{j(F)} - V_{k(F)} = Z_F I_{(F)}
\tag{6.42}
$$

From Equation (6.41) rows j and k, we can write

$$
\Delta V_j = V_{j(F)} - V_{j(o)} = (-Z_{jj} + Z_{jk})I_{(F)}
\tag{6.43a}
$$

and

$$
\Delta V_k = V_{k(F)} - V_{k(o)} = (-Z_{jk} + Z_{kk})I_{(F)}
\tag{6.43b}
$$

From Equations (6.42) and (6.43), the fault current is given by

$$
I_{(F)} = \frac{V_{j(o)} - V_{k(o)}}{Z_{jj} + Z_{kk} - 2Z_{jk} + Z_F}
\tag{6.44}
$$

Equations (6.42) and (6.44) are represented by the Thévenin's equivalent circuit shown in Figure 6.20(b). Having calculated the fault current, the nodal voltage changes are calculated using Equation (6.41). The actual voltages are then calculated using Equation (6.22a). The actual branch and machine currents are calculated using Equations (6.30) and (6.32), respectively.

6.6.3 General analysis of unbalanced short-circuit faults

Nodal sequence impedance matrices

In the case of a three-phase short circuit, only the PPS nodal impedance matrix is required, as presented in Section 6.6.2. For a two-phase unbalanced short circuit, the NPS nodal impedance matrix is also required. This can be efficiently derived from the NPS nodal admittance matrix, as we discussed for the PPS impedance matrix. From the early days of digital computer analysis of large power systems, an assumption has generally been made that the NPS and PPS impedance matrices are identical in order to remove the need to form, store and calculate the NPS nodal matrix. This is based on the assumption that the PPS and NPS models and parameters of all passive network elements are identical and that those for machines do not differ significantly. However, this assumption is only approximately valid when calculating the subtransient short-circuit current using machine subtransient impedances. It is not valid when calculating the short-circuit current some time after the instant of fault, e.g. the transient current using machine transient impedances, because these differ significantly from the machine NPS impedances. On the whole, with the huge and cost-effective increases in computing storage and speed, particularly since the early 1990s, this assumption is no longer needed.

In the case of a single-phase to earth and two-phase to earth faults, the nodal ZPS impedance matrix is also required in addition to the PPS and NPS impedance matrices. This matrix is radically different from the PPS and NPS impedance matrices both in connectivity due to e.g. transformer windings and impedance values of most network elements. Figure 6.21 shows the ZPS network model of Figure 6.15.

General fault impedance/admittance matrix for unbalanced short-circuit faults

It is useful to derive a single method of analysis that covers all unbalanced short-circuit faults, namely single-phase to earth, two-phase and two-phase to earth. Such a method needs to allow for both solid short circuits and those through fault impedances. Figure 6.22 shows a general fault connection node with four impedances or admittances, the values of which can be appropriately chosen to represent any type of unbalanced short circuit.

It can be shown that the phase voltages and currents at the fault location are related by the following equation:

$$
\begin{bmatrix} V_{R(F)} \\ V_{Y(F)} \\ V_{B(F)} \end{bmatrix} = \begin{bmatrix} Z_R + Z_E & Z_E & Z_E \\ Z_E & Z_Y + Z_E & Z_E \\ Z_E & Z_E & Z_B + Z_E \end{bmatrix} \begin{bmatrix} I_{R(F)} \\ I_{Y(F)} \\ I_{B(F)} \end{bmatrix} \tag{6.45a}
$$

and the fault impedance matrix is defined as

$$
\mathbf{Z}_F^{RYB} = \begin{bmatrix} Z_R + Z_E & Z_E & Z_E \\ Z_E & Z_Y + Z_E & Z_E \\ Z_E & Z_E & Z_B + Z_E \end{bmatrix} \tag{6.45b}
$$

For example, to represent a solid single-phase to earth fault, we set $Z_R = 0$, $Z_E = 0$, $Z_Y = \infty$ and $Z_B = \infty$. Similarly, we represent a solid two-phase (Y–B) short circuit by setting $Z_Y = 0$, $Z_B = 0$, $Z_R = \infty$ and $Z_E = \infty$. The sequence fault impedance matrix is calculated using $\mathbf{Z}_F^{PNZ} = \mathbf{H}^{-1} \mathbf{Z}_F^{RYB} \mathbf{H}$. The result is given by

$$
\mathbf{Z}_F^{PNZ} = \frac{1}{3} \begin{bmatrix} Z_R + Z_Y + Z_B & Z_R + h^2 Z_Y + h Z_B & Z_R + h Z_Y + h^2 Z_B \\ Z_R + h Z_Y + h^2 Z_B & Z_R + Z_Y + Z_B & Z_R + h^2 Z_Y + h Z_B \\ Z_R + h^2 Z_Y + h Z_B & Z_R + h Z_Y + h^2 Z_B & Z_R + Z_Y + Z_B + 9 Z_E \end{bmatrix} \tag{6.46}
$$

However, as we have already seen, some of the impedances will be infinite under certain fault conditions and as a result $\mathbf{Z}_F^{PNZ}$ will become undefined. This problem

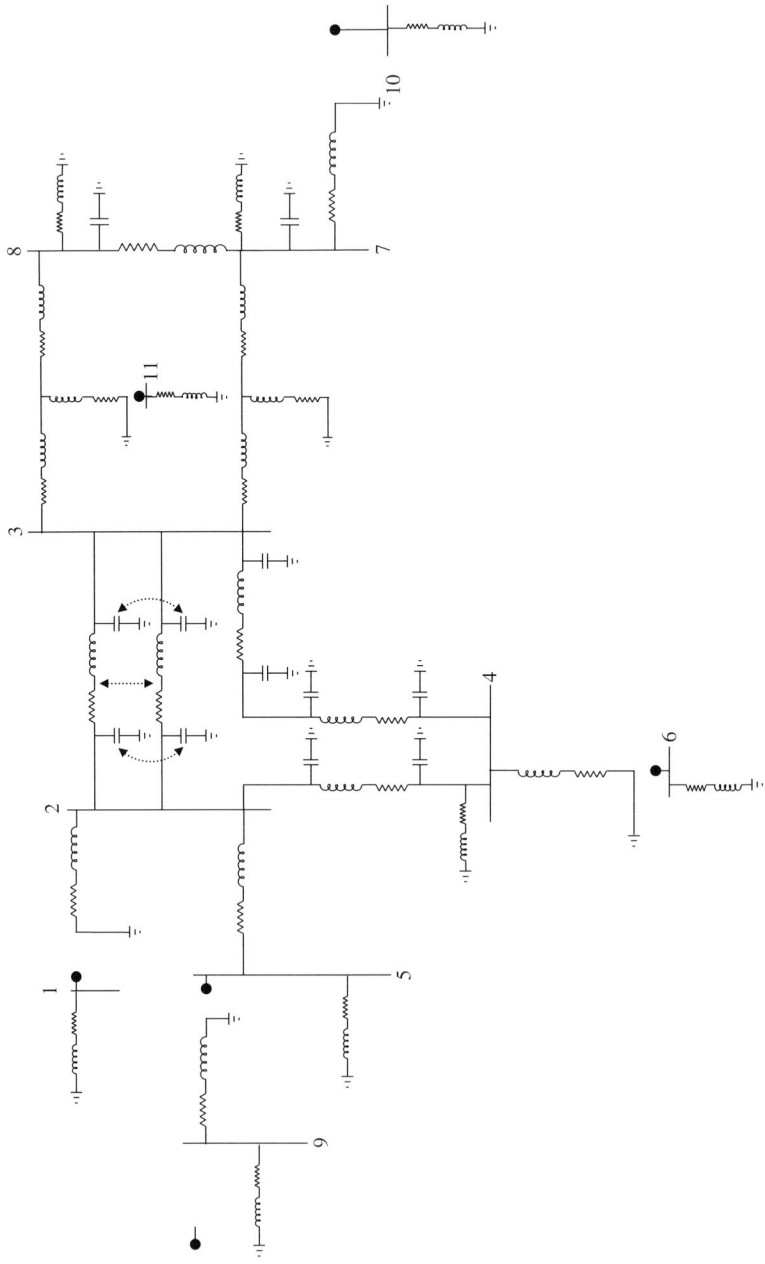

Figure 6.21 ZPS network model of Figure 6.15

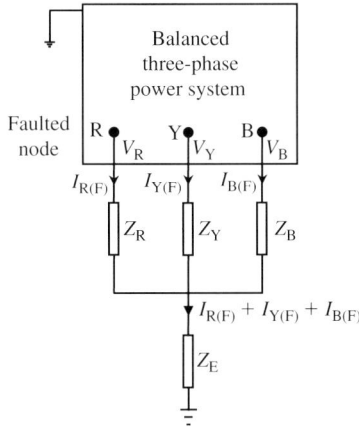

Figure 6.22   General fault impedance connection matrix

is solved by inverting $\mathbf{Z}_F^{PNZ}$ to obtain the sequence fault admittance matrix as follows:

$$\mathbf{Y}_F^{PNZ} = \frac{1}{3(Y_T + Y_E)} \begin{bmatrix} Y_E Y_T + 3Y_M & Y_E Y' - 3Y_N & Y_E Y'' \\ Y_E Y'' - 3Y_P & Y_E Y_T + 3Y_M & Y_E Y' \\ Y_E Y' & Y_E Y'' & Y_E Y_T \end{bmatrix} \quad (6.47)$$

where

$$Y_T = Y_R + Y_Y + Y_B \quad Y' = Y_R + h^2 Y_Y + h Y_B \quad Y'' = Y_R + h Y_Y + h^2 Y_B$$

$$Y_M = Y_R Y_Y + Y_Y Y_B + Y_R Y_B \quad Y_N = Y_Y Y_B + h Y_R Y_Y + h^2 Y_R Y_B$$

$$Y_P = Y_Y Y_B + h^2 Y_R Y_Y + h Y_R Y_B \quad Y_R = \frac{1}{Z_R} \quad Y_Y = \frac{1}{Z_Y} \quad Y_B = \frac{1}{Z_B} \quad Y_E = \frac{1}{Z_E}$$

Single-phase to earth short circuit through a fault impedance Z_F

As already outlined, this case is produced by setting $Z_R = Z_F$ or $Y_R = 1/Z_F$, $Z_E = 0$ or $Y_E = \infty$ and $Z_Y = Z_B = \infty$ or $Y_Y = Y_B = 0$. Therefore, the elements of Equation (6.46) are infinite and the sequence impedance matrix is thus undefined. However, Equation (6.47) results in indeterminate expressions which can be reduced to the following:

$$\mathbf{Y}_F^{PNZ} = \frac{1}{3Z_F} \begin{bmatrix} 1 & 1 & 1 \\ 1 & 1 & 1 \\ 1 & 1 & 1 \end{bmatrix} \quad (6.48)$$

Two-phase short circuit through a fault impedance $Z_F/2$

This case is produced by setting $Z_Y = Z_B = Z_F/2$ or $Y_Y = Y_B = 2/Z_F$, $Z_E = \infty$ or $Y_E = 0$ and $Z_R = \infty$ or $Y_R = 0$. Therefore, the elements of Equation (6.46) are infinite and the sequence impedance matrix is thus undefined. However,

Equation (6.47) results in the following useful matrix:

$$\mathbf{Y}_{F}^{PNZ} = \frac{1}{Z_F}\begin{bmatrix} 1 & -1 & 0 \\ -1 & 1 & 0 \\ 0 & 0 & 0 \end{bmatrix} \tag{6.49}$$

Two-phase to earth short circuit through a fault impedance Z_F and earth impedance Z_E

This case is produced by setting $Z_Y = Z_B = Z_F$ or $Y_Y = Y_B = 1/Z_F$, $Z_R = \infty$ or $Y_R = 0$. Where the earth impedance is equal to zero, $Z_E = 0$, $Y_E = \infty$. Therefore, the elements of Equation (6.46) are infinite and the sequence impedance matrix is consequently undefined. However, Equation (6.47) results in the following useful matrix:

$$\mathbf{Y}_{F}^{PNZ} = \frac{1}{3Z_F}\begin{bmatrix} 2 & -1 & -1 \\ -1 & 2 & -1 \\ -1 & -1 & 2 \end{bmatrix} \tag{6.50a}$$

where the earth impedance Z_E is finite, Equation (6.47) results in the following:

$$\mathbf{Y}_{F}^{PNZ} = \frac{1}{3Z_F(Z_F + 2Z_E)}\begin{bmatrix} 2Z_F + 3Z_E & -(Z_F + 3Z_E) & -Z_F \\ -(Z_F + 3Z_E) & 2Z_F + 3Z_E & -Z_F \\ -Z_F & -Z_F & 2Z_F \end{bmatrix} \tag{6.50b}$$

General mathematical analysis

Using Thévenin's theorem, and Equation (6.22b) the current and voltage changes in the PPS, NPS and ZPS networks, dropping the delta sign for currents for convenience, are given by

$$\Delta \mathbf{V}_{N\times 1}^{P} = \mathbf{Z}_{N\times N}^{P}\mathbf{I}_{N\times 1}^{P} \tag{6.51a}$$

$$\Delta \mathbf{V}_{N\times 1}^{N} = \mathbf{Z}_{N\times N}^{N}\mathbf{I}_{N\times 1}^{N} \tag{6.51b}$$

$$\Delta \mathbf{V}_{N\times 1}^{Z} = \mathbf{Z}_{N\times N}^{Z}\mathbf{I}_{N\times 1}^{Z} \tag{6.51c}$$

where $\mathbf{Z}_{N\times N}^{P}$, $\mathbf{Z}_{N\times N}^{N}$ and $\mathbf{Z}_{N\times N}^{Z}$ are the PPS, NPS and ZPS nodal impedance matrices, respectively, of the N node network. Prior to the short-circuit fault, the NPS and ZPS currents between all nodes, and the NPS and ZPS voltages at all nodes are zero because the three-phase system is assumed to be balanced. Where the system loading or operating condition is to be taken into account, the PPS voltages and currents are obtained from an initial loadflow study. Therefore, using the superposition principle, the actual nodal voltages during the fault are given by

$$\mathbf{V}_{N\times 1}^{P} = \mathbf{V}_{(o)N\times 1}^{P} + \mathbf{Z}_{N\times N}^{P}\mathbf{I}_{N\times 1}^{P} \tag{6.52a}$$

$$\mathbf{V}_{N\times 1}^{N} = \mathbf{Z}_{N\times N}^{N}\mathbf{I}_{N\times 1}^{N} \tag{6.52b}$$

$$\mathbf{V}_{N\times 1}^{Z} = \mathbf{Z}_{N\times N}^{Z}\mathbf{I}_{N\times 1}^{Z} \tag{6.52c}$$

where

$$\mathbf{V}^P_{N\times 1} = \begin{bmatrix} V^P_1 \\ \cdot \\ V^P_{(k)} \\ \cdot \\ V^P_N \end{bmatrix} \quad \mathbf{V}^N_{N\times 1} = \begin{bmatrix} V^N_1 \\ \cdot \\ V^N_{(k)} \\ \cdot \\ V^N_N \end{bmatrix} \quad \mathbf{V}^Z_{N\times 1} = \begin{bmatrix} V^Z_1 \\ \cdot \\ V^Z_{(k)} \\ \cdot \\ V^Z_N \end{bmatrix} \tag{6.53a}$$

$$\mathbf{I}^P_{N\times 1} = \begin{bmatrix} I^P_1 \\ \cdot \\ I^P_{(k)} \\ \cdot \\ I^P_N \end{bmatrix} \quad \mathbf{I}^N_{N\times 1} = \begin{bmatrix} I^N_1 \\ \cdot \\ I^N_{(k)} \\ \cdot \\ I^N_N \end{bmatrix} \quad \mathbf{I}^Z_{N\times 1} = \begin{bmatrix} I^Z_1 \\ \cdot \\ I^Z_{(k)} \\ \cdot \\ I^Z_N \end{bmatrix} \tag{6.53b}$$

The short-circuit fault is applied at node k. Thus, all injected sequence currents in the three sequence networks are zero except the sequence currents injected at the faulted node k. Equations (6.53a) and (6.53b) become

$$\mathbf{V}^P_{N\times 1} = \begin{bmatrix} V^P_1 \\ \cdot \\ V^P_{F(k)} \\ \cdot \\ V^P_N \end{bmatrix} \quad \mathbf{V}^N_{N\times 1} = \begin{bmatrix} V^N_1 \\ \cdot \\ V^N_{F(k)} \\ \cdot \\ V^N_N \end{bmatrix} \quad \mathbf{V}^Z_{N\times 1} = \begin{bmatrix} V^Z_1 \\ \cdot \\ V^Z_{F(k)} \\ \cdot \\ V^Z_N \end{bmatrix} \tag{6.54a}$$

$$\mathbf{I}^P_{N\times 1} = \begin{bmatrix} 0 \\ \cdot \\ -I^P_{F(k)} \\ \cdot \\ 0 \end{bmatrix} \quad \mathbf{I}^N_{N\times 1} = \begin{bmatrix} 0 \\ \cdot \\ -I^N_{F(k)} \\ \cdot \\ 0 \end{bmatrix} \quad \mathbf{I}^Z_{N\times 1} = \begin{bmatrix} 0 \\ \cdot \\ -I^Z_{F(k)} \\ \cdot \\ 0 \end{bmatrix} \tag{6.54b}$$

From Equations (6.52) to (6.54), we can extract the sequence voltage components of row k as follows:

$$\begin{bmatrix} V^P_{F(k)} \\ V^N_{F(k)} \\ V^Z_{F(k)} \end{bmatrix} = \begin{bmatrix} V^P_{k(o)} \\ 0 \\ 0 \end{bmatrix} - \begin{bmatrix} Z^P_{kk} & 0 & 0 \\ 0 & Z^N_{kk} & 0 \\ 0 & 0 & Z^Z_{kk} \end{bmatrix} \begin{bmatrix} I^P_{F(k)} \\ I^N_{F(k)} \\ I^Z_{F(k)} \end{bmatrix} \tag{6.55a}$$

or

$$\mathbf{V}^{PNZ}_{F(k)} = \mathbf{V}^{PNZ}_{k(o)} - \mathbf{Z}^{PNZ}_{kk}\mathbf{I}^{PNZ}_{F(k)} \tag{6.55b}$$

where

$$\mathbf{V}^{PNZ}_{F(k)} = \begin{bmatrix} V^P_{F(k)} \\ V^N_{F(k)} \\ V^Z_{F(k)} \end{bmatrix} \quad \mathbf{I}^{PNZ}_{F(k)} = \begin{bmatrix} I^P_{F(k)} \\ I^N_{F(k)} \\ I^Z_{F(k)} \end{bmatrix} \quad \mathbf{Z}^{PNZ}_{kk} = \begin{bmatrix} Z^P_{kk} & 0 & 0 \\ 0 & Z^N_{kk} & 0 \\ 0 & 0 & Z^Z_{kk} \end{bmatrix} \tag{6.55c}$$

The PPS, NPS and ZPS networks are interconnected at the fault location in a manner that depends on the fault type, as described in Chapter 2. For the general analysis of any type of unbalanced fault, in fact even balanced faults if desired, we need to describe the conditions at the fault location using a general method. One method is to use of a fault admittance matrix that is dependent on the fault type, as described in the previous section. Thus, the sequence fault currents can be written as

$$\mathbf{I}_{F(k)}^{PNZ} = \mathbf{Y}_{F}^{PNZ} \mathbf{V}_{F(k)}^{PNZ} \tag{6.56}$$

Substituting Equation (6.56) into Equation (6.55b), we obtain

$$\mathbf{V}_{F(k)}^{PNZ} = [\mathbf{U} + \mathbf{Z}_{kk}^{PNZ}\mathbf{Y}_{F}^{PNZ}]^{-1}\mathbf{V}_{k(o)}^{PNZ} \tag{6.57}$$

where $\mathbf{U}$ is the identity matrix. The sequence fault currents at the faulted node k are now obtained by substituting Equation (6.57) back into Equation (6.56), giving

$$\mathbf{I}_{F(k)}^{PNZ} = \mathbf{Y}_{F}^{PNZ}[\mathbf{U} + \mathbf{Z}_{kk}^{PNZ}\mathbf{Y}_{F}^{PNZ}]^{-1}\mathbf{V}_{k(o)}^{PNZ} \tag{6.58}$$

The sequence voltages at the faulted node k, $\mathbf{V}_{F(k)}^{PNZ}$, are calculated using Equation (6.57). The PPS, NPS and ZPS voltages throughout the PPS, NPS and ZPS networks can be calculated by substituting Equation (6.54) in Equation (6.52). Therefore, for a fault at node k, the sequence voltages at any node i are given by

$$\begin{bmatrix} V_i^P \\ V_i^N \\ V_i^Z \end{bmatrix} = \begin{bmatrix} V_{i(o)}^P \\ 0 \\ 0 \end{bmatrix} - \begin{bmatrix} Z_{ik}^P & 0 & 0 \\ 0 & Z_{ik}^N & 0 \\ 0 & 0 & Z_{ik}^Z \end{bmatrix} \begin{bmatrix} I_{F(k)}^P \\ I_{F(k)}^N \\ I_{F(k)}^Z \end{bmatrix} \tag{6.59}$$

The actual sequence currents flowing in the branches of the PPS, NPS and ZPS networks are calculated as follows:

$$\begin{bmatrix} I_{ij}^P \\ I_{ij}^N \\ I_{ij}^Z \end{bmatrix} = \begin{bmatrix} y_{ij}^P & 0 & 0 \\ 0 & y_{ij}^N & 0 \\ 0 & 0 & y_{ij}^Z \end{bmatrix} \begin{bmatrix} V_i^P - V_j^P \\ V_i^N - V_j^N \\ V_i^Z - V_j^Z \end{bmatrix} \quad i \neq j \tag{6.60}$$

The sequence current contributions of each machine are calculated as follows:

$$\begin{bmatrix} I_{i(m)}^P \\ I_{i(m)}^N \\ I_{i(m)}^Z \end{bmatrix} = \begin{bmatrix} I_{i(m)}^S \\ 0 \\ 0 \end{bmatrix} - \begin{bmatrix} y_{i(m)}^P & 0 & 0 \\ 0 & y_{i(m)}^N & 0 \\ 0 & 0 & y_{i(m)}^Z \end{bmatrix} \begin{bmatrix} V_i^P \\ V_i^N \\ V_i^Z \end{bmatrix} \tag{6.61}$$

Having calculated the sequence components of every nodal voltage and branch current, the corresponding phase voltages and currents are calculated using $\mathbf{V}^{RYB} = \mathbf{H}\mathbf{V}^{PNZ}$ and $\mathbf{I}^{RYB} = \mathbf{H}\mathbf{I}^{PNZ}$, where $\mathbf{H}$ is the transformation matrix given in Chapter 2. The reader should remember to modify the phase angles of the sequence currents and voltages in those parts of the network which are separated from the faulted node by star–delta transformers, as discussed in Chapter 4.

6.6.4 General analysis of open-circuit faults

One-phase open-circuit fault

The connection of the PPS, NPS and ZPS sequence networks was covered in Chapter 2, and shown in Figures 2.10 and 2.11 for one-phase and two-phase open-circuit faults, respectively. In these figures, the PPS, NPS and ZPS impedances are the Thévenin's impedances seen looking from the open-circuit fault points F and F'. We first illustrate how these impedances are calculated making use of the nodal impedance matrix and the Thévenin's impedance between two nodes using Figure 6.23.

In Figure 6.23, consider an open-circuit fault on a circuit connected between two nodes j and k and having sequence impedances Z^P, Z^N and Z^Z. In Equation (6.39), we derived the equivalent Thévenin's impedance between two nodes j and k and showed it in Figure 6.19(b). This Thévenin's impedance includes the circuit that is physically connected between nodes j and k. An open-circuit fault can be represented by removing the circuit between nodes j and k completely, i.e. all three phases, and then reinserting the appropriate sequence impedances Z^P, Z^N and Z^Z of the circuit. The disconnection of the circuit can be simulated by connecting $-Z^P$, $-Z^N$ and $-Z^Z$ between nodes j and k in the PPS, NPS and ZPS Thévenin's equivalents. Therefore, the equivalent Thévenin's impedance between the open-circuit points F and F' is calculated as illustrated in Figure 6.24 and is given by

$$Z^x_{(FF')} = \frac{-(Z^x)^2}{Z^x_{\text{Thév.}(jk)} - Z^x} \quad \text{where x = P, N, Z} \qquad (6.62a)$$

where, using Equation (6.39),

$$Z^x_{\text{Thév.}(jk)} = Z^x_{jj} + Z^x_{kk} - 2Z^x_{jk} \quad \text{where x = P, N, Z} \qquad (6.62b)$$

Nodal impedance matrix

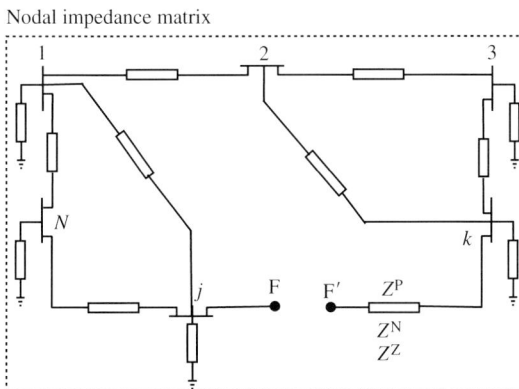

Figure 6.23 Open-circuit fault on a circuit connected between nodes j and k

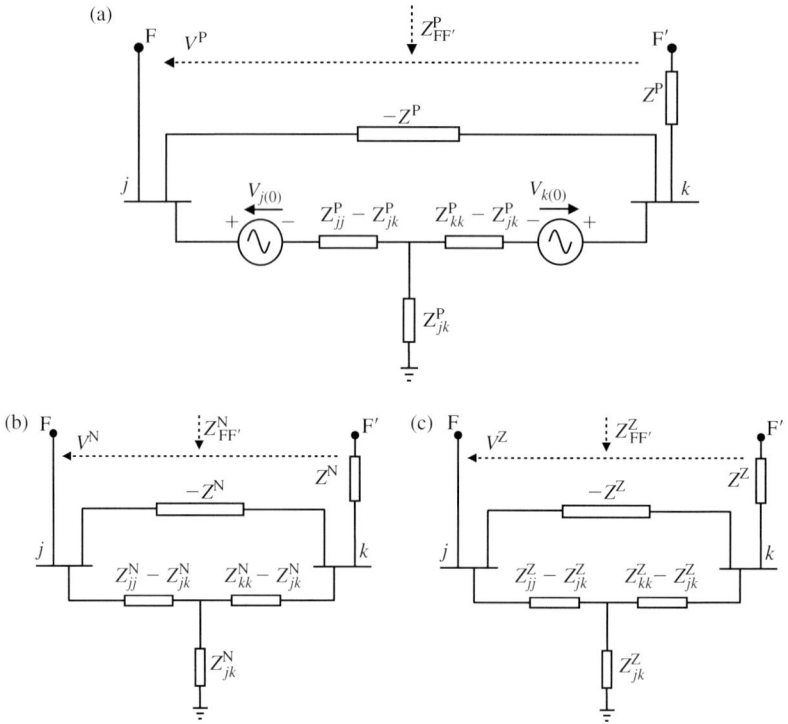

Figure 6.24 Calculation of (a) PPS, (b) NPS and (c) ZPS Thévenin's impedances between open-circuit points F and F′

The prefault current flowing between F and F′ at node j before the open-circuit fault occurs is a PPS current, and is calculated from the initial loadflow study using

$$I^P_{jk} = \frac{P_j - jQ_j}{V^*_j} \tag{6.63a}$$

Alternatively, an approximate value may be calculated as follows:

$$I^P_{jk} = \frac{V_j - V_k}{Z^P} \tag{6.63b}$$

Therefore, Equation (6.62) and the prefault current can be used in the equations derived in Chapter 2 for one-phase open circuit and two-phase open circuits to calculate the sequence currents and voltages at the fault location FF′.

To calculate the sequence voltage changes throughout the PPS, NPS and ZPS sequence networks using their nodal impedance matrices, we need to calculate the equivalent sequence currents to be injected at nodes j and k in each sequence network. These are easily calculated from Figure 6.24 by replacing each sequence voltage V^x in series with the circuit impedance Z^x (x = P, N or Z) with a current

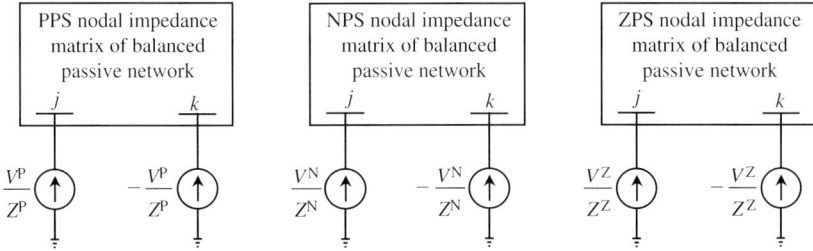

Figure 6.25   Representation of open-circuit faults by injection of PPS, NPS and ZPS current sources into respective sequence networks

source equal to V^x/Z^x in parallel with Z^x, the latter cancelling out when combined with $-Z^x$. This current source V^x/Z^x flows from node k to node j and can be considered as two separate current sources. The first is equal to V^x/Z^x and is injected into node j, and the second is equal to $-V^x/Z^x$ and is injected into node k. Figure 6.25 illustrates the sequence current sources injected into each sequence network. In summary, the sequence current sources injected into nodes j and k are given by

$$I^x_{s(j)} = \frac{V^x}{Z^x} \quad \text{and} \quad I^x_{s(k)} = -\frac{V^x}{Z^x} \quad x = P, N, Z \tag{6.64}$$

From Equations (6.51) and using $x = P$, N or Z, we have

$$
\begin{bmatrix} \Delta V^x_1 \\ \cdot \\ \Delta V^x_j \\ \Delta V^x_k \\ \cdot \\ \Delta V^x_N \end{bmatrix}
=
\begin{bmatrix}
Z^x_{11} & \cdot & Z^x_{1j} & Z^x_{1k} & \cdot & Z^x_{1N} \\
 & \cdot & & & & \\
Z^x_{j1} & \cdot & Z^x_{jj} & Z^x_{jk} & \cdot & Z^x_{jN} \\
Z^x_{k1} & \cdot & Z^x_{kj} & Z^x_{kk} & \cdot & Z^x_{kN} \\
 & \cdot & & & & \\
Z^x_{N1} & \cdot & Z^x_{Nj} & Z^x_{Nk} & \cdot & Z^x_{NN}
\end{bmatrix}
\begin{bmatrix} 0 \\ \cdot \\ I^x_{s(j)} \\ I^x_{s(k)} \\ \cdot \\ 0 \end{bmatrix}
\tag{6.65}
$$

Using Equation (6.64), we obtain

$$
\begin{bmatrix} \Delta V^x_1 \\ \cdot \\ \Delta V^x_j \\ \Delta V^x_k \\ \cdot \\ \Delta V^x_N \end{bmatrix}
=
\begin{bmatrix}
(Z^x_{1j} - Z^x_{1k})V^x/Z^x \\
\cdot \\
(Z^x_{jj} - Z^x_{jk})V^x/Z^x \\
(Z^x_{jk} - Z^x_{kk})V^x/Z^x \\
\cdot \\
(Z^x_{Nj} - Z^x_{Nk})V^x/Z^x
\end{bmatrix}
\quad x = P, N, Z \tag{6.66}
$$

The actual NPS and ZPS sequence voltages are equal to the changes calculated using Equation (6.66), as shown in Equations (6.52b) and (6.52c). The actual PPS voltages are calculated by adding the prefault voltages obtained from the initial loadflow study to the PPS voltage changes calculated using Equation (6.66), as shown in Equation (6.52a). The actual sequence currents flowing in the branches of the PPS, NPS and ZPS networks are calculated using Equation (6.60).

6.7 Three-phase short-circuit analysis of large-scale power systems in the phase frame of reference

6.7.1 Background

In some detailed short-circuit studies, fixed impedance analysis using three-phase modelling of the power system is used. One reason for this is to correctly include the inherent network unbalance, such as that due to untransposed overhead lines and cables. Another reason is the need to correctly calculate the earth return currents which requires the calculation of currents flowing through the earth wires of overhead lines and through cable sheaths. Such calculations require the explicit modelling of the earth wire and sheath conductors. The representation of all conductors of an overhead line or cable necessitates the use of three-phase models. Thus, three-phase modelling and analysis is then carried out in the phase frame of reference, since the symmetrical components sequence reference frame loses its advantages. In Chapter 3, we presented three-phase modelling of overhead lines and cables, and in Chapter 4, we presented three-phase modelling of transformers and other static power plant. In this section, we present a three-phase model of a current source, e.g. a synchronous machine and a general three-phase short-circuit analysis technique in the phase frame of reference that can be used in large-scale power systems.

6.7.2 Three-phase models of synchronous and induction machines

In Figure 6.14, we presented a PPS machine model using Thévenin's and Norton's equivalents. From Chapter 5, we know that an ac machine will also present an NPS impedance to the flow of NPS current. Whether a ZPS impedance is presented to the flow of ZPS currents depends on the stator winding connection and earthing method.

Synchronous machines

To maintain generality, we assume that, the synchronous machine NPS impedance is not equal to the PPS impedance. From the three independent sequence networks of the machine, shown in Figure 6.26, the following sequence equation can be written

$$
\begin{bmatrix} V^P \\ V^N \\ V^Z \end{bmatrix} = \begin{bmatrix} E^P \\ 0 \\ 0 \end{bmatrix} - \begin{bmatrix} Z^P & 0 & 0 \\ 0 & Z^N & 0 \\ 0 & 0 & Z^Z \end{bmatrix} \begin{bmatrix} I^P \\ I^N \\ I^Z \end{bmatrix}
\tag{6.67a}
$$

or

$$
\mathbf{V}^{PNZ} = \mathbf{E}^{PNZ} - \mathbf{Z}^{PNZ}\mathbf{I}^{PNZ}
\tag{6.67b}
$$

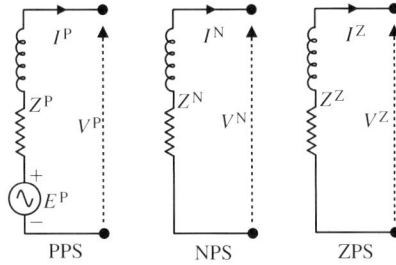

Figure 6.26 Independent machine sequence circuits

Before the short-circuit fault occurs, the machine is assumed to produce a balanced set of three-phase voltages having a PPS component only. However, the machine may be assumed to be connected to either a balanced or an unbalanced network. If the network is balanced, then only PPS currents flow, but if the network is unbalanced, then NPS and ZPS currents will also flow. Multiplying Equation (6.67b) by the sequence-to-phase transformation matrix $\mathbf{H}$, we obtain

$$\mathbf{V}^{RYB} = \mathbf{E}^{RYB} - \mathbf{Z}^{RYB}\mathbf{I}^{RYB} \tag{6.68a}$$

where

$$\mathbf{V}^{RYB} = \mathbf{H}\mathbf{V}^{PNZ} \quad \mathbf{I}^{RYB} = \mathbf{H}\mathbf{I}^{PNZ}$$

$$\mathbf{V}^{RYB} = [\, V^R \quad V^Y \quad V^B \,]^t \quad \text{and} \quad \mathbf{I}^{RYB} = [\, I^R \quad I^Y \quad I^B \,]^t$$

$$\mathbf{E}^{RYB} = \mathbf{H}\mathbf{E}^{PNZ} \quad \text{or} \quad [\, E^R \quad E^Y \quad E^B \,]^t = [\, E^P \quad h^2 E^P \quad h E^P \,]^t \tag{6.68b}$$

and

$$\mathbf{Z}^{RYB} = \mathbf{H}\mathbf{Z}^{PNZ}\mathbf{H}^{-1} \tag{6.68c}$$

From Equation (6.67), the machine sequence impedance matrix is given by

$$\mathbf{Z}^{PNZ} = \begin{bmatrix} Z^P & 0 & 0 \\ 0 & Z^N & 0 \\ 0 & 0 & Z^Z \end{bmatrix} \tag{6.69a}$$

After some algebra, Equation (6.68c) results in the following machine phase impedance matrix

$$\mathbf{Z}^{RYB} = \frac{1}{3}\begin{bmatrix} Z_S & Z_{M1} & Z_{M2} \\ Z_{M2} & Z_S & Z_{M1} \\ Z_{M1} & Z_{M2} & Z_S \end{bmatrix} \tag{6.69b}$$

where

$$Z_S = Z^Z + Z^N + Z^P \quad Z_{M1} = Z^Z + hZ^P + h^2 Z^N \quad Z_{M2} = Z^Z + h^2 Z^P + hZ^N \tag{6.69c}$$

The machine phase impedance matrix shows equal self-phase impedances for the three phases, full interphase but unequal mutual coupling between the three phases, and is non-symmetric, i.e. $Z_{12} \neq Z_{21}$. In cases where the machine's NPS impedance is assumed to be equal to the PPS impedance, i.e. $Z^N = Z^P$, then Equation (6.69b) simplifies to the following familiar symmetric phase impedance matrix

$$\mathbf{Z}^{RYB} = \begin{bmatrix} Z_S & Z_M & Z_M \\ Z_M & Z_S & Z_M \\ Z_M & Z_M & Z_S \end{bmatrix} \tag{6.70a}$$

where

$$Z_S = \frac{Z^Z + 2Z^P}{3} \quad Z_M = \frac{Z^Z - Z^P}{3} \tag{6.70b}$$

Since the machine's R^P and R^Z are usually taken to be equal, Z_M is mainly reactive, i.e. $Z_M \approx jX_M$, and it is worth noting that X_M is negative because $X^Z < X^P$.

From Equations (6.68a), (6.68b), (6.69b) and (6.70a), the three-phase Thévenin's equivalent circuit of the machine is as shown in Figure 6.27(a).

From Equation (6.68a), under a three-phase short-circuit fault at the machine terminals, the short-circuit current vector is given by $\mathbf{I}_S^{RYB} = \mathbf{Y}^{RYB}\mathbf{E}^{RYB}$. Substituting for $\mathbf{E}^{RYB}$ from Equation (6.68a) we obtain

$$\mathbf{I}_S^{RYB} = \mathbf{Y}^{RYB}\mathbf{V}^{RYB} + \mathbf{I}^{RYB} \tag{6.70c}$$

Equation (6.70c) represents the machine's three-phase Norton's equivalent circuit model shown in Figure 6.27(b).

Induction machines

A three-phase induction machine model can be derived in a similar way to a synchronous machine model, with one exception. Viewed from its terminals, an induction machine appears as an open circuit in the ZPS network where the stator winding is connected in either delta or star with an isolated neutral as is usually the case. Thus, the machine's ZPS impedance is infinite. Using the machine's sequence admittances $Y^P = 1/Z^P$, $Y^N = 1/Z^N$ and $Y^Z = 0$, as well as Equation (6.69a), the machine's sequence admittance matrix is given by

$$\mathbf{Y}^{PNZ} = \begin{bmatrix} Y^P & 0 & 0 \\ 0 & Y^N & 0 \\ 0 & 0 & 0 \end{bmatrix} \tag{6.71a}$$

Using Equation (6.68c), the induction machine phase admittance matrix is given by

$$\mathbf{Y}^{RYB} = \frac{1}{3} \begin{bmatrix} Y^P + Y^N & hY^P + h^2 Y^N & h^2 Y^P + hY^N \\ h^2 Y^P + hY^N & Y^P + Y^N & hY^P + h^2 Y^N \\ hY^P + h^2 Y^N & h^2 Y^P + hY^N & Y^P + Y^N \end{bmatrix} \tag{6.71b}$$

(a)

(b)

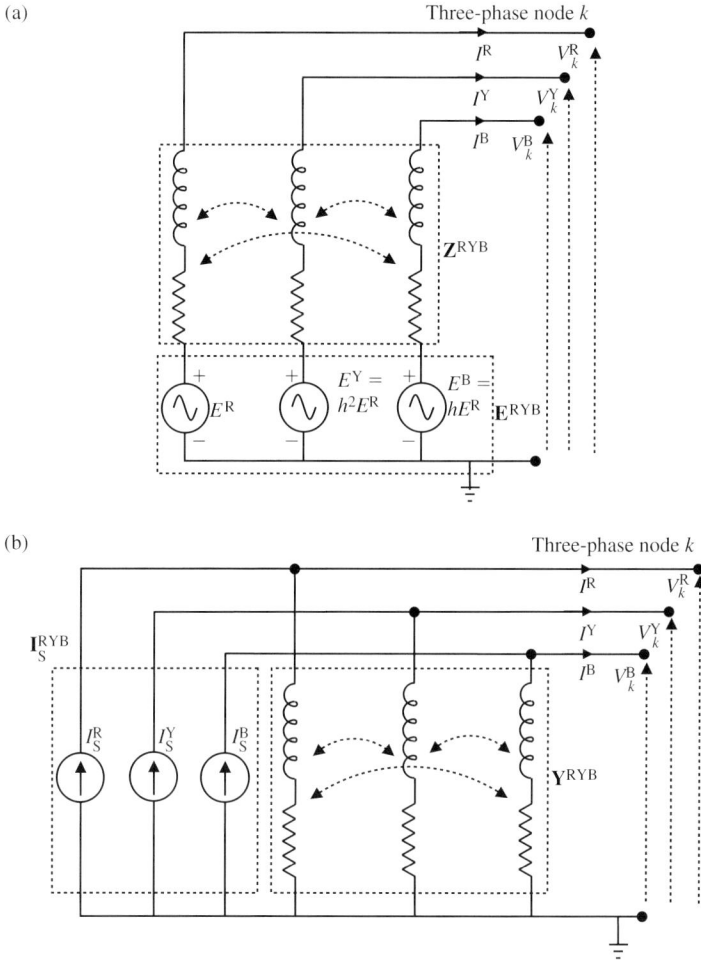

Figure 6.27   Three-phase (a) Thévenin's and (b) Norton's equivalent circuit models of machines

In the case where $Y^P \approx Y^N$, Equation (6.71b) reduces to

$$\mathbf{Y}^{RYB} = \frac{1}{3} \begin{bmatrix} 2Y^P & -Y^P & -Y^P \\ -Y^P & 2Y^P & -Y^P \\ -Y^P & -Y^P & 2Y^P \end{bmatrix} \qquad (6.71c)$$

6.7.3 Three-phase analysis of ac current in the phase frame of reference

In Chapter 3, we presented three-phase lumped admittance matrix models of overhead lines and cables, and in Chapter 4, similar three-phase models were presented for transformers and other static power plant. Where ac short-circuit analysis is being undertaken, the prefault loadflow study that sets up the initial conditions for the short-circuit study may be based on a balanced or an unbalanced loadflow study. The latter study produces unbalanced voltages and currents in the network, whereas the former produces PPS voltages and currents only. Figure 6.28 illustrates a power system containing balanced three-phase voltage or current sources, an unbalanced power network such as an untransposed overhead line and a three-phase node k where a balanced or unbalanced short-circuit fault may occur.

Similar to the collection of lumped shunt elements, shown in Figure 6.17 for single-phase elements, the three-phase self-admittance matrix at any three-phase node can be calculated as the sum of all the individual element self-admittance matrices at that node. The three-phase admittance matrix of the entire system can then be constructed, and the three-phase system admittance matrix equation is given as

$$
\begin{bmatrix} \mathbf{I}_1^{RYB} \\ \cdot \\ \mathbf{I}_k^{RYB} \\ \cdot \\ \mathbf{I}_N^{RYB} \end{bmatrix} = \begin{bmatrix} \mathbf{Y}_1^{RYB} & \cdot & \mathbf{Y}_{1k}^{RYB} & \cdot & \mathbf{Y}_{1N}^{RYB} \\ \cdot & & \cdot & & \cdot \\ \mathbf{Y}_{k1}^{RYB} & \cdot & \mathbf{Y}_{kk}^{RYB} & \cdot & \mathbf{Y}_{kN}^{RYB} \\ \cdot & & \cdot & & \cdot \\ \mathbf{Y}_{N1}^{RYB} & \cdot & \mathbf{Y}_{Nk}^{RYB} & \cdot & \mathbf{Y}_{NN}^{RYB} \end{bmatrix} \begin{bmatrix} \mathbf{V}_1^{RYB} \\ \cdot \\ \mathbf{V}_k^{RYB} \\ \cdot \\ \mathbf{V}_N^{RYB} \end{bmatrix} \qquad (6.72a)
$$

where

$$
\mathbf{Y}_{ii}^{RYB} = \sum_j \mathbf{y}_{ii}^{RYB} \qquad \mathbf{Y}_{ij}^{RYB} = -\mathbf{y}_{ij}^{RYB} \quad i \neq j
$$

and

$$
\mathbf{V}_k^{RYB} = [\, V_k^R \quad V_k^Y \quad V_k^B \,]^t \quad \mathbf{I}_k^{RYB} = [\, I_k^R \quad I_k^Y \quad I_k^B \,]^t
$$

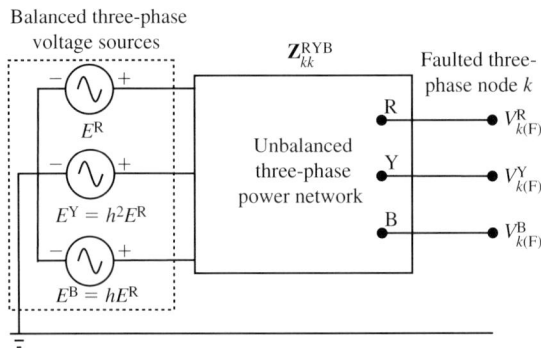

Figure 6.28   Illustration of a three-phase network for multiphase short-circuit analysis in the phase frame of reference

The elements of the three-phase admittance matrix of Equation (6.72a) are 3×3 matrices. Written in concise matrix form, Equation (6.72a) becomes

$$\mathbf{I}^{\text{RYB}} = \mathbf{Y}^{\text{RYB}} \mathbf{V}^{\text{RYB}} \tag{6.72b}$$

and

$$\mathbf{V}^{\text{RYB}} = (\mathbf{Y}^{\text{RYB}})^{-1} \mathbf{I}^{\text{RYB}} = \mathbf{Z}^{\text{RYB}} \mathbf{I}^{\text{RYB}} \tag{6.72c}$$

As described in Section 6.6, the matrix $\mathbf{Y}$ is sparse and generally symmetrical, and its inverse $\mathbf{Z}$ may be obtained by factorising $\mathbf{Y}$ into its upper triangle using Gaussian elimination and back substitution. Using the superposition theorem, the actual phase voltages at all nodes in the three-phase network are obtained using similar equations to those given in Equation (6.22), thus

$$\mathbf{V}^{\text{RYB}} = \mathbf{V}_0^{\text{RYB}} + \Delta\mathbf{V}^{\text{RYB}} \tag{6.73a}$$

The phase voltage changes $\Delta\mathbf{V}^{\text{RYB}}$ caused by the injected phase currents $\mathbf{I}^{\text{RYB}}$ are given by

$$\Delta\mathbf{V}^{\text{RYB}} = \mathbf{Z}^{\text{RYB}} \mathbf{I}^{\text{RYB}} \tag{6.73b}$$

For a general shunt short-circuit fault of any type at node k, the injected phase currents at all nodes in the system except node k are zero. Thus,

$$\mathbf{I}^{\text{RYB}} = [\,0 \quad \cdot \quad -\mathbf{I}_{\text{F}(k)}^{\text{RYB}} \quad \cdot \quad 0\,]^t \tag{6.73c}$$

where $\mathbf{I}_{\text{F}(k)}^{\text{RYB}} = [\,\mathbf{I}_{\text{F}(k)}^{\text{R}} \quad \mathbf{I}_{\text{F}(k)}^{\text{Y}} \quad \mathbf{I}_{\text{F}(k)}^{\text{B}}\,]^t$.

From Equations (6.73), the actual phase voltages at the faulted node k are given by

$$\mathbf{V}_{\text{F}(k)}^{\text{RYB}} = \mathbf{V}_{k(o)}^{\text{RYB}} - \mathbf{Z}_{kk}^{\text{RYB}} \mathbf{I}_{\text{F}(k)}^{\text{RYB}} \tag{6.74}$$

Using the general phase fault impedance matrix of Equation (6.45b), the phase fault admittance matrix is given by

$$\mathbf{Y}_{\text{F}}^{\text{RYB}} = \frac{1}{Y_{\text{T}}} \begin{bmatrix} Y^{\text{R}}(Y^{\text{Y}} + Y^{\text{B}} + Y_{\text{E}}) & -Y^{\text{R}}Y^{\text{Y}} & -Y^{\text{R}}Y^{\text{B}} \\ -Y^{\text{R}}Y^{\text{Y}} & Y^{\text{Y}}(Y^{\text{R}} + Y^{\text{B}} + Y_{\text{E}}) & -Y^{\text{Y}}Y^{\text{B}} \\ -Y^{\text{R}}Y^{\text{B}} & -Y^{\text{Y}}Y^{\text{B}} & Y^{\text{B}}(Y^{\text{R}} + Y^{\text{Y}} + Y_{\text{E}}) \end{bmatrix} \tag{6.75}$$

where

$$Y_{\text{T}} = Y^{\text{R}} + Y^{\text{Y}} + Y^{\text{B}} + Y_{\text{E}} \quad Y^{\text{R}} = \frac{1}{Z^{\text{R}}} \quad Y^{\text{Y}} = \frac{1}{Z^{\text{Y}}} \quad Y^{\text{B}} = \frac{1}{Z^{\text{B}}} \quad Y_{\text{E}} = \frac{1}{Z_{\text{E}}}$$

The phase fault admittance matrix for various unbalanced faults through a fault impedance Z_{F} and an earth impedance Z_{E} can be calculated. For a single-phase to earth short-circuit fault, the phase fault admittance matrix is given by

$$\mathbf{Y}_{\text{F}}^{\text{RYB}} = \begin{bmatrix} \dfrac{1}{Z_{\text{F}} + Z_{\text{E}}} & 0 & 0 \\ 0 & 0 & 0 \\ 0 & 0 & 0 \end{bmatrix} \tag{6.76a}$$

For a two-phase unearthed fault, the phase fault admittance matrix is given by

$$\mathbf{Y}_F^{RYB} = \begin{bmatrix} 0 & 0 & 0 \\ 0 & \dfrac{1}{Z_F} & \dfrac{-1}{Z_F} \\ 0 & \dfrac{-1}{Z_F} & \dfrac{1}{Z_F} \end{bmatrix} \tag{6.76b}$$

For a two-phase to earth short-circuit fault, the phase fault admittance matrix is

$$\mathbf{Y}_F^{RYB} = \frac{1}{Z_F(Z_F + 2Z_E)} \begin{bmatrix} 0 & 0 & 0 \\ 0 & Z_F + Z_E & -Z_E \\ 0 & -Z_E & Z_F + Z_E \end{bmatrix} \tag{6.76c}$$

For a three-phase earthed or unearthed short-circuit fault, the phase fault admittance matrix is given by

$$\mathbf{Y}_F^{RYB} = \frac{1}{mZ_F} \begin{bmatrix} 2 & -1 & -1 \\ -1 & 2 & -1 \\ -1 & -1 & 2 \end{bmatrix} \tag{6.76d}$$

where $m = 1$ for a three-phase unearthed fault and $m = 3$ for a three-phase to earth fault as presented in Chapter 2.

Using the fault admittance matrix, the actual phase voltages at the faulted node k are

$$\mathbf{I}_{F(k)}^{RYB} = \mathbf{Y}_F^{RYB} \mathbf{V}_{F(k)}^{RYB} \tag{6.77}$$

From Equations (6.74) and (6.77), we obtain the phase fault voltages as

$$\mathbf{V}_{F(k)}^{RYB} = [\mathbf{U} + \mathbf{Z}_{kk}^{RYB} \mathbf{Y}_F^{RYB}]^{-1} \mathbf{V}_{k(o)}^{RYB} \tag{6.78}$$

where $\mathbf{U}$ is the identity matrix. Substituting Equation (6.78) in Equation (6.77) the phase fault currents at the faulted node k are given by

$$\mathbf{I}_{F(k)}^{RYB} = \mathbf{Y}_F^{RYB} [\mathbf{U} + \mathbf{Z}_{kk}^{RYB} \mathbf{Y}_F^{RYB}]^{-1} \mathbf{V}_{k(o)}^{RYB} \tag{6.79}$$

The actual phase voltages at any node throughout the network, say node i, are calculated using Equation (6.73), thus

$$\mathbf{V}_i^{RYB} = \mathbf{V}_{i(o)}^{RYB} - \mathbf{Z}_{ik}^{RYB} \mathbf{I}_{F(k)}^{RYB} \tag{6.80}$$

The actual phase currents flowing in the network three-phase branches are given by

$$\mathbf{I}_{ij}^{RYB} = \mathbf{y}_{ij}^{RYB} [\mathbf{V}_i^{RYB} - \mathbf{V}_j^{RYB}] \tag{6.81}$$

The phase current contributions of each machine, say the machine connected at node i, are calculated using Equation (6.70c) as follows:

$$\mathbf{I}_{i(m)}^{RYB} = \mathbf{I}_{i(s)}^{RYB} - \mathbf{Y}_{i(m)}^{RYB} \mathbf{V}_i^{RYB} \tag{6.82}$$

Sequence currents and voltages, if required, can be calculated using the phase to sequence transformation $\mathbf{V}^{PNZ} = \mathbf{H}^{-1} \mathbf{V}^{RYB}$ and $\mathbf{I}^{PNZ} = \mathbf{H}^{-1} \mathbf{I}^{RYB}$.

6.7.4 Three-phase analysis and estimation of X/R ratio of fault current

General

Where the magnitude of the dc component of the short-circuit fault current is estimated from the X/R ratio at the fault location, the X/R ratio can be calculated using Thévenin's impedance matrix at the fault location $\mathbf{Z}_{kk}^{RYB}$ in accordance with IEC 60909 Method B, IEC 60909 Method C or IEEE C37.010 Method. These will be discussed in Chapter 7.

In general, the Thévenin's phase impedance matrix $\mathbf{Z}_{kk}^{RYB}$ representing the mutually coupled three-phase network is bilateral and the elements of $\mathbf{Z}_{kk}^{RYB}$ are reciprocal, i.e. $Z_{RY} = Z_{YR}$ except in special cases. This may be due to the generally unequal PPS and NPS impedances of machines and the non-symmetric admittance matrices of quadrature boosters and phase shifting transformers.

The effective X/R ratio at the fault location is dependent on the short-circuit fault type. To illustrate how the X/R ratio can be calculated, we use Figure 6.28 where the following equation can be written

$$
\begin{bmatrix} E^R \\ E^Y \\ E^B \end{bmatrix} - \begin{bmatrix} V_k^R \\ V_k^Y \\ V_k^B \end{bmatrix} = \begin{bmatrix} Z_{kk}^{RR} & Z_{kk}^{RY} & Z_{kk}^{RB} \\ Z_{kk}^{YR} & Z_{kk}^{YY} & Z_{kk}^{YB} \\ Z_{kk}^{BR} & Z_{kk}^{BY} & Z_{kk}^{BB} \end{bmatrix} \begin{bmatrix} I^R \\ I^Y \\ I^B \end{bmatrix} \tag{6.83}
$$

We will consider one-phase to earth and three-phase to earth short-circuit faults but the method is general and can be extended to cover other fault types.

One-phase to earth short-circuit fault

For a solid one-phase to earth short-circuit fault on phase R at node K, we have $I^R = I_F$, $I^Y = 0$, $I^B = 0$ and $V_k^R = 0$. Therefore, from the first row of Equation (6.83), we have

$$
E^R = Z_{kk}^{RR} I_F \tag{6.84}
$$

Similar equations can be written where the faulted phase is either phase Y or phase B. Therefore, in general, the X/R ratio of the phase fault current is calculated from the self-impedance of the faulted phase, or

$$
\left(\frac{X}{R} \right)_{\text{Phase x}} = \frac{\text{Imaginary}[Z_{kk}^{XX}]}{\text{Real}[Z_{kk}^{XX}]} \quad x = \text{RR, YY or BB} \tag{6.85}
$$

Alternatively, the X/R ratio can be calculated by transforming Equation (6.83) to the sequence reference frame using $\mathbf{V}^{PNZ} = \mathbf{H}^{-1}\mathbf{V}^{RYB}$, $\mathbf{I}^{PNZ} = \mathbf{H}^{-1}\mathbf{I}^{RYB}$ and $\mathbf{Z}_{kk}^{PNZ} = \mathbf{H}^{-1}\mathbf{Z}_{kk}^{RYB}\mathbf{H}$. The result is given by

$$
\begin{bmatrix} E^P \\ 0 \\ 0 \end{bmatrix} - \begin{bmatrix} V_k^P \\ V_k^N \\ V_k^Z \end{bmatrix} = \begin{bmatrix} Z_{kk}^{PP} & Z_{kk}^{PN} & Z_{kk}^{PZ} \\ Z_{kk}^{NP} & Z_{kk}^{NN} & Z_{kk}^{NZ} \\ Z_{kk}^{ZP} & Z_{kk}^{ZN} & Z_{kk}^{ZZ} \end{bmatrix} \begin{bmatrix} I^P \\ I^N \\ I^Z \end{bmatrix} \tag{6.86}
$$

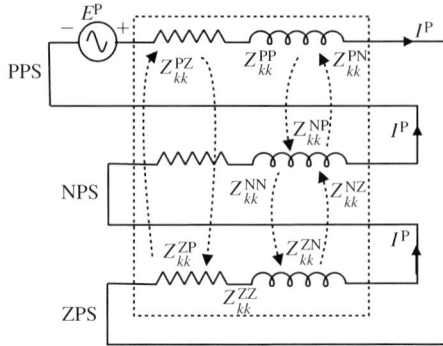

Figure 6.29 Calculation of X/R ratio of single-phase short-circuit fault current in an unbalanced three-phase network

As we discussed in Chapter 2, the boundary conditions $I^R = I_F$, $I^Y = 0$, $I^B = 0$ and $V_k^R = 0$ translate to $I^P = I^N = I^Z = \frac{1}{3}I_F$ and $V_k^P + V_k^N + V_k^Z = 0$. Using these sequence boundary conditions and Equation (6.86), the symmetrical component equivalent circuit is shown in Figure 6.29.

Now, summing the three rows of Equation (6.86), or using Figure 6.29, and using $I^P = \frac{1}{3}I^R$, the following equation can be written

$$E^P = \frac{(Z_{kk}^{PP} + Z_{kk}^{PN} + Z_{kk}^{PZ} + Z_{kk}^{NN} + Z_{kk}^{NP} + Z_{kk}^{NZ} + Z_{kk}^{ZZ} + Z_{kk}^{ZP} + Z_{kk}^{ZN})}{3}I_F \quad (6.87)$$

Therefore, the equivalent X/R ratio of the phase fault current I^R for a single-phase short-circuit fault on phase R is given by

$$\frac{X}{R} = \frac{\text{Imaginary}[Z_{kk}^{PP} + Z_{kk}^{PN} + Z_{kk}^{PZ} + Z_{kk}^{NN} + Z_{kk}^{NP} + Z_{kk}^{NZ} + Z_{kk}^{ZZ} + Z_{kk}^{ZP} + Z_{kk}^{ZN}]}{\text{Real}[Z_{kk}^{PP} + Z_{kk}^{PN} + Z_{kk}^{PZ} + Z_{kk}^{NN} + Z_{kk}^{NP} + Z_{kk}^{NZ} + Z_{kk}^{ZZ} + Z_{kk}^{ZP} + Z_{kk}^{ZN}]} \quad (6.88)$$

It is quite interesting to note that one third of the sum of the nine elements of the sequence impedance matrix for a one-phase fault on phase R is equal to the self-phase impedance of phase R that is Equations (6.84) and (6.87) are identical. Thus,

$$Z_{kk}^{RR} = \frac{(Z_{kk}^{PP} + Z_{kk}^{PN} + Z_{kk}^{PZ} + Z_{kk}^{NN} + Z_{kk}^{NP} + Z_{kk}^{NZ} + Z_{kk}^{ZZ} + Z_{kk}^{ZP} + Z_{kk}^{ZN})}{3} \quad (6.89)$$

Equation (6.89) can be obtained using Equations (2.25) and (2.26) from Chapter 2 even though Equation (2.26) was derived assuming a bilateral network.

Three-phase to earth short-circuit fault

For a three-phase to earth short-circuit fault, we have $V_k^R = V_k^Y = V_k^B = 0$. However, unlike the balanced network analysis presented in Chapter 2, because the network is now unbalanced, $I_F^R + I_F^Y + I_F^B = I_E \neq 0$. Therefore from Equation (6.83) we have

$$
\begin{bmatrix} E^R \\ E^Y \\ E^B \end{bmatrix} = \begin{bmatrix} Z_{kk}^{RR} & Z_{kk}^{RY} & Z_{kk}^{RB} \\ Z_{kk}^{YR} & Z_{kk}^{YY} & Z_{kk}^{YB} \\ Z_{kk}^{BR} & Z_{kk}^{BY} & Z_{kk}^{BB} \end{bmatrix} \begin{bmatrix} I_F^R \\ I_F^Y \\ I_F^B \end{bmatrix} \tag{6.90}
$$

or

$$
\begin{bmatrix} I_F^R \\ I_F^Y \\ I_F^B \end{bmatrix} = \begin{bmatrix} Y_{kk}^{RR} & Y_{kk}^{RY} & Y_{kk}^{RB} \\ Y_{kk}^{YR} & Y_{kk}^{YY} & Y_{kk}^{YB} \\ Y_{kk}^{BR} & Y_{kk}^{BY} & Y_{kk}^{BB} \end{bmatrix} \begin{bmatrix} E^R \\ E^Y \\ E^B \end{bmatrix} \tag{6.91}
$$

Therefore, using $E^Y = h^2 E^R$, $E^B = hE^R$ and expanding Equation (6.91), the effective impedance seen by phase fault current R is given by

$$
Z_{Eq}^{RR} = \frac{E^R}{I_F^R} = \frac{1}{Y_{kk}^{RR} + h^2 Y_{kk}^{RY} + h Y_{kk}^{RB}} \tag{6.92a}
$$

and the X/R ratio of phase R fault current is given by

$$
\left(\frac{X}{R} \right)_{\text{Phase R}} = \frac{\text{Imaginary}[Z_{Eq}^{RR}]}{\text{Real}[Z_{Eq}^{RR}]} \tag{6.92b}
$$

The effective impedance seen by phase fault current Y is given by

$$
Z_{Eq}^{YY} = \frac{E^Y}{I_F^Y} = \frac{1}{h Y_{kk}^{YR} + Y_{kk}^{YY} + h^2 Y_{kk}^{YB}} \tag{6.93a}
$$

and the X/R ratio of phase Y fault current is given by

$$
\left(\frac{X}{R} \right)_{\text{Phase Y}} = \frac{\text{Imaginary}[Z_{Eq}^{YY}]}{\text{Real}[Z_{Eq}^{YY}]} \tag{6.93b}
$$

The effective impedance seen by phase fault current B is given by

$$
Z_{Eq}^{BB} = \frac{E^B}{I_F^B} = \frac{1}{h^2 Y_{kk}^{BR} + h Y_{kk}^{BY} + Y_{kk}^{BB}} \tag{6.94a}
$$

and the X/R ratio of phase B fault current is given by

$$
\left(\frac{X}{R} \right)_{\text{Phase B}} = \frac{\text{Imaginary}[Z_{Eq}^{BB}]}{\text{Real}[Z_{Eq}^{BB}]} \tag{6.94b}
$$

It is important to note that the three-phase fault currents are generally unbalanced both in magnitude and phase and that the three X/R ratios are generally different. The differences being dependent on the degree of network unbalance.

In the author's experience, and depending on the type of study being carried out, it is often more advantageous in three-phase unbalanced analysis to use actual physical units, i.e. volts, amps and ohms, instead of per-unit quantities.

6.7.5 Example

Example 6.4 The three-phase Thévenin's impedance matrix at the fault location $\mathbf{Z}_{kk}^{RYB}$ of Figure 6.28 is given by

$$
\mathbf{Z}_{kk}^{RYB} = \begin{array}{c} R \\ Y \\ B \end{array} \begin{bmatrix} 0.1 + j1 & 0.05 + j0.3 & 0.04 + j0.32 \\ 0.05 + j0.3 & 0.12 + j0.9 & 0.04 + j0.28 \\ 0.04 + j0.32 & 0.04 + 0.28 & 0.09 + j1.1 \end{bmatrix}
$$

The balanced three-phase voltage sources have an rms value of unity. Calculate the fault currents and X/R ratios for a one-phase to earth fault on phase R, Y and B, and a three-phase to earth fault as presented in Section 6.7.4.

One-phase to earth fault

Using Equation (6.84), the current for a fault on phase R is equal to

$$
I_F^R = \frac{1}{0.1 + j1} \approx 1\angle -84.29°
$$

For a fault on phase Y

$$
I_F^Y = \frac{1}{0.12 + j0.9} = 1.1\angle -82.4°
$$

For a fault on phase B

$$
I_F^B = \frac{1}{0.09 + j1.1} = 0.906\angle -85.32°
$$

Using Equation (6.85) and for a fault on phase R, we have

$$
\left(\frac{X}{R}\right)_{\text{Phase R}} = \frac{1}{0.1} = 10
$$

For a fault on phase Y

$$
\left(\frac{X}{R}\right)_{\text{Phase Y}} = \frac{0.9}{0.12} = 7.5
$$

and for a fault on phase B

$$
\left(\frac{X}{R}\right)_{\text{Phase B}} = \frac{1.1}{0.09} = 12.2
$$

The sequence Thévenin's impedance matrix at the fault location $\mathbf{Z}_{kk}^{PNZ}$ is calculated as

$$
\mathbf{Z}_{kk}^{PNZ} = \begin{array}{c} P \\ N \\ Z \end{array} \begin{bmatrix} 0.06 + j0.7 & -0.051 - j0.023 & 0.064 + j0.022 \\ 0.041 - j0.017 & 0.06 + j0.7 & -0.064 - j1.547\times10^{-3} \\ -0.064 - j1.547\times10^{-3} & 0.064 + j0.022 & 0.19 + j1.6 \end{bmatrix}
$$

Using Equation (6.88), we have

$$\left(\frac{X}{R}\right)_{\text{Phase R}} = \frac{3}{0.3} = 10$$

Three-phase to earth fault

The Thévenin's phase admittance matrix at the fault location $\mathbf{Y}_{kk}^{RYB}$ is calculated as

$$Y_{kk}^{RYB} = \begin{array}{c} R \\ Y \\ B \end{array}\left[\begin{array}{ccc} 0.101 - j1.169 & -0.017 + j0.31 & -0.015 + j0.263 \\ -0.017 + j0.31 & 0.153 - j1.269 & -0.018 + j0.236 \\ -0.015 + j0.263 & -0.018 + j0.236 & 0.076 + j1.041 \end{array}\right]$$

Using Equation (6.91), we have

$$I_F^R = Y_{kk}^{RR} + h^2 Y_{kk}^{RY} + h Y_{kk}^{RB} = 1.46\angle -83.8°$$

$$I_F^Y = Y_{kk}^{YR} + h^2 Y_{kk}^{YY} + h Y_{kk}^{YB} = 1.54\angle 153.93° = 1.54\angle(-83.8 - 122.27)°$$

and

$$I_F^B = Y_{kk}^{BR} + h^2 Y_{kk}^{BY} + h Y_{kk}^{BB} = 1.3\angle 35.12° = 1.3\angle(-83.8 + 118.92)°$$

Using Equation (6.92a),

$$Z_{Eq}^{RR} = \frac{E^R}{I_F^R} = \frac{1}{Y_{kk}^{RR} + h^2 Y_{kk}^{RY} + h Y_{kk}^{RB}} = 0.074 + j0.681$$

and the X/R ratio of phase R fault current is

$$\left(\frac{X}{R}\right)_{\text{Phase R}} = \frac{0.681}{0.074} = 9.2$$

Using Equation (6.93a),

$$Z_{Eq}^{YY} = \frac{E^Y}{I_F^Y} = \frac{1}{h Y_{kk}^{YR} + Y_{kk}^{YY} + h^2 Y_{kk}^{YB}} = 0.0445 + j0.6478$$

and the X/R ratio of phase Y fault current is given by

$$\left(\frac{X}{R}\right)_{\text{Phase Y}} = \frac{0.6478}{0.0445} = 14.5$$

Using Equation (6.94a),

$$Z_{Eq}^{BB} = \frac{E^B}{I_F^B} = \frac{1}{h^2 Y_{kk}^{BR} + h Y_{kk}^{BY} + Y_{kk}^{BB}} = 0.0686 + j0.766$$

and the X/R ratio of phase B fault current is given by

$$\left(\frac{X}{R}\right)_{\text{Phase B}} = \frac{0.766}{0.0686} = 11.16$$

Further reading

Books

[1] Grainger, J. and Stevenson, W.D., *'Power System Analysis'*, 1994, ISBN 0701133380.
[2] Elgerd, O.I., *'Electric Energy Systems Theory'*, McGraw-Hill Int. Ed., 1983, ISBN 0-07-66273-8.
[3] Anderson, P.M., *'Analysis of Faulted Power Systems'*, Iowa State Press, Ames, IA, 1973.

Papers

[4] Nor, K.M., *et al.*, 'Improved three-phase power-flow methods using sequence components', *IEEE Transactions on Power Systems*, Vol. 20, No. 3, August 2005, 1389–1397.
[5] Berman, A., *et al.*, 'Analysis of faulted power systems by phase coordinates', *IEEE Transactions on Power Delivery*, Vol. 13, No. 2, April 1998, 587–595.

7

International standards for short-circuit analysis in ac power systems

7.1 General

Guidelines and standards for short-circuit analysis have been developed in some countries. These standards generally aim at producing consistency and repeatability of conservative results or results that are sufficiently accurate for their intended purpose. Three very popular and widely used approaches are the International Electro-technical Commission (IEC) 60909-0:2001 Standard, the American Institute for Electrical and Electronics Engineers (IEEE) C37.010:1999 Standard and the UK Engineering Recommendation ER G7/4 procedure. In this chapter, we give a brief introduction to the three approaches.

7.2 International Electro-technical Commission 60909-0 Standard

7.2.1 Background

In 1988, the International Electro-technical Commission published IEC Standard IEC 60909 entitled '*Short-Circuit Current Calculation in Three-Phase AC Systems*'. This was derived from the German Verband Deutscher Electrotechniker (VDE) 0102 Standard. The IEC Standard was subsequently updated in 2001 and is the only international standard recommending methods for calculating short-circuit currents in three-phase ac power systems. Its aim, when conceived, was to present a practical and concise procedure, which, if necessary, can be used to carry out hand calculations and which leads to conservative results with sufficient accuracy. However, non-conservative results were subsequently identified in particular applications and this objective was revised in the 2001 update to 'leading to results which are generally of acceptable accuracy'. Other methods, such as the superposition or ac method covered in Section 6.3.3, are not excluded if they give results of at least the same precision.

The Standard is applicable in low voltage and high voltage systems up to 550 kV nominal voltages and a nominal frequency of 50 or 60 Hz. The Standard deals with the calculation of maximum and minimum short-circuit currents and distinguishes between short circuits with and without ac current component decay corresponding to short circuits that are near to and far from generators.

7.2.2 Analysis technique and voltage source at the short-circuit location

The basis of the IEC 60909-0 Standard is the calculation of the initial symmetrical rms short-circuit current, I_k'', for any fault type by a method which is based on the use of an equivalent voltage source at the fault location such as the passive method covered in Section 6.3.2. The magnitude of this voltage source is equal to $cU_n/\sqrt{3}$ where U_n is the nominal system phase-to-phase voltage and c is the voltage factor. The voltage factor is intended to give a worst-case condition and is assumed to account for the variations in loadflow conditions, voltage levels between different locations and with time, transformer off-nominal tap ratios and internal subtransient voltage sources of generators and motors. Table 7.1 shows the recommended value of the c factor.

The calculations are based on the use of rated nominal impedances and neglecting capacitances and passive loads, except in the zero phase sequence (ZPS) networks in systems with isolated neutrals or resonant earth. The fault arc resistance is also neglected. The calculation method ignores the initial prefault system loading and operating conditions and is essentially a fixed impedance passive analysis method as described in Chapter 6. In calculating the maximum short-circuit currents, the resistances of lines and cables are calculated at 20°C whereas for minimum short-circuit current, higher conductor operating temperatures are used. A number of correction factors and scaling factors are used in order to satisfy

Table 7.1 Voltage factor c (IEC 60909-0)

	Voltage factor c for the calculation of	
Nominal system phase-to-phase voltage U_n	Maximum short-circuit currents (c_{max})	Minimum short-circuit currents (c_{min})
LV (100 V up to 1000 V)		
Upper voltage tolerance +6%	1.05	0.95
Upper voltage tolerance +10%	1.1	0.95
MV and HV (>1 kV up to 550 kV)	1.1	1

acceptable accuracy whilst maintaining simplicity. In addition to the voltage factor c, the following factors are used:

(a) Factor K_T for the correction of the impedance of network transformers.
(b) Factors K_G; K_S; $K_{G,S}$; $K_{T,S}$; K_{SO}; $K_{G,SO}$ and $K_{T,SO}$ for the correction of the impedance of generator and combined generator–transformer units.
(c) Factor κ for the calculation of the peak make current.
(d) Factor μ for the calculation of symmetrical short-circuit breaking current.
(e) Factor λ for the calculation of symmetrical steady state short-circuit current.
(f) Factor μq for the calculation of the symmetrical breaking current of asynchronous motors.

IEC 60909 provides three methods for the calculation of the equivalent system X/R ratio at the fault point used in the calculation of the peak make current as well as the dc break current component at the minimum time delay t_{min}. IEC 60909-0 defines t_{min} as the shortest time from the instant of short circuit to contact separation of the first pole to open of the switching device. This is taken as the sum of the minimum relay protection time and minimum circuit-breaker operating time. In calculating the initial symmetrical short-circuit current, I_k'', for a far from generator short circuit, the positive phase sequence (PPS) and negative phase sequence (NPS) generator impedances are assumed equal.

7.2.3 Impedance correction factors

Network transformers (factor K_T)

Impedance correction factors are applied to the nominal rated impedance values of two-winding and three-winding network transformers. The correction factor covers transformers with and without on-load tap-changers if the transformer's voltage ratio is different from the ratio of the base voltages of the network. The corrected PPS impedance is given by

$$Z_{TK} = K_T Z_T = K_T(R_T + jX_T)\Omega \tag{7.1a}$$

and the correction factor K_T is given by

$$K_T = 0.95 \times \frac{c_{max}}{1 + 0.6x_T} \tag{7.1b}$$

where $Z_T = R_T + jX_T$ is the uncorrected rated transformer PPS impedance in Ω and x_T is the rated transformer PPS reactance in pu. This correction factor is also applied to the transformer's NPS and ZPS impedances. For three-winding transformers, the correction factor is applied to the three reactances between windings. For example, if the three windings are denoted A, B and C and the pu reactance between windings A and B with C open is considered, the correction factor is given by

$$K_{T(AB)} = 0.95 \times \frac{c_{max}}{1 + 0.6x_{T(AB)}} \tag{7.2}$$

and similarly for the two other windings.

No correction is applied to the star winding neutral earthing impedance, where present. Similarly, this correction factor is not applied to generator step-up transformers.

Directly connected synchronous generators and compensators (factor K_G)

Where the short-circuit location is on a low or medium voltage network directly fed from generators without step-up transformers, such as in industrial power systems, a correction factor is applied to the PPS subtransient impedance of the generators. This correction is introduced to compensate for the use of the equivalent voltage source $cU_n/\sqrt{3}$ instead of the subtransient voltage E'' behind subtransient reactance calculated at rated operation. The corrected PPS subtransient impedance is given by

$$Z_{Gk} = K_G \, Z_G = K_G(R_G + jX_d'')\Omega \tag{7.3a}$$

where the correction factor K_G is given by

$$K_G = \frac{U_n}{U_{rG}} \times \frac{c_{max}}{1 + x_d'' \sin \varphi_{rG}} \tag{7.3b}$$

where $Z_G = R_G + jX_d''$ is the uncorrected generator subtransient impedance, U_{rG} is the rated phase–phase voltage of the generator, φ_{rG} is the phase angle between $U_{rG}/\sqrt{3}$ and the rated current I_{rG} and x_d'' is the generator subtransient reactance in pu. To calculate the peak make current using the factor κ (this is covered in Section 7.2.5), a fictitious generator resistance R_{Gf} is used such that:

$R_{Gf} = X_d''/20$ for $U_{rG} > 1\,\text{kV}$ and rated apparent power $S_{rG} \geq 100\,\text{MVA}$
$R_{Gf} = X_d''/14.286$ for $U_{rG} > 1\,\text{kV}$ and $S_{rG} < 100\,\text{MVA}$
$R_{Gf} = X_d''/6.667$ for $U_{rG} \leq 1\,\text{kV}$

The correction factor K_G is also applied to the NPS and ZPS generator impedances but not to the generator neutral earthing impedance, where present. If the terminal voltage of the generator is different from U_{rG} by a factor p_G, then U_{rG} in Equation (7.3b) is replaced by $U_{rG}(1 + p_G)$. Synchronous motors equipped with voltage regulation equipment are treated as synchronous generators.

Power station units with and without on-load tap-changers (factors K_S; $K_{G,S}$; $K_{T,S}$; and K_{SO}; $K_{G,SO}$; $K_{T,SO}$)

A power station unit is defined as the combined generator and its step-up transformer is shown in Figure 7.1. Two cases are considered: the first where the step-up transformer is equipped with an on-load tap-changer and the second where it is not.

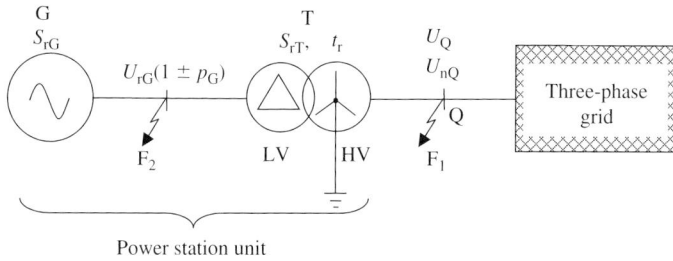

Figure 7.1 IEC 60909 power station unit for calculation of impedance correction factors for short-circuit faults at F1 and F2

Step-up transformer with an on-load tap-changer (factors K_S; $K_{G,S}$ and $K_{T,S}$)

For a short-circuit fault on the high voltage side of the step-up transformer, i.e. at F_1, the corrected impedance of the entire power station unit referred to the high voltage side of the step-up transformer is given by

$$Z_S = K_S \left[\frac{U_{rTHV}^2}{U_{rTLV}^2} \times Z_G + Z_{THV} \right] \Omega \tag{7.4a}$$

and the correction factor K_S is given by

$$K_S = \frac{U_{nQ}^2}{U_{rG}^2} \times \frac{U_{rTLV}^2}{U_{rTHV}^2} \times \frac{c_{max}}{1 + \left| x_d'' - x_T \right| \sin \varphi_{rG}} \tag{7.4b}$$

where Z_{THV} is the uncorrected transformer's impedance referred to the high voltage side, U_{nQ} is the nominal system voltage at the power station unit connection point, x_T is the transformer reactance in pu at nominal tap position and $t_r = U_{rTHV}/U_{rTLV}$ is the rated voltage ratio of the transformer. U_{nQ}^2 may be replaced with $U_{nQ}U_{Q\,min}$ if experience shows that the minimum operating voltage is such that $U_{Q\,min} > U_{nQ}$ and the worst-case current is not required. U_{rG} may be replaced with $U_{rG}(1 + p_G)$ if the actual generator voltage is always higher than the rated value. In Equation (7.4b), the absolute value of $(x_d'' - x_T)$ is taken.

For unbalanced short circuits, K_S is also applied to the NPS and ZPS impedances of the power station unit provided overexcited generator operation is assumed. The correction is not applied to the transformer's neutral impedance, if present. However, for underexcited generator operation, the application of K_S may give non-conservative results in which case the superposition method may be used. In calculating the current contribution of the entire power station unit on the transformer's high voltage side, it is not necessary to take into account the contribution of the motors within the auxiliary system.

When calculating the partial short-circuit currents from the generator and through the transformer for a fault between the generator and the transformer, i.e. at

F_2, the generator and transformer impedance correction factors are, respectively, given by

$$K_{G,S} = \frac{c_{max}}{1 + x_d'' \sin \varphi_{rG}} \tag{7.5a}$$

and

$$K_{T,S} = \frac{c_{max}}{1 - x_T \sin \varphi_{rG}} \tag{7.5b}$$

Step-up transformer without an on-load tap-changer (factors K_{SO}; $K_{G,SO}$ and $K_{T,SO}$)

For a short-circuit fault on the high voltage side of the step-up transformer, i.e. at F_1, the corrected impedance of the entire power station unit referred to the high voltage side of the step-up transformer is given by

$$Z_{SO} = K_{SO}(t_r^2 Z_G + Z_{THV})\Omega \tag{7.6a}$$

and the correction factor K_{SO} is given by

$$K_{SO} = \frac{U_{nQ}}{U_{rG}(1 + p_G)} \times \frac{U_{rTLV}}{U_{rTHV}} \times (1 \pm p_T) \times \frac{c_{max}}{1 + x_d'' \sin \varphi_{rG}} \tag{7.6b}$$

$1 \pm p_T$ is used if the transformer has off-load taps and one tap position is permanently used otherwise $p_T = 0$. If the highest current is required, then $1 - p_T$ is used.

For unbalanced short-circuit faults, K_{SO} is also applied to the NPS and ZPS impedances of the power station unit but not to the transformer's neutral earthing impedance, if present. In calculating the current contribution of the entire power station unit on the transformer's high voltage side, it is not necessary to take into account the contribution of the motors within the auxiliary system.

When calculating the partial short-circuit currents from the generator and through the transformer for a fault between the generator and the transformer, i.e. at F_2, the generator and transformer impedance correction factors are, respectively, given by

$$K_{G,SO} = \frac{1}{1 + p_G} \times \frac{c_{max}}{1 + x_d'' \sin \varphi_{rG}} \tag{7.7a}$$

and

$$K_{T,SO} = \frac{1}{1 + p_G} \times \frac{c_{max}}{1 - x_T \sin \varphi_{rG}} \tag{7.7b}$$

7.2.4 Asynchronous motors and static converter drives

Low voltage and medium voltage motors contribute to the initial symmetrical short-circuit current, the peak make current, the symmetrical breaking current and, for unbalanced faults, the steady state short-circuit current. The contribution of medium voltage motors is included in the calculation of maximum short-circuit currents. The contribution of low voltage motors in power station auxiliaries and industrial power systems, e.g., chemical plant, steel plant and pumping stations, etc., is taken into account.

Large numbers of low voltage motors, e.g. in industrial installations and power station auxiliaries, together with their connection cables, can be represented as an equivalent motor. The rated current of the equivalent motor $\sum_i I_{rM}$ is the sum of the rated currents of all motors. For asynchronous motors in low voltage supply systems directly connected to the fault location, i.e. without intervening transformers, the motor's contribution to the initial symmetrical short-circuit current may be neglected if

$$\sum_i I_{rM} \le 0.01 \times I_k'' \tag{7.8a}$$

where I_k'' is the total initial symmetrical short-circuit current at the fault location calculated without the contribution of the motors. If $I_{LR}/I_{rM} = 5$ where I_{LR} is the motor's locked rotor current, and $I_{kM}'' = I_{LR}$ where I_{kM}'' is the initial short-circuit current of the motor, Equation (7.8a) gives

$$\sum_i I_{kM}'' \le 0.05 \times I_k'' \tag{7.8b}$$

The short-circuit PPS, and NPS, motor impedance is calculated using

$$Z_M = R_M + jX_M = \frac{1}{I_{LR}/I_{rM}} \times \frac{U_{rM}^2}{S_{rM}} \Omega \tag{7.9a}$$

where S_{rM} and U_{rM} are the rated apparent power and rated voltage of the motor. With P_{rM}/p being the motor's rated active power per pair of poles, the recommended X_M/R_M ratios of the motor's short-circuit impedance are given by:

$X_M/R_M = 10$ for high voltage motors with $P_{rM}/p \ge 1$ MW.
$X_M/R_M = 6.67$ for high voltage motors with $P_{rM}/p < 1$ MW.
$X/R = 2.38$ for low voltage motor groups with connection cables.

Low voltage and medium voltage motors feeding through two-winding transformers to the short-circuit location may be neglected if

$$\frac{\sum_i P_{rM}}{\sum_i S_{rT}} \le \frac{0.8}{\left|\frac{100 \times c \times \sum_i S_{rT}}{\sqrt{3} U_{nQ} I_{kQ}''} - 0.3\right|} \tag{7.9b}$$

where $\sum_i P_{rM}$ is the sum of the rated powers of low voltage and medium voltage motors, $\sum_i S_{rT}$ is the sum of the rated apparent powers of all parallel transformers feeding the motors and I_{kQ}'' is the initial symmetrical short-circuit current at the fault location calculated without motors. The absolute value of the denominator should be used if it is negative.

Reversible static converter drives are considered for three-phase short circuits if the rotational masses of the motors and the static equipment provide reverse transfer of energy for deceleration at the time of the short circuit. They are considered to contribute to the initial symmetrical short-circuit current I_k'' and the peak make current but they do not contribute to the symmetrical short-circuit breaking current or the steady state short-circuit current. These drives are represented as an equivalent asynchronous motor with the equivalent PPS impedance as given in Equation (7.9a) with $I_{LR}/I_{rM} = 3$ and $X_M/R_M = 10$.

In the calculation of minimum short-circuit currents, the contribution of asynchronous motors is neglected.

7.2.5 Calculated short-circuit currents

Calculation of peak make current

From the initial rms symmetrical current I_k'', the peak make current for a single radial circuit is calculated using a peak factor κ and is given by

$$i_p = \kappa\sqrt{2}I_k'' \tag{7.10a}$$

where

$$\kappa = 1.02 + 0.98e^{-3/(X/R)} \tag{7.10b}$$

and (X/R) is the system X/R ratio at the fault point calculated using generator fictitious resistance R_{Gf} as described in Section 7.2.3. The variation of κ with the X/R ratio is shown in Figure 7.2.

For parallel independent radial circuits, the total peak make current is the sum of the individual make currents of each radial circuit. For meshed systems,

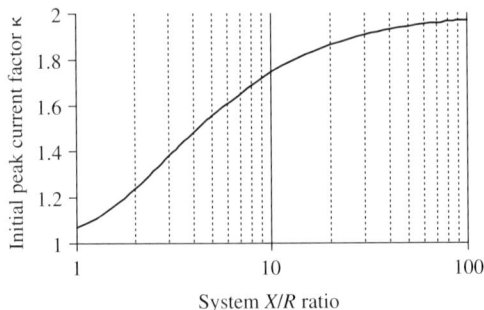

Figure 7.2　IEC 60909 factor κ for the calculation of initial peak short-circuit current

Equation (7.10a) is also used but Equation (7.10b) changes to

$$\kappa = 1.02 + 0.98e^{-3/(X/R)_i} \tag{7.11}$$

where $(X/R)_i$ is calculated in accordance with one of the following three methods:

Method A: Uniform X/R *ratio*
κ is calculated using the largest X/R ratio of all the network branches that feed partial short-circuit current into the fault location. This method can lead to significant errors.

Method B: Equivalent X/R *ratio at the fault point*
The X/R ratio at the fault point is calculated from the Thévenin's impedance obtained by network reduction of complex impedances. A safety factor of 1.15 is used to compensate for the underestimate in the dc current caused by the different X/R ratios of network branches and sources, i.e.

$$i_p = 1.15\kappa\sqrt{2}I_k'' \tag{7.12}$$

The factor $1.15\,\kappa$ should not exceed 1.8 and 2 in low voltage and high voltage systems, respectively. The 1.15 safety factor is not applied if the smallest X/R ratio in the network is greater than or equal to 3.33.

Method C: Equivalent frequency f_c
Initially, the magnitude of network reactances are scaled from their power frequency f values to a reduced frequency f_c then the Thévenin's impedance at the fault point is calculated. If the calculated PPS and ZPS impedances at f_c at the fault point are $Z_C^P = R_C^P + jX_C^P$ and $Z_C^Z = R_C^Z + jX_C^Z$, the X/R ratio at the fault point is calculated from

$$\frac{X}{R} = \begin{cases} \left(\dfrac{X_C^P}{R_C^P}\right) \times \dfrac{f}{f_c} & \text{for a three-phase short circuit} \\[3mm] \left(\dfrac{2X_C^P + X_C^Z}{2R_C^P + R_C^Z}\right) \times \dfrac{f}{f_c} & \text{for a single-phase to earth short circuit} \end{cases} \tag{7.13}$$

where f is nominal system frequency (50 or 60 Hz) and $f_c/f = 0.4$. This gives the equivalent frequency $f_c = 20$ Hz for $f = 50$ Hz and $f_c = 24$ Hz for $f = 60$ Hz. Equal PPS and NPS impedances are assumed. Similar equations can be derived for other fault types.

Calculation of dc current component

The maximum dc component of the short-circuit current is calculated using

$$i_{dc} = \sqrt{2}I_k'' \exp\left[\frac{-2\pi f t_{min}}{(X/R)_b}\right] \tag{7.14}$$

Table 7.2   Equivalent frequency f_c for the calculation of the break $(X/R)_b$ ratio used to estimate the magnitude of the break dc current component at t_{min}

$f \times t_{min}$ (cycles)		<1	<2.5	<5	<12.5
t_{min} (ms)	$f = 50\,\text{Hz}$	<20	<50	<100	<250
	$f = 60\,\text{Hz}$	<16.67	<41.67	<83.34	<208.34
f_c/f		0.27	0.15	0.092	0.055
f_c (Hz)	$f = 50\,\text{Hz}$	13.5	7.5	4.6	2.75
	$f = 60\,\text{Hz}$	16.2	9	5.52	3.3

where f is power frequency and $(X/R)_b$ is the break X/R ratio calculated using Method C equivalent frequency depending on the minimum time t_{min} from the instant of fault. In calculating $(X/R)_b$, the actual resistance of generators, R_G, is used. The equivalent frequency f_c to be used in calculating Z_C is determined as shown in Table 7.2.

Calculation of rms symmetrical breaking current

For a near to generator three-phase short-circuit fault, consider first either a singly fed short circuit or a non-meshed network consisting of parallel independent short-circuit sources. In the case of two sources namely a synchronous generator and an asynchronous motor, the symmetrical short-circuit breaking current is given by

$$I_b = I_{bG} + I_{bM} \tag{7.15a}$$

where

$$I_{bG} = \mu I''_{kG} \quad \text{and} \quad I_{bM} = (\mu q)I''_{kM} \tag{7.15b}$$

I''_{kG} and I''_{kM} are the synchronous generator and asynchronous motor initial symmetrical short-circuit current contributions measured at their terminals. Denoting I_{rG} and I_{rM} as the generator and motor rated currents, the multiplying factors μ and q are given by

$$\mu = \begin{cases} 0.84 + 0.26 \times \exp[-0.26 \times (I''_{kG}/I_{rG})] & \text{for } t_{min} = 20\,\text{ms} \\ 0.71 + 0.51 \times \exp[-0.30 \times (I''_{kG}/I_{rG})] & \text{for } t_{min} = 50\,\text{ms} \\ 0.62 + 0.72 \times \exp[-0.32 \times (I''_{kG}/I_{rG})] & \text{for } t_{min} = 100\,\text{ms} \\ 0.56 + 0.94 \times \exp[-0.38 \times (I''_{kG}/I_{rG})] & \text{for } t_{min} \geq 250\,\text{ms} \end{cases} \tag{7.16}$$

For asynchronous motors, I''_{kG}/I_{rG} in Equation (7.16) is replaced with I''_{kM}/I_{rM}. and

$$q = \begin{cases} 1.03 + 0.12 \times \ln(P_{rM}/p) & \text{for } t_{min} = 20\,\text{ms} \\ 0.79 + 0.12 \times \ln(P_{rM}/p) & \text{for } t_{min} = 50\,\text{ms} \\ 0.57 + 0.12 \times \ln(P_{rM}/p) & \text{for } t_{min} = 100\,\text{ms} \\ 0.26 + 0.10 \times \ln(P_{rM}/p) & \text{for } t_{min} \geq 250\,\text{ms} \end{cases} \tag{7.17}$$

where P_{rM} and p are the motor rated power in MW and number of pole pairs. The maximum value for q is unity. Equations (7.16) and (7.17) are plotted in Figures 7.3

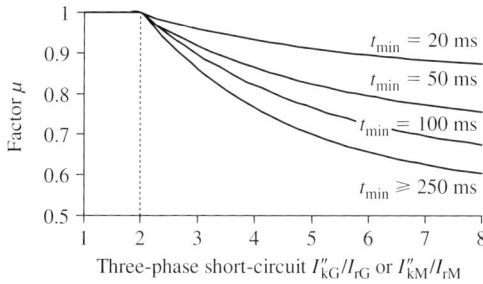

Figure 7.3   IEC 60909 factor μ used for calculation of short-circuit breaking current from synchronous generators

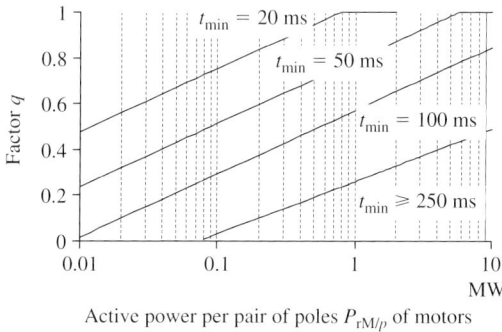

Figure 7.4   IEC 60909 factor q used with factor μ for calculation of short-circuit breaking current from asynchronous motors

and 7.4, respectively. Equation (7.16) for μ applies for synchronous generators employing rotating or static exciters provided, for the latter, that $t_{min} < 250\,\mathrm{ms}$ and the maximum excitation voltage is $<1.6\,\mathrm{pu}$ of the rated value. μ is set equal to unity if the excitation system is unknown.

For a far from generator three-phase short circuit, i.e. $I''_{kG}/I''_{rG} \leq 2$, or for unbalanced short circuits, $\mu = 1$ and $I_{bG} = I''_{kG}$. Similarly, for far from motor three-phase short circuit or for unbalanced short circuits $\mu q = 1$ and $I_{bM} = I''_{kM}$.

For meshed networks, the approximation $I_b = I''_k$ can be used. Alternatively, for improved accuracy, the decay in short-circuit currents from synchronous machines and asynchronous motors can be estimated from the partial step changes in voltage at the connection locations of the machines due to the fault and the following equation can be used:

$$
I_b = I''_k - \frac{1}{cU_n/\sqrt{3}} \left[\sum_i \Delta U''_{Gi}(1 - \mu_i)I''_{kGi} + \sum_j \Delta U''_{Mj}(1 - \mu_j q_j)I''_{kMj} \right]
$$

$$(7.18a)$$

where

$$\Delta U''_{Gi} = jX''_{diK} \times I''_{kGi} \quad \text{and} \quad \Delta U''_{Mj} = jX''_{Mj} \times I''_{kMj} \qquad (7.18b)$$

are the voltage drops at the terminals of the synchronous machine (i) and the asynchronous motor (j). X''_{diK} is the corrected subtransient reactance of synchronous machine (i), i.e. $X''_{diK} = K_\upsilon X''_{di}$ with $K_\upsilon = K_G, K_S$ or K_{SO} as appropriate and X''_{Mj} is asynchronous motor j subtransient or transient reactance. I''_{kGi} and I''_{kMj} are the initial symmetrical current contributions from synchronous machine (i) and asynchronous motor (j) as measured at their terminals. For a far from motor short circuit, $\mu_j = 1$ and $1 - \mu_j q_j = 0$ irrespective of the value of q_j.

Calculation of rms symmetrical steady state current

IEC 60909-0 recognises that the calculation of the steady state current I_k is less accurate than the calculation of I''_k because it depends on factors such as the generator initial operating point, machine saturation, excitation system characteristics and automatic voltage regulator action, etc. Worst-case assumptions are made to calculate maximum and minimum steady state short-circuit currents using factors expressed in terms of the generator rated current. For a single-fed near to generator three-phase short circuit, the maximum and minimum steady state short-circuit current is given by

$$I_{k\,max} = \lambda_{max} I_{rG} \qquad (7.19a)$$

$$I_{k\,min} = \lambda_{min} I_{rG} \qquad (7.19b)$$

λ_{max} is a factor that is dependent on the saturated direct axis synchronous reactance and for turbo-generators and salient-pole generators is obtained from IEC 60909-0 Figures 18(a and b) and 19(a and b), respectively. For turbo-generators, these figures apply for maximum excitation voltages of 1.3 and 1.6 pu whereas for salient-pole generators, they apply for maximum excitation voltages of 1.6 and 2 pu. The voltage factor $c = c_{max}$ is used. λ_{min} is calculated under constant no-load excitation, i.e. without voltage regulator action and the c factor used is $c = c_{min}$.

For meshed networks, $I_{k\,max} = I''_{k\,max}$ and $I_{k\,min} = I''_{k\,min}$ and similarly for unbalanced short-circuit faults.

7.2.6 Example

Example 7.1 In Figure 6.10, the dc current component was calculated using an accurate time domain solution, IEC 60909 Method B and IEEE C37.010 Method. Repeat the calculation using IEC 60909 Method C and compare the result against the accurate time domain solution assuming a 50 Hz power system.

The equivalent frequency used to calculate the peak make factor is 20 Hz and together with the equivalent frequencies shown in Table 7.2, the equivalent X/R ratios and corresponding time constants of the three parallel branches are

Table 7.3   Calculated equivalent X/R ratio and corresponding equivalent dc time constant

Time (ms)	f_c (Hz)	Equivalent X/R	Equivalent T_{dc} (ms)
$0 \leq t < 10$	20	15.9	50.7
$10 \leq t < 20$	13.5	16.6	52.9
$20 \leq t < 50$	7.5	18.3	58.1
$50 \leq t < 100$	4.6	20	63.7
$100 \leq t < 250$	2.75	22.5	71.8

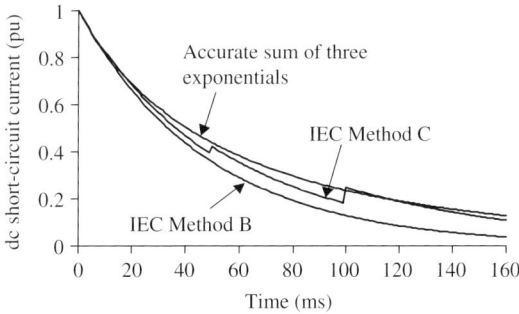

Figure 7.5   IEC 60909 dc current component of three parallel independent sources calculated using IEC Methods B and C and accurate time domain solution

calculated using Equation (7.13). The results are summarised in Table 7.3 and the dc current comparison is shown in Figure 7.5.

The equivalent frequency Method C is clearly far superior to Method B because it calculates new and higher X/R ratios and hence dc time constants as time elapses from the onset of the short-circuit fault. This process counteracts the increasing underestimate with time of Method B. In practice, it is found that Method C ensures that the degree of error is kept to a minimum.

7.3 UK Engineering Recommendation ER G7/4

7.3.1 Background

Following the publication of IEC 60909 in 1988, the UK electricity supply industry convened a group of power system engineers from across the industry to consider the application of IEC 60909 or a more appropriate alternative. As a result, Engineering Recommendation ER G7/4 '*Procedure to Meet the Requirements of IEC 60909 for the Calculation of Short-Circuit Currents in Three-Phase AC Power Systems*' was published in 1992. At that time, although most companies in the UK used computer methods for the calculation of short-circuit currents, a variety of approaches were used for both the modelling of power system elements and the techniques used in the calculation of short-circuit currents. As a result, different short-circuit current results would be obtained for the same power system characteristics.

UK ER G7/4 is a procedure that sets out 'good industry practice' and applies for the calculation of short-circuit currents used in the assessment of circuit-breaker make and break duties in ac power systems having a nominal voltage range of 380 V to 400 kV. It is also applicable for currents used in the calculation of protection relay settings and mechanical forces on conductors and busbars.

7.3.2 Representation of machines and passive load

Synchronous machines

Synchronous machines are represented by a voltage source behind an impedance model where the PPS reactance is time dependent and is given by

$$\frac{1}{X_G^P(t)} = \frac{1}{X_d} + \left(\frac{1}{X_d'} - \frac{1}{X_d}\right)e^{-t/T_d'} + \left(\frac{1}{X_d''} - \frac{1}{X_d'}\right)e^{-t/T_d''} \tag{7.20}$$

The time constants are affected by the external network between the machine and the point of fault, as discussed in Chapter 6. Where the machines are connected to the faulted network by step-up transformers having comparable reactances to the machine subtransient reactances, unsaturated subtransient and transient reactances are used. However, saturated reactances are used where the machines are directly connected to the faulted system. For extended fault clearance times, the effect of excitation control systems, and in particular those that have high speed response, may need to be considered.

Asynchronous machines

Asynchronous machines or groups of such machines are represented by a voltage source behind an impedance model where, for a single-cage or single-winding rotor, the time-dependent PPS reactance is given by

$$\frac{1}{X_m^P(t)} = \frac{1}{X'}e^{-t/T'} \tag{7.21}$$

where X' and T' are calculated from the machine parameters or tests as discussed in Chapter 5. Like synchronous machines, the time constant is modified by the external network impedance. Where neither the internal parameters of the machines nor its starting characteristics are known, the IEC 60909-0 approach of using the ratio of locked rotor current to rated current shown in Equation (7.9a) can be used to calculate the initial PPS impedance. If ac current decrement cannot be directly represented, the asynchronous machine's current adjustment factor (μq) of IEC 60909 of Equation (7.15b) is used to calculate the PPS reactance corresponding to any elapsed time t after the instant of fault as follows:

$$X_m^P(t) = \frac{X_m'}{\mu q} \tag{7.22}$$

X/R ratios for induction motors and static converter drives are considered as per IEC 60909-0 and is given in Section 7.2.4. The NPS impedance is assumed equal to the initial motor PPS impedance.

Small induction motors forming part of the general power system load

Numerous small induction motors that form part of the general load in the power system and hence are not individually identifiable are represented as an equivalent motor. The initial PPS impedance and time constant of the equivalent motor should be obtained by measurement. In the absence of measured data, indicative empirical estimates are suggested. For calculating the initial three-phase rms short-circuit current contribution at a 33 kV bulk supply substation busbar from induction motors in the general load supplied from that busbar the induction motor contribution is represented as follows:

(a) 1 MVA of short-circuit infeed per MVA of aggregate substation winter demand where the load is connected at low voltage, i.e. ≤ 1 kV.
(b) 2.6 MVA of short-circuit infeed per MVA of aggregate substation winter demand where the load is connected at medium voltage, i.e. >1 kV.

These short-circuit infeeds relate to a complete loss of supply voltage to the motors.
  The time variation of the PPS reactance of the equivalent motor is given by Equation (7.21) and the ac time constant is taken as $T' = 40$ ms. It is therefore assumed that the contribution to a three-phase short circuit from motors in the general load decays to a negligible value at around 120 ms. The effective X/R ratio of the equivalent motor at the 33 kV busbar is taken equal to 2.76.

Passive loads

Both the real and reactive power components of passive loads are represented in the analysis. In particular, ER G7/4 recognises that the reactive load may have an appreciable effect on short-circuit currents because of its effect on transformer tap positions and rotating plant internal voltages.

7.3.3 Analysis technique

IEC 60909 describes methods for the hand calculation of short-circuit currents that lead to results that are generally of acceptable accuracy for their intended purpose. However, other methods are not excluded provided they give results of at least the same precision. ER G/74 is essentially a computer-based short-circuit current calculation procedure that uses the superposition method and can be used as an alternative to IEC 60909 where higher precision is required. ER G7/4 is based on the following key principles:

(a) Credible prefault power system operating conditions giving rise to maximum short-circuit currents are first established.

(b) Prefault ac loadflow studies are carried out to determine the network voltage profile, internal voltages of generators and motors and transformer tap positions.

(c) The short-circuit contributions of all rotating plant are included. In addition to individually identifiable rotating plant, the aggregate effect of numerous small motors forming part of the general load in the power supply system are included using a suitable equivalent.

The loadflow results and the current changes due to the short circuit are combined using the superposition theorem as described in Chapter 6. A search is carried out for the worst-case prefault operating condition that leads to the maximum short-circuit current.

The above procedure is intended for use where calculations of improved accuracy are required. However, a simplified procedure that does not use a loadflow study may also be used where maximum currents are well within the rating of plant and equipment. Such a procedure will still include the contributions of all rotating plant including induction machines.

ER G7/4 does not specify the analysis methods to be used in calculating the ac and dc components of short-circuit currents or their decrements. The user may choose to use suitable time-step simulations or fixed impedance techniques similar to those described in Chapter 6.

7.3.4 Calculated short-circuit currents

The short-circuit currents calculated by ER G7/4 are the short-circuit making and breaking currents. The magnitude of each of these currents consists of an rms ac current component and a dc current component. Although the initial value of the dc component depends on the point on the voltage waveform at which the fault occurs, maximum dc current offset is assumed.

Short-circuit making current

This is the peak asymmetric current or the maximum instantaneous value of the prospective short-circuit current. If the analysis method used does not allow the initial peak current to be directly calculated, IEC 60909 Equation (7.10) is used.

Short-circuit breaking current

The short-circuit breaking current allows for both ac and dc current decay. Both the ac and dc currents are calculated at a specified break time t_b. The magnitude of the dc current component at $t = t_b$ is given by

$$i_{dc} = \sqrt{2}I_k''\exp\left[\frac{-2\pi f t_b}{(X/R)}\right] \tag{7.23}$$

To determine the ability of a circuit-breaker to interrupt these component currents or a combination of them, reference should be made to the test data of the circuit-breaker and the standards on which the test data is based, e.g. BS 116:1952, ESI 41-10:1987 or IEC 62271-100 Standard.

X/*R* ratio for estimation of the magnitude of the dc current component

The system X/R ratio at the fault point is calculated and used to estimate the magnitude of the required dc current component where explicit calculation of the asymmetric initial peak and break currents cannot be made. If some network elements have X/R ratios below 3.3, as may be the case in low voltage networks, then a 1.15 safety factor is used in calculating the peak asymmetric currents unless the equivalent frequency Method C of IEC 60909 is used to calculate the X/R ratio. Since the time constants of the dc current decrements for three-phase and earth faults, e.g. single-phase to earth fault, are different, different X/R ratios will need to be evaluated. Where a conservative estimate is adequate, the initial peak current can be assumed to be equal to $1.8\sqrt{2}I_k''$ for low voltage networks (i.e. 1 kV or below) and $2\sqrt{2}I_k''$ for networks operating above 1 kV thereby obviating the need to calculate the X/R ratio.

7.3.5 Implementation of ER G7/4 in the UK

In this section, we briefly describe the implementation of ER G7/4 in computer analysis methods in England and Wales. The analysis is based on the fixed impedance technique where synchronous and induction machines are represented by a voltage source behind a fixed impedance. Essentially, two rms symmetrical short-circuit currents are calculated corresponding to two time instants. The first is the initial symmetrical rms short-circuit current, I_k'', corresponding to the instant of fault, and the second is the symmetrical rms short circuit after the subtransient contribution has decayed to a negligible value, I_k'. In order to calculate these two PPS currents, two PPS networks are created: a subtransient network and a transient network using machine subtransient and transient reactances, respectively. Corresponding to these two time instants, two system X/R ratios are calculated using IEC 60909 equivalent frequency Method C and these are denoted $(X/R)_i$ and $(X/R)_b$ where i and b represent initial and break, respectively. The peak make current is calculated as follows:

$$i_p = \sqrt{2}I_k''[1.02 + 0.98 \times e^{-3/(X/R)_i}] \tag{7.24}$$

The rms subtransient current component at the fault location is assumed to have a single effective subtransient time constant of 40 ms, i.e. it is assumed that all subtransient components from across the system have decayed to a negligible value by 120 ms. The break time t_b is defined as the time from the instant of short-circuit fault to the time instant that corresponds to the peak of the major current loop just before current interruption at current zero. Thus, for a given break time t_b in ms, the rms ac break current component is given by

$$I_b = \begin{cases} I'_k + (I''_k - I'_k)e^{-t_b/40} & \text{for } t_b < 120 \text{ ms} \\ I'_k & \text{for } t_b \geq 120 \text{ ms} \end{cases} \tag{7.25}$$

An inherent assumption in Equation (7.25) is that the effect of even modern fast acting excitation control systems is negligible for $t_b < 120$ ms. However, after this time, the reduction in fault current due to increasing machine reactance is offset by the action of excitation control systems through field voltage forcing and the short-circuit current is assumed to remain constant at I'_k.

The magnitude of the dc current component at the break time t_b is calculated, assuming maximum offset, as follows:

$$i_{dc} = \sqrt{2}I''_k \exp\left[\frac{-2\pi f t_b}{(X/R)_b}\right] \tag{7.26}$$

Practical break times considered are those that correspond to the peaks of major asymmetrical current loops, i.e. approximately 30, 50, 70, 90 and 110 ms, etc. in the UK 50 Hz power system. If the amplitude of the peak asymmetric break current at $t = t_b$ is required, then this is calculated using Equations (7.25) and (7.26) as

$$i_b = \sqrt{2}I_b + i_{dc} \tag{7.27}$$

Alternatively, if the rms asymmetric breaking current of the total asymmetric current is required, as may be the case for the old British Standard BS 116 and IEEE C37.010 Standard, then this is calculated using Equations (7.25) and (7.26) as

$$I_{ab} = \sqrt{I_b^2 + i_{dc}^2} \tag{7.28}$$

For balanced faults, subtransient and transient PPS system admittance matrices are formed. In addition, NPS and ZPS system admittance matrices are formed in the case of unbalanced faults. The formation of the NPS matrix allows machines to have a different NPS reactance from machine PPS reactance as well as the correct modelling of plant such as quadrature boosters and phase shifters. Passive loads shunt admittances are included in the PPS and NPS short-circuit current calculations but not in the ZPS networks where loads are separated from the bulk distribution system by one or more delta connected transformer windings.

The initial and break system X/R ratios are calculated using the equivalent frequency Method C of IEC 60909. In the UK 50 Hz power system, 20 Hz is used in the calculation of $(X/R)_i$ and an appropriate equivalent frequency is selected from Table 7.2, depending on the break time t_b, in the calculation of $(X/R)_b$.

In addition to the PPS system X/R ratio, NPS and ZPS system X/R ratios are also calculated depending on the unbalanced fault being studied. In the calculation of the PPS X/R ratio, all passive shunts, e.g. line shunt capacitances and load impedances, etc., are excluded. However, the calculated NPS and ZPS X/R ratios generally include the network shunt elements.

7.4 American IEEE C37.010 Standard

7.4.1 Background

The IEEE method for the calculation of system short-circuit currents is described in IEEE Standard C37.010:1999 which is a revision of C37.010:1979. The Standard is entitled *IEEE Application Guide for AC High-Voltage Circuit Breakers Rated on a Symmetrical Current Basis*. Prior to 1964, the applicable American National Standard Institute (ANSI) Standards were based on 'Total Current Basis' as opposed to 'Symmetrical Current Basis'. The IEEE C37.010:1999 Standard is applicable to high voltage circuit-breakers above 1 kV for use in utility, industrial and commercial power systems operating at a nominal frequency of 60 Hz.

The IEEE C37.010 short-circuit calculation methodology was developed to provide conservative results for the determination of short-circuit duties and the sizing and selection of circuit-breaker ratings. The method is intended to provide a degree of accuracy that is within the practical limits of short-circuit calculation considering the accuracy of data that are usually available for such computations. The asymmetrical short-circuit current waveform is assumed to have maximum dc current offset. Other alternative methods, though not identified, are not excluded if they give results of at least the same precision. The Standard makes no distinction between radial or single-fed short circuits and those fed from meshed power systems. The Standard deals with the calculation of maximum short-circuit currents only and distinguishes between short circuits with significant and insignificant ac current component decay corresponding to short circuits that are local to or remote from generators, respectively.

Unlike IEC 60909 or ER G/74 which are short-circuit current calculation procedures only, the IEEE Standard is aimed at the sizing and selection of circuit-breakers. As a result, the IEEE short-circuit calculation methodology is included within a circuit-breaker application guide.

7.4.2 Representation of system and equipment

The IEEE Standard uses the concept of 'local' and 'remote' generation contribution to the short-circuit fault. If the external reactance between the generator terminals and the short-circuit location is such that $X_{\text{External}} > 1.5X_d''$, the generator is considered to be remote. Otherwise, it is considered to be local. The IEEE Standard recommends neglecting the shunt impedances of static loads as well as the PPS line capacitance. The NPS and PPS reactances of rotating plant are assumed equal and similarly for the NPS and PPS resistances.

Two separate and different impedance networks are created; the first is the closing & latching (momentary) impedance network and the second is the interrupting impedance network. From the first network, the first cycle symmetrical current and the system X/R ratio are calculated and these are used to calculate the closing and latching current duty. From the interrupting impedance network, the

Table 7.4 Machine reactance adjustment factors to account for ac short-circuit current decay according to IEEE C37.010 Standard

	PPS reactances (pu)	
Type of rotating machine	Closing and latching network	Interrupting network
All turbo-generators, all hydro-generators with damper windings and all condensers	$1.0X_d''$	$1.0X_d''$
Hydro-generators without damper windings	$0.75X_d'$	$0.75X_d'$
All synchronous motors	$1.0X_d''$	$1.5X_d''$
Induction motors		
Above 1000 hp at 1800 rpm or less	$1.0X_d''$	$1.5X_d''$
Above 250 hp at 3600 rpm	$1.0X_d''$	$1.5X_d''$
From 50 to 1000 hp at 1800 rpm or less	$1.2X_d''$	$3.0X_d''$
From 50 to 250 hp at 3600 rpm	$1.2X_d''$	$3.0X_d''$

Three-phase induction motors below 50 hp and all single-phase motors are neglected.
X_d'' and X_d' of synchronous machines are the rated voltage saturated values.

symmetrical interrupting current and the system X/R ratio are calculated and these are used to calculate the asymmetrical interrupting current duty. The Standard recommends machine reactance adjustment factors, i.e. multipliers to account for ac current decay and these are generally different for the closing and latching duty, and interrupting duty calculations as shown in Table 7.4.

7.4.3 Analysis technique

General

IEEE C37.010 is a fixed impedance short-circuit calculation method. The initial generator and motor prefault loading conditions are not modelled nor are the explicit decay rates of individual generators and motors. The Standard recommends the use of a voltage source at the short-circuit location whose magnitude is equal to the highest typical prefault operating voltage E at the fault location.

The system X/R ratio at the fault point is calculated from separate X and R networks. The standard assumes that the reduction of separate X and R networks tends to correct for the multiple time constants and associated decay rates due to short-circuit currents passing through multiple paths to the fault location. In practical cases, the resultant error is on the conservative side.

Both a simplified E/X method and a more accurate, semi-rigorous, E/X method that accounts for ac and dc current decrements are described. Where the use of impedances is preferred to the use of reactances in the calculation of the magnitude of short-circuit current, the E/X calculation may be replaced with E/Z calculation.

E/X simplified method

The E/X simplified method calculates the symmetrical short-circuit current and is very simple to use. The 'E' represents the prefault voltage at the short-circuit location. The 'X' is the equivalent system reactance at the fault point calculated with the resistances of network elements set to zero and machine reactances set to their subtransient values. The equivalent reactance at the fault point is equal to X^P for three-phase short circuits and $2X^P + X^Z$ for single-phase to earth short circuits. The calculated short-circuit currents are thus E/X^P for three-phase short circuits and $3E/(2X^P + X^Z)$ for single-phase short circuits. The Standard considers this simplified procedure to be sufficient provided that:

(a) The calculated three-phase short currents do not exceed 80% of the circuit-breaker symmetrical interrupting capability.
(b) The calculated single-phase to earth short-circuit currents do not exceed 70% of the circuit-breaker symmetrical interrupting capability for single-phase to earth short circuits.

Otherwise, the more accurate E/X procedure, described below, should be used.

E/X method with correction for ac and dc decrements

This more accurate procedure involves applying multiplying factors to the E/X calculated short-circuit results to account for ac and dc current decrements. The factors to be used depend on the calculated system X/R ratio seen at the fault point and on whether the fault is 'local' to or 'remote' from generation. The applicable factors are taken from three figures in the IEEE Standard namely 'Figures 8, 9 and 10'. The curves of 'Figures 8 and 9' give the K_{acdc} factor that includes the effects of both ac and dc decrements and applies for 'local' three-phase and single-phase to earth short circuits. 'Figure 8' gives a maximum value of K_{acdc} of 1.25 which supports the 80% criterion used in the application of the E/X simplified method for three-phase faults. 'Figure 9' gives a maximum multiplying factor of K_{acdc} of 1.41 which supports the 70% criterion used in the application of the E/X simplified method for one-phase faults. 'Figure 10' gives the K_{dc} factor that includes the effect of dc decrement only and applies for 'remote' three-phase and single-phase short-circuit faults. These three figures also include multipliers for longer circuit-breaker contact parting times than the minimum. The corrected E/X short-circuit current result should not exceed the symmetrical interrupting capability of the circuit-breaker. A more accurate multiplying factor can be determined by using the ratio of remote generation contributions to total fault current known as the no ac decrement or NACD ratio. The new multiplying factor $K_{\mathrm{Interpolated}}$ may be obtained by interpolation between K_{acdc} and K_{dc} using the NACD ratio as follows:

$$K_{\mathrm{Interpolated}} = K_{\mathrm{acdc}} + NACD \times (K_{\mathrm{dc}} - K_{\mathrm{acdc}}) \qquad (7.29)$$

7.4.4 Calculated short-circuit currents

First cycle symmetrical current

This is the half cycle rms current and is calculated using the closing and latching impedance network and the appropriate machine reactances shown in Table 7.4.

Closing and latching current

The closing and latching current is equivalent to the half cycle peak current. If the first cycle symmetrical current including motor infeed is below the symmetrical current rating of the circuit-breaker and the system X/R ratio is less than 17 at 60 Hz, then no further calculations are required. If the system X/R ratio is greater than 17, the closing and latching current, expressed as a peak current, may be calculated using a peak factor similar to the IEC 60909-0 peak factor κ of Equation (7.10) as follows:

$$\text{Peak factor} = \sqrt{2}\left[1 + \sin\{\tan^{-1}(X/R)\}\exp\left\{\frac{-[\pi/2 + \tan^{-1}(X/R)]}{(X/R)}\right\}\right] \quad (7.30)$$

where the X/R ratio is calculated from the closing and latching impedance network.

If the value of the closing and latching current duty is calculated as a total rms value of the asymmetrical peak current, a simplified rule using a factor equal to 1.6 times first cycle symmetrical current, which assumes an X/R ratio of 25 at 1/2 cycle, may be used. Alternatively, the total rms value may be calculated using a more accurate multiplying factor of the rms symmetrical current based on the calculated X/R at the fault location as follows:

$$I_{\text{rms total at 1/2 cycle (kA)}} = \sqrt{1 + 2\exp\left[-\frac{2\pi}{(X/R)}\right]} \times I_{\text{Sym at 1/2 cycle (kA)}} \quad (7.31)$$

Symmetrical interrupting current

This current is calculated using the interrupting impedance network and the appropriate machine reactances shown in Table 7.4.

Asymmetrical interrupting current

The asymmetrical interrupting current is calculated using the symmetrical interrupting current and an applicable multiplying factor which is a function of the circuit-breaker contact parting time and the system X/R ratio at the fault location. The factor is obtained from C37.010 'Figures 8, 9 and 10' depending on whether the fault is fed by local or remote sources and the short-circuit type. The asymmetrical interrupting current duty is calculated as an rms value of an asymmetrical current.

Time-delayed 30 cycle (steady state) current

The time-delayed 30 cycle (steady state) short-circuit current applies when transient effects have completely decayed. In calculating this current, C37.010 recommends the use of a network that comprises only generators represented by either their transient reactance or a higher reactance that takes into account the ac component decay. The dc component is assumed to have decayed to zero.

7.5 Example calculations using IEC 60909, UK ER G7/4 and IEEE C37.010

Example 7.2   Consider the simple example network shown in Figure 7.6 to illustrate the maximum current quantities supplied by a power station unit for a three-phase short-circuit fault on the high voltage side of the transformer.

At node Q, the power station unit supplies a static impedance load, the load voltage magnitude is 380 kV and the generator is operating at rated active and reactive power output. The calculations are carried out in accordance with IEC 60909, UK ER G7/4 and IEEE C37.010. The turbo-generator and transformer data on their MVA rating are:

$$Generator :\ \ S_{rG} = 588\,\text{MVA} \qquad U_{rG} = 23\,\text{kV} \qquad R_G = 0.18\%$$

$$x_d'' = 18\% \qquad\qquad x_d' = 24\% \qquad \cos \varphi_{rG} = 0.85$$

$$Transformer :\ \ S_{rT} = 588\,\text{MVA} \quad U_{rTHV} = 432\,\text{kV} \quad U_{rTLV} = 23\,\text{kV}$$

$$u_{Xr} = 15\% \qquad\qquad u_{Rr} = 0.2\% \qquad p_T = -16\% \text{ to } +2\%$$

$$System(load) :\ U_{nQ} = 400\,\text{kV}$$

IEC 60909 calculation

The transformer impedance in Ω referred to the high voltage side is calculated as

$$Z_{THV} = \left(\frac{0.2}{100} + j\frac{15}{100} \right) \times \frac{432^2}{588} = (0.63477 + j47.608)\Omega$$

The generator impedance in Ω is calculated as

$$Z_G = R_G + jX_d'' = \left(\frac{0.18}{100} + j\frac{18}{100} \right) \times \frac{23^2}{588} = (0.0016194 + j0.16194)\Omega$$

Using Equation (7.4b), the correction factor of the power station unit is given by

$$K_S = \frac{400^2}{23^2} \times \frac{23^2}{432^2} \times \frac{1.1}{1 + |0.18 - 0.15| \times 0.526783} = 0.9284$$

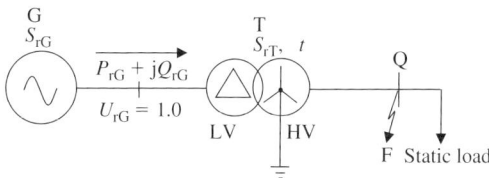

Figure 7.6   Example of a simple circuit for calculation of maximum short-circuit currents according to various standards

Using Equation (7.4a), the corrected impedance of the power station unit referred to the high voltage side is given by

$$Z_S = 0.9284 \left[\left(\frac{432}{23} \right)^2 (0.0016194 + j0.16194) + (0.63477 + j47.608) \right]$$

$$= (1.1197 + j97.239)\Omega = 97.245e^{j89.34} \, \Omega$$

The three-phase short-circuit current supplied by the power station unit is given by

$$I''_{kS} = \frac{1.1 \times 400/\sqrt{3}}{97.245e^{j89.34}} = 2.61232e^{-j89.34} \, kA$$

Since $U_{rG} > 1 \, kV$ and rated apparent power $S_{rG} > 100 \, MVA$, $R_{Gf} = X''_d/20$ and $R_{Gf} = 0.0081 \, \Omega$. Thus, the corrected impedance of the power station unit referred to the high voltage side is given by

$$Z_{Sf} = 0.9284 \left[\left(\frac{432}{23} \right)^2 (0.0081 + j0.16194) + (0.63477 + j47.608) \right]$$

$$= (3.2423 + j97.239)\Omega$$

Therefore, $X_{Sf}/R_{Sf} = 30$. From Equation (7.10b), the peak current factor is given by $\kappa = 1.02 + 0.98e^{-3/30} = 1.9067$. Using Equation (7.10a), the peak current is given by $i_p = 1.9067 \times \sqrt{2} \times 2.61232 = 7.044 \, kA$.

To calculate the rms breaking current, we need to calculate the factor μ which depends on the minimum time t_{min} and the ratio I''_{kG}/I_{rG}. The rated current is given by $I_{rG} = 588/(\sqrt{3} \times 23) = 14.76 \, kA$. The short-circuit current delivered by the generator at its terminals to the high voltage side fault is given by $I''_{kG} = 2.61232 \times 432/23 = 49.066 \, kA$. The required ratio is

$$\frac{I''_{kG}}{I_{rG}} = \frac{49.066}{14.76} = 3.324$$

Therefore, using Equation (7.16) and $t_{min} = 50 \, ms$, we have

$$\mu = 0.71 + 0.51 \times e^{-0.3 \times 3.324} = 0.898.$$

The rms short-circuit breaking current is given by

$$I_b = 0.898 \times 2.61232 = 2.346 \, kA$$

The magnitude of the dc breaking current at $t_{min} = 50 \, ms$ is calculated using Equation (7.14) and $(X/R)_b = 97.239/1.1197 = 86.84$, thus

$$i_{dc} = \sqrt{2} \times 2.61232 \times \exp\left[\frac{-2\pi \times 50 \times 0.05}{86.84} \right] = 3.083 \, kA$$

UK ER G7/4 calculation

The most realistic operating condition of the power station unit before the short circuit that maximises the short-circuit current is when the generator is producing rated power output and the magnitude of the high voltage is minimum. The rated generator power output in pu on rating is given by $S_{rG} = (0.85 + j0.526)$pu. With a transformer on-load tap-changer, $U_{rG} = 1.0$ pu. The generator load current is calculated from $S_{rG} = U_{rG}I_{rG}^*$ and is given by $I_{rG} = (0.85 - j0.526)$pu. Considering the voltage drop across the transformer rated impedance, we have

$$V_T = 1 - (0.002 + j0.15)(0.85 - j0.526) = 0.928 \times e^{-j7.83}$$

The transformer tap ratio required to achieve a high voltage voltage of 0.95 pu is equal to $t = 0.95/0.928 = 1.0237$.

For a real network, the above prefault quantities are calculated using a computer-based ac loadflow study.

The subtransient and transient internal voltages of the generator are given by

$$E_G'' = 1.0 + (0.0018 + j0.18)(0.85 - j0.526) = 1.107 \times e^{j7.89} \text{ pu}$$
$$E_G' = 1.0 + (0.0018 + j0.24)(0.85 - j0.526) = 1.146 \times e^{j10.2} \text{ pu}$$

The short-circuit subtransient current is calculated as

$$I_k'' = \frac{1.107 \times e^{j7.89}}{1.0237 \times [(0.0018 + j0.18) + (0.002 + j0.15)]} = 3.2767 \times e^{-j81.44} \text{ pu}$$

or

$$|I_k''| = 3.2767 \times \frac{588}{\sqrt{3} \times 400} = 2.78 \text{ kA}$$

The short-circuit transient current is calculated as

$$I_k' = \frac{1.146 \times e^{j10.2}}{1.0237 \times [(0.0018 + j0.24) + (0.002 + j0.15)]} = 2.870 \times e^{-j79.24} \text{ pu}$$

or

$$|I_k'| = 2.87 \times \frac{588}{\sqrt{3} \times 400} = 2.43 \text{ kA}$$

The system initial and break X/R ratios at the fault point are calculated as follows:

$$(X/R)_i = \frac{0.18 + 0.15}{0.0018 + 0.002} = 86.84 \quad (X/R)_b = \frac{0.24 + 0.15}{0.0018 + 0.002} = 102.6$$

The peak make short-circuit current is calculated using Equation (7.24)

$$i_p = \sqrt{2} \times 2.78 \times [1.02 + 0.98 \times e^{-3/86.84}] = 7.73 \text{ kA}$$

The short-circuit rms break current at 50 ms break time is calculated using Equation (7.25) or

$$I_b = 2.43 + (2.78 - 2.43)e^{-50/40} = 2.53 \text{ kA}$$

The magnitude of the dc break current at 50 ms break time is calculated using Equation (7.26) or

$$i_{dc} = \sqrt{2} \times 2.78 \times \exp\left[\frac{-2\pi \times 50 \times 0.05}{102.6}\right] = 3.37\,\text{kA}$$

The peak asymmetric break current at 50 ms break time is calculated using Equation (7.27) or

$$i_b = \sqrt{2} \times 2.53 + 3.37 = 6.95\,\text{kA}$$

The rms asymmetric break current at 50 ms break time is calculated using Equation (7.28) or

$$I_{ab} = \sqrt{2.53^2 + 3.37^2} = 4.21\,\text{kA}$$

IEEE C37.010 calculation

Simplified E/X method
The highest typical operating voltage on the transformer high voltage side is 1.03 pu. Using the subtransient reactance of the generator, the total subtransient reactance at the fault point is equal to $0.18 + 0.15 = 0.33$ pu. The short-circuit current is given by

$$E/X = 1.03/0.33 = 3.12\,\text{pu} \quad \text{or} \quad 3.12 \times \frac{588}{\sqrt{3} \times 400} = 2.65\,\text{kA}$$

No further calculations are required if this current is smaller than 80% of the symmetrical interrupting capability of the circuit-breaker. Otherwise, the more accurate method is used.

E/X method with correction for ac and dc decrements
The first cycle rms current is calculated using the 'closing and latching' network and is given by

$$E/X = 1.03/0.33 = 3.12\,\text{pu} \quad \text{or} \quad 3.12 \times \frac{588}{\sqrt{3} \times 400} = 2.65\,\text{kA}$$

Using Equation (7.30), the closing and latching duty peak current is calculated as

$$\sqrt{2} \times 2.65 \times \left[1 + \sin\{\tan^{-1}(86.84)\}\exp\left\{\frac{-[\frac{\pi}{2} + \tan^{-1}(86.84)]}{86.84}\right\}\right]$$

$$= 2.778 \times 2.65 = 7.36\,\text{kA}$$

This peak current duty must not exceed the corresponding breaker capability of 2.6 times the rated symmetrical short-circuit current. Alternatively, if the total rms current is required, this can be calculated using Equation (7.31) as follows:

$$\sqrt{1 + 2\exp\left[-\frac{2\pi}{86.84}\right]} \times 2.65 = 1.69 \times 2.65 = 4.48\,\text{kA}$$

For a 5-cycle breaker and a 3-cycle contact parting time, the multiplying factor for $X/R = 86.84$ is obtained from 'Figure 8' in IEEE C37.010 and is equal to 1.225. Therefore, the product $1.225 \times 2.65 = 3.25\,\text{kA}$ must not exceed the symmetrical interrupting capability of the circuit-breaker.

Example 7.3   The above example for IEEE C37.010 analysis is trivial. Therefore, consider the network shown in Figure 7.7 that provides a more realistic example of the IEEE C37.010 calculation method. All quantities shown in Figure 7.7 are in % on 100 MVA base and F is the short-circuit fault location. The circuit-breaker being considered is denoted as breaker A and is a 132 kV, 16 kA, 5-cycle breaker with a 3-cycle contact parting time.

Case 1: three-phase short-circuit fault
The reader may draw the PPS impedance network model of the system to help in the calculation of the Thévenin's impedance at the fault point. The Thévenin's equivalent PPS reactance (all resistances set to zero) and PPS resistance (all reactances set to zero) at the fault point are calculated as

$$X^{\text{P}} = \frac{\frac{(0.18+0.1)}{2} \times (0.025 + 0.02)}{\frac{(0.18+0.1)}{2} + (0.025 + 0.02)} = 0.0340\,\text{pu}$$

and

$$R^{\text{P}} = \frac{\frac{(0.0018+0.002)}{2} \times (0.01 + 0.0005)}{\frac{(0.0018+0.002)}{2} + (0.01 + 0.0005)} = 0.00160\,\text{pu}$$

The highest typical operating voltage at the circuit-breaker location is 1.03 pu. The first cycle symmetrical short-circuit current is equal to

$$\frac{1.03}{0.034} = 30.29\,\text{pu} \quad \text{or} \quad 30.29 \times \frac{100}{\sqrt{3} \times 132} = 13.25\,\text{kA}$$

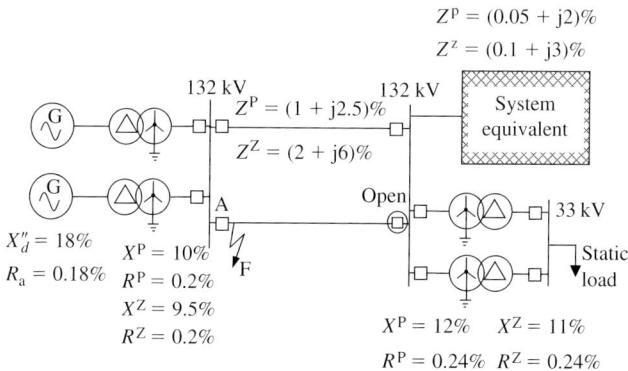

Figure 7.7 Example of a simple system for the calculation of maximum short-circuit currents according to IEEE C37.010 Standard

The X/R ratio is given by $0.034/0.0016 = 21.25$. Using Equation (7.30), the closing and latching duty peak current is calculated as follows:

$$\sqrt{2} \times 13.25 \times \left[1 + \sin\{\tan^{-1}(21.25)\}\exp\left\{ \frac{-[\frac{\pi}{2} + \tan^{-1}(21.25)]}{21.25} \right\} \right]$$

$$= 2.63 \times 13.25 = 34.9\,\text{kA}$$

This peak closing and latching duty is lower than the circuit-breaker peak current capability of $2.6 \times 16 = 41.6\,\text{kA}$. Alternatively, if the rms value of the total asymmetric current is required, this can be calculated using Equation (7.31) as follows:

$$\sqrt{1 + 2\exp\left[-\frac{2\pi}{21.25} \right]} \times 13.25 = 1.58 \times 13.25 = 20.9\,\text{kA}$$

For this 5-cycle circuit-breaker with a 3-cycle contact parting time, the interrupting duty to be compared with the symmetrical interrupting capability of $16\,\text{kA}$ and is equal to $1.0 \times 13.25 = 13.25\,\text{kA}$ where the multiplying factor of 1.0 is obtained from 'Figure 8' of IEEE C37.010 at $X/R = 21.25$.

Case 2: single-phase to earth short-circuit fault
Again, the reader may draw the ZPS impedance network model of the system. The Thévenin's equivalent ZPS reactance (all resistances set to zero) and ZPS resistance (all reactances set to zero) at the fault point are calculated as

$$X^Z = \frac{\frac{(0.095)}{2} \times \left(0.06 + \frac{\frac{0.11}{2} \times 0.03}{\frac{0.11}{2} + 0.03} \right)}{\frac{(0.095)}{2} + \left(0.06 + \frac{\frac{0.11}{2} \times 0.03}{\frac{0.11}{2} + 0.03} \right)} = 0.0297\,\text{pu}$$

and

$$R^Z = \frac{\frac{(0.002)}{2} \times \left(0.02 + \frac{\frac{0.0024}{2} \times 0.001}{\frac{0.0024}{2} + 0.001} \right)}{\frac{(0.002)}{2} + \left(0.02 + \frac{\frac{0.0024}{2} \times 0.001}{\frac{0.0024}{2} + 0.001} \right)} = 0.000954\,\text{pu}$$

The first cycle symmetrical short-circuit current is equal to

$$\frac{3 \times 1.03}{2 \times 0.034 + 0.0297} = 31.93\,\text{pu} \quad \text{or} \quad 31.93 \times \frac{100}{\sqrt{3} \times 132} = 14\,\text{kA}$$

The X/R ratio is equal to

$$\frac{2 \times 0.034 + 0.0297}{2 \times 0.0016 + 0.00095} = 23.6$$

Using Equation (7.30), the closing and latching duty peak current is calculated as

$$\sqrt{2} \times 14 \times \left[1 + \sin\{\tan^{-1}(23.6)\}\exp\left\{ \frac{-[\frac{\pi}{2} + \tan^{-1}(23.6)]}{23.6} \right\} \right]$$

$$= 2.65 \times 14 = 37.1\,\text{kA}$$

This peak closing and latching duty is lower than the circuit-breaker peak current capability of $2.6 \times 16 = 41.6\,\text{kA}$. If the rms value of the total asymmetric current is required, this can be calculated using Equation (7.31) as

$$\sqrt{1 + 2\exp\left[-\frac{2\pi}{23.6}\right]} \times 14 = 1.59 \times 14 = 22.3\,\text{kA}$$

For this 5-cycle circuit-breaker with a 3-cycle contact parting time, the interrupting duty to be compared with the symmetrical interrupting capability of $16\,\text{kA}$ is equal to $1.06 \times 14 = 14.84\,\text{kA}$ where the multiplying factor of 1.06 is obtained from 'Figure 9' of IEEE C37.010 at $X/R = 23.6$.

7.6 IEC 62271-100-2001 and IEEE C37.04-1999 circuit-breaker standards

7.6.1 Short-circuit ratings

Breaking or interrupting rating

IEC 62271-100 defines a rated short-circuit breaking current in kA and this is characterised by two values; the rms value of its ac component and the percentage dc component. Standard values of the ac component are selected from the $R10$ series 6.3, 8, 10, 12.5, 16, 20, 25, 31.5, 40, 50, 63, 80 and 100 kA.

The IEEE Standard defines a symmetrical interrupting capability equal to rated short-circuit current $\times K$ where $K = 1$ for most modern circuit-breakers. For some older circuit-breakers, $K =$ rated maximum voltage/operating voltage except that for single-phase to earth faults, the symmetrical interrupting capability is 15% higher. The IEEE rated short-circuit current in amperes is equivalent to the IEC rms ac component of the rated short-circuit breaking current.

dc current rating

In both IEC and IEEE Standards, the value of the percentage dc component can be determined from an envelope that consists of a single decaying exponential having a time constant T_{dc}. The percentage value of the dc component (%dc) in per cent of the peak ac current component or peak symmetrical interrupting capability is given by

$$\%I_{\text{dc}} = 100 \times e^{-(T_{\text{r}} + T_{\text{cb}})/T_{\text{dc}}} \tag{7.32a}$$

where

$$T_{\text{dc}} = \frac{(X/R)}{2\pi f} \tag{7.32b}$$

and $(T_{\text{r}} + T_{\text{cb}})$ is the circuit-breaker contact parting time measured from the fault instant. T_{cb} is the minimum opening time of the circuit-breaker and T_{r} is the protection relay time and is equal to half cycle of power frequency (10 ms at 50 Hz and 8.33ms at 60 Hz). T_r is equal to zero for self-tripping circuit-breakers.

If the rms component of the short-circuit breaking current or the symmetrical current component is denoted I_{Sym} in kA, the dc current component in kA is given by

$$I_{dc(kA)} = \frac{\%I_{dc}}{100} \sqrt{2} I_{Sym(kA)} \tag{7.33}$$

Both IEC and IEEE specify a standard dc time constant $T_{dc} = 45$ ms. In addition, IEC provides special case time constants related to the rated voltage of the circuit-breaker. These recognise the higher time constants that may be encountered at locations close to generating stations, transformer or series reactor fed faults and extra high voltage transmission lines, etc. The special case time constants are:

(a) $T_{dc} = 120$ ms for rated voltages up to and including 52 kV.
(b) $T_{dc} = 60$ ms for rated voltages from 72.5 kV up to and including 420 kV.
(c) $T_{dc} = 75$ ms for rated voltages 550 kV and above.

The dc current envelope rating is shown in Figure 7.8.

From Equation (7.32b), a dc time constant of 45 ms is equivalent to a system X/R ratio of 14.14 at 50 Hz and 17 at 60 Hz. For IEEE Standard, the dc current value is taken at the time of primary arcing contact parting.

Asymmetrical ratings

IEC specifies a rated short-circuit making current (peak current at half cycle) of 2.5 times the rms value of the ac component of the rated short-circuit breaking current for $T_{dc} = 45$ ms and a nominal frequency of 50 Hz. The factor is 2.6 for a nominal frequency of 60 Hz. For all special case time constants, the factor is 2.7 irrespective of the nominal frequency.

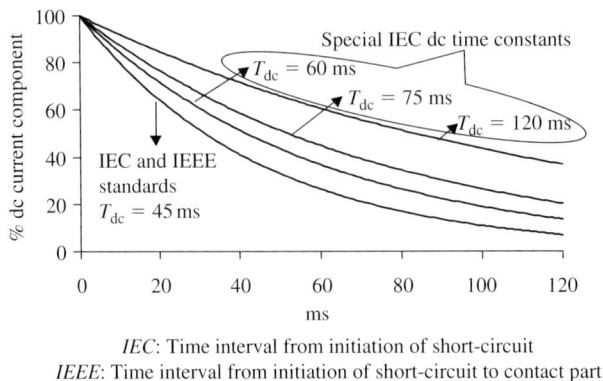

IEC: Time interval from initiation of short-circuit
IEEE: Time interval from initiation of short-circuit to contact part

Figure 7.8 Percentage dc current component for standard IEC and IEEE dc time constants, and special IEC cases

IEEE specifies a peak current capability at half cycle of 2.6 times the rated rms symmetrical short-circuit current. In addition, IEEE defines a total rms (asymmetric) fault current as follows:

$$I_{rms(total)} = \sqrt{I_{rms(symmetrical)}^2 + I_{dc}^2} \qquad (7.34a)$$

and using Equations (7.31a) and (7.32), we obtain

$$I_{rms(total)} = I_{rms(symmetrical)}\sqrt{1 + 2\exp\left[-\frac{4\pi c}{(X/R)}\right]} \qquad (7.34b)$$

where c is the circuit-breaker contact parting time in cycles.

7.6.2 Assessment of circuit-breakers short-circuit duties against ratings

For IEC circuit-breakers, the peak make current duty calculated using Equations (7.10a) and (7.11) should not exceed the rated short-circuit making current of 2.5, 2.6 or 2.7 times the rms value of the ac component of the rated short-circuit breaking current. For IEEE circuit-breakers, the peak make current should not exceed 2.6 times the symmetrical interrupting current.

The assessment of circuit-breaker interrupting duty against rating is not straightforward where the calculated duty dc time constant is significantly greater than the circuit-breaker rated dc time constant. Figure 7.9 illustrates two current waveforms. One for a circuit-breaker having a rated symmetrical rms short-circuit current of 50 kA and a rated dc time constant of 45 ms. The other waveform represents a duty of 41 kA symmetrical rms current and 100 ms duty dc time constant. At the first, second and third current peaks, the amplitude of the duty is lower than the

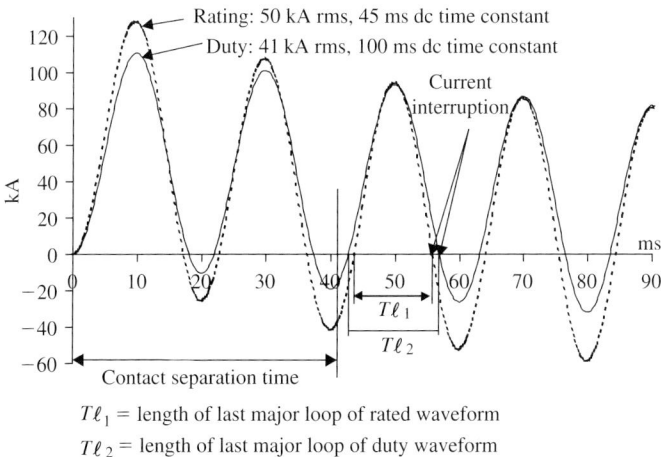

Figure 7.9   Illustration of the effect of large duty dc time constant on circuit-breaker short-circuit rating

corresponding rating. For the fourth and subsequent peaks, the reverse is true. This reversal is caused by the higher duty dc time constant. From the circuit-breaker's viewpoint, the larger duty dc time constant shown in this example results in a duty current waveform with the following characteristics:

(a) It produces a slower rate of current decay and larger dc current component at contact separation.
(b) The duration of each major current loop of the duty current waveform is longer than the corresponding rating, e.g. $T_{\ell_2} > T_{\ell_1}$.
(c) The arcing time duration measured from contact separation to arc extinction (current interruption) is increased.
(d) The arc energy is increased. This may be measured either from contact separation to arc extinction or over the major current loop lengths.
(e) The minor current loops have reduced in both amplitude and duration.
(f) The rate of change of current at current interruption (zero current) is decreased and, as a result, the peak of the transient recovery voltage (TRV) could also be decreased.

The arc energy is defined as

$$E = \int_{t_1}^{t_2} v(t)i(t)\mathrm{d}t \tag{7.35a}$$

where

$$v(t) = i(t)R(t) \tag{7.35b}$$

and $v(t)$ and $i(t)$ are instantaneous arc voltage and instantaneous arc current, respectively. $t_2 - t_1$ is the arcing time duration or, more usually, the duration of the last major current loop before interruption. If available, a circuit-breaker arc model may be used since this accurately represents the time-dependent non-linear variation of the arc resistance and hence arc voltage. Where such a model is not available, two simplified options exist. The first is to assume that the arc voltage is constant and the arc energy is therefore proportional to

$$E \propto \int_{t_1}^{t_2} |i(t)|\mathrm{d}t \quad \text{for constant arc voltage} \tag{7.36a}$$

The second is to assume that the arc resistance is constant and the arc energy is therefore proportional to

$$E \propto \int_{t_1}^{t_2} i^2(t)\mathrm{d}t \quad \text{for constant arc resistance} \tag{7.36b}$$

Current international consensus in both IEC and IEEE Standards is that where the symmetrical short-circuit current duty is less than 80% of the rated symmetrical short-circuit current, then a circuit-breaker tested with a time constant of 45 ms may be considered adequate for any higher time constant except where current zeros are delayed for several cycles. For short-circuit duties above 80%, IEEE allows the use of the derating (correction) method described in Section 7.4.3 provided that the time constant does not exceed 120 ms ($X/R = 45.2$ at 60 Hz).

For short-circuit duties above 80%, IEC 62271-308:2002 *Guide for Asymmetrical Short-Circuit Breaking Test Duty T100a* includes a criterion when the duty time constant exceeds the standard test time constant of the circuit-breaker. The criterion is a measure of the arc energy and is based on the product $\hat{I} \times \Delta T$ where $\hat{I}$ and ΔT are the peak and duration of the last short-circuit major current loop, before interruption, respectively. The IEC Guide may be applied such that if the energy represented by the $\hat{I} \times \Delta T$ area under the duty current curve is less than that of the rated current curve, the duty is judged to be within rating and current interruption is deemed successful. For example, from Figure 7.9, and considering the third major current loop, we have

$$\hat{I}_{\text{Duty}} \times T_{\ell_2(\text{Duty})} = 93.16\,\text{kA} \times 14.19\,\text{ms} = 1.322\,\text{kA s}$$

The corresponding circuit-breaker rating is equal to

$$\hat{I}_{\text{Rating}} \times T_{\ell_1(\text{Rating})} = 94.06\,\text{kA} \times 12.15\,\text{ms} = 1.142\,\text{kA s}$$

Therefore, the (50 kA rms, $T_{\text{dc}} = 45$ ms) circuit-breaker is not capable of breaking a duty of (41 kA rms, $T_{\text{dc}} = 100$ ms). It is interesting to note the corresponding duty and rating values for the second major current loop. These are equal to

$$(\hat{I} \times T)_{\text{Duty}} = 100.96\,\text{kA} \times 15.39\,\text{ms} = 1.554\,\text{kA s}$$

and

$$(\hat{I} \times T)_{\text{Rating}} = 107.1\,\text{kA} \times 13.55\,\text{ms} = 1.454\,\text{kA s}$$

An alternative criterion is to evaluate the total arc energy that corresponds to the last major loop of both the rated and duty current waveforms using Equation (7.36). A trade-off between the ac and dc components of the total asymmetric instantaneous currents can then be calculated so that the total arc energy of the major loop duty current is equal to that of the rated or tested current. In addition, the increase in the last major loop arc duration may be conservatively limited to a small amount of 10–15% compared to the major loop arc duration of the rated current. This method enables the calculation of effective derating factors that may be applied to the rated symmetrical short-circuit current of the circuit-breaker.

Further reading

Books

[1] IEC 60909-0:2001–2007, *Short-Circuit Currents in Three-Phase AC Systems – Part 0: Calculation of Currents*, 1st end.
[2] IEC 60909-1:2002, *Short-Circuit Currents in Three-Phase AC Systems – Part 1: Factors for the Calculation of Short-Circuit Currents*.

[3] Engineering Recommendation ER G 7/4, *Procedure to Meet IEC 60909 for the Calculation of Short-Circuit Currents in Three-Phase AC Power Systems*, Electricity Network Association, UK, 1992.

[4] IEEE Standard C37.010:1999, *IEEE Application Guide for AC High-Voltage Circuit Breakers Rated on Symmetrical Current Basis*.

[5] British Standard BSEN 62271-100:2001, *High-Voltage Switchgear and Controlgear – Part 100: High-Voltage Alternating-Current Circuit-Breakers*.

[6] IEC TR 62271-308:2002, *High-Voltage Switchgear and Controller – Part 308: Guide for Asymmetrical Short-Circuit Breaking Test Duty T100a*.

Papers

[7] Bridger, B., All Amperes are not created equal: a comparison of current ratings of high-voltage circuit breakers rated according to ANSI and IEC Standards, *IEEE Transactions on Industry Applications*, Vol. 29, No. 1, January/February 1983, 195–201.

[8] Hartman, C.N., Understanding asymmetry, *IEEE Transactions on Industry Applications*, Vol. IA-21, No. 4, July/August 1985, 842–848.

[9] Knight, G., *et al.*, Comparison of ANSI and IEC 909 short-circuit current calculation procedures, *IEEE Transactions on Industry Applications*, Vol. 29, No. 3, May/June 1993, 625–630.

[10] Rodolakis, A., A comparison of North American (ANSI) and European (IEC) fault calculation guidelines, *IEEE Transactions on Industry Applications*, Vol. 29, No. 3, May/June 1993, 515–521.

[11] Roennspiess, O.E., *et al.*, A comparison of static and dynamic short circuit analysis procedures, *IEEE Transactions on Industry Applications*, Vol. 26, No. 3, May/June 1990, 463–475.

[12] Berizzi, A., *et al.*, Short-circuit calculation: a comparison between IEC and ANSI Standards using dynamic simulation as reference, *IEEE Transactions on Industry Applications*, Vol. 30, No. 4, July/August 1994, 1099–1106.

[13] Cooper, C.B., *et al.*, Application of test results to the calculation of short-circuit levels in large industrial systems with concentrated induction-motor loads, *Proceedings IEE*, Vol. 116, No. 11, November 1969, 1900–1906.

[14] Cooper, C.B., *et al.*, Contribution of single-phase motors to system fault level, *IEE Proceedings*, Vol. 139, No. 4, July 1992, 359–364.

[15] Calculation of electric power system short-circuits during the first few cycles, AIEE Committee Report, *AIEE Winter General Meeting*, New York, January/February 1956, pp. 120–127.

[16] Calculated symmetrical and asymmetrical short-circuit current decrement rates on typical power systems, AIEE Committee Report, *AIEE Winter General Meeting*, New York, January/February 1956, pp. 274–285.

[17] Skuderna , J.E., The *X/R* method of applying power circuit breakers, *AIEE Winter General Meeting*, New York, February 1959, pp. 328–338.

[18] Satou, T., *et al.*, Influence of the time constant of dc. component on interrupting duty of gas circuit breakers, *IEEE*, 2001, 300–305.

[19] Smeets, R.P.P., *et al.*, Economy motivated increase of dc time constants in power systems and consequences for fault current interruption, *IEEE*, 2001, 462–467.

[20] Shinato, T., *et al.*, Evaluation of interruption capability of gas circuit breakers on large time constants of dc component of fault current, CIGRE, 2002, 13–104.

Network equivalents and practical short-circuit current assessments in large-scale ac power systems

8.1 General

In this chapter, we discuss some of the practical issues faced by power system engineers engaged in short-circuit current assessment. We will present methods for the derivation of practical power system equivalents and describe how generation plant, power transmission and distribution networks may be represented in large-scale system studies. Some factors that maximise the calculated short-circuit current magnitudes are discussed as well as the range of uncertainties faced when making such calculations.

An introduction to the subject of probabilistic analysis of ac and dc short-circuit currents is presented, and finally an introduction to the theory of quantified risk assessment that may be used in the support of practical decisions concerning health and safety is given.

8.2 Power system equivalents for large-scale system studies

8.2.1 Theory of static network reduction

A power system equivalent is a mathematical model intended to represent a large portion of a power system network that may contain generation plant, network

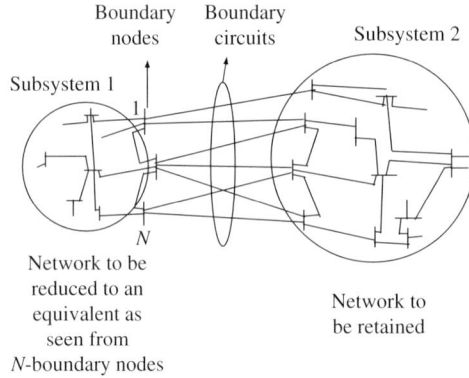

Figure 8.1   Illustration of static network reduction terminology used in the calculation of power system equivalents

branches, load or any combination of these. Figure 8.1 illustrates the general case of a large interconnected network divided into two parts or subsystems where one part is to be reduced to a small equivalent network whereas the second part is to be retained in full. The two parts are connected by a number of boundary circuits. The derivation of the power system equivalent is carried out by a network reduction process and Figure 8.1 shows the terminology used.

Network reduction is effectively a mathematical process that combines and eliminates network nodes and branches, so that the reduced equivalent contains smaller number of nodes and branches than the original network. The equivalent must be capable of reproducing the electrical behaviour of the original network at the boundary interface or nodes with sufficient accuracy. There are several types of power system equivalents the characteristics of which depend on the type of power system analysis being carried out. Equivalents may generally be classed as static or dynamic equivalents. The former may be used in dc and ac loadflow analysis, ac short-circuit analysis and harmonic analysis etc. The latter are used in power system stability and control analysis. In this chapter, we are concerned with the derivation of power system equivalents by network reduction for use in power system short-circuit analysis.

To illustrate the theory, subsystem 1 in Figure 8.1 is the power system network whose reduced equivalent is to be calculated as seen from nodes 1 to N which are termed the boundary nodes. All internal nodes and branches in subsystem 1 are to be eliminated in deriving the equivalent seen 'looking' from the boundary nodes into subsystem 1. The equivalent, as seen from the boundary nodes 1 to N consists of these nodes only, self (shunt) impedances to the reference node, usually the zero voltage node, and transfer impedances between all boundary nodes. All internal nodes and their interconnecting elements, e.g. lines, cables, transformers, etc., within subsystem 1 are to be eliminated by the reduction process. The reduced equivalent is illustrated in Figure 8.2. Both the boundary circuits and subsystem 2 are excluded from the reduction process.

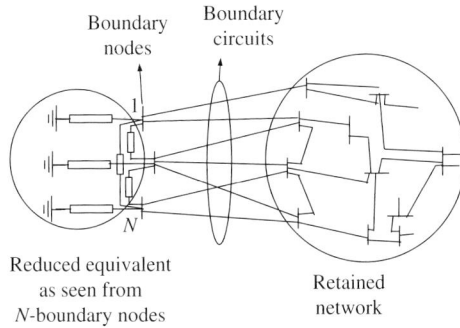

Figure 8.2 Illustration of the power system equivalent produced by the network reduction process of subsystem 1 of Figure 8.1

The number of boundary nodes should not be very large otherwise a very large number of transfer or series branches will be generated between all boundary nodes in the equivalent and this is equal to $N(N-1)/2$, where N is the number of boundary nodes. For example, if the original full network of subsystem 1 consists of 500 nodes and 1000 series branches, a 50-boundary node equivalent of subsystem 1 will result in a 1225 series branches!

8.2.2 Need for power system equivalents

During the early days of power system network analysis by digital computers, the need for network equivalents arose in order to allow the analysis of large-scale power systems to be practically possible given the limited computer storage and speed at that time. However, due to the huge developments in computer technology, this constraint is no longer applicable and very large power systems can nowadays be analysed quickly and cost effectively on personal computers.

However, power system equivalents are still required in industry. Practical examples include a distribution network company that owns and operates a distribution network and obtains its bulk power needs from a transmission company from one or several substations. The transmission network may be very large and an equivalent at one or more boundary nodes that are judiciously selected is usually sufficient for the short-circuit analysis needs of the distribution company. Similarly, the transmission company itself would not need to represent the distribution network in its entirety and an appropriate equivalent at one or more boundary nodes is usually sufficient. The added practical advantages are that the exchange and processing of data between companies is simplified, extraneous networks are removed from studies and the volume of data to be exchanged is reduced thus minimising the scope for errors in data handling. In general and in order to maintain sufficient accuracy of calculated short-circuit currents, experience shows that the network voltage level being studied should be represented in full as well as individual transformers stepping up or down to the next voltage level. For example, consider a network consisting of the following voltage levels: 400 kV transmission; 132 kV

subtransmission; 33, 11 and 0.43 kV distribution. If the voltage level being studied is 132 kV, then the 132 kV network should be modelled in its entirety (all
lines, cables, series reactors, etc.) as well as all 132 kV/400 kV autotransformers,
132 kV/33 kV and 132 kV/11 kV transformers. Power system equivalents may be
placed at 400, 33 or 11 kV voltage levels if required as this will generally have a
negligible effect on the precision of the results of short-circuit currents calculated
on the 132 kV network.

Further, the ongoing liberalisation, restructuring and privatisation of electricity
supply industries around the world since the 1990s have created the need for the
calculation of equivalents. Generally, new electricity trading markets and new market participants are established, e.g. independent generating companies (Gencos),
independent transmission system owners and/or operators (TSOs) and independent
distribution system owners and/or operators (DSOs). Figure 8.3 illustrates some
typical exchange requirements of either detailed network data or power system
equivalents among the main participants in a typical electricity market. The actual
flow of data may differ from one market to another depending on whether network
owners are also operators of networks owned by other companies.

Normally, network codes are also established that include legal obligations for
the exchange of planning and operational data between various companies and
much of the detailed technical data is usually classed as confidential. As a result,
using Figure 8.3, TSOs may be required to provide equivalents to DSOs and
vice versa. TSOs may also exchange equivalents among themselves but where
the accuracy of these equivalents is of critical importance in terms of operating
safe and secure interconnected transmission networks, full network data may be
exchanged. Gencos and large industrial consumers are usually required to provide

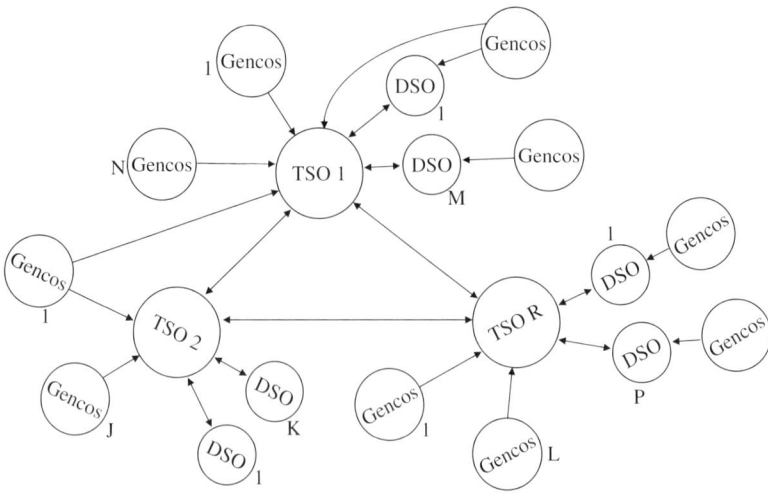

Figure 8.3 Illustration of exchange requirements of detailed network data or power system equivalents
in a liberalised electricity market

full network data to their host network operators who may not be allowed to pass it on to other network users due to confidentiality.

8.2.3 Mathematical derivation of power system equivalents

Conventional bus impedance or admittance matrices

Various mathematical methods can be used to carry out the network reduction process and derive the required power system equivalent. Most methods use either the bus impedance matrix or the bus admittance matrix of the entire network. For a W node network, let the bus impedance matrix be given by

$$
\mathbf{Z} =
\begin{bmatrix}
Z_{11} & Z_{12} & . & Z_{1j} & Z_{1k} & Z_{1\ell} & . & Z_{1W} \\
Z_{21} & Z_{22} & . & Z_{2j} & Z_{2k} & Z_{2\ell} & . & Z_{2W} \\
. & . & . & . & . & . & . & . \\
Z_{j1} & Z_{j2} & . & Z_{jj} & Z_{jk} & Z_{j\ell} & . & Z_{jW} \\
Z_{k1} & Z_{k2} & . & Z_{kj} & Z_{kk} & Z_{k\ell} & . & Z_{kW} \\
Z_{\ell1} & Z_{\ell2} & . & Z_{\ell j} & Z_{\ell k} & Z_{\ell\ell} & . & Z_{\ell W} \\
. & . & . & . & . & . & . & . \\
Z_{W1} & Z_{W2} & . & Z_{Wj} & Z_{Wk} & Z_{W\ell} & . & Z_{WW}
\end{bmatrix}
\tag{8.1}
$$

For a given set of boundary nodes N, an impedance submatrix $\mathbf{Z}_{\text{Bound}}$, whose diagonal terms correspond to the boundary nodes, can be extracted from the full network impedance matrix of Equation (8.1). For example, if the boundary nodes are nodes 2, 5 and 9, then $\mathbf{Z}_{\text{Bound}}$ is given by

$$
\mathbf{Z}_{\text{Bound}} =
\begin{array}{c}
\\ 2 \\ 5 \\ 9
\end{array}
\begin{array}{c}
\;2\quad\; 5\quad\; 9\;\\
\begin{bmatrix}
Z_{22} & Z_{25} & Z_{29} \\
Z_{52} & Z_{55} & Z_{59} \\
Z_{92} & Z_{95} & Z_{99}
\end{bmatrix}
\end{array}
\tag{8.2}
$$

In the general case of N-boundary nodes, and renumbering the nodes 1 to N for convenience, $\mathbf{Z}_{\text{Bound}}$ is given by

$$
\mathbf{Z}_{\text{Bound}} =
\begin{bmatrix}
Z_{11} & Z_{12} & Z_{13} & . & Z_{1N} \\
Z_{21} & Z_{22} & Z_{23} & . & Z_{2N} \\
. & . & . & . & . \\
Z_{N1} & Z_{N2} & Z_{N3} & . & Z_{NN}
\end{bmatrix}
\tag{8.3}
$$

The terms of $\mathbf{Z}_{\text{Bound}}$ are directly extracted from the full network bus impedance matrix. Therefore, the diagonal terms represent the driving point or Thévenin's impedances seen at the boundary nodes and the off-diagonal terms represent the transfer impedances between the boundary nodes. The nodal admittance matrix of the boundary nodes $\mathbf{Y}_{Bound}$ is obtained by inverting $\mathbf{Z}_{Bound}$ as follows:

$$\mathbf{Y}_{Bound} = \mathbf{Z}_{Bound}^{-1} = \begin{bmatrix} Y_{11} & -y_{12} & -y_{13} & . & -y_{1N} \\ -y_{21} & Y_{22} & -y_{23} & . & -y_{2N} \\ -y_{31} & -y_{32} & Y_{33} & . & . \\ . & . & . & . & . \\ -y_{N1} & -y_{N2} & -y_{N3} & . & Y_{NN} \end{bmatrix} \tag{8.4a}$$

where

$$Y_{ii} = y_{ii} + \sum_{\substack{j=2 \\ j \neq i}}^{n} y_{ij} \tag{8.4b}$$

The diagonal element of node i, Y_{ii} is the sum of all admittances connected to node i including the shunt admittance y_{ii} to the reference node. Therefore, the shunt admittance connected at boundary node i is calculated by summing all the elements of row i of Equation (8.4a). An off-diagonal element in the nodal admittance matrix of Equation (8.4a) is equal to the negative of the admittance connecting the boundary nodes i and j.

Using Equation (8.3), the calculation of a single-node equivalent results in a single equivalent impedance admittance, as shown in Figure 8.4, as follows:

$$V_1 = Z_{11} \times I_1 \quad \text{or} \quad I_1 = Y_{11} \times V_1 \tag{8.5}$$

Similarly, the calculation of a two-node equivalent using Equation (8.3) results in the following nodal impedance and nodal admittance matrices:

$$\begin{bmatrix} V_1 \\ V_2 \end{bmatrix} = \begin{bmatrix} Z_{11} & Z_{12} \\ Z_{21} & Z_{22} \end{bmatrix} \begin{bmatrix} I_1 \\ I_2 \end{bmatrix} \tag{8.6a}$$

and its inverse, using Equation (8.4) is expressed as

$$\begin{bmatrix} I_1 \\ I_2 \end{bmatrix} = \begin{bmatrix} y_{11} + y_{12} & -y_{12} \\ -y_{21} & y_{21} + y_{22} \end{bmatrix} \begin{bmatrix} V_1 \\ V_2 \end{bmatrix} \tag{8.6b}$$

Equation (8.6) is general and allows for a situation where these matrices are non-symmetric, i.e. $Z_{12} \neq Z_{21}$ and $y_{12} \neq y_{21}$ such as in the case of quadrature boosters or phase shifting transformers as discussed in Chapter 4. For symmetric matrices,

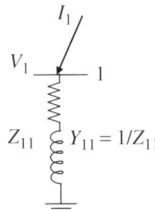

Figure 8.4   Power system equivalent of a single-boundary node

the two-node equivalent can be represented as a star equivalent impedance or a π equivalent admittance circuit as shown in Figure 8.5.

Again, assuming symmetric matrices, the calculation of a three-node equivalent results in the following matrices:

$$\begin{bmatrix} V_1 \\ V_2 \\ V_3 \end{bmatrix} = \begin{bmatrix} Z_{11} & Z_{12} & Z_{13} \\ Z_{12} & Z_{22} & Z_{23} \\ Z_{13} & Z_{23} & Z_{33} \end{bmatrix} \begin{bmatrix} I_1 \\ I_2 \\ I_3 \end{bmatrix} \tag{8.7a}$$

and

$$\begin{bmatrix} I_1 \\ I_2 \\ I_3 \end{bmatrix} = \begin{bmatrix} y_{11} + y_{12} + y_{13} & -y_{12} & -y_{13} \\ -y_{12} & y_{22} + y_{12} + y_{23} & -y_{23} \\ -y_{13} & -y_{23} & y_{33} + y_{13} + y_{23} \end{bmatrix} \begin{bmatrix} V_1 \\ V_2 \\ V_3 \end{bmatrix} \tag{8.7b}$$

The three-node equivalent can be represented as a star equivalent impedance or π equivalent admittance circuit as shown in Figure 8.6.

It is interesting to note that two-boundary nodes are described by a single-π equivalent circuit whereas for three-boundary nodes, three-π equivalents are required. Therefore, it is to be expected that the calculation of a multiple-node equivalent results in multiple-π equivalent networks. Although a one-node increase in the number of boundary nodes adds one shunt impedance only, it, however, adds transfer admittances/impedances to all the original boundary nodes. Therefore, the number of transfer admittances/impedances rises rapidly with the number of boundary nodes. In practice, the transfer admittance/impedance between any two

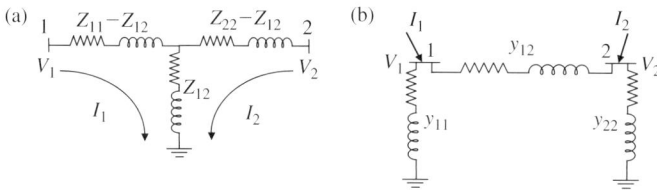

Figure 8.5 Power system equivalent for two-boundary nodes: (a) star impedance equivalent and (b) π admittance equivalent

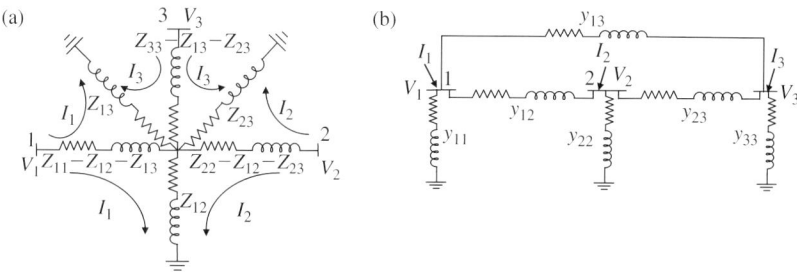

Figure 8.6 Power system equivalent for three-boundary nodes: (a) star impedance equivalent and (b) π admittance equivalent

nodes is equal to the apparent impedance between these nodes in the original unre-duced network. Thus, some transfer impedances will be very large (admittances very close to zero) where nodes are electrically quite remote from each other and hence these branches may be neglected.

Direct derivation of admittance matrix of power system equivalents

The conventional method described above requires the derivation of the entire net-work bus impedance matrix using techniques such as formulating and indirectly inverting the bus admittance matrix or by a step-by-step impedance matrix building process. An alternative is to derive the admittance matrix of the reduced equiva-lent directly from the full network admittance matrix given a specified number of boundary nodes. Figure 8.7(a) shows an interconnected passive linear network with N-boundary nodes and it is required to reduce the entire network to an equivalent as seen from these boundary nodes. All passive elements, e.g. lines, cables and trans-formers, are represented by their appropriate impedances. Rotating machines, e.g. synchronous generators are represented by an appropriate positive phase sequence (PPS) impedance such as subtransient or transient impedance, and their voltage sources are short-circuited.

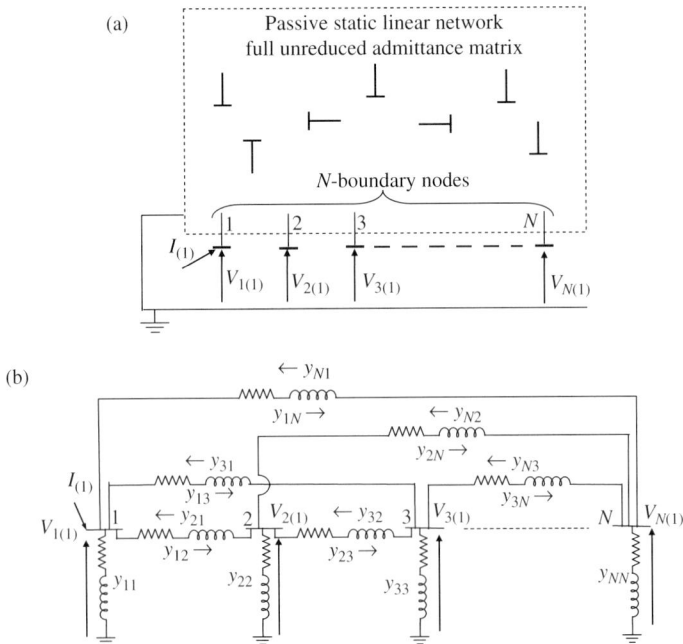

Figure 8.7 Direct derivation of a power system equivalent of N-boundary nodes: (a) network to be reduced as seen from N-boundary nodes and (b) multiple-π equivalent admittance circuit

In the calculation of negative phase sequence (NPS) and zero phase sequence (ZPS) equivalents, the machines are represented by their NPS and ZPS impedances to earth. The technique is based on the connection of a voltage source at each boundary node in turn, measuring the injected current into the boundary node and the resultant voltages at all other boundary nodes. The injected voltage source at each boundary node must have the same magnitude and phase angle, if any, though the latter may be zero. The magnitude of the voltage source is arbitrary, but a sufficiently large value may be used to improve the precision of calculated parameters.

The objective is to derive the equivalent nodal admittance matrix of the N-boundary nodes which is represented by the multiple-π circuit of Figure 8.7(b). A current $I_{(1)}$ injected into node 1 will produce currents and voltages throughout the network including voltages at the boundary nodes 1, 2, 3, ..., N. From Figure 8.7(b), we can write

$$I_{(1)} = y_{11}V_{1(1)} + y_{12}[V_{1(1)} - V_{2(1)}] + y_{13}[V_{1(1)} - V_{3(1)}] + \cdots + y_{1N}[V_{1(1)} - V_{N(1)}]$$
(8.8)

where y_{11} is the shunt admittance between node 1 and the reference node, y_{12} is the transfer admittance between nodes 1 and 2, $V_{1(1)}$ is the voltage at node 1 due to injected current $I_{(1)}$, $V_{2(1)}$ is the voltage at node 2 due to injected current $I_{(1)}$, etc. Rearranging Equation (8.8), we obtain

$$I_{(1)} = Y_{11}V_{1(1)} - y_{12}V_{2(1)} - y_{13}V_{3(1)} - \cdots - y_{1N}V_{N(1)}$$
(8.9a)

where

$$Y_{11} = y_{11} + y_{12} + y_{13} + \cdots + y_{1N}$$
(8.9b)

At boundary node 2, where the injected current is 0, we can write

$$0 = y_{22}V_{2(1)} + y_{21}[V_{2(1)} - V_{1(1)}] + y_{23}[V_{2(1)} - V_{3(1)}] + \cdots + y_{2N}[V_{2(1)} - V_{N(1)}]$$

or

$$0 = -y_{12}V_{1(1)} + Y_{22}V_{2(1)} - y_{23}V_{3(1)} - \cdots - y_{2N}V_{N(1)}$$
(8.10a)

where

$$Y_{22} = y_{21} + y_{22} + y_{23} + \cdots + y_{2N}$$
(8.10b)

and similarly for nodes 3, 4, 5, ..., N. For node N, we obtain

$$0 = -y_{N1}V_{1(1)} - y_{N2}V_{2(1)} - y_{N3}V_{3(1)} - \cdots + Y_{NN}V_{N(1)}$$
(8.11a)

where

$$Y_{NN} = y_{N1} + y_{N2} + y_{N3} + \cdots + y_{NN}$$
(8.11b)

Equations (8.9a) and (8.10a) to (8.11a) can be written in matrix form as follows:

$$
\begin{bmatrix} I_{(1)} \\ 0 \\ 0 \\ 0 \\ 0 \end{bmatrix} = \begin{bmatrix} Y_{11} & -y_{12} & -y_{13} & . & -y_{1N} \\ -y_{21} & Y_{22} & -y_{23} & . & -y_{2N} \\ -y_{31} & -y_{32} & Y_{33} & . & -y_{3N} \\ . & . & . & . & . \\ -y_{N1} & -y_{N2} & -y_{N3} & . & Y_{NN} \end{bmatrix} \begin{bmatrix} V_{1(1)} \\ V_{2(1)} \\ V_{3(1)} \\ . \\ V_{N(1)} \end{bmatrix}
\tag{8.12}
$$

The process is now repeated for boundary node 2. A current $I_{(2)}$ is now injected into node 2 and this will produce currents and voltages throughout the network including voltages at the boundary nodes $1, 2, 3, \ldots, N$. Using a similar derivation process to that of Equation (8.12), we obtain

$$
\begin{bmatrix} 0 \\ I_{(2)} \\ 0 \\ 0 \\ 0 \end{bmatrix} = \begin{bmatrix} Y_{11} & -y_{12} & -y_{13} & . & -y_{1N} \\ -y_{21} & Y_{22} & -y_{23} & . & -y_{2N} \\ -y_{31} & -y_{32} & Y_{33} & . & -y_{3N} \\ . & . & . & . & . \\ -y_{N1} & -y_{N2} & -y_{N3} & . & Y_{NN} \end{bmatrix} \begin{bmatrix} V_{1(2)} \\ V_{2(2)} \\ V_{3(2)} \\ . \\ V_{N(2)} \end{bmatrix}
\tag{8.13}
$$

Continuing with the above current injection process, i.e. to boundary nodes $3, 4, \ldots, N$, writing the derived equations in matrix form in each case, we can collect the individual matrix Equations (8.12), (8.13), etc., into the following system of matrix equations:

$$
\begin{bmatrix} I_{(1)} & 0 & 0 & . & 0 \\ 0 & I_{(2)} & 0 & . & 0 \\ 0 & 0 & I_{(3)} & . & 0 \\ . & . & . & . & . \\ 0 & 0 & 0 & . & I_{(N)} \end{bmatrix} = \begin{bmatrix} Y_{11} & -y_{12} & -y_{13} & . & -y_{1N} \\ -y_{21} & Y_{22} & -y_{23} & . & -y_{2N} \\ -y_{31} & -y_{32} & Y_{33} & . & -y_{3N} \\ . & . & . & . & . \\ -y_{N1} & -y_{N2} & -y_{N3} & . & Y_{NN} \end{bmatrix}
$$

$$
\times \begin{bmatrix} V_{1(1)} & V_{1(2)} & V_{1(3)} & . & V_{1(N)} \\ V_{2(1)} & V_{2(2)} & V_{2(3)} & . & V_{2(N)} \\ V_{3(1)} & V_{3(2)} & V_{3(3)} & . & V_{3(N)} \\ . & . & . & . & . \\ V_{N(1)} & V_{N(2)} & V_{N(3)} & . & V_{N(N)} \end{bmatrix}
\tag{8.14}
$$

or in concise matrix form

$$
\mathbf{I}_{N \times N} = \mathbf{Y}_{N \times N} \times \mathbf{V}_{N \times N}
\tag{8.15}
$$

where $\mathbf{I}_{N \times N}$ is a diagonal $N \times N$ matrix of known injected currents and $\mathbf{V}_{N \times N}$ is a known $N \times N$ non-symmetric matrix of resultant voltages, i.e. $V_{1(2)} \neq V_{2(1)}$ in the general case. Knowing the injected currents and voltages into the boundary nodes, the equivalent admittance matrix of the N-boundary nodes can be calculated as follows:

$$
\mathbf{Y}_{N \times N} = \mathbf{I}_{N \times N} \times \mathbf{V}_{N \times N}^{-1}
\tag{8.16}
$$

where $\mathbf{Y}_{N \times N}$ is a general non-symmetric admittance matrix. In most practical system networks, $\mathbf{Y}_{N \times N}$ is actually a symmetric matrix. The admittance connecting node i to any other boundary node, e.g. y_{ij}, is equal to the negative of the corresponding off-diagonal element of $\mathbf{Y}_{N \times N}$. The shunt admittance at boundary node i shown in Figure 8.7(b) is given by

$$y_{ii} = Y_{ii} - \sum_{\substack{j=2 \\ j \neq i}}^{N} y_{ij} \tag{8.17}$$

The impedances of the admittance branches of Figure 8.7(b) are calculated as follows:

$$z_{ij} = \frac{1}{y_{ij}} \quad j = i+1, i+2, \ldots, N \ \text{for} \ i = 1, 2, \ldots, N-1 \tag{8.18a}$$

and

$$z_{ii} = \frac{1}{y_{ii}} \quad i = 1, 2, \ldots, N \tag{8.18b}$$

In practical applications, the technique described above can be used to calculate more than one PPS equivalent such as subtransient and transient equivalents corresponding to subtransient and transient reactances of machines. In deriving the PPS equivalent, and depending on the national approach followed, line and cable shunt susceptance and shunt admittance of static loads may be excluded. In addition, the technique can also be directly applied to the calculation of NPS and ZPS equivalents. The shunt elements that may be excluded in the PPS equivalent calculation may also be excluded in the NPS equivalent calculation but they are generally included in the calculation of the ZPS equivalent.

Generalised time-dependent power system equivalents

Power system PPS, NPS and ZPS equivalents are calculated and included in PPS, NPS and ZPS models of the power system network used to calculate both balanced and unbalanced short-circuit fault currents. A generalised PPS time-dependent power system equivalent for a single node consists of a voltage source behind a series PPS impedance. The source voltage is the PPS Thévenin's or open-circuit voltage and the PPS impedance consists of a resistance in series with a time-dependent reactance. In the general case where the equivalent contains active sources, a generator model can be used to represent the equivalent whose PPS reactance at any point in time is given by

$$\frac{1}{X_{\mathrm{Eq}}^{\mathrm{P}}(t)} = \frac{1}{X_{\mathrm{Eq}}} + \left(\frac{1}{X_{\mathrm{Eq}}'} - \frac{1}{X_{\mathrm{Eq}}} \right) \exp\left(\frac{-t}{T_{\mathrm{Eq}}'} \right) + \left(\frac{1}{X_{\mathrm{Eq}}''} - \frac{1}{X_{\mathrm{Eq}}'} \right) \exp\left(\frac{-t}{T_{\mathrm{Eq}}''} \right)$$
$$\tag{8.19}$$

where X_{Eq}, X_{Eq}' and X_{Eq}'' are the steady state, transient and subtransient PPS reactances of the equivalent at the boundary node. T_{Eq}' and T_{Eq}'' are the transient and subtransient time constants of the equivalent at the boundary node. The NPS and ZPS equivalent impedances at the boundary node, denoted 1, are $Z_{11(\mathrm{Eq})}^{\mathrm{N}}$ and $Z_{11(\mathrm{Eq})}^{\mathrm{Z}}$, respectively. Figure 8.8 illustrates the PPS, NPS and ZPS equivalents of a single-boundary node.

The sequence representation of a single-node equivalent can be extended to equivalents of two or more boundary nodes. The shunt branch connected at each

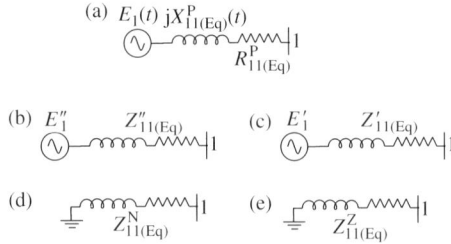

Figure 8.8 Sequence power system equivalents for a single-boundary node: (a) PPS time-dependent equivalent, (b) PPS subtransient equivalent, (c) PPS transient equivalent, (d) NPS equivalent and (e) ZPS equivalent

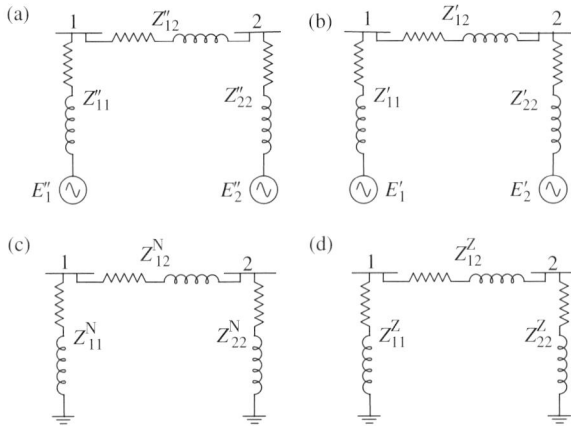

Figure 8.9 Sequence power system equivalents for two-boundary nodes: (a) PPS subtransient π equivalent, (b) PPS transient π equivalent, (c) NPS π equivalent and (d) ZPS π equivalent

boundary node in the π or multiple-π equivalents represents an equivalent source of short-circuit current, i.e. rotating plant. Figure 8.9 illustrates the PPS, NPS and ZPS π admittance circuit equivalents for two-boundary nodes.

8.3 Representation of power systems in large-scale studies

8.3.1 Representation of power generating stations

In a practical power system, the amount of installed generation plant exceeds the system peak demand by an appropriate margin, typically 20%, to allow for generation unavailability and demand forecasting errors. In order to maximise the magnitude of calculated short-circuit currents in planning and design studies, all generation plant is usually assumed operating but appropriately scaled to match system demand plus power losses. For off-peak demand year round operational planning studies, usually only scheduled generation dispatched to meet system demand, system frequency control and operating reserves may be used in the calculation of short-circuit currents.

In Chapter 5 , we discussed the modelling of individual generators for short-circuit analysis purposes. In a power station comprising a number of generators that may be connected to the power system network being studied either directly or through their own step-up transformers, the precision in the calculated short-circuit currents is generally improved if each generator is represented individually. However, the representation of power stations as equivalents is generally adequate if they are electrically quite remote from the locations being studied.

As presented in Chapter 5, power stations driven by renewable energy sources such as wind are being connected in increasing numbers and sizes both to distribution and transmission networks. A single wind farm power station may comprise, tens to hundreds of generators rated at, say, 2 to 5 MW each. The large number of small generators in such a wind farm may be comparable to the total number of large generators individually modelled in the entire network of an average size utility. Figure 8.10(a) illustrates a typical layout of a 128 MW offshore wind farm consisting of eight rows of eight wind turbines each and each wind turbine generator is rated at 2 MW. Each turbine generator has its own transformer that steps up to medium voltage, typically 20–33 kV, and all are connected by cables to a single collector busbar.

The wind farm may be modelled in its entirety including all 64 generators for local wind farm connection design studies. However, for studies at electrically remote locations on the wider host distribution, subtransmission or transmission systems, a reduced short-circuit equivalent for the wind farm generators, their transformers and cables at the collector busbar is usually sufficient. PPS, NPS and ZPS equivalents would be required to completely describe the equivalent as illustrated in Figure 8.10(b). The ZPS equivalent impedance is usually infinite because the transformer is usually delta–star connected.

8.3.2 Representation of transmission, distribution and industrial networks

For short-circuit analysis on transmission systems, transmission networks are usually modelled in their entirety in order to avoid loss of accuracy of calculated short-circuit currents. Generators connected to transmission networks are modelled in full and distribution networks supplied from a transmission network are modelled in part with appropriate equivalents for the remainder.

For short-circuit analysis on distribution systems, distribution networks may be modelled in their entirety as well as generators connected to these networks. Transmission networks supplying these distribution networks may be modelled using appropriate equivalents.

For short-circuit analysis in power station auxiliary systems or industrial power systems where a significant number of motors of different sizes may be used, a full representation of the entire station auxiliary and industrial networks is usually

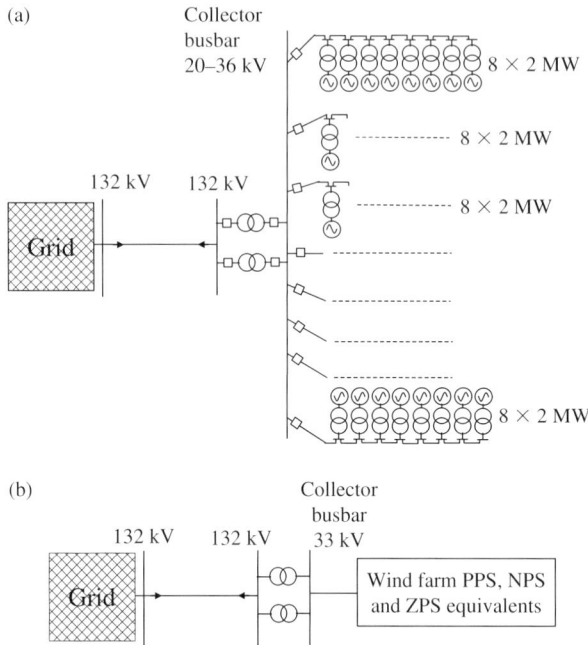

Figure 8.10 Offshore wind farm and sequence equivalents: (a) typical layout of a 128 MW wind farm (64 generators) and (b) sequence equivalents at collector busbar (boundary node(s))

required. However, groups of parallel low voltage, e.g. 0.415 kV may be lumped together to form a single equivalent model for each. The modelling of motors above 1 kV individually improves calculation accuracy.

Where ac superposition analysis is used, general power system static load at bulk supply substations in transmission and distribution networks is usually modelled as shunt PPS and NPS impedances derived from the prefault load voltage, MW and MVAr demand. The load ZPS impedance is derived from the network and transformer impedances that provide a path for ZPS currents. Where required by national practice, small induction motors forming part of the general substation load may be modelled as an equivalent motor connected at the substation.

8.4 Practical analysis to maximise short-circuit current predictions

8.4.1 Superposition analysis and initial ac loadflow operating conditions

The magnitude of short-circuit current is primarily determined by the amount of connected generation plant and network topology. Where the superposition analysis technique is used, such as that described by the UK ER G7/4 guide, and although of secondary influence, the effect of initial network and generating plant operating conditions is nonetheless important when assessing the short-circuit duties on existing switchgear. A loadflow operating condition that gives a particular network voltage profile, transformer tap positions and machine internal voltages that maximise the short-circuit current at one location, will not necessarily do so at other locations in the network. For example, short-circuit currents at locations remote from generation plant will be higher for higher prefault voltages at these locations. However, at generating stations with generator–transformers equipped with on-load tap-changers, the short-circuit currents delivered on the high voltage side of the transformers will be higher for lower prefault voltage on the transformer high voltage side. This is because the fault currents delivered from this power station unit depend on the ratio of the generator impedance to transformer impedance and on the transformer tap position. Similarly, a lower voltage profile on a subtransmission or distribution system, e.g. 132 kV, will generally cause an increase in the short-circuit current delivered through autotransformers from the transmission system, e.g. 275 or 400 kV. Also, a lower voltage at 132 kV together with a higher voltage at 33 or 11 kV will cause higher short-circuit currents at 33 and 11 kV delivered through the 132 kV/33 kV and 132 kV/11 kV transformers which are equipped with high voltage winding on-load tap-changers. Therefore, in calculating maximum short-circuit currents at various locations in a power system, there may be conflicting requirements in the loadflow study, so that a single study cannot be established to calculate maximum short-circuit currents at various locations.

8.4.2 Effect of mutual coupling between overhead line circuits

In Chapter 3, we showed that for double-circuit overhead lines, there will be like sequence PPS, NPS and ZPS couplings between the two circuits where six-phase transpositions are assumed and ZPS coupling only where ideal nine-phase transpositions are assumed. The PPS/NPS inter-circuit mutual impedances are generally small and typically only a few per cent of the PPS/NPS self-circuit impedances. However, the ZPS inter-circuit mutual impedance can be significant in comparison with the ZPS self-circuit impedance. To illustrate the effect of ZPS mutual coupling on the magnitude of the ZPS short-circuit current, consider the ZPS representation of a double-circuit overhead line shown in Figure 8.11(a) where the two circuits are denoted A and B.

Using Equation (3.89a) from Chapter 3, the series voltage drop across circuits A and B are $\Delta V_A = Z_A^Z I_A^Z + Z_{AB}^Z I_B^Z$ and $\Delta V_B = Z_{AB}^Z I_A^Z + Z_B^Z I_B^Z$. The effective ZPS impedance of each circuit is given by $Z_{A(\text{effective})} = Z_A^Z + Z_{AB}^Z$ and $Z_{B(\text{effective})} = Z_B^Z + Z_{AB}^Z$ if the ZPS currents I_A^Z and I_B^Z are equal and flow in the same direction. However, if the ZPS currents are equal but flow in the opposite direction, $Z_{A(\text{effective})} = Z_A^Z - Z_{AB}^Z$ and $Z_{B(\text{effective})} = Z_B^Z - Z_{AB}^Z$. Therefore, if the

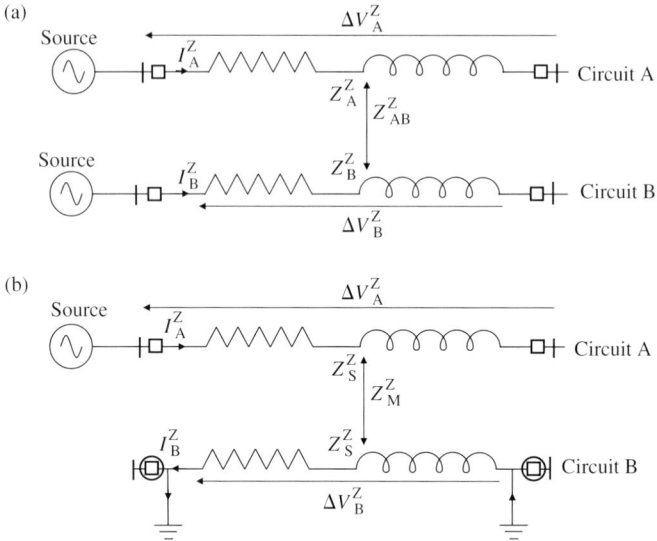

Figure 8.11   Effect of ZPS mutual coupling between two overhead line circuits: (a) ZPS inter-circuit coupling with both circuits in service; (b) ZPS inter-circuit coupling with one circuit out of service and earthed at both ends

ZPS mutual impedance between the two circuits is neglected in the modelling of the line, the short-circuit current calculated may be either an overestimate or an underestimate depending on the direction of ZPS currents flowing in each circuit. Where the ZPS currents flow in the same direction, there will be an overestimate but where the ZPS currents in each circuit flow in opposite directions, there will be an underestimate.

Figure 8.11(b) shows another practical case of a double-circuit overhead line with two identical circuits and with one circuit assumed of service and earthed at both ends for safety reasons. The voltage drops across each circuit are given by $\Delta V_A = Z_S^Z I_A^Z + Z_M^Z I_B^Z$ and $\Delta V_B = Z_M^Z I_A^Z + Z_S^Z I_B^Z$. Since circuit B is earthed at both ends,

$$\Delta V_B = 0 \quad \text{and} \quad I_B^Z = -\frac{Z_M^Z}{Z_S^Z} I_A^Z$$

Therefore, the effective impedance of the in-service circuit A is given by

$$Z_{A(\text{effective})}^Z = \frac{\Delta V_A^Z}{I_A^Z} = Z_S^Z - \frac{(Z_M^Z)^2}{Z_S^Z}$$

For example, for a typical 400 kV double-circuit line used in England and Wales with four subconductor bundle per phase, the ZPS impedance per circuit and the

ZPS mutual impedance between the two circuits are $Z_S^Z = (0.103 + j0.788)\Omega/\text{km}$ and $Z_M^Z = (0.085 + j0.420)\Omega/\text{km}$, respectively. Therefore, the effective impedance of one circuit with the second circuit earthed at both ends is equal to $Z_{A(\text{effective})}^Z = (0.0415 + j0.565)\Omega/\text{km}$. This represents a 28.7% reduction in the magnitude of the series impedance and an increase in its effective X/R ratio from 7.6 to 13.7. The increase in X/R ratio occurs because the phase conductors of the earthed circuit effectively act as short-circuited secondary turns where a significant proportion of the return short-circuit current flows via these 'turns' and much less returns via the single conductor earth wire.

8.4.3 Severity of fault types and substation configuration

Figure 8.12 shows typical air and gas insulated substations that include several circuit-breakers. For the assessment of the making and breaking (interrupting) duties of existing circuit-breakers, the selection of new circuit-breakers or the assessment of substation infrastructure integrity, both three-phase and single-phase short-circuit fault currents are usually calculated. In isolated or high impedance earthed systems, three-phase fault calculations are sufficient but for solidly earthed systems such as 132 kV and above in the UK, single-phase fault currents have to be calculated as well. Single-phase short-circuit currents will be higher than three-phase short-circuit currents at locations where the ZPS equivalent impedance at the fault point is lower than the PPS/NPS impedance. Figure 8.13(a) shows the direction of short-circuit currents for a close-up fault at F before fault clearance in a double-busbar or transfer bus substation layout. The maximum short-circuit breaking (interrupting) duty on circuit-breaker A is calculated assuming it is the last circuit-breaker to open to clear the fault, i.e. all circuit-breakers B, C and D have already opened. This duty can also occur under automatic circuit-breaker reclosure onto a persistent short-circuit fault. The two duty conditions are depicted in Figure 8.13(b).

Bus coupler and bus section circuit-breakers, shown as A and B in Figure 8.14(a) and (b) are required to make and interrupt the maximum fault current associated with short-circuit infeeds from all connected circuits when energising a section of busbar which is still inadvertently earthed. The resultant direction of current flow in the substation is also indicated. The making onto such a fault also imposes maximum duty on substation infrastructure equipment.

Figure 8.15(a) illustrates the direction of short-circuit current flows for a close-up fault on an outgoing line in a 1 and ½ circuit-breaker substation where circuit-breaker A or B being the last to open will be required to clear the fault. Figure 8.15(b) illustrates a busbar short-circuit fault on the same substation configuration, cleared by opening of circuit-breakers A, B, C and D with the last to open seeing the duty.

(a)

(b)

Figure 8.12 Air and gas insulated substations: (a) 400 kV outdoor air insulated substation and (b) 275 kV outdoor gas insulated substation

(a) Double-bus (UK) or transfer bus (America) substation

Four switch mesh (UK) or ring bus (America) substation

(b) Double-bus (UK) or transfer bus (America) substation

Four switch mesh (UK) or ring bus (America) substation

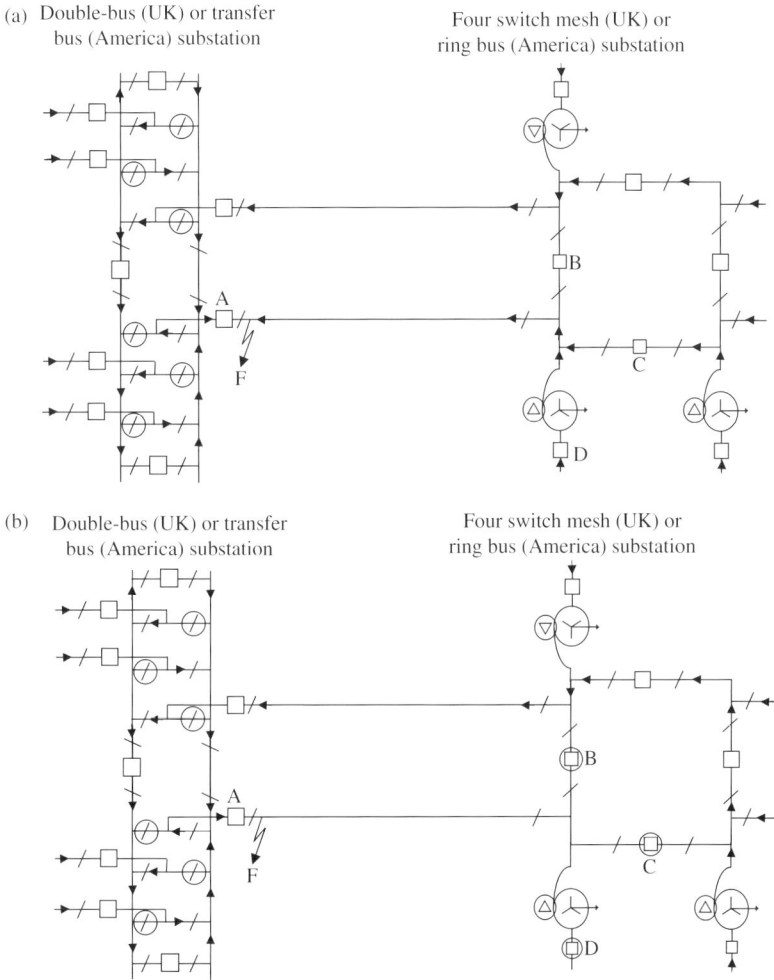

Figure 8.13   Illustration of the effect of substation configuration on maximum short-circuit currents: (a) direction of current flows for a short-circuit fault at F before fault clearance and (b) A is last circuit-breaker to clear short-circuit fault F or A recloses onto a persistent fault F by automatic circuit reclosure

8.5 Uncertainties in short-circuit current calculations: precision versus accuracy

The confidence that can be assigned to the calculated short-circuit currents depends on several factors which can be generally classified as follows:

(a) Confidence in the power system generation, network and load data used in the calculations.

(b) Confidence in the mathematical models used for power system plant.

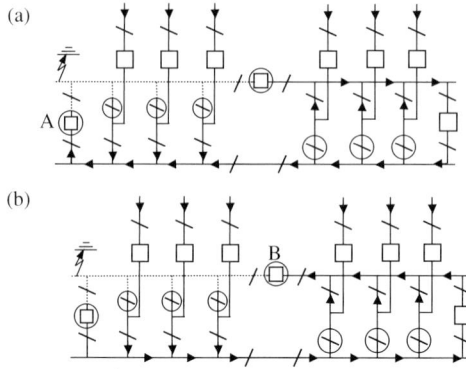

Figure 8.14 Short-circuit currents seen by bus section and bus coupler circuit-breakers in double-busbar substations: (a) inadvertent energising of a 'faulted' busbar section using bus coupler circuit-breaker A and (b) inadvertent energising of a 'faulted' busbar section using bus section circuit-breaker B

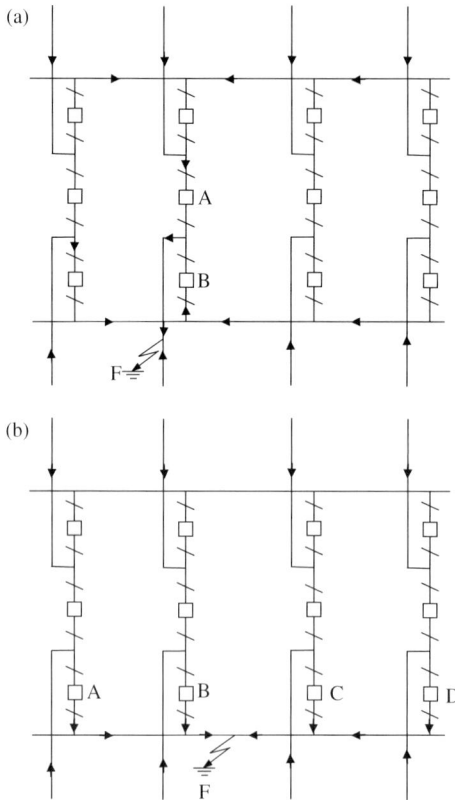

Figure 8.15 Short-circuit currents in '1 and ½' circuit-breaker substations: (a) direction of current flows for a close-up short-circuit fault on a line at F before fault clearance and (b) direction of current flows for a short-circuit busbar fault at F before fault clearance

(c) Confidence in the accuracy of analysis techniques used to calculate the ac and dc short-circuit currents.
(d) Confidence in the ability of engineers to correctly manage, create and use the system data, network topology diagram and network model.
(e) Confidence in the engineers' ability to understand and correctly apply the relevant national or international engineering standard being used, e.g. IEC 60909, UK ER G7/4 or IEEE C37.010.
(f) Confidence in the engineers' ability to make correct interpretation of short-circuit current results and to make appropriate engineering judgement.
(g) Confidence in the rated data or declared capabilities of circuit-breakers and substation infrastructure equipment.

At the system planning and design stage, actual data for new plant is usually not available and typical generic data is generally used as informed by similar plant family types already in use. Uncertainty in plant data is defined as the interval of design tolerances associated with plant generic data and these should be known and factored into the analysis. For example, design tolerances are $\pm 15\%$ on subtransient and transient reactances for synchronous machines manufactured to IEC 60034 Standard. Actual data for some existing small generation plant connected to distribution networks may not be known due to age of plant or change of ownership, etc. In such situations, typical parameters from similar plant may be used. Design tolerances for impedances of generator–transformers and network transformers manufactured to IEC 60076 Standard are $\pm 10\%$ and $\pm 15\%$ for two-winding and three-winding transformers, respectively. For autotransformers, the design tolerance is $\pm 15\%$. Factory test certificates for transformers and generators, when and where available, are usually the best-quality data. For overhead lines, PPS/NPS and ZPS impedances (and susceptances) are usually calculated based on line constructional data as presented in Chapter 3. Various assumptions are made such as uniform conductor height above ground, transposition, uniform earth resistivity (ZPS impedance varies with the logarithm of earth resistivity) for the entire line length, etc. Generally, the PPS impedances may have small tolerances of a few per cent, but the tolerances for the ZPS impedances are generally larger and may possibly reach 20%. There may also be a tolerance of a few per cent, on the length of installed lines which affects the total line impedance. For underground cables, the sequence impedances are normally measured at commissioning and in practice this is the only way of obtaining a good snapshot estimate of the ZPS impedance. There may also be a significant tolerance in the short-circuit infeed from induction motors due to unknown quantity and/or parameters. In addition, there is a significant uncertainty of the short-circuit infeed from small single-phase induction motors forming part of the general power system load.

The actual tolerances associated with various plant may be positive for some plant and negative for others so that some may cancel others out. The effect of impedance tolerances on the magnitude of short-circuit current depends on the location of the fault and the relative short-circuit contribution through the plant. For example, consider a generator–transformer unit with machine subtransient reactance and transformer reactance of 15% on MVA rating each. A 15% underestimate in the machine subtransient reactance will result in 8% overestimate in the short-circuit current delivered on the transformer's high voltage side. However, if the contribution of the generator–transformer unit is 20% of the total short-circuit current, the effect of the 15% underestimate in the generator subtransient impedance will only be 1.6% overestimate in the total short-circuit current.

Mathematical models used in steady state power frequency analysis for lines, cables, transformers, etc. are quite accurate and together with the analysis technique used, should contribute very little error to the ac component of short-circuit current at a fixed point in time, e.g. at the instant of fault. However, the estimation of the time variation of the ac component in fixed impedance analysis techniques is necessarily approximate and depending on the approximation method used this will lead to some error. The error in the dc component of short-circuit current depends on the analysis technique used. If this component is estimated from the system X/R ratio at the fault point, then, in the author's experience, the range of error may be up to $\pm 5\%$ if IEC 60909 Method C is used or up to $+30\%$ if IEEE C37.010 Method is used. IEC Method B can give substantial underestimates and its use is not recommended.

In most power system networks, there is a very large volume of data to be managed; that of the existing plant and network where some plant may be 40 or 50 years old, and that of the future network that includes modifications, extensions and decommissioning. Errors can be reduced if there is a single source of technical data such as a single database that is continuously kept complete and accurate. This is not a trivial task! The problem may be compounded where those responsible for the data in such a database are not the ones that use the data in network analysis or are not even engineers. In the author's experience, technical data can be the biggest single source of error in short-circuit analysis!

Standardisation of the short-circuit work undertaken through documented quality control procedures will contribute to precision in the calculations. Uncertainties can be reduced by the use of sensitivity studies on assumptions made in order to provide a measure of the possible range of errors. However, precision does not guarantee accuracy! Precision is a measure of consistency or repeatability and relates to variations between individual calculations of the same quantity. Accuracy, however, is the degree of closeness that the calculated value approaches the true value.

Given the discussion above, is there a need, in practice, for a safety margin to be applied to the calculated maximum short-circuit current results? The answer obviously depends on the views taken regarding all the above factors and individual practices. For example, an approach such as that of IEC 60909 that relies on acceptable accuracy for the intended purpose may be deemed not to require additional safety margin. Equally, an approach such as that of the UK ER G7/4 that relies on the ability of power system engineers to search for and find the worst-case realistic short-circuit currents may be deemed not to require an additional safety margin. In practice, the application or otherwise of safety margins depend on the philosophy of individual companies including attitude to safety, investment, asset risk management and risk aversion.

8.6 Probabilistic short-circuit analysis

8.6.1 Background

The international standards of short-circuit current calculation, IEC 60909 and IEEE C37.010 are essentially deterministic in approach. The former is aimed at calculating results with acceptable accuracy for their intended purpose and the latter is generally aimed at calculating sufficiently conservative results. The UK ER G7/4 approach is also deterministic although it is aimed at the calculation of worst-case realistic results. These approaches include various assumptions aimed at calculating maximum values of short-circuit currents. On the other hand, probabilistic short-circuit analysis techniques are generally aimed at calculating a probability distribution of short-circuit current magnitudes at various locations in the system. This can provide information on the probability of short-circuit currents exceeding certain values or falling below certain values.

8.6.2 Probabilistic analysis of ac short-circuit current component

Probabilistic analysis techniques of ac short-circuit current component recognise that the magnitude of short-circuit current at a given location is primarily determined by the following three random statistical factors:

(a) the nature of the fault type e.g. single-phase, three-phase, etc
(b) the location of the fault e.g. on busbars, on outgoing lines from a substation, etc
(c) the state of the power system in terms of generation plant, network state and demand level at the time of the occurrence of the fault

The deterministic short-circuit analysis approach considers (a) and (b) above although not faults some distance away from substations except where tower currents are being calculated. However, for network design studies, this approach does not consider (c) but instead it usually assumes that all installed generation plant is available and connected to the network and can contribute short-circuit current. The network elements e.g. lines and transformers etc are all assumed to be 100% available. In network operational planning studies, 100% availability of generation plant and network may be assumed in the area close to the faulted locations but not on a system wide basis.

The probabilistic short-circuit analysis approach aims at avoiding the compounding of safety factors or the assumption of simultaneous coincident worst case conditions. Thus, many power system states are considered and analysed at the time of occurrence of the fault and a fault current is calculated for each state.

(a) Probability (% exceeding abscissa)

(b) Number of three-phase faults
    per substation per annum

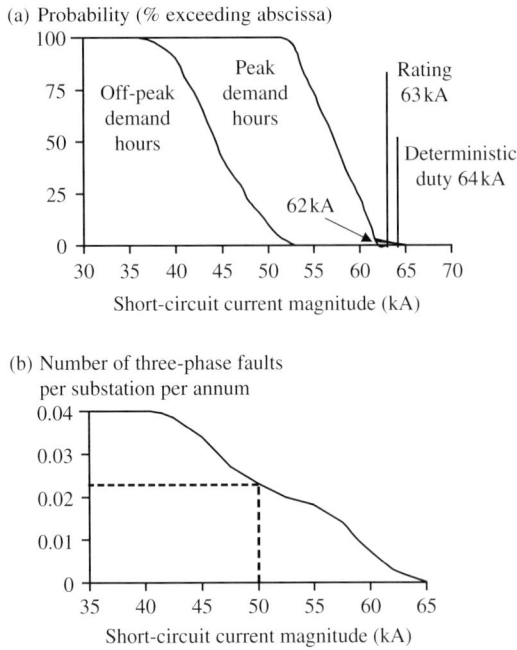

Figure 8.16   Illustration of probabilistic analysis results of short-circuit current ac component in a 400 kV transmission system: (a) probability distribution of ac short-circuit current magnitude and (b) risk of three-phase short-circuit faults exceeding abscissa

The probability distributions of the calculated short-circuit current magnitudes may be compared to the deterministic or probabilistic rating of circuit-breakers or busbars, as required, as illustrated in Figure 8.16(a) in a typical 400 kV system.

The current magnitude calculated by the deterministic approach is 64 kA and is greater than the circuit-breaker rating of 63 kA. The maximum current calculated by the probabilistic approach for peak demand hours is 65 kA. Assuming a 1 kA safety margin is used, there is a 2% probability of exceeding 62 kA. These results may be combined with an assumed uniform probability of three-phase fault occurrence throughout the year per substation, in the absence of better information. The outcome is illustrated in Figure 8.16(b). This shows that 0.023 faults per annum exceed 50 kA or a risk of one fault in 43.5 years exceeding 50 kA. Also, this shows that 0.04 faults per annum exceed 62 kA or a risk of one fault in 250 years exceeding 62 kA. The off-peak demand hours show much lower short-circuit currents due primarily to the reduced amount of connected generation plant.

Monte Carlo simulations may be used to simulate the behaviour of the power system under fault conditions by applying faults at random times during a year of system operation and at random locations in the network. However, the description of such simulations is outside the scope of this book. Probability distributions of subtransient, transient or other intermediate short-circuit current magnitudes may be calculated.

8.6.3 Probabilistic analysis of dc short-circuit current component

Assumptions of international standards

The international standards of short-circuit current calculation, IEC 60909, IEEE C37.010 and UK ER G7/4 make similar assumptions regarding the dc component of short-circuit fault current. The standards generally assume that the short-circuit fault results in maximum dc offset on the faulted phase for a single-phase short-circuit fault or one of the phases for a three-phase short-circuit fault. IEEE C37.010 considers that this corresponds to full 100% asymmetry except at generator bus-bars where asymmetry greater than 100% may occur. IEC 60909 makes similar assumption of full asymmetry and notes that the dc current component may exceed the peak value of the ac component for some near to generator faults. Similarly, ER G7/4 assumes full asymmetry of dc current component.

Factors affecting dc short-circuit current magnitude

In IEC 60909 and as implied in IEEE C37.010 and ER G7/4, evolving short circuit faults are not considered. Therefore, a three-phase short circuit is assumed to occur simultaneously on all three phases. This is important because sequential short-circuit faults can lead to dc current asymmetry greater than 100% but this is not dealt with in this book.

Recalling Equation (1.17) derived in Chapter 1 for the dc short-circuit current component following a simultaneous three-phase short-circuit fault, we have

$$i_{i(dc)}(t) = -\sqrt{2} I_{rms} \sin\left[\varphi_i - \tan^{-1}\left(\frac{\omega L}{R}\right)\right] \times \exp\left[\frac{-t}{\left(\frac{L}{R}\right)}\right] \qquad (8.20)$$

where

$$I_{rms} = \frac{V_{rms}}{\sqrt{R^2 + (\omega L)^2}} \qquad (8.21)$$

and

$$i = r, y, b \quad \varphi_y = \varphi_r - \frac{2\pi}{3} \quad \varphi_b = \varphi_r + \frac{2\pi}{3} \qquad (8.22)$$

The magnitude of the dc current component in any phase depends on the magnitude of the initial or subtransient ac current component I_{rms} and the instant on the voltage waveform φ_r when the short circuit occurs. The rate of decay of the dc current component depends on the time constant L/R or initial X/R ratio at the fault point ($X/R = \omega L/R$). The X/R ratio will vary with the fault location and fault type as well as the state of the system at the time of short circuit. The system state is determined by the availability of generation plant and network elements as

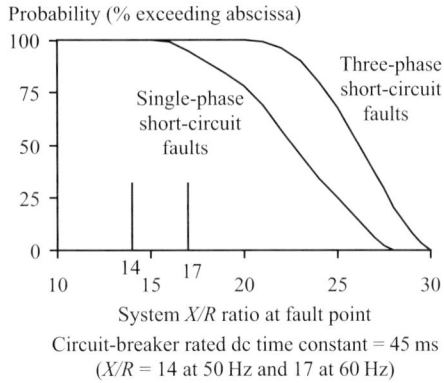

Probability (% exceeding abscissa)

System X/R ratio at fault point
Circuit-breaker rated dc time constant = 45 ms
(X/R = 14 at 50 Hz and 17 at 60 Hz)

Figure 8.17   Illustration of a typical probability distribution of X/R ratio of dc short-circuit current

well as the availability of any induction motor short-circuit infeed. The probability distributions of initial three-phase and single-phase X/R ratios at a 400 kV location comprising a large power station, a number of transmission circuits and a local distribution load are illustrated in Figure 8.17. The X/R ratios are calculated using IEC 60909 Method C. The results shows that the rated X/R ratios are almost always exceeded at such a location.

From Equation (8.20), the initial magnitude of the dc current component in any phase at the instant of short circuit, $t = 0$, expressed in per cent of the peak ac short-circuit current component is given by

$$\left[\frac{i_{i(dc)}(t=0)}{\sqrt{2}I_{rms}} \right]\% = -100 \times \sin\left[\varphi_i - \tan^{-1}\left(\frac{\omega L}{R} \right) \right] \quad i = r, y, b \qquad (8.23)$$

Maximum dc current offset or full asymmetry assumed by the deterministic IEC, IEEE and ER G7/4 calculations occurs for $\varphi_i - \tan^{-1}(\omega L/R) = -\pi/2$ or

$$\varphi_i = \tan^{-1}\left(\frac{\omega L}{R} \right) - \frac{\pi}{2} \qquad (8.24)$$

At any value of phase R angle φ_r, one or two dc current components will be negative. However, since we assumed a simultaneous three-phase fault condition, it is instructive to plot the absolute value of Equation (8.23) and this is shown in Figure 8.18. The initial dc current is plotted against the voltage phase angle φ_r for the IEC 62271 and IEEE C37.04 circuit dc time constant of 45 ms as well as the additional maximum quoted IEC 62271 circuit dc time constant of 120 ms. It is shown that for any value of φ_r, one of the three phases will have a dc current component of at least 86.6%. The effect of the dc time constant on the initial magnitude of the dc current is, as expected, negligible.

The above analysis assumed a simultaneous three-phase short-circuit fault. In the case of a circuit-breaker being inadvertently closed onto a network where a three-phase short-circuit fault already exists, as discussed in Section 8.4.3, the above analysis will equally apply if it is assumed that the three poles of the circuit-breaker

Initial magnitude of dc current component
(% of peak ac component)

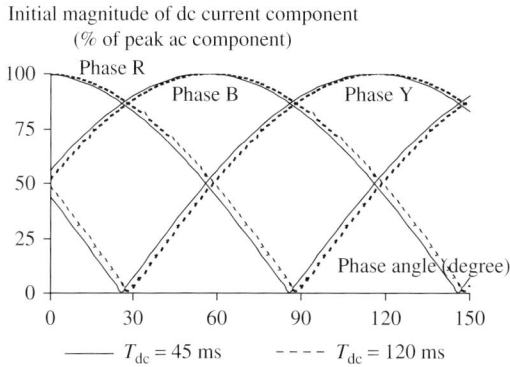

Figure 8.18   Absolute initial magnitude of dc current component in the three phases under a simultaneous three phase short-circuit fault

close simultaneously. In practice, the three poles do not close mechanically at the same time due to pole stagger or span which, according to IEC 62271 Standard, should not exceed half cycle of power frequency (10 ms for 50 Hz systems and 8.33 ms for 60 Hz systems). Also, electrical closure or prestrike may occur before mechanical closure and this is discussed in the next section.

Probabilistic analysis of voltage phase angle φ

In practice, the deterministic assumption of maximum dc offset has a substantial impact on the sizing of circuit-breakers and substation infrastructure such as busbars particularly with the increasing X/R ratios in power systems. The assumption of near to zero voltage phase angle φ_r results in maximum dc offset so the question is: For non-simultaneous closure of the three circuit-breaker poles when closing onto a fault, what is the probability of one of the poles closing at a voltage phase angle near zero?

Now, we describe the factors that can give rise to the dispersion between the three circuit-breaker poles. First, we note that there are two distinct cases relating to whether the circuit-breaker is independent pole operated, i.e. has one closing mechanism per pole, or three-pole operated, i.e. has one closing mechanism driving all three poles. The latter results in near simultaneous mechanical closure so that the mechanical dispersion is very small, controllable and fixed. However, for circuit-breakers with independent pole operation, larger and variable dispersion can occur since there is no mechanical coupling between the three poles. Further, for any pole, using Figure 8.19, there is a dispersion in the time lag between the closing order T_{co} and the instant of actuation of the closing mechanism T_a. Second, there is a dispersion in the operating time between actuation and instant of mechanical closure when contacts touch T_{mc}. Also, electrical closures due to electrical breakdown or prestrikes between the approaching contacts, at T_{ec}, may occur. These parameters are illustrated in Figure 8.19 for a single phase or pole

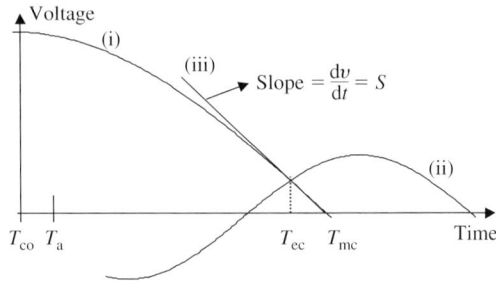

Figure 8.19   Illustration of circuit-breaker electrical closure, or prestrike, and mechanical closure

of the circuit-breaker. Curve (i) is a cosinusoid of period $4 \times T_{mc}$ and indicates the time variation of pole 1 gap withstand voltage during closing, assuming the dielectric strength of the insulation medium to be constant. Curve (ii) indicates the system voltage across pole 1 gap of the circuit-breaker. As the withstand voltage of pole 1 contact gap falls to the level of voltage across the gap, an electrical closure or prestrike occurs at T_{ec} before the contacts mechanically touch. Similar process may occur for poles 2 and 3 of the circuit-breaker. The latter part of curve (i) can be approximated as a straight line and indicated by curve (iii) whose negative slope is defined as S where $S = dv/dt$ = average rate of decay of dielectric strength of the gap during circuit-breaker closing in kV/s = (dielectric strength of the insulation medium in kV/cm) × (relative speed of approach of the contacts at impact in cm/s).

To derive an expression for the probability density function of the voltage phase angle φ, we neglect the voltage polarity effect on the electrical strength of the gap, i.e. $v(t) = \sqrt{2}V_{rms}|\sin(\omega t)|$. Figure 8.20(a) illustrates the mechanical closing instants and the corresponding range of electrical closing instants or angles for two different slopes S_1 and S_2 and Figure 8.20(b) shows an expanded section of a small element on the voltage curve.

Using Figure 8.20(b), consider two closing curves of the same slope S separated by a small time interval Δt, T_{mc1} and T_{mc2} are their respective mechanical closing times and T_{ec1} and T_{ec2} are their respective electrical closing instants. On the voltage waveform, the electrical phase angles φ_{e1} and φ_{e2} correspond to the time instants T_{ec1} and T_{ec2}, respectively. The probability element of electrical closing between φ_{e1} and φ_{e2} is $F(\varphi)\Delta\varphi$ where $F(\varphi)$ is the probability density function of angle φ. The probability element between the mechanical closure instants T_{mc1} and T_{mc2} is $F(t)\Delta t$ where $F(t)$ is the probability density function of mechanical closure with time t. Therefore, with φ and t being two functionally connected stochastic variables, we can write

$$F(\varphi)\Delta\varphi = F(t)\Delta t \tag{8.25}$$

or

$$F(\varphi) = \frac{F(t)\Delta t}{\Delta\varphi} \tag{8.26}$$

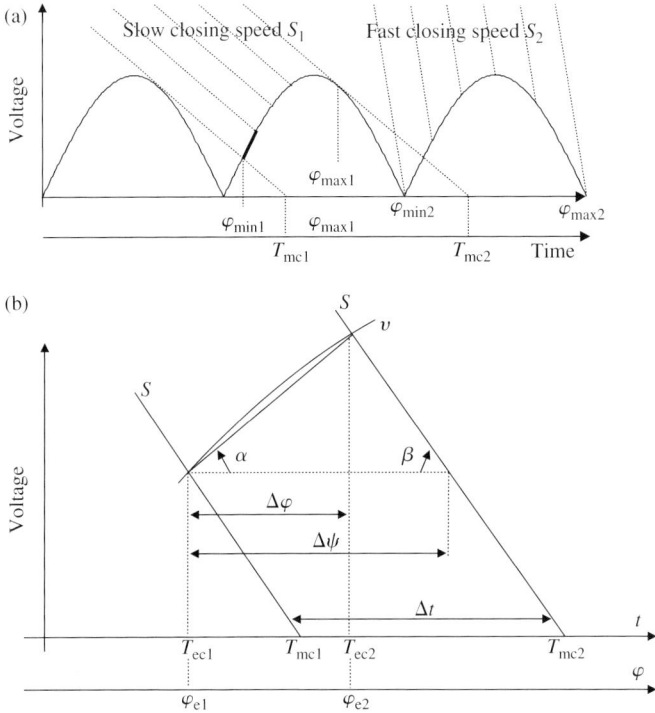

Figure 8.20 Probabilistic analysis of voltage phase angle of circuit-breaker closure onto a fault: (a) range of electrical closing angle due to a circuit-breaker prestrike and (b) an expanded section of a small element on the voltage curve of Figure 8.20(a)

The instant when the circuit-breaker closing order is given is random and is independent of the voltage phase angle φ and the breaker closes independently of time. Thus, the mechanical closing instant is uniformly distributed and has a constant probability density $F(t)$ between the minimum and the maximum closing angles of 0 and π as shown in Figure 8.20(a). Thus,

$$\int_0^{\pi/\omega} F(t)\mathrm{d}t = 1 \quad \text{or} \quad F(t) = \frac{\omega}{\pi} \tag{8.27}$$

However, the instant at which electrical closure, or prestrike, occurs is correlated with the voltage across the circuit-breaker contact gap. Thus, the distribution of the electrical closing instant, or angle, will not be uniform. From the geometry of Figure 8.21(b), we have

$$\Delta\varphi \tan\alpha = -(\Delta\psi - \Delta\varphi)\tan\beta \quad \Delta\psi = \omega\Delta t \quad \tan\beta = S \tag{8.28}$$

where S is the slope of the closing curve and is negative, $\tan\alpha$ is the slope of the source voltage $v(t) = \sqrt{2}V_{\text{rms}}\sin(\omega t)$, i.e. $\tan\alpha = \sqrt{2}V_{\text{rms}}\omega\cos\varphi$ and $\varphi = \omega t$.

From these relations and Equation (8.28), we obtain

$$\frac{\Delta t}{\Delta \varphi} = \frac{1}{\omega} \left(1 - \frac{\sqrt{2} V_{rms} \omega \cos \varphi}{S} \right) \qquad (8.29)$$

Therefore, using Equations (8.27) and (8.29) in Equation (8.26), we obtain

$$F(\varphi) = \frac{1}{\pi} \left(1 + \frac{\sqrt{2} V_{rms} \omega \cos \varphi}{S} \right) \qquad \varphi_{min} \leq \varphi \leq \varphi_{max} \qquad (8.30)$$

The probability distribution $f(\varphi)$ of the closing angle being equal to or greater than φ is equal to

$$f(\varphi) = \int_{\varphi}^{\varphi_{max}} F(\varphi) d\varphi$$

and is given by

$$f(\varphi) = \frac{1}{\pi} \left[(\varphi_{max} - \varphi) - \frac{\sqrt{2} V_{rms} \omega}{S} (\sin \varphi_{max} - \sin \varphi) \right] \qquad (8.31)$$

Besides probabilistic design and operational planning studies, the calculated probability equations may also be used in quantified risk assessments where calculated short-circuit currents may potentially exceed the rating of circuit-breakers. This is discussed in the next section.

For other causes of short-circuit currents, e.g. a lightning strike on top of a tower, then three scenarios may be considered. The first is where the magnitude of the lightning current is so large that the induced tower top voltage will cause back-flashover to the three phases of the circuit irrespective of the actual instantaneous value of the voltage on each phase. This is similar to the simultaneous three-phase short circuit where one of the phases will have an initial dc current component of at least 86.6%. The second scenario is where a substantial lightning strike directly hits one of the phases of the circuit and causes a short circuit. In this case, since the short circuit is effectively independent of the voltage magnitude or phase angle, the probability of the short circuit occurring between voltage zero and voltage peak is uniform. The probability of maximum dc current offset requires a short circuit at or near to voltage zero and if this is assumed to be within $0 \pm 10°$, then this equates to 40° over a full power frequency cycle of 360°. Therefore, the probability of maximum dc current offset is approximately equal to $(40°/360°) \times 100 \approx 11\%$. The third scenario is where the magnitude of the lightning induced voltage alone is insufficient but together with that of the actual instantaneous value of the voltage on any one phase will cause back-flashover and short circuit. Under this mechanism, the short circuit is more likely to occur when the phase voltage is significantly away from voltage zero. In this case, the probability of maximum dc current offset arising may be assumed to be negligible.

It is noted that the variable S used in Equations (8.30) and (8.31) is an important circuit-breaker parameter for the practical application of point-on-wave closing technique of circuit-breakers. This application is aimed at reducing transient over-voltages such as when energising shunt capacitor banks. To achieve this, the technique aims at closing the three poles at or very close to voltage zero on the source voltage waveform which requires a high value of S. Therefore, a circuit-breaker equipped with point-on-wave closing will produce maximum dc offset in each phase if it were inadvertently closed onto an existing three-phase to earth short-circuit fault. The effect of the parameter S on the probability of closing at various voltage phase angles is illustrated in an example in the next section.

8.6.4 Example

Example 8.1 Derive and plot the probability density and distribution functions of the electrical closing angle for a 400 kV circuit-breaker having eight gaps in series. The power system has a nominal frequency of 50 Hz.

Using Equations (8.30) and (8.31), and above a certain limiting circuit-breaker closing speed, i.e. when $|S| \geq S_{\text{limit}}$, $\varphi_{\min} = 0$ and $\varphi_{\max} = \pi$. Also, at $\varphi = \varphi_{\max} = \pi$, $F(\varphi) = 0$ and Equation (8.30) gives $S = -S_{\text{limit}} = -\sqrt{2}V_{\text{rms}}\omega$. Thus,

$$S_{\text{limit}} = \sqrt{2} \times \frac{400\,\text{kV}}{\sqrt{3} \times 8} \times 2\pi \times 50\,\text{rad/s} = 12.825\,\text{kV/ms}$$

For $|S| < S_{\text{limit}}$, we have $F(\varphi_{\max}) = 0$ giving

$$\varphi_{\max} = \cos^{-1}\left(\frac{S}{S_{\text{limit}}}\right)$$

Also $\varphi_{\min}$ can be determined from

$$\int_{\varphi_{\min}}^{\varphi_{\max}} F(\varphi)\mathrm{d}\varphi = 1 \quad \text{or} \quad (\varphi_{\max} - \varphi_{\min}) - \frac{S_{\text{limit}}}{S}(\sin\varphi_{\max} - \sin\varphi_{\min}) - \pi = 0$$

It can be easily shown that for $[|S| \geq S_{\text{limit}}, \varphi_{\min} = 0, \varphi_{\max} = \pi]$; $[S = -\frac{1}{2}S_{\text{limit}}, \varphi_{\min} = 13.2°, \varphi_{\max} = 120°]$; $[S = -\frac{1}{4}S_{\text{limit}}, \varphi_{\min} = 30.5°, \varphi_{\max} = 104.47°]$; $[S = -\frac{1}{8}S_{\text{limit}}, \varphi_{\min} = 45.5°, \varphi_{\max} = 97.18°]$. The results of the probability functions are shown in Figure 8.21(a) and (b).

At low circuit-breaker closing speed, e.g. $S = -\frac{1}{8}S_{\text{limit}}$ the electrical closing angles are higher and their range is more restricted. Therefore, the assumption of maximum 100% dc current offset is unrealistic. Even the worst case of $\varphi_{\min} = 45.5°$ gives an initial dc current magnitude, using Equation (8.27), of 65% (for 45 ms circuit dc time constant) and 68% (for 120 ms circuit dc time constant). However, for modern circuit-breakers with high closing speeds, all closing angles are likely but it is generally more likely that a circuit-breaker will close at smaller angles.

(a) Probability density

(b) Probability exceeding abscissa

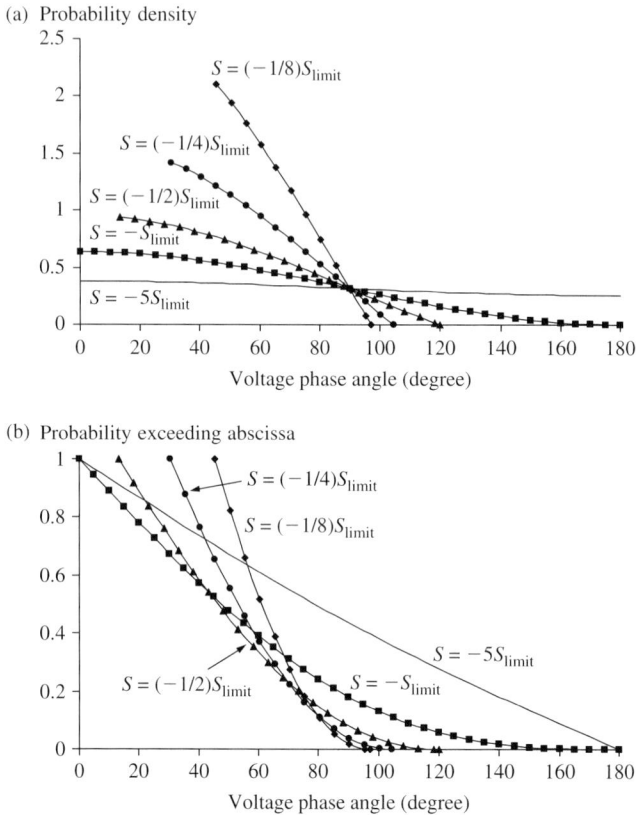

Figure 8.21 Probability density and distribution of electrical closing angle of a 400 kV circuit-breaker: (a) probability density of electrical closing angles and (b) probability of electrical closing angles

At $|S| \geq 5S_{limit}$, there is almost equal probability of closing at any angle between $0°$ and $180°$, i.e. the closure instant is independent of the voltage phase angle. In other words, there is equal probability of 100% and zero dc current offset.

8.7 Risk assessment and safety considerations

8.7.1 Background

Short-circuit studies carried out in long-term network design short-term network operational timescales or real time may identify short-circuit duties on existing switchgear in excess of ratings. The condition is usually defined as switchgear that is potentially overstressed. In the unlikely event of switchgear being identified to be potentially permanently overstressed, and in the absence of operational solutions (this is discussed in the next chapter) to remove the condition, it is usual to replace such switchgear with higher rated equipment. In network operational planning timescales, temporary overstressing may be identified such as may occur during switching operations to reconfigure the network. Depending on operating practices, substation configurations and circuit definitions, the duration of such a potential temporary overstressing may extend from a few minutes to possibly an hour.

8.7.2 Relevant UK legislation

When the health and safety of employees and the general public is affected, there may be legal duties which must be observed. With reference to the UK, the relevant legislation can be summarised as follows:

- *The Health and Safety at Work Act 1974 (Sections 2 and 3)*: 'An employer has a duty to ensure the health and safety of all his employees and to carry on his business in such a way that persons in his employment are not exposed to danger'.
- *The Electricity at Work Regulations 1989*:
  - *Regulation 5*: 'No electrical equipment shall be put into use where its strength and capability may be exceeded in such a way as may give rise to danger'.
  - *Regulation 29*: 'Provided a company takes all reasonable steps and exercises all due diligence to avoid danger arising, a defence can be formed by the implementation of risk management procedures'.
- *The Management of Health and Safety at Work Regulations 1999 (Section 3 Risk Assessments)*: 'Every employer shall make an assessment of the risks to the health and safety of his employees whilst at work and the risks to persons not in his employment in connection with his undertaking'.
- *The Electricity Safety, Quality and Continuity Regulations 2002*: Part 1 places general duties on employers to prevent danger so far as is reasonably practicable, and to ensure their equipment is sufficient for the purpose in which it is used.

Breaches of legislation in the UK are criminal offences and can lead to prosecution. Switchgear should not be overstressed if this gives rise to danger but if an employer takes all reasonably practicable steps and exercises all due diligence to avoid danger, a defence can be formed by the implementation of a suitable risk management procedure such as a quantified risk assessment.

8.7.3 Theory of quantified risk assessment

The objectives of quantified risk assessment are to identify the risk of fatality to an individual associated with equipment catastrophic failure and to reduce these individual risks to a level which is As Low As Reasonably Practicable (ALARP). The UK Health and Safety at Work Act 1974 includes the ALARP concept and the Health and Safety Executive have interpreted this concept in a risk diagram as shown in Figure 8.22.

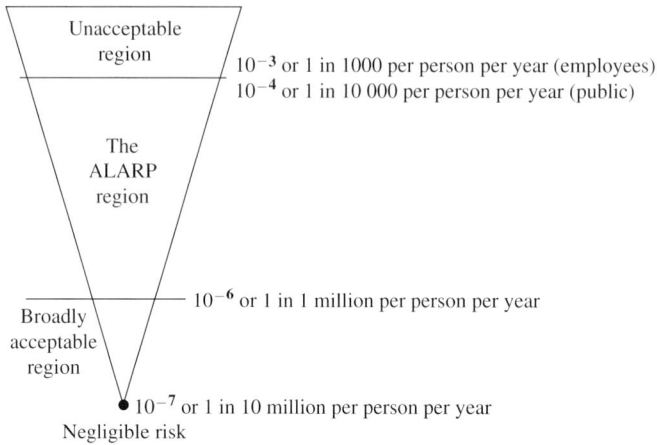

Figure 8.22 The 'ALARP' principle of the UK's Health and Safety at Work Act

The diagram consists of three regions. The bottom 'broadly acceptable region' is where the individual risk is so small that no further measures are necessary. The top 'unacceptable region' is where the individual risk is so large that it cannot be tolerated. The middle 'ALARP region' is where the individual risk must be 'As Low As Reasonably Practicable' considering the benefits of accepting the risk against the costs incurred of any further risk reduction measures. The UK Health and Safety Executive suggests certain numerical levels of risks. A risk of 10^{-7} or 1 in 10 million per person per year represents negligible risk. A risk level at the lower limit of the ALARP region is 10^{-6} or 1 in 1 million per person per year. Also, the suggested level of risk at the upper limit of the ALARP region is 10^{-4} or 1 in 10 000 per person per year for the public and 10^{-3} or 1 in 1000 per person per year for employees, respectively. The maximum tolerable level of risk to the public, 1 in 10 000 per person per year, is based on the proportion of people killed in road-traffic-related accidents in the UK per year. At the bottom of the ALARP region, the risk may be tolerated if the cost of reducing it would exceed the improvement gained. However, at the top of the ALARP region, the risk must be reduced unless the cost of reduction, which would exceed the improvement gained, is grossly disproportionate to the improvement gained.

8.7.4 Methodology of quantified risk assessment

A quantified risk assessment method consists of two steps: (a) risk analysis and (b) risk evaluation. Risk analysis may use an event tree risk analysis model in order to quantify the risks. This is a method originally devised to assess protective systems reliability and safety in nuclear power plants. Risk analysis consists of the following steps:

(a) Analysis of the site layout and configuration, maximum short-circuit current and critical location of the fault, and identification of potentially overstressed switchgear and its rating, type, age, maintenance record, etc. An estimation of the frequency of short-circuit fault f and proportion of time per year when the plant is subject to potential overstressing P_1 are made.

(b) Analysis of the consequences of plant failure on people. Estimates of the probability of catastrophic plant failure P_2, the proportion of time the individual is exposed to the hazard P_3 and the probability of fatal injury P_4.

The individual risk (IR) is estimated as follows:

$$IR = f \times P_1 \times P_2 \times P_3 \times P_4 \text{ per person per year} \tag{8.32}$$

Risk evaluation consists of comparing the estimated IR against the criteria for risk unacceptability, acceptability and tolerability as described in Section 8.7.3 and shown in Figure 8.22.

Further reading

Books

[1] Brown, H., 'Solution of Large Networks by Matrix Methods', 2nd edn, John Wiley & Sons, 1985, ISBN 0-471-80074-0.

Papers

[2] El-Kady, M.A., 'Probabilistic short-circuit analysis by Monte Carlo simulations', IEEE Transactions on PAS, Vol. PAS-102, No. 5, May 1983, 1308–1316.

[3] Srivastava, K.D., et al., 'The probabilistic approach to substation bus short-circuit design', Electric Power Systems Research, Vol. 4, No. 1, 1981, 191–200.

[4] Ford, G.L., et al., 'An advanced probabilistic short-circuit program', IEEE Transactions, on PAS, Vol. PAS-102, No. 5, May 1983, 1240–1248.

[5] Tleis, N., 'Computation of multiple prestriking transients in three-phase cable-motor systems', PhD Thesis, The University of Manchester, England, UK, November 1989.

[6] Svensen, O.H., 'The influence of prestrike on the peak values of energisation transients' IEEE Transactions on PAS, Vol. PAS-95, No. 2, March/April 1976, 711–719.

[7] Fukuda, S., et al., 'Switching surge reduction with circuit-breaker resistors in extra-high voltage systems', CIGRE, Session 1970, August/September, 13-04.

9

Control and limitation of high short-circuit currents

9.1 General

Many power system networks can be subject to high potential short-circuit fault currents. Some of the reasons for this include the connection of new generation plant to transmission and distribution networks, the strengthening of the power networks by the addition of new parallel routes, the use of low impedance equipment to improve voltage and reactive power control and system transient stability, the connection of many induction motors in industrial networks, and others. In this chapter, we describe some of the practical methods and techniques that may be used for the control and limitation of the magnitude of short-circuit currents in both power system operational and design timescales. Various types of established and emerging state of the art short-circuit fault current limiters that may be used to control and limit fault currents to safe values as well as their modelling techniques will be described.

9.2 Limitation of short-circuit currents in power system operation

9.2.1 Background

Depending on national or even particular utility practices, power system operation may extend from real time up to typically 6 months to 2 years ahead. The objectives are to ensure that the power system is operated in a safe, secure, reliable and economic manner, whilst planned maintenance and construction outages can be carried out as necessary. In both real time and operational planning timescales, the only measures that are usually available to control and limit the magnitudes of potential short-circuit currents are those that utilise existing equipment in the existing system. Some of these measures are described in the next sections.

9.2.2 Re-certification of existing plant short-circuit rating

This method may be used when a short-circuit current duty is predicted to be in excess of the nominally declared short-circuit capability of the circuit-breaker or substation infrastructure equipment. The method consists of re-examination of existing equipment's test certificates for any additional inherent capability that may be available over and above that specified in initial tender documents. If found available, this additional capability is usually re-certified with the equipment manufacturer before being released for use.

9.2.3 Substation splitting and use of circuit-breaker autoclosing

For substations that are normally run solid, and where potentially excessive short-circuit currents are predicted, the substation may be split in order to reduce the short-circuit currents for faults on either side of the split. The reduction is caused by the increase of the effective impedance between the fault location and some of the short-circuit current sources in the system. Substation splitting, in its simplest form, in single- and double-busbar substations, is accomplished by opening bus section or bus coupler circuit-breakers and operating them in a normally open position as shown in Figure 9.1.

Unfortunately, substation splitting reduces the degree of interconnectivity of the substation and in some cases may equate, electrically, to the removal of circuit(s) out of service. The reduction in reliability is illustrated in Figure 9.1(a) in the case of a single-busbar transformer fed substation as may be found in typical distribution systems. A fault outage on transformer T1 results in the loss of supply to busbar section 1 and all demand supplied from the distribution feeders connected to it.

Since the operation of the three transformers in parallel causes excessive short-circuit currents in the single-busbar substation, it is required to improve the security of supply to busbar section 1 and limit the potential short-circuit currents to within circuit-breaker ratings. These objectives may be met by using one of the schemes shown in Figure 9.2.

In Figure 9.2(a), the bus section circuit-breaker is normally run open with an autoclosing facility so that fault currents on either busbar sections are safely limited. In the event of loss of transformer T1, the normally open bus section circuit-breaker is automatically closed so that transformers T2 and T3 supply the entire substation load. Sometimes autotripping may also be used in order to automatically reopen the bus section breaker when T1 is returned to service. Figure 9.2(b) shows an alternative arrangement where the bus section circuit-breaker is run normally closed and either T2 or T3 is operated on hot standby. This means that the transformer is

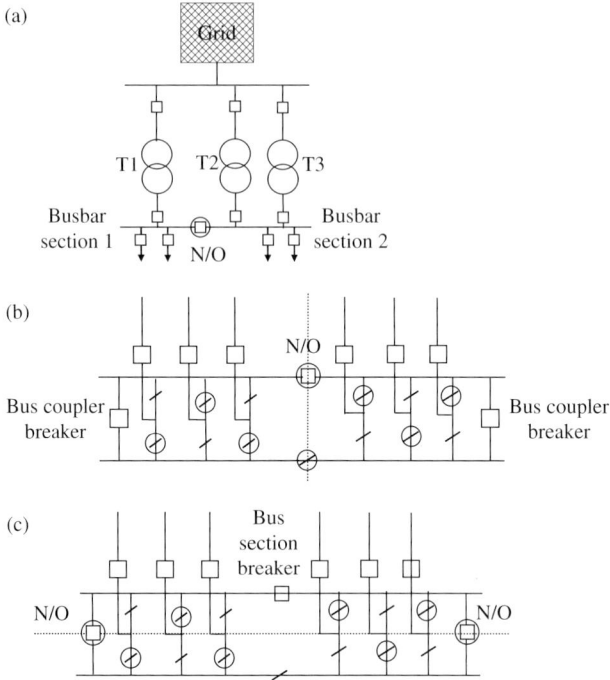

Figure 9.1 Substation splitting for short-circuit current limitation: (a) Single-busbar substations, bus section breaker normally open (N/O), (b) vertical split in double-busbar substations and (c) horizontal split in double-busbar substations

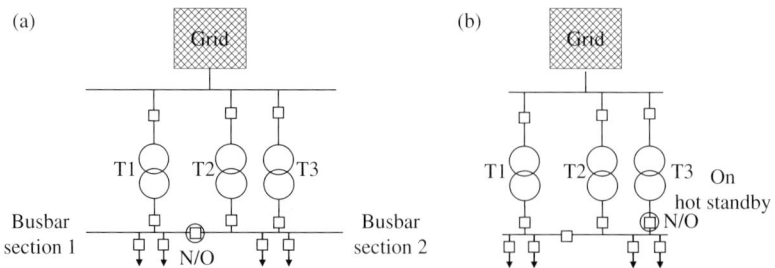

Figure 9.2 Use of autoclose schemes in short-circuit current limitation: (a) normally open bus section circuit-breaker with autoclosing and (b) normally open breaker of a hot standby transformer with autoclosing

permanently energised from its primary but its secondary breaker is normally run open with an autoclosing facility. This breaker is automatically closed in the event of loss of transformer T1 to prevent overloading of transformer T2. Autotripping may also be used.

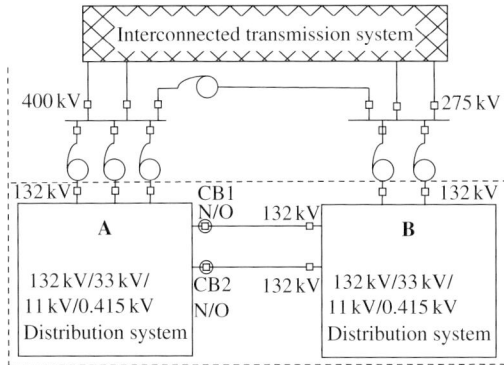

Figure 9.3 Illustration of subtransmission/distribution network splitting for short-circuit current control

The principle of substation splitting can also be applied in industrial power systems where generally single-busbar substations are used and these are interconnected by busbar sectionalisers which may be a cable circuit controlled by two circuit-breakers. In this case, one of these circuit-breakers is normally operated open.

9.2.4 Network splitting and reduced system parallelism

Transmission systems operating at different voltage levels, e.g. 400 and 275 kV (or 220 kV), are normally operated in parallel. In addition, distribution systems operating at, say, 132 or 110 kV may in some cases be operated in parallel with transmission systems. The latter is illustrated in Figure 9.3 where the 132 kV distribution system is supplied from two substations that may be many kilometres apart from each other and the entire distribution system is normally operated interconnected.

Sources of short-circuit currents in the transmission and distribution systems can contribute to faults at substations A or B. However, if the interconnected distribution network is split by operating circuit-breakers CB1 and CB2 normally open, the short-circuit sources in distribution subnetwork A become electrically remote from faults at substation B and vice versa. The magnitude of short-circuit current reduction depends on the specific characteristics of the distribution system, e.g. the degree of interconnection and amount of connected generation plant as well as the relative proportions of currents supplied from the transmission and distribution systems. In general, reductions of up to 30% might be obtained. As in Section 9.2.3, it should be noted that the interconnection is planned in the first place to increase the reliability of the distribution network and the effect of network splitting on demand security would need to be considered. Although more difficult in this case, autoclosing may also be considered as discussed in Section 9.2.3.

9.2.5 Sequential disconnection of healthy then faulted equipment

This method of short-circuit current limitation is only suitable when the circuit-breaker breaking or interruption duty is the limit rather than the making duty which is within equipment rating. The method is illustrated using Figure 9.4.

Consider a short-circuit fault on the line side of circuit-breaker F that results in a short-circuit current that exceeds the rating of circuit-breaker F. All other circuit-breakers in the substation may also be overstressed. The short-circuit breaking current at F with the remote circuit-breaker F_r open is given by $I_F = \sum_{j=1}^{4} I_j$. In this method of current limitation, another upstream circuit-breaker feeding the short-circuit fault, e.g. CB1, is opened first then circuit-breaker F is opened to safely interrupt the reduced short-circuit current. This method can introduce significant disadvantages and risks to the reliability and safety of power systems and these have to be thoroughly evaluated. The fault clearance time is delayed by the upstream breaker operation time. Also, a fault at F would normally constitute a $(N-1)$ condition but in this method, another healthy circuit is deliberately disconnected giving a $(N-2)$ contingency. The term $(N-x)$ represents the total system N less x circuits. Where automatic circuit reclosure on the faulted circuit is not employed, i.e. circuit-breaker F will not reclose after clearing the fault, circuit-breaker CB1 can be reclosed immediately once circuit-breaker F has opened and cleared the fault. However, if automatic reclosure is employed, then the decision to reclose CB1 should consider the two possibilities of the fault being transient or permanent. In practice, the engineering of such schemes may be very complex. In addition, from a safety point of view, in-service breaker F is potentially overstressed, but its safe operation is dependent on the success of a sequential tripping scheme that must ensure the opening of breaker CB1 first. If for some reason this scheme fails, there may be significant adverse system reliability and safety consequences.

9.2.6 Increasing short-circuit fault clearance time

This method consists of delaying the current interruption time of the circuit-breaker from, say, 50 to 100 ms. The delay is used in order to benefit from the decrease with

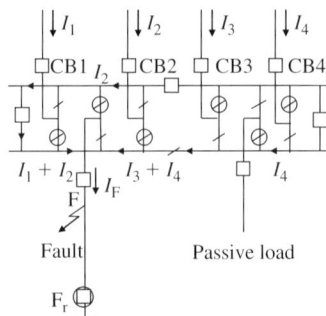

Figure 9.4 Fault current limitation by opening healthy equipment (CB1) then faulted equipment (CBF)

time of both the ac and the dc components of the short-circuit current. However, the effect of any delay in the fault clearance time on the stability of the power system will need to be considered. The method depends on national practice in terms of the standard used for the calculation of short-circuit currents since some ignore the decay of short-circuit currents with time altogether.

9.2.7 De-loading circuits

This method of fault current limitation consists of opening one circuit-breaker at one end of a circuit but keeping the circuit energised from the other end(s) in order to benefit from its reactive power gain generated by its susceptance. The opened circuit-breaker is operated normally open to reduce the short-circuit infeed into the substation. An autoclose arrangement may be used in the event of a loss of another infeeding circuit into the substation. The method is conceptually similar to operating transformers on hot standby with autoclosing as described in Section 9.2.3.

9.2.8 Last resort generation disconnection

In real-time system operation, the disconnection of generation plant that can feed short-circuit currents into a local substation for fault current limitation is highly unusual. Such action can be very costly in liberalised electricity markets so that even if the disconnection is required only once and for a few hours, the costs that may be incurred may be similar to or exceed the cost of purchasing and installing a new extra high voltage circuit-breaker. Some of the other techniques already described are examined first and usually the splitting of the substation or the de-loading of one feeder circuit may be used.

9.2.9 Example

Example 9.1   Figure 9.5 shows a 400 kV transmission substation supplying a 132 kV substation through four 240 MVA 400 kV/132 kV autotransformers.

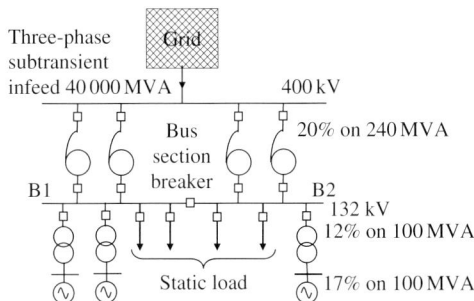

Figure 9.5   System for Example 9.1

The 132 kV substation is of a double-busbar design and the 132 kV network is radial. Three 100 MVA generators are connected to the 132 kV busbar as shown. The 132 kV circuit-breakers and substation infrastructure three-phase short-circuit rating is 5000 MA corresponding to 21.9 kA three-phase short-circuit current. Calculate the 132 kV three-phase busbar short-circuit current when the 132 kV substation is operated solid and split using the busbar section circuit-breaker. For quicker hand calculation, all network resistances are ignored and the 132 kV feeders are assumed to supply a static load only. The generator's subtransient reactances are given. The initial ac currents are to be calculated.

Solution

132 kV substation solid
The positive phase sequence (PPS) subtransient reactance of the system infeed on a 100 MVA base is calculated as

$$X_{\text{Thév.}} = \frac{100\,\text{MVA}}{40\,000\,\text{MVA}} \times 100 = 0.25\%$$

The autotransformer PPS reactance on a 100 MVA base is equal to

$$X_T = 20\% \times \frac{100\,\text{MVA}}{240\,\text{MVA}} = 8.34\%$$

The Thévenin's equivalent impedance 'seen' at the faulted 132 kV busbar is calculated as

$$X_{\text{Thév.}} = \frac{1}{\frac{1}{0.0025 + 0.0834/4} + \frac{3}{0.17 + 0.12}} = 0.0188\,\text{pu}$$

The 132 kV busbar three-phase fault level is equal to 100 MVA/0.0188 pu = 5319 MVA and this corresponds to a three-phase fault current of 5319 MVA/($\sqrt{3} \times 132$ kV) = 23.3 kA which exceeds the 21.9 kA rating. Clearly, the bus section and feeder circuit-breakers are all potentially overstressed. Also, the generator–transformer breakers are overstressed for a fault on the transformer side of the breaker since each generator contribution to the fault is equal to 1.5 kA. The reader is encouraged to show that the 132 kV breakers of the autotransformers are not overstressed since the fault current for a fault on the transformer side with the 400 kV breaker open is only 19 kA. It is noted that under the busbar fault condition, the current supplied through each autotransformer is 4.7 kA, but the duty on the autotransformer 132 kV breaker is not equal to 23.3 kA–4.7 kA = 18.6 kA!

132 kV substation split with bus section breaker normally open
The Thévenin's equivalent impedance 'seen' at the fault point for a fault on busbar section B1 is calculated as

$$X_{\text{Thév.}} = \frac{1}{\frac{2}{0.29} + \cfrac{1}{0.0834/2 + \cfrac{1}{\cfrac{1}{0.17 + 0.12 + 0.0834/2} + \cfrac{1}{0.0025}}}} = 0.03386\,\text{pu}$$

The 132 kV busbar three-phase fault level is given by $X_{Thév.} = 100\,MVA/$ $0.03386\,pu = 2953\,MVA$ and this corresponds to a three-phase fault current of $2953\,MVA/(\sqrt{3} \times 132\,kV) = 12.9\,kA$ which is substantially less than the $21.9\,kA$ rating. The three-phase fault current at busbar B2 is equal to $11.4\,kA$. Therefore, the short-circuit current duties are well within the ratings of all 132 kV circuit-breakers.

This huge reduction in short-circuit current magnitude is due to the assumption that the 132 kV network emanating from the 132 kV substation is radial. Where this network is interconnected, an equivalent interconnecting impedance would still exist between the two 132 kV busbar sections when the 132 kV busbar is split. The effect of this is that the reduction in the magnitude of short-circuit current may be significantly reduced.

9.3 Limitation of short-circuit currents in power system design and planning

9.3.1 Background

Depending on national or even particular utility practices, power system design and planning may extend from typically 6 months ahead up to 7 years ahead or more. The objectives are generally to design the connection of new generation plant and demand centres to the existing network as well as the network infrastructure reinforcements required to ensure secure and reliable power system into the future. In future network planning and design timescales, many competing measures are available to the designer to secure the network and ensure safe short-circuit fault level management including the planning and engineering of the installation of new equipment.

In general, some of the methods described in Section 9.2 can also be used at the network design stage, e.g. re-certification of existing plant short-circuit capability, substation splitting and use of circuit-breaker autoclosing, network splitting and reduced system parallelism, increasing circuit-breaker fault clearance time and de-loading of circuits. In the next sections, we will describe some of the additional measures generally available to the system design engineer.

9.3.2 Opening of unloaded delta-connected transformer tertiary windings

Many transformers used in power systems have an additional unloaded third or tertiary winding. As we discussed in Chapter 4, the delta tertiary winding presents a low zero phase sequence (ZPS) impedance path to the flow of ZPS currents. Therefore, when opened, no ZPS current can circulate inside the winding. This has the effect of increasing the shunt ZPS impedance to earth to a very large value depending on the transformer core construction. This increase is greatest for 5-limb and shell-type core transformers. However, experience shows that the overall effect on the reduction of the magnitude of single-phase to earth short-circuit current is relatively small and local and hence this technique, which is very attractive economically, is potentially useful in marginal overstressing at specific locations. The delta tertiary winding also provides a path for the circulation of third harmonic currents which arise mainly from the transformer magnetising current. The opening of the delta winding causes third harmonic ZPS currents to circulate further into the system which may cause unacceptable voltages and interference in communication networks. Therefore, the use of this technique requires careful consideration.

9.3.3 Specifying higher leakage impedance for new transformers

A short-circuit overstressing problem may be caused when a substation is being reinforced through the addition of one or more new transformers. The short-circuit currents fed through the transformers to the fault location can be reduced by specifying higher than average transformer leakage impedance, e.g. by up to 40% or 50%. However, the increased transformer impedance causes an increase in the transformer reactive power losses and voltage drop and hence the provision of the required supply voltage quality may require a tap-changer to have a wider tapping range. Nonetheless, appreciable reduction in the magnitude of short-circuit current may be obtained.

9.3.4 Upgrading to higher nominal system voltage levels

The short-circuit current magnitude in kA for a given short-circuit fault level in MVA is inversely proportional to the nominal system voltage level. In most practical situations, however, electrical power systems and their voltage levels already exist and the construction of higher voltage systems for short-circuit limitations only may not be economically justifiable. To illustrate the effect on the short-circuit current magnitude of connecting a power station unit (a generator and its step-up transformer) at two different voltage levels, consider a 100 MVA generator having a 20% subtransient reactance and a step-up transformer having a 15% leakage reactance, both on 100 MVA. With the generator unloaded and a 1 pu prefault voltage, the initial magnitude of the short-circuit current for a three-phase short-circuit fault on the transformer's high voltage side is equal to

$$\frac{1}{0.2 + 0.15} \times \frac{100\,\text{MVA}}{\sqrt{3} \times \text{V}_{\text{line}}}$$

Therefore, the magnitude of the short-circuit current delivered from the power station unit will be equal to 1.25 and 0.75 kA for connecting to 132 and 220 kV systems, respectively. In practice, the reduction in short-circuit current for connecting to 220 kV will be a little higher because the transformer's leakage reactance increases with winding voltage. In general, the benefits of such reductions will need to be balanced against the increased costs of connecting at a higher voltage level.

9.3.5 Uprating and replacement of switchgear and other substation equipment

When a short-circuit duty is predicted to be in excess of the available rating of existing equipment, it is usually the case that more than one circuit-breaker is found to be potentially overstressed. Depending on the particular technology and design of the circuit-breaker, some uprating or replacement of some components that provides higher short-circuit capability may be possible. Alternatively, the circuit-breakers are replaced with new circuit-breakers with higher short-circuit rating. The costs of such replacements are usually quite high and the potential short-circuit current duty is not reduced.

9.3.6 Wholesale replacement of switchgear and other substation equipment

This is a very expensive and usually last resort design option if the existing substation is still relatively new although the new equipment may be specified to enable sufficient future growth in short-circuit duty. However, consideration of wholesale replacement may be affected by other factors. These include: if the substation equipment is approaching the end of its nominal life, if significant equipment unreliability is identified, environmental or safety issues, installation, and system outage and access requirements.

9.3.7 Use of short-circuit fault current limiters

A fault current limiter is a power system device that, when installed at specific location(s) in the power system, is capable of appreciably reducing the short-circuit current magnitude very quickly following the instant of the short-circuit fault. Some limiters may also interrupt the short-circuit current. Various types of fault current limiters and their modelling requirements are presented in Section 9.4.

9.3.8 Examples

Example 9.2 Using Example 9.1 system and data shown in Figure 9.5, calculate the initial three-phase short-circuit current for a solid 132 kV busbar, but with the reactances of autotransformers now 30% higher.

The Thévenin's equivalent impedance 'seen' at the fault point is given by

$$X_{\text{Thév.}} = \frac{1}{\frac{1}{0.0025 + (0.0834 \times 1.3)/4} + \frac{3}{0.17 + 0.12}} = 0.02266\,\text{pu}$$

The 132 kV busbar three-phase fault level is equal to 100 MVA/0.02266 pu = 4412 MVA and this corresponds to a three-phase fault current of 4412 MVA/ $(\sqrt{3} \times 132\,\text{kV}) = 19.3\,\text{kA}$ which is within the 21.9 kA rating. If the reactance of each autotransformer is 50% higher, the three-phase fault current would be equal to 17.5 kA.

Example 9.3   Again, we will use the system and data of Figure 9.5. In addition, the ZPS system infeed at 400 kV is assumed equal to the PPS infeed. The auto-transformers have unloaded, closed 13 kV delta-connected tertiary windings and their equivalent 400, 132 and 13 kV windings ZPS reactances on 240 MVA base are 19.2%, 0% and 24%, respectively. The generator–transformers' wind-ings are star–delta connected and the ZPS reactance is 11% on 100 MVA. Calculate the single-phase short-circuit fault current at the solid 132 kV busbar under the following conditions:

(a) Normal condition with autotransformer's delta windings closed.
(b) Autotransformers's delta windings are opened. The core construction is 3-limb. The effective equivalent 400 kV, 132 kV and neutral reactances are -4%, 12% and 100% on 100 MVA base, respectively.

Delta windings closed
From Example 9.1, the PPS/NPS (negative phase sequence) Thévenin's equiv-alent impedance 'seen' at the fault point is equal to 0.0188 pu. Also, the ZPS Thévenin's equivalent impedance 'seen' at the fault point is calculated as

$$X_{\text{Thév.}}^{Z} = \frac{1}{\frac{1}{0.0025 + 0.08/4} + \frac{3}{0.11} + \frac{1}{0.1/4}} = 0.00895\,\text{pu}$$

The single-phase fault current is equal to

$$\frac{3}{2 \times 0.0188 + 0.00895} \times \frac{100\,\text{MVA}}{\sqrt{3} \times 132\,\text{kV}} = 28\,\text{kA}$$

As expected, in this example, the single-phase fault current is 20% higher than the three-phase fault current.

Delta windings opened
The opening of the autotransformer delta windings for 3-limb cores would produce changes in the ZPS equivalent reactances of the autotransformers, as discussed in Chapter 4. The Thévenin's PPS/NPS reactance is unchanged and is equal to 0.0188 pu. The ZPS Thévenin's equivalent impedance 'seen' at the

fault point is amenable for hand calculation but requires one simple star-to-delta transformation. It is easily shown that $X_{Th\acute{e}v.}^{Z} = 0.0324\,pu$. The single-phase fault current is equal to

$$\frac{3}{2 \times 0.0188 + 0.0324} \times \frac{100\,MVA}{\sqrt{3} \times 132\,kV} = 18.7\,kA$$

It is interesting to note that although opening the delta windings increases the ZPS Thévenin's impedance by a factor of $0.0324/0.00895 = 3.6$, the single-phase fault current is reduced by 33% which in this example is quite significant.

Where the autotransformers are of 5-limb core or shell-type construction, then as we discussed in Chapter 4, the opening of the delta winding will cause the ZPS shunt neutral impedance to become very large. Values may range from 3000% to 5000% on 100 MVA base. The reader may wish to repeat the calculation of single-phase fault current for a 5-limb or shell-type autotransformer and compare with questions (a) and (b) above.

9.4 Types of short-circuit fault current limiters

9.4.1 Background

There are several types of short-circuit fault current limiters for application in low, medium, high and extra high voltage systems. Some, like the current limiting series reactor, have been used for decades. Others, like superconducting and solid state limiters, are undergoing extensive research and development and some are now being prototyped and are expected to be ready for commercial applications within the next few years.

Besides their main benefits of avoiding the otherwise unnecessary replacement of switchgear, fault current limiters introduce additional benefits caused by the reduction of stresses on equipment during short-circuit faults. A number of concomitant economic benefits include increased power plant life by limiting let-through current, lower thermal, mechanical and electrodynamical stresses of equipment, improved network capacity by allowing several transformers to operate in parallel and possibly lower need for spares. In this section, we briefly describe established and emerging fault current limiter technologies as well as their modelling needs.

9.4.2 Earthing resistor or reactor connected to transformer neutral

An earthing resistor or reactor may be connected to the neutral of a transformer star connected winding in order to limit earth fault currents, i.e. single-phase and two-phase to earth. Usually, neutral earthing reactors are used on autotransformers such as those connecting 400–132 kV networks and 380 kV/365 kV to 110 kV networks. Neutral earthing resistors are usually connected to the neutral of the lower voltage winding of distribution transformers such as on the 11 kV neutral of a star-connected winding of a 132 kV/11 kV transformer.

The modelling of earthing reactors and resistors was covered in Chapter 4. The use of earthing reactors in autotransformers with a tertiary winding is modelled as shown in Figure 4.23 where the effects are to increase the shunt ZPS impedance branch, increase the net ZPS series impedance and move the shunt ZPS impedance towards the high voltage side of the transformer. As a result, the main benefit of this method is to reduce the single-phase to earth short-circuit current during a fault on the low voltage substation side of the transformer. The ohmic value of the earthing reactor for use with existing transformers is usually a compromise between the desire to increase the reactance as much as possible to limit the short-circuit current and the need to limit the voltage at the transformer neutral point to within the insulation level of the winding.

For two-winding and three-winding transformers, the effect of the neutral reactor is to increase the transformer's ZPS leakage impedances as shown in Figures 4.12 and 4.19 of Chapter 4 respectively.

9.4.3 Pyrotechnic-based fault current limiters

These fault current limiters, also known as 'I_S-limiter', are widely used in low and medium voltage systems and industrial power systems at nominal voltages up to 40.5 kV with interrupting currents up to 210 kA symmetrical rms. The I_S-limiter is a device that consists of two parallel conductors as shown in Figure 9.6(a).

The first is the main conductor that carries the load current under normal unfaulted system operating conditions. The second is a parallel fuse with a high breaking capacity that limits the short-circuit current at the first current's rise and interrupts it at the next current zero. Figure 9.6(b) shows the effect of I_S-limiter operation where the two sources of fault currents are characterised by an rms current equal to 31.5 kA and a peak factor equal to 1.8 giving an initial peak asymmetric current of 80 kA. The I_S-limiter is activated by a small charge that opens the main conductor and diverts the short-circuit current to the fuse. The normal instantaneous current through the I_S-limiter is monitored by an electronic measuring and tripping circuit. Both the current and its rate of rise are continuously evaluated and compared to selected set points. If both set points are simultaneously reached, the I_S-limiter trips and the total operating time to current interruption is 5–10 ms. After an operation, the I_S-limiter insert (one unit per phase), i.e. the main conductor, the parallel fuse and the charge need to be replaced. These limiters can be used in a variety of applications such as to couple substation busbars, connection of two separate subsystems, connection in series with generator feeders or in parallel with a current limiting reactor, etc.

Since the current is generally limited and interrupted before the half cycle asymmetric peak is reached, as illustrated in Figure 9.6(b), the I_S-limiter current will not contribute to this initial half cycle peak current. Therefore, the limiter can be

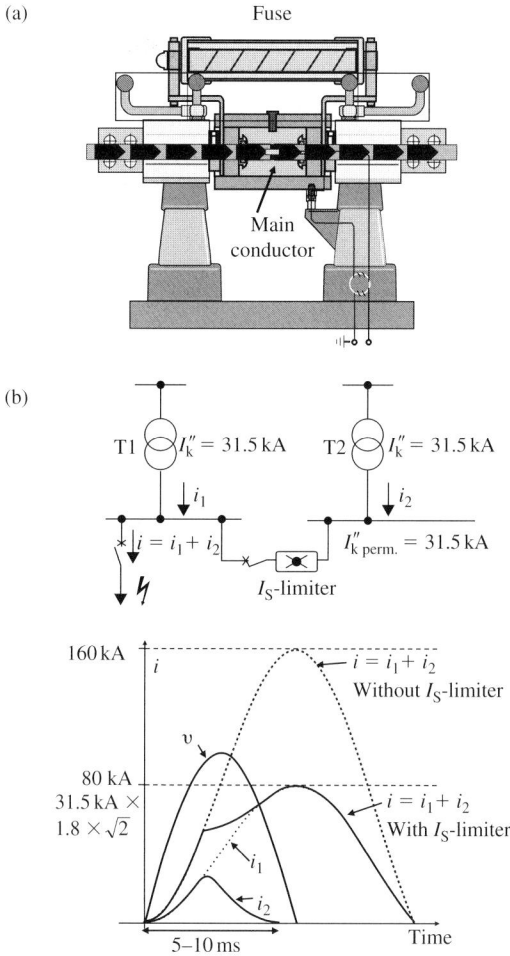

Figure 9.6 The I_S-limiter (*Source*: ABB Germany): (a) insert and insert holder and (b) performance

modelled as a zero admittance (phase, PPS, NPS and ZPS) element in short-circuit analysis studies.

9.4.4 Permanently inserted current limiting series reactor

Short-circuit current limiting series reactors have been used in power systems at almost all voltage levels for decades. They introduce a leakage impedance into the current path which limits the fault current. Because they are permanently inserted in the system, i.e. under normal unfaulted system conditions, they have several disadvantages. These include: they introduce a voltage drop, have active

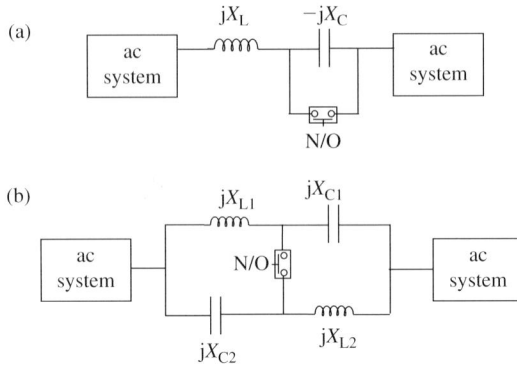

Figure 9.7   Series resonant fault current limiters using a bypass switch: (a) tuned series resonant fault limiter design (i) and (b) tuned series resonant fault limiter design (ii)

and reactive power losses, may adversely affect the optimum distribution of power flow in interconnected networks and may also adversely affect the transient and dynamic stability performance of power systems. Series reactors are modelled using their leakage impedance as discussed in Chapter 4.

9.4.5 Series resonant current limiters using a bypass switch

To eliminate the active and reactive power losses in bus tie applications using single reactor component, an alternative is to use a power frequency tuned resonant series inductor and capacitor combination. The combination has a zero impedance under normal operating conditions. To limit the fault current, the capacitor element is shorted out using a fast closing bypass switch. A variant of this approach is to use a series/parallel arrangement of tuned inductors and capacitors in order to create two parallel tuned elements in series by the operation of a single bridging switch at the circuit mid-point. Both circuits are shown in Figure 9.7. Some of the disadvantages of these circuits are that a number of large components are mounted at line potential and there may be a risk of sub-synchronous resonance which may damage turbo-generator shafts. In the second circuit, if all the inductive and capacitive reactances are chosen to be equal, the impedance switches from zero to almost an infinite value, and while this is very attractive, may be impractical because of its significant adverse effect on downstream protection. Also, the shorting switch may have to close near zero voltage across the capacitor to avoid a very large transient current duty.

9.4.6 Limiters using magnetically coupled circuits

Figure 9.8 shows two configurations of magnetically coupled circuits. Two windings, a primary and a secondary, having equal number of turns are shown. The two

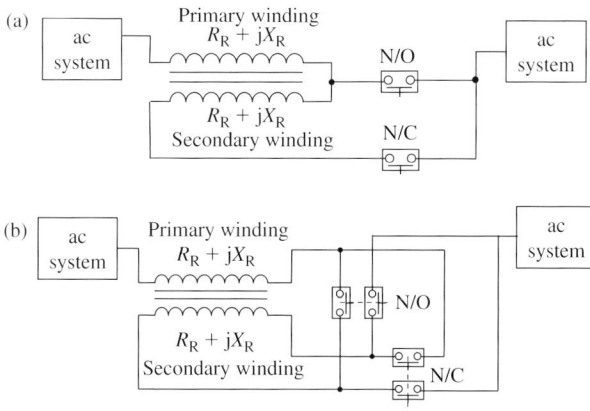

Figure 9.8 Fault current limiters using magnetically coupled circuits: flux cancelling limiter with (a) reverse secondary winding bypass and (b) secondary winding polarity changeover

windings are connected so that the same current flows through them and the primary winding MMF balances that of the secondary winding.

During normal or unfaulted operation, the voltage induced across the self-inductance of the primary winding is cancelled by the voltage induced in the same primary winding due to the current flowing in the secondary winding, and vice versa. Therefore, the net voltage across each winding is zero. Since close mutual coupling between the two windings is required as well as safe clearances where high voltages are involved, an iron core would be required. The device would result in resistive losses. Because under normal unfaulted operation, the secondary MMF cancels out the primary MMF, this limiter is termed a flux cancelling limiter. A sensing and control circuit is required to detect the onset of short-circuit current and switch the device into the limiting mode.

Flux cancelling limiter with reverse secondary winding bypass

This is shown in Figure 9.8(a). When an external short-circuit fault occurs, the coupling between the two windings is broken by immediately closing the normally open switch and opening the normally closed switch. This effectively open circuits the secondary winding and causes the primary winding to appear as a series reactor with an impedance Z_R that acts to limit the short-circuit current. The actions of the switches make the circuit appears as 1:1 transformer with an open-circuited secondary winding. The voltage rating of the normally closed switch is equal to the voltage across the primary winding under short-circuit conditions. This switch only carries the normal load current. The current rating of the normally open switch is equal to the limited short-circuit current and under unfaulted system conditions, the voltage across this switch is zero because it is equal to the voltage across the secondary winding.

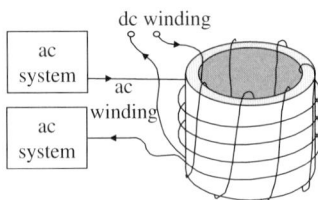

Figure 9.9   Orthogonally wound saturable reactor fault current limiter with dc control winding

Flux cancelling limiter with reverse parallel to series reconnection

This is shown in Figure 9.8(b). When an external short-circuit fault occurs, the direction of the secondary winding is reversed so that the induced voltage adds to the voltage across the self-inductance of each winding, instead of subtracting, and hence doubles the effective impedance in series with the network. The normally open switches carry the limited short-circuit current whereas the normally closed switches carry the load current.

9.4.7 Saturable reactor limiters

Saturable reactors have a non-linear voltage–current characteristic so that when the voltage across the reactor rises above a certain threshold, the reactor current increases disproportionately due to core saturation. As a result, its effective impedance reduces below that without saturation. Figure 9.9 illustrates a saturable reactor with a dc and an ac windings that are wound orthogonally on an iron core.

The amount of saturation is controlled by injecting a variable dc current into the dc control winding. With the exception for the core saturation, the current in the main ac winding is not affected by that in the dc control winding. Because the two windings are orthogonal, the mutual coupling between them is negligible. The device operates in the saturation region under normal unfaulted system conditions and must be quickly taken out of saturation upon the onset of short-circuit current. The increase in impedance offered by this limiter is generally small and may not be sufficient if appreciable current limitation is required. A conventional series iron core saturable reactor may also be used where saturation is obtained by injecting a dc current into the main winding.

9.4.8 Passive damped resonant limiter

Figure 9.10 illustrates one phase of a three-phase damped resonant limiter circuit using only passive components.

The limiter consists of an isolation transformer whose primary winding is connected in series with the system and a capacitor is connected across its secondary

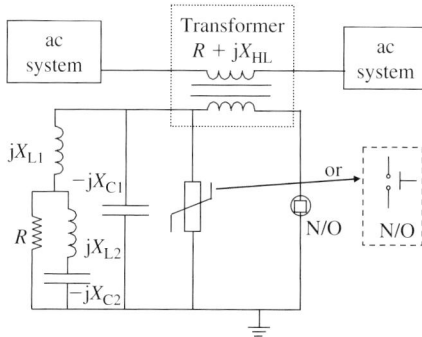

Figure 9.10   Passive damped resonant fault current limiter

winding. A non-linear resistor, e.g. a varistor, or a fast closing triggered switch, is connected in parallel with the capacitor and a damped tuned filter is connected in parallel with the varistor. Under normal unfaulted system condition, the secondary circuit appears as a capacitor at 50 Hz that when transferred to the primary of the transformer, is equal to and hence cancels out the transformer's leakage reactance. Therefore, at 50 Hz, the limiter appears as a short circuit except for the resistance of the transformer. When a short circuit fault occurs in the power system, the increase in current flowing in the transformer's primary winding causes a corresponding increase in the secondary winding's current. The rise in voltage across the secondary winding impedance would cause the varistor or switch to conduct and short circuit the capacitor. Thus, the effective impedance that remains in the primary circuit is the transformer's leakage impedance and this acts to limit the fault current to the desired value. A circuit-breaker in parallel with the varistor or switch may be used to short circuit the secondary and reduce the energy absorption duty on the varistor.

The fault current limiting described can be obtained without the circuitry involving X_{L1}, R, X_{L2} and X_{C2}. However, with series capacitors in transmission systems, sub-synchronous resonance phenomenon has occurred and caused damage to turbo-generator shafts. Thus, the purpose of this circuitry is to provide damping at sub-synchronous frequencies. The circuit consists of a reactance X_{L1}, and a 'C' filter whose X_{L2} and X_{C2} are tuned at power frequency (50 or 60 Hz), so that the damping resistor R is short-circuited at this frequency. X_{L1}, is used because the filter can not be connected across the main capacitor X_{C1} to avoid short-circuiting this capacitor. The value of X_{L1} is chosen so that at power frequency, the parallel combination of X_{L1} and X_{C1} appears as a capacitive reactance whose value is equal to the leakage reactance of the series transformer.

In summary, the limiter has a negligible impedance under normal unfaulted system operation and a fault limiting reactance equal to the transformer's leakage reactance under short-circuit conditions. A damping circuit is included to reduce the risk of sub-synchronous resonance. The modelling of this limiter as a leakage impedance in short-circuit studies is straightforward.

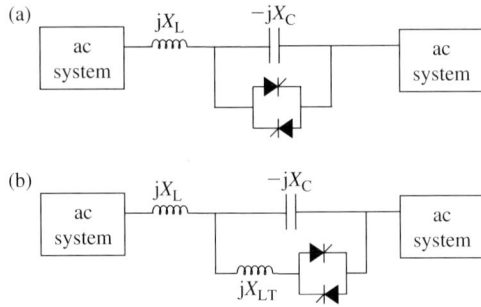

Figure 9.11 Fault current limiters using power electronic switches: (a) thyristor protected series capacitor and (b) thyristor controlled series capacitor

9.4.9 Solid state limiters using power electronic switches

Series resonant limiter using thyristor protected series capacitor

Figure 9.11(a) illustrates this type of limiter which is based on a flexible ac transmission system series compensation device called the thyristor protected series capacitor. In parallel with the capacitor is a power electronic switch that consists of anti-parallel thyristors. Under normal unfaulted system operation, the thyristor switches are not conducting current, and the series inductance and capacitance are tuned at power frequency. Therefore, only the inductor's resistance remains in the circuit and this incurs active power losses. Under system fault conditions, the thyristor switches are made to conduct the fault current and bypass the capacitor thus inserting the reactor impedance into the fault path within a few milliseconds. The reactor acts to limit the fault current.

Series resonant limiter using thyristor controlled series capacitor

The circuit is illustrated in Figure 9.11(b) and is essentially a flexible ac transmission system series compensation device employing thyristor controlled series capacitor used for improving the power transfer capability of the network. Under normal system operating conditions, the thyristor conduction angles are small (firing angles are large), so that most of the current flows through the series capacitor. The effective impedance of the parallel inductor/capacitor combination is capacitive. However, when a short-circuit fault occurs, the firing angle of the thyristors is quickly reduced, so that the parallel inductor/capacitor combination changes to inductive thus limiting the fault current. Fault current limitation is generally an attractive by-product of these controlled series capacitor devices.

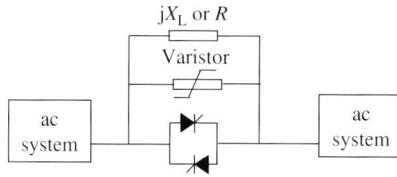

Figure 9.12   Solid state fault current limiter using normally conducting power electronics switches

Solid state limiter using normally conducting power electronics switches

This type of fault limiter is illustrated in Figure 9.12.

It consists of a power electronics switch that is connected in series in the ac system. A fault limiting resistor or reactor is connected in parallel with the switch. Under normal unfaulted system operation, the switch conducts and carries the normal load current and the fault limiting resistor/reactor is short-circuited. When a system short-circuit fault occurs, the rising fault current is detected and the switch is turned off thus diverting the fault current to the parallel resistor/reactor which acts to limit the fault current. The fault current limiting is accomplished within a few milliseconds. The use of a current limiting resistor brings the useful benefit of reducing the phase angle difference between the voltage and current and quickly removing the dc fault current component. A varistor may be connected in parallel with the switch to limit the transient voltage that appears across it. The switches may accomplish the current switch off using gate turn off thyristors or modern alternatives. Because they are normally conducting, the switches incur on-state active power losses and these are typically 0.1–0.2% of the throughput power.

9.4.10 Superconducting fault current limiters

Background

A superconductor is a wire or a coil such as Bismuth-2233 or $YBa_2Cu_3O_7$ that, when cooled, acts as a perfect conductor that has a zero resistance. Beyond a certain critical limit of a particular property, the conductor loses its superconductivity state and transition to a normal high resistive state. The property referred to is the conductor critical temperature, critical current or current density, or magnetic field. Figure 9.13 illustrates the magnetic field and resistance/current, properties of a superconductor. Cryogenic cooling equipment is used to cool the superconductor using liquid nitrogen and for high temperature superconductors, the cooling temperature is less than $-100°C$. The 'high temperature' term refers to material that becomes superconductors at temperatures that are typically above $30°K$. To the author's knowledge, $100°K$ has so far been achieved.

The transition from the superconducting to the normal resistive state occurs automatically within a few milliseconds. No special devices or circuitry is needed to detect and trigger the fault limiting action since the transition in the state material

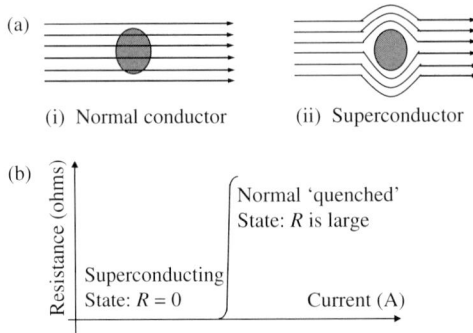

(a)

(i) Normal conductor   (ii) Superconductor

(b)

Normal 'quenched'
State: R is large

Superconducting
State: $R = 0$ Current (A)

Resistance (ohms)

Figure 9.13 Basic electrical property of a superconductor: (a) magnetic flux lines and (b) resistance/current characteristics of a high temperature superconductor

is only dependent on the magnitude of the current flowing through it. In practical installations, recovery of the superconducting property should occur quickly in order to limit and withstand subsequent faults in the power system within a short period of time.

At the time of writing, superconducting fault current limiters remain mainly laboratory prototypes although the resistive type (see below) is nearing commercial use at low distribution voltages. Currently, there are four types of superconducting fault current limiters: (a) resistive superconducting limiter, (b) shielded inductance superconducting limiter, (c) saturated inductance superconducting limiter and (d) air-gap superconducting limiter. These are briefly described in the next section.

Resistive superconducting limiter

This limiter uses a superconductor connected in series in the system. Under normal unfaulted system operation, the limiter is in the superconducting state and has a zero resistance. When a system short-circuit fault occurs, the current or current density in the superconductor increases beyond a certain critical limit. The material loses its superconducting state and effectively becomes a high series resistance which limits the fault current. Figure 9.14 illustrates two connections: a series connection and a shunt connection.

The main disadvantage of the series connection is that fast reclosing operation cannot be achieved in a cost-effective design because recovery of the limiter in readiness for a second limiting operation may take several minutes. This is caused by the heat generated in the conductor in the resistive state which must be dissipated to avoid conductor damage. In the shunt connection, a resistor or a reactor is connected in parallel with the limiter. Under normal unfaulted system operation, the current flows through the superconductor that has a zero resistance. Under faulted system conditions, the resistance of the limiter increases and the fault current is shared between the limiter and the resistor/reactor. If the design enables

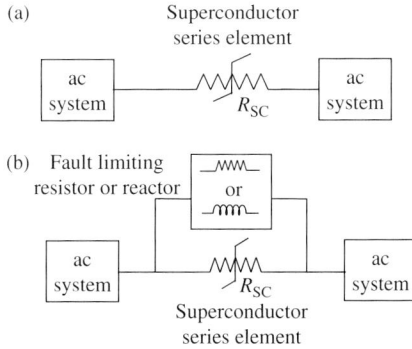

Figure 9.14 (a) Resistive superconducting fault current limiter and (b) resistive superconducting fault current limiter with a parallel current limiting resistor/reactor

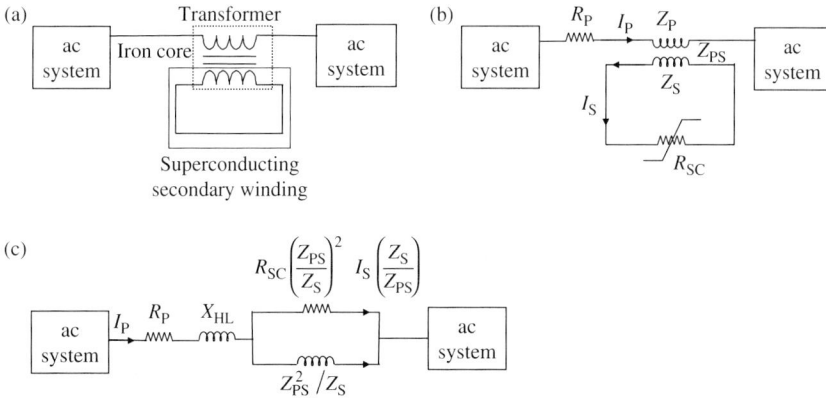

Figure 9.15 Shielded inductance superconducting fault current limiter: (a) circuit schematic, (b) equivalent circuit and (c) equivalent circuit referred to primary side

most of the fault current to flow through the resistor/reactor, multiple operation in a short period of time may be possible.

Shielded inductance superconducting limiter

In this type of limiter, a series transformer whose primary winding is connected in series in the power system is used. The secondary winding of the transformer is a single turn superconductor as illustrated in Figure 9.15(a).

During normal unfaulted system operation, the superconducting secondary winding is effectively short-circuited and its MMF balances that of the primary winding so that almost no flux penetrates the iron core. This introduces the transformer leakage reactance into the primary circuit as illustrated in Figure 9.15(b) and (c). During faulted system operation, the superconductor current increases beyond the critical value and becomes highly resistive. In order to achieve MMF

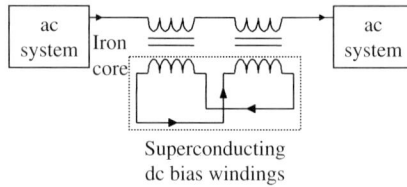

Figure 9.16   Saturated inductance superconducting fault current limiter

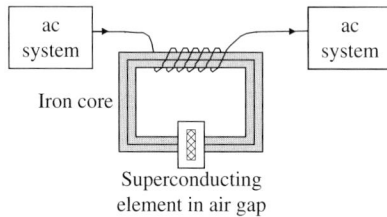

Figure 9.17   Air-gap superconducting current fault limiter

balance, flux penetrates the iron core and the effective impedance appearing in the primary circuit is sharply increased thus limiting the fault current. The mass of iron and copper, cryogenics, civil engineering, overall size and cost can significantly exceed that of the resistive limiter so that practical applications may be unlikely unless new breakthroughs are made.

Saturated inductance superconducting limiter

In this type of limiter, illustrated in Figure 9.16, the line current flows through a series winding around an iron-cored reactor.

During normal unfaulted system operation, the iron of this reactor is held in saturation by a second dc superconducting winding and the effective reactance of the device is quite small. Under faulted system operation, the high fault current drives the core out of saturation and the effective reactance increases to the air core value. Two such devices are connected in series to provide limiting action for both polarities. The use of superconducting dc winding reduces steady state losses and because it remains in the superconducting state during system fault conditions, it enables instantaneous recovery. Another advantage is the improvement in the ratio of faulted to unfaulted impedance. The total mass of iron and copper required, as well as overall size, for a practical three-phase high voltage device are significant.

Air-gap superconducting limiter

This limiter is illustrated in Figure 9.17 and is similar to the shielded inductance limiter except that the secondary winding is replaced by a strip of superconductor inserted into an air gap in the iron core.

During normal unfaulted system operation, the superconductor expels the flux from the gap causing a high reluctance and a low primary inductance. However, during a faulted system condition, the superconductor can no longer support the currents necessary to expel the flux which now exceeds the superconductor critical limit and the magnetic reluctance drops causing the primary inductance to rise. The ratio of faulted to unfaulted limiter impedance may not be significant.

9.4.11 The ideal fault current limiter

The ideal fault current limiter should be invisible to the system in which it is installed except the short-circuit conditions it is designed to operate under. The ideal limiter should have the following characteristics:

(a) It should have a zero impedance and zero active and reactive power losses under normal unfaulted system conditions.
(b) It should detect, discriminate and respond to all types of short-circuit faults in less than 1 or 2 ms.
(c) When it responds, it should insert a very high limiting impedance to limit current flow. A mainly inductive impedance may be preferred in higher voltage networks although resistive limiters have the benefit of quickly suppressing the dc fault current component.
(d) It should automatically and quickly recover once the fault has been removed ready for another current limiting operation.
(e) It should be capable of performing successive current limiting operations without replacement.
(f) It should cause no unacceptable overvoltages or harmonics in the power system.
(g) It should have no adverse impact on power system protection performance.
(h) It should be highly reliable and fail-safe.

In practical installations, such an ideal limiter is not achievable and various design compromises have to be made.

9.4.12 Applications of fault current limiters

There are many different possible applications of fault current limiters in low, medium, high and extra high voltage power networks as well as in industrial power systems. The most efficient, in terms of fault current reduction, and economic method is chosen depending on network and substation specific factors. A brief summary of the main applications is given below.

Figure 9.18 illustrates a case where the solid coupling of the 132 kV (110 kV) busbar is not possible due the short-circuit ratings of connected circuit-breakers being exceeded.

The busbar is split into two sections which are connected by a fault current limiter. The limiter's impedance is chosen in order to reduce the short-circuit infeed from one side of the limiter for a fault on the other side, and vice versa,

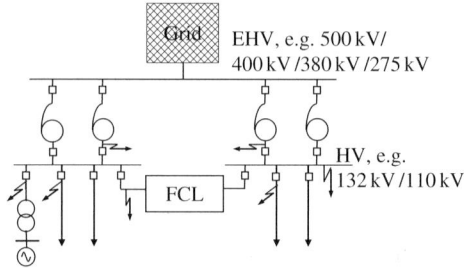

Figure 9.18   Fault current limiter used to couple substation busbars at 132 kV (110 kV)

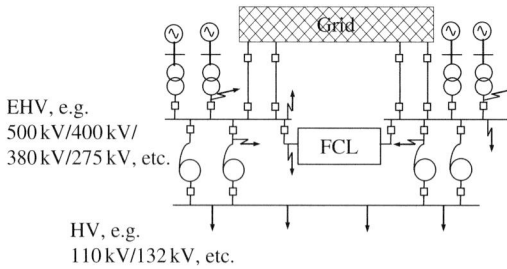

Figure 9.19   Fault current limiter used to couple substation busbars at extra high voltage levels

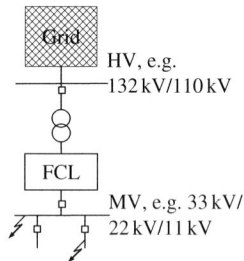

Figure 9.20   Fault current limiter in high-to-medium voltage transformer circuit

to well within switchgear ratings. For sufficiently large limiter impedance, the majority of the fault current on either side of the limiter is supplied through the transformers from the higher voltage network.

Figure 9.19 illustrates an extra high voltage substation with a significant amount of connected generation where the substation cannot be operated solid because the available short-circuit currents will exceed various switchgear and substation infrastructure ratings.

Busbar splitting through a fault current limiter is quite effective in limiting the short-circuit current magnitude to well within switchgear and substation infrastructure ratings.

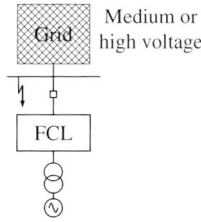

Figure 9.21 Fault current limiter in series with a generator–transformer circuit

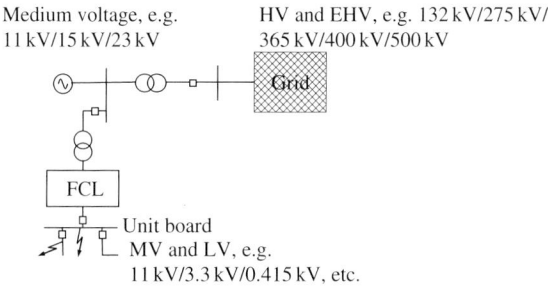

Figure 9.22 Fault current limiter in series with unit transformer supplying power station auxiliaries

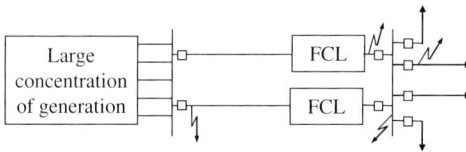

Figure 9.23 Fault current limiter in series with exporting circuits connecting large concentration of generation

Figure 9.20 illustrates a fault current limiter application in series with a transformer in a high-to-medium voltage substation. The limiter acts to limit transfomer fed short-circuit currents for faults on the medium voltage substation.

Figure 9.21 illustrates a fault current limiter application in series with a generator–transformer circuit. The limiter acts to limit the short-circuit current infeed from the generator for faults on the high voltage substation and the generator terminals.

Figure 9.22 illustrates a fault current limiter application in series with a generator–unit transformer that supplies the power station auxiliaries. The limiter acts to limit the short-circuit infeed from the generator and the grid via the unit transformer for faults on the unit board and lower voltages within the auxiliary system.

Figure 9.23 illustrates a fault current limiter application in series with circuit feeders that export power from a large concentration of generation plant located at one or several locations.

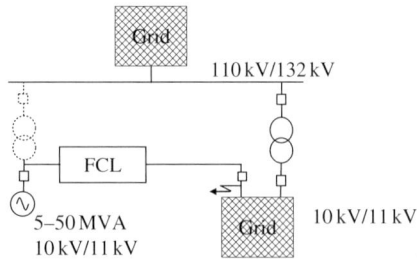

Figure 9.24   Fault current limiter facilitating the connection of a generator to a local grid

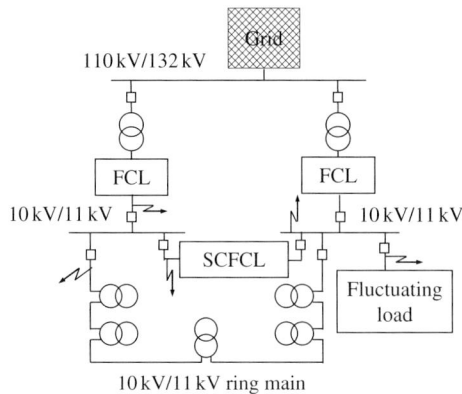

Figure 9.25   Fault current limiters that limit fault currents and improve supply voltage quality

Figure 9.24 illustrates the connection of a generator through a fault current limiter to a medium voltage grid to limit the generator's short-circuit contribution to faults within this grid. In this example, the limiter serves as an alternative to a transformer (and switchgear) connection to the high voltage grid.

Figure 9.25 illustrates fault current limiters in series with low impedance transformers and medium voltage substation busbars coupled via a resistive superconducting fault current limiter.

Operating the medium voltage busbars split reduces the short-circuit fault level at each busbar and reduces the supply voltage quality to customers supplied from the busbar section that supplies a fluctuating load such as an arc furnace load. Coupling the busbar sections through a superconducting fault current limiter, or ideally one that has a very low unfaulted impedance, increases the available short-circuit level at the busbar supplying the fluctuating load and improves the supply voltage quality.

9.4.13 Examples

Example 9.4   Using Example 9.1 system and data, calculate the short-circuit current for a single-phase short-circuit fault at the solid 132 kV busbar if each

400 kV/132 kV autotransformer is equipped with a neutral earthing reactor having a reactance of 10 Ω.

The value of the neutral earthing reactor in per cent on 100 MVA is equal to

$$X_{\text{NER}} = \frac{10\,\Omega}{\frac{(132)^2}{(100)}} \times 100 = 5.739\%$$

The changes in the transformer ZPS T equivalent reactances and the new 400, 132 and 13 kV reactances are calculated as

$$\Delta X_{400\,\text{kV}}^Z = \frac{-3 \times \left(\frac{400}{132} - 1\right)}{\left(\frac{400}{132}\right)^2} \times 5.739\% = -3.8\%$$

thus

$$X_{400\,\text{kV}}^Z = -3.8\% + 8\% = 4.2\%$$

$$\Delta X_{132\,\text{kV}}^Z = \frac{3 \times \left(\frac{400}{132} - 1\right)}{\left(\frac{400}{132}\right)} \times 5.739\% = 11.5\%$$

thus

$$X_{132\,\text{kV}}^Z = 11.5\% + 0\% = 11.5\%$$

$$\Delta X_{13\,\text{kV}}^Z = \frac{3}{\left(\frac{400}{132}\right)} \times 5.739\% = 5.7\%$$

thus

$$X_{13\,\text{kV}}^Z = 5.68\% + 10\% = 15.68\%$$

The ZPS Thévenin's equivalent impedance 'seen' at the 132 kV fault point is calculated using a simple star-to-delta transformation. Thus, $X_{\text{Thév.}}^Z = 0.01877$ pu. It is interesting to compare this with the value obtained in Example 9.3 of 0.00895 pu. The effect of the neutral earthing reactor is to double the ZPS Thévenin's impedance at the 132 kV faulted busbar. It is also worth noting that the opening of the delta-connected tertiary windings is more effective in increasing the ZPS Thévenin's impedance by a factor of $0.0324/0.01877 = 1.72$.

The single-phase fault current is equal to

$$\frac{3}{2 \times 0.0188 + 0.01877} \times \frac{100\,\text{MVA}}{\sqrt{3} \times 132\,\text{kV}} = 23.3\,\text{kA}$$

Example 9.5   Figure 9.26(a) shows a power station having N operational but unloaded identical generators connected to a high voltage busbar through identical transformers. An interbus series reactor short-circuit limiter is used as shown in Figure 9.26(b). Derive a general expression relating the ratio of the three-phase short-circuit current with and without the reactor. Assume that the generators and transformers have equal MVA rating.

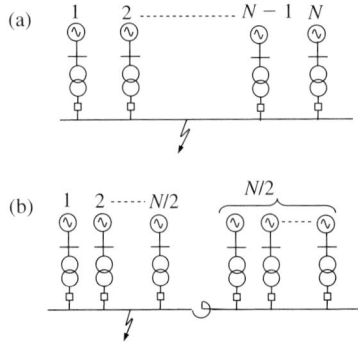

Figure 9.26 Systems for Example 9.5

Let the PPS reactance of each generator and transformer be denoted X_G and X_T in pu on rating and let the series reactor's PPS reactance on the generator's rating be denoted X_R. The three-phase short-circuit current for a fault on the high voltage busbar of Figure 9.26(a) is given by

$$I_F = \frac{N}{X} \quad \text{where } X = X_G + X_T$$

Using a short-circuit limiting series reactor with $N/2$ generators arranged on either side, it can be easily shown that the three-phase short-circuit current for a fault on either side of the high voltage busbars is given by

$$I_{F(R)} = \frac{\frac{4X}{N} + X_R}{\frac{2X}{N}\left(\frac{2X}{N} + X_R\right)}$$

Therefore, using the above two equations, we obtain

$$\frac{I_{F(R)}}{I_F} = \frac{1 + \frac{4}{N}\frac{X}{X_R}}{2 + \frac{4}{N}\frac{X}{X_R}}$$

Clearly, the limiting value is found by setting $N \to \infty$ giving a fault current of 50% of that without the reactor. However, it is more instructive to examine a practical situation where $X_G \approx X_T$ and $X_R \approx X_G$. Thus,

$$\frac{I_{F(R)}}{I_F} = \frac{1 + \frac{8}{N}}{2 + \frac{8}{N}}$$

and for $N = 2, 4, 6, 8$, and 10, $I_{F(R)}/I_F = 0.83, 0.75, 0.7, 0.67$ and 0.64, respectively. Small reductions in fault current occur as N increases and the current reduces asymptotically towards 0.5.

Further reading

Papers

[1] Fernandez, P.C., *et al.*, Brazilian successful experience in the usage of current limiting rectors for short-circuit limitation, *International Conference on Power System Transients* (IPST 05) in Montreal, Canada, 19–23 June 2005, Paper No. IPST05-215.

[2] Parton, K.C., A new power system fault limiter, *Electrical Review International*, Vol. 202, No. 5, February 1978, 63–65.

[3] Salim, K.M., *et al.*, Preliminary experiments on saturated DC reactor type fault current limiter, *Preparation of MT-17*, Geneva, Switzerland, 24–28 September 2001.

[4] Ihara, S., *et al.*, Design options for passive fault current limiter, 1 River Road-Schenectady, NY 12345, USA.

[5] Tleis, N., *et al.*, Design of a 400 kV passive fault current limiter, *International Conference on Switchgear and Substations*, Tokyo, Japan, 2000.

[6] Putrus, G.A., *et al.*, *A Static Fault Current Limiter and Circuit Breaker Employing GTO Thyristor* School of Engineering, University of Northmbria, Newcastle.

[7] Weller, R.A., *et al.*, *Computer Modelling of Superconductivity*, Cambridge, UKCB3 0HE.

[8] Raju, B.P., *et al.*, Fault current limiting reactor with superconducting DC bias winding, *International Conference on Large High Voltage Electric Systems*, 1–9 September 1982 session, Paris.

[9] Power, A.J., An overview of transmission fault current limiters, *IEE Colloquium on Fault Current Limitus – A Look at Tomorrow*, Digest, 1995/026, IEE, London, 1995.

[10] *IEE Colloquium on Fault Current Limiters, a Look at Tomorrow*, Digest, 1995/026, IEE, London, 1995.

[11] Gubser, D.U. Superconductivity: an emerging power-dense energy-efficient technology, *IEEE Transactions on Applied Superconductivity*, Vol. 14, No. 4, December 2004, 71–75.

An introduction to the analysis of short-circuit earth return current, rise of earth potential and electrical interference

10.1 Background

We outlined in Chapter 1, one area where short-circuit current calculations are made to calculate rise of earth potential at substations and at overhead transmission line towers. During an unbalanced single-phase to earth or two-phase to earth short-circuit fault, either the total earth fault current or a proportion of it will flow through the general mass of the earth and return to the source. The earth current will return to the neutral point(s) of the supply transformers or sources supplying the fault current through a substation earth electrode system that has a finite impedance. The product of the earth return current and the substation earthing impedance causes a rise in the potential of the substation earthing system and of the earth in the vicinity of the substation with respect to the potential of the general mass of the earth. Excessive rise of earth potential may be dangerous to people in the vicinity or may even cause damage to plant.

The flow of current in earth through a substation earthing system will produce potential gradients or contours within and around the substation. However, the detailed analysis of these contours using electromagnetic field theory is outside the scope of this book. Here, we present the calculation of rise of earth potential only at the location where the current enters earth.

The calculation of rise of earth potential requires the calculation of the earth return current or the proportion of earth fault current that returns through earth. We will present methods for the calculation of such currents that are generally accurate and sufficient in the majority of practical applications except where several long cable circuits emanate from the substation. In these cases, detailed multiphase modelling of connected equipment is recommended.

At substations, an earthing conductor is used to connect to earth exposed metal surfaces of various substation equipment. Because of its huge volume, the overall body of the earth is used as a reference, i.e. it has zero voltage. Earthing conductors do not normally carry current and should have a zero voltage, i.e. their voltage is identical to that of remote earth except under short-circuit conditions. An earth electrode is a part of the earthing system that is in contact with the earth and may comprise copper conductor, cast iron plates or steel piles, etc. Earthing of neutral points in a power system is made by the connection to the earthing electrode system of: substation equipment, metalwork, cladding, overhead line earth wires, cable sheaths and armouring terminating at the substation. In North America, the term grounding is used and is analogous to earthing.

10.2 Electric shock and tolerance of the human body to ac currents

10.2.1 Step, touch, mesh and transferred potentials

Electric shock may occur when a person touches an earthed structure during a short-circuit fault or walks within the vicinity of an earthing system during a fault or touches two separately earthed structures. Figure 10.1 illustrates a person in the vicinity of an ac substation subject to step, touch, mesh and transferred potentials. Step potential is the surface potential difference between the 2 ft of a person that are assumed to be 1 m apart. Touch potential is the potential difference between the hands and feet of a person when they are standing 1 m away from an earthed

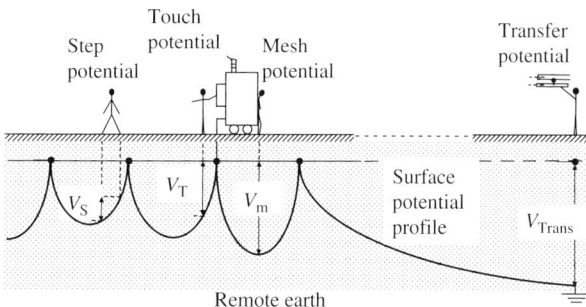

Figure 10.1   Illustration of persons subject to step, touch, mesh and transferred potentials

object they are touching with their hands. Mesh potential is the maximum touch potential to be found within a mesh of the substation earthing system. Transferred potential is a case of a touch potential in a remote location where the shock voltage may be approaching the full rise of earth potential of an earth electrode.

10.2.2 Electrical resistance of the human body

The human body presents a resistance (impedance) to the flow of ac power frequency current which decreases non-linearly with the applied touch voltage. Various parts of the human body such as skin, blood, muscles, tissues and joints, tend to oppose current flow. Broadly speaking, the human body presents an impedance that consists of two parts. The internal impedance which is mainly resistive, and the impedance of the skin at the point of contact which consists of a resistance and capacitance in parallel. Above about 200 V touch voltage, the dc resistance and ac impedance of the human body are generally equal.

Figure 10.2 shows a simplified representation of the resistances of the human body where the following equations can be written

$$R_{\text{Hand–Hand}} = 2R_{\text{Arm}} \quad R_{\text{Foot–Foot}} = 2R_{\text{Leg}} \quad R_{\text{Hand–Foot}} = R_{\text{Arm}} + R_{\text{Leg}} \quad (10.1)$$

Measurements of the hand-to-hand and hand-to-foot body resistances suggest that the arm and leg resistances are related by $R_{\text{Leg}} \approx 0.6 \times R_{\text{Arm}}$. Thus,

$$R_{\text{Hand–Foot}} = 0.8 \times R_{\text{Hand–Hand}} \quad (10.2)$$

$$R_{\text{One hand–Both feet}} = 0.65 \times R_{\text{Hand–Hand}} \quad (10.3)$$

and if both hands and both feet are in parallel

$$R_{\text{Both hands–Both feet}} = 0.4 \times R_{\text{Hand–Hand}} \quad (10.4)$$

and

$$R_{\text{Both hands–Body centre}} = 0.25 \times R_{\text{Hand–Hand}} \quad (10.5)$$

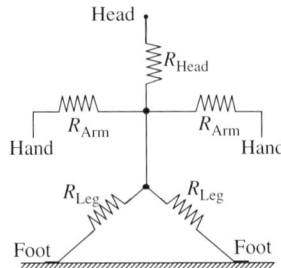

Figure 10.2 Representation of the electrical resistances of the human body

The above equations indicate that the effective resistance of interest is between extremities, e.g. from one hand to the other, from one hand to both feet or from one foot to the other. The resistance of internal body tissues is around $300\,\Omega$ excluding skin resistance. Including skin, the total resistance from hand to hand is typically between 600 and $3300\,\Omega$. The total body resistance reduces with increasing touch voltage because of the considerable reduction in the skin resistance part due to damage or puncture. For example, for $R_{\text{Hand–Hand}} = 1725\,\Omega$ at a 100 V touch voltage, $R_{\text{Arm}} = 862.5\,\Omega$, $R_{\text{Leg}} = 517.5\,\Omega$, $R_{\text{Hand–Foot}} = 1380\,\Omega$, $R_{\text{One hand–Both feet}} = 1121\,\Omega$, $R_{\text{Both hands–Both feet}} = 690\,\Omega$, $R_{\text{Both hands–Body centre}} = 431\,\Omega$ and $R_{\text{Foot–Foot}} = 1035\,\Omega$. The most dangerous currents in terms of ventricular fibrillation of the heart (this is discussed in the next section) are those that flow from left hand to one or both feet. For other paths through the body, IEC 60479-1:2005 provide a rough estimation of the current magnitude required to present an equivalent danger of ventricular fibrillation. For example, for a left foot to a right foot path, the magnitude of the current required is 25 times higher. According to American IEEE Standard 80:2000, the total body resistance from one hand to both feet, hand to hand and foot to foot is generally taken equal to $1000\,\Omega$.

10.2.3 Effects of ac current on the human body

The effects of an ac electric current passing through the vital parts of a human body depend on magnitude, frequency and duration of this current. The most common physiological effects of electric current on the human body, in order of increasing current magnitude, are perception, muscular contraction, unconsciousness, ventricular fibrillation, respiratory nerve blockage and burning. Accordingly, a number of touch current thresholds are defined such as perception, reaction or muscular contraction, let-go and ventricular fibrillation. The threshold of reaction is usually taken as 0.5 mA independent of duration. The threshold of let-go current is the maximum touch current at which a person can let-go of the electrodes they are holding. This threshold is assumed in IEC 60479-1:2005 to be equal to 10 mA for adult males and 5 mA for all population of adult males, females and children. Ventricular fibrillation is a condition of incoordinate action of the main chambers (ventricles) of the heart resulting in immediate arrest of blood circulation. Humans are vulnerable to the effects of power frequency currents and values as low as tens of mA can be lethal. In comparison with power frequency current, it is generally accepted that the human body can tolerate a slightly higher current at 25 Hz and approximately five times larger current at zero frequency i.e. dc current. Fortunately, the human body may tolerate hundreds of amperes in the case of lightning surges because of their very short microseconds duration. The threshold of perception is generally taken as 1 mA. Currents from 0.5 to 5 mA, usually termed let-go currents, do not impair the ability of a person holding an energised object to control his muscles and release it. From 5 mA with a duration of 7 s up to 200 mA with a duration at 10 ms, currents are quite painful. They can result in strong involuntary muscular contraction, difficulty in breathing and immobilisation but do not usually cause organic damage and the effects are temporary and would disappear if the current is interrupted. It is generally not until a current magnitude of around 100 mA that ventricular fibrillation, stoppage of the heart or inhibition of respiration might occur and cause injury or death. In such cases, resuscitation does not work and the only known remedy is controlled electric shock. Industrial practice usually attempts to ensure that shock currents are kept below the ventricular fibrillation threshold in order to prevent death or injury. According to American IEEE Standard 80:2000 that provides guidance on safe earthing practices in ac substations, the non-fibrillating 50–60 Hz current is related to the energy absorbed by the human body and is given by

$$I_B = \frac{0.116}{\sqrt{t_s}} \text{ A} \quad 0.03 \text{ s} \leq t_s \leq 3 \text{ s} \tag{10.6}$$

where I_B is the tolerable body current and t_s is the duration of shock current in seconds. The constant 0.116 is based on a body weight of 50 kg and a probability of ventricular fibrillation of 0.5% in a large population of adult males. Equation (10.6) gives tolerable body currents of 367 and 116 mA for 0.1 and 1 s, respectively. Based on the permissible body current, American IEEE Standard 80:2000 provides criteria for the permissible touch and step voltages for a 50 kg person as follows:

$$V_{\text{Step 50 kg}} = (1000 + 6C_S \times \rho_S)I_B \text{ V} \tag{10.7}$$

$$V_{\text{Touch 50 kg}} = (1000 + 1.5C_S \times \rho_S)I_B \text{ V} \tag{10.8}$$

where ρ_S is the resistivity of earth surface material in Ωm, I_B is the permissible shock current given in Equation (10.6), the number 1000 is the body resistance in Ω and C_S is a derating factor of the earth's surface layer given as

$$C_S = 1 + \frac{8}{\pi r^2} \sum_{n=1}^{\infty} K^n \int_0^r \sin^{-1}\left(\frac{2r}{R_1 + R_2}\right) x \, dx \text{ m}^{-1} \tag{10.9a}$$

and

$$R_1 = \sqrt{(r+x)^2 - (2nh_S)^2} \quad R_2 = \sqrt{(r+x)^2 + (2nh_S)^2} \tag{10.9b}$$

and $K = (\rho - \rho_S)/(\rho + \rho_S)$ is the reflection coefficient between the surface layer having a resistivity of ρ_S and the soil beneath having a resistivity of ρ, r, in m, is the equivalent radius of the foot that is assumed a conducting plate on the earth's surface and usually taken equal to 0.08 m. h_S is the thickness of the surface layer in m.

10.3 Substation earth electrode system

10.3.1 Functions of substation earth electrode system

Substation earth electrode systems provide a means to carry and dissipate electric currents into earth under both normal and short-circuit conditions. Also, they are designed to provide a degree of safety for persons working or walking in the vicinity of earthed equipment so that they are not exposed to the danger of critical electric shock. In providing an earthing point for various equipment associated with the substation, the substation earth electrode system must fulfil the following requirements:

(a) The resistance of the earth electrode system to a remote earth must be sufficiently low to ensure operation of protection relays for earth faults at the substation and along lines and cables connected to it.
(b) The earth potential gradient within and near to the substation should be such that under earth faults, 'step' and 'touch' voltages are limited to safe levels.
(c) The earth electrode system must be isolated from services entering the substation so that any rise in substation earth voltage, which can be kilovolts at times, is not 'transferred' to telephones, water mains, railway sidings, etc.
(d) The earth electrode system should be such that non-current carrying parts of electrical equipment, e.g. sheaths and armouring of low voltage power and control cables, do not suffer heavy fault currents.
(e) The earth electrode system should be capable of carrying the maximum earth fault currents without overheating, mechanical damage or unduly drying out the soil around buried earth electrodes and conductors.

10.3.2 Equivalent resistance to remote earth

General

In electrical power systems, various shapes of buried earth electrode systems are used and approximate formulae are available to enable the calculation of the equivalent resistance to remote earth of common types of electrode systems. At power frequency, the inductive component of the impedance of the earth electrode system is very small and is usually neglected. The earth electrode system is designed to have a resistance, so that the voltage rise with respect to remote earth is limited to a certain value. For example, if the maximum earth fault current flowing through this resistance is $10\,\text{kA}$ and the earth electrode potential rise is to be limited to $1\,\text{kV}$, the earth electrode should be designed to have a resistance of less than or equal to $0.1\,\Omega$. Approximate formulae for the equivalent resistance to remote earth of common types of electrode systems are given below.

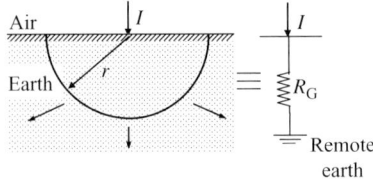

Figure 10.3 Buried hemisphere earth electrode

Buried hemisphere

The equivalent resistance to remote earth of a buried hemisphere, shown in Figure 10.3, is given by

$$R_G = \frac{\rho_e}{2\pi r} \; \Omega \qquad (10.10a)$$

where ρ_e is average earth resistivity in Ωm and r is hemisphere radius. The arrows represent the diffusion of current through the earth. For example, for $\rho_e = 100 \, \Omega$m and $r = 1$ m, $R_G \approx 15.9 \, \Omega$.

If the sphere is buried at a depth of h in m which is large compared to twice its radius, the resistance to earth is given by

$$R_G = \frac{\rho_e}{4\pi} \left(\frac{1}{r} + \frac{1}{2h} \right) \Omega \qquad (10.10b)$$

One driven vertical rod

Driven earth rods are generally solid and circular and are effective in small area substations or where low soil resistivity strata, into which the rod can penetrate, lies beneath a layer of high soil resistivity. The resistance to earth of a single isolated earth rod driven vertically into earth, shown in Figure 10.4, is given by

$$R_G = \frac{\rho_e}{2\pi L} \left[\log_e \left(\frac{4L}{r} \right) - 1 \right] \Omega \quad L \gg r \qquad (10.11)$$

where L and r are the length and radius of the rod in m, respectively. For example, for a driven rod in an 11 kV substation with $L = 3$ m, $r = 0.01$ m and $\rho_e = 100 \, \Omega$m,

$$R_G = \frac{100}{2\pi \times 3} \left[\log_e \left(\frac{4 \times 3}{0.01} \right) - 1 \right] = 32.3 \, \Omega$$

Where the vertical electrode is surrounded by an infill material such as semi-conductive concrete or cement, the earth resistance is given approximately by

$$R_G = \frac{1}{2\pi L} \left\{ \rho_e \left[\log_e \left(\frac{4L}{r_i} \right) - 1 \right] + \rho_i \log_e \left(\frac{r_i}{r_{rod}} \right) \right\} \Omega \qquad (10.12)$$

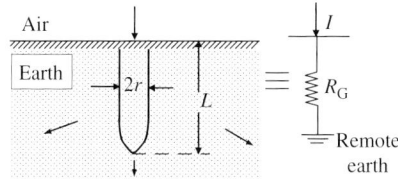

Figure 10.4   An isolated driven vertical rod earth electrode

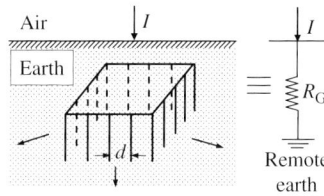

Figure 10.5   Multiple driven vertical rods in a hollow square earth electrode

where ρ_i is resistivity of infill material in Ωm, r_i is radius of shell of infill material in m and r_{rod} is the radius of the rod in m. The resistivity of concrete infill material lie in the range of 30–90 Ωm.

Multiple driven vertical rods in a hollow square

Figure 10.5 shows multiple driven vertical rods in parallel arranged in a hollow square with adjacent rods being equally spaced. The equivalent resistance to earth of the combined rods is given by

$$R_{G(N)} = \frac{1}{N}\left[R_G + \frac{\rho_e}{2\pi d} \times \lambda(N)\right] \Omega \qquad (10.13)$$

where R_G is the resistance of a single isolated rod given in Equation (10.11), N is the number of rods along each side of the square, d is the distance between adjacent rods in m and $\lambda(N)$ is a factor that depends on the number of rods along each side of the hollow square. British Standard BS EN 7430:1998 Table 3 gives the value of λ as a function of N where $\lambda = 2.71$ for $N = 2$, $\lambda = 7.9$ for $N = 10$ and $\lambda = 9.4$ for $N = 20$. For example, using the previous example rod data, $N = 20$ and $d = 10$ m, we obtain

$$R_{G(N)} = \frac{1}{20}\left[32.3 + \frac{100}{2\pi \times 10} \times 9.4\right] = 2.36 \, \Omega$$

Buried horizontal strip or wire

Horizontal strip or round wire conductor electrodes are very useful where high resistivity earth layer underlies shallow surface layers of low resistivity. The resistance to earth of a single strip or round wire conductor, shown in Figure 10.6,

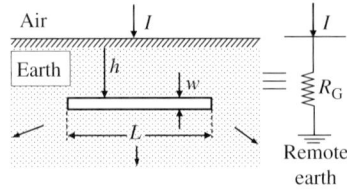

Figure 10.6   Buried horizontal strip or round wire earth electrode

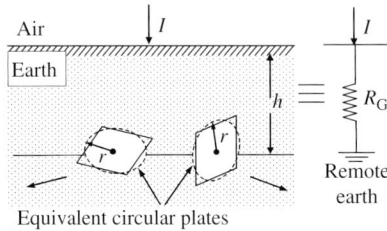

Equivalent circular plates

Figure 10.7   Buried horizontal or vertical flat plate earth electrode

is given by

$$R_G = \frac{\rho_e}{2\pi L}\left[\log_e\left(\frac{2L^2}{wh}\right) - k\right] \Omega \tag{10.14}$$

where $k = 1$ for a strip conductor and $k = 1.3$ for a round wire conductor, L is length of strip or wire in m, h is depth in m and w is width of strip or diameter of round wire in m. For example, for a round wire conductor with $L = 50$ m, $h = 0.5$ m, $w = 0.02$ m and $\rho_e = 100\ \Omega$m, we obtain

$$R_G = \frac{100}{2\pi \times 50}\left[\log_e\left(\frac{2 \times (50)^2}{0.02 \times 0.5}\right) - 1.3\right] = 3.76\ \Omega$$

Buried vertical or horizontal flat plate

Vertical or horizontal flat plate electrodes have large surface area and are used in soils where it is difficult to drive rods or where soil resistivity is very high. The resistance to earth of a buried vertical or horizontal flat plate, shown in Figure 10.7, is given by

$$R_G = \frac{\rho_e}{8r}\left(1 + \frac{r}{2.5h + r}\right) \Omega \tag{10.15a}$$

where h is depth to plate centre in m, r is the radius of an equivalent circular plate in m and is given by

$$r = \sqrt{\frac{A}{\pi}}\ \text{m} \tag{10.15b}$$

where A is area of plate in m^2.

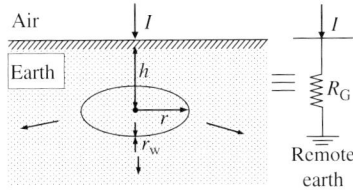

Figure 10.8   Buried horizontal ring of wire earth electrode

For example, for a 1.1 m × 1.1 m plate buried at a depth of 1 m in a soil having

$$\rho_e = 50\,\Omega m \quad r = \sqrt{\frac{1.1 \times 1.1}{\pi}} = 0.62\,\text{m}$$

$$R_G = \frac{50}{8 \times 0.62}\left(1 + \frac{0.62}{2.5 \times 1 + 0.62}\right) = 12.1\,\Omega$$

If the plate is placed on the surface of the earth, $h = 0$ and $R_G = 50/(4 \times 0.62) = 20.2\,\Omega$.

Buried horizontal ring of wire

The resistance to earth of a buried horizontal ring, shown in Figure 10.8, is given by

$$R_G = \frac{\rho_e}{4\pi^2 r}\log_e\left(\frac{32r^2}{hr_w}\right)\Omega \tag{10.16}$$

where r is radius of ring in m, h is depth in m and r_w is radius of wire in m. For example, for $\rho_e = 100\,\Omega m$, $r = 0.5$ m, $h = 0.5$ m and $r_w = 0.01$ m,

$$R_G = \frac{100}{4\pi^2 \times 0.5}\log_e\left(\frac{32 \times 0.5^2}{0.5 \times 0.01}\right) = 37.4\,\Omega$$

Buried horizontal grid or mesh

Figure 10.9 illustrates a horizontal grid or mesh system where R_G represents the equivalent resistance to remote earth and R_m represents the local conductor resistance within the mesh.

The mesh is generally buried at a depth of 0.3–1 m. The resistance of the metal conductors forming the mesh is generally negligible compared to the resistance of the volume of earth in which the mesh is buried. The resistance to earth of a buried grid or mesh, is given by

$$R_{G(\text{mesh})} = \rho_e\left(\frac{1}{4r} + \frac{1}{\sum \ell}\right)\Omega \tag{10.17}$$

where $r = \sqrt{A/\pi}$ in m and A is area of mesh in m². $\sum \ell$ is sum of lengths of buried conductors. For example, for a large 400 kV/132 kV substation with

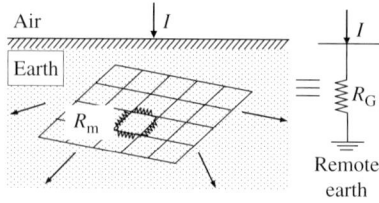

Figure 10.9   Buried horizontal grid or mesh earth electrode

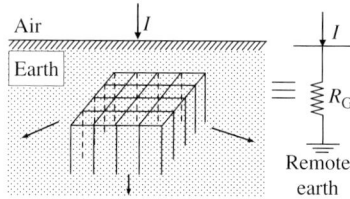

Figure 10.10   Combined horizontal mesh and driven vertical rods around periphery earth electrode

a 171 m × 171 m square mesh that consists of 20×20 conductor rods at 9 m intervals buried in a soil having $\rho_e = 100\,\Omega m$, $r = \sqrt{(171 \times 171)/\pi} = 96.5$ m, $\ell = (20 \times 171) \times 2 = 6840$ m and

$$R_{G(mesh)} = 100 \times \left(\frac{1}{4 \times 96.5} + \frac{1}{6840} \right) \approx 0.274\,\Omega$$

Combined horizontal mesh with driven vertical rods around periphery

Using Figure 10.10, the total equivalent resistance of an earthing system that consists of a combination of horizontal mesh and vertical rods electrodes is given by

$$R_{Eq} = \frac{R_{G(mesh)}R_{G(N)} - R_{Mut}^2}{R_{G(mesh)} + R_{G(N)} - 2R_{Mut}}\,\Omega \tag{10.18}$$

where $R_{G(mesh)}$ and $R_{G(N)}$ are as given in Equations (10.17) and (10.13), respectively, R_{Mut} is a mutual resistance and is given by

$$R_{Mut} = R_{G(mesh)} - \frac{\rho_e}{\pi \ell}\left[\log_e \left(\frac{\pi L}{w} \right) - 1 \right]\,\Omega \tag{10.19}$$

where L is length of driven vertical rod in m and w is width of strap in m. For example, from the previous mesh and multiple vertical rods, $R_{G(mesh)} = 0.274\,\Omega$ and $R_{G(N)} = 2.36\,\Omega$. For $w = 0.05$ m, the mutual resistance is

$$R_{Mut} = 0.274 - \frac{100}{\pi \times 6840}\left[\log_e \left(\frac{\pi \times 3}{0.05} \right) - 1 \right] = 0.274 - 0.0197 = 0.254\,\Omega$$

Therefore, the total equivalent resistance is

$$R_{Eq} = \frac{0.274 \times 2.36 - 0.254^2}{0.274 + 2.36 - 2 \times 0.254} = 0.2738 \, \Omega$$

In this example, R_{Eq} is practically equal to $R_{G(mesh)}$. Thus, the buried horizontal mesh is by far the most effective earth electrode and the additional vertical rods have introduced negligible improvement.

10.4 Overhead line earthing network

10.4.1 Overhead line earth wire and towers earthing network

Most high voltage lines and generally all extra high voltage lines have one or more earth wires. The earth wire is usually bonded to the tower top and connected to the earth meshes of substations where the line terminates. Terminal towers of lattice steel structures and poles of metallic construction are usually bonded to the substation earth electrode system. The line's earth wire, as seen from the sub-station earthing system, usefully serves to extend this earthing system beyond the substation due to the tower connections to earth via the tower footing resistances. Figure 10.11 illustrates a single-circuit overhead line with a single earth wire where Z_S is the self-series impedance of the earth wire per span with earth return. A span is the distance between two adjacent towers and Z_S can be calculated using Equation (3.19a). R_T is the tower earthing or footing resistance with the impedance of the tower itself being negligible at power frequency. For homogeneous earth resistivity ρ_e and with only earth wires present, i.e. no counterpoise conductor, the tower earthing or footing resistance R_T is given by

$$R_T = \frac{\rho_e}{2\pi r} \, \Omega \tag{10.20}$$

where r is the radius of an equivalent hemispherical electrode having the same earth resistance as the tower footing. This radius is in the order of 1–2 m for smaller towers and may be 6–10 m for large towers with multiple legs. For example, for a typical 275 kV tower used in England and Wales having an equivalent radius of 2.8 m, the tower footing resistance is equal to 5.7 Ω for a 100 Ωm earth resistivity.

10.4.2 Equivalent earthing network impedance of an infinite overhead line

In order to calculate the equivalent impedance of the combination of the earth wire and tower footing resistances, the line or number of spans is assumed infinite. Using Figure 10.11, the earth wire per span and tower footing resistance can be

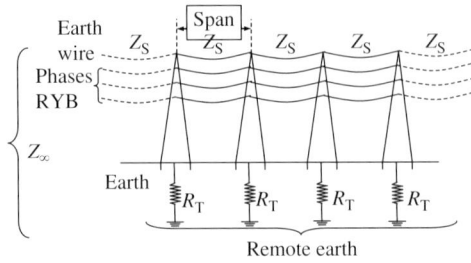

Figure 10.11 Illustration of overhead line earth wire and tower footing resistance network of a single-circuit overhead line

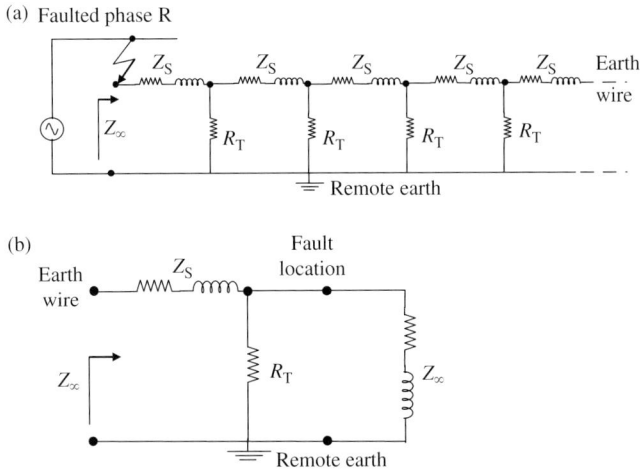

Figure 10.12 Equivalent earthing impedance of the infinitely long line shown in Figure 10.11: (a) ladder network representation of the earth wire and tower footing resistances and (b) derivation of equivalent impedance of a ladder network

represented as a ladder network that consists of an infinite number of series–parallel elements as shown in Figure 10.12(a).

Along the entire line length, the span lengths are assumed identical and the footing resistances of each tower are assumed equal. The tower at the fault location is not included in the equivalent circuit of the earth wire/tower footing. Assume that the driving point impedance of this ladder network as seen from the fault location is Z_∞. Since the line is assumed infinitely long, the addition of one more series–parallel element does not change Z_∞. This is illustrated in Figure 10.12(b) where

$$Z_\infty = Z_S + \frac{Z_\infty R_T}{Z_\infty + R_T}$$

Solving this equation for Z_∞, the practical solution is given by

$$Z_\infty = \frac{Z_S}{2} + \sqrt{\frac{Z_S^2}{4} + Z_S R_T} = \frac{Z_S}{2}\left(1 + \sqrt{1 + \frac{4R_T}{Z_S}}\right) \Omega \qquad (10.21)$$

It is interesting to note that if $Z_S = R_T$ then $Z_\infty/R_T = 1.6180$ a number known since medieval times and related to the Fibonacci numbers which were first published in the *Liber Abaci* in 1202 AD. Equation (10.21) shows that the driving point impedance of this ladder network depends on the earth wire characteristics, e.g. material and size, tower footing resistance and earth resistivity along the line route. For example, consider a 132 kV single-circuit overhead line with a single aluminium conductor steel reinforced (ACSR) earth wire that has three spans per km and a self-impedance of $(0.18 + j0.72)\Omega$/km. The tower footing resistance $R_T = 10\,\Omega$. The driving point or ladder equivalent impedance is equal to

$$Z_\infty = \frac{0.06 + j0.24}{2}\left(1 + \sqrt{1 + \frac{4 \times 10}{0.06 + j0.24}}\right) = (1.265 + j1.09)\Omega$$

$$= 1.67\,\Omega\angle40.75°$$

In practice, the line length beyond which the driving point impedance remains nearly constant is not infinite. This is discussed in the next section.

10.5 Analysis of earth fault ZPS current distribution in overhead line earth wire, towers and in earth

In this analysis, we use the single-circuit overhead line of Figure 10.11 and consider a one-phase to earth short-circuit fault at the terminal tower, numbered 0, as shown in Figure 10.13(a). We recall that Z_S is the earth wire impedance in Ω/span.

The fault current I_F will divide between the terminal tower and the first span of the earth wire. The latter current will then flow into tower 1 and second span, and so on. Writing Kirchoff's current law for towers or nodes n and $(n-1)$, we have

$$I_n = i_n - i_{n+1} \qquad (10.22a)$$

and

$$I_{n-1} = i_{n-1} - i_n \qquad (10.22b)$$

Writing Kirchoff's voltage law for the nth earth wire span, we have

$$i_n Z_S - I_F Z_m + I_n R_T - I_{n-1} R_T = 0 \qquad (10.23)$$

In Section 10.9, we will introduce the concept of a screening factor k for an overhead line with an earth wire but for now we will use the formula

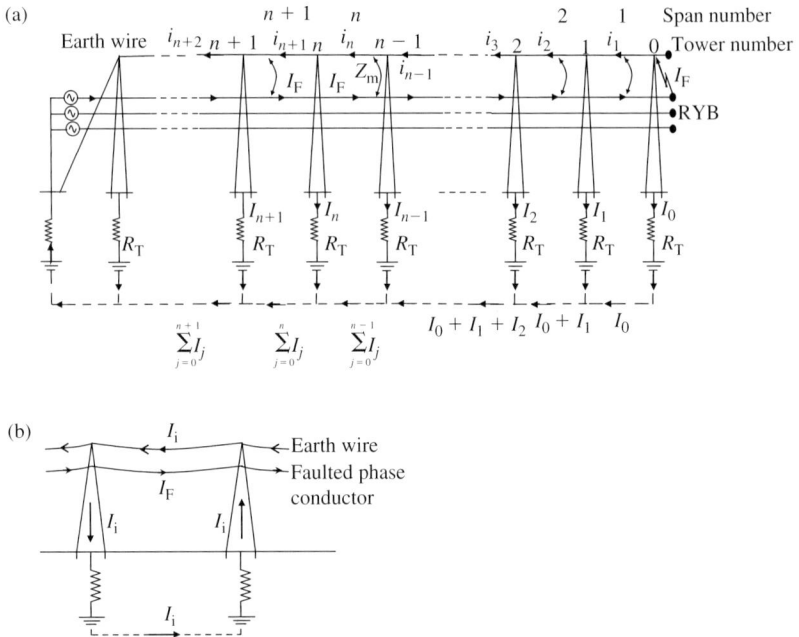

Figure 10.13 Earth fault current distribution in earth wire, towers and earth of single-circuit overhead line: (a) earth fault current distribution in earth wire, towers and earth under a one-phase to earth short-circuit fault and (b) induced current circulating between towers

$Z_m = (1 - k)Z_S$ in Equation (10.23) to obtain

$$i_n = \frac{R_T}{Z_S}(I_{n-1} - I_n) + (1 - k)I_F = I_{C(n)} + I_i \qquad (10.24a)$$

Similarly for the $(n + 1)$th earth wire span, we can write

$$i_{n+1} = \frac{R_T}{Z_S}(I_n - I_{n+1}) + (1 - k)I_F = I_{C(n+1)} + I_i \qquad (10.24b)$$

where

$$I_i = (1 - k)I_F \quad I_{C(n)} = \frac{R_T}{Z_S}(I_{n-1} - I_n) \quad I_{C(n+1)} = \frac{R_T}{Z_S}(I_n - I_{n+1}) \quad (10.24c)$$

Equation (10.24) suggests that the earth wire current i consists of two components: a variable component I_C and a constant component I_i. The former is termed a conductive current that reduces with distance from the fault location as it progressively dissipates through the towers into earth. I_i is termed an inductive component because it is induced in the earth wire by the fault current I_F flowing through the faulted phase conductor through the mutual impedance Z_m between the faulted phase conductor and the earth wire. If I_F is constant throughout the length of the line, I_i will also be constant throughout the length of the line as it effectively circulates within each span as illustrated in Figure 10.13(b). Now, substituting Equation (10.24) into Equation (10.22) and rearranging, we obtain

$$\frac{Z_S}{R_T}I_n = \Delta^2 I_n \tag{10.25a}$$

Also, substituting Equation (10.22) into Equation (10.24a), we obtain

$$\frac{Z_S}{R_T}i_n = \Delta^2 i_n + (1-k)\frac{Z_s}{R_T}I_F \tag{10.25b}$$

where

$$\Delta^2 f_n = (f_{n+1} - f_n) - (f_n - f_{n-1}) \tag{10.25c}$$

Equations (10.25) are forward finite difference equations with a constant term in Equation (10.25b). It can be shown that the solution is given by

$$I_n = Ae^{\gamma n} + Be^{-\gamma n} \tag{10.26a}$$

and

$$i_n = ae^{\gamma n} + be^{-\gamma n} + (1-k)I_F \tag{10.26b}$$

where A, B, a and b are constants and γ is the propagation constant and is given by

$$\gamma = \sqrt{\frac{Z_S}{R_T}} = \alpha + j\beta \text{ per span} \tag{10.26c}$$

By substituting Equations (10.26) into Equation (10.22), we obtain

$$a = \frac{A}{1 - e^{\gamma}} \quad \text{and} \quad b = \frac{B}{1 - e^{-\gamma}} \tag{10.27}$$

If the line is sufficiently long, the conductive current in the earth wire progressively dissipates through the towers to earth and becomes negligible at, say, tower n away from the fault location. Thus, substituting $A = a = 0$ in the varying terms of Equation (10.26a) and (10.26b), we obtain

$$I_n = Be^{-\gamma n} \tag{10.28a}$$

$$i_n = \frac{B}{1 - e^{-\gamma}}e^{-\gamma n} + (1-k)I_F \tag{10.28b}$$

Therefore, at the faulted terminal tower, $n = 0$, and for the first span, $n = 1$,

$$I_0 = B \quad \text{and} \quad i_1 = \frac{B}{1 - e^{-\gamma}}e^{-\gamma} + (1-k)I_F \tag{10.29a}$$

and the fault current at the faulted terminal tower, tower 0, is equal to

$$I_F = I_0 + i_1 \tag{10.29b}$$

From Equations (10.29a) and (10.29b), we obtain

$$B = kI_F(1 - e^{-\gamma}) \tag{10.30}$$

Substituting Equation (10.30) into Equation (10.28), we obtain

$$I_n = [kI_F(1 - e^{-\gamma})]e^{-\gamma n} \quad n \geq 0 \tag{10.31a}$$

$$i_n = kI_F e^{-\gamma n} + (1 - k)I_F \quad n \geq 1 \tag{10.31b}$$

Equation (10.31) can be used to calculate the currents in any tower or earth wire span.

Using Equations (10.28a) and (10.29a), the conductive current in tower n is given by

$$I_n = I_0 e^{-\gamma n} \tag{10.32a}$$

where I_0 is the conductive current in terminal tower 0. The tower number n, where the magnitude of the conductive current decays to a negligible value, say, 1.83% of the magnitude of I_0 can be determined using Equation (10.32a). Thus,

$$n = \frac{4}{\alpha} = \frac{4}{\text{Real}(\gamma)} = \frac{4}{\text{Real}\left(\sqrt{\frac{Z_S}{R_T}}\right)} \tag{10.32b}$$

If the remaining conductive current in tower n were chosen as 5% of I_0, the constant '4' in Equation (10.32b) changes to '3'. It is also of practical interest to determine the tower number n, or span number, and corresponding line length, where the tower conductive current drops to a negligible value of the sum of all tower conductive currents, denoted I_C. Substituting $n = 0, 1, 2, \ldots, n$ in Equation (10.32a), and summing up all resulting tower currents, we obtain

$$I_C = I_0 + I_0 e^{-\gamma} + I_0 e^{-2\gamma} + I_0 e^{-3\gamma} + \cdots + I_0 e^{-n\gamma}$$

which can be rewritten as

$$I_C = I_0 \left[\frac{1 - e^{-\gamma(n+1)}}{1 - e^{-\gamma}}\right] \tag{10.33a}$$

Therefore, using Equation (10.33a), it can be shown that the tower number, or span number, n, where the tower current $I_n = I_0 e^{-\gamma n}$ drops to 1.83% of I_C is given by

$$n \cong \frac{\log_e(1 + e^{-2\alpha} - 2e^{-\alpha}\cos\beta) + 8}{1.8\alpha} \tag{10.33b}$$

where $\alpha = \text{Real}(\gamma)$, $\beta = \text{Imaginary}(\gamma)$ and γ is given in Equation (10.26c). Z_S is in Ω/span and R_T is in Ω. The use of 1.83% is arbitrary but such a small value ensures a significant reduction in the nth tower current in per cent of I_C. The line length that corresponds to such n towers or spans, given in Equation (10.33b), is given by

$$\ell_n = n \times S_p \, \text{km} \tag{10.33c}$$

where $S_p = 1/N_p$ is the span length in km and N_p is the number of spans per km.

To illustrate how the above equations may be used, consider the 132 kV overhead line of Section 10.4.2 that has three spans per km and $Z_S = (0.18 + j0.72)\Omega/\text{km}$. It is required to determine n and the corresponding line length where the conductive

Table 10.1 Overhead line earthing network: propagation constant, tower number and line length where tower conductive current drops to 1.83% of total conductive current

	$Z_S = (0.06 + \text{j}0.24)\Omega/\text{span}$				
$R_T(\Omega)$	1	10	25	50	100
$\gamma = \sqrt{Z_S/R_T}$ $= \alpha + \text{j}\beta/\text{span}$	0.392037 +j0.30609	0.1239729 +j0.09679	0.0784074 +j0.06121	0.0554424 +j0.04328	0.0392037 +j0.03061
n	9	19	23	26	28
ℓ_n (km)	3	6.3	7.8	8.8	9.3

current drops to 1.83% of the total conductive current. Use a range of tower footing resistances of $R_T = 1, 10, 25, 50$ and $100\,\Omega$. For $R_T = 1\,\Omega$, we have

$$\gamma = \sqrt{\frac{(0.18 + \text{j}0.72)/3}{1}} = (0.392037 + \text{j}0.30609)/\text{span}$$

$$n = \frac{\log_e(1 + e^{-2\times0.392037} - 2e^{-0.392037}\cos 0.30609) + 8}{1.8 \times 0.392037} \approx 9 \quad \text{and}$$

$$\ell_n = 9 \times 0.334 = 3\,\text{km}$$

Similar calculations can be made for the remaining values of R_T and the results are summarised in Table 10.1.

For the given value of Z_S, n and ℓ_n saturate at around 28 and 9.3 km, respectively, as R_T reaches $100\,\Omega$. It is informative for the reader to calculate the variations in n and ℓ_n for $Z_S = (0.5 + \text{j}0.075)\Omega/\text{span}$.

10.6 Cable earthing system impedance

Various methods of cable sheath/armour earthing arrangements were described in Chapter 3 and these depend on the cable rated voltage and the length of the cable. Like an overhead line's earth wire, where the sheath/armour of cables are bonded to the substation earth electrode system, this serves to extend the effective area of the substation earthing system. A major portion of the earth fault current will normally return via the sheaths/armours of the cable supplying the fault current depending on the effectiveness of the sheath/armour earthing. A similar approach to that for an overhead line may be used to derive an equivalent earthing impedance for the cable sheath/armour where this is earthed allowing current circulation.

10.7 Overall substation earthing system and its equivalent impedance

The overall substation earthing system impedance consists of three main components. These are the substation earth electrode system, e.g. earth mesh or mat, the earth wire and tower footing network of overhead lines entering the

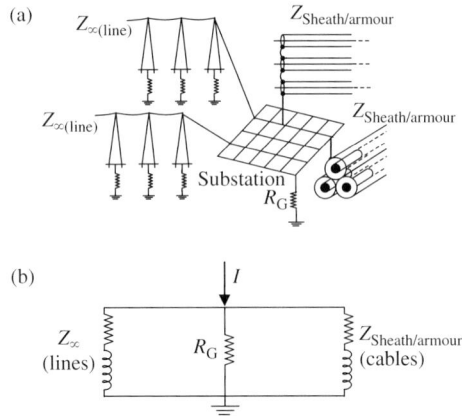

Figure 10.14 General extended substation earthing system: (a) components of substation earthing system and (b) substation equivalent earthing system impedance

substation, and the sheath/armour of cables emanating from the substation. In practice, where electrodes are spaced sufficiently far apart, it is normally sufficient to consider the component impedances to be in parallel as 'seen' from the substation as shown in Figure 10.14.

Therefore, the overall substation earthing impedance with respect to remote earth is given by

$$Z_E = \frac{1}{\frac{1}{R_G} + \sum_i \frac{1}{Z_{\infty(i)-\text{lines}}} + \sum_j \frac{1}{Z_{\text{Sheath/armour}(j)-\text{cables}}}} \ \Omega \quad (10.34)$$

This impedance may be obtained by calculation at the design stage and confirmed by measurement.

10.8 Effect of system earthing methods on earth fault current magnitude

An earth fault current is a current that flows to earth and has a magnitude that depends on the method of system earthing. In solidly earthed and low impedance earthed systems, high levels of earth fault current result. However, earth fault currents are intentionally limited to very low levels in high impedance earthed systems. These systems may be earthed through a high impedance resistor or a high impedance reactor connected to transformer neutral points, or through the connection of an earthing or zig-zag transformer, sometimes with a neutral resistor, to systems supplied through delta-connected transformer windings. Industrial power systems tend to use low impedance earthing. Resonant earthing is where the system is earthed through a high impedance reactor called a Peterson coil, or arc suppression coil, i.e. usually connected to the neutral of distribution or zig-zag transformers. The variable impedance reactor is tuned to the overall system phase to earth capacitive reactance. In isolated systems, no intentional connection to earth is made but these systems are effectively earthed through the distributed capacitance of lines, cables and transformers windings. In both resonant earthed and isolated systems, the magnitude of earth fault current is very small. In effectively earthed systems, earth fault currents may be higher than three-phase fault currents.

10.9 Screening factors for overhead lines

We recall Figure 10.13 and Equation (10.31). The conductive current flowing through the first earth wire span away from the faulty terminal tower is considerably higher than those flowing in the spans further away from the fault. The conductive current flowing through the earth wire diminishes to zero as it dissipates through the towers to earth. The final value of the earth wire current i_{n+1} in span $(n+1)$ is equal to the constant inductive current I_i obtained from Equation (10.31b) by setting $n \to \infty$,

$$i_{n+1} = I_i = (1 - k)I_F \qquad (10.35)$$

It is important to note that this current, illustrated in Figure 10.13(b), does not enter earth and hence does not contribute to the rise of earth potential as will be seen later. From Figure 10.13(a) and starting from the faulted tower, we can write

$$I_F = I_0 + i_1 = I_0 + I_1 + i_2 = I_0 + I_1 + I_2 + i_3 = I_0 + I_1 + I_2 + I_3 + \cdots + I_n + i_{n+1}$$

or

$$I_F = I_C + (1 - k)I_F \qquad (10.36a)$$

where $I_C = \sum_{j=0}^{n} I_j$ is the sum of all conductive or tower currents given in Equation (10.33a). From Figure 10.13, we observe that this current dissipates through the towers and return to the source through the earth. This current is therefore termed the earth return current and is given by

$$I_{ER} = I_C = \sum_{j=0}^{n} I_j \qquad (10.36b)$$

Thus, from Equations (10.35) and (10.36), we have

$$I_F = I_{ER} + I_i \qquad (10.37a)$$

and

$$I_{ER} = kI_F \qquad (10.37b)$$

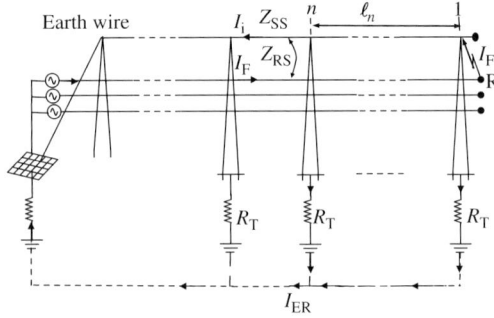

Figure 10.15 Calculation of screening factor for a single-circuit overhead line

We are now ready to define the line's screening factor using Equation (10.37b). This is defined as the ratio of the earth return current I_{ER} to the total fault current entering the earthing system at the fault location, i.e.

$$k = \frac{I_{ER}}{I_F} = \frac{I_{ER}}{3I^Z} \tag{10.38}$$

where the earth fault current entering the earthing system is $I_F = 3I^Z$ and I^Z is its zero phase sequence (ZPS) component. To express the line's screening factor in terms of the line's impedances, we assume that the line is sufficiently long so that the total earth wire impedance is large compared to the terminal resistances and we neglect the line shunt susceptance.

Figure 10.15 shows a line fed by a three-phase voltage source at the sending end and is open-circuited at the receiving end. When a single-phase short circuit occurs on phase R at the receiving end, the series voltage drop across one span of the faulted phase conductor and earth wire of length S in km beyond tower n is given by

$$\begin{bmatrix} \Delta V_R \\ \Delta V_Y \\ \Delta V_B \\ 0 \end{bmatrix} = S \begin{bmatrix} Z_{RR} & Z_{RY} & Z_{RB} & Z_{RS} \\ Z_{RY} & Z_{YY} & Z_{YB} & Z_{YS} \\ Z_{RB} & Z_{YB} & Z_{BB} & Z_{BS} \\ Z_{RS} & Z_{YS} & Z_{BS} & Z_{SS} \end{bmatrix} \begin{bmatrix} I_F \\ 0 \\ 0 \\ -I_i \end{bmatrix} \tag{10.39}$$

Therefore, from Equation (10.39), $0 = Z_{RS}I_F - Z_{SS}I_i$ or $I_i = (Z_{RS}/Z_{SS})/I_F$. Therefore, using Equations (10.37a) and (10.38), the line's screening factor is given by

$$k = \frac{I_{ER}}{3I^Z} = 1 - \frac{Z_{RS}}{Z_{SS}} \tag{10.40}$$

where Z_{SS} is the self-impedance of the earth wire with earth return, and Z_{RS} is the mutual impedance between the faulted phase conductor R and the earth wire with earth return. It is noted that Equation (10.40) can be rewritten as $Z_{RS} = (1 - k)Z_{SS}$ which is the equation we used for Z_m in Equation (10.23). The self and mutual impedances can be calculated using Equations (3.19a) and (3.20a), respectively, and are in Ω/km. From Equations (10.38) and (10.40), the earth return current I_{ER} and the earth wire inductive current I_i are given by

$$I_{ER} = k \times (3I^Z) = \left(1 - \frac{Z_{RS}}{Z_{SS}}\right) \times (3I^Z) \qquad (10.41a)$$

$$I_i = (1 - k) \times (3I^Z) = \frac{Z_{RS}}{Z_{SS}} \times (3I^Z) \qquad (10.41b)$$

Equation (10.41a) shows that the screening factor k determines the proportion of the line's earth fault current returning through earth at a distance greater than or equal to ℓ_n where ℓ_n is given in Equation (10.33c). In addition, Equation (10.41a) suggests that in the absence of coupling between the faulted phase conductor and the earth wire, or in the absence of an earth wire, all the earth fault current returns to the source through the earth. The above analysis is also applicable to double-circuit overhead lines with one earth wire where Z_{RS} now represents the mutual impedance between the faulted conductor phase R on one of the two circuits. The analysis method presented can be extended to calculate the screening factor for circuits with two earth wires. In England and Wales, for most 132, 275 and 400 kV single-circuit and double-circuit overhead lines with one earth wire, the magnitude of k ranges from 0.60 to 0.75 i.e. 60–75% of the line's phase fault current returns through earth and only 25–40% returns through the earth wire.

10.10 Screening factors for cables

10.10.1 General

The current flowing in a single-core cable or the flow of unbalanced currents in the cores of a three-core cable induces voltages in the metallic sheath/armour of the cable. Under an earthed fault condition, and if the sheath/armour is earthed at each end, a current will circulate in the sheath/armour earth loop. This current is part of the earth fault current and the remaining part is the earth return current that returns through earth. The cable's screening factor is dependent on the cable's layout, earthing method, sheath/armour material, terminal earthing resistances and earth resistivity.

10.10.2 Single-phase cable with metallic sheath

Consider a single-phase cable with a metallic sheath but no armour having a length L in km. The cable is fed from a single-phase source at end 1 and subjected to a short-circuit fault at end 2 as shown in Figure 10.16(a). The cable core is denoted C and the sheath is denoted S and the latter is earthed at each end. The sheath terminal earthing resistances are denoted R_1 and R_2. Figure 10.16(b) shows the cable sheath earthing equivalent circuit. I_F, I_S and I_{ER} are the conductor fault current, sheath current and earth return current, respectively.

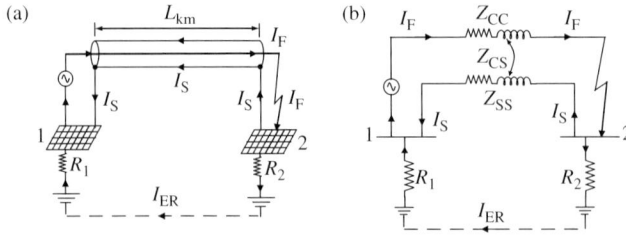

Figure 10.16 Calculation of screening factor for a single-phase cable with metallic sheath earthed at each end: (a) current distribution during a one-phase to earth short-circuit fault and (b) equivalent cable sheath circuit

The series voltage drop across the length of the conductor and sheath is given by

$$\begin{bmatrix} \Delta V_C \\ \Delta V_S \end{bmatrix} = L \begin{bmatrix} Z_{CC} & Z_{CS} \\ Z_{CS} & Z_{SS} \end{bmatrix} \begin{bmatrix} I_F \\ -I_S \end{bmatrix} \tag{10.42}$$

From Equation (10.42) and Figure 10.16, we have

$$\Delta V_S = L(Z_{CS}I_F - Z_{SS}I_S) = -I_{ER}(R_1 + R_2) \quad \text{and} \quad I_F = I_{ER} + I_S$$

Therefore, the cable's screening factor is given by

$$k = \frac{I_{ER}}{I_F} = \frac{Z_{SS} - Z_{CS}}{Z_{SS} + \frac{R_1 + R_2}{L}} \tag{10.43}$$

The earth return current and sheath current are given by

$$I_{ER} = k \times I_F \quad I_S = (1 - k) \times I_F \tag{10.44}$$

The sheath current can be expressed as follows:

$$I_S = \frac{Z_{CS} + \frac{R_1 + R_2}{L}}{Z_{SS} + \frac{R_1 + R_2}{L}} \times I_F = I_C + I_i \tag{10.45}$$

As in the case of an overhead line, the current returning through the cable sheath consists of a conductive component and an inductive component given by

$$I_C = \frac{\frac{R_1 + R_2}{L}}{Z_{SS} + \frac{R_1 + R_2}{L}} \times I_F \tag{10.46a}$$

$$I_i = \frac{Z_{CS}}{Z_{SS} + \frac{R_1 + R_2}{L}} \times I_F \tag{10.46b}$$

I_i is the inductive component of the sheath current due to inductive coupling between the core conductor and the sheath through the mutual impedance Z_{CS}. I_C is the conductive component of the sheath current due to conduction between the cable's earthing terminals 2 and 1.

Equation (10.43) of the cable's screening factor k includes the effect of the terminal earthing resistances. However, it is sometimes required to represent the effect of the cable parameters only. Therefore, with the terminal resistances set to zero, the cable's screening factor is given by

$$k = \frac{I_{ER}}{I_F} = 1 - \frac{Z_{CS}}{Z_{SS}} \tag{10.47}$$

Equation (10.47) can be rewritten using Equations (3.101a) and (3.103) as follows:

$$k = \frac{I_{ER}}{I_F} \simeq \frac{R_{S(ac)}}{Z_{SS}} \tag{10.48}$$

where $R_{S(ac)}$ is the sheath's ac resistance and Z_{SS} is given in Equation (3.101a).

It is interesting to note that the self-impedance of the faulted core conductor Z_{CC} does not have any effect on the cable's screening factor and hence on the current that returns through the sheath and also on the earth return current. Using $I_F = 3I^Z$, the earth return current and sheath current are given by

$$I_{ER} = k \times (3I^Z) \tag{10.49a}$$

$$I_i = (1 - k) \times (3I^Z) \tag{10.49b}$$

10.10.3 Three-phase cable with metallic sheaths

We presented in Chapter 3, a wide variety of three-phase cable circuit layouts and earthing arrangements. We will now illustrate the technique of calculating the cable's screening factor for a general three-phase cable layout since the technique can be extended and applied to other layouts including armoured cables. Figure 10.17 illustrates a three-phase single-core cable having a length L in km and metallic sheaths, but no armour, and laid out in a flat symmetrical arrangement. At both ends of the cable, the sheaths are solidly bonded and connected to earth electrode where the terminal earthing resistances are denoted R_1 and R_2.

The cable is fed by a three-phase voltage source at end 1 and is open-circuited at end 2, the cable cores are denoted C1, C2 and C3, and the sheaths are denoted S1, S2 and S3. Under a single-phase to earth short-circuit fault on phase R, or conductor C1, at end 2, Figure 10.17(a) shows the resultant currents returning through the three sheaths and the current returning through earth. Figure 10.17(b) shows an equivalent circuit representation of the cable under this short-circuit fault condition. Using Equation (3.86), the series voltage drop across the cable, is given by

$$\begin{bmatrix} \Delta V_{C1} \\ \Delta V_{C2} \\ \Delta V_{C3} \\ \Delta V_{S1} \\ \Delta V_{S2} \\ \Delta V_{S3} \end{bmatrix} = L \begin{bmatrix} Z_{C1C1} & Z_{C1C2} & Z_{C1C3} & Z_{C1S1} & Z_{C1S2} & Z_{C1S3} \\ Z_{C1C2} & Z_{C2C2} & Z_{C2C3} & Z_{C2S1} & Z_{C2S2} & Z_{C2S3} \\ Z_{C1C3} & Z_{C2C3} & Z_{C3C3} & Z_{C3S1} & Z_{C3S2} & Z_{C3S3} \\ Z_{C1S1} & Z_{C2S1} & Z_{C3S1} & Z_{S1S1} & Z_{S1S2} & Z_{S1S3} \\ Z_{C1S2} & Z_{C2S2} & Z_{C3S2} & Z_{S1S2} & Z_{S2S2} & Z_{S2S3} \\ Z_{C1S3} & Z_{C2S3} & Z_{C3S3} & Z_{S1S3} & Z_{S2S3} & Z_{S3S3} \end{bmatrix} \begin{bmatrix} I_F \\ 0 \\ 0 \\ -I_{S1} \\ -I_{S2} \\ -I_{S3} \end{bmatrix}$$

$$\tag{10.50a}$$

and

$$\Delta V_{S1} = \Delta V_{S2} = \Delta V_{S3} = -I_{ER}(R_1 + R_2) \tag{10.50b}$$

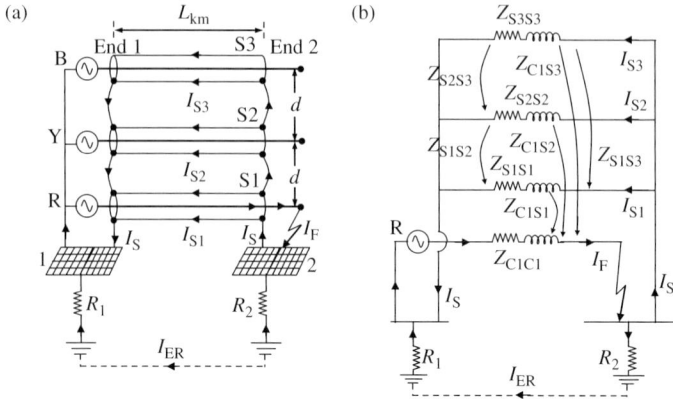

Figure 10.17   Calculation of screening factor for a three-phase single-core cable having a symmetrical flat arrangement and solidly bonded earthed sheaths: (a) current distribution during a one-phase to earth short circuit and (b) equivalent circuit of faulted cable conductor and sheaths

From Equation (10.50) and the symmetrical cable layout shown in Figure 10.17(a), we have $Z_{S1S1} = Z_{S2S2} = Z_{S3S3}$ and $Z_{S1S2} = Z_{S2S3}$. Thus, we can write

$$\begin{bmatrix} \Delta V_{S1} \\ \Delta V_{S2} \\ \Delta V_{S3} \end{bmatrix} = L \begin{bmatrix} Z_{C1S1}I_F \\ Z_{C1S2}I_F \\ Z_{C1S3}I_F \end{bmatrix} - L \begin{bmatrix} Z_{S1S1} & Z_{S1S2} & Z_{S1S3} \\ Z_{S1S2} & Z_{S1S1} & Z_{S1S2} \\ Z_{S1S3} & Z_{S1S2} & Z_{S1S1} \end{bmatrix} \begin{bmatrix} I_{S1} \\ I_{S2} \\ I_{S3} \end{bmatrix} = \begin{bmatrix} -I_{ER}(R_1 + R_2) \\ -I_{ER}(R_1 + R_2) \\ -I_{ER}(R_1 + R_2) \end{bmatrix}$$

$$(10.51a)$$

and

$$I_S = I_{S1} + I_{S2} + I_{S3} \quad I_F = I_S + I_{ER} \tag{10.51b}$$

After some algebra, it can be shown that the cable's screening factor is given by

$$k = \frac{I_{ER}}{I_F} = \frac{\left[\begin{array}{c} Z_{S1S1}^2 - 2Z_{S1S2}^2 + Z_{S1S1}Z_{S1S3} + (Z_{S1S2} - Z_{S1S3}) \\ \times Z_{C1S1} - (Z_{S1S1} - Z_{S1S2})(Z_{C1S1} + Z_{C1S2} + Z_{C1S3}) \end{array} \right]}{Z_{S1S1}^2 - 2Z_{S1S2}^2 + Z_{S1S1}Z_{S1S3} + (3Z_{S1S1} + Z_{S1S3} - 4Z_{S1S2})\frac{(R_1+R_2)}{L}}$$

$$(10.52)$$

Z_{S1S1} is the self-impedance of sheath S1 with earth return, Z_{S1S2} is the mutual impedance between sheaths S1 and S2 with earth return, and Z_{S1S3} is the mutual impedance between sheaths S1 and S3 with earth return. Z_{C1S1}, Z_{C1S2} and Z_{C1S3} are the mutual impedances between the faulted phase conductor R and sheaths S1, S2 and S3, with earth return, respectively. All impedances are in Ω/km. The self-impedance Z_{S1S1} can be calculated using Equation (3.101a) and all mutual impedances can be calculated using Equation (3.103).

For three-core cables and three single-core cables in a touching trefoil or equilateral layouts, the screening factor can be derived from Equation (10.52) by setting

$Z_{S1S2} = Z_{S1S3} = Z_{C1S2} = Z_{C1S3}$. Thus,

$$k = \frac{I_{ER}}{I_F} = \frac{Z_{S1S1} - Z_{C1S1}}{Z_{S1S1} + 2Z_{S1S2} + 3\frac{(R_1+R_2)}{L}} \tag{10.53}$$

Substituting Equations (3.101a) and (3.103) into Equation (10.53), it can be shown that the cable's screening factor is given by

$$k = \frac{I_{ER}}{I_F} \cong \frac{R_{S(ac)}}{R_{S(ac)} + 3\pi^2 10^{-4}f + j4\pi 10^{-4}f \log_e\left[\frac{D_{erc}^3}{d^2(r_{is}+r_{os})/2}\right] + 3\frac{(R_1+R_2)}{L}} \tag{10.54}$$

where $R_{S(ac)}$ is the sheath ac power frequency resistance, f is the nominal power frequency, d is the distance between the centres of cable phases, r_{is} and r_{os} are sheath inner and outer radii, and D_{erc} is the depth of equivalent earth return conductor given in Equation (3.15).

Equations (10.52) and (10.53) of the cable's screening factor k include the effect of the terminal earthing resistances R_1 and R_2, but if it is required to represent the effect of the cable parameters only, these resistances can be set to zero. However, the effect of the terminating resistances is important since they act to reduce the screening factor and hence the earth return current and increase the current that returns through the cable's sheaths. Using $I_F = 3I^Z$, the earth return current and total sheath current are given by

$$I_{ER} = k(3I^Z) \tag{10.55a}$$

$$I_S = (1 - k)(3I^Z) \tag{10.55b}$$

As in the case of a single-phase cable, the sheath current I_S consists of an inductive and a conductive component but we leave this simple derivation for the reader.

It is to note that in the case of three single-core cables of symmetrical flat formation, two different screening factors can be obtained. One would correspond to a short-circuit fault on the cable conductor laid in the central position and the other for a short-circuit fault on one of the cable conductors occupying an outer position. In our analysis here, we have considered the latter case. The derivation of the screening factor for the former case is straightforward using the method we have presented. Further, the method can be used to derive screening factors for cables having other physical layouts as well as armoured cables.

In most practical cable installations in medium, high and extra high voltage systems, and unlike overhead lines, the majority of the earth fault current for the fault conditions presented in this section returns through the cable sheaths and the rest through the earth.

10.11 Analysis of earth return currents for short-circuits in substations

Earth return currents are calculated under both single-phase and two-phase to earth short-circuit faults in substations. The calculation procedure and the current distribution under a solid one-phase to earth short-circuit fault in a substation is presented using a simple network of two substations separated by a single-circuit overhead power line with a single earth wire, as shown in Figure 10.18. The neutrals of the transformer star windings at the two substations are solidly connected to their respective substation earth electrode system. The overhead line earth wire is bonded to the earth mesh at both substations.

At substation 1, the initial rms phase fault current is the sum of the incoming phase fault currents on the line and the transformer and is given by

$$I_F'' = I_{F(trans)}'' + I_{F(line)}'' \text{ A} \tag{10.56a}$$

Since the phase fault current under a one-phase to earth fault is equal to three times the ZPS current, we have

$$I_F'' = 3I_{(trans)}^Z + 3I_{(line)}^Z \text{ A} \tag{10.56b}$$

At the short-circuit location, the short-circuit current supplied through the transformer returns to the substation transformer neutral, i.e. it circulates within the transformer. This current does not enter earth and as a result does not contribute to the earth return current and substation rise of earth potential. Based on the analysis presented in Section 10.5, part of the incoming earth fault current on the overhead line's faulted phase will flow through the line's earth wire and completely dissipates through the towers to earth at distance ℓ_n from the fault location. At this distance, the current returning through the earth wire is the constant inductive current and that returning through earth is the earth return current. Using the line's screening factor presented in Section 10.9, and $I_{F(line)}'' = 3I^Z$, the earth return

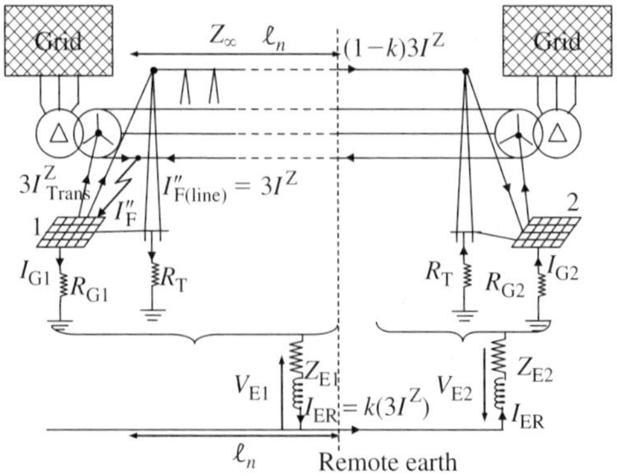

Figure 10.18   Analysis of earth return currents for short circuit in substations

current at distance ℓ_n is given by

$$I_{ER} = k \times (3I^Z)A \qquad (10.57a)$$

and the earth wire inductive current is given by

$$I_S = (1 - k) \times (3I^Z)A \qquad (10.57b)$$

where k is as given in Equation (10.40) and ℓ_n is as given in Equation (10.33c).

If the line length is shorter than ℓ_n, the earth return current I_{ER} calculated by Equation (10.57a) will be an overestimate because in this case not all the conductive current returning through the earth wire would have dissipated through the towers to earth. Using Equation (10.34), the overall earthing impedance at the faulted substation with respect to remote earth consists of the parallel combination of the substation earth electrode resistance and the line's earth wire/tower footing earthing impedance and is given by

$$Z_{E1} = \frac{1}{\frac{1}{R_{G1}} + \frac{1}{Z_\infty}} \Omega \qquad (10.58)$$

10.12 Analysis of earth return currents for short circuits on overhead line towers

The calculation procedure of earth return current due to a short-circuit earth fault at a tower and the resultant current distribution in earth and line's earth wire is presented using the simple network shown in Figure 10.19.

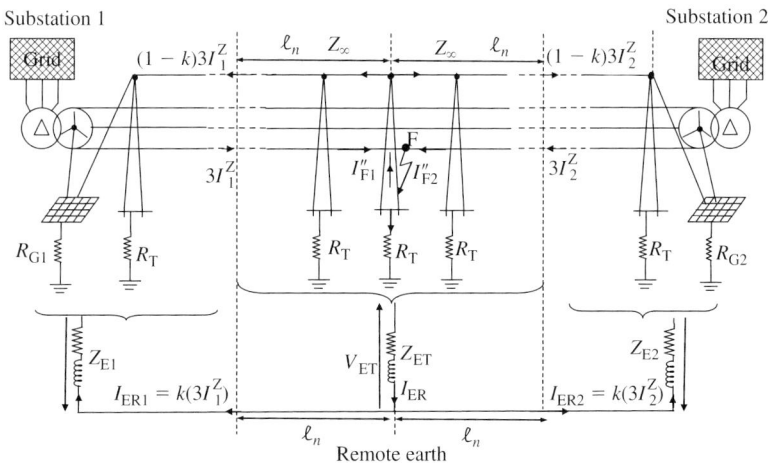

Figure 10.19 Analysis of earth return currents for a short circuit on an overhead line tower

The initial rms phase fault current is the sum of the incoming phase fault currents on both sides of the tower and is given by

$$I''_F = I''_{F1} + I''_{F2} \text{A} \tag{10.59a}$$

or rewritten in terms of their respective ZPS currents

$$I''_F = 3I^Z_1 + 3I^Z_2 \text{A} \tag{10.59b}$$

At the faulted tower, part of the fault current flows into the earth through the faulted tower and its footing resistance. The other part flows into the earth wire in both directions away from the faulted tower. As in the analysis presented in Section 10.5, the currents carried by the earth wires of the first spans on both sides of the faulted tower are considerably higher than those flowing in the spans further away from the fault. After a distance equal to ℓ_n in both directions, each earth wire current becomes constant and equal to the respective inductive current. Therefore, using the line's screening factor k, the earth wire currents I_{S1} and I_{S2} and the earth return currents I_{ER1} and I_{ER2} at distance ℓ_n from the faulted tower are given by

$$I_{S1} = (1 - k) \times (3I^Z_1)\text{A} \tag{10.60a}$$

$$I_{S2} = (1 - k) \times (3I^Z_2)\text{A} \tag{10.60b}$$

and

$$I_{ER1} = k \times (3I^Z_1)\text{A} \tag{10.61a}$$

$$I_{ER2} = k \times (3I^Z_2)\text{A} \tag{10.61b}$$

For a short-circuit fault on the tower, the total earth return current that will cause a rise of earth potential at the tower base is given by

$$I_{ER} = I_{ER1} + I_{ER2} = k \times [(3I^Z_1) + (3I^Z_2)]\text{A} \tag{10.62}$$

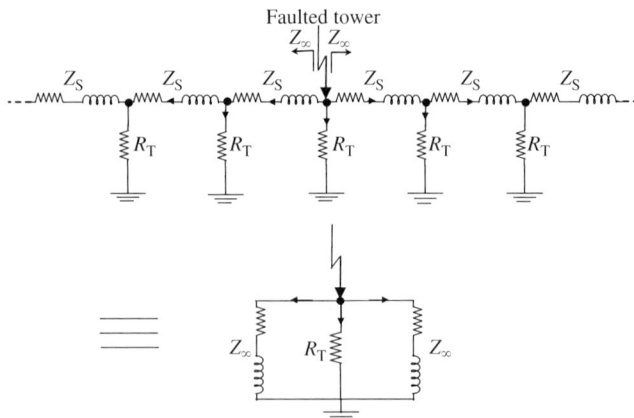

Figure 10.20   Equivalent earthing network of Figure 10.19 as seen from the faulted tower

The overall earthing impedance at the faulted tower with respect to remote earth consists of the faulted tower footing resistance and the two earth wire ladder impedances in parallel as shown in Figure 10.20. Using Equation (10.34), the overall earthing impedance is given by

$$Z_{ET} = \frac{1}{\frac{1}{R_T} + \frac{2}{Z_\infty}} \ \Omega \tag{10.63}$$

10.13 Calculation of rise of earth potential

Under normal unfaulted system conditions, the substation earthing system may have a potential that is slightly above that of remote earth of 0 V because on naturally unbalanced overhead lines, ZPS currents flow and return to the source via the substation earthing system. Also, the load in low voltage distribution networks may not be perfectly balanced. However, the design of the line and the degree of load unbalance usually ensure that these ZPS currents are kept low, so that the rise of earth potential with respect to remote earth is low and at safe levels.

During a single-phase to earth or two-phase to earth short circuits within a substation, large ZPS current flows through the earthing system in solidly earthed and low impedance earthed systems and this may cause a significant rise in earth potential depending on the magnitude of the effective substation earthing system impedance. In general, the electrode rise of earth potential with respect to remote earth system is given by

$$V_E = Z_E \times I_{ER} \ V \tag{10.64a}$$

where Z_E is given in Equation (10.34) and I_{ER} is the earth return current. In the case of an earth fault in a substation, as shown in Figure 10.18, the rise of earth potential at substation 1 with respect to remote earth, using Equation (10.58), is given by

$$V_{E1} = \frac{1}{\frac{1}{R_{G1}} + \frac{1}{Z_\infty}} \times I_{ER} \ V \tag{10.64b}$$

For an earth fault on a tower, as shown in Figure 10.19, the rise of earth potential at the tower with respect to remote earth, using Equation (10.63), is given by

$$V_{ET} = \frac{1}{\frac{1}{R_T} + \frac{2}{Z_\infty}} \times I_{ER} \ V \tag{10.65a}$$

Alternatively, using the tower fault current I_T, the rise of earth potential at the tower with respect to remote earth can be calculated using Equation (10.20) as follows:

$$V_{ET} = R_T \times I_T = \frac{\rho_e}{2\pi r} I_T \ V \tag{10.65b}$$

Where the tower fault current enters earth at the tower earthing electrode, the distribution of the current in the earth is non-uniform. However, the distribution becomes more uniform and radial after a small distance from the tower outside its earthing electrode. Therefore, the voltage on the surface of the earth with respect to remote earth at a distance x m from the edge of the tower earth electrode is given by the following approximate equation:

$$V(x) \cong \frac{1}{1 + \frac{2\pi R_T}{\rho_e} x} V_{ET} \quad x > 2r \tag{10.66}$$

where V_{ET} is given in Equation (10.65), r is the equivalent radius of a sphere representing the tower earth electrode. Equation (10.66) can be used where the distance x is greater than twice the equivalent radius r.

10.14 Examples

Example 10.1   Consider the simple system of Figure 10.18. It is required to calculate the short-circuit earth return current and rise of earth potential at substation 1 for a single-phase to earth short-circuit fault in the substation.

The data for this system is as follows: a 50 Hz system and a 25 km 400 kV single-circuit overhead line with a single earth wire between substations 1 and 2.

Substation 1
PPS (NPS) subtransient short-circuit infeed at 400 kV	15 000 MVA
ZPS short-circuit infeed at 400 kV	10 000 MVA
PPS and ZPS X/R ratios are assumed infinite for convenience in hand calculation	
Resistance of earth mesh electrode system	$R_1 = 0.1\,\Omega$

Substation 2
PPS (NPS) subtransient short-circuit infeed at 400 kV	32 000 MVA
ZPS short-circuit infeed at 400 kV	35 000 MVA
PPS and ZPS X/R ratios are assumed infinite	

Overhead line
PPS impedance	$(0.017 + j0.3)/\Omega/km$
ZPS impedance	$(0.12 + j0.82)/\Omega/km$
Earth wire self-impedance	$(0.21 + j0.7)/\Omega/km$
Mutual impedance between faulted phase conductor and earth wire	$(0.04 + j0.19)/\Omega/km$
Tower footing resistance	$10\,\Omega$
Span length	$0.366\,km$

Substation 1

$$\text{PPS/NPS subtransient impedance} = j\frac{400^2}{15\,000} = j10.67\,\Omega$$

ZPS impedance $= j\dfrac{400^2}{10\,000} = j16\,\Omega$

Substation 2

PPS/NPS subtransient impedance $= j\dfrac{400^2}{32\,000} = j5\,\Omega$

ZPS impedance $= j\dfrac{400^2}{35\,000} = j4.6\,\Omega$

Overhead line

Screening factor $k = \dfrac{(0.21 + j0.7) - (0.04 + j0.19)}{0.21 + j0.7} = 0.735\angle{-1.73°}$

$$= 0.735 - j0.022 \approx 0.735$$

Earth wire self-impedance per span $= 0.366\,\text{km} \times (0.21 + j0.7)\,\Omega/\text{km}$
$$= (0.077 + j0.256)\,\Omega \text{ per span}$$

Driving point impedance of infinite line

$$Z_\infty = \frac{0.077 + j0.256}{2} \times \left(1 + \sqrt{1 + \frac{4 \times 10}{0.077 + j0.256}}\right)$$

$$= (1.35 + j1.1)\,\Omega = 1.74\,\Omega\angle 39.17°$$

Substation 1 overall earthing impedance

$$Z_{E1} = \frac{1}{\frac{1}{0.1} + \frac{1}{1.35 + j1.1}} = (0.095 + j0.0033)\,\Omega$$

The propagation constant is equal to

$$\gamma = \alpha + j\beta = \sqrt{\frac{0.077 + j0.256}{10}} = 0.1312115 + j0.0975524 \text{ per span}$$

The tower number or number of line spans where the conductive current drops to 1.8% of the earth return current is equal to

$$n = \frac{\log_e(1 + e^{-2 \times 0.1312115} - 2e^{-0.1312115} \cos 0.0975524) + 8}{1.8 \times 0.1312115} = 18$$

The corresponding line length from the faulted substation is equal to $\ell_n = 18 \times 0.366 = 6.6\,\text{km}$.

To calculate the subtransient earth fault current, the Thévenin's impedances at the point of fault are given by

PPS/NPS subtransient impedance $= \dfrac{1}{\frac{1}{j10.67} + \frac{1}{(0.425 + j7.5) + j5}} = (0.09 + j5.76)\,\Omega$

ZPS impedance $= \dfrac{1}{\frac{1}{j16} + \frac{1}{(3 + j20.5) + j4.6}} = (0.45 + j9.8)\,\Omega$

The earth fault current, using IEC-60909 method and a voltage factor $c = 1.1$, is calculated as

$$I''_F = \frac{3 \times (400\,\text{kV}/\sqrt{3}) \times 1.1}{2 \times (0.09 + \text{j}5.76) + (0.45 + \text{j}9.8)}$$

$$= 35.7\,\text{kA}\angle{-}88.3° = (1.05 - \text{j}35.7)\text{kA}$$

The ZPS component of the earth fault current is calculated as

$$I^Z = \frac{1.05 - \text{j}35.7}{3} = (0.35 - \text{j}11.9)\text{kA}$$

The ZPS current component supplied through the line is calculated as

$$I^Z_{\text{Line}} = \frac{\text{j}16}{3 + \text{j}25.1} \times (0.35 - \text{j}11.9) = 1.11 - \text{j}7.45 = 7.5\,\text{kA}\angle{-}81.5°$$

The earth return current at a distance of 6.6 km from the fault location is given by

$$I_{\text{ER}} = 0.735 \times [3 \times (1.11 - \text{j}7.45)] = (2.45 - \text{j}16.4) = 16.6\,\text{kA}\angle{-}81.5°$$

The earth wire current at a distance of 6.6 km from the fault location is calculated as

$$I_S = (1 - 0.735) \times [3 \times (1.11 - \text{j}7.45)] = (0.88 - \text{j}5.92) = 5.98\,\text{kA}\angle{-}81.5°$$

The rise of earth potential at substation 1 is calculated as

$$V_E = 16.6\angle{-}81.5° \times (0.095 + \text{j}0.0033) = 1.57\,\text{kV}\angle{-}79.5°$$

Example 10.2   Consider the simple system of Figure 10.19. It is required to calculate the short-circuit earth return current for a single-phase to earth short-circuit fault at a tower midway through the line. Calculate the rise of earth potentials at the tower and at 5 and 10 m distance from the edge of the tower earth electrode. Use the same data as in Example 10.1 and assume an earth resistivity of 100 Ωm.

To calculate the subtransient earth fault current, the Thévenin's impedances at the faulted tower are calculated as follows:

$$\text{PPS/NPS subtransient impedance} = \frac{1}{\dfrac{1}{\text{j}10.67 + (0.2125 + \text{j}3.75)} + \dfrac{1}{\text{j}5 + (0.2125 + \text{j}3.75)}}$$

$$= (0.1126 + \text{j}5.45)\Omega$$

$$\text{ZPS impedance} = \frac{1}{\dfrac{1}{\text{j}16 + (1.5 + \text{j}10.25)} + \dfrac{1}{\text{j}4.6 + (1.5 + \text{j}10.25)}}$$

$$= (0.8 + \text{j}9.5)\Omega$$

Using IEC 60909 voltage factor $c = 1.1$, the tower earth fault current is equal to

$$I_F'' = \frac{3 \times (400\,\text{kV}/\sqrt{3}) \times 1.1}{2 \times (0.1126 + j5.45) + (0.8 + j9.5)} = 37.3\,\text{kA}\angle{-87.1°}$$
$$= (1.87 - j37.2)\text{kA}$$

The ZPS component of the tower earth fault current is calculated as

$$I^Z = \frac{1.87 - j37.2}{3} = (0.62 - j12.4)\text{kA}$$

The ZPS current component supplied through the line from substation 1 is calculated as

$$I_{\text{Line}(1)}^Z = \frac{1.5 + j14.85}{3 + j41.1} \times (0.62 - j12.4) = 0.1 - j4.5 = 4.5\,\text{kA}\angle{-88.7°}$$

The ZPS current component supplied through the line from substation 2 is calculated as

$$I_{\text{Line}(2)}^Z = (0.62 - j12.4) - (0.1 - j4.5) = 0.52 - j7.9 = 7.91\,\text{kA}\angle{-86.2°}$$

The earth return current flowing towards substation 1 calculated as

$$I_{\text{ER}(1)} = 0.735 \times [3 \times (0.1 - j4.5)] = (0.22 - j9.92) = 9.92\,\text{kA}\angle{-88.7°}$$

The earth return current flowing towards substation 2 is calculated as

$$I_{\text{ER}(2)} = 0.735 \times [3 \times (0.52 - j7.9)] = (1.146 - j17.4) = 17.45\,\text{kA}\angle{-86.2°}$$

The earth wire current at a distance of 6.6 km from the faulted tower on the side of substation 1 is calculated as

$$I_{S(1)} = (1 - 0.735) \times [3 \times (0.1 - j4.5)] = (0.0795 - j3.57)$$
$$= 3.57\,\text{kA}\angle{-88.7°}$$

The earth wire current at a distance of 6.6 km from the faulted tower on the side of substation 2 is calculated as

$$I_{S(2)} = (1 - 0.735) \times [3 \times (0.52 - j7.9)] = (0.413 - j6.28)$$
$$= 6.29\,\text{kA}\angle{-86.2°}$$

The total earth return current is calculated as

$$I_{\text{ER}} = I_{\text{ER}(1)} + I_{\text{ER}(2)} = (0.22 - j9.92) + (1.146 - j17.4)$$
$$= 27.35\,\text{kA}\angle{-87.1°}$$

The overall earthing impedance at the faulted tower is calculated as

$$Z_{\text{ET}} = \frac{1}{\frac{1}{10} + \frac{2}{1.35 + j1.1}} = (0.65 + j0.48)\Omega = 0.8\angle{36.4°}\,\Omega$$

The rise of earth potential at the faulted tower is calculated as

$$V_{\text{ET}} = 27.35\angle{-87.1°} \times 0.8\angle{36.4°} = 21.88\,\text{kV}\angle{-50.7°}$$

The equivalent radius for the tower's earth electrode represented as a hemisphere is given by

$$r = \frac{100}{2\pi \times 10} = 1.6\,\text{m}$$

Therefore, the required distances of 5 and 10 m are greater than $2r$. The voltage on the earth's surface at 5 m from the edge of the spherical earth electrode is

$$V(x = 5) \cong \frac{1}{1 + \frac{2\pi \times 10}{10} \times 5} \times 21.88 = 5.28\,\text{kV}$$

The voltage at a distance of 10 m reduces to 3 kV.

10.15 Electrical interference from overhead power lines to metal pipelines

10.15.1 Background

The proximity of overhead power lines, underground cables or traction lines to adjacent structures that have metallic parts such as communication cables, fences, surface or underground pipelines can produce harmful voltages in these structures. Our focus in this book is on the voltages produced in metal pipelines by overhead power lines. Metal pipelines are usually formed of steel tubes that are welded together and used for the transportation of various substances such as crude oil, natural gases, water, liquefied petroleum gases and sewage. The length of pipelines may range from several kilometres to hundreds or even thousands of kilometres. Most pipelines are usually buried at low depth although some may be installed above ground. Since the soil is electrolytic, ac corrosion of buried pipelines may occur when ac current is exchanged between the pipe and the earth. Therefore, buried pipelines have a few millimetre thick coating that insulates the metal from the surrounding earth and provides the primary protection against corrosion. New coatings include polyethylene and epoxy but for old pipelines, bitumen and glass cloth were used. Cathodic protection systems provide additional protection against ac corrosion. The technique consists of applying a low dc voltage along the pipe that negatively polarizes the metal with respect to earth thereby helping to minimise electrochemical corrosion of the metal by the soil. The risk of ac corrosion of pipelines begins at a much lower ac pipeline voltage than that which endangers safety. Generally, the ac voltage between the pipeline and a reference electrode above the pipeline should be less than 10 V if earth resistivity is greater than 25 Ωm but should not exceed 4 V if earth resistivity is less than or equal to 25 Ωm.

   Metal pipelines are conductors that are generally insulated from earth. In proximity to overhead power lines, pipelines may be exposed to electrical interference for part of their length and this causes voltages to appear on the pipeline. Many countries specify maximum permissible touch voltages to protect pipeline workers. Under permanent or steady state conditions, maximum permissible touch voltages tend to range from 15 to 65 V. Under short duration fault conditions, the range is generally from 300 to 1500 V depending on short-circuit fault clearance times. In some countries, the limits are reduced if the public has access to the pipelines.

There are three types of electrical interference from power lines to pipelines. These are generally termed electrostatic or capacitive, electromagnetic or inductive, and resistive or conductive couplings.

10.15.2 Electrostatic or capacitive coupling from power lines to pipelines

Buried pipelines in proximity to overhead lines are not exposed to capacitive coupling from the power line because the earth acts as an electrostatic shield. Only pipelines installed above earth are subject to capacitive coupling from the conductors of overhead lines as illustrated in Figure 10.21(a). Where the pipeline runs physically in parallel with the conductors of the power line, the parallel exposure of the pipeline and power line is termed a parallelism. This is illustrated in Figure 10.21(b).

The coupling occurs under both normal and faulted power system conditions. The coupling causes voltages to appear on the insulated pipeline metal with respect to earth or currents to circulate in an earthed pipeline through the earthing connection. In general, voltage problems caused by capacitive coupling can be easily solved by earthing the pipeline. The pipeline voltages for a given pipeline exposure with the power line can be calculated using matrix analysis techniques as presented in Chapter 3. The self and mutual potential coefficients of the power line conductors are given in Equations (3.2). Equation (3.2b) can also be used to calculate the mutual potential coefficients between the pipeline and the power line's conductors where these form a parallelism. The self-potential coefficient of

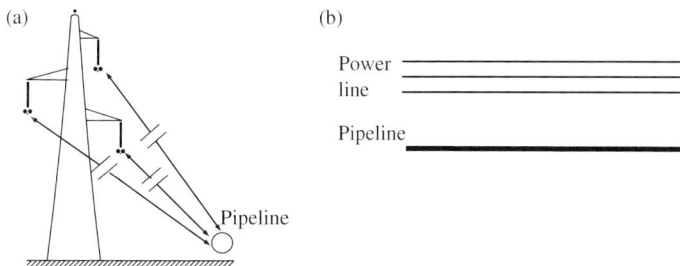

Figure 10.21   Illustration of electrostatic or capacitive coupling interference from a power line to a pipeline: (a) capacitive coupling and (b) parallel exposure

a pipeline close to the earth is given by

$$P_p = 17.975109 \times 10^6 \times \log_e \left[\frac{h_p + \sqrt{h_p^2 - r_p^2}}{r_p} \right] \text{km/F} \qquad (10.67)$$

where h_p is the pipeline's height above ground measured from the pipe's centre and r_p is the pipeline's radius, both are in m. Recalling Equation (3.1) and writing it in partitioned matrix form for a multi-conductor system that consists of power lines and pipelines, we can write

$$\begin{bmatrix} \mathbf{V}_C \\ \mathbf{V}_p \\ \mathbf{V}_E \end{bmatrix} = \begin{matrix} C \\ p \\ E \end{matrix} \begin{bmatrix} \mathbf{P}_C & \mathbf{P}_{Cp} & \mathbf{P}_{CE} \\ \mathbf{P}_{pC} & \mathbf{P}_p & \mathbf{P}_{pE} \\ \mathbf{P}_{EC} & \mathbf{P}_{Ep} & \mathbf{P}_E \end{bmatrix} \begin{bmatrix} \mathbf{Q}_C \\ \mathbf{Q}_p \\ \mathbf{Q}_E \end{bmatrix} \mathbf{V} \qquad (10.68)$$

where C, p and E represent the power lines' phase conductors, pipelines and power lines' earth wires, respectively. The equation is general and allows for the presence of more than one power line with more than one earth wire and more than one pipeline. All potential coefficients in Equation (10.68) are matrices. The earthed earth wires are now eliminated by substituting $\mathbf{V}_E = 0$ in Equation (10.68), giving

$$\begin{bmatrix} \mathbf{V}_C \\ \mathbf{V}_p \end{bmatrix} = \begin{matrix} C \\ p \end{matrix} \begin{bmatrix} \mathbf{P}'_C & \mathbf{P}'_{Cp} \\ \mathbf{P}'_{pC} & \mathbf{P}'_p \end{bmatrix} \begin{bmatrix} \mathbf{Q}_C \\ \mathbf{Q}_p \end{bmatrix} \mathbf{V} \qquad (10.69a)$$

where

$$\mathbf{P}'_C = \mathbf{P}_C - \mathbf{P}_{CE}\mathbf{P}_E^{-1}\mathbf{P}_{EC} \quad \mathbf{P}'_{Cp} = \mathbf{P}_{Cp} - \mathbf{P}_{CE}\mathbf{P}_E^{-1}\mathbf{P}_{Ep}$$

$$\mathbf{P}'_{pC} = \mathbf{P}_{pC} - \mathbf{P}_{pE}\mathbf{P}_E^{-1}\mathbf{P}_{EC} \quad \mathbf{P}'_p = \mathbf{P}_p - \mathbf{P}_{pE}\mathbf{P}_E^{-1}\mathbf{P}_{Ep} \qquad (10.69b)$$

The next step is to apply the pipelines' earthing constraint to Equation (10.69a). For an insulated pipeline, $\mathbf{Q}_p = 0$ and, from Equation (10.69a), the pipelines' voltages to earth due to capacitive coupling with the power lines are given by

$$\mathbf{V}_p = \mathbf{P}'_{pC}\mathbf{P}'^{-1}_C\mathbf{V}_C \, \mathbf{V} \qquad (10.70a)$$

where $\mathbf{V}_C$ are the known phase voltages to earth of the power lines. If a person touches pipeline i whose voltage is $V_{p(i)}$, the current that would flow through the person's body is determined by the series combination of his contact resistance to earth and the pipeline's capacitive reactance. In practice, the latter is much greater than the person's resistance and therefore the discharge current is given by

$$I_{p(i)} = j2\pi f 10^3 C_{p(i)} L_i V_{p(i)} \text{ mA} \qquad (10.70b)$$

where

$$C_{p(i)} = \frac{1}{P_{p(i)}} \text{ F/km} \qquad (10.70c)$$

and L_i is length of pipeline exposed to capacitive coupling in km. If the pipelines are solidly earthed or earthed through a very low impedance, $\mathbf{V}_p = 0$ and, from Equation (10.69a), we obtain

$$\mathbf{Q}_p = -\mathbf{P}'^{-1}_p\mathbf{P}'_{pC}(\mathbf{P}'_C - \mathbf{P}'_{Cp}\mathbf{P}'^{-1}_p\mathbf{P}'_{pC})^{-1}\mathbf{V}_C \, \text{C/km} \qquad (10.71a)$$

Since the phasor equivalent of the current $i = dq/dt$ is $I = j\omega Q$, the pipelines' charging currents are given by

$$\mathbf{I}_p = j2\pi f \mathbf{Q}_p \text{ A/km} \qquad (10.71b)$$

The discharge current through the body of a person that touches pipeline i is given by

$$I_{p(i)} = -j2\pi f \, 10^3 L_i Q_{p(i)} \text{ mA} \qquad (10.71c)$$

If the pipeline or some of its sections are not parallel to the power line, the distance between the pipeline, or a section of the pipeline, and the various power line conductors is no longer constant. Two such situations are illustrated in Figure 10.22. In both cases, the non-parallel pipeline exposure can be converted to a parallelism where the pipeline is parallel to the power line and is at an equivalent distance from the power line given by

$$x_{Eq} = \sqrt{x_1 x_2} \text{ m} \qquad (10.72a)$$

and

$$1/3 \leq x_1/x_2 \leq 3 \qquad (10.72b)$$

where x_{Eq} is the geometric mean distance to the power line and x_1 and x_2 are the minimum and maximum distances of the pipeline to the power line. The constraint of Equation (10.72b) is applied in order to maintain sufficient accuracy in calculating the mutual parameters between the pipeline and the power line. This constraint effectively places a limit on the length of a non-parallel pipeline section which necessitates dividing the pipeline into a number of sections each of which are converted to a parallel section to the power line.

For an insulated pipeline having a number of sections of both parallel and non-parallel exposures, the total pipeline voltage to earth can be calculated as the mean of the voltages in each section weighted by its length to the pipeline's total length

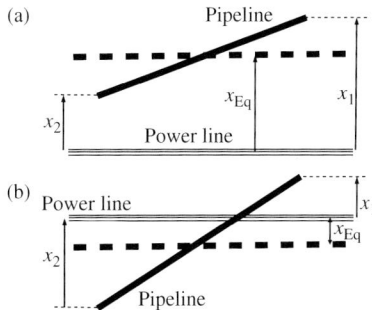

Figure 10.22   Conversion of non-parallel exposures to parallel exposures between a power line and a pipeline: (a) oblique exposure near power line and (b) oblique exposure crossing power line

as follows:

$$V_p = \frac{1}{L} \sum_{j=1}^{N} V_{p(j)} L_j \text{ V} \tag{10.73}$$

This voltage can be used to calculate the current that would flow through the body of a person that touches or comes into contact with the pipeline. The matrix analysis technique presented above can be extended and applied to double-circuit power lines.

10.15.3 Electromagnetic or inductive coupling from power lines to pipelines

General

Since the earth does not act as an electromagnetic shield, both above ground and buried pipelines in proximity to overhead lines are exposed to inductive coupling from the power line. The coupling exists under both normal steady state and faulted power system conditions and induces longitudinal voltages or electromotive forces (EMFs) on the pipeline. These EMFs produce voltage stresses on the pipeline and can also cause currents to circulate in the pipeline. These voltages (to earth) can reach several tens of volts under steady state conditions and a few kilovolts under fault conditions. The latter can damage the pipeline's insulation coating and cathodic protection systems. The longitudinal induced EMF depends on the distance between the power line and pipeline and the length of parallelism. The inductive zone of influence of a power line increases with the earth's resistivity and is generally taken as $y = 200\sqrt{\rho_e}$ where y is in metres. For example, for

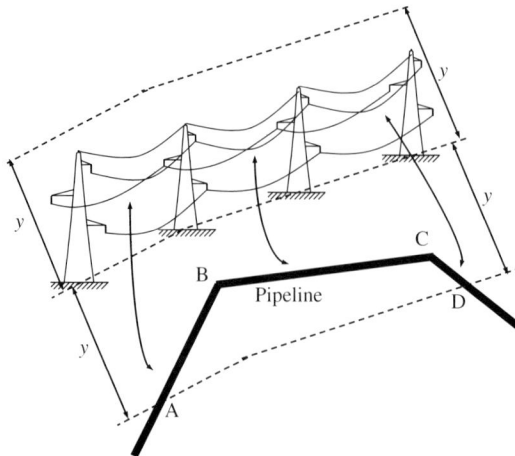

Figure 10.23   Illustration of electromagnetic or inductive coupling from power line to pipeline and power line's inductive zone of influence

$\rho_e = 50\,\Omega m$, $y = 1.42\,km$. The inductive coupling mechanism and zone of influence are illustrated in Figure 10.23 where the sections of the pipeline that fall within the zone of influence are AB, BC and CD. Each pipeline section presents an oblique exposure that can be converted to a parallelism using Equation (10.72).

Analysis of longitudinal induced EMFs on the pipeline

Mutual impedance between power line and pipeline

The EMF induced on a pipeline can be calculated under either normal steady state or faulted power system operating conditions. In both cases, the mutual impedances between the pipeline and the relevant power line conductors are required. Figure 10.24(a) illustrates the mutual inductive coupling between a single-circuit overhead power line with one earth wire and a buried pipeline parallel to the power line within the zone of influence.

The mutual impedance between the pipeline p and an overhead line phase or earth wire conductor j, with earth return, is calculated using Equation (3.20a) as follows:

$$Z_{pj} = \pi^2 10^{-4} f + j4\pi 10^{-4} f \log_e\left(\frac{D_{erc}}{d_{pj}}\right) \Omega/km \quad j = R, Y, B \text{ or } E \quad (10.74)$$

where d_{pj} is the distance between the centres of the pipeline and line conductor j, f is system frequency and D_{erc} is the depth of equivalent earth return given in Equation (3.15). If the distances between the pipeline and the power line's conductors are greater than the distance given in Equation (3.20b), i.e. $d_{pj} \leq \frac{D_{erc}}{7.32}$, Equation (10.74) results in an error that generally increases with distance particularly at low earth resistivity. The following equation gives an alternative formula for the mutual impedance and is valid for any distance between the pipeline and

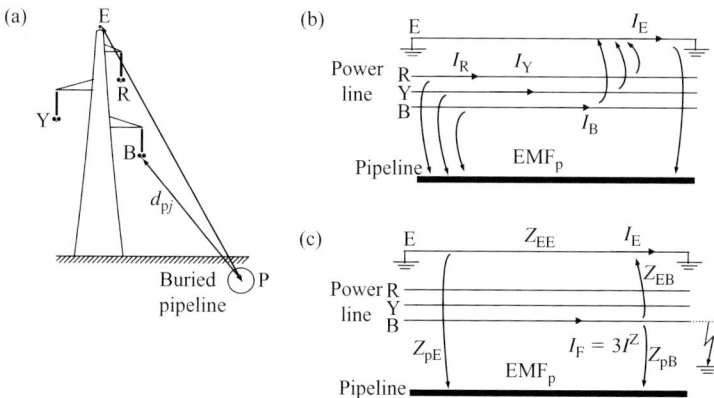

Figure 10.24   Calculation of induced EMF on a pipeline from a power line: (a) illustration of mutual coupling between pipeline and power line, (b) induced EMF under normal operation and (c) induced EMF under fault condition

the power line:

$$Z_{pj} = \pi^2 10^{-4} f + j4\pi f \, 10^{-4} \sqrt{\frac{\left[\log_e \left(1 + 1.382 \frac{D_{erc}^2}{d_{pj}^2} \right) \right]^2}{4} - \frac{\pi^2}{16}} \ \Omega/\text{km} \quad (10.75)$$

The use of Equation (10.75) may result in a safe upper overestimate in the magnitude of the mutual impedance that is always lower than 8% provided that Equation (10.72) is used.

Induced EMFs during steady state unfaulted system conditions

In Figure 10.24(a), we consider a general case of a single-circuit overhead line with one earth wire. However, the analysis approach is general and applicable for more than one earth wire, double-circuit lines and underground cables. The phase and earth wire currents are illustrated in Figure 10.24(b). Each current induces a voltage on the pipeline through the appropriate mutual impedance between the pipeline and the conductor. Therefore, the total longitudinal EMF induced on the pipeline due to the three-phase currents and the earth wire current is given by

$$-\text{EMF}_p = (Z_{pR} I_R + Z_{pY} I_Y + Z_{pB} I_B) + Z_{pE} I_E \ \text{V/km} \quad (10.76)$$

where the currents are in amps. The series voltage drop across the earth wire conductor is given by $\Delta V_E = Z_{ER} I_R + Z_{EY} I_Y + Z_{EB} I_B + Z_{EE} I_E = 0$ giving

$$I_E = \frac{-1}{Z_{EE}} (Z_{ER} I_R + Z_{EY} I_Y + Z_{EB} I_B)$$

Substituting I_E into Equation (10.76), we obtain

$$-\text{EMF}_p = (Z_{pR} I_R + Z_{pY} I_Y + Z_{pB} I_B) - \frac{Z_{pE}}{Z_{EE}} (Z_{ER} I_R + Z_{EY} I_Y + Z_{EB} I_B) \ \text{V/km}$$
$$(10.77)$$

Equation (10.77) can be used whether the phase currents I_R, I_Y and I_B are balanced or not. If the power line has no earth wire, the $Z_{pE} I_E$ term in Equation (10.76) disappears and the induced EMF is given by

$$-\text{EMF}_p = Z_{pR} I_R + Z_{pY} I_Y + Z_{pB} I_B \ \text{V/km} \quad (10.78)$$

Induced EMFs during a short-circuit fault in the power system

In many practical power systems, balanced three-phase short-circuit faults on high voltage power systems result in fault currents having a small degree of unbalance caused by the unbalanced network such as due to untransposed overhead lines. Therefore, the vector sum of the induced EMFs on an adjacent pipeline results in a small net induced EMF on the pipeline because the phase angle displacements between the three currents are not significantly different from 120°. The highest induced EMF on the pipeline will, therefore, occur under an unbalanced one-phase to earth short-circuit fault on the power system. The high ZPS current that flows on the power line induces a high EMF on a nearby pipeline depending on its proximity to the power line. Figure 10.24(c) illustrates a one-phase to earth short-circuit fault on phase conductor B, i.e. the nearest phase to the pipeline. The fault location is assumed at one extremity of the parallelism. For a fault within the length of the parallelism, the net EMF is calculated as the vector sum of the EMFs due to the opposing fault currents within the parallelism. From Figure 10.24(c), the series voltage drop across the earth wire is given by $\Delta V_E = Z_{EB}I_F + Z_{EE}I_E = 0$ or $I_E/I_F = -Z_{EB}/Z_{EE}$. The EMF induced on the pipeline is given by $-EMF_p = Z_{pB}I_F + Z_{pE}I_E$ and using $I_E/I_F = -Z_{EB}/Z_{EE}$, we obtain

$$-EMF_p = Z_{pB}I_F k \text{ V/km}, \quad I_F \text{ is in amps} \tag{10.79a}$$

where

$$k = 1 - \frac{Z_{EB}}{Z_{EE}} \times \frac{Z_{pE}}{Z_{pB}} \tag{10.79b}$$

and k is the screening factor. Equation (10.79b) is similar to Equation (10.40) except that the screening factor now includes the effect of both the earth wire and pipeline. Equation (10.79a) shows that since the fault current is constant within the parallelism, the longitudinal EMF per km induced on the pipeline will also be constant.

Analysis of pipeline voltages caused by inductive coupling

The calculation of pipeline voltages caused by the EMFs produced by inductive coupling from the power line, where the pipeline is within an overhead line's zone of influence, requires the modelling of the pipeline and determination of its electrical characteristics. Like the conductors of an overhead transmission line, a metallic pipeline can be considered as a long lossy transmission conductor of known geometrical dimensions and physical characteristics. The magnetic coupling from the power line to the pipeline may be represented as distributed induced EMF sources on the pipeline. The equivalent circuit of a distributed pipeline section of length L in parallel with the conductors of an overhead power line is shown in Figure 10.25. $E(x)$ is the induced EMF increment from the power line per km, z and y are the pipeline's series impedance and shunt admittance per km and Z_A and Z_B are the impedances of the pipeline seen from ends A and B, respectively. From Figure 10.25(b), we can write

$$V(x) - z \, dx \, I(x) + E(x) \, dx - [V(x) + dV(x)] = 0 \tag{10.80a}$$

or

$$\frac{dV(x)}{dx} = -zI(x) + E(x) \tag{10.80b}$$

and

$$I(x) - dI(x) = y \, dx \, V(x) + I(x) \tag{10.81a}$$

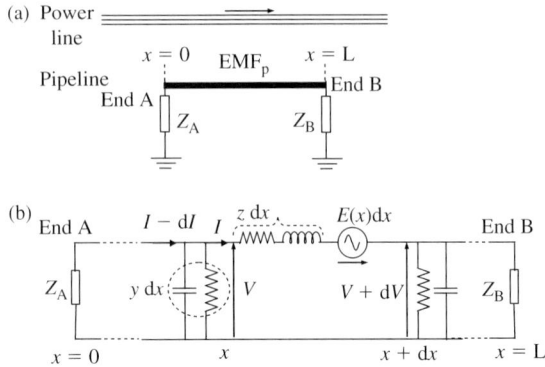

Figure 10.25 Equivalent circuit of a distributed metal pipeline inductively coupled to a power line: (a) pipeline section parallel to a power line and (b) pipeline distributed parameter equivalent circuit

or

$$\frac{dI(x)}{dx} = -yV(x) \tag{10.81b}$$

Differentiating Equations (10.80b) and (10.81b) with respect to x and using the original equations, we obtain

$$\frac{d^2V(x)}{dx^2} = \gamma^2 V(x) + \frac{dE(x)}{dx} \tag{10.82a}$$

$$\frac{d^2I(x)}{dx^2} = \gamma^2 I(x) - yE(x) \tag{10.82b}$$

where

$$\gamma = \sqrt{zy} \ \text{km}^{-1} \tag{10.82c}$$

γ is the propagation constant of the lossy pipeline. Under normal unfaulted power system operation, the load currents influencing the parallel pipeline section are constant. The current will also be constant under a one-phase to earth fault condition on the power line beyond the section parallel to the pipeline. Under both conditions, the induced EMF is constant, i.e. $E(x) = E_o$. It can be shown that the solutions of Equations (10.82a) and (10.82b) are given by

$$V(x) = \frac{E_o}{\gamma} \left\{ \frac{[Z_B(Z_A - Z_o) - Z_A(Z_B + Z_o)e^{\gamma L}]e^{-\gamma x}}{-[Z_A(Z_B - Z_o) - Z_B(Z_A + Z_o)e^{\gamma L}]e^{-\gamma(L-x)}}{(Z_A + Z_o)(Z_B + Z_o)e^{\gamma L} - (Z_A - Z_o)(Z_B - Z_o)e^{-\gamma L}} \right\} \text{V} \tag{10.83a}$$

$$I(x) = \frac{E_o}{\gamma Z_o} \left\{ 1 + \frac{\begin{array}{c}[Z_B(Z_A - Z_o) - Z_A(Z_B + Z_o)e^{\gamma L}]e^{-\gamma x} \\ +[Z_A(Z_B - Z_o) - Z_B(Z_A + Z_o)e^{\gamma L}]e^{-\gamma(L-x)}\end{array}}{(Z_A + Z_o)(Z_B + Z_o)e^{\gamma L} - (Z_A - Z_o)(Z_B - Z_o)e^{-\gamma L}} \right\} V$$

(10.83b)

where

$$Z_o = \sqrt{z/y} \ \Omega$$

(10.83c)

Z_o is the characteristic impedance of the lossy pipeline. Equation (10.83a) can be used to calculate the pipeline voltage to earth along the pipeline section, i.e. at any value of x between zero and L and Equation (10.83b) can be used to calculate the pipeline current. The terminal impedances Z_A and Z_B are chosen to represent the pipeline's electrical characteristics at ends A and B of the pipeline section. For a pipeline section that continues for a few kilometres beyond end A, the electrical continuity of the pipeline persists and the pipeline's impedance seen from point A may be taken as equal to the pipeline's characteristic impedance or $Z_A = Z_o$. If the pipeline is solidly earthed at end A then $Z_A = 0$ but if the pipeline has insulating joints and unearthed at end A, then $Z_A = \infty$. These characteristics apply equally to end B. The magnitude of the pipeline voltage will obviously vary from zero, where the pipeline is earthed to a maximum where the pipeline is insulated and unearthed. For the case where the pipeline parallel section is insulated and unearthed at end B but extends beyond end A, i.e. $Z_A = Z_o$ and $Z_B = \infty$, Equations (10.83a) and (10.83b) simplify to the following:

$$V(x) = \frac{E_o}{2\gamma}[(2e^{-\gamma L} - e^{-2\gamma L})e^{\gamma x} - e^{-\gamma x}] V$$

(10.84a)

$$I(x) = \frac{E_o}{2\gamma Z_o}[2 + (e^{-2\gamma L} - 2e^{-\gamma L})e^{\gamma x} - e^{-\gamma x}] A$$

(10.84b)

The maximum pipeline voltage to earth occurs at end B, i.e. $x = L$ and is given by

$$V(x = L) = \frac{E_o}{\gamma}(1 - e^{-\gamma L}) V$$

(10.85a)

and the current at end B is $I(x = L) = 0$. At end A, i.e. $x = 0$, Equation (10.84a) gives

$$V(x = 0) = \frac{E_o}{2\gamma}(2e^{-\gamma L} - e^{-2\gamma L} - 1) V$$

(10.85b)

Beyond end A where the pipeline extends away perpendicular to the power line, the attenuation of the pipeline voltage can be evaluated to ensure that appropriate safety measures are taken to protect personnel working on the pipeline even many kilometres away. Using Equation (10.85b) that gives the pipeline voltage at end A, the pipeline voltage along a new axis u at a distance u beyond end A is given by

$$V(u) = \frac{E_o}{2\gamma}[2e^{-\gamma L} - e^{-2\gamma L} - 1]e^{-\gamma u} V$$

(10.86)

Where the pipeline extends beyond end A and remains inductively coupled to the power line, this can be considered a new pipeline section having an oblique exposure and converted to a parallelism using Equation (10.72).

Electrical characteristics of metal pipelines

Series impedance with earth return

In calculating the pipeline voltage to earth, we represented the pipeline as a long lossy distributed parameter conductor or transmission line. The pipeline was characterised by its propagation constant and characteristic impedance given in Equations (10.82c) and (10.83c), respectively, where z and y are the pipeline's series impedance with earth return and shunt admittance. The series impedance consists of the internal impedance and external impedance with earth return. For pipelines installed above ground, the series impedance with earth return is given by

$$z = \frac{\sqrt{\rho_p \mu_p f}}{3.163 r_p} + \pi^2 10^{-4} f + j \left[\frac{\sqrt{\rho_p \mu_p f}}{3.163 r_p} + 4\pi 10^{-4} f \log_e \left(\frac{D_{erc}}{r_p} \right) \right] \ \Omega/\text{km} \quad (10.87)$$

where r_p is the pipeline's outer radius, μ_p is the relative permeability of the pipeline's metal, typically about 300 for steel, ρ_p is the pipeline's resistivity, f is the power frequency and D_{erc} is as given in Equation (3.15). For buried pipelines, the external impedance part of the series impedance is dependent on the soil's electrical characteristics and is a complicated function of the propagation constant γ. For computer-based calculations, the series impedance of a buried pipeline is given by

$$z = \frac{\sqrt{\rho_p \mu_p f}}{3.163 r_p} + \pi^2 10^{-4} f + j \left[\frac{\sqrt{\rho_p \mu_p f}}{3.163 r_p} + 4\pi 10^{-4} f \right.$$

$$\left. \times \log_e \left(\frac{1.85}{r_{p(Eq)} \sqrt{\gamma^2 + 8\pi^2 f 10^{-7} (1/\rho_e + j2\pi f \varepsilon_e)}} \right) \right] \ \Omega/\text{km} \quad (10.88)$$

where $r_{p(Eq)} = \sqrt{r_p^2 + 4d_p^2}$, in m, is the equivalent radius of the pipeline, d_p is the depth of the pipeline, ρ_e and ε_e are earth's resistivity and permitivity, respectively. In Equation (10.88), γ should be in m^{-1}. Generally, for modern buried coated pipelines, Equation (10.87) is found to be sufficiently accurate.

Shunt admittance

For an uncoated pipeline installed above ground at a height h_p in m, the shunt admittance $y = j2\pi f C_p$ is calculated using Equations (10.67) and (10.70c). Thus

$$y = j \frac{0.349549 \times f \times 10^{-6}}{\log_e \left[\frac{h_p + \sqrt{h_p^2 - r_p^2}}{r_p} \right]} \ \text{S/km} \quad (10.89)$$

where h_p is measured from the pipe's centre. For a coated and buried pipeline, its equivalent shunt admittance y consists of the coating's admittance y_c in series with the external earth admittance y_e and is given by

$$y = \frac{1}{\frac{1}{y_c} + \frac{1}{y_e}} \text{ S/km} \tag{10.90}$$

where

$$y_c = \frac{2000\pi r_p}{\rho_c t_c} + j\frac{\pi r_p f \varepsilon_c 10^{-6}}{9 t_c} \text{ S/km} \tag{10.91a}$$

and

$$y_e = \frac{1000\pi(1/\rho_e + j2\pi f \varepsilon_e)}{\log_e\left(\frac{1.12}{\gamma r_{p(Eq)}}\right)} \text{ S/km} \tag{10.91b}$$

where ρ_c is the resistivity of the pipeline's coating in Ωm, ε_c is the coating's relative permittivity and t_c is the coating's thickness in m. For coated pipelines with usually high coating resistivity, y_e is much greater than y_c and Equation (10.90) gives

$$y = y_c = \frac{2000\pi r_p}{\rho_c t_c} + j\frac{\pi r_p f \varepsilon_c 10^{-6}}{9 t_c} \text{ S/km} \tag{10.92}$$

10.15.4 Resistive or conductive coupling from power systems to pipelines

In Sections (10.11)–(10.13), we discussed how ZPS earth return currents due to short-circuit earth faults in substations and on overhead line towers cause a rise of earth potential with respect to remote earth over a given area. If a buried pipeline is located in the zone of influence, i.e. earth potential rise, irrespective of whether the pipeline is parallel to the power line or not, the coating insulation will be exposed to a voltage stress since the pipeline's metal remains at virtually earth potential. If this voltage stress is greater than the dielectric strength of the insulation coating, it may puncture the coating and damage cathodic protection systems. If the pipeline passes through the earth electrode systems of substations or overhead power line towers, the voltage stress may be so high that it may puncture the coating. The intense leakage current may damage the pipeline's metal. If the pipeline is earthed and connected to the earth electrode of a substation or tower, the rise of earth potential of the substation or tower will be transferred to the pipeline and may endanger safety.

10.15.5 Examples

Example 10.3   Consider a single-circuit 400 kV overhead transmission line with one earth wire and an above ground insulated metal pipeline in the vicinity as shown in Figure 10.26.

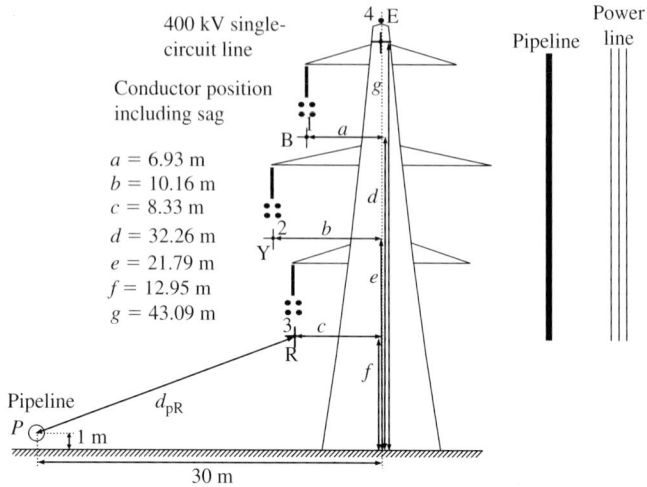

Figure 10.26 Power line and above ground pipeline for calculation of capacitive coupling for Example 10.3

The geometrical and physical data of the overhead line circuit is identical to that used in Example 3.4 and system frequency is 50 Hz. The pipeline is parallel to the axis of the power line at a distance of 30 m, has an outer radius of 0.35 m and its height above ground is 1 m. The length of parallel exposure of the pipeline and power line is 4 km. The three-phase voltages on the power line are in a balanced steady state condition. Calculate the touch voltage to earth on the pipeline induced by capacitive coupling with the power line and the discharge current that can flow through a person's body when touching the pipeline.

The pipeline's self-potential coefficient and mutual potential coefficients with the power line are

$$P_p = 17.975109 \times 10^6 \times \log_e \left[\frac{1 + \sqrt{1 - 0.35^2}}{0.35} \right] = 30.75 \times 10^6 \text{ km/F}$$

$$P_{pR} = 17.975109 \times 10^6 \times \log_e \left[\frac{\sqrt{(30 - 8.33)^2 + (12.95 + 1)^2}}{\sqrt{(30 - 8.33)^2 + (12.95 - 1)^2}} \right]$$

$$= 0.72978 \times 10^6 \text{ km/F}$$

$$P_{pY} = 0.90175 \times 10^6 \text{ km/F} \quad P_{pB} = 0.73726 \times 10^6 \text{ km/F}$$

$$P_{pE} = 0.56191 \times 10^6 \text{ km/F}$$

The potential coefficient matrix of the power line and pipeline coupled system is

$$
\mathbf{P} = \begin{array}{c} R \\ Y \\ B \\ p \\ E \end{array}
\left[
\begin{array}{ccc|c|c}
85.366 & 22.165 & 15.253 & 0.72978 & 10.46 \\
22.165 & 94.72 & 28.719 & 0.90175 & 18.0 \\
15.253 & 28.719 & 101.773 & 0.73726 & 30.985 \\
\hline
0.72978 & 0.90175 & 0.73726 & 30.75 & 0.56191 \\
\hline
10.46 & 18.0 & 30.985 & 0.56191 & 156.833
\end{array}
\right] \times 10^6 \, \text{km/F}
$$

Thus

$$
\mathbf{P}_C = \begin{bmatrix} 85.366 & 22.165 & 15.253 \\ 22.165 & 94.72 & 28.719 \\ 15.253 & 28.719 & 101.773 \end{bmatrix} \times 10^6 \, \text{km/F}
$$

$$
\mathbf{P}_{Cp} = \begin{bmatrix} 0.72978 \\ 0.90175 \\ 0.73726 \end{bmatrix} \times 10^6 \, \text{km/F}
$$

$$
\mathbf{P}_{CE} = \begin{bmatrix} 10.46 \\ 18.0 \\ 30.985 \end{bmatrix} \times 10^6 \, \text{km/F} \quad \mathbf{P}_{pC} = \mathbf{P}_{Cp}^T \quad \mathbf{P}_{EC} = \mathbf{P}_{CE}^T
$$

$$
\mathbf{P}_p = 30.75 \times 10^6 \, \text{F/km}
$$

$$
\mathbf{P}_{pE} = \mathbf{P}_{Ep} = 0.56191 \times 10^6 \, \text{km/F} \quad \mathbf{P}_E = 156.83 \times 10^6 \, \text{F/km}
$$

$$
\mathbf{P}'_C = \mathbf{P}_C - \mathbf{P}_{CE}\mathbf{P}_E^{-1}\mathbf{P}_{EC} = \begin{bmatrix} 84.668 & 20.964 & 13.186 \\ 20.964 & 92.654 & 25.163 \\ 13.186 & 25.163 & 95.651 \end{bmatrix} \times 10^6 \, \text{km/F}
$$

$$
\mathbf{P}'_{Cp} = \mathbf{P}_{Cp} - \mathbf{P}_{CE}\mathbf{P}_E^{-1}\mathbf{P}_{Ep} = \begin{bmatrix} 0.692 \\ 0.837 \\ 0.626 \end{bmatrix} \times 10^6 \, \text{km/F}
$$

$$
\mathbf{P}'_{pC} = \mathbf{P}_{pC} - \mathbf{P}_{pE}\mathbf{P}_E^{-1}\mathbf{P}_{EC} = \mathbf{P}_{Cp}'^T
$$

$$
\mathbf{P}'_p = \mathbf{P}_p - \mathbf{P}_{pE}\mathbf{P}_E^{-1}\mathbf{P}_{Ep} = 30.748 \times 10^6 \, \text{km/F}
$$

$$
\mathbf{V}_C = \frac{1}{\sqrt{3}} \begin{bmatrix} 400000 \\ h^2 400000 \\ h 400000 \end{bmatrix} \text{V}
$$

Therefore,

$$V_p = \mathbf{P}'_{pC}\mathbf{P}_C^{-1}\mathbf{V}_C = (142.012 - j524.367)V = 543.26\ V\angle{-74.85°}$$

$$I_p = 2\pi f 10^3 \frac{1}{P_p} \times L \times V_p = 2\pi \times 50 \times \frac{1}{30.75 \times 10^6} \times 4 \times (142.012$$
$$- j524.367) \times 1000 = (5.803 - j21.429)mA = 22.2\ mA\angle{-74.85°}$$

$$Q_p = -\mathbf{P}'^{-1}_p\mathbf{P}'_{pC}(\mathbf{P}'_C - \mathbf{P}'_{Cp}\mathbf{P}'^{-1}_p\mathbf{P}'_{pC})^{-1}\mathbf{V}_C = (-0.462 + j1.706) \times 10^{-5}\ C/km$$
$$I_p = -j2\pi f 10^3 L Q_p = -j2\pi \times 50 \times 10^3 \times 4 \times (-0.462 + j1.706) \times 10^{-5}$$
$$= (5.806 - j21.439)mA = 22.2\ mA\angle{-74.85°}$$

As discussed in Section 10.2.3, a steady state current higher than about 5 mA would generally be considered unacceptable from personnel safety viewpoint. Therefore, the pipeline in this example would generally be earthed through an appropriate resistance typically of the order of a few hundred ohms.

Example 10.4   Consider a single-circuit overhead power line and a parallel metal pipeline in the vicinity. The power line carries a balanced set of three-phase steady state currents. Derive an expression for the induced EMF on the pipeline in terms of the distances between the pipeline and the power line's phase conductors. Consider a power line with and without one earth wire.

We start with case where the power line has no earth wire. The balanced three-phase currents are PPS only thus $I_R = I^P$, $I_Y = h^2 I^P$ and $I_B = h I^P$. Therefore, from Equation (10.78), we have $-EMF_p = (Z_{pR} + h^2 Z_{pY} + h Z_{pB})I^P$. Using Equation (10.74),

$$Z_{pj} = \pi^2 10^{-4} f + j4\pi 10^{-4} f \log_e\left(\frac{D_{erc}}{d_{pj}}\right) \quad \text{for} \quad j = R, Y, B$$

$$h = -\frac{1}{2} + j\frac{\sqrt{3}}{2} \quad \text{and} \quad h^2 = -\frac{1}{2} - j\frac{\sqrt{3}}{2}$$

we have

$$-EMF_p = I^P \left\{ \pi^2 10^{-4} f + j4\pi 10^{-4} f \log_e\left(\frac{D_{erc}}{d_{pR}}\right) + \left(\frac{-1}{2} - j\frac{\sqrt{3}}{2}\right) \right.$$
$$\times \left[\pi^2 10^{-4} f + j4\pi 10^{-4} f \log_e\left(\frac{D_{erc}}{d_{pY}}\right)\right]$$
$$\left. + \left(\frac{-1}{2} + j\frac{\sqrt{3}}{2}\right)\left[\pi^2 10^{-4} f + j4\pi 10^{-4} f \log_e\left(\frac{D_{erc}}{d_{pB}}\right)\right] \right\}$$

or after a little algebra

$$-\text{EMF}_p = 4\pi 10^{-4} f I^P \left\{ \frac{\sqrt{3}}{2} \log_e \left(\frac{d_{pB}}{d_{pY}} \right) + j \log_e \left(\frac{\sqrt{d_{pY} d_{pB}}}{d_{pR}} \right) \right\} \text{V/km}$$

where the mutual impedances are in Ω/km and $I_R = I^P$ is in amperes.

If the line has one earth wire, using Equation (10.77), and by analogy with the above equation, we obtain

$$-\text{EMF}_p = 4\pi 10^{-4} f I^P \left\{ \frac{\sqrt{3}}{2} \log_e \left(\frac{d_{pB}}{d_{pY}} \right) + j \log_e \left(\frac{\sqrt{d_{pY} d_{pB}}}{d_{pR}} \right) \right.$$

$$\left. - \frac{Z_{pE}}{Z_{EE}} \left[\frac{\sqrt{3}}{2} \log_e \left(\frac{d_{EB}}{d_{EY}} \right) + j \log_e \left(\frac{\sqrt{d_{EY} d_{EB}}}{d_{ER}} \right) \right] \right\} \text{V/km}$$

Example 10.5 Consider the single-circuit 400 kV overhead transmission line of Example 10.3 and a non-parallel buried coated steel pipeline in the vicinity as shown in Figure 10.27. The pipeline is insulated and unearthed at end B and extends perpendicular to the overhead power line at end A.

The geometrical and physical data of the pipeline is as follows:

Outer radius $= 0.25$ m, depth $= 1$ m, resistivity and relative permeability of steel are $0.138 \, \mu\Omega$m and 300, respectively. Insulation coating is polyethylene with thickness $= 5$ mm, resistivity $= 25 \times 10^6 \, \Omega$m and relative permittivity $= 5$.

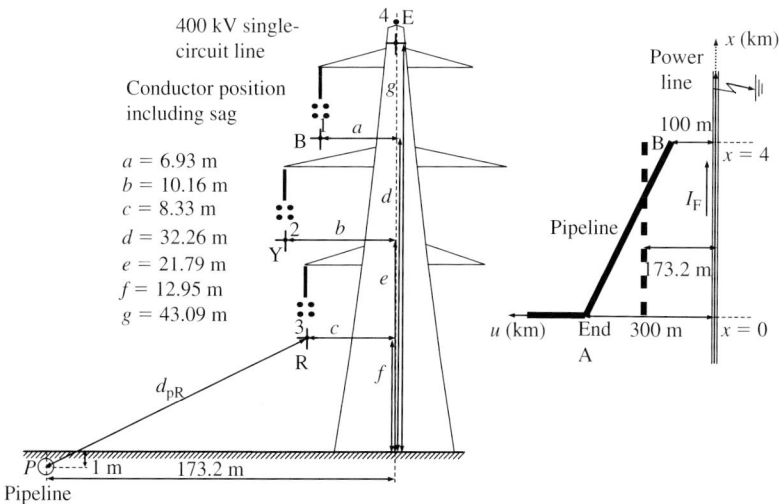

Figure 10.27 Power line and buried coated steel pipeline for calculation of inductive coupling for Example 10.5

Earth resistivity is $100\,\Omega$m. System frequency is $50\,$Hz.
Balanced steady state currents on the power line of $2000\,$A rms. One-phase to earth short-circuit fault current on phase R at the edge of zone of influence is $13\,000\,$A rms.

Calculation of mutual impedances
The non-parallel pipeline is converted to a parallel pipeline at a geometric mean distance from the power line equal to $d_{\text{Eq}} = \sqrt{100 \times 300} = 173.2\,$m. The zone of influence of the power line is $y = 200\sqrt{100} = 2000\,$m. Hence, the pipeline is well within the zone. The depth of earth return is $D_{\text{erc}} = 658.87\sqrt{100/50} = 931.78\,$m. The distances between the pipeline and the power line's phase and earth wire conductors are $d_{\text{pR}} = \sqrt{(12.95 + 1)^2 + (173.2 - 8.33)^2} = 165.46\,$m, similarly $d_{\text{pY}} = 164.63\,$m, $d_{\text{pB}} = 169.56\,$m and $d_{\text{pE}} = 178.72\,$m. The mutual impedances with earth return between the pipeline and the power line's conductors are

$$Z_{\text{pR}} = \pi^2 10^{-4} \times 50 + \text{j}4\pi 50 \times 10^{-4} \sqrt{\frac{\left[\log_e\left(1 + 1.382\frac{931.78^2}{165.46^2}\right)\right]^2}{4} - \frac{\pi^2}{16}}$$

$$= (0.0493 + \text{j}0.1088)\Omega/\text{km}$$

Similarly

$$Z_{\text{pY}} = (0.0493 + \text{j}0.1091)\Omega/\text{km}$$

$$Z_{\text{pB}} = (0.0493 + \text{j}0.1071)\Omega/\text{km}$$

and

$$Z_{\text{pE}} = (0.0493 + \text{j}0.1036)\Omega/\text{km}$$

From Example 3.4, we have

$$Z_{\text{ER}} = (0.0493 + \text{j}0.1609)\Omega/\text{km} \quad Z_{\text{EY}} = (0.0493 + \text{j}0.178)\Omega/\text{km},$$

$$Z_{\text{EB}} = (0.0493 + \text{j}0.2147)\Omega/\text{km} \quad \text{and} \quad Z_{\text{EE}} = (0.1136 + \text{j}0.6988)\Omega/\text{km}.$$

Electrical characteristics of the pipeline
The pipeline's shunt admittance is

$$y = \frac{2000\pi \times 0.25}{25 \times 10^6 \times 5 \times 10^{-3}} + \text{j}\frac{\pi \times 0.25 \times 50 \times 5 \times 10^{-6}}{9 \times 5 \times 10^{-3}}$$

$$= (0.01256 + \text{j}0.00436)\text{S}/\text{km}$$

The series impedance of the pipeline with earth return is

$$z = \frac{\sqrt{0.138 \times 10^{-6} \times 300 \times 50}}{3.163 \times 0.25} + \pi^2 10^{-4} \times 50$$

$$+ j \left[\frac{\sqrt{0.138 \times 10^{-6} \times 300 \times 50}}{3.163 \times 0.25} + 4\pi 10^{-4} \times 50 \times \log_e \left(\frac{931.78}{0.25} \right) \right]$$

$$= (0.10688 + j0.5167)\Omega/\text{km}$$

The pipeline's propagation constant is

$$\gamma = \sqrt{(0.10688 + j0.5167) \times (0.01256 + j0.00436)}$$

$$= (0.05525 + j0.06295)\text{km}^{-1}$$

The pipeline's characteristic impedance is

$$Z_o = \sqrt{\frac{0.10688 + j0.5167}{0.01256 + j0.00436}} = 5.478 + j3.11 = 6.3\angle 29.58° \, \Omega$$

Steady state unfaulted power system operation

Induced EMF
The balanced three-phase currents are PPS only thus $I_R = 2000\,\text{A}$, $I_Y = 2000\,\text{Ae}^{-j2\pi/3}$ and $I_B = 2000\,\text{Ae}^{j2\pi/3}$. The induced EMF on the pipeline is

$$-\text{EMF}_p = (0.0493 + j0.1088)(2000) + (0.0493 + j0.1091)(2000e^{-j2\pi/3})$$

$$+ (0.0493 + j0.1071)(2000e^{j2\pi/3}) - \frac{(0.0493 + j0.1036)}{(0.1136 + j0.6988)}$$

$$\times [(0.0493 + j0.1609)(2000) + (0.0493 + j0.178)(2000e^{-j2\pi/3})$$

$$+ (0.0493 + j0.2147)(2000e^{j2\pi/3})] = (3.57 + j0.182)$$

$$- (-13.13 - j8.14) = 16.7 + j8.32\,\text{V/km}$$

Therefore, $\text{EMF}_p = -16.7 - j8.32 = 18.66\angle -153.5°$ V/km. If the overhead power line had no earth wire, the induced EMF is $\text{EMF}_p = -3.57 - j0.182 = 3.57\angle -177.1°$ V/km. It is interesting to note that the effect of the power line's earth wire under steady state conditions is to increase the induced voltage on the pipeline. This is opposite to the effect under fault conditions as shown in Equation (10.79).

Pipeline voltage to earth
The pipeline voltage to earth along the pipeline between ends A and B as well as beyond end A can be calculated using Equation (10.84a) between ends A and B and Equation (10.86) beyond end A where $E_o = (-16.7 - j8.32)$V/km and $\gamma = (0.05525 + j0.06295)$km^{-1}. The maximum pipeline voltage at end B at $x = 4$ km is equal to $-62.95 - j22.4 = 66.8\angle-160.4°$ V and the voltage at end A at $x = 0$ is equal to $4.82 + j8.83 = 10.06\angle61.37°$ V.

One-phase to earth short-circuit fault condition

Induced EMF
   The overhead line's screening factor is

$$k = 1 - \frac{Z_{ER}}{Z_{EE}} \times \frac{Z_{pE}}{Z_{pR}} = 1 - \frac{(0.0493 + j0.1609)}{(0.1136 + j0.6988)} \times \frac{(0.0493 + j0.1036)}{(0.0493 + j0.1088)}$$

$$= 1 - (0.23769\angle-7.8°) \times (0.96048\angle-1.07°) = 0.7744 + j0.0352$$

$$= 0.775\angle2.6°.$$

The induced EMF on the pipeline is $-EMF_p = Z_{pR}I_F k$ or

$$-EMF_p = (0.0493 + j0.1088) \times 13\,000 \times 0.775\angle2.6° = -446.5 - j1117.5$$

$$= 1203.4\angle-111.78° \text{ V/km}$$

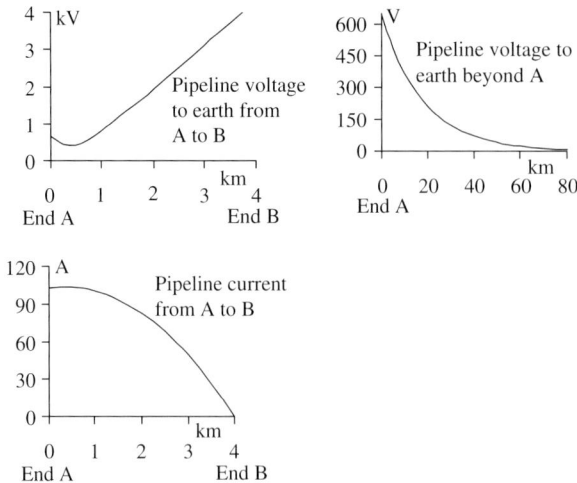

Figure 10.28   Calculated voltages to earth and current profiles on a steel pipeline inductively coupled to a 400 kV power line under a one-phase to earth fault condition. Corresponding to Figure 10.27

Pipeline voltage to earth and pipeline current

We now calculate the profile of the pipeline voltage to earth along the pipeline between ends A and B as well as the attenuation beyond end A using Equations (10.84a) and (10.86) respectively. We also calculate the pipeline current along the pipeline between ends A and B using Equation (10.84b). In these Equations $E_o = (-446.5 - j1117.5)$V/km, $\gamma = (0.05525 + j0.06295)$km^{-1} and $Z_o = (5.478 + j3.11)\Omega$. The results of $|V(x)|$ and $|I(x)|$ are shown in Figure 10.28.

Further reading

Books

[1] Sunde, E.D., *Earth Conduction Effects in Transmission Systems*, Van Nostrand, Dover Publications, New York, 1968.
[2] Tagg, G.T., *Earth Resistance*, G. Newnes, London, 1964.
[3] DD IEC/TS 60479-1:2005, *Effects of Current on Human Beings and Livestock – Part 1: General Aspects*.
[4] British Standard BS 7430:1998, *Code of Practice for Earthing*.
[5] Engineering Recommendation S34, *A Guide for Assessing the Rise of Earth Potential at Substation Sites*, Electricity Network Association, London, 1986.
[6] IEEE Standard 80:2001, *Design of Power System Grounding*.
[7] British Standard BS EN 12954:2001, *Cathodic Protection of Buried or Immersed Metallic Structures–General Principles and Applications for Pipelines*.

Papers

[8] Thapar, B., *et al.*, Evaluation of ground resistance of a grounding grid of any shape, *IEEE Transaction on Power Delivery*, Vol. 6, No. 2, April 1991, 640–647.
[9] Nahman, J., *et al.*, Earth fault currents and potential distribution in composite systems, *IEE Proceedings Generation Transmission and Distribution*, Vol. 142, No. 2, March 1995, 135–142.
[10] Sverak, J.G., *et al.*, Safe substation grounding part 1, *IEEE Transactions on PAS*, Vol. PAS-100, No. 9, September 1981, 4281–4290.
[11] Sverak, J.G., *et al.*, Safe substation grounding part II, *IEEE Transactions on PAS*, Vol. PAS-101, No. 10, October 1982, 4006–4023.
[12] Endrenyi, J., Analysis of transmission tower potentials during ground faults, *IEEE Transactions on PAS*, Vol. PAS-86, No. 10, October 1967, 1274–1283.
[13] Verma, R. and Mukhedkar, D., Ground fault current distribution in substation, towers and ground wire, *IEEE Transactions on PAS*, Vol. PAS-98, No. 3, May/June 1979, 724–730.
[14] Dawalibi, F.P., Ground fault current distribution between soil and neutral conductors, *IEEE Transaction on PAS*, Vol. PAS-99, March/April 1980, 452–461.
[15] Popovic, L.M., Determination of reduction factor for cable lines consisting of three single-core cables, *IEEE Transactions on PD-18*, No. 3, 2003, 736–743.

[16] Taflove, A., *et al.*, Prediction method for buried pipeline voltages due to 60 Hz AC inductive coupling, Part I: Analysis, *IEEE Transactions on PAS*, Vol. PAS-18, No. 3, May/June 1979, 780–794.

[17] Djogo, G. and Salama, M.M.A., Calculation of inductive coupling from power lines to multiple pipelines and buried conductors, *Electric Power Systems Research*, Vol. 41, 1997, 75–84.

[18] Dawalibi, F.P., *et al.*, Analysis of electrical interference from power lines to gas pipelines – Part I: Computation methods, *IEEE Transactions on Power Delivery*, Vol. 4, No. 3, July 1989, 1840–1846.

[19] Kouteynikoff, P., Results of an international survey on the rules limiting interference coupled into metallic pipelines by high voltage power systems, *Electra*, No. 110, January 1987, 55–66.

[20] Favez, B. and Gougeuil, J.C., Contribution to studies on problems resulting from the proximity of overhead lines with underground metal pipelines, Report No. 336, CIGRE, Session 1966.

[21] Wait, J.R., Mutual coupling between grounded circuits and the effect of thin vertical conductor in the earth, *IEEE Transaction on Antennas and Propagation*, Vol. AP-31, No. 4, July 1983, 640–644.

[22] Carson, J.R., Wave propagation in overhead wires with ground return, *Bell System Technical Journal*, Vol. 5, 1926, 539–554.

[23] Pollaczek, F., Ueber das feld einer unendlich langen wechsel stromdurchflossenen einfachleitung, *Electrishe Nachrichten Technik*, Vol. 3, No. 9, 1926, 339–359.

[24] Schelkunoff, S.A., The electromagnetic theory of coaxial transmission lines and cylindrical shells, *Bell System Technical Journal*, Vol. XIII, 1934, 532–578.

[25] Haberland, G., Theorie der Leitung von Wechselstrom durch die Erde, Zeitschrifi fir Angewandte Mathematik und Mechanik, No. 6, 1926, 366.

Appendices

A.1 Theory and analysis of distributed multi-conductor lines and cables

In order to describe accurately the performance of multi-conductor transmission lines, we used in Chapter 3 two-port nodal admittance equations that included the line's series impedance and shunt admittance matrices. We will now briefly present the theory of distributed multi-conductor lines. The theory also applies to cables.

Consider a multi-conductor line that consists of N conductors. Each conductor, say conductor k, has a self-impedance Z_{kk}, with earth return, and is mutually coupled to all other conductors through mutual impedances Z_{kj}, with earth return. In addition, conductor k has a self-admittance Y_{kk} and is mutually coupled to all other conductors through mutual admittances Y_{kj}. Taking an element of infinitesimal length Δx on the line, the series voltage developed in length Δx on conductor k when currents flow in all conductors, $1, 2, \ldots, k, \ldots, N$, is given by

$$\Delta V_{k_{(k=1,\ldots,N)}} = -\sum_{j=1}^{N} Z_{kj} I_j \Delta x \qquad \text{(A.1a)}$$

The minus sign is because the change in voltage is negative with increasing x.

Similarly, the shunt displacement current due to the voltage applied to conductor k and voltages on all other conductors, $1, 2, \ldots, N$, is given by

$$\Delta I_{k_{(k=1,\ldots,N)}} = -\left(-Y_{kk} V_k + \sum_{j=1\neq k}^{N} Y_{kj} V_j \right) \Delta x \qquad \text{(A.1b)}$$

Writing Equations (A.1) in matrix form for $\Delta x \to 0$, we have

$$\frac{\mathrm{d}\mathbf{V}(x)}{\mathrm{d}x} = -\mathbf{Z}\mathbf{I}(x) \qquad \text{(A.2a)}$$

$$\frac{\mathrm{d}\mathbf{I}(x)}{\mathrm{d}x} = -\mathbf{Y}\mathbf{V}(x) \qquad \text{(A.2b)}$$

where $\mathbf{Z}$ and $\mathbf{Y}$ are the series phase impedance and shunt phase admittance matrices of the line in per-unit length. Differentiating Equations (A.2a) and (A.2b) with respect to x, we obtain

$$\frac{d^2\mathbf{V}(x)}{dx^2} = \gamma^2\mathbf{V}(x) \tag{A.3a}$$

$$\frac{d^2\mathbf{I}(x)}{dx^2} = (\gamma^T)^2\mathbf{I}(x) \tag{A.3b}$$

where

$$\gamma = \sqrt{\mathbf{ZY}} \quad \text{and} \quad \gamma^T = \sqrt{\mathbf{YZ}} \tag{A.3c}$$

γ is the line's propagation coefficient matrix and γ^T is its transpose noting that $\mathbf{Z}$ and $\mathbf{Y}$ are symmetric matrices.

The general solutions to Equations (A.3a) and (A.3b) are given by

$$\mathbf{V}(x) = e^{-\gamma x}\mathbf{V}_i + e^{\gamma x}\mathbf{V}_r \tag{A.4a}$$

$$\mathbf{I}(x) = e^{-\gamma^T x}\mathbf{I}_i + e^{\gamma^T x}\mathbf{I}_r \tag{A.4b}$$

where $\mathbf{V}_i$, $\mathbf{V}_r$, $\mathbf{I}_i$ and $\mathbf{I}_r$ are column matrices that satisfy the boundary conditions of the line. Differentiating Equation (A.4a) with respect to x and substituting into Equation (A.2a), we obtain a different form for Equation (A.4b) as follows:

$$\mathbf{I}(x) = \mathbf{Y}_o(e^{-\gamma x}\mathbf{V}_i - e^{\gamma x}\mathbf{V}_r) \tag{A.5a}$$

where

$$\mathbf{Y}_o = \mathbf{Z}^{-1}\gamma \tag{A.5b}$$

$\mathbf{Y}_o$ is the characteristic admittance matrix of the line and its inverse, the characteristic impedance matrix is given by

$$\mathbf{Z}_o = \gamma^{-1}\mathbf{Z} \tag{A.5c}$$

Also, differentiating Equation (A.5a) and substituting in Equation (A.2b), we obtain another formulation for $\mathbf{Y}_o$ and $\mathbf{Z}_o$ as follows:

$$\mathbf{Y}_o = \mathbf{Y}\gamma^{-1} \quad \text{and} \quad \mathbf{Z}_o = \gamma\mathbf{Y}^{-1} \tag{A.5d}$$

Equations (A.4a) and (A.5a) are general and provide the voltage and current solutions at any point along the line. However, if we are only interested in the line's terminal conditions, i.e. at $x = 0$ and at $x = \ell$, we can derive an equivalent two-port π admittance model. Substituting $x = 0$ and at $x = \ell$ in Equations (A.4a) and (A.5a), and using $\mathbf{V}(x=0) = \mathbf{V}_S$, $\mathbf{V}(x=\ell) = \mathbf{V}_R$, $\mathbf{I}(x=0) = \mathbf{I}_S$ and $\mathbf{I}(x=\ell) = \mathbf{I}_R$, we have

For $x = 0$:
$$\mathbf{V}_S = \mathbf{V}_i + \mathbf{V}_r \tag{A.6a}$$

$$\mathbf{I}_S = \mathbf{Y}_o(\mathbf{V}_i - \mathbf{V}_r) \tag{A.6b}$$

For $x = \ell$:
$$\mathbf{V_R} = e^{-\boldsymbol{\gamma}\ell}\mathbf{V_i} + e^{\boldsymbol{\gamma}\ell}\mathbf{V_r} \tag{A.7a}$$

$$\mathbf{I_R} = \mathbf{Y_o}(e^{-\boldsymbol{\gamma}\ell}\mathbf{V_i} - e^{\boldsymbol{\gamma}\ell}\mathbf{V_r}) \tag{A.7b}$$

From Equation (A.6), we obtain

$$\mathbf{V_i} = \frac{1}{2}(\mathbf{V_S} + \mathbf{Y_o^{-1}I_S}) \tag{A.8a}$$

$$\mathbf{V_r} = \frac{1}{2}(\mathbf{V_S} - \mathbf{Y_o^{-1}I_S}) \tag{A.8b}$$

Substituting Equation (A.8) into Equation (A.7), and after some matrix and hyperbolic functions manipulations, we obtain

$$\mathbf{I_S} = \mathbf{AV_S} - \mathbf{BV_R} \tag{A.9a}$$

$$\mathbf{I_R} = -\mathbf{BV_S} + \mathbf{AV_R} \tag{A.9b}$$

where

$$\mathbf{A} = \mathbf{Y_o}\coth(\boldsymbol{\gamma}\ell) \tag{A.10a}$$

$$\mathbf{B} = \mathbf{Y_o}\operatorname{cosech}(\boldsymbol{\gamma}\ell) \tag{A.10b}$$

Substituting Equation (A.3c), $\boldsymbol{\gamma} = \sqrt{\mathbf{ZY}}$, and using Equation (A.5d), $\mathbf{Y_o} = \mathbf{Y}\boldsymbol{\gamma}^{-1}$, in Equation (A.10), we obtain

$$\mathbf{A} = \mathbf{Y}(\sqrt{\mathbf{ZY}})^{-1}\coth(\sqrt{\mathbf{ZY}}\ell) \tag{A.11a}$$

$$\mathbf{B} = \mathbf{Y}(\sqrt{\mathbf{ZY}})^{-1}\operatorname{cosech}(\sqrt{\mathbf{ZY}}\ell) \tag{A.11b}$$

Using Equations (A.9), the π equivalent admittance model of a multi-conductor line or cable is shown in Figure A.1.

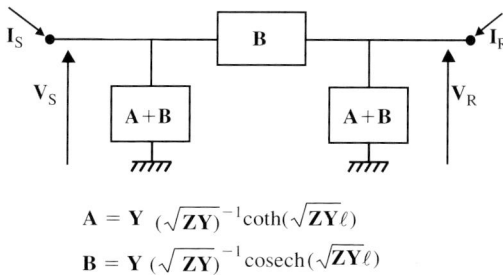

$$\mathbf{A} = \mathbf{Y}\,(\sqrt{\mathbf{ZY}})^{-1}\coth(\sqrt{\mathbf{ZY}}\ell)$$
$$\mathbf{B} = \mathbf{Y}\,(\sqrt{\mathbf{ZY}})^{-1}\operatorname{cosech}(\sqrt{\mathbf{ZY}}\ell)$$

Figure A.1 π equivalent circuit of a multi-conductor overhead line or a multi-conductor cable where **A**, **B**, **Z** and **Y** are square matrices, and **V** and **I** are column matrices

A.2 Typical data of power system equipment

A.2.1 General

In this section, typical data and parameters for various power systems equipment is assembled and presented in tabular format. Throughout, the parameters used are defined as follows:

For overhead lines, cables, reactors and capacitors

$Z^P/R^P/X^P$: positive phase sequence (PPS) impedance/resistance/reactance of a three-phase circuit

$Z^Z/R^Z/X^Z$: zero phase sequence (ZPS) impedance/resistance/reactance of a three-phase circuit

B^P/B^Z: PPS/ZPS susceptance of a three-phase circuit

$$Z^P = R^P + jX^P \quad \text{and} \quad Z^Z = R^Z + jX^Z$$

Z^P_{12}/Z^Z_{12}: PPS/ZPS mutual impedance between two adjacent three-phase circuits

B^P_{12}/B^Z_{12}: PPS/ZPS mutual susceptance between two adjacent three-phase circuits

$$Z^P_{12} = R^P_{12} + jX^P_{12} \quad \text{and} \quad Z^Z_{12} = R^Z_{12} + jX^Z_{12}$$

For transformers

$Z/R/X$ is leakage impedance/resistance/reactance.

$$Z^P_{HL} = R^P_{HL} + jX^P_{HL} \quad \text{and} \quad Z^Z_{HL} = R^Z_{HL} + jX^Z_{HL}$$
$$Z^P_{HT} = R^P_{HT} + jX^P_{HT} \quad \text{and} \quad Z^Z_{HT} = R^Z_{HT} + jX^Z_{HT}$$
$$Z^P_{LT} = R^P_{LT} + jX^P_{LT} \quad \text{and} \quad Z^Z_{LT} = R^Z_{LT} + jX^Z_{LT}$$

HL represents high voltage to low voltage leakage parameter, HT represents high voltage to low voltage leakage parameter and LT represents low voltage to tertiary voltage leakage parameter.

For synchronous machines

$X''_d/X'_d/X_d$: d-axis subtransient/transient/synchronous reactance and X''_q is q-axis subtransient reactance. X^Z is ZPS reactance and R_a is armature ac resistance

T''_d/T'_d: d-axis subtransient/transient short-circuit time constant

T_a: armature or dc short-circuit time constant. H is inertia constant.

For induction machines

Double-cage or deep-bar: $X''/X'/X$ is subtransient/transient/steady state reactance

T''/T': subtransient/transient short-circuit time constant

T_a: armature or dc short-circuit time constant

Single-cage: X'/X is transient/steady state reactance

T': transient short-circuit time constant

T_a: armature or dc short-circuit time constant

A.2.2 Data

Table A.1 Typical parameters of synchronous generators

	Rated MVA	R_a % on rated MVA	X''_d % on rated MVA	X'_q % on rated MVA	X'_d % on rated MVA	X_d % on rated MVA	X^Z % on rated MVA	T''_d ms	T'_d ms	T_a ms	H MW × s/MVA
Turbo-generators	5	0.39	15	27	25	290	8	23	550	180	1
	50	0.3	17	19	24	225	9	20	634	192	1.8
	110	0.16	14	14.6	18	200	8	19	662	283	5.1
	150	0.2	20	21	29	182	9	22	651	336	3.2
	200	0.16	17.5	19	28	194	8	31	908	367	4.5
	300	0.15	17.4	18.2	23	214	9	25	1140	381	5.8
	500	0.21	25	26.4	34	224	12	32	1010	395	6.2
	800	0.20	23	24.2	32	219	14	51	942	344	3.9
	1280	0.26	24	24.1	36	155	11	28	1406	293	4.5
	1990	0.3	35	37	51	270	15	17	1010	384	8.5
Hydro-generators	90	0.17	16.6	21	22	128	5	31	834	352	3.2
	320	0.12	14	19	22	116	5	79	1100	456	4.6

Table A.2 Typical parameters of induction generators

Rated voltage kV	Rated power MVA/MW	Stator resistance Ω	Stator reactance Ω	Magnetising reactance Ω	Rotor resistance, running/starting Ω	Rotor reactance, running/starting Ω
Doubly fed induction generators						
0.69	0.9/0.8	0.003	0.030	1.2	0.0022/0.003	0.056/0.052
0.69	2.0/1.8	0.0014	0.021	0.81	0.0016/0.0019	0.022/0.021
0.69	2.0.1/2	0.0016	0.018	0.78	0.0021/0.0025	0.026/0.023
0.69	3.0/2.85	0.0011	0.01	0.6	0.0008/0.001	0.014/0.013
1	3.1/3	0.0023	0.019	0.9	0.0015/0.007	0.052/0.042
Squirrel-cage induction generators						
0.69	0.68/0.6	0.004	0.028	1.6	0.0032/0.0048	0.041/0.033
0.69	1.6/1.43	0.0022	0.045	1.4	0.0021/0.0036	0.035/0.029
0.69	2.56/2.3	0.0018	0.03	0.7	0.0019/0.003	0.012/0.01

Table A.3 Typical parameters of induction motors

Rated voltage kV	Rated power kW	Rated current A	Ratio of locked rotor current to rated current	Rated power factor	dc time constant in ms
0.415	45	74	7	0.89	15
6	500	60	5.7	0.84	52
6	3000	362	5.3	0.9	60
6.6	10 000	993	6	0.92	98

Rated voltage kV	Rated kVA	Stator resistance Ω	Stator reactance Ω	Magnetising reactance Ω	Rotor resistance (running) Ω	Rotor reactance (running) Ω
3.3	900	0.0968	1.331	38.7	0.0726	0.847

Generic data in pu on rated apparent power for small and large three-phase induction motors

	Stator resistance pu	Stator reactance pu	Magnetising reactance pu	Rotor resistance pu	Rotor reactance pu
Small induction motors	0.031	0.1	3.2	0.018	0.18
Large industrial induction motors	0.013	0.067	3.8	0.009	0.17

Measured data of a 4.2 MVA double-cage induction motor, reactances in pu on 4.2 MVA

Rated voltage kV	Rated power MW	Subtransient reactance X'' pu	Transient reactance X' pu	Steady state reactance X pu	Subtransient short-circuit time constant T'' ms	Transient short-circuit time constant T' ms	dc short-circuit time constant T_a ms
6.6	3.46	0.16	0.2	3.73	20	80	40

Table A.4 Typical sequence parameters of single-circuit and double-circuit overhead lines

(a) Single-circuit line parameters (50 Hz)

Positive and zero phase sequence parameter	400 kV, 4×400 mm² ACSR, 1×400 mm² ACSR earth wire	400 kV, 2×500 mm², ACAR, 1×175 mm² ACSR earth wire	275 kV, 2×300 mm², AAAC, 1×160 mm² ACSR earth wire	275 kV, 2×175 mm², ACSR, 2×70 mm² ACSR earth wires	132 kV, 1×400 mm², ACSR, 1×175 mm² ACSR earth wire	132 kV, 2×175 mm², ACSR, 1×175 mm² ACSR earth wire	66 kV, 1×96.8 mm², ACSR, 1×64.5 mm² ACSR earth wire	33 kV, 1×64.5 mm², ACSR, 1×64.5 mm² earth wire	11 kV, 1×32.25 mm², ACSR, no earth wires
Z^P (Ω/km)	$0.017+j0.278$	$0.03+j0.30$	$0.044+j0.31$	$0.08+j0.346$	$0.069+j0.38$	$0.08+j0.3$	$0.183+j0.376$	$0.30+j0.44$	$0.537+j0.376$
Z^Z (Ω/km)	$0.1+j0.78$	$0.14+j0.8$	$0.113+j0.78$	$0.246+j0.706$	$0.185+j0.95$	$0.19+j0.86$	$0.42+j1.1$	$0.51+j1.04$	$0.81+j1.3$
B^P (μS/km)	4.1	3.6	3.7	3.323	3.04	3.9	2.0	–	–
B^Z (μS/km)	2.28	2.05	2.1	2.467	1.66	2.0	0.6	–	–

(b) Double-circuit line parameters (50 Hz)

Positive and zero phase sequence parameter	400 kV, 4×400 mm² ACSR, 1×400 mm² ACSR earth wire	400 kV, 2×400 mm² ACSR, 1×400 mm² ACSR earth wire	275 kV, 2×300 mm² AAAC, 1×160 mm² ACSR earth wire	132 kV, 2×175 mm² ACSR, 1×175 mm² ACSR earth wire	66 kV, 1×100 mm² ACSR, 1×30 mm² ACSR earth wire
Z^P (Ω/km)	$0.017+j0.252$	$0.034+j0.32$	$0.044+j0.289$	$0.08+j0.303$	$0.187+j0.375$
Z^Z (Ω/km)	$0.188+j1.20$	$0.205+j1.26$	$0.276+j1.28$	$0.30+j1.33$	$0.450+j1.147$
Z^P_{12} (Ω/km)	$-0.0003-j0.0130$	$-0.0003-j0.0130$	$-0.00043-j0.0107$	$-0.0004-j0.0133$	$-0.00036-j0.0110$
Z^Z_{12} (Ω/km)	$0.085+j0.420$	$0.085+j0.42$	$0.116+j0.446$	$0.111+j0.478$	$0.254+j0.646$
B^P (μS/km)	4.46	3.5	3.96	3.72	2
B^Z (μS/km)	1.9	1.6	1.75	1.47	0.8
B^P_{12} (μS/km)	0.126	0.084	0.076	0.097	0.02
B^Z_{12} (μS/km)	-0.72	-0.544	-0.63	-0.76	-0.3

(Continued)

(c) Single-circuit line parameters (60 Hz)

Positive and zero phase sequence parameter	230 kV	345 kV	500 kV	765 kV	1200 kV
Z^P (Ω/km)	0.051 + j0.50	0.04 + j0.39	0.026 + j0.31	0.012 + j0.3	0.010 + j0.27
Z^Z (Ω/km)	0.13 + j1.25	0.112 + j0.92	0.09 + j0.86	0.07 + j0.80	0.05 + j0.70
B^P (μS/km)	3.36	4.5	5.1	5.0	4.5
B^Z (μS/km)	2.0	2.7	3.2	2.9	3.4

Table A.5 Typical sequence parameters of single-circuit three-phase cables

(a) Cable parameters (50 Hz)

Positive and zero phase sequence parameter	400 kV 2500 mm² copper conductor, copper screen/aluminium sheath, XLPE, in tunnel	400 kV 2000 mm² copper conductor, aluminium sheath, flat formation, cross-bonded, oil filled	275 kV 1600 mm² copper conductor, aluminium sheath, flat formation, cross-bonded, oil filled	220 kV 1000 mm² copper conductor, aluminium sheath, XLPE, three core	132 kV 1000 mm² copper conductor, aluminium sheath, XLPE, three core	132 kV 1000 mm² copper conductor, aluminium sheath, flat formation, single core, oil filled	66 kV 630 mm² copper conductor, aluminium sheath, flat formation, oil filled, single core	33 kV 400 mm² aluminium conductor, sheath, three core, oil filled	33 kV 300 mm² aluminium conductor, copper conductor, XLPE, lead sheath, flat single core	11 kV 120 mm² aluminium/ copper conductor, XLPE, lead sheath, trefoil single core
Z^P (Ω/km)	0.011 + j0.19	0.0143 + j0.14	0.016 + j0.175	0.027 + j0.122	0.027 + j0.11	0.025 + j0.18	0.04 + j0.174	0.1 + j0.08	0.12/0.075 + j0.16	0.3/0.19 + j0.1
Z^Z (Ω/km)	0.048 + j0.057	0.065 + j0.057	0.073 + j0.054	0.10 + j0.05	0.1 + j0.04	0.114 + j0.061	0.57 + j0.076	0.35 + j0.055	–	–
B^P (μS/km)	76	121	138	56	75	193	262	254	85	150
B^Z (μS/km)	76	121	138	56	75	193	262	254	85	150

(b) Cable parameters (60 Hz)

Positive and zero phase sequence parameter	500 kV 3000 MCM copper conductor, aluminium sheath, flat formation	345 kV 2500 MCM copper conductor, lead sheath, pipe-type high pressure oil filled cable	230 kV 2000 MCM copper conductor, paper-insulated lead sheath, trefoil formation	230 kV 1600 MCM copper conductor, lead sheath, pipe-type high pressure oil filled cable	115 kV copper conductor, paper-insulated lead sheath	115 kV copper conductor, lead sheath, pipe-type high pressure gas filled cable
Z^P (Ω/km)	0.011 + j0.136	0.028 + j0.176	0.04 + j0.128	0.34 + j0.155	0.06 + j0.31	0.038 + j0.13
Z^Z (Ω/km)	0.058 + j0.06	0.378 + j0.512 at 5 kA	0.171 + j0.083	0.40 + j0.36 at 5 kA	0.256 + j0.20	0.38 + j0.30 at 5 kA
B^P (μS/km)	108	97	70	135	190	80
B^Z (μS/km)	108	97	70	135	190	80

Table A.6 Typical sequence parameters of transformers

(a) Two-winding transformers

HV winding terminals (kV)	LV winding terminals (kV)	Vector group	Rated MVA	R^P_{HL} % on rated MVA	X^P_{HL} % on rated MVA	R^Z_{HL} % on rated MVA	X^Z_{HL} % on rated MVA
275	33	YD	75	0.49	19.4	0.42	16.5
400	33	YD	150	0.34	22.2	0.32	20.8
275	66	YD	120	0.38	19	0.32	16.1
400	66	YD	180	0.38	24.3	0.32	20.7
132	66	YY	30	0.8	10.5	0.78	9.7
132	11	YY	55	0.6	15.5	0.6	14
132	33	YD	90	0.6	22	0.5	18.8
132	11	YY	30	1	24	1	20.7
132	25	1 Phase	18	0.48	12	0.48	12
23.5	11	DY	32.5	0.15	6.75	0.15	5.75
66	33	YY	60	0.36	13	0.32	11.8
66	11	DY	30	0.52	13	0.48	11.6
33	11	DY	37.5	0.5	12.5	0.47	11
33	11	DY	10	0.67	9.2	0.6	8.4
33	0.69	DY	2.1	0.75	8	0.75	7.5
11	0.43	DY	0.1	0.6	4.75	0.6	4.3
20	0.43	DY	1	0.5	5	0.5	4.2
20	0.43	DY	2	0.6	6	0.6	5.3

(b) Autotransformers and three-winding transformer

HV winding terminals (kV)	LV winding terminals (kV)	Tertiary winding terminals (kV)	Vector group	Rated MVA	Z^P_{HL} % on rated MVA	Z^P_{HT} % on rated MVA	Z^P_{LT} % on rated MVA	Z^Z_{HL} % on rated MVA	Z^Z_{HT} % on rated MVA	Z^Z_{LT} % on rated MVA
400	132	13	Y0y0d1	240	20.3	54	32	19.4	51.9	32.4
275	132	13	Y0y0d1	240	21	46.5	22.6	16	38.5	25
275	66	11	YYD	120	21.8	30.5	6.3	17.4	29.4	6.5

(c) Positive phase sequence impedance variation with tap position

(i) Autotransformer, low voltage winding line-end tapped

Rating	240 MVA		
No-load voltage ratio	275 kV/132 kV		
Tap range	±15% in 18 steps of 1.67%		
Tap position	1	10	19
Tap position (%)	+15	0	−15
Impedance (% on rating)	14.8	19.7	29

(ii) Two-winding transformer, high voltage winding tapped

Rating	30 MVA		
No-load voltage ratio	132 kV/11 kV		
Tap range	−20% to +10% in 18 steps of 1.67% on HV winding		
Tap position	1	7	19
Tap position (%)	+10	0	−20
Impedance (% on rating)	22.8	21.5	18.9

Table A.7   Typical sequence parameters of quadrature boosters and phase shifters

Rated voltage kV	Rated throughput MVA	Rated quadrature voltage in %	Rated phase shift degrees	Apparent positive phase sequence impedance at zero phase shift % on rated MVA	Zero phase sequence impedance % on rated MVA
400	2500/2000	−20/+20	±11.3	8	See Chapter 4 for detailed zero sequence data of two typical plant
275	1000/750	−20/+20	±11.3	5	
235	300	−72.8/+72.8	±40	12	
132	250/150	−53.6/+53.6	±30	4	

Table A.8   Typical sequence parameters of series reactors

Rated voltage kV	Rated MVA	Positive phase sequence resistance R^P % on rated MVA	Positive phase sequence reactance X^P % on rated MVA	Zero phase sequence resistance R^Z % on rated MVA	Zero phase sequence reactance X^Z % on rated MVA
500	2000	0.1	22	0.1	22
400	2000/1320	0.085/0.13	20/26.4	0.085/0.13	20/26.4
275	750	0.06	10.5	0.65	10.5
132	90	0.20	18	0.20	18
66	60	0.23	15.25	0.23	15.25
33	50	0.1515	12	0.1515	12
11	12.5	0.3	20	0.3	20
4.2	7.5	0.16	8	0.16	8

Table A.9   Typical sequence parameters of shunt reactors

Rated voltage kV	Rating MVAr	Positive phase sequence susceptance B^P % on 100 MVA	Zero phase sequence susceptance B^Z % on 100 MVA
500	300/200	−300/−200	−300/−200
400	200/150/100	−200/−150/−100	−200/−150/−100
275	150/100	−150/−100	−150/−100
132	60/45/30	−60/−45/−30	−60/−45/−30
33	30/20	−30/−20	−30/−20
11	10/5	−10/−5	−10/−5

Table A.10   Typical sequence parameters of shunt capacitors

Rated voltage kV	Rating in MVAr	Positive phase sequence susceptance B^P % on 100 MVA base	Zero phase sequence susceptance B^Z % on 100 MVA base
500	300/200	300/200	300/200
400	200/150/100	200/150/100	200/150/100
275	150/100	150/100	150/100
132	60/45/30	60/45/30	60/45/30
33	30/20	30/20	30/20
11	10/5	10/5	10/5

Table A.11   Typical positive phase sequence X/R ratios and dc time constants of power system equipment

(a) Synchronous generators

Rated MVA	X_d'' (% on rated MVA)	R_a (% on rated MVA)	Typical X/R ratio	T_{dc} (50 Hz), ms	T_{dc} (60 Hz), ms
5	15	0.39	38.5	122.4	102.1
110	14	0.16	87.5	278.5	232.1
500	25	0.21	119	379	315.6

(b) Generator transformers

Rated MVA	Typical X/R ratio	T_{dc} (50 Hz), ms	T_{dc} (60 Hz), ms
5	25	79.5	66.3
110	40	127	106.1
500	70	223	185.7

(c) Network transformers

Voltage ratio	Rated MVA	Typical X/R ratio	T_{dc} (50 Hz), ms	T_{dc} (60 Hz), ms
500 kV/235 kV	750	90	286.5	238.7
400 kV/275 kV	500	70	222.8	185.7
380 kV/150 kV	450	85	270.5	225.5
400 kV/132 kV	240	55	175	145.9
150 kV/110 kV	63	50	159	132.6
132 kV/33 kV	90	40	127.3	106.1
132 kV/11 kV	30	25	79.6	66.3
33 kV/11 kV	10	15	47.7	39.8
11 kV/0.43 kV	1	8	25.5	21.2

(d) Series reactors

Nominal voltage (kV)	Rated MVA	Typical X/R ratio	T_{dc} (50 Hz), ms	T_{dc} (60 Hz), ms
500	2000	220	700	583.6
400	1320	200	636.6	530.5
275	750	175	557	464.2
132	90	90	286.5	238.7
33	50	79	251.5	209.5
10	45	75	238.7	200
4.2	7.5	50	159	132.6

(e) Overhead lines

Voltage (kV)	11	33	66	115	132	230	275	345	400	500	765	1200
Typical X/R ratio	0.7	1.5	2	4	5	7	9	10	12	15	25	27
T_{dc} (50 Hz), ms	2.2	4.7	6.4	12.7	15.9	22.3	28.6	31.8	38.2	47.7	79.6	86
T_{dc} (60 Hz), ms	1.85	4	5.3	10.6	13.3	18.5	23.9	26.5	31.8	39.8	66.3	71.6

(Continued)

Table A.11   (Continued)

(f) Cables

Voltage (kV)	11	33	66	115	132	230		275	345	400	500
Typical X/R ratio	0.5	1.5	3	4	7	1 (pipe type)		10	8	12	13
T_{dc} (50 Hz), ms	1.6	4.7	9.5	12.7	22.3	3.2		31.8	25.5	38.2	41.4
T_{dc} (60 Hz), ms	1.3	4	8	10.6	18.5	2.6		26.5	21.2	31.8	34.5

Note: The typical ratios given for overhead lines and cables correspond to one particular design in each case. However, the X/R ratio for, say, 11 kV cable, may vary from 0.4 to 2.5 depending on material, geometry and layout of the conductors. Similarly, the X/R ratio for, say a 132 kV overhead line, may vary from 2.5 to 7 depending on material, physical dimensions and spacing of the conductors.

Index

American IEEE C37.010 standard, 469
 ac decrement factor, 471
 asymmetrical interrupting current, 472
 closing and latching current, 472
 closing and latching impedance
  network, 469
 dc decrement factor, 471
 E/X method with correction for ac and
  dc decrement, 471
 E/X simplified method, 470
 first cycle symmetrical current, 469, 471
 interrupting impedance network, 469
 no ac decrement (NACD) ratio, 471
 reactance adjustment factors, 470
 representation of:
  local generation, 469, 471
  remote generation, 469, 471
 symmetrical interrupting current, 469,
  472
 time-delayed 30 cycle current, 472
 X/R ratio, 470
Arc resistance, 7
As Low As Reasonably Practicable, 517
Autotransformers, 232
 PPS/NPS equivalent circuits, 232
 ZPS equivalent circuit, 237

Back-flashover, 514
Balanced three-phase impedances, 39
Boundary circuit, 486
Boundary node, 486, 489, 493
Bundled conductors, 76
Bus admittance matrix, 489
Bus impedance matrix, 489

Cables:
 ac resistance, 150
 aerial, 140
 belted, 141
 capacitance, 146, 148

cross bonded, 143
earth continuity conductor, 143
earthing impedance, 567
full shunt susceptance matrix, 164, 166
insulation, 140
layouts, 154
multi-conductor, 145
parameters, 145
permittivity, 147
phase impedance matrix, 157, 159,
 162, 169
phase susceptance matrix, 165, 167
pipe-type, 141, 147
proximity effect factor, 150, 151
resistivity, 151
sea return impedance, 152
self and mutual impedances, 149, 150,
 152, 153
self-contained, 141
sequence impedance matrix, 157,
 160, 162
sequence susceptane matrix, 165,
 166, 167
sheath voltage limiters, 142
single-core, 141
single-point bonded, 142
skin effect factor, 150, 151
solidly bonded, 142
submarine, 140, 141
susceptance, 146, 147
temperature coefficient, 151
three-core, 141
transposed, 143
underground, 140
Capacitance matrix, 94, 107
Circuit-breakers, 3, 10, 501
 arcing time, 10
 breakdown, 511

Circuit-breakers (*continued*)
  breaking duty, 501
  bus coupler, 501
  bus section, 501
  contact separation, 10
  current interruption, 10
  electrical closure, 511
  dielectric strength, 512
  IEC62271 time constant, 510
  IEEE C37.04 time constant, 510
  independent pole, 511
  making duty, 501
  persistent fault, 501
  point-on-wave closing, 514, 515
  pole stagger, 511
  prestrike, 511
  standards ratings, 479
    arc energy, 482, 483
    arc extinction, 482
    arcing time, 482
    asymmetrical ratings, 480, 481
    dc current rating, 479
    derating factor, 483
    major current loop, 482
    percentage dc current, 479
    short-circuit rating, 479
    standard dc time constant, 480
  substation:
    1 and $^1/_2$ switch, 501
    double-bus, 501
    transfer bus, 501
  three-pole operated, 511
  trip circuit, 10

Data exchange, 487
Depth of earth return, 83
Driving point impedance, 489

Earth electrode, 551
Earth fault current distribution, 563
  cables, 576
  conductive current, 564, 572
  earth wire current, 564
  inductive current, 564, 572
  line length, 566
  overhead line, 563
  propagation constant, 565
  sheath current, 572, 573
  towers, 577
Earth fault factor, 49
Earth resistivity, 83
Earth return current, 550, 569, 576
Earth return path impedance, 81

Earthing conductor, 551
Earthing methods, 568
Earthing network:
  cable sheath, 571
  earth wire, 561, 563
  infinite overhead line, 561
    earthing impedance, 561
    ladder network, 562
  tower footing resistance, 561
Earthing transformers, 242
Electric shock, 551
  heart fibrillation, 553, 554
  let-go current, 553
  muscular contraction, 553
  non-fibrillating current, 554
  resistance of human body, 552
  ventricular fibrillation, 553
Electrical interference:
  capacitive coupling, 585
    discharge current, 587
    pipeline charging current, 587
    pipeline voltage to earth, 586
  conductive coupling, 595
  inductive coupling, 588
    induced longitudinal voltage, 588
    inductive zone of influence, 588
    induced EMF, steady-state, 590
    induced EMF, faults, 590
    pipeline voltage to earth, 591, 592
    pipeline current, 593
    propagation constant, 592
    characteristic impedance, 593
    terminal impedances, 593
  oblique exposure, 587
  parallel exposure, 585
  resistive coupling, 595
Electrical parameters, 76
Electricity at Work Regulations, 517
Electromagnetic forces, 12
Physical units, 15
Equipment failure, 6
Equivalent, 485
  direct derivation, 492, 493, 494, 495
  distribution, 497
  industrial systems, 497
  multiple π equivalent, 491
  multiple-node, 491
  NPS equivalent, 493
  power station, 497
  power system, 485, 489
  PPS equivalent, 495
  single-node, 490

subtransient, 495
three-node, 491
time-dependent, 495
transient, 495
transmission, 497
two-node, 490
ZPS equivalent, 493
External impedance, 321

Fault current limiter, 529
current limiting reactor, 533
flux cancelling limiter, 535, 536
ideal fault current limiter, 543
magnetically coupled limiter, 534
neutral earthing resistor and reactor, 531
passive damped resonant limiter, 536
pyrotechnic limiter, 532
saturable reactor limiter, 536
series resonant current limiter, 534
superconducting fault current limiter,
539
cryogenic cooling, 539
high temperature superconductor, 539
resistive state, 539
resistive superconducting limiter, 540
shielded inductance limiter, 541
saturated inductance limiter, 542
air-gap superconducting limiter, 542
thyristor controlled series capacitor, 538
thyristor protected series capacitor, 538
Faults:
back flashover, 5, 75
balanced, 4
causes, 5
cross-country, 5, 65
impedance, 7
mechanical effects, 12
nature, 4
severity, 503
shunt, 43
simultaneous, 63
statistics, 6
thermal effects, 12
types, 4
unbalanced, 4
weather related, 5

General analysis of three-phase faults, 417,
422
fault between nodes, 427
fault impedance/admittance matrix, 429
sequence admittance matrices, 419,
428, 429

sequence impedance matrices, 422,
428, 429
simultaneous faults, 425
General analysis of unbalanced faults,
432
Geometric mean radius, 85
Grounding, 551

Health and safety, 517
Human error, 6

IEC 60909-0 Standard:
analysis technique, 452
asynchronous motors, 456
dc current, 459
equivalent frequency f_c, 460
generators, 454
impedance correction factors, 453
network transformers, 453
peak make current, 458
peak factor k, 458
power station units, 454
static converter drives, 458
symmetrical breaking current, 460
μ factor for generators, 460
μq factor for asynchronous motors,
460
symmetrical steady-state current, 462
VDE, 451
voltage factor c, 452
X/R ratio, 459
method A, 459
method B, 459
method C, 459
Individual risk, 519
Induction motor modelling, 357
deep-bar rotor, 365
double cage rotor, 365
dq frame reference, 362
initial motor loading, 377
operator reactance, 364, 366
phase frame reference, 358
single cage rotor, 363
single-phase short circuit current, 375
NPS reactance, 376, 377
NPS resistance, 376, 377
NPS slip, 377
ZPS reactance, 377
ZPS resistance, 377
slip speed, 358, 359
subtransient reactance, 367
time constants, 364, 366, 367, 368
three-phase fault current, 368, 370,
371, 372

Induction motor modelling (*continued*)
  PPS reactance, 373, 374
    external impedance, 374
  transient reactance, 364
Induction motor tests, 378
  locked rotor test, 379
  no load test, 381
  open-circuit voltage, 382
  stator dc resistance, 378
  sudden short-circuit test, 382

Leakage impedance variation, 245
  leakage flux variation, 245
  nominal tap position, 245
  tap position, 245
Lightning strike, 5, 6
Limitation of short-circuit current,
    520
  autoreclosing, 524
  autotripping, 521
  de-loading circuits, 525
  fault clearance time, 524
  higher system voltage, 528
  hot standby, 521
  network splitting, 523
  opening delta tertiary winding, 527
  re-certification, 521
  sequential disconnection, 524
  substation splitting, 521
Load modelling, 285
Loss of synchronism, 3

Machine internal voltage:
  AVRs, 339
    Effect on short-circuit current, 339
    field voltage, 340
    static exciter, 340
Machine overheating, 6
Measurement of:
  cable:
    dc resistance, 197
    earth loop impedance, 196
    PPS/NPS impedance, 195
    susceptance, 196
    ZPS impedance, 195
  distributed multi-conductor line, 190,
      607
    self and mutual phase impedance,
      187
    self and mutual shunt susceptance,
      188
  double-circuit line parameters, 192, 193
  QB and PS impedances, 268
  single-circuit line parameters, 187

synchronous machine:
  currents, 348
  d-axis reactances, 348, 349
  resistances, 351
  short-circuit time constants, 349, 350,
    351
transformers, 249
  three-winding leakage impedance,
    253
  autotransformer leakage impedance,
    254
  copper loss, 250
  iron loss, 249
  magnetising reactance, 249
  no-load current, 249, 250
  short-circuit PPS impedance, 250, 251
  short-circuit ZPS leakage impedance,
    252
  two-winding leakage impedance test,
    253
Multiple-circuit lines, 123
  three-circuit lines, 124
  four-circuit lines, 126
Mutual impedance, 81, 589

Network codes, 487, 488
Network reduction, 486, 489
  dynamic, 486
  static, 486

Overhead line, 74
  counterpoise, 75, 76
  double-circuit, 75
  earth wire, 75
  shielding, 75
  single-circuit, 74
Overstressed, 516, 517

Peak current envelope, 324
Peak make current, 11
Per-unit analysis, 15
  actual quantity, 19
  base admittance, 17
  base impedance, 17
  base quantity, 19
  change of base, 18
  mutual admittance, 23
  mutual impedance, 20
  single-phase systems, 15
  three-phase systems, 19
Phase sequence networks, 34
  negative, 34
  positive, 34
  zero, 34

Phase shifters, 262
  NPS equivalent circuit, 266
  PPS equivalent circuit, 263
  ZPS equivalent circuit, 267
PI (π) equivalent circuit, 174
  multiphase, 182, 607
  sequence models, 175, 177, 179, 181
Pipeline, 584
  ac corrosion, 584
  cathodic protection, 584
  coating, 584
  distributed parameter, 592, 594
  earthing, 585
  equivalent circuit, 591
  potential coefficient, 586
  series impedance with earth return,
      594
  shunt admittance, 594
  touch voltage, 585
Potential:
  mesh, 552
  step, 551
  touch, 551, 552
  transferred, 552
Potential coefficients, 78
Potential coefficient matrix, 94
Power quality, 4
Probabilistic analysis, 507
  ac short-circuit current, 508
  deterministic analysis, 507, 511
  maximum dc offset, 509
  Monte Carlo, 508
  probability distribution, 508, 514
  probability distribution of X/R ratio, 510
  probability of maximum dc offset, 514
  random factors, 507
  risk, 508
  sequential short-circuit, 509
  simultaneous three-phase short-circuit,
      509
  voltage phase angle, 512
  worst-case, 507
Protection relay, 3

Quadrature boosters, 262
  NPS equivalent circuit, 266
  PPS equivalent circuit, 263
  ZPS equivalent circuit, 267

Reduced network, 486
Remote earth, 551
Renewable, 2
  wind turbine, 2

Retained network, 486
  power system, 486
Rise of earth potential, 3, 550, 551, 579
Risk assessment, 518
  risk analysis, 518, 519
  risk evaluation, 518, 519
Risk management, 517
Risk of fatality, 517, 518

Safety, 2
Screening factor:
  cables, 571, 572, 573, 574
  overhead lines, 570
Self impedance, 79
  complex propagation constant, 80
  external reactance, 81
  internal impedance, 79
  skin depth, 80
  tubular conductor, 80
Series:
  capacitors, 273
    modelling of, 278, 279
    types of schemes, 278
  reactors, 272
    modelling of, 273
Series phase:
  impedance matrix, 92, 105, 110, 114, 126
Series
  sequence impedance matrix, 93, 116,
      127
Short-circuit analysis, 402
  ac analysis, 403
  fixed impedance, 402
  general analysis in phase-frame, 438,
      441, 445
  passive analysis, 402
  time domain analysis, 404
  time variation of ac current, 405, 409,
      412
  time variation of dc current, 405, 410,
      415
  X/R ratio, 403
Shunt:
  phase susceptance matrix, 95, 105, 107,
      117
  sequence susceptance matrix, 95, 117
Shunt capacitors, 272
  modelling of, 275
Shunt reactors, 272
  modelling of, 275
Single-phase transformer, 202
  equivalent circuit, 203, 205
  pi (π) equivalent circuit, 209
Static variable compensators, 283

Substation earth electrode system:
  function, 555
  potential gradient, 555
  resistance to remote earth, 555
    buried flat plate, 558
    buried grid or mesh, 559
    buried hemisphere, 556
    buried ring of wire, 559
    mesh and vertical rods, 560
    multiple driven vertical rods, 557
    one driven vertical rod, 556
Substation earthing impedance, 567
Superposition theorem, 398
  open-circuit fault, 400
  short-circuit fault, 398
Surge impedance, 5
Symmetrical components, 31
Synchronous machine modelling, 302
  $dq0$ modelling:
  operator reactance analysis, 308
    d-axis, 309
    q-axis, 308
  Park's transformation, 305
  per-unit system, 306
  Single-phase to earth short-circuit fault,
    328
    short-circuit current, 329, 332, 338
    time constants, 329
    ZPS reactance, 331
    ZPS resistance, 331
  stator, rotor, field, damper windings, 302
  subtransient and transient time
      constants:
    d-axis, 312
    q-axis, 310, 323
  subtransient, transient reactances:
    d-axis, 309, 311
    q-axis, 310, 311, 312
  three-phase ryb modelling, 302
  three-phase sequence equivalent circuit,
    314
  three-phase short circuit faults, 315, 323
    current components, 317, 318
    external impedance, 321
    peak current envelope, 324
    PPS reactance and resistance, 318
    short-circuit current, 318, 338
    stator/armature time constant, 317
  time varying inductance, 304
  transformation matrix ryb to $dq0$, 305
  two-phase short-circuit fault, 324
    NPS reactance, 325, 326
    NPS resistance, 328

  short-circuit current, 325, 328, 338
    time constants, 327
  two-phase to earth short-circuit fault,
    332
    NPS reactance, 334
    NPS resistance, 328
    short-circuit current, 333, 336
    time constants, 334
Synchronous motor, 342
  condenser, 342
  compensator, 342

Temporary overstressing, 516, 517
Thermal heating of conductors, 12
Thévenin's impedance, 399, 422, 489
Thévenin's theorem, 398
  open-circuit fault, 400
  short-circuit fault, 398
Three-phase modelling of:
  induction machines, 440
  reactors and capacitors, 286, 287
  phase shifting transformers, 297
  quadrature boosters, 445
  static load, 299
  synchronous machines, 438
  transformers, 287, 290, 293
Three-phase π models:
  double-circuit, 184
  single-circuit, 183
Three-phase three-winding transformers,
    224
  double-secondary windings, 224
  harmonic currents, 224
  PPS/NPS equivalent circuits, 225
  PPS/NPS π equivalent circuits, 229
  ZPS equivalent circuit 229
Three-phase two-winding transformers,
    213
  PPS/NPS equivalent circuits, 214
  phase shift, 218, 222
  ZPS admittance matrix, 220
  ZPS equivalent circuit, 215, 217
Three-phase voltages:
  complex instantaneous, 29
  complex phasor, 29
  real instantaneous, 29
Time constant, 10
Traction transformers, 243
Transfer impedance, 489
Transformation matrix, 33
Transposition(s), 96
  double-circuit lines, 108
  ideal, 122, 125
  independent circuit, 108

inter-circuit impedance matrix, 108
matrix, 98, 119
nine, 122
perfect, 96
semi-transposed, 101
single-circuit lines, 96
six, 114, 118

Unbalance, 28, 76
  external sources, 28
  internal sources, 28
Unbalanced three-phase impedances, 39
Uncertainty in short-circuit current, 504
  accuracy, 505, 506
  confidence, 504
  data volume, 506
  database, 506
  design tolerance, 505
  factory test certificates, 505
  generic data, 505
  precision, 506
  risk management, 507
  safety margin, 506, 507
  time-variation of current, 506
UK Engineering Recommendation ER
    G7/4, 463
  break time, 467
  breaking current, 466, 468
  dc current, 466, 468
  good industry practice, 464
  making current, 466, 467
  principles, 465
  representation of, 464
    passive load, 465
    small induction motors, 465
    synchronous machines, 464
    asynchronous machines, 464
  rms asymmetric breaking current, 468
  superposition method, 465
X/R ratio, 467

Untransposed:
  double-circuit lines, 102

Voltage transformers, 201

Windfarm, 497
  PPS, NPS and ZPS equivalents, 497
Wind turbine generators, 2, 385
  modelling for short-circuit analysis, 388
    'fixed' speed induction generators,
      388
    small speed range induction
      generators, 388, 389
    variable speed doubly fed induction
      generators, 389, 390, 391, 392
    series-converter connected generator,
      393, 394, 395
  types, 385
    'fixed' speed induction generators,
      386
    small speed range induction
      generators, 386
    doubly fed induction generators, 387
    series-converter connected generator,
      387

X/R ratio, 10

Zig-zag transformers, 242
ZPS magnetising reactance:
  core magnetic reluctance, 246
  effect of core construction, 246
  five-limb and shell-type cores, 247
  impedance of transformers, 246
  three single-phase bank, 247
  three-limb cores, 248